7th International Brazing and Soldering Conference 2018

AA004954

New Orleans, Louisiana, USA
15-18 April 2018

Editors:

Robin Gourley **Charles Walker**

ISBN: 978-1-5108-7844-0

Printed from e-media with permission by:

Curran Associates, Inc.
57 Morehouse Lane
Red Hook, NY 12571

Some format issues inherent in the e-media version may also appear in this print version.

Printed by Curran Associates, Inc. (2019)

American Welding Society
8669 NW 36 Street, #130
Miami, Florida 33166-6672
USA

Phone: 800-443-9353
305-443-9353

www.aws.org

Additional copies of this publication are available from:

Curran Associates, Inc.
57 Morehouse Lane
Red Hook, NY 12571 USA
Phone: 845-758-0400
Fax: 845-758-2633
Email: curran@proceedings.com
Web: www.proceedings.com

Brazing and Soldering

Proceedings of the 7th International Brazing and Soldering Conference

April 15-18, 2018
Double Tree by Hilton
New Orleans, LA

Edited by:
Robin Gourley
Charles Walker

Sponsored and Published by the American Welding Society

Miami, FL 33166-6672

www.aws.org

TABLE OF CONTENTS

American Welding Society®

aws.org

IBSC2018

INTERNATIONAL BRAZING AND SOLDERING CONFERENCE

Thank You

2018 IBSC SPONSORS

GOLD

Aimtek Quality and Service

LucasMilhaupt® Global Brazing Solutions®

PRINCE & IZANT COMPANY — CUSTOMER FOCUSED. SOLUTION DRIVEN.

SILVER

Ajax Tocco

Alexy Metals

OXY-GON INDUSTRIES, INC

BRONZE

DANKAY

HUA GUANG

Metglas®, Inc.

LANYARD SPONSORSHIP

HARRIS — A LINCOLN ELECTRIC COMPANY

PEN SPONSORSHIP

SOLAR ATMOSPHERES

REFRESHMENT BREAK SPONSORSHIP

AIM MRO

REGISTRATION SPONSORSHIP

WF

Wetting of bulk Cu, Cu_3Sn and Cu_6Sn_5 by liquid Sn at different temperatures

O. Y. Liashenko[1,2], *F. Hodaj*[1]
[1]*Univ. Grenoble Alpes, CNRS, Grenoble INP, SIMAP, F-38000 Grenoble, France*
[2]*Cherkasy National University, Cherkasy, Ukraine*

Abstract

The mechanical properties of joints in electronics depend on the adhesive properties of the interfaces as well as on the microstructure of the solder joint [1]. The importance of wettability in adhesion is well-known: a good wetting is a necessary condition for a good adhesion [2]. The microstructure of Sn-based solder after the reflow process can strongly depend on the nucleation of tin that takes place at preferred sites for heterogeneous nucleation which could be inclusions or not dissolved intermetallics (IMCs) in the solder bulk such as Cu_6Sn_5 and Ag_3Sn or IMC reaction layer(s) formed at solder/substrate interface (Cu_6Sn_5 and Cu_3Sn for Cu substrate or Ni_3Sn_4 for Ni substrate) [3]. Thus, the knowledge of the contact angles of liquid Sn on different phases involved in the joint is of paramount importance for predicting the temporal evolution of the joint system. For this purpose, we have performed wetting experiments of bulk Cu, Cu_3Sn, and Cu_6Sn_5 substrates at 250-600°C. The experiments are performed using a dispensed drop set-up under high vacuum which allows to separate processes of melting and spreading and to realize an isothermal spreading. The role of the oxidation of the substrate on the contact angles is also discussed.

Introduction

Fabrication of integrated circuits includes numerous reflows of solder-copper contacts [1]. During a reflow, several processes occur such as melting of the solder, *wetting of the substrate by the liquid solder, interfacial reaction at the solder/substrate interface and solidification of unreacted solder* [4-5]. Wetting and spreading of liquid Sn on the unreacted Cu substrate as well as on the reaction products formed at the Sn/Cu interface (i.e. Cu_6Sn_5 and Cu_3Sn intermetallics), can play a crucial role on the subsequent evolution of the interfacial system, especially in the framework of packaging challenges during miniaturization [6-8]. *Thus, the wetting properties of Cu, Cu_6Sn_5 and Cu_3Sn phases by liquid Sn are of great importance from both theoretical and practical point of view.*

Wetting of liquid Sn-based solder alloys on Cu substrate have been extensively studied and the observed contact angles vary from ~30 to 60° (see for example Ref. [9]). On the contrary, only one study is reported on the wetting of Cu_3Sn and Cu_6Sn_5 bulk substrates [10]. Note however that in these studies the wetting experiments have been performed by the sessile drop technique and thus the melting interferes with the spreading process leading to non-isothermal spreading. *Thus, the principal aim of this work is to study the wetting of liquid Sn on the bulk Cu substrate as well as on the reaction products formed at liquid Sn/Cu interface, i.e. on the bulk η-Cu_6Sn_5 and ε-Cu_3Sn substrates by the dispensed drop technique which allows to separate processes of melting and spreading and to realize an isothermal spreading.*

Experimental procedure

Bulk η-Cu_6Sn_5 compound was prepared from Sn and Cu pieces sealed in a quartz tube under Ar, maintained first for 24h at 1000°C, quenched in water and then annealed for 20 days at 350°C. XRD analysis of cross-sectioned surface of the sample showed only peaks of expected monoclinic Cu_6Sn_5 phase.

Bulk ε-Cu_3Sn compound was prepared from Cu and Sn pieces placed in an alumina crucible maintained first for 30 min at 800°C, then for 4 hours at 630°C under high vacuum ($1.6\cdot10^{-4}$ Pa). XRD analysis of powders issued from the sample revealed the presence of the only orthorhombic Cu_3Sn phase. Cu_6Sn_5, Cu_3Sn and Cu substrates were polished with diamond paste down to 0.25 μm.

The average roughness of the samples, measured by high resolution optical profilometry on 350 μm×260 μm areas, was 5-10 nm.

Wetting experiments consist in heating pure Sn in an alumina crucible placed above the substrate under high vacuum (~10^{-5} Pa). Afterwards, the molten metal is extruded from the crucible through a capillary, forming droplets 1-2 mm in diameter. The wetting process is filmed by a CDD camera (25 frames.s^{-1}) and the *contact angle θ* and the *drop base diameter d* were determined with an accuracy of ±2° and ±2% respectively. After cooling, specimens were prepared for scanning electron microscopy (SEM) analysis. Note that only characterization of some selected samples will be presented here.

Results and discussion

1. Wetting of Cu substrate by liquid Sn or Sn-Cu alloys

Wetting of Cu by liquid Sn was performed at temperatures between 250°C and 600°C. In order to study the influence of the surface oxidation of the substrates on the wetting process two types of wetting experiments were performed:

(i) The deposition of the liquid droplet is performed once the experimental temperature is attained. These experiments are called *"without preheating of substrate"*.
(ii) The substrate is firstly heated to high temperature and maintained at this temperature for a given time under high vacuum before cooling down to the experimental temperature and performing the wetting experiment. These experiments are called *"with preheating of substrate"*.

Experiments without preheating of Cu substrate
Fig. 1 gives the temperature-time profile of different samples for the wetting experiment. An isothermal holding for 10 min was made after the deposition of each droplet.

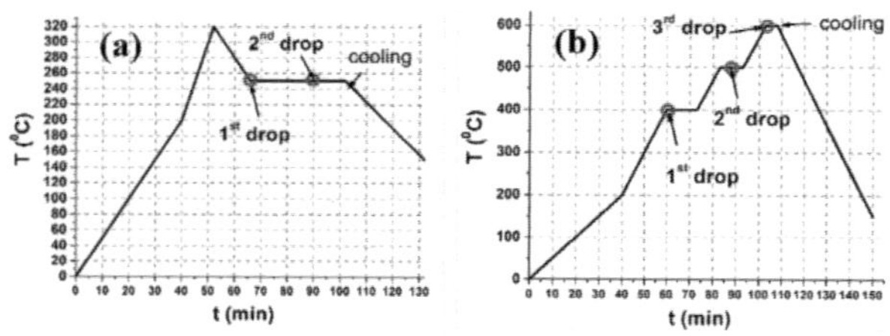

Figure 1 Temperature-time profile during wetting of Cu substrate by liquid Sn (experiments without preheating of the substrate). (a) T = 250°C (2 droplets); (b) T = 400, 500 and 600°C (3 droplets).

For wetting experiments performed at 250-500°C, the contact angle θ, measured to be in the range 26-35°C as can be seen from Fig.2, remains constant for 10 min and is equal to the initially measured contact angle (40 ms after the drop deposition). The drop base diameter d also remains constant. On the contrary, for T = 600°C, the apparent contact angle decreases from 27 to ~0° in about 15 min (see Fig. 2). These experiments show that the initial measured contact angle is lower than 35° and it decreases with the temperature.

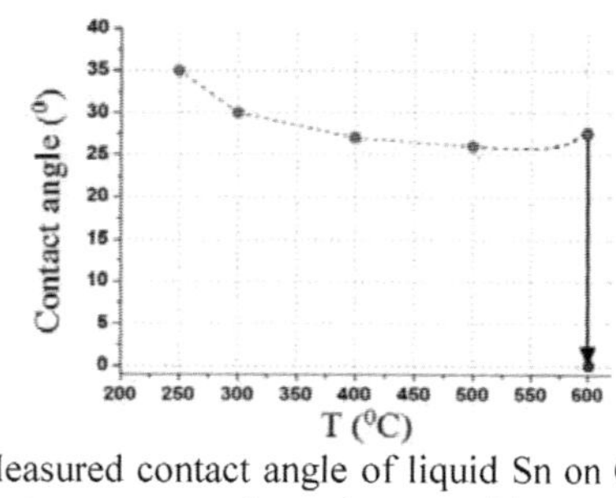

Figure 2 Measured contact angle of liquid Sn on Cu substrate at different temperatures (experiments without preheating of substrate).

In fact, when a droplet of Sn is put in contact with a Cu substrate, the dissolution of Cu in liquid Sn will be followed by the intermetallics formation at the interface: Cu_6Sn_5 on the Sn side and Cu_3Sn on the Cu side (see Fig. 3). Thus it appears astonishing that no spreading is observed for T < 500°C despite the fact that the solubility of Cu in liquid Sn is significant (about 13 at% at 400°C and 25at% at 500°C). Note however that observation of droplets deposited at $T \leq$ 500°C showed that during their subsequent heating and isothermal holding at 600°C (see Fig. 1b) all droplets started to spread after about 10 min of maintaining at 600°C. Once the spreading process starts at 600°C it is very rapid, and the contact angle becomes lower than 10° in less than 4 min. This spreading process at 600°C is undoubtedly linked with the interfacial interaction (dissolution + reaction) between Cu and liquid alloy.
In order to verify if the oxidation of Cu surface (native oxide) has any effect on the wetting process, we have performed wetting experiments with preheating of Cu substrate.

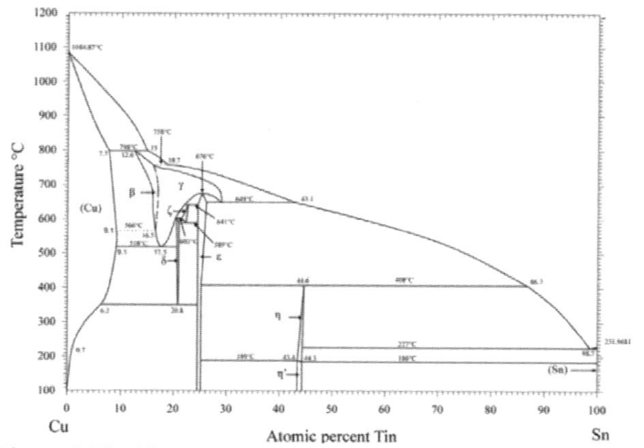

Figure 3 The binary Cu-Sn phase diagram [11].

Experiments with preheating of the Cu substrate
This series of experiments was performed by annealing of the copper substrate at 800°C for one hour prior to the deposition of the liquid tin droplets. Fig. 4 gives the temperature-time profile for two sets of experiments during which 8 droplet depositions were performed at temperatures from 250 to 800°C.
Droplets were deposited during the cooling stage. The duration of the observation of each deposited droplet was only 3 min because, in all cases, no evolution of the contact angle was observed after 3 min of contact.

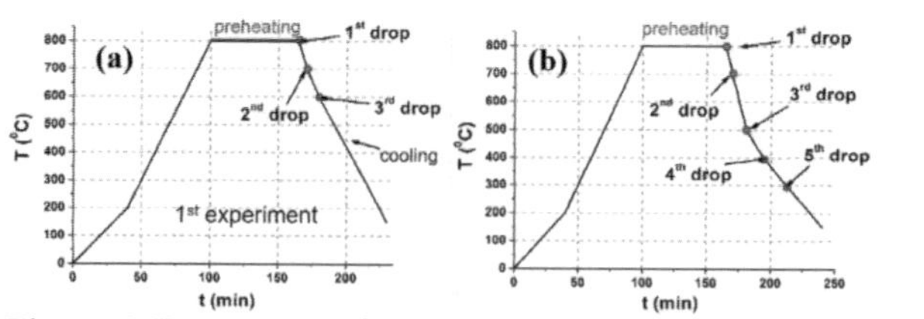

Figure 4 Temperature-time profile of experiments performed with preheating at 800°C for 1 hour. The deposition of Sn droplets on Cu substrate was made after an isothermal holding of about 3 min at experimental temperature. Droplet deposition is marked by o.

Fig. 5 and Table 1 summarize the values of the initial and final measured contact angles at different temperature. Note that for low (300°C) and high temperature (700 and 800°C) experiments, the spreading of the drop was not observed (i.e. the initial and final measured contact angles are the same). As it can be seen from the Table 1, the switch to the longer deoxidation duration at 800°C (experiment at 500°C marked by *) does not lead to much lower measured contact angle. This result suggests that 1 hour of deoxidation at 800°C is sufficient to overcome the oxidation of the copper substrate surface.

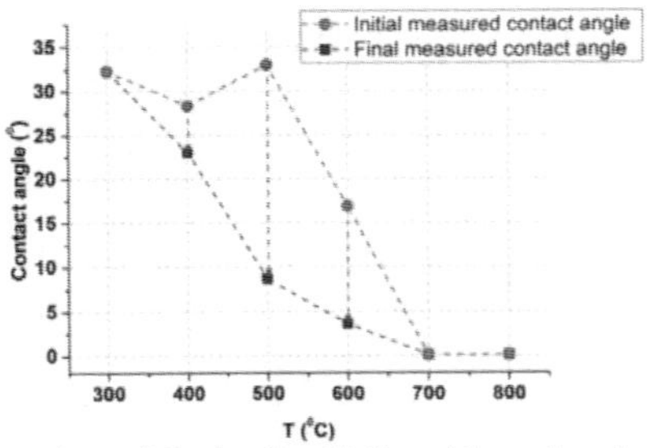

Figure 5 Wetting of Cu by liquid Sn with preheating of the Cu at 800°C for 1 hour. Variation of the initial measured contact angle θ_i (in 40 ms) and final measured contact angle θ_f with temperature (for the spreading time see Table 1).

T, °C	300	400	500	500*	600	700	800
θ_i, °	32	28	33	32	17	0	0
θ_f, °	32	23	9	14	4	0	0
t_s (s)	#	4	1	3	0.1	#	#

Table 1 Wetting data in Cu/Sn couple at different temperature: θ_i = initial measured contact angle (in 40 ms); θ_f = final measured contact angle. Preheating of the substrate for 1 h at 800°C except for experiments marked by (*) for which the preheating time was 2 h at 800°C. (#): not measured.

Fig. 6 gives the variation of the contact angle and droplet base diameter (*d*) with spreading time for experimental temperatures 400, 500 and 600°C. It shows that spreading from the initial to the final measured contact angle is very rapid. It should be emphasized that in experiments without deoxidation procedure we have not observed any variation of the contact angle at 400 and 500°C. These observations point out the strong correlation between contact angle measurements and oxidation extent of both copper substrate and liquid tin surfaces even under high vacuum.

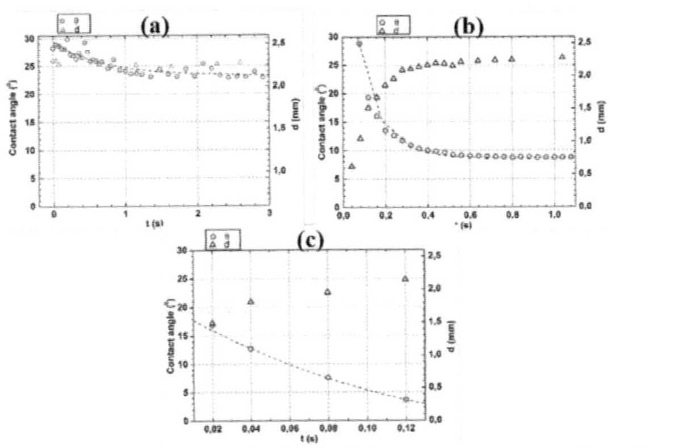

Figure 6 Variation with time of the contact angle (o) and drop base diameter (Δ) of the droplet at 400°C (a), 500°C (b) and 600°C (c) in Cu/Sn couple with preheating of Cu substrate.

The comparison of the wetting experiments of liquid Sn on Cu substrate without preheating and with preheating (see Figs. 2 and 5) shows that:

(i) If the wetting experiments are performed at low temperature range 250-400°C, the preheating process at 800°C has no influence on the initial measured contact angle or even on the final contact angle. This suggests that even if deoxidation process occurs at T = 800°C, during cooling at low T (with a cooling rate around 5°C/min - see Fig. 4) an oxidation of the Cu substrate is possible at low T.

(ii) For experiments performed at 500°C, even if the initial measured contact angle seems to be almost the same, as in the case of the Cu substrate preheated at 800°C, a rapid spreading occurs from θ_i = 32-33° to θ_f = 9-14 °C whereas no spreading is observed with substrate not preheated at 800°C. Moreover, the initial measured contact angles for experiments performed at high temperatures (700 and 800°C - see Fig. 5) are close to 0°.

Characterization of Cu/Sn sample

Fig. 7 gives a cross-section analysis of the sample after the experiment at T = 400°C (sample marked as 4th droplet in Fig. 4b). It shows the formation of a planar layer close to the Cu substrate that is the ε-Cu_3Sn compound (~ 6 μm thick) and a thick scallop-like layer (~ 12 μm thick) on the Sn side that is identified as η-Cu_6Sn_5 phase as it is predicted by the binary Cu-Sn phase diagram (Fig. 3).

This sample revealed a strong dissolution process that happened during the stage of spreading (that lasts about 5 s - see Fig. 6a), but also and above all during the interaction for times longer than 5 s where no significant spreading occurs. Indeed, for this experiment, during 35 min down to the solidification temperature the triple line advances only about 150 μm. The important dissolution of Cu in liquid Sn is testified by the existence of numerous large Cu_6Sn_5 precipitates inside the drop bulk, which are formed during cooling of Sn-Cu liquid alloy. It is interesting to note that the Cu/droplet interface in the region of about 200 μm close to the triple line is not planar (Fig. 7) and has a profile similar with the dissolutive wetting profile described in Ref. [2].

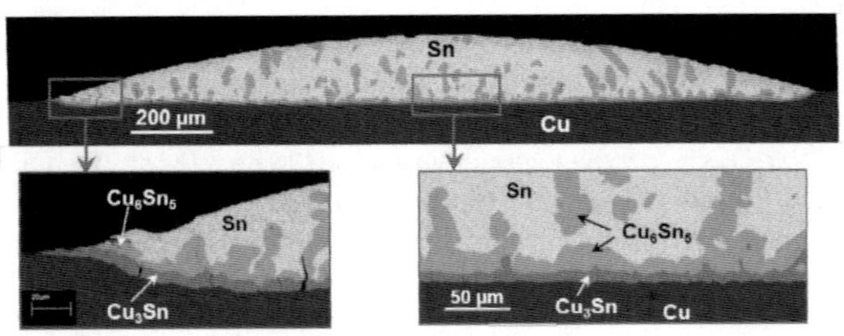

Figure 7 Cross-section of the interface between tin droplet and copper substrate at 400°C after wetting experiment with preheating of Cu substrate. The drop was maintained for 3 min at 400°C and then cooled (see Fig. 4b).

Summary of wetting experiments in Cu/Sn couple

Wetting experiments of preheated Cu substrates by liquid Sn at temperatures from 300 to 600°C show that the first stage of spreading down to a contact angle of about 30° in less than 40 ms can be attributed to the non-reactive wetting of liquid Sn on the Cu surface [12]. Given the fact that these wetting experiments are performed with Cu surface that could be not exempted completely of oxide nanoparticles it can be concluded that the value of the equilibrium contact angle of liquid Sn on a non-reacted and clean Cu surface could be less than 30°.

The very rapid and non-reactive spreading stage can be followed by the reactive spreading stage that correspond to the dissolution of Cu substrate on the liquid drop and/or to the formation of reactive Cu_3Sn and Cu_6Sn_5 layers at Cu/Sn interface [2, 12]. Indeed, for contact time longer than 40 ms and experimental temperature $T \geq 400°C$ a second spreading stage is observed during some tenth of seconds to some seconds according to the temperature (see Fig. 6).

2. Wetting of Cu_3Sn by liquid Sn

As in the case of Cu substrate, the wetting of bulk Cu_3Sn was studied on the non-preheated or preheated Cu_3Sn substrate. The temperature of preheating process under high vacuum was chosen to be equal to 600°C, lower than the decomposition temperature of Cu_3Sn compound (see Fig. 3).

Experiments without preheating of the Cu_3Sn substrate

Fig. 8 gives the temperature-time profile for two sets of experiments during which 7 wetting experiments were performed at temperatures from 250 to 600°C. For T = 250 and 300°C two wetting experiments were performed (1st and 2nd drop on Fig. 8a and 3rd, 4th drops on Fig. 8b).

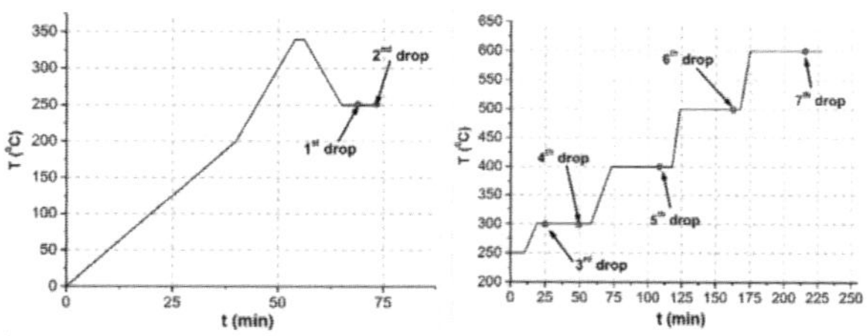

Figure 8 Temperature-time profile during two sets of experiments. Wetting of Cu_3Sn substrate by liquid Sn at different temperature without preheating.

Fig. 9 gives the spreading curves corresponding to the experiments performed at 250, 300 and 350°C. These curves show that the first measured contact angle values (in 40 ms) are very high for low temperatures: θ_0 =125 and 76° at 250 and 300°C respectively. Afterwards, the contact angle decreases rapidly to about 40° in about 10s. Finally, a slow spreading stage is observed at 300°C down to θ = 30° in some minutes.

The very high values of the first measured contact angle at 250°C (θ_0 = 125°) corresponds undoubtedly to the equilibrium contact angle of liquid tin on an oxide surface. Indeed, it is well known that the equilibrium contact angles of non-reactive liquid metals on oxide surfaces are as high as 120-140° [2].

The rapid decrease of the contact angle with time corresponds to the destruction of the native oxide layer by interaction with the liquid phase. The same comments are valid for the experiment performed at 300°C.

At T = 400°C, the first measured contact angle in 40 ms is 37°. This value is close to those obtained at 250 and 300°C after some seconds of interaction between liquid Sn and substrate. Afterwards, the contact angle decreases from 37° to about 25° in about 5 min.

Note that at T = 250 to 400°C, liquid Sn reacts with Cu_3Sn substrate and forms the interfacial reaction compound Cu_6Sn_5. While for T > 415°C, no reactive compound is formed between liquid Sn and Cu_3Sn (see Fig. 7), only dissolution of Cu_3Sn in liquid Sn occurs.

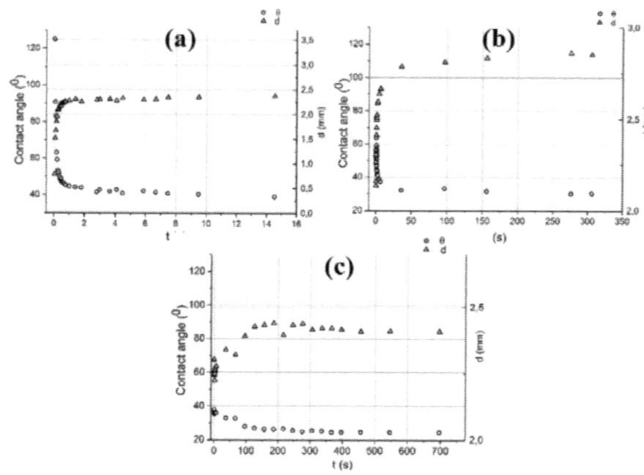

Figure 9 Variation of the contact angle (o) and drop base diameter (Δ) with time during wetting of Cu_3Sn by liquid Sn at T = 250, 300 and 400°C. Experiments performed without preheating.

Contrary to the experiments performed at temperature lower than 400°C, the high temperature experiments show very low values of the first measured contact angle in 40 ms (θ_0 = 18° at 500°C and θ_0 = 8° at 600°C) and no spreading kinetics was observed in about 5 min.

Table 2 summarizes the first measured contact angle in 40 ms (θ_0) and the "final" measured contact angle (noted θ^*) after 5 min of contact between liquid Sn and Cu_3Sn substrate.

T(°C)	250		300		400	500	600
θ_0(°)	120	125	76	45	37	18	8
θ^*(°)	40	40	30	33	25	18	8

Table 2 Variation with temperature of the first measured contact angle in 40 ms (θ_0) and "final" contact angle θ^* measured after 5 minutes of contact of Sn droplet with Cu_3Sn substrate (without preheating).

Experiments with preheating of the Cu_3Sn substrate

Fig. 10 gives the temperature-time profile for wetting experiments with Cu_3Sn substrate preheated at 600°C under vacuum for 1 h.

For all wetting experiments performed at 400, 500 and 600°C the spreading was instantaneous (in 40 ms) and no evolution of the contact angle was observed during 5 min. These first measured values of contact angles (in 40 ms) at different temperatures are summarized in Table 3.

Comparison of Table 2 with Table 3 shows that the observed contact angles on the preheated substrate are significantly lower than those observed on the non-preheated Cu_3Sn substrate. These results clearly indicate the role of the preheating of the Cu_3Sn substrate on the wetting kinetics as well as on the values of the final contact angles and thus the role of the surface deoxidation.

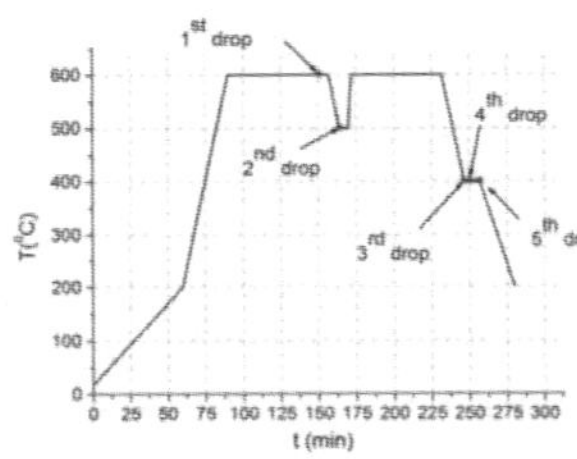

Figure 10 Temperature-time profile of wetting experiments with the Cu_3Sn substrate (with preheating at 600°C).

T(°C)	400			500	600
θ_0(°)	20	9	20	21	7

Table 3 First measured contact angles in 40 ms (θ_0) versus temperature for the Sn droplets deposited on Cu_3Sn substrate preheated at 600°C under vacuum.

Characterization of Cu_3Sn/Sn sample

Fig. 11 gives some SEM micrographs of the sample corresponding to the wetting experiment performed at 300°C. This sample corresponds to the 5th drop of Fig. 10, the time of reaction at 300°C is 5 min and the time during cooling from 300 to 230°C is evaluated to be about 20 min. The deposited droplet has a circular form and a Cu_6Sn_5 reaction layer of thickness about 20 μm is formed at the interface.

From Fig. 11b it can be seen that Cu_3Sn substrate was dissolved in liquid Sn which is testified by the presence of large particles of Cu_6Sn_5 phase inside the drop bulk.

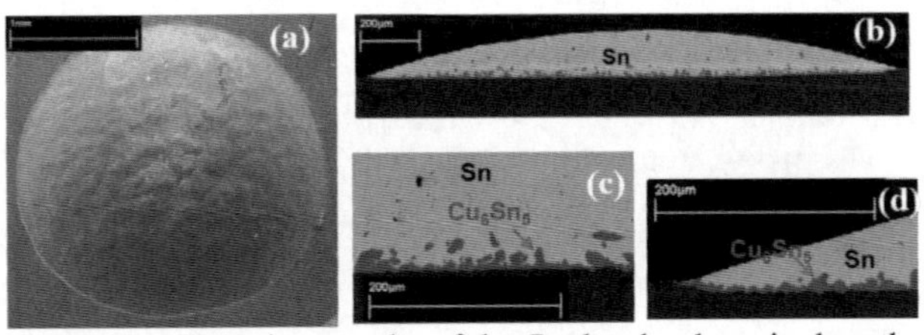

Figure 11 SEM micrographs of the Sn droplet deposited on the bulk Cu_3Sn surface at 300°C.

Summary of wetting experiments in Cu_3Sn/Sn couple

Investigations of wetting of Cu_3Sn by liquid Sn showed that for not preheated substrate, the first measured contact angle values (in 40 ms) are very high for low temperatures: θ_o = 125° at 250°C and θ_o = 76° at 300°C. θ = 125° corresponds to the equilibrium contact angle of liquid Sn on an oxide surface whereas θ = 76° corresponds to the equilibrium contact angle of liquid Sn on a composite surface Cu_3Sn + oxide. Afterwards, in both cases the contact angle decreased rapidly to about 40° in about 10s. The decrease of the contact angle corresponds to the destruction of the oxide native layer by interaction with liquid phase that depends on the thickness of the native oxide layer and the experimental temperature. Thus, the final contact angle of liquid Sn on Cu_3Sn at 250, 300 and 400°C is about 40°, 30° and 25°, which points out the fact that the surface deoxidation is more efficient with increasing experimental temperature.

For preheated Cu_3Sn substrate, the first contact angle at 400°C (in 40 ms) is 20° much lower than the first measured contact angle at the same temperature obtained in the case without heat treatment (37°) meaning that, starting from 400°C, the heat treatment under vacuum of Cu_3Sn at 600°C has an impact on its surface deoxidation.

3. Wetting of Cu_6Sn_5 substrates

Given the fact that the Cu_6Sn_5 phase is not stable at $T \geq 415$°C (at $T = 415$°C it undergoes a peritectic decomposition: $Cu_6Sn_5 \rightarrow Cu_3Sn$ + liquid - see Fig. 3), the wetting experiments should be performed at $T < 415$°C. Thus, we choose a relatively high temperature (390°C) in order to deoxidize under vacuum (if possible) the Cu_6Sn_5 surface. In this case we implemented the preheating procedure for 2 hours at 390°C and droplet deposition at the same temperature.

It was observed that the contact angle of the droplet remains almost constant (about 20°) for wetting time lower than 0.4 s but afterwards it starts to decrease continuously. Afterwards the height of the drop decreases with the time while the drop base diameter remains constant and after 30 min of contact the drop was not visible.

In order to understand this phenomenon, we have performed SEM analysis of top views and cross sections of the sample (Fig. 12). As it was expected from camera recording, the droplet is not detected on the cross-section of the sample and Sn particles are detected inside the bulk Cu_6Sn_5 substrate. The Cu_6Sn_5 grain-boundaries are wetted by liquid Sn not only underneath the drop but also all over the substrate. They clearly show that the liquid alloy is infiltrated inside the substrate to different extent and the "infiltration paths" are the grain-boundaries of the Cu_6Sn_5 substrate. Note that infiltration can also take place by a combination of infiltration of opened porosities and wetting of grain boundaries. However, the examination of Figs. 12b strongly suggests that the infiltration takes place at least via the Cu_6Sn_5 grain boundary wetting.

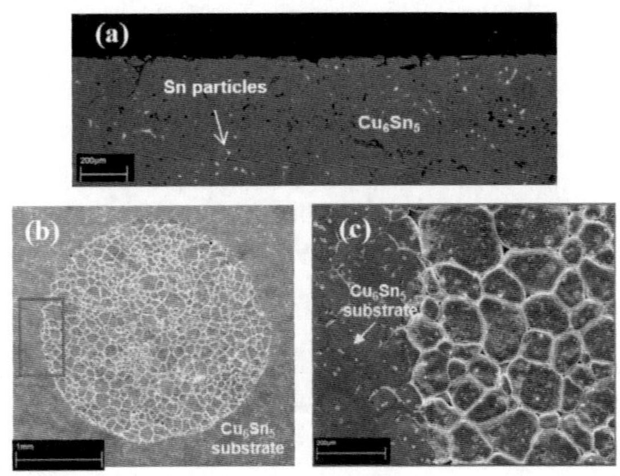

Figure 12 SEM micrographs of the sample liquid Sn/Cu_6Sn_5 substrate held at 390°C for 20 minutes. (a) cross-section, (b) top-view, (c) enlargement of the zone indicated by a red rectangle in (b).

Summary of wetting experiments in Cu_6Sn_5/Sn couple

Wetting experiments of liquid Sn on flat bulk Cu_6Sn_5 substrate at 390°C showed that the first measured contact angle (in 40 ms) is about 20°. This value is lower than that obtained on Cu surface and close to that obtained on Cu_3Sn surface at the same temperature. During the contact between the liquid Sn and Cu_6Sn_5 substrate a significant decrease in the drop height without any noticeable change in the drop base diameter of the droplet is observed due to a rapid infiltration of open porosities of the Cu_6Sn_5 substrate and/or to grain-boundary infiltration.

Conclusions

The wettability of bulk Cu, Cu_3Sn and Cu_6Sn_5 by liquid Sn was studied by implementing the dispensed drop technique in the high-vacuum furnace. It was found that the oxidation extent of the substrate can be altered by prolonged annealing at high temperature under the vacuum.

Wetting experiments of liquid Sn on Cu substrate allow us to study, in a qualitative way, the influence of the annealing of the Cu substrate under high vacuum at 800°C on its wetting behavior: If the wetting experiments are performed at low temperature range 250-400°C, such annealing has practically no influence on the first and final measured contact angles. On the contrary, for higher temperature wetting experiments (500-600°C) such annealing leads to a small decrease in the first measured contact angles (in 40 ms) and a significant increase in the spreading kinetics.

The first measured (in 40 ms) contact angles of liquid Sn on a Cu_3Sn surface preheated at 600°C depend on the experimental temperature: they decrease from 23 to 10° in the temperature range 300-500°C, which should be mainly due to the oxidation extent of the surface of the substrate. During wetting experiments of the bulk Cu_6Sn_5 compound at 390°C by liquid Sn low values of the first measured (in 40 ms) contact angle are obtained (~20°). It allows us to conclude that the contact angles of liquid Sn on two main Cu-Sn intermetallics are very low and could be lower than that on Cu. The pronounced infiltration of Sn into the Cu_6Sn_5 bulk is another feature observed during our experiments. It was found that complete infiltration of millimetric metallic droplet take about 30 min and proceeds through the wetted grain-boundaries of the Cu_6Sn_5 and/or open porosities of the intermetallic bulk.

From experimental results of wetting of Cu and Cu_6Sn_5 substrates it can be concluded that wetting of Cu substrate should proceed in two stages: (i) a rapid *non-reactive stage* during which the contact angle decreases down to 30° very rapidly and (ii) a slow *reactive stage* during which the contact angle decreases from 30 to about 20° (contact angle on Cu_6Sn_5 phase).

References

[1] Tu K. N, Solder Joint Technology: Materials, Properties, and Reliability, Springer (New York, 2007).

[2] Eustathopoulos N. *et al*, Wettability at High Temperatures, Pergamon Materials Series, Volume 3 (Cambridge, 1999).

[3] Liashenko O. *et al*, "Spectrum of heterogeneous nucleation modes in crystallization of Sn-0.7wt%Cu solder: experimental results versus theoretical model calculations," *J. Mater. Sci.: Mater. Electron.*, Vol. 26, No. 20 (2015), pp. 8464-8477.

[4] Bertheau J. *et al*, "Effect of intermetallic compound thickness on shear strength of 25 μm diameter Cu-pillars," *Intermetallics*, Vol. 51 (2014), pp. 37-47.

[5] Taneja D. *et al*, "Further insight into interfacial interactions in nickel/liquid Sn-Ag solder system at 230-350°C," *J Mater Sci. Mater Electron.*, Vol. 28, No. 24 (2017), pp. 18366-18378.

[6] Liashenko O. Y. *et al*, "Phase growth competition in solid/liquid reactions between copper or Cu_3Sn compound and liquid tin-based solder," *J. Mater. Sci. Mater. Electron.*, Vol. 25, No. 10 (2014), pp. 4664-4672.

[7] Liashenko O. Y. and Hodaj F., "Differences in the interfacial reaction between Cu substrate and metastable supercooled liquid Sn–Cu solder or solid Sn-Cu solder at 222° C: Experimental results versus theoretical model calculations," *Acta Mater.*, Vol. 99 (2015), pp. 106-118.

[8] Liashenko O. Y. *et al*, "On the initial stages of phase formation at the solid Cu/liquid Sn-based solder interface," *Acta Mater.*, Vol. 117 (2016), pp. 216-227.

[9] Matsumoto T. and Nogi K., "Wetting in soldering and microelectronics," *Annu. Rev. Mater. Res.*, Vol. 38 (2008), pp. 251-273.

[10] Boettinger W. J. *et al*, The Metal Science of Joining, ed. M.J Cieslak, J.H. Perepezko, S.Kang, M.J. Glicksman, The Minerals, Metals and Materials Society, Warrendale, (Pennsylvania, 1992).

[11] Binary Alloy Phase Diagrams, 2nd Ed. Editor: T.B. Massalski, H. Okamoto, P.R. Subramanian, L. Kacprzak, ASM International, Materials Park (Ohio, 1990).

[12] Protsenko P. *et al*, "Thermodynamics and kinetics of dissolutive wetting of Si by liquid Cu," *Acta Mater.*, Vol. 58, No. 20 (2010), pp. 6565-6574.

Voids formation&extinction in SAC305 paste at the beginning of melting stages using rotational viscometer

Tomoaki Yuasa, Hiroyuki Nakayama , Takahisa Yamazaki
Tokyo Institute of Technology, 2-12-1, O-okayama, Meguro ward, Tokyo, 152-8550, Japan
E-mail: yuasa.t.aa@m.titech.ac.jp, Tel: +81-3-5734-3236

Abstract

Sn-3.0Ag-0.5Cu (SAC305) solder has been used in packaging of electronic components. However, in reflow soldering using SAC305 solder paste, there are residual voids at the joint. We attempted to measure the behavior of the solder paste and voids from the observation of melting process as torque by shear stress generated between the solder and the surface of the electrode substrate by using a cone-plate rotational type viscometer. In the case of Ni cone, the poor wettability in the reflow process was observed because it was difficult to measure the torque at the beginning of melt and it turned out that segregation of oxygen and copper in the solder alloy on the oxide layer at the Ni cone substrate occurred by the EPMA. It was considered this segregation became the starting point of voids residue at the beginning of melt. Then, we used Ni cone covered with Au coating layer to prevent void formation at the interface. The wettability of the paste was improved and it turned out to be effective when the thickness of Au layer was 50nm and 100nm but the effect was reduced when the thickness was 20nm by detecting torque.

Introduction

Assembling in reflow soldering using lead-free solder has been selected as a general packaging method of electronic components since RoHS was enforced in 2006. Currently in Japan, Sn-3.0Ag-0.5Cu alloy solder (known as SAC 305 alloy solder) recommended by the Japan Electronics and Information Technology Industries Association is used. Although, solder paste generates voids and residual voids at the joint decrease the thermal conductivity and the electrical conductivity at the interface, then the performance of the electronic component deteriorates [1][2]. In soldering, copper is used for the substrate. Then, copper and tin form intermetallic compounds Cu_6Sn_5 with high volume thermal expansion rate [3]. Joint strength decreases. nickel is plated on the copper substrate to prevent the inter metallic compound growth. But in reflow soldering, an interface with poor wetting is formed on the nickel plate surface. Therefore, the behavior of the solder paste and voids from the melting stage to the completely molten state was investigated to measure a torque by shear resistance generated between the solder and the surface of the electrode substrate by using a cone-plate type rotational viscometer in order to propose a method to reduce voids from joint.

Experimental procedure

Flow characteristics of the melting solder paste would be changed with the void formation, flux in the paste and the formation of IMC. A cone-plate type rotational viscometer was used because this type of viscometer can measure regardless of flow characteristics and change the contact interface with the solder paste by various materials of the cone.
Figure 1 shows the schematic diagram of a cone-plate type rotational viscometer. This viscometer consists of motor, torque driver, photoelectric sensor, time interval meter, heat resistant Petri dish and cone. The torque generated between the cone and the fluid by shear resistance of the fluid can be measured by reading the phase difference between the up and down rotation of the torque driver. Table 1 shows the properties of a solder paste used in this viscosity measurement. The solder paste used is J3-8600-KF manufactured by Ishikawa Metal Co., Ltd. The dimension of the cone with slope of 3° for this viscometer is shown in Figure 3. We prepared two types of nickel cone, they were a normal nickel surface and a gold-deposited nickel cone. Gold was coated on the nickel surface using a magnetron sputtering system and the thickness of the gold layer was set to 100 nm, 50nm and 20nm. In the case of using the normal nickel cone, approximately 15 g of solder paste was placed on a petri dish and heated. When the temperature of the hot plate reached 300 ℃, the rotation of the cone was started as viscosity measurement. We observed the wetting behavior when the solder is in cream state, from the beginning of melting stages to the completely molten state.
In the case of using the gold-deposited nickel cone, approximately 6.5 g of solder paste was placed on a petri dish and heated. When the temperature of the solder reached to 218 ℃, viscosity measurement was started. In the case using a normal nickel surface cone, the rotational speed of the cone was set to several rotational speeds and the viscosity was measured. On the other hand, in the case of using the gold-deposited nickel cone, the rotational speed of the cone was set to 2.97 rad/s constant.

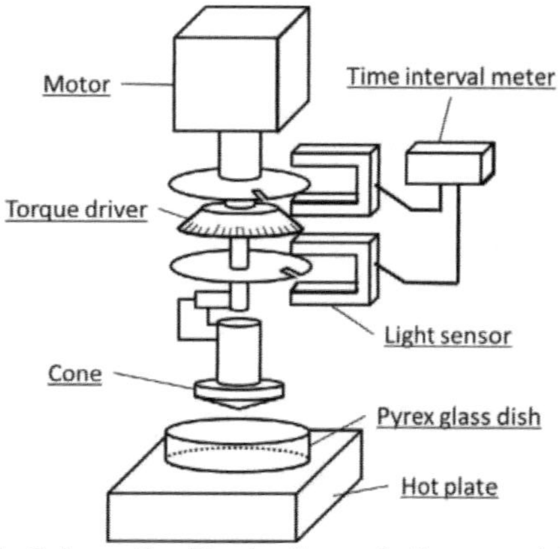

Figure 1 Schematic illustration of the cone/plate-type rotational viscometer.

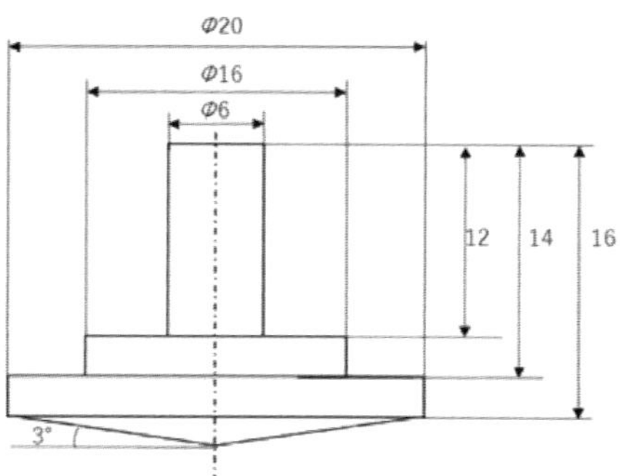

Figure 2 Specifications of the cone.

Table 1 Physical properties of a solder paste.

Compositions	Sn-3.0Ag-0.5Cu
Liquidus temperature, °C	220
Solidus temperature, °C	217
Particle size, μm	25~38
The flax content, mass%	11.0
Viscosity of paste, Pa・s	220

Experimental results

Results and discussion of experiment with normal surface

The viscosity of the solder in the lower-speed region in the cream state was 117 Pa·s, and it was confirmed that the solder was extruded to the outside by the cone. In the high-speed region, the solder and the cone were dissociated. When the hot plate temperature was 250 ℃, the solder started to melt and changed the form with metallic luster. The solder in solid-liquid mixed state gathered at the center of the cone, and it was impossible to measure the viscosity. When the hot plate temperature reached 300 ℃, the solder became completely molten state. In this state, the viscosity of the solder was 114 mPa·s. The flow curve is shown in Figure 3.

Figure 4 and Figure 5 show the state of deposits of the solder on the cone after the viscosity test. Figure 4 shows the state of deposits on the cone at the beginning of melting stages. Figure 5 shows the state of deposits on the cone at the completely molten state. In Figure 4 and Figure 5, element mapping by EPMA was performed on cross sections of these deposits and the nickel cone. The analysis results are shown in Figure 6 and 7. It was confirmed that nickel dissolved in the solder to form a compound layer composed of nickel, copper and tin at the interface when the solder reached the completely molten state. On the other hand, it was confirmed that copper segregated at the interface before nickel dissolved at the beginning of melting. Therefore, in initial stage of the formation of the interface between the SAC305 solder and nickel, it was suggested that segregation of copper bonded to voids to oxide on nickel surface formed a boundary with poor wetting, which is a starting point of the voids residue.

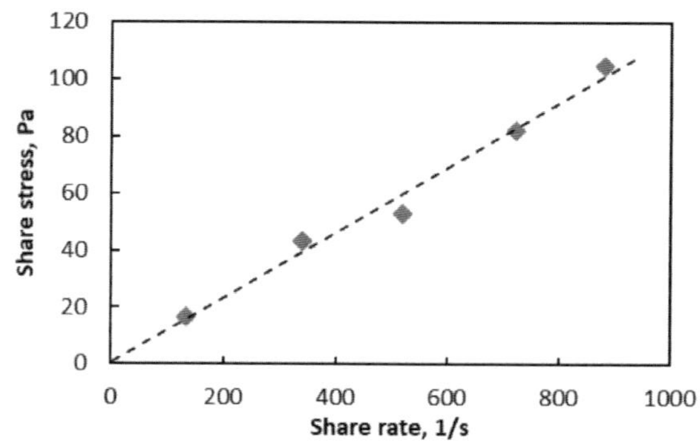

Figure 3 Flow curve of SAC305 at the completely molten state at 300℃.

Figure 4 Deposits on the cone at the beginning of melting.

Figure 5 Deposits on the cone at the completely molten state.

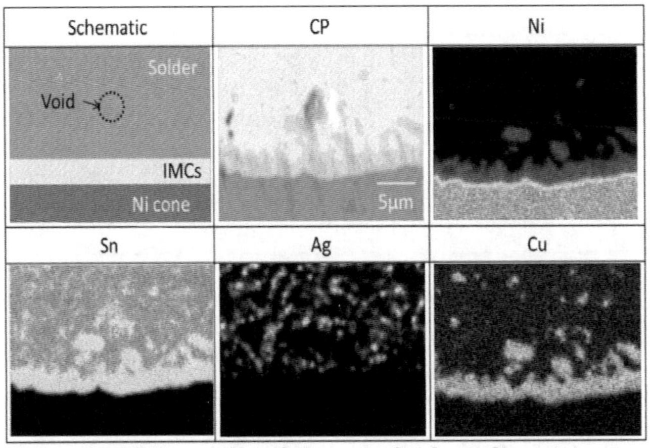

Figure 6 EPMA results of the completely molten state at solder/nickel cone interface.

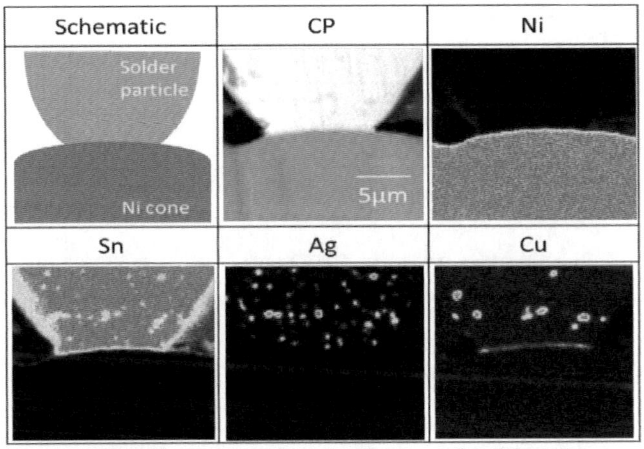

Figure 7 EPMA results of the beginning of melting stages at solder/nickel cone interface.

Results and discussion of torque measurement using gold-deposited nickel cone in the viscometer

Experiments using nickel cone suggested that copper bonded to voids in the solder segregated to the oxide on the nickel surface. Therefore, these experiments were conducted using a nickel cone in which a chemically stable gold layer was deposited on the nickel surface in order to prevent the influence of oxides on the nickel surface.

Changes in torque for each gold layer thickness are shown in Figures 8 to 10. In order to compare with the measured torque, the value of the torque converted from the viscosity between the completely melted solder and the nickel cone is indicated by a red dotted line in each graph. The conversion equation for the viscosity and the torque generated for the cone is shown below

$$T = \frac{2\pi R^3 \omega}{3 sin\theta} \cdot \eta \qquad (1)$$

Where R is the radius of the cone (m), ω is the angular velocity ($\mathrm{rad/s}$), θ is the cone slope, η is the viscosity of fluid [4]. Since the viscosity of the solder measured using the nickel cone was 114 $\mathrm{mPa \cdot s}$, the torque value was converted to 0.0136 $\mathrm{mN \cdot m}$ when the angular velocity was set to 2.97 $\mathrm{rad/s}$.

Regardless of the thickness of the gold layer, it was confirmed that the solder was in contact with the entire surface of the cone, and the viscosity could be measured. When the thickness of the gold layer was 20 nm, the change in torque was not stabilized, and the measured torque was smaller than the torque converted from viscosity measured by the experiment using nickel cone. On the other hand, when the thickness of the gold layer was 50 nm and 100 nm, the measured torque stably showed a value more than the torque converted from viscosity measured by the experiment using nickel cone. In order to observe the state of the cone surface at the point P in Figure 10, the experiment conditions were made equal and the cone was pulled out from solder in 20 seconds after the cone dipping. The state of the cone surface is shown in Figure 11. Then, in the region A in Figure 11, it was observed that the solder was solidified on the cone surface and the solder was good wetted on the gold coated cone surface. The cone was cut along the dotted line in Figure 11, and Figure 12 shows the EPMA results of solder the cross section of the interface between the solder and the cone in the region A. From Figure 12 shows that the gold layer was dispersed in the solder in approximately 20 seconds after the cone dipping, and a copper-rich compound layer was formed on the nickel surface. From the torque change shown in Figs. 8 to 10 and the result of EPMA mapping, the following consideration is given. When gold-deposited cone is dipped in solder, gold and solder are in contact with each other as shown in Figure 13, and an interface with good wettability is formed regardless of the thickness of the gold layer. However, the gold layer begins to disperse immediately after cone dipping as shown in the schematic diagram of Figure 14. In this state, nickel and solder are in contact. If the solder is still in a solid-liquid mixing state, it is considered that an interface with poor wettability is formed on the nickel surface. When a part of the gold layer is dispersed in the solder, it is considered that the change in torque stabilizes at a value close to the viscosity measurement result using the nickel cone as shown in Figures 9 and 10 if the solder is already in a completely molten state. In other words, it was suggested that it is possible to suppress the residue of voids by depositing a gold layer of 100 nm or 50 nm on the nickel surface. On the other hand, if the solder is still in a solid-liquid mixed state because the nickel surface and the solder slip, it is considered that the change in torque is not stabilized as shown in Figure 8 because the nickel surface and the solder slipped. Then, the interface between the nickel surface and the solder with poor wettability is considered to be the starting point of the residual void.

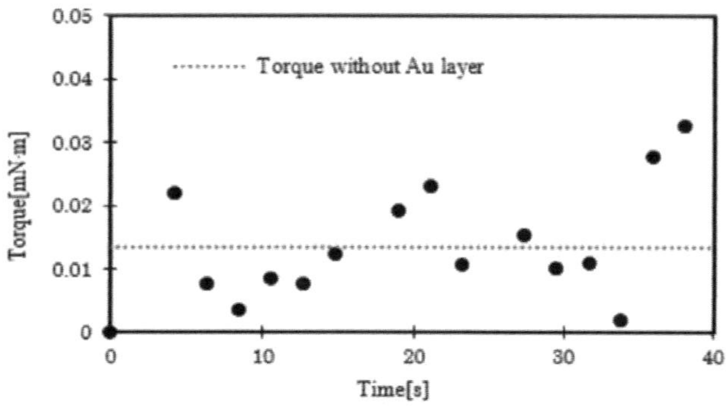

Figure 8 Torque change during rotation in gold layer 20 nm.

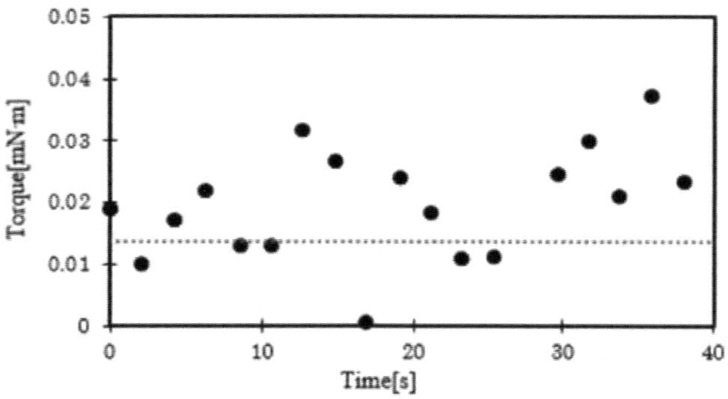

Figure 9 Torque change during rotation in gold layer 50 nm.

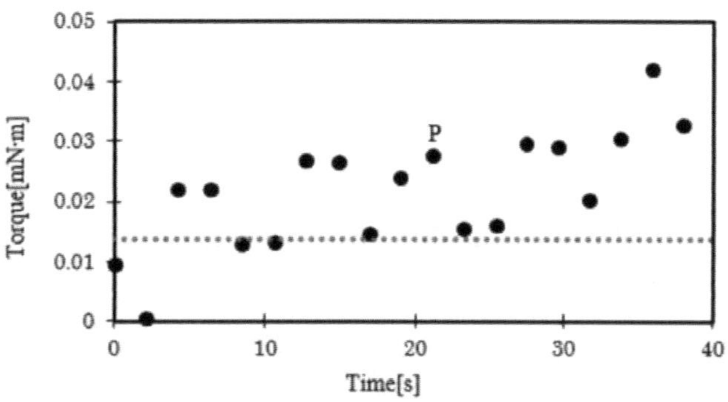

Figure 10 Torque change during rotation in gold layer 100 nm.

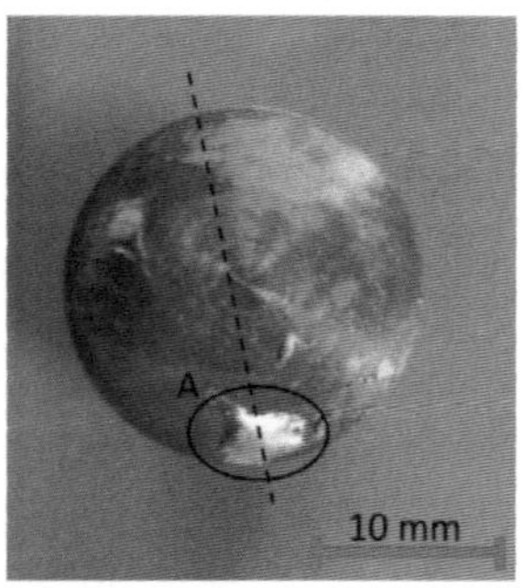

Figure 11 Appearance of the cone surface when the cone was pulled out from the solder at point P.

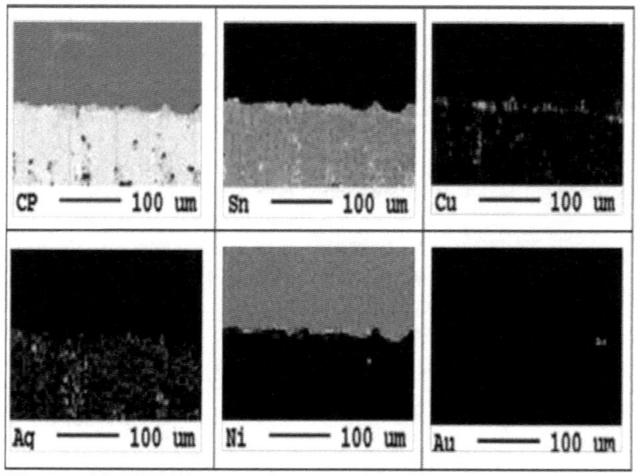

Figure 12 EPMA mapping at solder/gold/nickel interface when the cone was pulled out from the solder at point P.

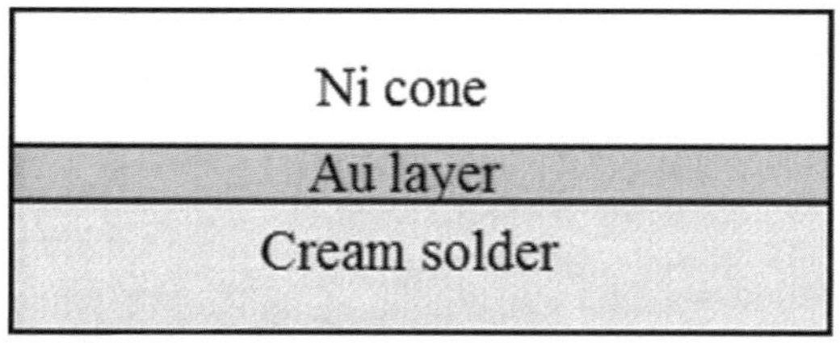

Figure 13 Schematic diagram of interface immediately after cone dipping.

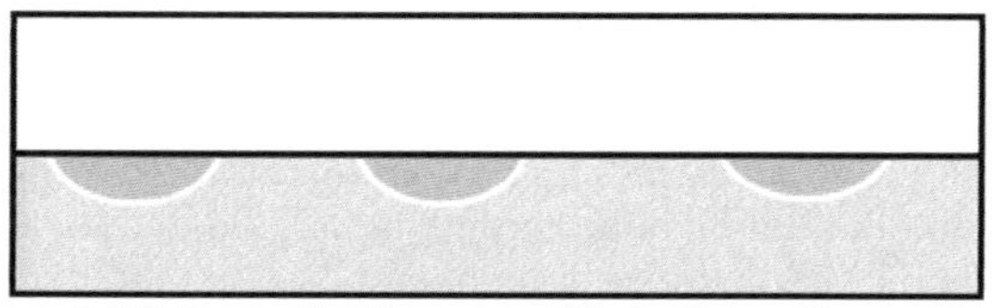

Figure 14 Schematic diagram of interface after dispersion of gold layer.

Simulation on behavior of void

Simulation model

We simulated the behavior of voids and solid copper in the SAC 305 solder which reached the solidus line in the process of gradually increasing the temperature. The lattice Monte Carlo method was used for numerical simulation. It was assumed that void and solid copper move by movement of tin atoms in liquid metal. Tin cells, void cells, and copper cells were placed on two-dimensional lattice points. For each step, one cell is randomly selected and the arrangement of cells was changed by exchanging the position of the selected cell and the cell next to the selected cell. Then, if the surface energy of the whole system was smaller than that before the cell arrangement change, the cell arrangement was updated. Here, it is assumed that the surface energy is generated only between the nearest neighbor cells. Moreover, even when the total of the surface energy of the whole system became larger, the exchange of the cells is admitted if the number of the second closest cells increased in order to enable the cluster of cells to move with shear deformation when the void cell and void cell, void cell and copper cell, copper cell and copper cell are coupled. Figure 15 shows a schematic diagram of cell cluster movement with shear deformation. When the number of lattice points is N, the step at the time when the number of exchange of the cell reaches N/2 times is set as 1 Monte Carlo Step (1 MCS). There were 400 lattice points (20×20) in total. The distance between lattice points was set at 0.01 μm. The number of particles, the temperature and volume of the system were assumed to be constant. When the void moves through the liquid metal, the solder try to maintain the shape of the sphere to lower the surface energy, so voids that had reached the left or right ends were defined as being discharged. We made the voids which reached to the left or right ends unmovable because the number of cells is kept constant in this simulation. The copper cell and the void cell which reached the upper and lower stages of the lattice point were set so as to be repelled without adhering.

Next, the time width per 1 MCS was obtained by choosing one of the tin cells on the lattice and tracing the movement. When the distance from the starting point of the tin cell is x, the time is t, and the self-diffusion coefficient is D, the following expression holds.

$$x^2 = 2Dt \qquad (2)$$

Assuming that the time width in 1 MCS is Δt,

$$x^2 = 2D\Delta t \cdot \mathrm{MCS} \qquad (3)$$

Figure 16 shows the relationship between MCS and x^2. The inclination of the straight line obtained by linear approximating the plotted points by the least-squares method is the value of $2Dt$ in equation (3), and its value was 0.0001. The diffusion coefficient in liquid metal can be expressed as follows using Einstein's equation and Sutherland's equation,

$$D = \frac{kT}{4\pi\mu a} \qquad (4)$$

where, k is the Boltzmann constant, T is the temperature, μ is the viscosity, and a is the atomic radius. D was calculated when the temperature of the system was 500 K, the viscosity of pure tin was 1.86 $\mathrm{mPa \cdot s}$ [5] and a was 140 pm. As a result of calculation, Δt was found to be $\Delta t = 1.7 \times 10^{-6}$ s.

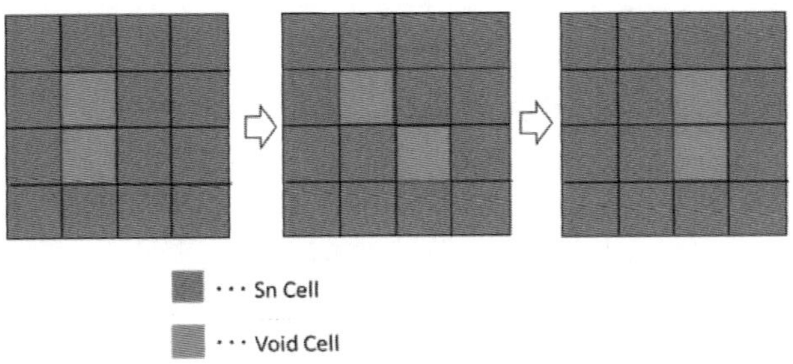

Figure 15 schematic diagram of an example of cell cluster movement with shear deformation.

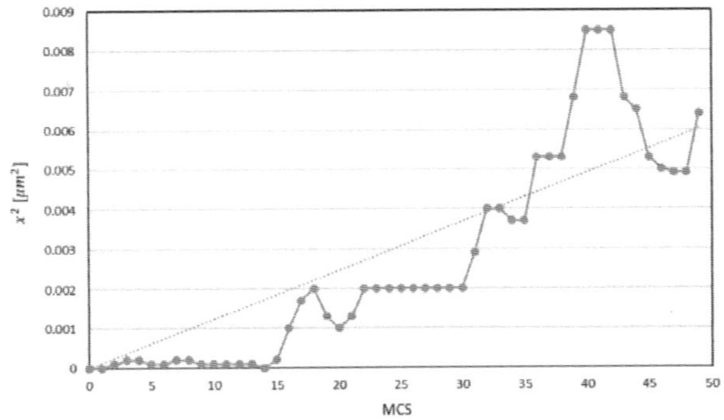

Figure 16 Relation between MCS and x^2 and approximate straight line.

Table 2 The parameter of the surface energy between each cell.

Cell	Surface Energy 10^{-17} J
Void/Void	0
Void/Sn	5.0
Void/Cu	0
Cu/Sn	0.90
Sn/Sn	0

Simulation results and discussion

The simulation results are shown in Figure 17. It was observed that the void cell and the copper cell were moving in a coupled state when about 50 MCS had passed. After about 400 MCS, void nucleus was formed at the center of the solder. Void cluster were formed at the center in about 1200 MCS, and this cluster did not move at all. From the above, it was simulated that voids were formed in the solder before 800 MCS (this step is converted to 1.3×10^{-3} s in terms of time) elapsed, and the concentration of copper around the void was observed.

In this simulation, the movement of the void due to the flow of the solder caused by the rotation of the cone is not considered. In Figure 17, the voids staying in the center of the solder may move due to the flow of the solder and be discharged to the outside. Copper segregated on the nickel surface is considered to be a starting point for the formation of voids on the substrate surface if the solder is in contact with the nickel surface. It is expected that this void will not be emitted by solder flow. In other words, it is considered that it is effective to prevent the segregation of copper on nickel surface in the process of void formation in solder at the beginning of the melting stages in order to prevent void residue.

Conclusions

In this study, behavior at the interface of SAC 305 alloy solder from the melting stage to the completely molten state was investigated using a cone-plate type rotational viscometer. As a result of experiments, it was suggested that a poor wetting interface was formed because copper in the solder segregated

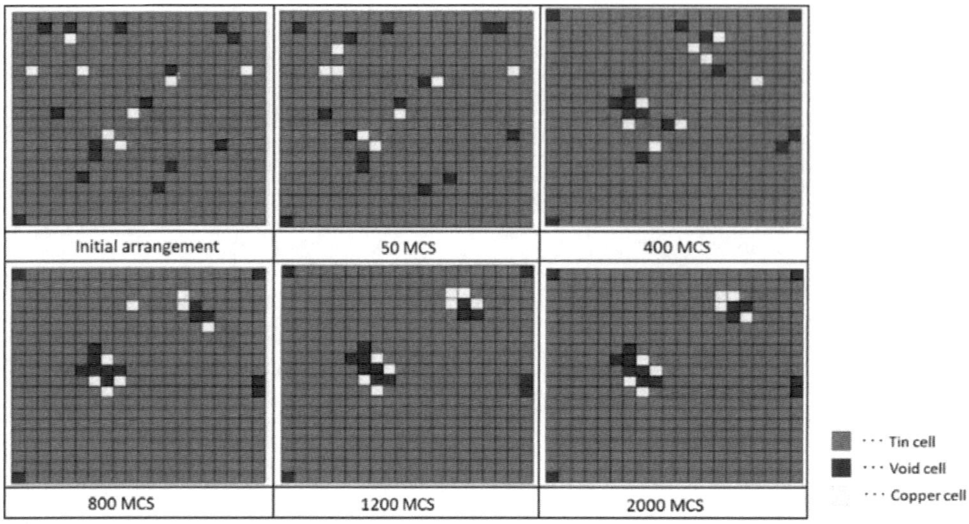

Figure 17 Behavior of void cell and copper cell in the solder at the beginning of the melting stages.

on the nickel surface in the solid-liquid mixing state in the process of increasing the temperature of the solder and this segregation of copper became the starting point of the void residue. It was clarified that this copper segregation could be prevented by depositing a gold layer on the nickel surface. It was suggested that the thickness of the deposited gold layer was effective for 50 nm or 100 nm and it was not very effective in the case of 20 nm. The behavior of void and solid copper in the solder at the beginning of the melting stages was simulated by calculating the surface energy of the whole system. As a result of the simulation, the void was formed in the center of the solder in about 1.3×10^{-3} s, and the copper concentrated around the void. The formed void did not move and stayed in the center of the solder.

References

[1] Weicheng Lin, "The Void-free Reflow Soldering of BGA with Vacuum", in Electronic Packaging Technology, 2007. ICEPT 2007. 8th International Conference, Shanhai, pp. 1-5 (2007)

[2] Hailong Li, Chunqing Wang, Meng Yang, Ningning Wang, Rong An and Yanjun Xu, "The effect of voids on thermal conductivity of solder joints", in Electronic Packaging Technology and High Density Packaging (ICEPT-HDP), 2012 13th International Conference, Guilin, pp. 1061-1064 (2012)

[3] Nan Jiang, J.A. Clum, R.R. Chromik, and E.J. Cotts, "THERMAL EXPANSIONOF SEVERAL Sri-BASED INTERMETALLIC COMPOUNDS", Scripta Msterialia, Vol. 37, No. 12, pp.1851-1854 (1997)

[4] Takahisa Yamazaki, Shinya Oishi, Hirosato Gamou, Toshi-Taka Ikeshoji and Akio Suzumura, "Flow curves of Sn and Sn-3.5Ag obtained by rotational viscometry using a stainless steel cone", IOP Conference Series: Materials Science and Engineering, vol. 61 (2013)

[5] Marc J. Assael, Agni E. Kalyva and Konstantinos D. Antoniadis, "Reference Data for the Density and Viscosity of Liquid Copper and Liquid Tin", Journal of Physical and Chemical Reference Data, vol. 39 (2010)

Effect of Rare Earth Elements on Microstructures and Properties of Lead-free Solders

Wei Zhou and Wengang Zhai
School of Mechanical and Aerospace Engineering, Nanyang Technological University, Singapore 639798, Singapore
wzhou@cantab.net

Xi Wang, Xianfen Li, and Yucheng Wu
School of Materials Science and Engineering, Hefei University of Technology, Anhui Province 230009, P. R, China

Introduction

Soldering is an indispensable manufacturing process during fabrication of modern microelectronic devices. Sn-Pb solders possess advantages of low melting point, excellent performance and low cost. However, due to toxicity of lead, environmental regulations around the world have mandated elimination of lead-containing solders in electronic assemblies.

Sn-0.7Cu, Sn-3.8Ag, Sn-9Zn, Sn-3.8Ag-0.7Cu and many other lead-free solders have already been developed; however, their solderability and mechanical properties require further improvement. Considerable effort has been made in the past decade to use trace amounts of rare earth elements to alloy the lead-free solders with the objective of improving the microstructures and properties [1-22].

The published papers tend to highlight positive effect of rare earth elements, for example, La [2-6], Ce [4, 7-17] and mixed rare earth elements [18-22] have all been reported to help to refine microstructures of various types of lead-free solders. However, it should be noted that the refinement of microstructures does not necessarily lead to improvement in mechanical properties.

We aim to obtain a clearer picture on how rare earth elements affect microstructures and properties of lead-free solders by examining the literature carefully and also by performing our own experimental work. Attention is mainly focused on two light rare earth metals La and Ce since they are the most commonly used in lead-free solders, perhaps because of their relatively low price, as shown in Table 1. Heavy rare earth metals are much more expensive than La and Ce, as shown in Table 2.

Table 1 Price of light rare earth metals (≥99%) in Dec 2017 (USD/kg) [23]

Metal	La	Ce	Sm	Nd	Pr
Price	7	7	7	60	85

Table 2 Price of heavy rare earth metals (≥99%) in Dec 2017 (USD/kg) [23]

Metal	Y	Gd	Er	Dy	Tb	Sc
Price	35	55	95	350	550	15000

Prices of typical metals used in solders are listed in Table 3 for the purpose of comparison. Rare earth elements are usually perceived to be very expensive, but it can be seen from Tables 1 and 3 that both La and Ce are in fact much cheaper than Sn. Therefore, it would make much economic sense to add La and Ce to lead-free solders if they could be proven to be very beneficial.

Table 3 Price of other metals used in solders (≥99%) in Dec 2017 (USD/kg) [23]

Metal	Sn	Ag	Cu	Zn	Pb
Price	20	555	6	1.8	1.9

Effect of La on Microstructures and Properties

Dudek et al. investigated the effect of La additions (0.1 wt% and 0.5 wt%) on microstructure and shear strength of a Sn-3.9Ag-0.7Cu alloy [2, 3, 4]. With La addition, $LaSn_3$ phase formed in the eutectic and along the boundaries of the α-Sn dendrites, leading to refinement of the Sn matrix microstructure. It is noted that an addition of 0.5 wt% La reduced the length of Sn dendrite by as much as 60%. However, unfortunately, the shear strength decreased with the La addition in spite of the microstructural refinement.

Zhang et al. added various amounts of La (0.02, 0.04, 0.06, 0.08, 0.1, 0.2 and 0.5 wt%) into a Sn-9Zn lead-free solder and found that the solder's wettability was improved and the microstructures were refined [5]. The strength was found to be generally improved but the shear strength decreased when the La concentration reached 0.5 wt%. Major results of these previous researches are summarized in Table 4.

Furthermore, Pei and Qu studied SnAg lead-free solders and found that up to 0.25% La doping tended to improve creep resistance and the 0.1% La addition was the optimal amount for improving the fatigue life [6].

We carried out extensive experiments to add various amounts of La (0.05, 0.1, 0.2, 0.5, 1.0, 1.5 wt.%) to a Sn-3.8Ag-0.7Cu solder and studied the rare earth element's effect on the microstructures, wettability and shear strength of the soldered joints. La with purity of 99.99 wt.% was used. The solder alloys were melted in a furnace and covered with B_2O_3 to protect the melt from oxidation. The melt was stirred with a tungsten filament to ensure the homogenization of chemical composition.

Table 4 Summary of previous researches on effect of La

Solder	La (wt%)	Tensile (T) / Shear (S) Strength	Microstructure
Sn-3.9Ag-0.7Cu [2, 3, 4]	0.1	S ↓ 12.7%	Refined
	0.5	S ↓ 35.4%	Refined
Sn-9Zn [5]	0.06	T ↑ 31.0% S ↑ 28.8%	Refined
	0.1	T ↑ 17.2% S ↑ 3.4%	Refined
	0.5	T ↑ 13.6% S ↓ 9.1%	Refined

After solidification, the solder samples were ground on progressively finer SiC paper, polished with diamond paste, and then etched in a solution of 5 mL HNO_3, 2 mL HCl and 93 mL CH_4O for about 2 s. The etched samples were observed under the optical microscope and scanning electron microscope. Typical microstructures are shown in Figs. 1-3.

Addition of La was found generally to refine the microstructures but the refining effect decreased when the amount of La was high at 1 and 1.5 wt%.

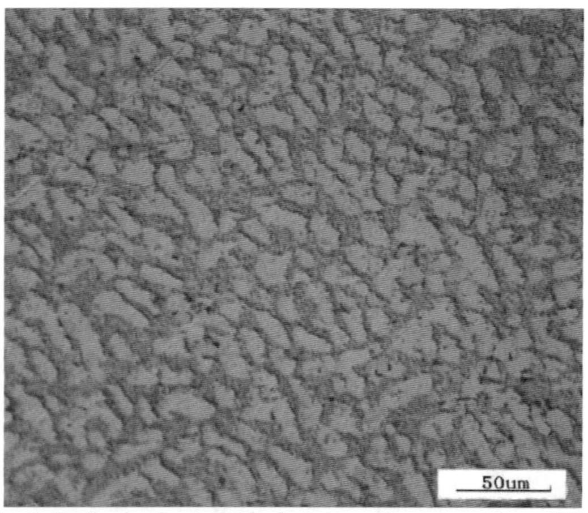

Fig. 1 Optical micrograph showing microstructures of Sn-3.8Ag-0.7Cu lead-free solder without La addition.

It can be seen from La-Sn binary phase diagram [24] that La and Sn form intermetallic $LaSn_3$. When small amounts of La are added, tiny $LaSn_3$ particles form and they can promote heterogeneous nucleation during solidification to refine the microstructures, but excessive La produces large amounts of $LaSn_3$ particles with small inter-particle distance. This may result in coarsening of $LaSn_3$ particles.

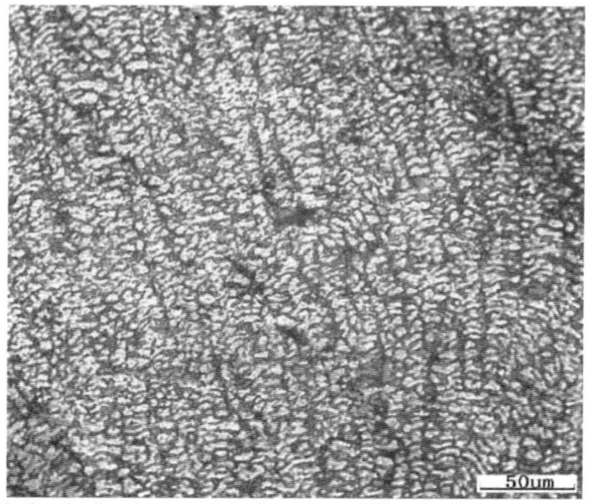

Fig. 2 Optical micrograph showing refined microstructures of Sn-3.8Ag-0.7Cu lead-free solder with 0.5% La addition.

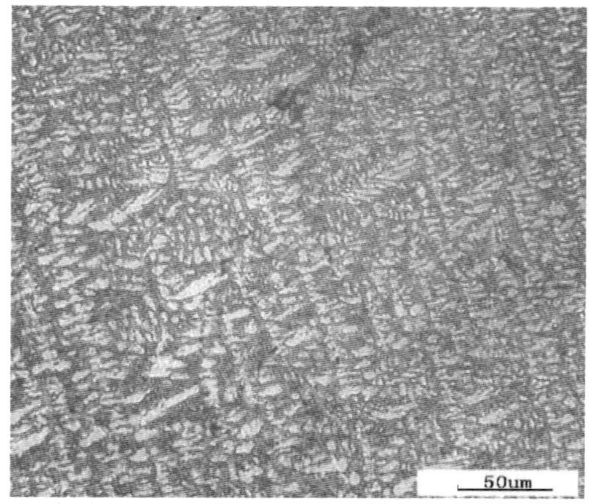

Fig. 3 Optical micrograph showing less refined microstructures of Sn-3.8Ag-0.7Cu lead-free solder with 1% La addition.

Wettability experiment was carried out according to GB11364-89 standard. Copper plate was used as substrate, and 0.2 g of solder alloy was placed on the copper plate and covered by solder paste. Wettability samples were heated at 260 °C in a furnace for 5 min. Photographs were taken and the wetting areas were calculated after cleaning.

As shown in Fig. 4, the wetting area increased considerably (by about 35%) when only 0.05 wt% La was added. However, the increase in wettability gets less when the La concentration increases to 0.1 and 0.2 wt%. After the addition of La reaches 0.5 wt%, the wettability of Sn-3.8Ag-0.7Cu-xLa show little difference from that of the solder without La addition.

Shear strength was tested at a constant displacement rate of 2 mm/min using two pieces of 40mmx20mmx2mm copper plates with lap length of 5 mm. The results are presented in Fig. 5.

Clearly, adding La to the solder leads to some improvement in shear strength (by about 10%) only when the amount of La added is the lowest at 0.05 wt%. It should be especially noted that adding 0.2 wt% or more La to the solder results in drastic reduction of the shear strength from the original 33 MPa to around 25 MPa (i.e., by about 25%).

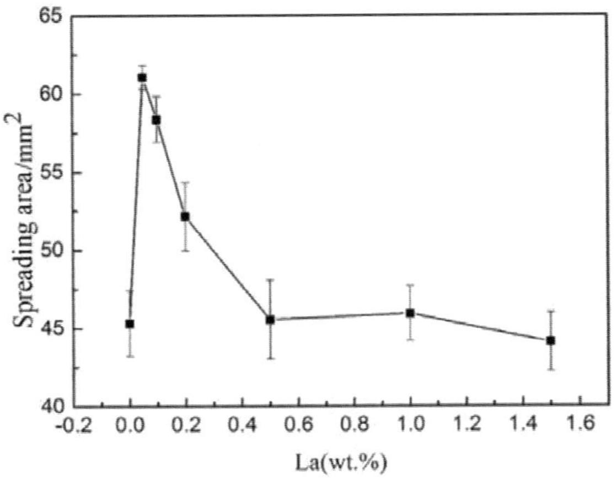

Fig. 4 Wetting areas for Sn-3.8Ag-0.7Cu solder containing various amounts of La.

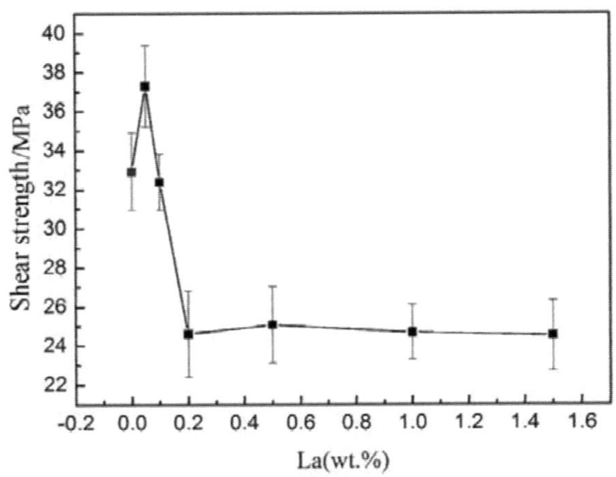

Fig. 5 Shear strength for Sn-3.8Ag-0.7Cu solder containing various amounts of La.

Our experimental results seem to be in strong contrast to those obtained by Zhang et al. [5] (refer to Table 4), but both sets of results point to the same fact that very small amounts of La are better than larger amounts of La in terms of mechanical properties.

Our experimental results are in reasonable agreement with those obtained by Dudek et al. [2, 3, 4] (refer to Table 4), who studied a very similar solder (Sn-3.9Ag-0.7Cu vs Sn-3.8Ag-0.7Cu); however, unfortunately they did not test solders with La concentration below 0.1 wt% and thus did not observe the beneficial effect of adding very small amounts of La.

Effect of Ce on Microstructures and Properties

Dudek et al. studied the effect of La, Ce and Y additions in Sn-3.9Ag-0.7Cu solder alloy [4]. New intermetallic compounds were detected as $RESn_3$. With the three kinds of RE elements added, the shear strength of Sn-3.9Ag-0.7Cu-0.5RE solder alloy dropped while the elongation increased. Similar performance of shear strength of Sn-3.9Ag-0.7Cu-0.5Ce was reported by Xie et al. [7].

The ductility enhancement by Ce addition was attributed to $CeSn_3$ particles serving as microscopic fracture and void nucleation sites. The decrease of ultimate tensile strength was attributed to the formation of CeO_2 particles in Sn matrix [8].

Boh et al. found trace amount of Ce addition (0.1, 0.3 and 0.5 wt%) to Sn-1.0Ag lead-free solder could increase the microhardness and tensile strength [9]. Meanwhile, the microstructure was refined and the wettability was enhanced by Ce addition.

Table 5 Summary of previous researches on effect of Ce

Solder	Ce (wt%)	Tensile (T) / Shear (S) Strength	Micro-structure
Sn-3.8Ag-0.7Cu [10]	Matrix: 13.5 N		Refined
	0.01	↑ 1.1%	
	0.03	↑ 4.4%	
	0.10	↑ 0.9%	
Sn-0.5Cu-0.05Ni [10]	Matrix: 12.7 N		Refined
	0.01	↑ 4.7%	
	0.03	↑ 5.5%	
	0.10	↑ 0.8%	
Sn-3.8Ag-0.7Cu [13]	Matrix: 7.1 N		-
	0.03	↑ 11.3%	
Sn-1Ag [9]	0.1	22 MPa	Refined
	0.3	T ↑ 13.6%	
	0.5	T ↑ 18.2%	
Sn-3.9Ag-0.7Cu [4]	Matrix: 26.8 MPa		-
	0.1	S ↓ 10.8%	
	0.5	S ↓ 17.9%	
Sn-3.8Ag-0.7Cu [8]	Matrix: 39.0 MPa		Refined
	0.006	T ↑ 12.6%	
	0.02	T ↑ 9.2%	
	0.17	T ↓ 15.9%	
Sn-0.7Cu [8]	Matrix: 27.8 MPa		
	0.006	T ↑ 4.3%	
	0.02	T ↑ 1.4%	
	0.17	T ↓ 4.7%	
Sn-58Bi [16]	Matrix: 71MPa		Refined
	0.1	↑ 1.4%	
	0.5	↑ 4.2%	
	1.0	↑ 4.0%	
Sn-3.0Ag-0.5Cu [17]	Matrix: 34.8MPa		Refined
	0.02	↑ 1.5%	
	0.15	↑ 9.0%	
	0.5	↑ 5.4%	

Xue et al. and Zhang et al. systematically studied the effect of rare earth Ce on microstructure, solderability and mechanical properties [10-14]. They found 0.03%-0.05% Ce addition is beneficial to the wettability and mechanical properties.

Chriastel et al. added small amounts of Ce to Sn-3.5Ag-0.7Cu, Sn-0.7Cu and Sn-3.5Ag lead-free solders to improve their properties [15]. Results showed that the microstructure was remarkably refined by Ce addition. Meanwhile, the mean thickness of Cu_6Sn_5 was decreased with Ce addition.

The effect of Ce on microstructures and properties of lead-free solders is summarized in Table 5. All the studies have shown consistently that addition of various amounts of Ce helps to refine the microstructures. However, the results are not so consistent with regard to the element's effect on tensile or shear strength. Therefore, further studies are warranted.

We have added various amounts of Ce to Sn-3.8Ag-0.7Cu and Sn-5Bi-0.7Cu solders in order to study the element's effect on the microstructures and properties. The work is going on. Some preliminary results on Sn-3.8Ag-0.7Cu solder are shown in Figs. 6 and 7, and the results on Sn-5Bi-0.7Cu solder will be published elsewhere.

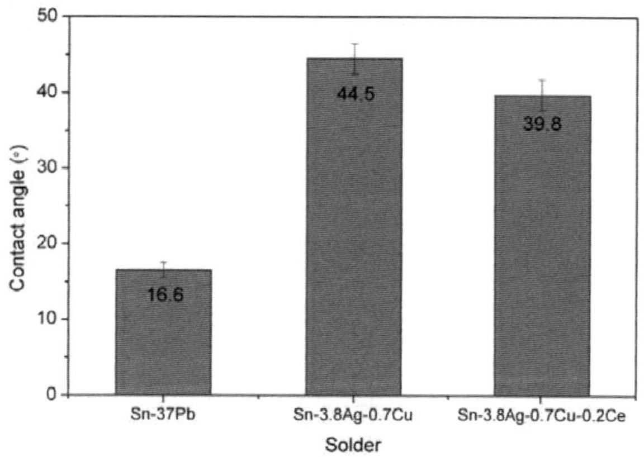

Fig. 6 Effect of 0.2 wt% Ce on contact angle of Sn-3.8Ag-0.7Cu solder.

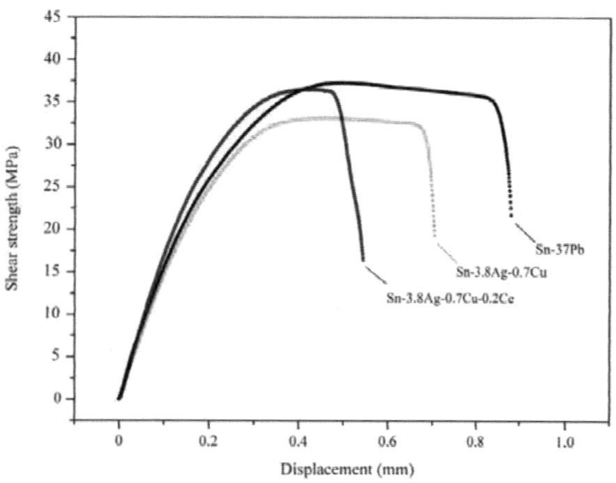

Fig. 7 Effect of 0.2 wt% Ce on shear strength of Sn-3.8Ag-0.7Cu solder.

In Figs. 6 and 7, the results for Sn-37Pb are included as benchmarks for comparison. It can be seen from the figures that addition of 0.2 wt% Ce to Sn-3.8Ag-0.7Cu solder improves the wettability a bit and also increases shear strength but at the expense of ductility. The Ce-alloyed solder still cannot match the properties of the conventional Sn-37Pb solder.

Conclusions

Great effort has been made to obtain a clearer picture on how light rare earth elements La and Ce affect microstructures and properties of lead-free solders. It has been shown consistently that adding the rare earth elements helps to refine microstructures of various types of lead-free solders.

However, it must be noted that the microstructural refinement does not necessarily lead to improvement of the wettability or mechanical properties.

Overall beneficial effects are shown for addition of very small amounts of La (e.g., at 0.05 wt%), but adding larger amounts of La (0.2, 0.5, 1.0, 1.5 wt.%) to Sn-3.8Ag-0.7Cu solder is found to result in drastic decrease of strength by about 25%.

Acknowledgments

Wei Zhou would like to acknowledge financial support from Nanyang Technological University and A*Star.

References

[1] Wu C. M. L., Yu D. Q., Law C. M. T., and Wang L., "Properties of Lead-free Solder Alloys with Rare Earth Element Additions," *Mater Sci Eng R*, Vol. 44, No. 1 (2004), pp. 1-44.

[2] Dudek M. A., Sidhu R. S., and Chawla N., "Novel Rare-Earth-Containing Lead-free Solders with Enhanced Ductility," *JOM*, Vol. 58, No. 6 (2006), pp. 57-62.

[3] Dudek M. A., Sidhu R. S., Chawla N., and Renavikar M., "Microstructure and Mechanical Behavior of Novel Rare Earth-Containing Pb-Free Solders," *J Electronic Mater*, Vol. 35, No. 12 (2006), pp. 2088-2097.

[4] Dudek M. A. and Chawla N., "Effect of Rare-Earth (La, Ce, and Y) Additions on the Microstructure and Mechanical Behavior of Sn-3.9Ag-0.7Cu Solder Alloy." *Metall Mater Trans A*, Vol. 41, (2010), pp. 610-620.

[5] Zhang L., Han J., Guo Y., and He C., "Microstructures and Properties of SnZn Lead-free Solder Joints Bearing La for Electronic Packaging," *IEEE Trans Electron Devices*, Vol. 59, No. 12 (2012), pp. 3269-3272.

[6] Pei M. and Qu J., "Creep and Fatigue Behavior of SnAg Solders with Lanthanum Doping," *IEEE T Compon Pack Tech*, Vol. 31, No. 3 (2008), pp.712-718.

[7] Xie H. X., Chawla N., and Shen Y. L., "Mechanisms of Deformation in High-ductility Ce-containing Sn-Ag-Cu Solder Alloys," *Microelectron Reliab*, Vol. 51, No. 6 (2011), pp. 1142-1147.

[8] Drienovsky M., Trnkova R. L., Martinkovic M., Ozvold M., Cernickova I., Palcut M., and Janovec J., "Influence of

Cerium Addition on Microstructure and Properties of Sn-Cu-(Ag) Solder Alloys," *Mater Sci Eng: A*, Vol. 623, No. 19 (2015), pp. 83-91.

[9] Noh B., Choi J., Yoon J., and Jung S., "Effects of Cerium Content on Wettability, Microstructure and Mechanical Properties of Sn-Ag-Ce Solder Alloys," *J Alloy Compound*, Vol. 499, No. 2 (2010), pp. 154-159.

[10] Wang, J. X., Xue S. B., Han Z. J., Yu S. L., Chen Y., Shi Y. P., and Wang H., "Effects of Rare Earth Ce on Microstructures, Solderability of Sn-Ag-Cu and Sn-Cu-Ni Solders as well as Mechanical Properties of Soldered Joints," *J Alloy Compound*, Vol. 467, No. 1 (2009), pp.219-226.

[11] Zhang L., Xue, S. B., Gao L. L., Dai W., Ji F., Chen Y., and Yu S. L., "Microstructure Characterization of SnAgCu Solder Bearing Ce for Electronic Packaging," *Microelectron Eng*, Vol. 88, No. 9 (2011), pp. 2848-2851.

[12] Zhang L., Xue, S. B., Gao L. L., Sheng Z., Yu S. L., Chen Y., Dai W., Ji F., and Guang Z., "Reliability Study of Sn-Ag-Cu-Ce Soldered Joints in Quad Fflat Packages," *Microelectron Reliab*, Vol. 50, No. 12 (2010), 2071-2077.

[13] Zhang L., Xue, S. B., Gao L. L., Sheng Z., Guang Z., Chen Y., and Yu S. L., "Properties of SnAgCu/SnAgCuCe Soldered Joints for Electronic Packaging," *J Mater Sci: Mater Electron*, Vol. 21, No. 6 (2010), pp. 635-642.

[14] Zhang L., Xue, S. B., Chen Y., Han Z. J., Wang J. X., Yu S. L., and Lu F. Y., "Effects of Cerium on Sn-Ag-Cu Alloys Based on Finite Element Simulation and Experiments," *J Rare Earth*, Vol. 27, No. 1 (2009), pp. 138-144.

[15] Chriastel J., Trnkova R. L., Dimova P. K., and Ozvold M., "Reaction of Liquid Sn-Ag-Cu-Ce Solders with Solid Copper," *J Electron Mater*, Vol. 40, No. 9 (2011), pp. 1956-1961.

[16] Bang J. O., Lee H. S., and Jung S. B., "Fabrication of Sn-58Bi-xCe Solder Paste Using Electrical Wire Explosion Method and Its Mechanical Property," *J Nanosci Nanotech*, Vol. 17, No. 5 (2017), pp. 3033-3037.

[17] Tu X., Yi D., Wu J., Wang B., "Influence of Ce Addition on Sn-3.0Ag-0.5Cu Solder Joints Thermal Behavior, Microstructure and Mechanical Properties," *J Alloy Compound*, Vol. 698, (2017), pp. 317-328.

[18] Wu C. M. L., Yu, D. Q., Law C. M. T., and Wang L., "Microstructure and Mechanical Properties of New Lead-Free Sn-Cu-RE Solder Alloys," *J Electron Mater*, Vol.31, No.9, (2002), pp. 928-932.

[19] Wu C. M. L., Yu, D. Q., Law C. M. T., and Wang L., "Improvements of Microstructure, Wettability, Tensile and Creep Strength of Eutectic Sn-Ag Alloy by Doping with Rare-earth Elements," *J Mater Res*, Vol. 17, No. 12 (2002), pp. 3146-3154.

[20] Wu C. M. L., Yu, D. Q., Law C. M. T., and Wang L., "The Properties of Sn-9Zn Lead-Free Solder Alloys Doped with Trace Rare Earth Elements," *J Electron Mater*, Vol. 31, No. 9 (2002), pp. 921-927.

[21] Wang L., Yu D.Q., Zhao J., and Huang M. L., "Improvement of Wettability and Tensile Property in Sn-Ag-RE Lead-free Solder Alloy," *Mater. Lett.*, Vol. 56, No. 6 (2002), pp. 1039-1042.

[22] Yu D. Q., Zhao J., and Wang L., "Improvement on the Microstructure Stability, Mechanical and Wetting Properties of Sn-Ag-Cu Lead-free Solder with the Addition of Rare Earth Elements." *J Alloy Compound*, Vol. 346, No. 1-2 (2004), pp. 170-175.

[23] http://mineralprices.com/default.aspx#rar

[24] Baker H, Okamoto H. Alloy Phase Diagrams (ASME International, ASME Handbook), 1992, pp. 1065-1081.

Stress-strain response of lead-free solder joints in MEMS device using finite element code

Liang Zhang
School of Mechatronic Engineering, Jiangsu Normal University, Xuzhou 221116, China;
Zhi-quan Liu
Institute of Metal Research, Chinese Academy of Sciences, Shenyang110016, China.
Yu-tong Ji
School of Information and Art, Jiangsu Vocational Institute of Architectural Technology, Xuzhou 221116, China.

Abstract

Based on finite element simulation, the stress-strain response of Sn3.8Ag0.7Cu0.03Ce, Sn3.8Ag0.7Cu, Sn0.7Cu solder joints in MEMS device after reflow soldering under thermal cycles loading (-55℃~125℃). It is found that stress-strain can be influenced by the solder materials obviously, the Sn3.8Ag0.7Cu0.03Ce solder joints represents priority property among the three solder materials, and the weakest solder joint can be found in the solder joints array. Based on fatigue life prediction model, combing the data of finite element simulation, the fatigue life of Sn3.8Ag0.7Cu0.03Ce, Sn3.8Ag0.7Cu, Sn0.7Cu solder joints in MEMS device can be calculated, and the fatigue life of Sn3.8Ag0.7Cu0.03Ce solder joint is demonstrated to be the highest of all.

Introduction

Rapid growth of microelectronics, photonics, biomedical, micro-electromechanical systems (MEMS), and nano-electromechanical systems (NEMS) technologies have facilitated the development of new micro-interconnection materials, which play a significant role in reliability of these systems[1]. In electronic industry, the lead-free solders can be utilized as interconnection materials for mechanical support and electrical connection[2], among series of lead-free solders, SnAgCu solder are proposed as the one of the best alternative lead-free alloys for the traditional SnPb alloys because of their good soldering and wetting behavior on several substrate materials[3].

In order to investigate the reliability of lead-free solder joints in electronic device, finite element simulation was utilized by lots of researchers. Mustafa[4] simulated the stress and strain in the Sn1.0Ag0.5Cu solder layer, based on the finite element simulation, the suitable layer thickness can be obtained. Chen[5] analyzed the reliability of Sn3.8Ag0.7Cu solder joints in QFP device under random vibration test, the results shown that solder fatigue failure is more critical than Cu lead failure for the PQFP assembly. Zhang[6] investigated the thermos-mechanical reliability of SnAgCu, SnPb, SnAgCuAl solder joints in FCBGA device using finite element simulation, which can be used to determine the effect of different control factors on the reliability of solder joints, it was demonstrated that the solder materials was the most important factor among the control factors. Nourani[7] studied the effect of processing parameters on fracture toughness of lead-free solder joints (Sn3.0Ag0.5Cu, Sn3.8Ag0.7Cu, Sn0.3Ag0.7Cu), the effects of solder thickness and local end geometry were found to be significant only at the higher strain rates. Chen[8] demonstrated that doping 0.02wt.% Ni in Sn1.0Ag0.5Cu solder can increase the fatigue resistance and make it has a comparable fatigue life with the high Ag content SnAgCu solders.

In this work, Sn3.8Ag0.7Cu0.03Ce, Sn3.8Ag0.7Cu, Sn0.7Cu solders were selected as interconnection ball in MEMS device, the thermal fatigue reliability of the three kinds of solder joints was analyzed by finite element code, which can help to find the appropriate solder for MEMS device.

FEM analysis

The a quarter finite element model for 9 × 9 BGA assembly is shown in Fig.1, in this model, chip, solder joints, Cu pads and FR-4 were contained and established. Considering the symmetry of the whole structure of BGA structure, only 1/4 finite element model was established in this study. In the modeling, the SOLID186 elements (56149) and sweep meshing method were used in this structure for the five materials. In order to reduce the solution time, each solder interconnection can be roughly meshed in the global model while the finer mesh can be applied for the diagonal solder joints, especially for the corner solder joints.

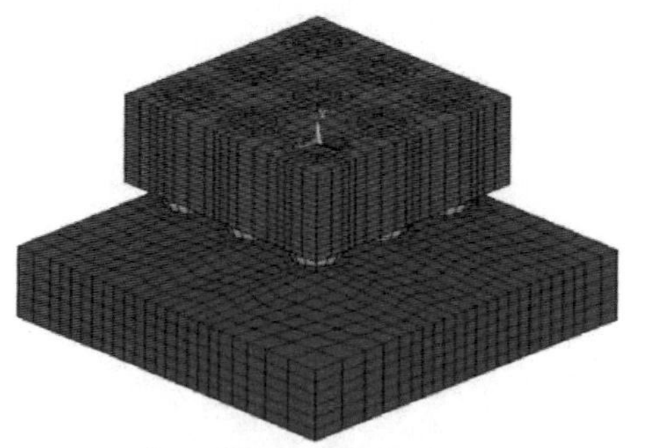

Fig.1 1/4 Finite element model

In this work, due to the high homologous temperature, Sn-0.7Cu, Sn-3.8Ag-0.7Cu and Sn-3.8Ag-0.7Cu-0.03Ce solders are assumed to be creep deformation, and the solder creep behavior is temperature and time dependent. Creep is considered as the primary damage mechanism for solders during thermal cycling and is solely used to simulate the material behavior[9]. Other materials (chip, Cu pads and FR-4 substrate) in BGA structure are assumed to be isotropic, linearly elastic and independent of temperature. For the steady-state creep behavior of solders, Garofalo-Arrhenius model has been proposed as a good choice to describe the creep behavior of solders as Eq.(1) [10].

$$\frac{d\gamma}{dt} = C\left(\frac{G}{T}\right)\left[\sinh\left(\omega\frac{\tau}{G}\right)\right]^n \exp\left(-\frac{Q}{RT}\right) \qquad (1)$$

where $\frac{d\gamma}{dt}$ is the creep shear strain rate, C is the materials constant , τ is the shear stress, G is the shear modulus, T is the absolute temperature, ω is the stress level, n is the stress exponent, R is the gas constant.

If the Sn-0.7Cu, Sn-3.8Ag-0.7Cu and Sn-3.8Ag-0.7Cu-0.03Ce solders obey the von Mises criterion, the creep equation can be rearranged and expressed as follows:

$$\dot{\varepsilon} = C_1\left[\sinh(C_2\sigma)\right]^{C_3} \exp\left(-\frac{C_4}{T}\right) \qquad (2)$$

where $C_1 \sim C_4$ are the materials constants, $\dot{\varepsilon}$ is the equivalent creep strain rate, σ is the equivalent stress, T is the absolute temperature. After curve-fitting, the parameters for Sn-3.8Ag-0.7Cu-0.03Ce, Sn3.8Ag-0.7Cu and Sn-0.7Cu solders can be determined[11,12], as shown in follows:

Sn-3.8Ag-0.7Cu-0.03Ce:

$$\dot{\varepsilon}=284000\left[\sinh(0.02432\sigma)\right]^{6.1} \exp\left(-\frac{6400}{T}\right) \qquad (3)$$

Sn-3.8Ag-0.7Cu:

$$\dot{\varepsilon}=501.3\left[\sinh(0.316\sigma)\right]^{4.96} \exp\left(-\frac{5433.5}{T}\right) \qquad (4)$$

Sn-0.7Cu:

$$\dot{\varepsilon}=4.35\left[\sinh(0.096\sigma)\right]^{3.7} \exp\left(-\frac{4835.2}{T}\right) \qquad (5)$$

The entire structure was subjected to a time dependent temperature distribution to reflect thermal cycling performed in the finite element simulation, which can show the response of stress-strain of solder joints in electronic device. The cyclic temperature loading (218K-398K) as shown in Fig.2 was proposed to evaluate the reliability of lead-free solder joints in device[13], and the thermal cyclic loading was selected to be imposed on the BGA assembly. The temperatures ranging from 218K to 398K was carried out for the testing, and the reference temperature (initial temperature) is 298K, dwell time at all peak temperature is 15min, the rates of descend and ascend temperature are 12 K/min. Moreover, it is assumed that in the finite element simulation no stress occurs at the initial temperature. For the thermal cycle loading, at all nodes in the BGA structure were loaded a uniform temperature distribution.

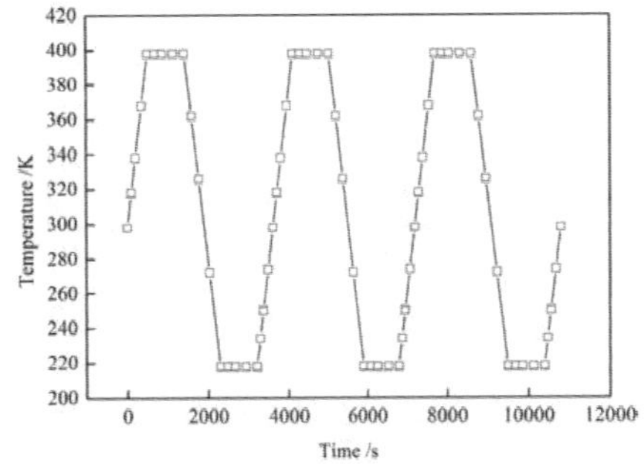

Fig.2 Cyclic temperature loading

Results and Discussion

The most critical solder joint is the outermost corner joint, as indicated in Fig. 3~Fig.5, the von Mises stress, equivalent creep strain and creep strain energy density occur at the edge of the same solder joint that located on the diagonal of the BGA structure. It is found that the stress-strain distribution in the Sn-0.7Cu solder joints are not uniform, and the interface between the top of chip substrate and solder joint is the high stress–strain region, and the maximum stress/stain occur at the edge area of the solder joint which is farthest from the centre of the BGA structure, and this solder joint is then defined as the key solder joint that is the easiest to be failed under thermal cycles. And there is no essential difference in the stress distribution between the Sn-0.7Cu, Sn-3.8Ag-0.7Cu and Sn-3.8Ag-0.7Cu-0.03Ce solder joints, only the stress value differs.

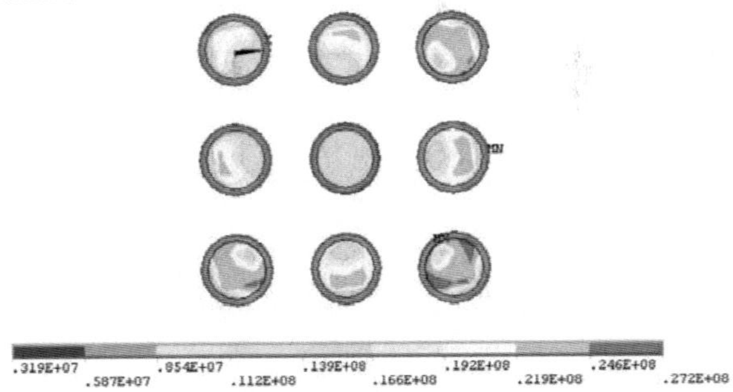

Fig.3 von Mises stress distribution of solder joints

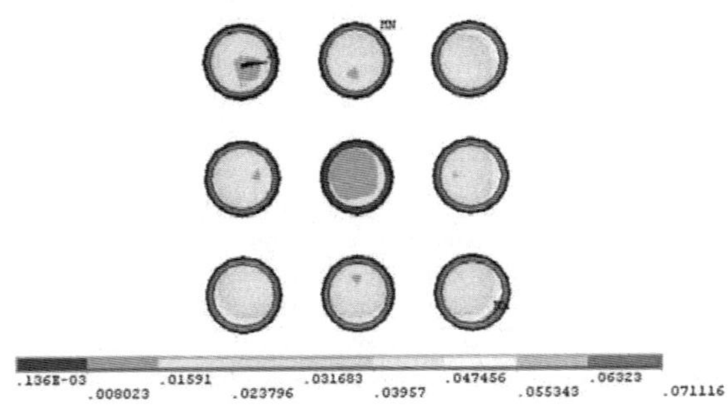

Fig.4 Creep strain distribution of solder joints

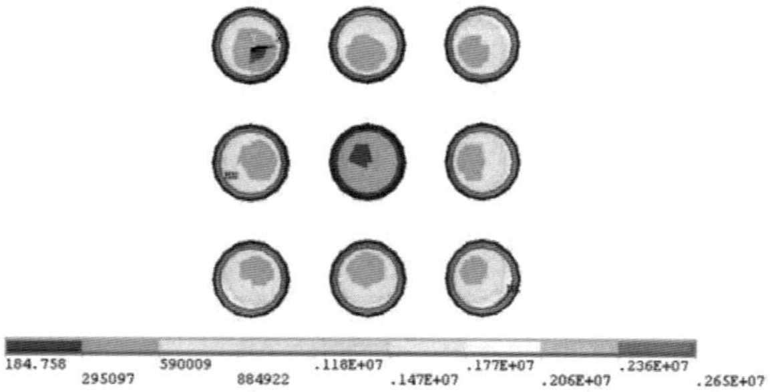

Fig.5 Creep strain energy density distribution of solder joints

In service of thermal cycles loading, the stress will change with the increase of loading time. Fig.6 plots the history of von Mises stress at the most critical locations of all studied BGA assembly based on creep model for Sn-0.7Cu, Sn-3.8Ag-0.7Cu and Sn-3.8-Ag-0.7Cu-0.03Ce solder joints after three cycles, respectively. It is found that the tendency of curves based on Garofalo-Arrhenius is similar to that of curves based on Anand equations [14]: stress relax appears at the initial stage of temperature loading, periodic variety can also be found in the figure. In addition, it is observed that the first cycle behavior is different from others, which indicates that the initial stress-strain response of solder joints is not stable. Sn-3.8Ag-0.7Cu-0.03Ce solder joints in BGA device show the smallest von Mises stress among the lead-free solder joints, moreover, the von Mises stress of Sn-3.8Ag0.7Cu solder joints is lower than that of Sn-0.7Cu solder joints, which can be found that the addition of 3.8Ag and 0.03Ce can decrease the von Mises stress of solder joints during thermal cycling loading.

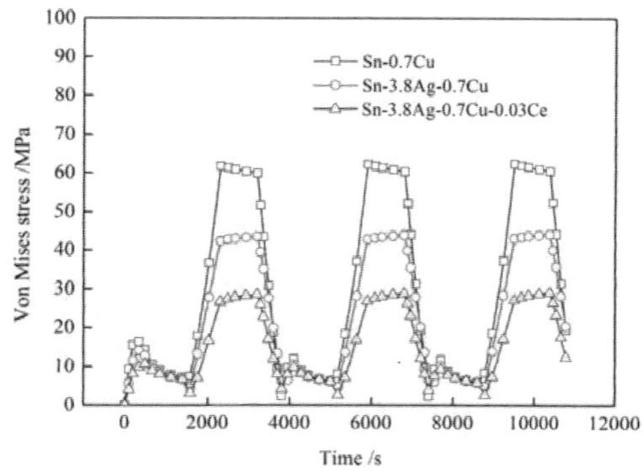

Fig.6 von Mises stress history of different solder joints

Fig.7 plots the equivalent creep strain of different solder joints in BGA device. It is observed that the equivalent creep strain of all lead-free solder joints increase as the temperature cycle proceeds, and the thermally induced cyclic creep deformation tends to accumulate as the number of the thermal cycle increase. The Sn-3.8Ag-0.7Cu-0.03Ce lead-free solder joints show the more creep resistant than Sn-3.8Ag-0.7Cu and Sn-0.7Cu solder joints, from the creep strain history response, the Sn-3.8Ag-0.7Cu-0.03Ce solder joints represent superiority of creep resistance among the three lead-free solder joints. Fig.8 shows the creep strain energy density of solder joints, the similar phenomenon can be found with equivalent creep strain. Therefore, it is safe to conclude that the creep resistant of Sn-0.7Cu solder joints are indeed improved by adding minute amounts of Ag and Ce.

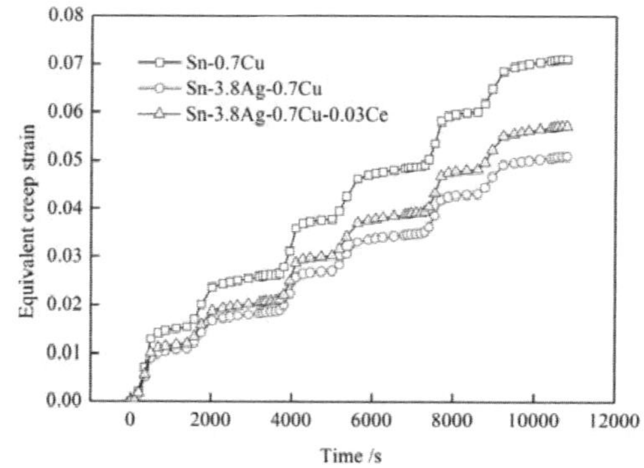

Fig.7 Equivalent creep strain history of different solder joints

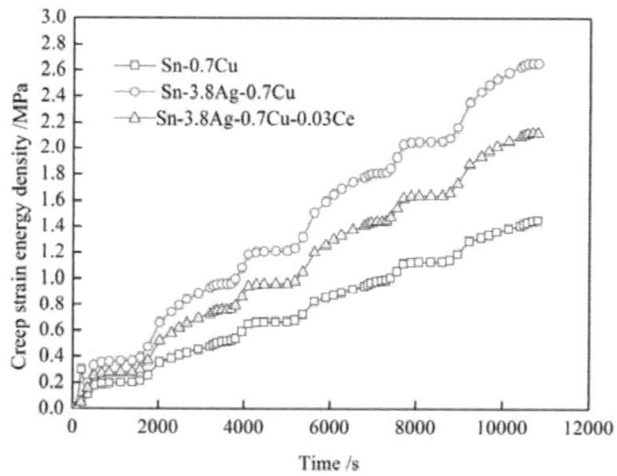

Fig.8 Creep strain energy density history of different solder joints

In order to calculate the fatigue life of Sn-0.7Cu, Sn-3.8Ag-0.7Cu and Sn-3.8Ag-0.7Cu-0.03Ce solder joints in MEMS device with BGA structure, the fatigue life model of solder joints calculated based on accumulated creep strain per cycles is shown as Eq. (6), for traditional SnPb solder, the θ is 0.69, C is -0.18, the fatigue life of SnPb solder joints can be calculated based on Eq.(6), combing with Eq.(7), the fatigue life of Sn-0.7Cu, Sn-3.8Ag-0.7Cu and Sn-3.8Ag-0.7Cu-0.03Ce solder joints can be determined, are 1460, 1630, 1853 cycles, respectively, which the addition of 3.8Ag can enhance the fatigue life of Sn-0.7Cu solder joints with 11.6% amplification, and the addition of 0.03Ce can increase the fatigue life of Sn3.8Ag-0.7Cu solder joints with 13.7% amplification.

$$N_f = \theta\left(\varepsilon_{acc}\right)^C \quad (6)$$

where N_f is the fatigue life of solder joints, C is accumulated creep strain per cycles, θ is the material constant.

$$\frac{N(lead-free)}{N(Sn37Pb)} = \left[\frac{\Delta\varepsilon(lead-free)}{\Delta\varepsilon(Sn37Pb)}\right]^C \quad (7)$$

Conclusions

Finite element simulation was utilized to analyze the stress-strain response of Sn-0.7Cu, Sn-3.8Ag-0.7Cu and Sn-3.8Ag-0.7Cu-0.03Ce solder joints with BGA structure in MEMS device, the results indicates that the Sn-3.8Ag-0.7Cu-0.03Ce lead-free solder joints show lower von Mises stress and equivalent creep strain, which result in the high fatigue life under thermal cyclic loading, increasing in the order Sn-0.7Cu, Sn-3.8Ag -0.7Cu and Sn-3.8Ag-0.7Cu-0.03Ce.

Acknowledgments

This work was supported by the National Natural Science Foundation of China (No. 51475220), the Qing Lan Project, the China Postdoctoral Science Foundation funded project(2016M591464) and Six talent peaks project in Jiangsu Province(XCL-022).

References

[1] Tsukamoto, H., Nishimura, T., Suenaga, S., McDonald, S, D., Sweatman, K, W., Nogita, K., "The influence of solder composition on the impact strength of lead-free solder ball grid array joints," *Microelectronics Reliability*, Vol. 51, No.3(2011), pp. 657-667.

[2] Zhang, L., Han, J, G., Guo, Y, H., Sun, L., "Properties and microstructures of SnAgCu-xEu alloys for concentrator silicon solar cells solder layer," *Solar Energy Materials and Solar Cells*, Vol.130 (2014), pp.397-400.

[3] Zhang, L., Liu, Z, Q., Yang, F., Zhong, S, J., "Cu_6Sn_5 whiskers precipitated in Sn3.0Ag0.5Cu/Cu interconnection in concentrator silicon solar cells solder layer," *Materials*, Vol. 10, No. 4(2017), pp.327.

[4] Mustafa, M., Suhling, J, C., Lall, P., "Experimental determination of fatigue behavior of lead-free solder joints in microelectronic packaging subjected to isothermal aging," *Microelectronics Reliability*, Vol. 56(2016), pp.136-147.

[5] Chen, F, X., Pang, J, H, L., "Study on reliability of PQFP assembly with lead free solder joints under random vibration test," *Microelectronics Reliability*, Vol.55, No.12 (2015), pp.2769-2776.

[6] Zhang, L., Sun, L., Zhong, S, J., Ma, J., Bao, L., "Reliability of Pb-free solder joints in FCBGA using finite element simulation and Taguchi method," *16th International Conference on Electronic Packaging Technology*, 2015, pp.197-200.

[7] Nourani, A., Spelt, J, K., "Effect of processing parameters on fracture toughness of lead-free solder joints," *Engineering Fracture Mechanics*, Vol.142(2015), pp.64-78.

[8] Chen, F, X., "Materials characterization and low cycle fatigue model of low Ag content lead-free solder," *IEEE Transactions on device and Materials Reliability*, Vol.17, No.2, pp.298-306.

[9] Li, X, Y., Wang, Z, S., "Thermo-fatigue life evaluation of SnAgCu solder joints in flip chip assemblies," *Journal of Materials Processing Technology*, Vol.183, No.1(2007), pp.6-12.

[10] Lau, J, H., Lee, S, W, R., "Modeling and analysis of 96.5Sn-3.5Ag lead-free solder joints of wafer level chip scale package on buildup microvia printed circuit board," *IEEE Transactions on Electronics Packaging Manufacturing*, Vol.25, No.1(2002), pp.51-57.

[11] Zhang, L., Xue, S, B., Gao, L, L., Zeng, G., Chen, Y., Yu, S. L., Sheng, Z., "Creep behavior of SnAgCu solders with rare earth Ce doping," *Transactions of Nonferrous Metals Society of China*, Vol.20, No.3, pp.412-417.

[12] Pang, J, H, L., "Lead-free solder," *Springer*, 2012.

[13] Zhang, L., Sun, L., Guo, Y, H., He, C, W., "Reliability of lead-free solder joints in CSP device under thermal cycling," *Journal of Materials Science: Materials in Electronics*, Vol.25, No.3(2014), pp.1209-1213.

[14] Zhang, L., Cui, J, H., Han, J, G., He, C, W., Guo, Y, H., Yuan, J, M., "Finite element analysis of SnAgCu (Zn, Co, Fe) lead-free solder joints for electronic packaging," *International Journal of Nonlinear Sciences and Numerical Simulation*, Vol.15, No3-4(2014), pp.197-206.

[15] Hong, B, Z., "Thermal fatigue analysis of a CBGA package with lead-free solder fillets," *Intersociety Conference on Thermal Phenomena*, 1998, pp.205-211.

Wetting kinetics and spreading phenomena of Sn-35Bi-1Ag solder on the T2 substrate

Yulong Li, Zhiliang Wang, Weifeng Long*
Key Lab for Robot &Welding Automation of Jiangxi Province,Mechanical & Electrical Engineering School, Nanchang University, Nanchang, 330031, China

Abstract

Wetting kinetics of Sn-35Bi-1Ag solder on randomly polished Cu plate was studied. Two temperatures of 473K and 553K with a same dwell time of 1min were selected as the processing parameter. In the experiments, the solder was put on a heating platform installed in an atmosphere controlled chamber, two sets of cameras were used to *insitu*/on time monitor the spreading process from the side and the top of the melting sessile drop. After processing the videos, the contact angle, spreading area and associated equivalent radius of the melting solder can be acquired, simultaneously. Results showed that the increase of test temperature obviously promoted the spreading of Sn-35Bi-1Ag solder on the T2 Cu substrate, although the processes were similarly characterized into four stages: i) an initial stage, ii) a rapid spreading stage, iii) a limited spreading stage and iv) an asymptotic stage. According to the analysis, the kinetics of the rapid spreading stage may be determined by the chemical reaction rate of the Cu and Sn characterized by the $R^n \sim t$ model with a n value of ~1. For the limited spreading stage, the kinetics can also be characterized by the $R^n \sim t$ model with a n value of ~4 at the test temperature of 473K, in which the diffusion rate may determine the process; while at the test temperature of 553K, it may be controlled by the combined effects of dissolving driven wetting and interficial reaction, $n \sim 3$. The evolution of the dynamic contact angle θ with the spreading time t can be expressed with the equation, $\cos\theta_F - \cos\theta = (\cos\theta_F - \cos\theta)e^{(-kt)}$, in which θ_F, θ_0 were the final coantact angle and initial contact angle, k was the constant, respectively. With the increase of test temperature, the constant k increased, and the spreading rate went faster.
Keyword: Sn-35Bi-1Ag; soldering; kinetics

Introduction

Sn-Pb based filler alloys are the dominant electronic solders widely used in electronic packaging because of their unique combination of material properties and low cost [1]. In spite of having these properties, recently increased environmental and human health concerns over the toxicity of lead and lead-containing compounds have banned the use of Sn-Pb solder alloys [2]. Therefore, the development of new lead-free solder alloys has become quite important to meet its growing demand for modern technology [3]. Moreover, the trends toward miniaturizationand higher packing density of environmental friendly microelectronic have also motivated to develop lead-free alloys to fulfill the requirements. Recently, several research groups have concentrated on developing binary and ternary Sn-based lead-free solders such as Sn-9Zn (199 °C), Sn-8Zn-3Bi (189 °C), Sn-3.5Ag (221 °C), Sn-0.7Cu (227 °C) and Sn-3.0Ag-0.5Cu (217 °C) for the application in electronic packaging systems [4-6]. However, most of the solder alloys have higher melting temperature that can damage components as well as printed circuit boards (PCBs) during soldering process. Among them, the melting temperature of eutectic Sn-9Zn based solder alloys is as same as eutectic Sn-Pb solders. However, their poor oxidation resistance and dross formations inhibit their applications in modern packaging systems [7-8]. Among lead-free solder alloys, Sn-35Bi-1Ag solder alloys is a potential substitute, owing to the low-melting point (178 °C), which is close to that of Sn-Pb (183 °C), and there is no Pb [9]. 1% of Ag in Sn-35Bi-1Ag solder can refine the microstructure [10], with excellent mechanical properties and wettability [11].

For the soldering process, the wetting phenomenon of the solder on the substrate is very important. However, the soldering process of the solder on the substrate is accompanied by the reaction, diffusion and dissolution of the interface [12]. Therefore, it is difficult to establish a comprehensive theory to describe the wetting phenomenon. In spite of the difficulties, the wetting and spreading of the solder on different substrates can provide guidance for the actual production process. In the existing literatures, (i) recording the melting sessile drop from the side for analyzing the dynamic contact angle verus time and (ii) recording the melting sessile drop from the top for analyzing the spreading area or equivalent radius verus time are often used to characterize the dynamic wetting processes. However, few studies have studied the above two characteristics of a wetting process, simultaneously. If the dynamic contact angle and the spreading area/ equivalent radius were monitored at the same time, the researchers will have a much better understanding of the spreading process. Therefore, an experimental system was assembled, in wich a heating platform was installed in an atmosphere controlled chamber, two sets of cameras were used to real-time monitor the melting sessile drop from the side and the top, simultaneously. After experiments, analysis the videos, the contact angle and equivalent radius verus time can be acquired, simultanuously. Then, the wetting kinetics of Sn-35Bi-1Ag solder on the pure Cu substrate at different temperatures were preliminarily studied.

Materials and experimental procedure

The assemled experimental system, featuring a heating platform assistted with two sets of cameras can be used to monitor the triple line movement from the side and the top of the melting sessile drop, simultaneously, as shown in Fig. 1. The lead-free solder of Sn-35Bi-1Ag solder was in the form of a paste, mixturing of the solder powders and rosin fulx. The substrate was the random polished T2 Cu plate with a geometric size of 20mm × 20mm × 2mm.

Wetting and spreading experiments were performed on the heating platform in a controlled chamber filling with high-purity argon atmosphere (99.999 % purity). Two temperatures, 473K and 553K with a same dwell time of 1min, were chosen to preliminarily study the effects of the temperature on the spreading process. For each temperature, three times of the experiments were repeated. During the wetting and spreading, the melting sessile drop movement were recorded from the beginning of the solder melting to the solidification.

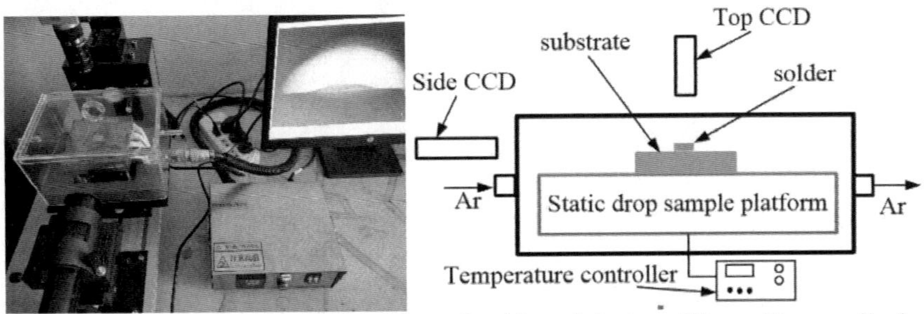

Fig. 1 The assemled experimental system, taking the video from the side and the top of the melting sessile drop for simultaneously acquiring the contact angle and the equivalent radius

Results and discussion

Spreading phenomena and data

To investigate the effects of the temperature on the spreading process of the Sn-35Bi-1Ag solder over the randomly polished Cu substrate, two soldering temperatures of 473K and 553K were employed. Figure 2 shows a series of instantaneous frames from a top CCD camera video of the Sn-35Bi-1Ag solder over the Cu substrate at 473K and 553K. As shown in Fig. 2, before the soldering alloys melted, the solder is in the form of a paste, mixturing of the solder and rosin flux. As the temperature rises to the melting point of the flux, the flux begins to spread and the equivalent radius of the paste increases slightly. As the temperature rises near the melting point of the solder, it has a slight shrinkage at the time of initial melting. Upon further increases in the temperature, the solder is advanced at a significant spreading rate until the maximum spreading limit is achieved. Ultimately, the solder gradually cooled and solidified. In addition, as seen in Fig. 2 a and b, it can be clearly observed that the equivalent radius increases and the spreading time decreases with increasing soldering temperature.

In order to more clearly characterize the spreading phenomenon of the Sn-35Bi-1Ag solder over the Cu substrate at different temperatures. The contact angle is recorded in real time by using a side CCD camera video, as can be seen in Fig. 3. Before the soldering alloys melted, the contact angle of the solder is approximately 90°. As the temperature rises to the melting point of the flux, the flux begins to spread and the contact angle decreases gradually. When the temperature rises near the melting point of the solder, the solder quickly form a metallic liquid and slightly shrinks. Upon further increases in the temperature, the molten solder continues to spread forward until the minimum contact angle is achieved. Meanwhile, the similar conclusions can be obtained from Fig. 2 a and b, the final The video captured from the top was processed for acquiring a curve of equivalent radius verus time; additionlly, synchronized video taken form the side was processed for acquiring the the dymanic contact angel verus time. The wetting and spreading characteristics of the solder on the T2 Cu substrate can be analyzed from the dymanic contact angel and the equivalent radius. Scanning electron microscopy (SEM, FEIQuanta 200F) and energy spectrum analysis (EDS, IE250X Max50) were used for observation the the soldered interface cross-section.

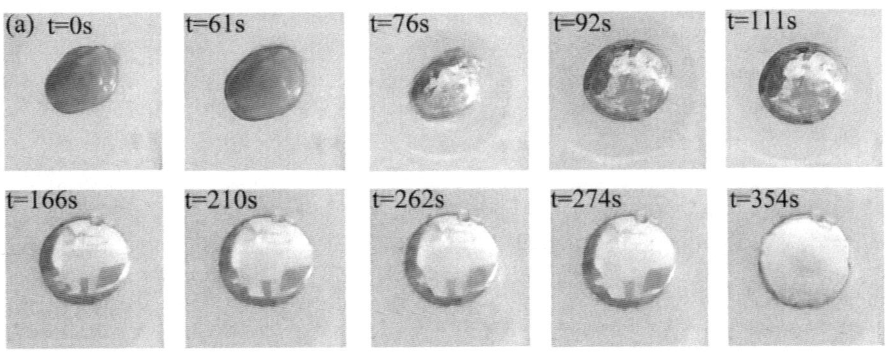

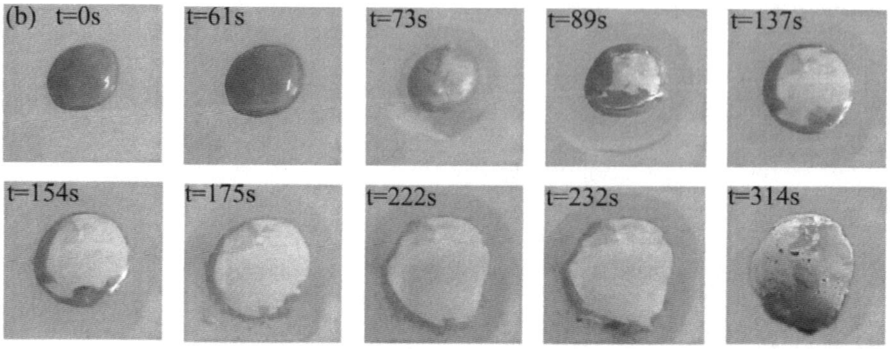

Fig. 2 A sequence of instantaneous frames during the Sn-35Bi-1Ag solder on Cu substrate with different soldering temperatures: (a) 473K, (b) 553K

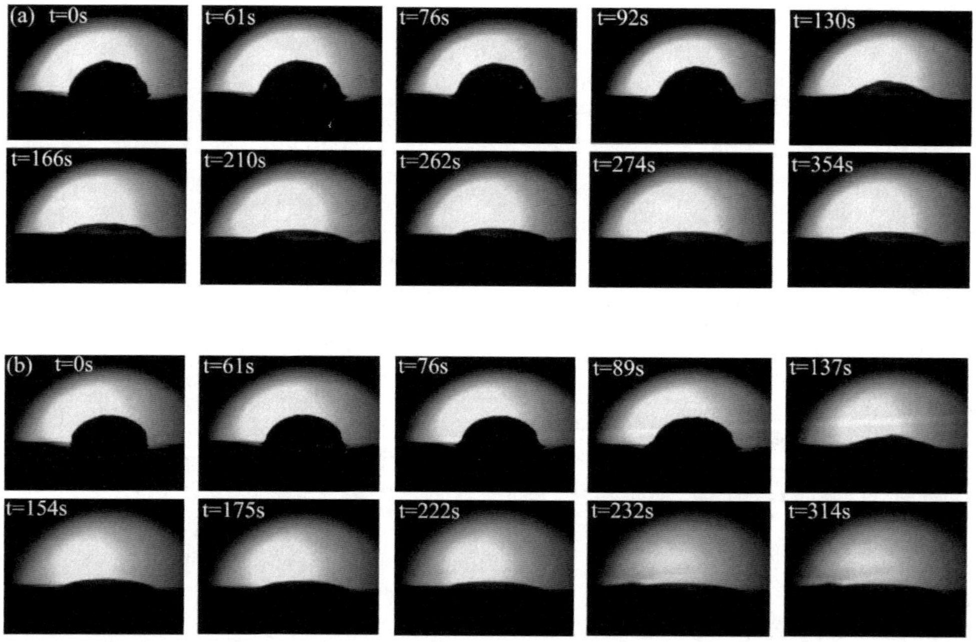

Fig. 3 A sequence of instantaneous frames during the Sn-35Bi-1Ag solder on Cu substrate with different soldering temperatures: (a) 473K, (b) 553K

Figure 4 presents the curve of the equivalent wetting radius and the contact angle versus time for the Sn-35Bi-1Ag solder on Cu substrate at 473K and 553K. As shown in Fig. 4, it can be noticed that the spearding process of the Sn-35Bi-1Ag solder on the Cu substrate are similar at test temperatures of 473K and 553K. The entire spreading process can be divided into four stages: (i) an initial stage, as the temperature rises to the melting temperature of the flux, the flux begins to spread. The active agent in the flux reacts with the oxide film on the substrate. Meanwhile, with the spreading of the flux, the solder has a slight flow. As the temperature further increases near to the melting point of the solder, the solder quickly forms a metallic liquid and slightly shrinked. The contraction process can not be clearly observed from the contact angle, but it can be clearly represented in the top camera video. The duration of the contraction process was approximately $10^0 \sim 2 \times 10^1$ s. The reasons for this phenomenon are as follows: a) there was an oxide film on the surface of the substrate. b) the cohesion force of Sn-35Bi-1Ag induced by the free energy of the liquid-solid interfacial tension LV near the melting point is greater than that of the adhesion force of the liquid-solid phase [13]. (ii) a rapid spreading stage, it is characterized by a rapid increase of a high spreading rate. The transition from poor initial spreading to rapid spreading is mainly attributed to the Cu-Sn phase formed by the reaction of Sn and Cu at the interface, resulting in a decrease of the interfacial tension and interface reaction of the release of free energy on the promotion of wetting and spreading [14]. (iii) a limited spreading stage, at this stage, there is not enough liquid solder in direct contact with the substrate to maintain the dissolution of the element and the chemical reaction [15]. the spreading rate is significantly reduced from the last stage. (iv) an asymptotic stage, as Sn content in Sn-35Bi-1Ag solder is further reduced, the reaction between the substrate and the Sn-35Bi-1Ag filler metal is terminated. As a result, the spreading of the solder gradually terminated and the largest spreading area and the smallest contact angle were obtained.

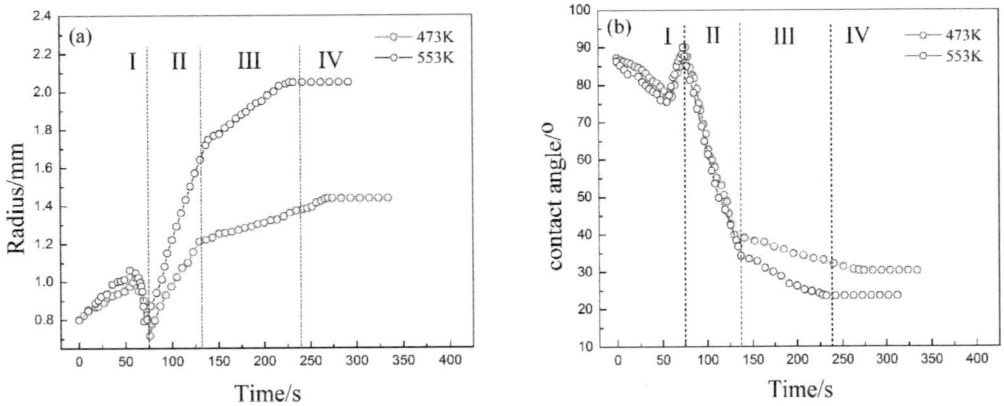

Fig. 4 Equivalent radius-time and contact angel-time evolution of the triple line for the Sn-35Bi-1Ag solder spreading on Cu substrate with different soldering temperature: (a) equivalent radius, (b) contact angel

Microstructure analysis at the interface

Figure 5 shows the cross-sectional structure of the Sn-35Bi-1Ag on Cu substrates after spreading at 473K and 553K, respectively. The interfacial structures at the temperatures of 473K and 553K in Fig. 8 (a) and (b) are composed of residual Sn-rich phase (gray), Bi-rich phase (white), Cu-Sn phase and Ag-Sn phase. The Cu_6Sn_5, Cu_3Sn and Ag_3Sn phase can be identified by using EDS. With the increase of temperature, the interfacial reaction continues. the thickness of Cu_3Sn gradually increases, and the morphologies of Ag_3Sn phase changed from acicular to spherical [17]. Near the solder side of the interface Cu_6Sn_5 in some parts of the solder grow to the interior. With the increases of temperature, the thickness of IMCs layer gradually increases.

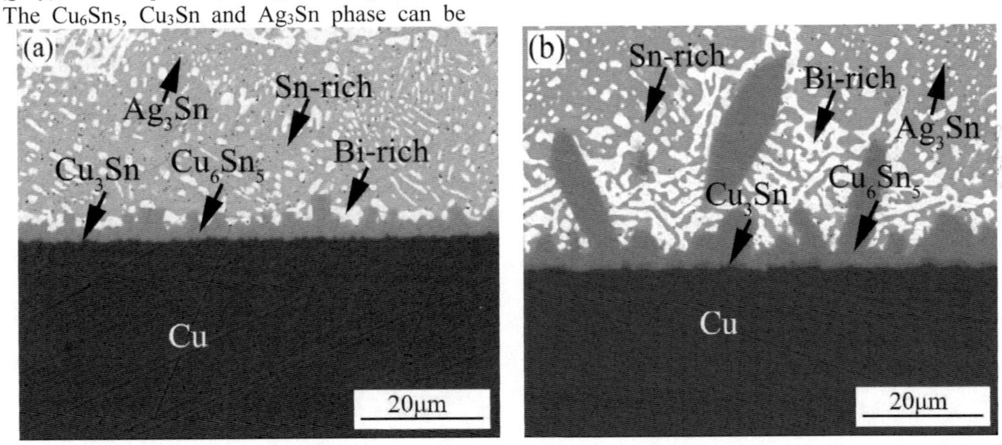

Fig. 5 Microstructure of the cross-section of Sn-35Bi-1Ag/Cu with different temperatures: (a) 473K, (b) 553K

The evolution mechanism of the interfacial reaction layer could be explained as follows: the solder spreads on the Cu substrate. First, mutual diffusion occurs between the molten solder and the substrate. With the increases of temperature, the diffusion speed of the atoms increases significantly. Then, when Cu and Sn reached a certain concentration, Cu and Sn reacted to form Cu_6Sn_5 intermetallic compounds. With the increases of time, the interfacial reaction continues and the thickness of Cu6Sn5 increases. The Cu_6Sn_5 and Cu matrix reaction in the Cu matrix side of the formation of Cu_3Sn, according to previous studies [18], Cu_6Sn_5 transforms into Cu_3Sn by the following reaction:

$$Cu_6Sn5 + 9Cu \rightarrow 5Cu_3Sn \quad (1)$$

as the temperature further increases, the consumption rate of Cu_6Sn_5 increases and the thickness of Cu_3Sn gradually increases [19].

Wetting kinetics

In order to further reveal the wetting mechanism of the Sn-35Bi-1Ag solder on Cu substrate, the change of the equivalent radius and the contact angle are measured and analyzed. Figure 6 shows the logarithm of the equivalent radius versus time. In general, the wetting and spreading kinetics is expressed by the following empirical power model (2) [20]:

$$R^n \sim t \quad (2)$$

In the equation, R is the equivalent radius at time t, and n is the exponent. In non-reactive wetting liquid/solid systems, the wetting kinetics have been thoroughly investigated by Tanner and Gennes *et al.* [21-22]. In their studies, the exponent n was calculated as 10 for the sessile drop case. However,

Eustathopoulos *et al.* reported that the spreading time in reactive wetting systems was in the range of 10^1-10^4 s [23]. This is several orders of magnitude higher than the time observed for the non-reactive wetting systems, which implies that the spreading rate for the reactive wetting systems was dominated by chemical reactions rather than viscous dissipation.

As shown in Fig. 6 (a) and (b), at 473K, the values of n for the rapid spreading stage and the limited spreading stage are calculated as 1 and 4, respectively. an interfacial reaction control model was proposed to describe reactive wetting in the previous study [24]: when the diffusion rate of an element is higher than the reaction rate at the triple line, the wetting kinetics are reaction-limited (R~t) [25], but when the reaction rate is higher than the diffusion rate of the active solute, diffusion is the dominant rate-limiting factor(R^4~t) [26]. Based on this correlation, it is deduced that the rapid spreading stage (ii) and the limited spreading stage (iii) in the Sn-35Bi-1Ag/Cu system at 473K was determined by the reaction rate ($n=1$) and the diffusion rate ($n=4$), respectively. On the other hand, for stages ii and iii of spreading over Cu substrates at 553K, the values of $n=1$ and 3, respectively. It is deduced that stage (ii) in the Sn-35Bi-1Ag/Cu system at 553K was determined by the reaction rate ($n=1$). As for stage (iii), $n=3$, which is probably because that the temperature increases, substrate and solder mutual dissolution faster, the interface of the chemical reaction rate has become significantly stronger, the wetting kinetics are controlled by the dual effects of dissolution-driven wetting and chemical reaction.

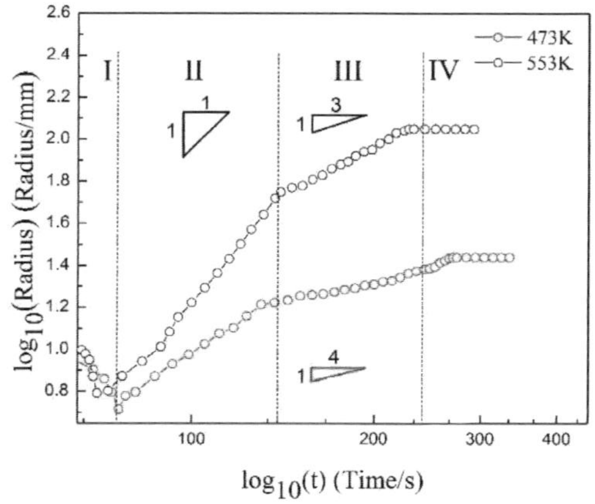

Fig. 6 Logarithm of the radius of the solder on different substrates with different soldering temperatures

Measuring and analyzing the change of contact angle, Dezellus got the coupling equation about the near-linear spreading behavior [27]:

$$\cos\theta_F - \cos\theta = (\cos\theta_F - \cos\theta_o)e^{(-kt)} \quad \textbf{(3)}$$

Where θ was the contact angle at the corresponding time t, θ_F was the contact angle formed between the solder and the interface at the stop of spreading, θ_0 was the contact angle with the interface at the start of spreading, *ie*, the initial contact angle, k was the kinetics constant, Dezellus gave a physical explanation of k [27]. The relationship $ln(cos\theta_F\text{-}cos\theta)$ versus t is made of the solder on the Cu substrate with different soldering temperatures. The linear analysis of the stage (ii) and the stage (iii) of wetting and spreading is carried out, which obtaines the k value at different temperatures.

As shown in Fig. 7, for the soldering temperatures of 473K and 553K, the k values obtaine by the fitting in the stages (ii) and (iii) of spreading of the solder on the Cu substrate are -0.121, -0.239, respectively.

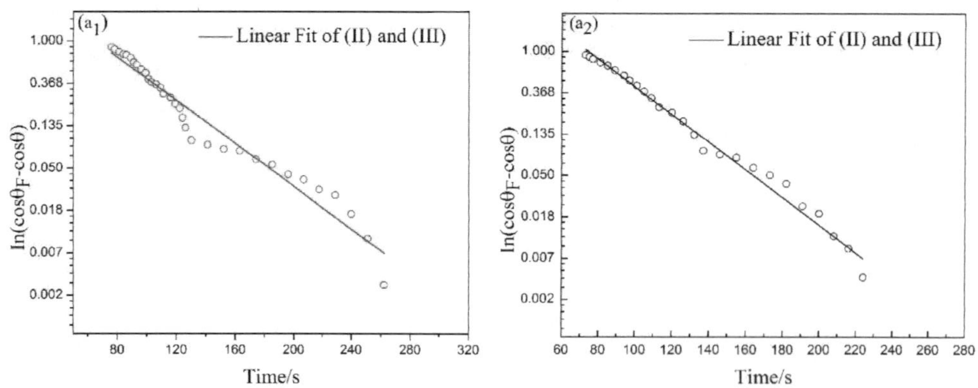

Fig. 7 Curve of $ln(cos\theta_F\text{-}cos\theta)$ and time t of the solder on Cu substrate with different soldering temperatures: $\mathbf{a_1}$ a polished Cu substrate at 473K, $\mathbf{a_2}$ a polished Cu substrate at 553K

From the comparison of k value and contrast shown in Fig. 7, it can be seen that in the (ii) and (iii) stages, as the temperature increases, the absolute value of k increases correspondingly

Conclusions

The assemled experimental system, featuring a heating platform assistted with two sets of cameras can be used to monitor the triple line movement from the side and the top of the melting sessile drop, simultaneously. Experiments were carried out on the spreading of Sn-35Bi-1Ag solder over the Cu substrate in a high-purity argon atmosphere. With the increase of soldering temperature, the ultimate spreading radius increases, the final contact angle decreases, and the spreading time decreases. The entire spreading process can be divided into four stages. The layered structure of both 473k and 553K are composed of residual Sn-rich /Birich/Ag_3Sn/Cu_6Sn_5/Cu_3Sn/Cu substrate. Variation in the microstructure suggested that the spreading mechanism changed from a reaction-limited and diffusion limited mechanism at 473K and to one that was controlled by chemical reactions, dissolution, and chemical reactions into solid at 553K.

Acknowledgments

This work was sponsored by the National Natural Science Foundation of China (No. 51665038), Key Project of the Nature Science Foundation of Jiangxi Province (20171ACB21011), Postgraduate Innovative Special Foundation of Nanchang University (cx2016085).

References

[1] Wang, F.J., *et al*, "Intermetallic compound formation at Sn-3.0 Ag-0.5 Cu-1.0 Zn lead-free solder alloy/Cu interface during as-soldered and as-aged conditions," *J ALLOY COMPD*, Vol. 438, No. 1 (2007), pp. 110-115. [J]

[2] Li, X, *et al*, "Effects of the melt state on the microstructure of a Sn-3.5% Ag solder at different cooling rates," *APPL SURF SCI*, Vol. 258, No. 15 (2012), pp. 5677-5682. [J]

[3] Billah, M. M, *et al*, "Sharif A. Effect of micron size Ni particle addition in Sn-8Zn-3Bi lead-free solder alloy on the microstructure, thermal and mechanical properties," *J ALLOY COMPD*, Vol. 585, (2014), pp. 32-39. [J]

[4] Zhang, L, *et al,* "Structure and properties of lead-free solders bearing micro and nano particles," *ADV MATER SCI ENG: R: Reports*, Vol. 82, (2014), pp. 1-32. [J]

[5] Fouzder, T, *et al*, "Effect of nano Al 2 O 3 additions on the microstructure, hardness and shear strength of eutectic Sn-9Zn solder on Au/Ni metallized Cu pads,". *MICROELECTRON RELIAB*, Vol. 50, No. 12 (2010), pp. 2051-2058. [J]

[6] Gain, A. K, *et al*, "Thermal aging effects on microstructures and mechanical properties of an environmentally friendly eutectic tin-Cu solder alloy," *MATER DESIGN*, Vol. 110, (2016), pp. 275-283. [J]

[7] Gain, A. K, *et al*, "Effect of nano Ni additions on the structure and properties of Sn-9Zn and Sn-Zn-3Bi solders in Au/Ni/Cu ball grid array packages," *ADV MATER SCI ENG*, Vol. 162, No. 2 (2009), pp. 92-98. [J]

[8] Lee, J. E, *et al*, "Effects of Ag and Cu addition on microstructural properties and oxidation resistance of Sn-Zn eutectic alloy," *J ALLOY COMPD*, Vol. 454, No. 1 (2008), pp. 310-320. [J]

[9] Fouzder, T, *et al*, "Microstructure, wetting characteristics and hardness of tin-bismuth-silver (Sn-Bi-Ag) solders on silver (Ag)-surface finished Cu (Cu) substrates," *J MATER SCI-MATER EL*, Vol. 28, No. 22 (2017), pp. 16921-16931. [J]

[10] PENG, C, *et al*, "Influence of minor Ag nano-particles additions on the microstructure of Sn30Bi0.5Cu solder reacted with a Cu substrate," *J MATER SCI-MATER EL*, Vol. 22, No. 7 (2011), pp. 797-806. [J]

[11] ZHANG, L, *et al*, "Microstructures and properties of Sn58Bi, Sn35Bi0.3Ag, Sn35Bi1.0Ag solder and solder joints," *J MATER SCI-MATER EL*, Vol. 26, No. 10 (2015), pp. 7629-7634. [J]

[12] Li. Y. L, *et al*, "Wetting kinetics of theAgCuTi filler metal on pure molybdenum substrate," *SURF INTERFACE ANAL*, Vol. 47, No. 8 (2015), pp. 838. [J]

[13] Dezellus, O, *et al*, "Fundamental issues of reactive wetting by liquid metals," *J MATER SCI*, Vol. 45, No. 16 (2010), pp. 4256-4264. [J]

[14] Kritsalis, P, *et al*, "Contribution to the study of reactive wetting in the CuTi/Al2O3 system," *J MATER SCI*, Vol. 26, No. 12 (1991), pp. 3400-3408. [J]

[15] Landry, K, *et al*, "Wettability of carbon by aluminum and aluminum alloys," *ADV MATER SCI ENG* , Vol. 254, No. 1 (1998),pp. 99-111. [J]

[16] Kumar, S, *et al*, "Mechanical and electronic properties of Ag3Sn intermetallic compound in lead free solders using ab initio atomistic calculation," *ADV MATER SCI ENG* , Vol. 178, No. 1 (2013), pp. 10-21. [J]

[17] Landry, K, *et al*, "Wettability of carbon by aluminum and aluminum alloys,". *ADV MATER SCI ENG*, Vol. 254, No. 1 (1998), pp. 99-111. [J]

[18] Hu, X. W, *et al*, "Interfacial reaction and IMC growth between Bi-containing Sn0. 7Cu solders and Cu substrate during soldering and aging," *J ALLOY COMPD*, Vol. 582, (2014), pp. 341-347. [J]

[19] Lim, G. T, *et al*, "Temperature effect on intermetallic compound growth kinetics of Cu pillar/Sn bumps," *J ELECTRON MATER*, Vol. 38, No. 11 (2009), pp. 2228-2233. [J]

[20] Zhao, H, *et al*, "Spreading kinetics of liquid solders over an intermetallic solid surface. part 1: eutectic lead solder," *J ELECTRON MATER*, Vol. 38, No. 9 (2009), pp. 1838-1845. [J]

[21] Tanner, L. H., "The spreading of silicone oil drops on horizontal surfaces," *EUR PHYS J-APPL PHYS*, Vol. 12, No. 9 (1979), pp. 1473. [J]

[22] De, Gennes. P. G. "Wetting: statics and dynamics," *REV MOD PHYS*, Vol. 57, No. 3 (1985), pp. 827. [J]

[23] Landry, K, *et al*, "Dynamics of wetting in reactive metal/ceramic systems: linear spreading," *ACTA MATER*, Vol. 44, No. 10 (1996), pp. 3923-3932. [J]

[24] Dezellus, O, *et al*, "Fundamental issues of reactive wetting by liquid metals," *J MATER SCI-MATER EL*, Vol. 45, No. 16 (2010), pp. 4256-4264. [J]

[25] Landry, K, *et al*, "Dynamics of wetting in reactive metal/ceramic systems: linear spreading," *ACTA MATER*, Vol. 44, No. 10 (1996), pp. 3923-3932. [J]

[26] Mortensen, A, *et al*, "Kinetics of diffusion-limited spreading of sessile drops in reactive wetting," *SCRIPTA MATER*, Vol. 36, No. 6 (1997), pp. 645-651. [J]

[27] Dezellus, O, *et al*, "Chemical reaction-limited spreading: the triple line velocity versus contact angle relation,". *ACTA MATER*, Vol. 50, No. 19 (2002), pp. 4741-4753. [J]

[28] Dezellus, O, *et al*, "Progress in modelling of chemical-reaction limited wetting," *J EUR CERAM SOC*, Vol. 23, No. 15 (2003), pp. 2797-2803. [J]

Capillary Flow on an Aluminum Brazing Sheet in a Gravity Field: A Phenomenological Study

Cheng-Nien Yu, Yangyang Wu, and Dusan P. Sekulic
Department of Mechanical Engineering, University of Kentucky, Lexington, Kentucky, USA
sekulicd@engr.uky.edu

Abstract

The presence of micrometeoroids and space debris in open space raises the question of a mitigation of the damage caused by a possible impact on the man-made structures. Moreover, an anticipated need for a bonding of complex structures in space has inspired a consideration of brazing of aluminum alloys in space. The study to be discussed in this lecture addresses an issue of filling the micro holes by liquid aluminum braze under terrestrial brazing of aluminum. It is observed that a hole with a diameter of the order of magnitude of up to a few hundreds of micrometers, for a given set of process and materials parameters, can be self-filled during the brazing process. Results indicate that a threshold diameter exists for the given set of core and clad dimensions. Samples with hole sizes below the threshold diameter will be filled during the brazing process. The holes with sizes larger than the threshold diameter will not be filled. The phenomenology of the process is studied. This research has been funded by NASA's Physical Sciences Research Program.

Introduction

Micrometeroids and orbital debris (MMOD) have been a considerable risk if striking the outer shell of satellites and/or the International Space Station [1-3]. For a long-term manned station orbiting the earth, as well as for extending the mission duration of the International Space Station the prolonged exposure to such a negative impact is significant, in particular keeping in mind that the components' design targets 30 and more year's life [2]. Since the total number of micrometeoroids and space debris in low earth orbit has been increasing over time, it is indeed a continuous and evolving challenge to mitigate their impact [4-6]. An example from NASA, Fig. 1, presents a hole of a diameter of 0.57 mm on the Ammonia Tank Assembly of the International Space Station [7]. This damage is inflicted by space debris.

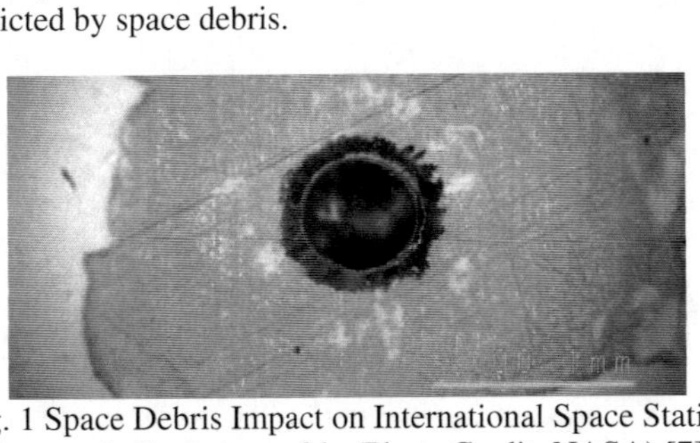

Fig. 1 Space Debris Impact on International Space Station Ammonia Tank Assembly (Photo Credit: NASA) [7]

Featuring low weight, aluminum is the primary material for building module shells at the International Space Station [8-9]. A habitable module wall was made of aluminum alloy 2219-T87 and a baseline bumper was made of AA6061-T6 [2]. A technical development for sealing punctures on the module walls was reported in 1999, promoting what is known as KERMIt (Kit for External Repair of Module Impacts) [10]. The repair kit features both mechanical fastening and an epoxy based adhesive [11]. Methods to form metallic filling with less manual treatment would be potentially beneficial to future developments of hole repair in space.

Aluminum brazing process features bonding or filling capillary clearances without melting mating aluminum surfaces which maintains geometrical similarity to base components [12]. In addition, the surface tension driven liquid aluminum capillary flow offers a possibility to fill irregular or unreachable configurations of gaps without tremendous efforts in precise heating manuplation. However, molten aluminum flow can be impeded by tenacious aluminum oxide layers, which may be aggravated by increasing temperature, humidity, or oxygen levels, and these factors lead to poor joint formation [13-16]. In order to disrupt aluminum oxide layers on the surface, substances such as K_x-Al_y-F_z flux, Mg, Ni, etc., are required depending on the brazing environment.

It has been recently shown by the authors that the flux embedded TRILLIUM™ brazing sheets were able to feature Al-Si liquid aluminum flow to fill a hole on an aluminum brazing sheet (Al-Si/Al-Mn), without the need of adding additional flux in a nitrogen protected environment after a brazing cycle under a benchmark terrestrial condition [17]. Without additional fluxing procedure there might be a possibility to braze (repair) under a stringent working environment. We hypotesize that the same would be possible to facilitate in space.

The development for manufacturing processes in space is one of the important aspects of NASA's strategic plan [18]. Vacuum brazing has been reported successful by an Electron Beam gun in space and considered as a potential manufacturing process in space [19]. To advance toward the ultimate goal of repairing holes on aluminum surfaces in space, it is important to understand fundamental mechanisms of the hole filling process of liquid aluminum on a solid aluminum alloy. In this study, standard AA4045/AA3003, AA4343/AA3003, and TRILLIUM™ brazing sheets were selected for a preliminary benchmark studies. The liquid Al-Si molten clad spreading over an Al-Mn surface featuring a hole will be phenomenologically discussed. Threshold diameters for filling/non-filling of a hole on various brazing sheets will be determined.

Experimental Setup

In this section, materials and methods of study will be discussed in a sequence of sub-sections: Materials, Sample Design, Brazing Facility, and Brazing Conditions. Test results will be given after this section.

Materials

Aluminum brazing sheets typically consist of at least two different layers of aluminum alloys: core and clad layers. As an example, Fig. 2 shows that a thin clad layer of aluminum silicon alloy was rolled onto a core layer of e.g., aluminum manganese substrate, forming a uniform brazing sheet. When heated, the Al-Si clad layer melts first and forms a liquid metal layer, while the core layer of Al-Mn away from the interface plane remains solid. The center portion of the brazing sheet used in this study features a cylindrical hole machined by calibrated SiC drill bits.

Material compositions of each layer used for this study are summarized in Table 1. Clad materials have: (i) Al-Si alloy nominally designated as AA4045, with an embedded potassium fluoroaluminate flux, (ii) AA4045 without flux, or (iii) AA4343 without flux. Brazing sheet thickness ranges between 0.3 mm and 0.5 mm. Clad layer thicknesses are between 25 µm and 36 µm. The clad layer may be on one side of the core layer, such as Material #3, or on both sides of the core layer, such as Material #4 and Material #6. In all cases, AA3003 has been used as a core alloy, rather than, e.g., AA2219, to establish a phenomenology of the surface tension driven filling a hole as a benchmark.

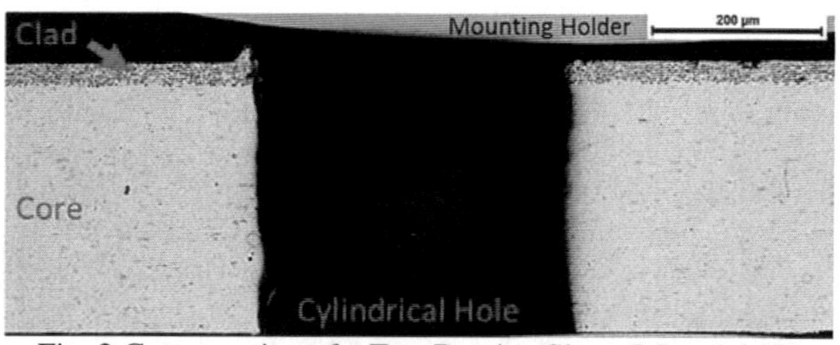

Fig. 2 Cross-section of a Test Brazing Sheet (Material #3) prior to a Brazing Test

Table 1 Materials

Material	Base Alloy	Cladding Alloy	Clad Gauge (µm)	Sheet Gauge (mm)
#3	AA3003	AA4045 (with embedded K_x-Al_y-F_z flux)	1 x 25	0.3
#4	AA3003	AA4343	2 x 30	0.4
#6	AA3003	AA4045	2 x 36	0.5

Sample Design

As shown in Fig. 3, test samples were cut or punched out of a sheet of a selected gauge to the required dimensions (4 mm square or disk shapes, to fit the heating platform in the hot stage microscopy facility chamber. Each sample was machined by various sizes of SiC drill bits, depending on the size selected for the hole. In this study, diameters of the hole were produced between 200 µm and 900 µm. The machined samples were cleaned in a sequence of a soapy water, water, and 95% ethanol using an ultrasonic cleaner. For brazing aluminum in controlled atmosphere other than the TRILLIUM™ sheets (Material #3), a flux is required to disrupt aluminum oxide layers, which naturally form in a contact with air. Therefore, after cleaning, a potassium fluoroaluminate flux ~15g/m^2 was loaded onto each sample to ensure sufficient amount of flux in each brazing process (i.e. additional flux was also applied on Material #3 to keep similar flux levels to Materials #4 and #6). Note that the industrial standard is well known to be around 5 g/m^2 or less, but the flux density in this study was intentionally selected to be higher to ensure enhanced oxide removal for the benchmark studies.

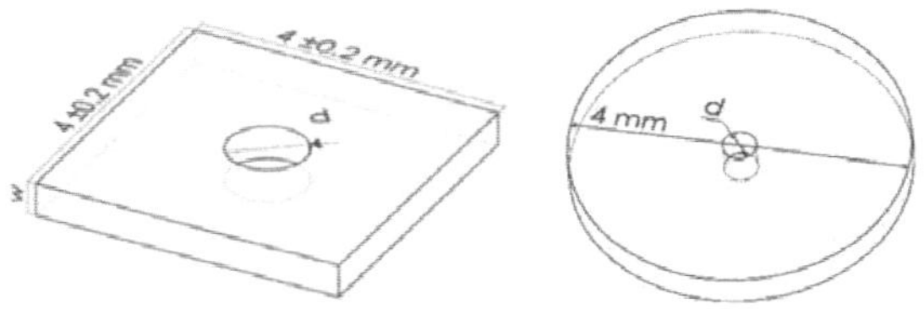

Fig. 3 Dimensions of Test Samples

Brazing Facility

In each brazing test, a test sample was placed on top of the heating element located at the center of the hot stage chamber. Through the quartz glass window at the lid of the hotstage, the sample surface heating, melting, surface tension driven flow and solidification of clad/filler were recorded by the digital camera installed on the microscope. The hot stage chamber allows gas input, controlled by a flowmeter. The chamber atmosphere was monitored by an oxygen analyzer at the gas outlet. Auxiliary lighting was provided from sides of the stage window to ensure proper visualization. Temperature of the sample's substrate was monitored and recorded by the heating unit of the hotstage and the control system. See Fig. 4 for details.

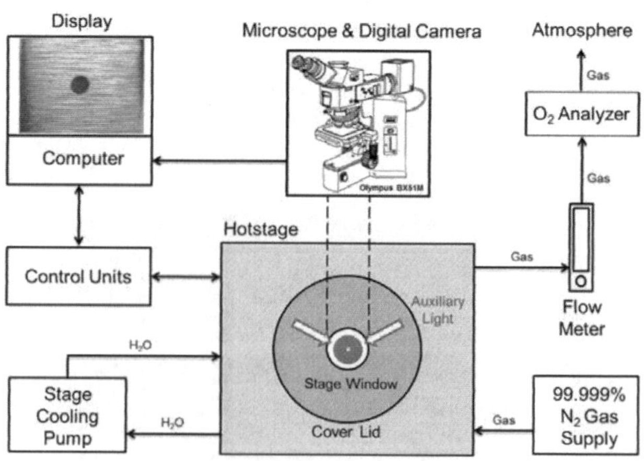

Fig. 4 Hostage Microscopy Facility

Brazing Conditions

Temperature of the sample was measured and controlled by the heating element and temperature sensors within the hotstage, the water cooling system, and the control unit. Brazing temperature history was programmed as follows: heating from the room temperature to 600°C by adding 100°C per minute, dwell at 600°C for 2 minutes, and cooling from 600°C to 40°C by 100°C per minute. Note that the measured temperature was at the heating assembly located at the bottom of the sapphire substrate, see Fig. 8, so that sample surface temperature is expected to be lower than registered temperature. The temperature history is shown in Fig. 5.

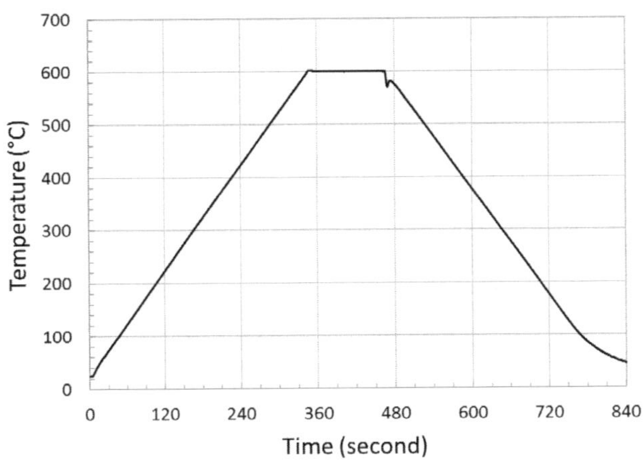

Fig. 5 Temperature History at the Heating Element in a Brazing Cycle

Interfacial test configurations were varied by having a sapphire glass covering: the bottom side of the sample (Set 1), both sides of the sample (Set 2), or on neither side, i.e., solely at the edge to support the sample (Set 3). See Fig. 6, Fig. 7, and Fig. 8, respectively, for a graphic presentation. For each test sample (Materials #3, #4, and #6 in Table 1), a potassium fluoro-aluminate flux powder was deposited and covered the surfaces of a test brazing sheet, see Figs. 6 through 8. The surrounding atmosphere was a 99.999% N_2 gas. Note that the sapphire substrate always exists in each set. It was used to prevent direct contact of test specimens components and the heating ceramic block, and to protect the heating element from liquid flux or metals if produced excessively and overflowed.

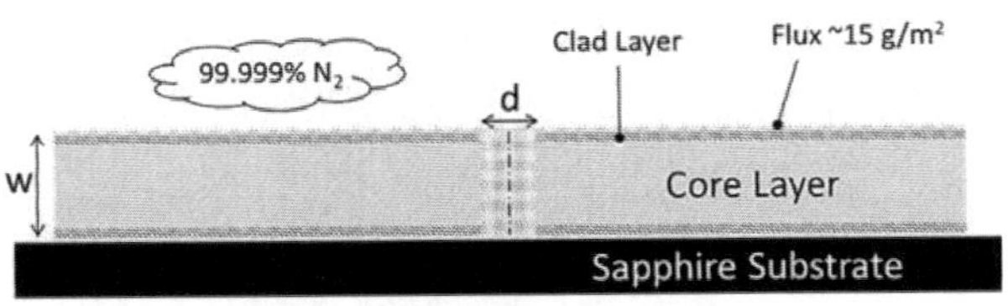

Fig. 6 Test Sample with the Bottom Side Covered by a Sapphire Glass (Set 1)

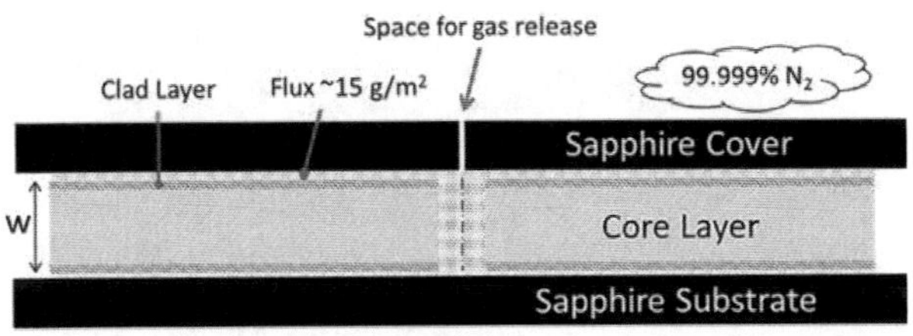

Fig. 7 Test Sample with Both Sides Covered by Sapphire Glasses (Set 2)

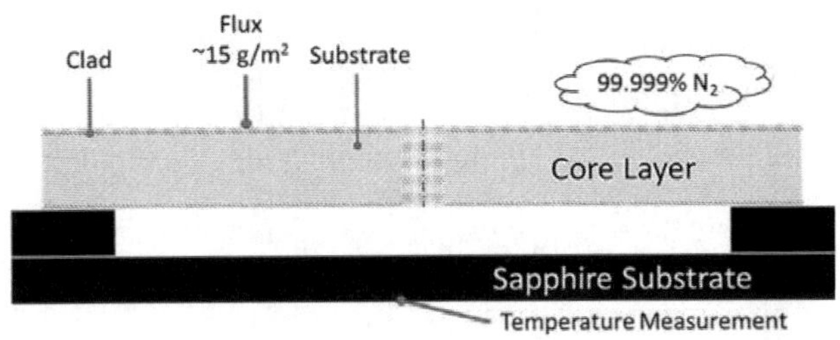

Fig. 8 Test Sample with Both Sides Open to Surrounding Atmosphere (Set 3)

Results

In this section, results will be discussed in a sequence of sub-sections: Hole-filled Case, Hole-non-filled Case, Gas Bubbles, Microstructure of the Braze, and Threshold Diameters. In the first two sub-sections, each of the hole-filled and hole-non-filled tests was studied to demonstrate phenomenological features of a hole filling process. Samples were video recorded in real time *in-situ*, through the hotstage window. Selected frames are presented below. Recording speed was 31 frames per second (32 milliseconds per frame).

Hole-filled Case

In Fig. 9, a series of featured frames taken by the hotstage microscopy facility was presented. The diameter of the hole prior to brazing was 0.69 mm. The test interfacial configurations are as shown in Fig. 6 (Set 1). In Frame 0 (an onset of observation, well before melting), the clad surface has the flux coverage, both (clad and flux) in solid state, prior to reaching the melting temperatures. The hole was also filled with the potassium fluoroaluminate flux. When melted, see Frame 717, the flux became transparent and only at the onset of melting occupied the hole. The molten clad appears as a bright-silver-like colored liquid, as shown in Frame 784. Later in the process, molten flux does not fill the hole, see Frame 871 and Frame 1015.

As heating continued, molten clad starts to form a thin layer of liquid pinned down around the edge of the hole, see Frame 1015 and Frame 1119. Liquid metal continuously flows toward the center of the hole and eventually fills the hole, see Frame 1151 through Frame 1154. During filling, liquid metal pushes remaining liquid flux and residual gas (if any) out of the hole and completes the filling, see Frame 1153 and Frame 1154. When quenched, the molten metal and flux residue solidifies, see Frame 4490.

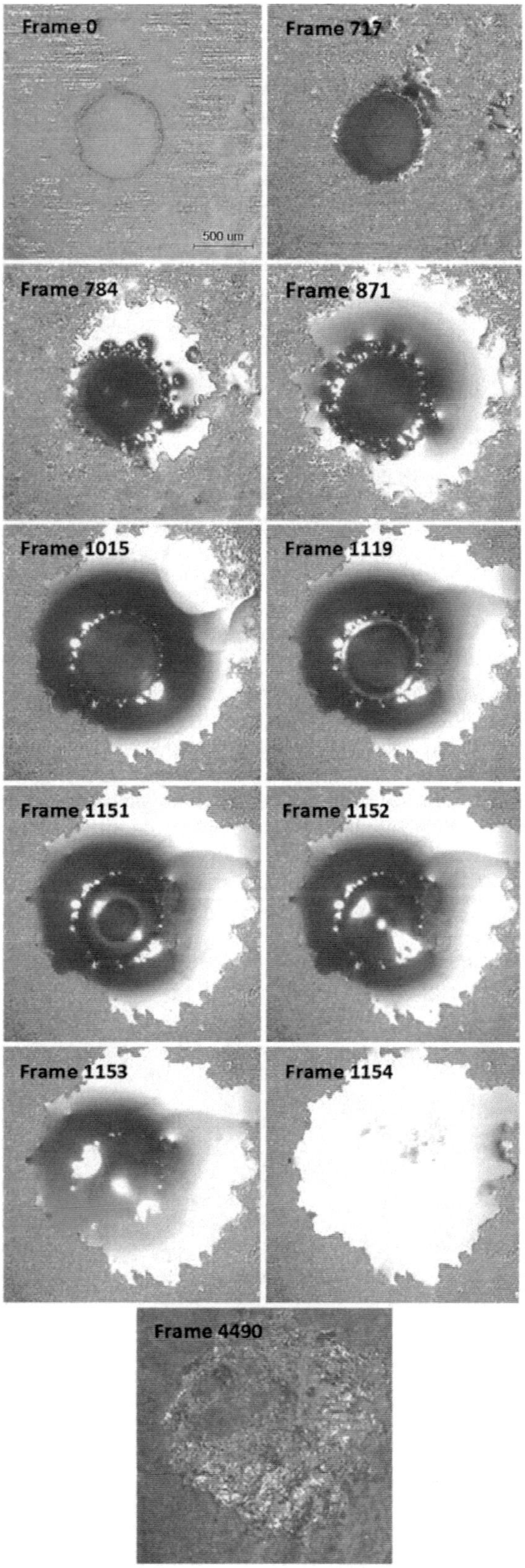

Fig. 9 Featured Frames of a Hole-filled Case
(Material #6, Set 1, Square Sample)

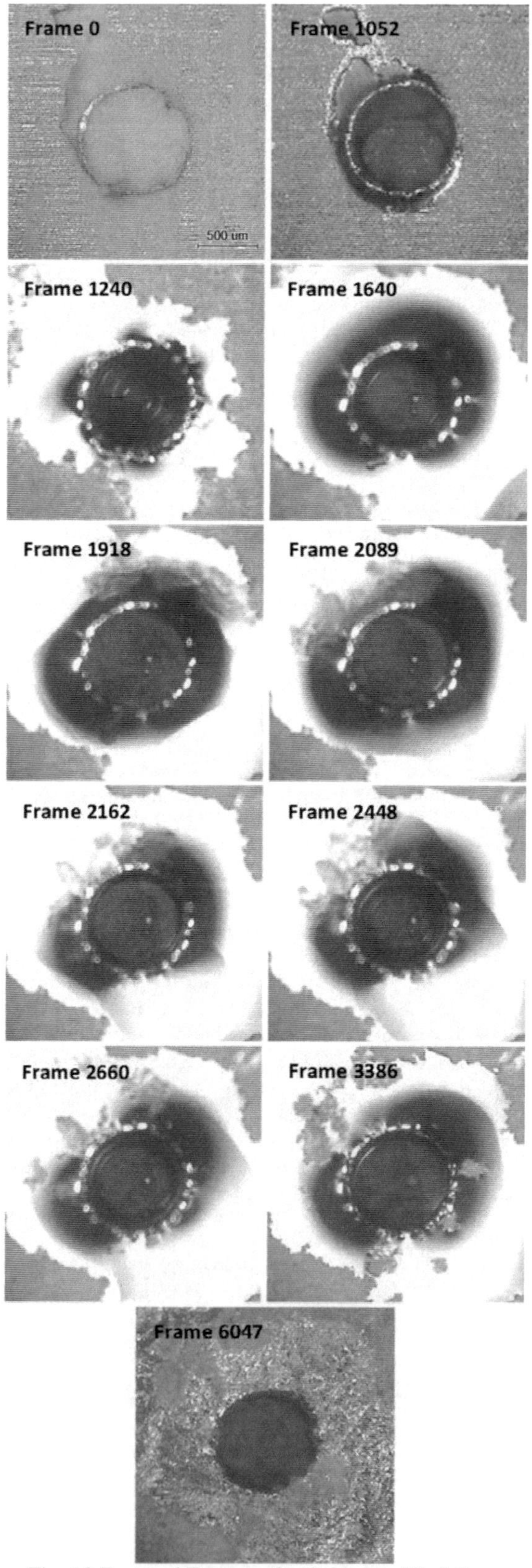

Fig. 10 Featured Frames of a Hole-non-filled Case
(Material #6, Set 1, Square Sample)

It is observed that a hole filling process starts from the peripheral side of the hole. The initial state of the filling process was preceded by a formation of a liquid flux and/or metal ring around the hole and it increases toward the center of the hole rather quickly. In this case, to fill a 0.69 mm diameter hole, the duration between the formation of a liquid ring around the hole and the instant of a completed filling was 4.5 seconds.

Hole-non-filled Case

For the test sample presented in Fig. 10, the diameter of the hole was 0.87 mm. The interfacial configuration of the test sample was Set 1, as shown in Fig. 6. In Frame 0, the clad layer of the test sample and the covering flux were both in solid state, prior to reaching the melting temperature. The hole was filled with the same potassium fluoro-aluminate flux powder as in the Hole-filled Case. When melted, the flux becomes transparent and occupies the hole, see Frame 1052. The molten clad has a silver-like color, see Frame 1240. Later in the process, the molten flux appears to leave the hole, compare Frames 1240 through 1918.

As heating continues molten clad starts to form a thin toroidal layer of liquid around the edge of the hole, see Frame 2089 and Frame 2162. Liquid metal continues to grow a thicker ring layer around the hole and the transparent molten flux grows even more toward the center of the hole, see Frame 2448. However, both liquid domains cease growing and instead start reducing the size of the liquid layer ring around the hole, compare Frame 2448 through Frame 3386. No further major topographic change was noticed until the solidification in Frame 6047.

For a hole-non-filled case, it is observed that a hole with a larger diameter features a liquid layer formation around the edge and it grows toward the center of the hole to some extent, however, it is unable to succeed in reaching the center of the hole. The amount of liquid around the hole is reduced but some portion of liquid remains to be pinned at the edge of the hole.

Gas Bubbles during Brazing Processes

During a brazing process, there are significant quantities of gas phase formation, i.e. bubbles penetration of the liquid driven by buoyancy in the gravity field and retarded by surface tension, see Fig. 11.

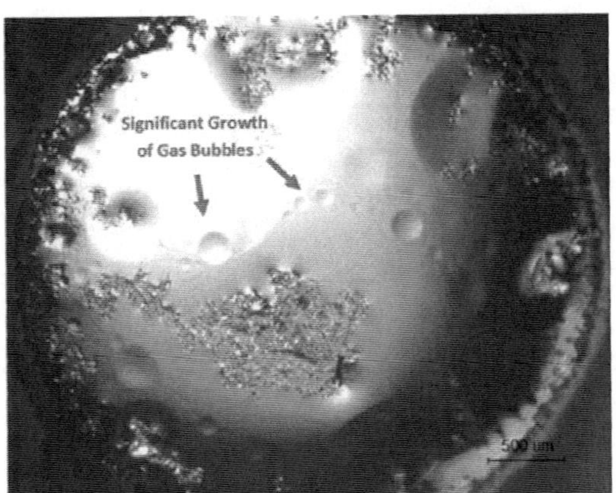

Fig. 11 Gas Bubble Growth during Brazing Processes (Material #3, Set 1, Disk Sample)

This terrestrial observation has a great importance for the process consideration in the microgravity environment. The gas phase bubbles are expected to be trapped within the molten metal thus forming porosities when solidified. The trapped gas bubbles will form voids and impact the mechanical integrity. Similar situation was reported in a soldering process in reduced gravity environment [20].

Microstructure of the Braze in Hole-filled and Hole-non-filled Cases

Brazed samples were cut, mounted, grinded, polished, and etched at the hole area to reveal the solidified melt microstructure and more importantly the topographical features in non-filled cases. It is shown in Fig. 12 that the pre-existed open hole has now being closed by a re-solidified liquid aluminum that was originally produced by melting the surface clad layer. The majority of the filled hole features the re-solidified eutectic Al-Si phase (needle like). The α-Al phase (white color) forms at solid-liquid aluminum interfacial regions.

A thin domain of α-Al phase is also visible at the bottom of the core layer, originally not cladded. It constitutes a clear evidence that the molten clad spreads in a small quantity to the bottom side of the core layer and subsequently re-solidifies. It should be concluded that the molten clad travels along the cylindrical wall of the hole to the bottom side of the core layer.

In Fig. 13, the hole cross-section provides an evidence of a liquid metal layer re-solidification around the hole and formation of the eutectic Al-Si phase. It also shows that although very thin, the molten clad flows toward the bottom side of the brazing sheet. Observe the bottom side of the micrograph for a thin layer of α-Al and eutectic Al-Si phases.

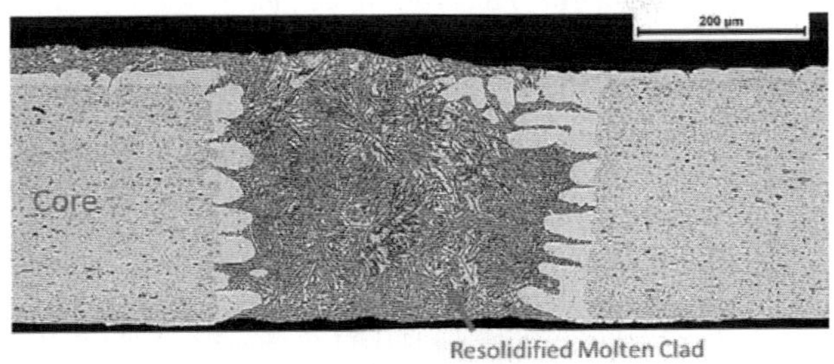

Fig. 12 Cross-section of a Filled Brazed (Material #3, Set 3, Disk Sample)

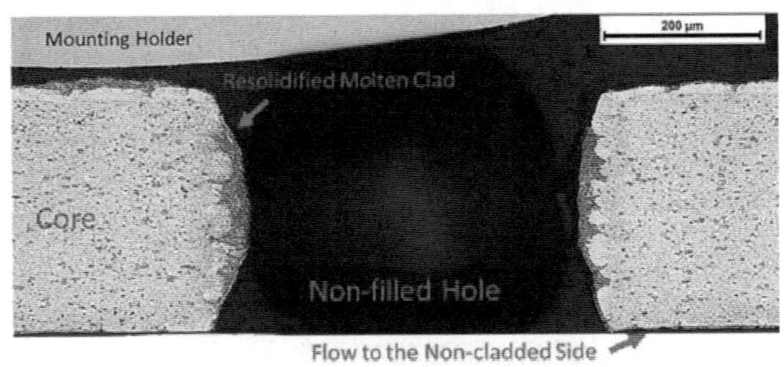

Fig. 13 Cross-section of a Non-filled Brazed (Material #3, Set 3, Disk Sample)

Threshold Diameters

Three brazing materials were used for three different brazing configurations, Figs. 6 through 8. Refer to "Materials" and "Brazing Conditions" sections for details. In Figs. 14 through 16, experimental results for various combinations of materials (brazing sheet/clad thicknesses), hole diameters, and interfacial configurations are presented. Each data point is a result of either having a molten clad filled hole or non-filled hole, marked by a solid dot or an empty dot, respectively. Note that in Fig. 16 (Set 3) there are several samples brazed at 610°C, marked by square symbols.

It can be clearly observed that by increasing the hole diameter, the hole which originally can be filled will cease to be filled at a threshold limit of its diameter – "threshold diameter." For example, in Fig. 14 the threshold diameter of a 0.5 mm thick brazing sheet is located between 0.69 mm and 0.85 mm. There may be some outliers in tests close to threshold diameters. For example, in Fig. 14 for the 0.4 mm thick brazing sheet, the sample of 0.55 mm hole diameter was filled, but several holes smaller than 0.55 mm were not filled. In this case, it can still be conservatively argued (depending on the level of statistical confidence) that the threshold diameter should be located between 0.46 mm and 0.58 mm.

Tests from Set 2 (Fig. 7) and Set 3 (Fig. 8) offer an insight into an interfacial impact on hole-filling processes. Test samples in Set 2 (both sides of the brazing sheet are facing sapphire glasses) have smaller threshold diameters than Set 1 (Fig. 6). The causes of this phenomenological observation are under investigation. There may be several possible reasons for the reduction in threshold diameters: (i) Non-wetting sapphire glass surface impedes the wettability of the molten clad. (ii) The weight of the sapphire glass affects the shape of liquid metal free surface around the hole. (iii) In addition to reacting with aluminum surface oxide layers, the potassium fluoroaluminate flux may possibly react with sapphire glasses (a sapphire is also an aluminum oxide, Al_2O_3) and may lead to a less activity. Without a non-wetting interface at the bottom, results from Set 3 (Fig. 8) were expected to have an improved threshold diameter compared to Set 1 (Fig. 6). However, the results show similar threshold diameters as for Set 1. This implies a limited involvement of the non-wetting sapphire interface at the bottom of the hole. Further study of interfacial impacts on the hole-filling processes is needed.

It has been noticed that a hole on each of the brazing sheets in each of the interfacial configurations can be filled if the hole diameter reduces or the thickness of the brazing sheet increases (compare solid dots and empty dots in Figs. 14 through 16). Note that for larger brazing sheet thickness in this study also means thicker clad layer. Whether the increase of threshold diameter is the result of increasing total brazing sheet thickness or the clad layer thickness has to be explored.

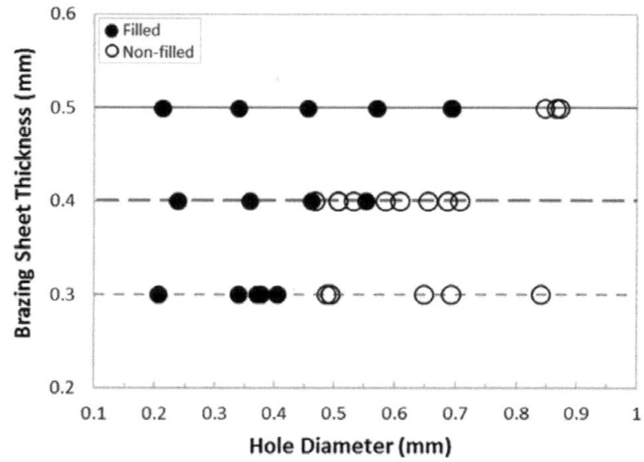

Fig. 14 Test Results for Top N_2 & Bottom Sapphire Interfaces (Set 1, Square Samples)

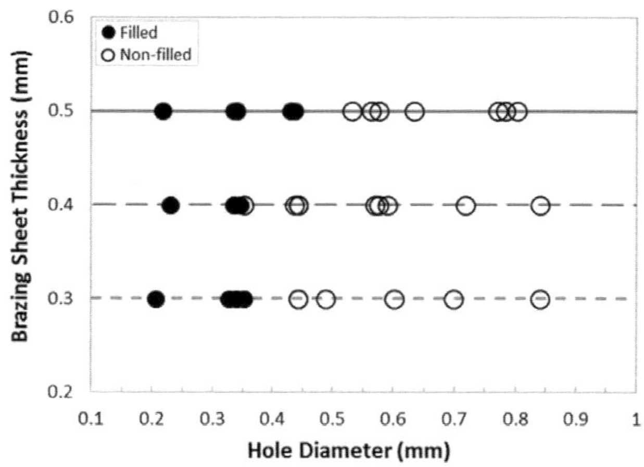

Fig. 15 Test Results for Top & Bottom Sapphire Interfaces (Set 2, Square Samples)

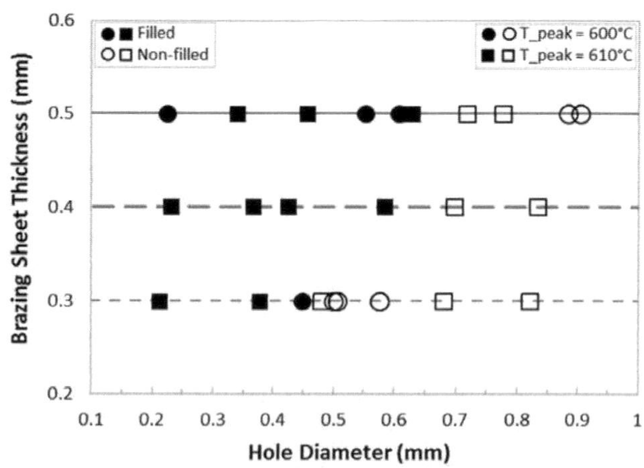

Fig. 16 Test Results for Top & Bottom N_2 Interfaces (Set 3, Sqaure Samples)

Conclusions

This study explored a phenomenology of hole-in-brazing sheet filling in a controlled atmosphere brazing. It is established that a "self-healing" (filling the hole) would be achieved if the hole diameter and/or thickness of the sheet/clad have beforehand established dimensions. Oppositely, the liquid formation pinned down around the hole instead of growing toward the center and filling the hole.

Threshold diameter of a hole generally appears to exist for a given material system. Samples with a larger sheet thickness (also a larger clad thickness) have larger threshold diameters. When the hole size is close to the size of the threshold diameter, it is more difficult to predict whether the hole will or will not be filled. For hole diameters much less or much larger than the threshold diameter, the result is more predictable with a higher statistical confidence.

Acknowledgments

This work has been funded by NASA's Physical Sciences Research Program, Grant# NNX17AB52G. Materials and the early stage of this work were supported by Gränges AB, Finspång, Sweden. TRILLIUM™ Technology is a registered trademark of Gränges AB, protected by United States Patent No. 8871356 and corresponding pending patent applications in other major countries. CY expresses sincere gratitude to Dr. Doug Hawksworth, Diomedea Inc., Canada, for discussions regarding the subject. The research team acknowledges Mr. Adam Price from the Institute for Sustainable Manufacturing, University of Kentucky for technical support. The team thanks Dr. Fiqiri Hodaj, Grenoble Institute of Technology, France, for discussion during his visit to the brazing laboratory at the University of Kentucky, as well as Dr. Paul Steen from Cornell University, USA for bringing attention to an early study of a hole formation in a liquid phase.

References

[1] Laurance, M. R. and Brownlee, D. E., "The Flux of Meteoroids and Orbital Space Debris Striking Satellites in Low Earth Orbit", *Nature*, Vol. 323, No. 11 (Sept. 1986), pp. 136-138.

[2] Christiansen, E. L. *et al*, "Space Station Freedom Debris Protection Techniques", *Advances in Space Research*, Vol. 13, No.8 (Aug. 1993), pp. 191-200.

[3] Lear, D. M. *et al*, "Investigation of Shuttle Radiator Micro-Meteoroid & Orbital Debris Damage", *50th AIAA/ASME/ASCE/AHS/ASC Structures, Structural Dynamics, and Materials Conference*, Palm Springs, CA, USA, May 4-7, 2009. Article Number 2009-2361.

[4] Liou, J.-C. and Johnson, N. L., "Risks in Space from Orbiting Debris", *Science*, Vol. 311, No. 5759 (Jan. 2006), pp. 340–341.

[5] Ebisuzaki, T. *et al*, "Demonstration Designs for the Remediation of Space Debris from the International Space Station", *Acta Astronautica*, Vol. 112 (July 2015), pp. 102-113.

[6] Christiansen, E. L. *et al*, "Handbook for Designing MMOD Protection", NASA Technical Report Server TM2009-214785, NASA Johnson Space Center, Huston, TX (June 2009).

[7] Online Source: "ATA Docking Target (Impact 16)", NASA ARES Hypervelocity Impact Technology's International Space Station Impact Gallery (Website https://ares.jsc.nasa.gov/orbital_debris/hvit/impact-galleries/iss-impact-gallery.html). Accessed in Jan. 2018.

[8] Dunn, B., Materials and Processes for Spacecraft and High Reliability Applications, Springer International Publishing AG (Cham, Switzerland, 2016).

[9] Online Source: Barry, P. L., "Home, Space Home", NASA Science News, Mar. 14, 2001 (Website https://science.nasa.gov/science-news/science-at-nasa/2001/ast14mar_1). Accessed in Jan. 2018.

[10] Hall, S. B., "Marshall Researchers Developing Patch Kit to Mitigate ISS Impact Damage", *The Orbital Debris Quarterly News*, Vol. 4, No. 4, The Orbital Debris Program Office, NASA Johnson Space Center, Houston, Texas, USA (Oct. 1999), pp. 1-3.

[11] Chaudhary, R. I. *et al*, "Variability Reduction and Design of Experiments Techniques for Improving System Performance of the International Space Station Kit for External Repair of Module Impacts", Defense Technical Information Center, Fort Belvoir, VA, USA (Feb. 2000). Accession# ADA373716.

[12] Hampston, G. and Jacobson, D. M., Principles of Soldering and Brazing, ASM International (Materials Park, OH, USA 1993).

[13] Hunter, M. S. and Fowle, P., "Natural and Thermally Formed Oxide Films on Aluminum", *Journal of the Electrochemical Society*, Vol. 103, No. 9 (Sept. 1956), pp. 482-485.

[14] Swidersky, H.-W., "Aluminium Brazing with Non-corrosive Fluxes - State of the Art and Trends in NOCOLOK Flux Technology", *DVS-Berichte 212*, Dusseldorf, Germany, 2001, pp. 164-169.

[15] Zahr, J. *et al*, "Correlations Between the Atmosphere Conditions on the Natural Aluminum Oxide Layer and the Brazeability", *5th International Brazing and Soldering Conference*, Las Vegas, NV, USA, Apr. 22-25, 2012, pp. 275-280.

[16] Yu, C.-N. *et al*, "Impact of the Background Atmosphere on the Aluminum Brazing Joint Formation", *ASME 2013 International Mechanical Engineering Congress & Exposition*, San Diego, California, USA, Nov. 15-21, 2013. Paper# IMECE2013-64776.

[17] Hawksworth, D. K. *et al*, "A Mechanistic Study of Aluminium Controlled Atmosphere Brazing Processes", *12th Vehicle Thermal Management Systems Conference and Exhibition*, Nottingham, U.K., May 10-13, 2015.

[18] Terrier, D. *et al*, "NASA Strategic Technology Investment Plan 2017", Office of the Chief Technologist, NASA (Sept. 2017).

[19] Flom, Y., "Brazing in Space: Enabling the Next Frontier", *Welding Journal*, Vol. 84, No. 10 (Oct. 2005), pp. 25-29.

[20] Pettegrew, R. D. *et al*, "Experimental Methods in Reduced-gravity Soldering Research", NASA Technical Report Server TM-2002-211993, NASA Glenn Research Center, Cleveland, OH, USA (Dec. 2002).

A new procedure for determining the spreading and flow behavior of brazing fillers using flux

T. Uhlig, B. Sattler and G. Wagner
Institute of Materials Science and Engineering, Chemnitz University of Technology, Chemnitz, Germany
thomas.uhlig@mb.tu-chemnitz.de

S. Weis
University of Applied Sciences Zwickau, Zwickau, Germany
sebastian.weis@fh-zwickau.de

Abstract

The development of new brazing fillers requires knowledge about their properties in comparison to well-established fillers. Therefor the determination of the spreading and flow behavior is necessary. By now, different measuring techniques are combined in order to reach a comprehensive understanding of the mechanisms. Especially in case of flux brazing, conventional measuring techniques cannot be used. Due to the industrial demand of application-related experiments, the intended setup should be simple and easily to arrange. This paper deals with the development of a new procedure to determine the spreading and flow behavior of brazing fillers employing flux. The experimental setup, which consists of a commercially available induction heater, equipped with a temperature controller, is mounted inside a gas and vacuum tight chamber. A CCD camera system is used to record the spreading and flow behavior on surfaces as well as inside narrow gaps. The results of experiments using different fillers and base materials are presented. The measurements of the spreading area and the flow velocity under defined conditions allow a quantitative evaluation of the properties of the fillers. Thus, the comparison of different fillers among each other is possible.

Introduction

To evaluate the processing properties of brazing fillers, some conventional methods are well established. One of the most used properties is the wetting angle of the brazing filler on a distinct base material. The Young-Dupré' equation describes fundamental relations between contact angle and surface energies respectively interfacial tensions [2]. Considering this basic calculation assuming idealized conditions, further equations by Wenzel or Cassie are more useful in practice. Even these equations only partly represent what actually happens [3, 4]. The spreading and flow of brazing fillers is a sophisticated process that cannot be described easily due to the variety of mutually parameters, as already stated by Prabhu and Shibata [5, 6]. Only a few methods are based on dynamic tests to examine the properties of the fillers. For example the wetting method in v-grooves after Yost et al. [7] or tests on capillary gaps by Hall et al. [8]. Admittedly, these and similar methods are very specialized by means of the test conditions. So the significance for most practical applications is not given. Additionally there is a demand from the industry to carry out fast and easy measurements. Other important subjects are repeatability, comparability of results, time need for analysis and the capturing of flow and spreading phenomena in real time. In contrast to previous methods, the procedure may be carried out using fluxes.

Experimental Procedure

The basic principle is running optic observation during testing. A simple CCD-camera captures single images of the base material surface in time intervals of 100 ms. For the implementation of this procedure, a special test bed was designed and built, which allows different process atmospheres, Fig. 1. Therefore, the sample is arrangement in the center of a gas and vacuum tight chamber.

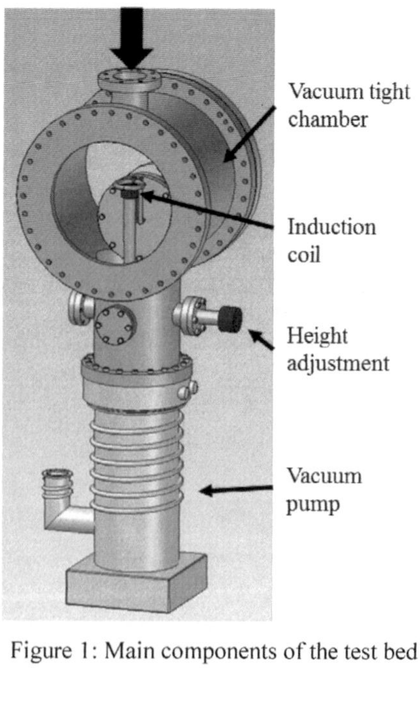

Figure 1: Main components of the test bed

Only the visible radiation emitted by the sample due to it's high temperature is recorded. The saved image sequences are used for a quantitative evaluation of the spreading and flow behavior as described later in detail. The sample is heated up by an induction coil. The temperature measurement as well as the automatic control of the power output is carried out by thermocouples and an adapted controller. Thus, the heating rate and the brazing temperature can be set easily. The arrangement of base material and brazing filler is placed on a height adjustable carrier made of graphite. Thereby it is possible to change the samples position within the focus of the camera and the inductive field. The base materials are cylindrically shaped. During induction heating, this offers the advantage of a uniform temperature distribution. The samples can be manufactured easily what allows a simple use of the test in industrial practice. In all employed tests, the sample arrangement is placed horizontally to annihilate gravity effects. Two types of sample geometries are used for the investigations, Fig. 2. Sample "type I" provides a drilled hole in the center of the base material. The brazing filler in form of kinked wire is fixed in a well-defined position thereby. This arrangement is used to examine the flow characteristics on a free surface. After melting the filler in running test, a sufficient wetting area is provided for observation. The other geometry "type II" is used for testing the flow behavior in gaps. The gap is fabricated by cutting the cylindrical sample in half and welding the parts together with defined spacers in between.

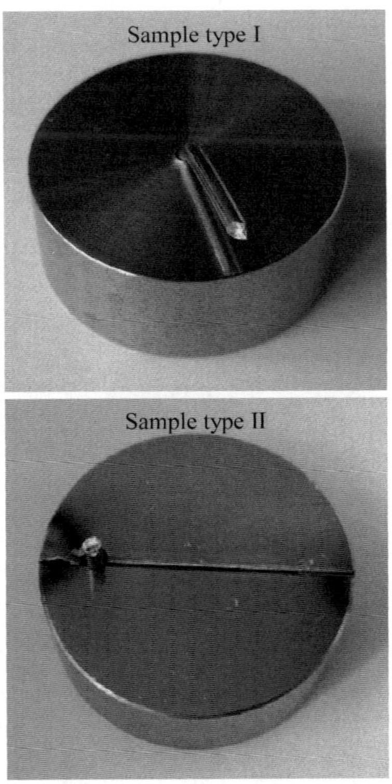

Figure 2: Geometries of samples: "type I" for investigations on a free surface, "type II" for investigations on gap

The measuring principle is tested on Ag/Cu- and Cu-based fillers, Tab. 1. They are used to validate the repeatability and differentiability of the results.

Table 1: Investigated brazing fillers

filler ISO 3677	min. brazing temperature [°C]
B-Ag72Cu-780	780
B-Ag60CuSn-600/730	720
B-Ag45CuZnSn-640/680	670
B-Ag35CuZnMnSnSi-680/700	700
B-Cu94P-710/890	760
B-Cu60Zn(Si)(Mn)-870/900	900

For preparing the test procedure, it is important to know the substantial influencing parameters. The effect of the temperature, which is one of the most important factors, is determined by several preliminary tests. A certain amount of brazing filler is applied on mild steel in combination with flux. Then the arrangements are heated up until the liquidus temperature of the brazing filler as well as temperatures above the liquidus. Figure 3 shows a comparison of images during the tests at different temperatures and after the tests. It appears that temperatures above the liquidus do not lead to an essential difference in spreading, even when the temperature exceeds 110% of liquidus. This represents an advantage of the procedure, because it is not too sensitive when a slight overheating happens.

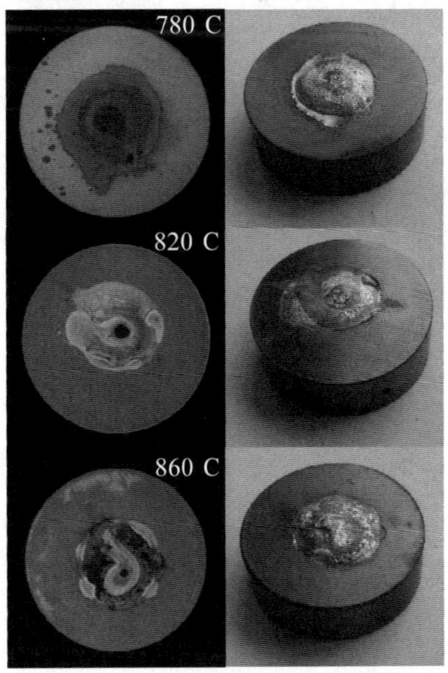

Figure 3: Influence of temperature on wetting of B-Ag72Cu-780 on mild steel

Another factor is the surface treatment of the base material. This includes cleaning conditions as well as structures caused by the manufacturing of the samples. To show this effect, base material samples with grinding grooves are used. Figure 4 points out, how the spreading and flow behavior of the brazing filler are influenced by different surface conditions. The orientation of the grooves is tilted in steps of 45° in relation to the centerline of an applied filler wire. It is visible, that the direction of the grooves leads to preferred paths of flow. To eliminate this detrimental influence, the samples that are used for further experiments are finished by precision turning. This leads to nearly radially symmetric structures. Hence, the results are not influenced by the positioning of the brazing filler and the base material to each other.

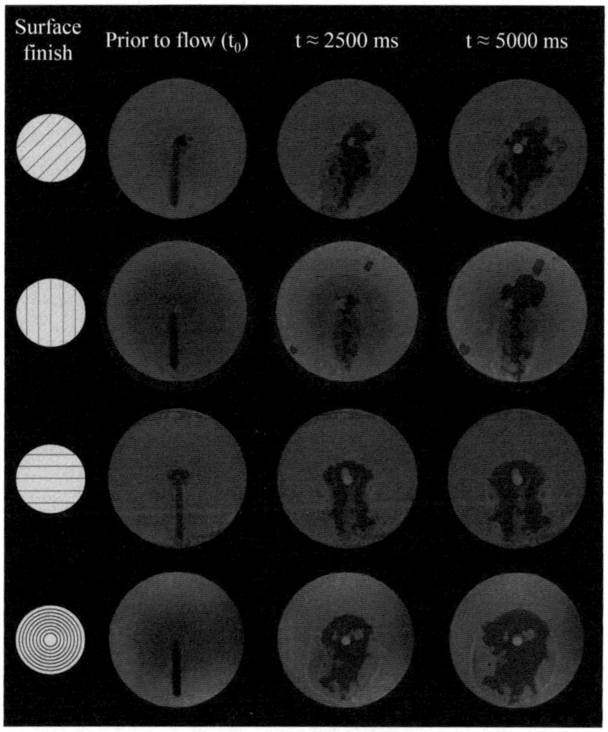

Figure 4: Influence of the surface finish on wetting of B-Ag72Cu-780 on mild steel

Results

First, the evaluation of the procedure will be explained using "type I" samples. A constant amount of brazing filler in form of wire is placed on the radially symmetric substrate surface. Under this conditions, the width of the spreading area is regarded. To comprehend the dynamic of this process, a parameter of time is needed. So, when the brazing filler begins to melt, a distinct period is set and afterwards the width of the spreading area is measured. Basically, the period can be freely chosen. In the performed investigations a period of 5 s is set. On one hand, 5 s of flow time lead up a proper distinction of results and on the other hand it is short enough to avoid wetting of the whole surface. For this reason, it is important to set the limit. An example for this type of measurement is shown in Fig. 5.

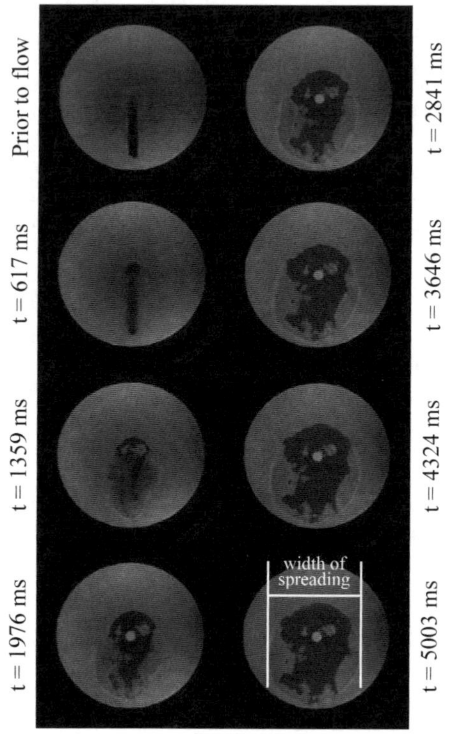

Figure 5: Measurement of the spreading width using sample "type I"

The testing procedure is evaluated with the two fillers B-Ag72Cu-780 and B-Ag35CuZnMnSnSi-680/700 on four different base materials using flux. The measured spreading widths are shown in Fig. 6. It can be seen that, the wetted areas of the two fillers do not differ significantly on mild steel substrates. In contrast to that, on stainless steel and copper, B-Ag35CuZnMnSnSi-680/700 spreads about twice as wide as B-Ag72Cu-780. An opposite effect occurs on brass CuZn37.

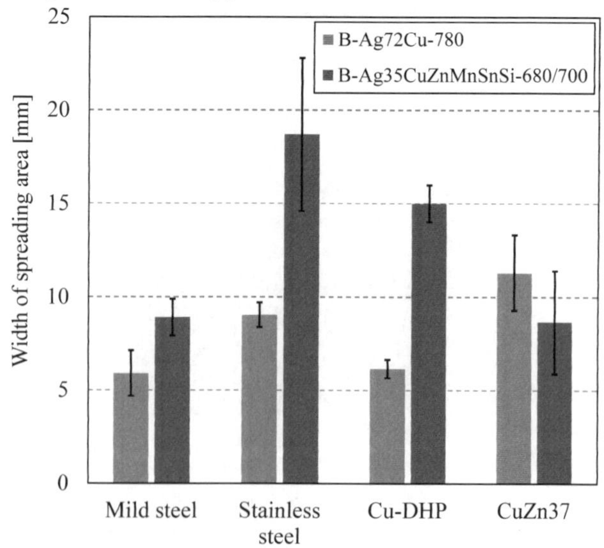

Figure 6: Measured width of spreading areas of two fillers on four different base materials

The experimental setup for the investigation of the flow behavior in gaps using sample "type II" is presented in Fig. 7. In this case the flow velocity of the liquid brazing filler inside the gap is of interest. The values are determined using single frames of the CCD-camera. First tests were carried out using gap widths of 50 μm and 100 μm. A width of 150 μm leads to accurately evaluable images. Due to the reduction of the flow velocity inside the wider gap, the results can be easily distinguished. Apart from that, the flow front is better detectable. At first, the possible flow distance respectively length of the gap is measured. After that, considering the capture time of every single frame is known, the duration of the flow process can be determined. With that information, an average velocity can be calculated. In Fig. 7, an example of a sequence is shown. The arrows indicate the different locations of the flow front inside the gap during the test.

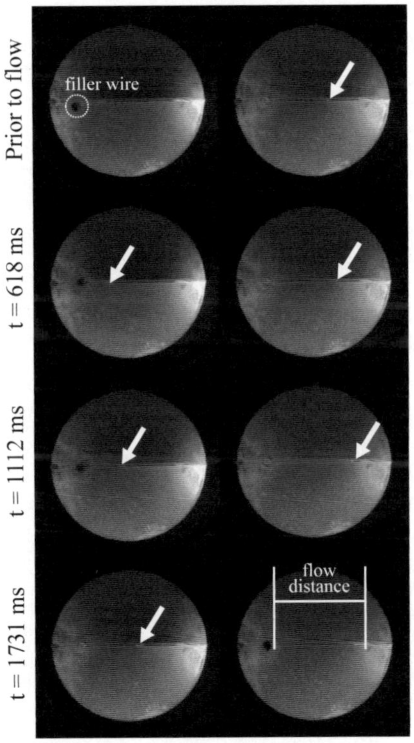

Figure 7: Measurement of the flow velocity in gaps using sample "type II"

The experiments are carried out using four brazing fillers on two different base materials. The calculated values of the flow velocities range between 1 mm/s and 15 mm/s. The results are summarized in Fig. 8. A standard deviation of about 20% of the mean value occurs in each sample series. The brazing fillers B-Ag72Cu-780, B-Ag35CuZnMnSnSi-680/700 and B-Cu60Zn(Si)(Mn)-870/900 come up with significantly higher flow velocities than B-Ag60CuSn-600/730. Reference data on flow velocities tend to noticeably higher values, see Yost et al. [7] for example. The lower values in this investigation are probably caused by the flow resistance that the molten flux generates when it is displaced by the liquid brazing filler.

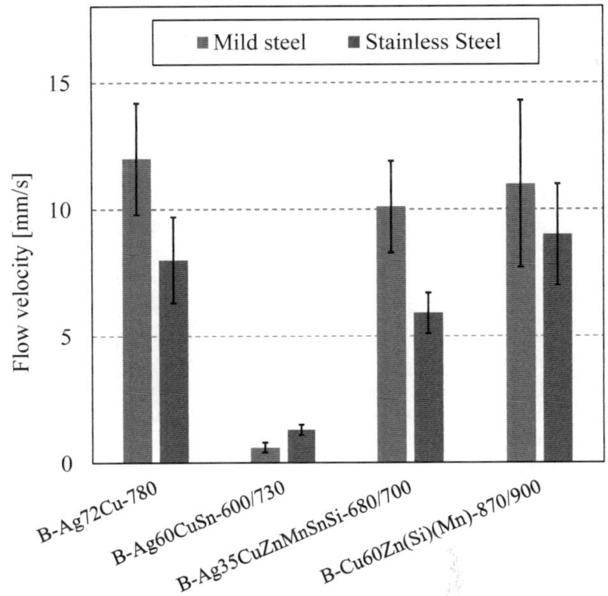

Figure 6: Calculated flow velocities of four fillers on two different base materials

Conclusions

In the present study, a new procedure was investigated, to characterize spreading and flow behavior considering fluxes by using an in-situ test. One of the main objectives was using a simple, but practice-oriented experimental setup. Furthermore, the test should be expeditious and easy to establish. The properties of different brazing fillers can be compared to each other regarding the width of spreading areas and the flow velocity inside gaps. The evaluation generates quantitative results on these properties for different combinations of brazing fillers and base materials. There is a scatter in the results, which is caused by the dynamics of the process. The mean values show clear tendencies, which allow a distinction of the different properties of the fillers. Further research will investigate the feasibility of using fillers in form of pastes for the developed procedure.

Acknowledgments

Part of the research was carried out within the scope of the project # 18705 BR funded by the Bundesministerium für Wirtschaft und Energie (BMWi) via the Arbeitsgemeinschaft industrieller Forschungsvereinigungen "Otto von Guericke" e.V. The authors thank the BMWi for financing this project.

References

[1] Weirauch, D.A. *et al*, "The spreading kinetics of Ag-28Cu(L) on nickel(s): Part I. Area of spread test on nickel foil" *Journal of Material Research,* Vol. 11 (1996), pp.1897-1916

[2] Zaremba, P., Hart- und Hochtemperaturlöten, DVS-Verlag (Düsseldorf 1988)

[3] Warren, J.A. *et al,* "Effect of phase change and solute diffusion on spreading on a dissolving substrate" *Acta Materialia,* Vol. 57 (2009), pp. 6022-6036

[4] Yost, F.G., "The triple line in reactive wetting" *Scripta Materialia,* Vol. 38 (1998), pp. 1225-1228

[5] Prabhu, K.N. *et al,* "Review of non-reactive and reactive wetting of liquids on surfaces" *Advances in Colloid and Interface Science,* Vol. 133 (2007), pp. 61-89

[6] Shibata, H. *et al,* "The Contact Angle between Liquid Iron and a Single-Crystal Magnesium Oxide Substrate at 1873 K" *Metallurgical and Materials Transactions B,* Vol. 35 (2004), pp. 179-181

[7] Yost, F.G. *et al,* "Solder wetting kinetics in narrow v-grooves" *Acta Materialia,* Vol. 45 (1997), pp. 5337-5345

[8] Hall, A.C. *et al,* "Visual observations of liquid filler metal flow within braze gap" *Science and Technology of Welding and Joining,* Vol. 9 (2004), pp. 95-102

Diffusion Fields in TRILLIUM™ Brazing Sheet under Different Heating Rates

Y. Wu[1], H. Fu[2], D. K. Hawksworth[3], C.-N. Yu[1], D. P. Sekulic[1]

[1]*Department of Mechanical Engineering, College of Engineering, University of Kentucky, Lexington, KY, 40506, USA*
sekulicd@engr.uky.edu

[2]*Corning Inc., Harrodsburg, KY, 40330, USA*

[3]*Diomedea Inc., Oakville, ON, L6J 2V8, Canada*

Abstract

The influence of heating rate on the diffusion of elements across the clad-core interface of brazing sheet clad with TRILLIUM™ composite was investigated. Migration of elements in such a system is responsible for brazeability performance, and hence understanding the impact of the heating history on residue formation after melting is of paramount importance. A series of tests have been performed exposing test specimens to different heating rates. The objective was to establish the volume of filler metal available on melting for flow into the joint and to predict the resultant fillet size. For this purpose, TRILLIUM™ brazing sheet samples were heated at heating rates of 1°C/min, 5°C/min, 10°C/min, 15°C/min, 20°C/min and 40°C/min. The distribution of Si, K, Cu, Ti, Fe, Mn, Al was determined using Glow Discharge Optical Emission Spectroscopy (GDOES). It was demonstrated that Si diffuses more prominently from the clad to core when the heating rate decreases. The Si depleted α-Al zone increases and the core alloy has more localized erosion at the clad-core interface at lower heating rates, however, the main clad/core interface position remains the same. Cu has the highest mobility and distributes evenly at different heating rates. Joint size is the smallest for heating rate of 1°C/min. As heating rate increases, the joint size increases correspondingly. When heating rate reaches 10°C/min, joint size does not change markedly with further increase of heating rate.

Introduction

Aluminum brazing is widely used in automotive heat exchanger manufacturing, aerospace and variety of other applications. It is often performed by furnace brazing [1]. The characteristics of a brazed joint depend on critical process parameters including surface preparation, furnace atmosphere, preheat conditions, heating rate, peak temperatures, dwell time above the melting point and cooling rates.

The impact of the peak temperature, dwell time, and atmosphere, for example, have already been widely discussed [2, 3, 4, 5]. However, the influence of heating rate is not widely reported. If the heating rate is too high, filler metal will not be uniformly heated and some filler metal may not fully melt, and thus will affect the quantity that can flow into a joint. In addition, thermal stresses would be larger at a higher heating rate, clearly detrimental to joint mechanical properties. On the other hand, if the heating rate is too low, Si diffusion from the clad to the core will be more significant, and thus less Al-Si eutectic would be available for brazed joint formation.

Moreover, although the Si diffusion in an aluminum brazing sheet has been frequently studied [2, 3, 4, 6], the migration of other elements (K, Cu, Ti, etc.) have not been reported in detail. This study will give an insight into diffusion fields in brazing sheet under different heating rates.

Materials and experimental procedures

Trillium™ brazing sheet with a total thickness of 317 μm and a 25 μm single clad layer (8% of the total thickness) was used in this study (see Fig. 1). The brazing sheet includes approximately 5 wt.% potassium-aluminum-fluoride flux embedded in the clad [7, 8, 9]. Chemical composition as declared by the manufacturer of this brazing sheet is listed in Table 1. The clad is AA4045 and the core is AA3003. Dimensions of the samples used in the study were 21 × 26 × 0.32 mm^3.

Table 1. Chemical composition (nominal) of Trillium™ brazing sheet (wt.%)

	Si	Fe	Cu	Mn	Mg	Zn	Ti
Clad	10.6	0.08	0.001	0.001	0.007	0.002	0.006
Core	0.068	0.252	0.344	1.435	0.002	0.003	0.144

The samples were ultrasonically cleaned first in a soapy water, rinsed in water and finally in ethanol. Then the samples were exposed to different heating rates in the controlled atmosphere transparent hot zone furnace (Centorr model #: 2091). Furnace chamber was purged using 99.999% ultra high purity nitrogen at 2 scfh (944 ccm) for 2 hours. During the test, nitrogen is continuously supplied at the same rate as during purging. The series of test runs at different heating rates were executed as follows: samples were heated first from room temperature to 540°C at the largest heating rate of the furnace system (around 42°C/min), followed by a dwell time of 1 min at that temperature level. Subsequently, the samples were heated to 600°C under various heating rates. The selected rates include:

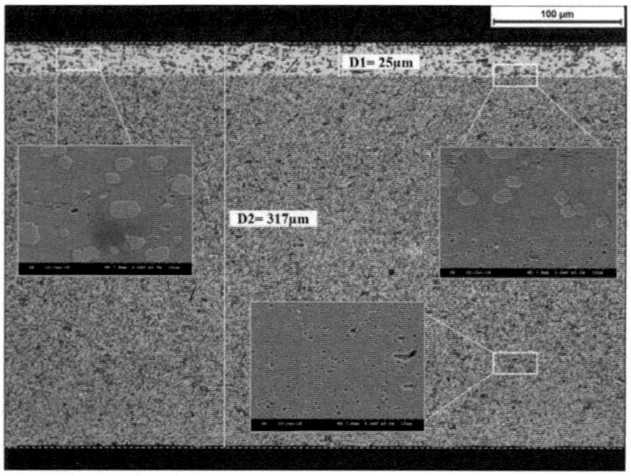

Fig. 1 Trillium™ brazing sheet with clad thickness and total thickness measurement using Image-Pro Plus 6.0. D1 represents the clad thickness; D2 represents the total thickness of the brazing sheet. Three SEM image inserts show details of the clad, the clad-core interface and the core, respectively.

40°C/min, 20°C/min, 15°C/min, 10°C/min, 5°C/min and 1°C/min. After reaching the peak of 600°C, another dwell lasting 2 min was imposed, followed by a natural cooling. The temperature profile is shown in Fig. 2. It was previously reported that the diffusion of Si becomes quite rapid just before reaching the solidus temperature for an Al-Si alloy, i.e., in the range between, say, 540°C to 574°C [2]. That is the reason why the intermediate temperature of 540°C was chosen.

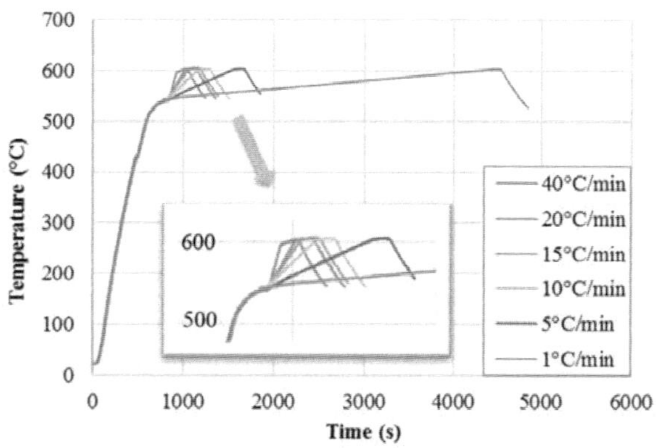

Fig. 2 Temperature profile for different heating rates

Analysis using glow-discharge optical emission spectroscopy (GDOES) was performed at randomly selected locations on the re-solidified clad. GDOES is a surface analysis technique for the direct depth profiling of thin films and industrial coatings [10]. During GDOES experiments, the samples are sputtered by Ar+ ions and accelerated neutral species with very low energies (<50 eV). The sputtered atoms are then excited by the plasma and de-excited by emitting photons with a characteristic wavelength, enabling element discrimination [11, 12]. The analysis was duplicated to verify the reproducibility of data and independence of profiles obtained on the selected location.

Elements concentration along the depth of the solidified clad and the substrate near the interface were collected for each sample. The samples were further processed by cutting-off a cross-section 5 mm away from the end of the longer edge of each sample and then imbedded in resin by cold mounting. Subsequently, the samples were polished in a sequence of steps using #220, #500, #800, #1200 SiC papers followed by 9 µm, 3 µm and OPU suspensions and then etched by Keller's reagent.

Table 2. Designation labels for samples exposed to various heating rates

B0: Unbrazed sample 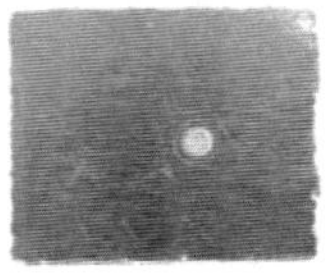 One test location was made.	
B40C: Brazed sample at 40°C/min 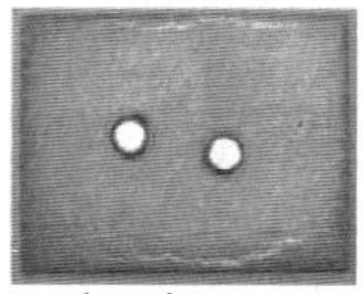 Two test locations were made.	B20C: Brazed sample at 20°C/min 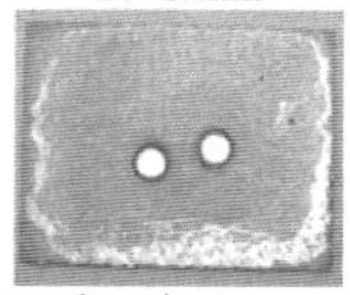 Two test locations were made.
B15C: Brazed sample at 15°C/min 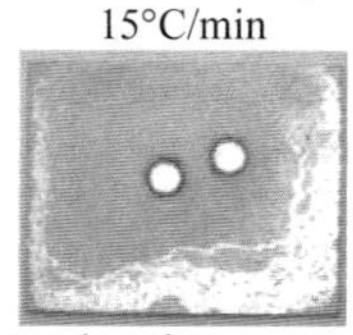 Two test locations were made.	B10C: Brazed sample at 10°C/min 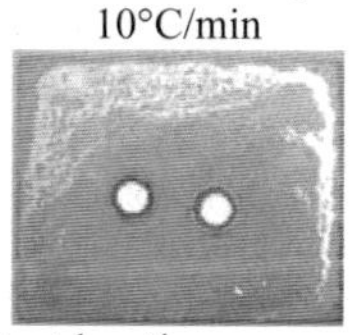 Two test locations were made.
B5C: Brazed sample at 5°C/min 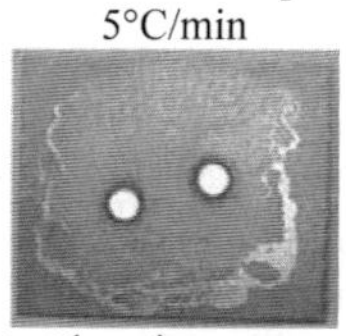 Two test locations were made.	B1C: Brazed sample at 1°C/min 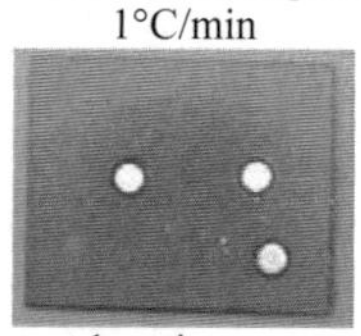 Three test locations were made.

Results and discussion

Figure 3 shows a comprehensive presentation of the distribution of elements across the interfaces of the studied specimens. Each plot shows the concentration distribution for all major elements obtained for (i) non-brazed and (ii) brazed coupons. These distributions indicate migration of elements for samples exposed to different heating rates. Designation labels for samples exposed to various heating rates in Fig. 3, are listed in Table 2.

After comparing the nominal concentrations given in Table 1 and the concentrations of the same elements obtained by

GDOES in Fig. 3(a) for the non-brazed samples (B0), one can draw the conclusion that the values for Si, Fe, Cu, and all the other elements agree quite well (see Table 3).

Table 3. Comparison of elements' concentration between the nominal and the GDOES measured data

	Clad		Core	
	Nominal (%)	Measured (%)	Nominal (%)	Measured (%)
Si	10.6	10.5	0.068	0.1
Fe	0.08	0.08	0.252	0.25
Cu	0.001	0.003	0.344	0.34
Mn	0.001	0	1.435	1.4
Ti	0.006	0.005	0.144	0.15
Al	89.30	88	97.75	97

Diffusion profiles – the impact of a heating rate change

From Fig. 3 (a), one can see that for the unbrazed sample B0, the elements distribution features a smooth transition at the interface of the clad and the core (at the depth between 20–28 μm from the top of the clad layer). The diffusion has already taken place when the clad and core were hot rolled together (notice the variable scaling factors for presented concentration distributions).

One may hypothesize that the hot rolling process leads to a local concentration increase at the very top of the brazing sheet at the depth of ~1 μm, for Si, Fe, Cu, K. For example, for B0 (which is the original Trillium™ sample), the concentration of Si in the clad is about 10.6%. However, at the top surface interface, the concentration of Si can reach about 13%. Due to the low concentration of Mn, Ti, these two elements do not feature a visible concentration change artefact.

The same distributions of elements are re-plotted next, with each plot offering concentration distributions of a single element but for all coupons on the same plot (refer to Fig. 4), i.e., each plot is presenting the concentration distributions of the selected element (e.g., Si) for various heating rates during brazing including the distribution of the same element in the virgin sample (before brazing).

The listed conclusions follow from the study of the diffusion plots.

1) Si diffuses from the clad layer (nominally declared as 10.6%, here measured as 10.5%) into the core alloy (nominally declared as 0.068%, here measured as 0.1%) where it stabilizes at a concentration of about 1.5 %, which corresponds to the solid solubility of Si in Al [2].

   A concentration peak of Si at a depth around 1μm is an artefact related to the conditions following the manufacturing of a brazing sheet by hot rolling. The concentration peak can reach as much as 18.5%. The additional peculiarity is that the peak concentrations appear to be higher in a heated sample than that in the virgin state (unbrazed B0) sample. It is hypothesized that when heated, the eutectic liquid's Si floats to the surface, hence more Si will be located at the top surface. This is further demonstrated by the images in Table 4, where one can see more Al-Si eutectic at the top for the heated samples.

   The heating rates decrease prompts more Si to diffuse into the core of the substrate. From the diffusion curves in Fig. 4(a), one can see that the slope of the Si distribution is steep for an unbrazed case (compared to the cases of 40 and 20°C/min) while the distribution gradually changes for further lowering the heating rates. This means that the concentration of Si changes more drastically for the lower heating rates within the first 5 μm.

2) Potassium has a high concentration at the surface of the re-solidified clad layer (up to 1 μm of depth), leaving the rest of the re-solidified residue potassium depleted. This means the flux has floated to the top surface and acted upon the oxide layer of the aluminum brazing sheet, promoting wettability of the liquid metal over its substrate.

   However, one can see that the potassium concentration is visibly higher than zero at depths of ~15–20 μm for B1C and B5C samples. It can be hypothesized that at low heating rates, when a thinner surface layer of molten filler metal is produced, molten flux remains trapped within the unmolten residual clad layer and does not segregate to the surface as previously observed.

   An entrapment of the potassium-aluminum-fluoride salt within the re-solidified clad layer with a very low heating rate may explain the "oxide skinning" observed on the surface after brazing (refer to sample B1C in Table 2).

3) Cu diffuses from the core alloy (nominally declared as 0.344%, here measured as 0.34%) into the clad layer (nominally declared as 0.001%, here measured as 0.003%) stabilizing at a concentration of approximately 0.2%.

   The concentration of copper in the core and in the clad is generally at the level between 0.20% and 0.25% except for the B1C sample, which means that Cu has good mobility and becomes evenly distributed when the sample is heated. For B1C, more Cu moved to the clad, leaving the concentration larger than that in the core.

   What is unusual (unless an artefact of the performed measurements) is that sudden concentration decreases may be observed at a roughly 1 μm of depth for all the heated samples, hence opposite from the distribution of other elements.

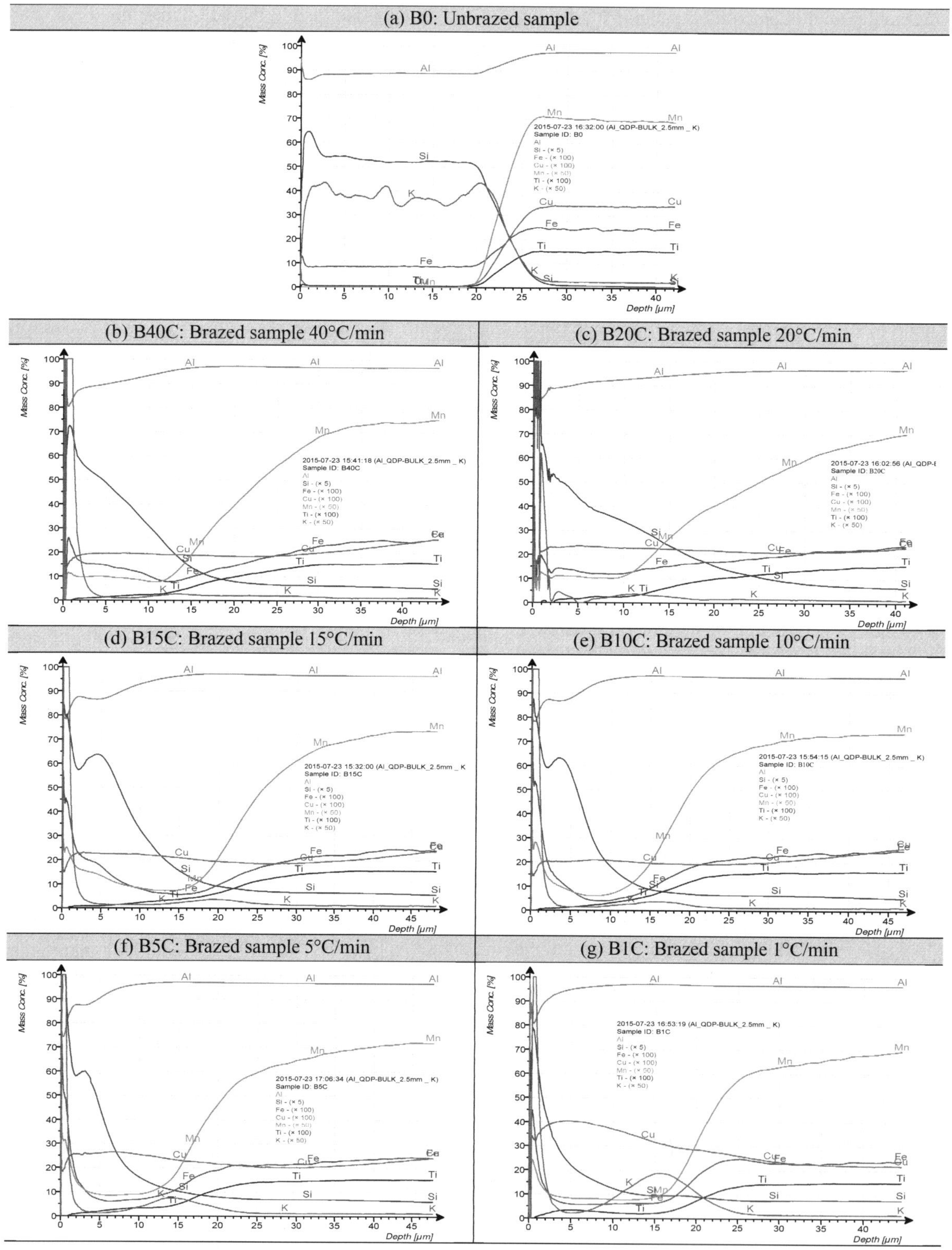

Fig. 3 GDOES data of elements' concentration profile along depth for different TRILLIUM™ brazing sheet samples. (a) B0, (b) B40C, (c) B20C, (d) B15C, (e) B10C, (f) B5C, (g) B1C

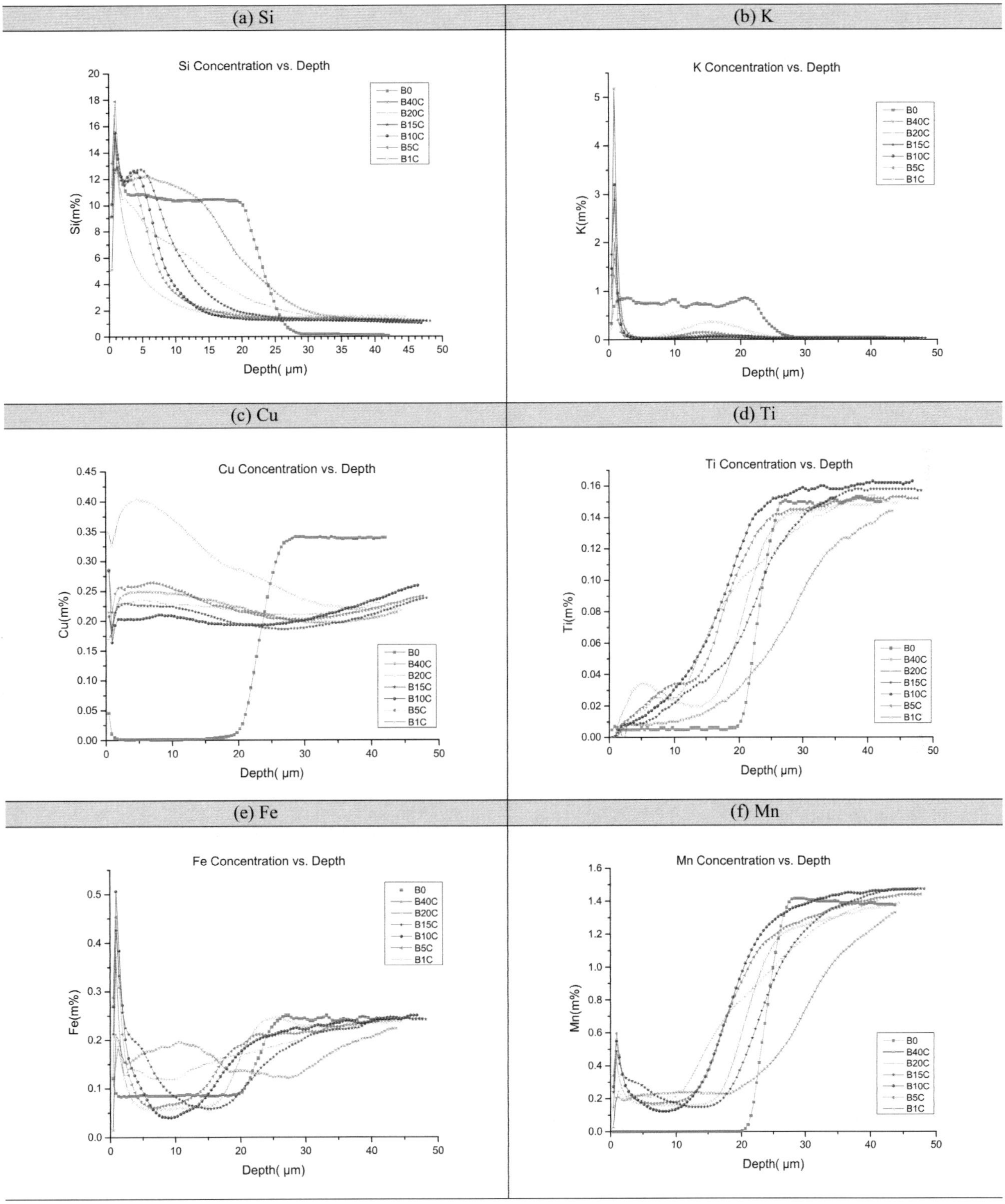

Fig. 4 Elements concentration of Si, Fe, Mn, K, Cu, Ti along depth for different samples obtained under different heating rates: (a) Si, (b) K, (c) Cu, (d) Ti, (e) Fe, (f) Mn

Cu can form Al-Si-Cu eutectic in Al-Si alloy. This ternary eutectic has a lower melting temperature (solidus: 525°C) than Al-Si alloy. Thus, adding Cu into Al-Si alloy can lower the melting temperature of the filler metal [13]. As a result, with more Cu diffused to the clad, melting temperature may decrease to some extent. In addition, corrosion resistance is increasing when Cu is increasing from 0% to 5% [14]. However, Cu can also form $CuAl_2(\theta)$ intermetallic in Al-Si alloy, which is brittle and harmful for the joint strength. With the heating rate decrease to 1°C/min, Cu concentration dramatically increases in the clad layer, which may help lower the melting temperature, and improve the corrosion resistance. On the other hand, this may also promote the growth of $CuAl_2(\theta)$ intermetallic.

4) Ti diffuses from the core alloy (nominally declared as 0.144%, here measured as 0.15%) into the clad layer (nominally declared as 0.006%, here measured as 0.005%). Heating rate will not change significantly the Ti concentration distribution, which means that Ti has low mobility in the brazing sheet. However, as heating rate decreases, Ti is prone to move to the clad from the core but at a very low rate (refer to B1C).

   The effect of Ti on Al-Si alloy is not clear, but it was reported that for Zn-Al filler metal, the microstructure of a Zn-Al alloy can be refined remarkably with the trace addition of 0.01 wt.% to 0.05 wt.% Ti. The Al_3Ti particles contribute to the improvement of the microstructure. The wettability of Zn-Al is significantly improved when the quantity of Ti added in the alloy is 0.03 wt.%. [15]

5) Fe diffuses from the core alloy (nominally declared as 0.252%, here measured as 0.25%) into the clad layer (nominally declared as 0.08%, here measured as 0.08%). It becomes more evenly distributed between the core and the clad when the heating rate increases. There are also concentration local extrema at a depth around 1 μm for the heated samples. The highest concentration of Fe can reach as much as 0.55%.

6) Mn diffuses from the core alloy (nominally declared as 1.435%, here measured as 1.4%) into the clad layer (nominally declared as 0.001%, here measured as 0%) stabilizing at a concentration of approximately 0.2%. Concentration peaks for heated samples at a depth around 1 μm was also observed, and the highest concentration of Mn can reach as much as 0.6%.

Joint size comparison

From the micrographs (Table 4), one can learn that the core is eroded in each case. With the heating rate decreasing, the eroded zone increases. Excluding the random erosion intrusions into the core, the main clad/core interface location remains the same after heating at different rates and following the re-solidification (compared to the original coupon, refer to the middle dash line, the right column in Table 4). At each eroded location, one can observe that Al-Si aggregation at the top, forming a triangular shape. However, at places that are not eroded, Al-Si eutectic is in smaller quantity and it is uniformly distributed. Due to an inhomogeneous solidification, the re-solidified clad thickness for higher heating rates can be larger than for the original unbrazed sample, refer to the top dash line for samples B40C, B20C, B15C, right column in Table 4. However, the re-solidified clad layer thickness is thinner than the original for B10C, B5C, B1C samples.

Silicon depleted alpha phase zone is increasing when the heating rate decreases. It means that the volume of molten filler metal layer that forms when the temperature exceeds the eutectic temperature is reduced with an increased diffusion (and reduced heating rate).

This was further demonstrated by the measured data using Image-Pro Plus; the Si depleted α-Al area can be measured (see Fig. 5). This zone is marked horizontally, so that the equivalent Si depletion α-Al domain thickness can be estimated (see Table 5). Results in Table 5 were averaged by repeated measurements of at least two different locations for a given sample. One can see that for a heating rate of 1°C/min, the Si depletion zone is 24.17 μm, which is very close to the original clad thickness (25 μm). We can also observe that the core is more intensely eroded at low heating rates, which again, contributes to the increase of the Si depletion zone thickness. These data are important for predicting the joint size.

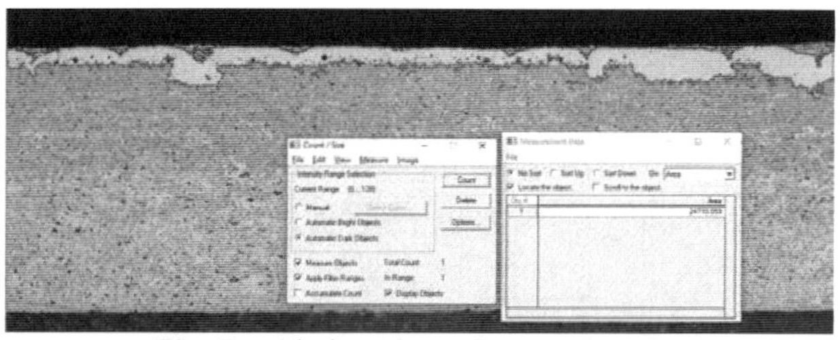

Fig. 5 α-Al phase layer for sample B1C.

Table 5. Equivalent Si depleted α-Al phase thickness for different heating rates.

Heating rate (°C/min)	Equivalent Si depleted α-Al phase thickness (μm)	Standard deviation (STD) (μm)
40	14.8	0.87
20	15.9	1.91
15	17.2	1.11
10	17.5	0.53
5	17.6	2.91
1	24.2	1.55

When heating rate decreases, the available liquid metal is decreasing, thus one may predict that the brazed joint size will be smaller at lower heating rate. Fig. 6 illustrates the joint size for heating rates of 1°C/min and 10°C/min when braze

Table. 4 Metallurgical graphics for different heating rates (left: 100×, right: 200×)
(a) B0, (b) B40C, (c) B20C, (d) B15C, (e) B10C, (f) B5C, (g) B1C
On the right column: the topmost dash line indicates the original (unbrazed) top surface of the clad;
the middle dash line indicates the original (unbrazed) clad/core interface;
the bottom dash line indicates the original (unbrazed) core bottom surface.

(a) B0: Unbrazed

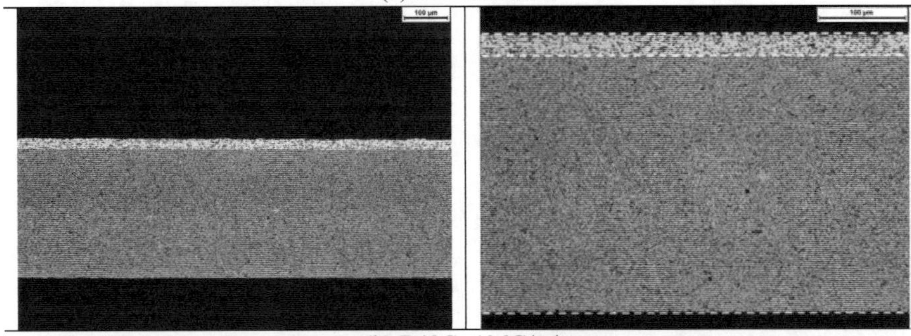

(b) B40C: 40 ℃/min

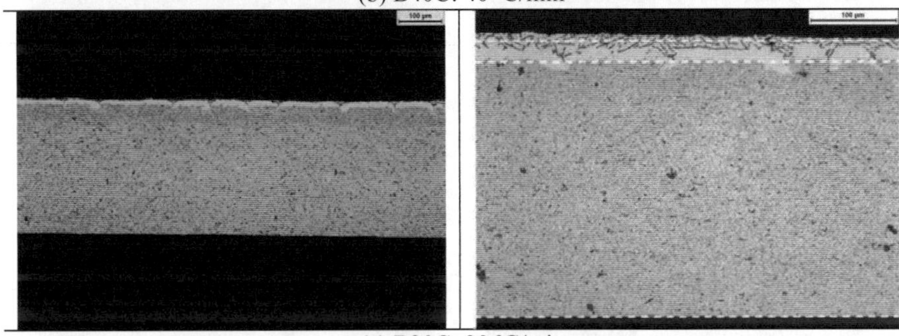

(c) B20C: 20 ℃/min

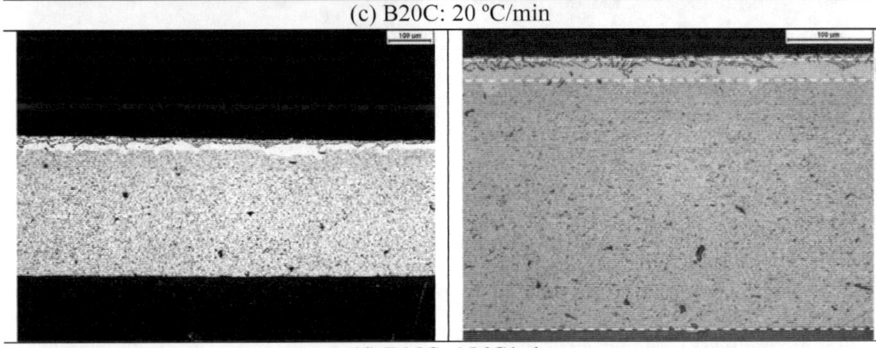

(d) B15C: 15 ℃/min

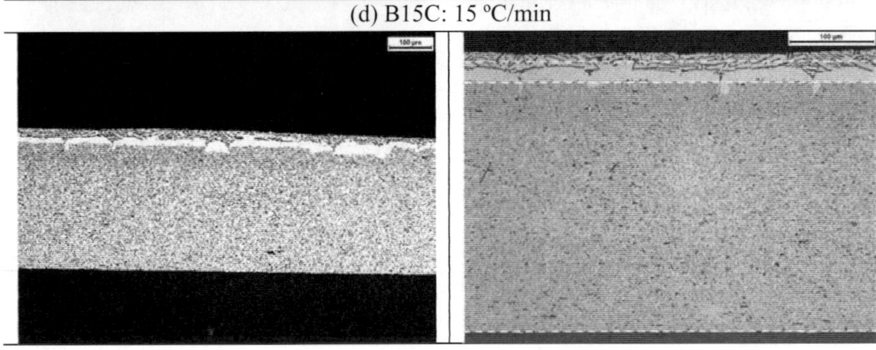

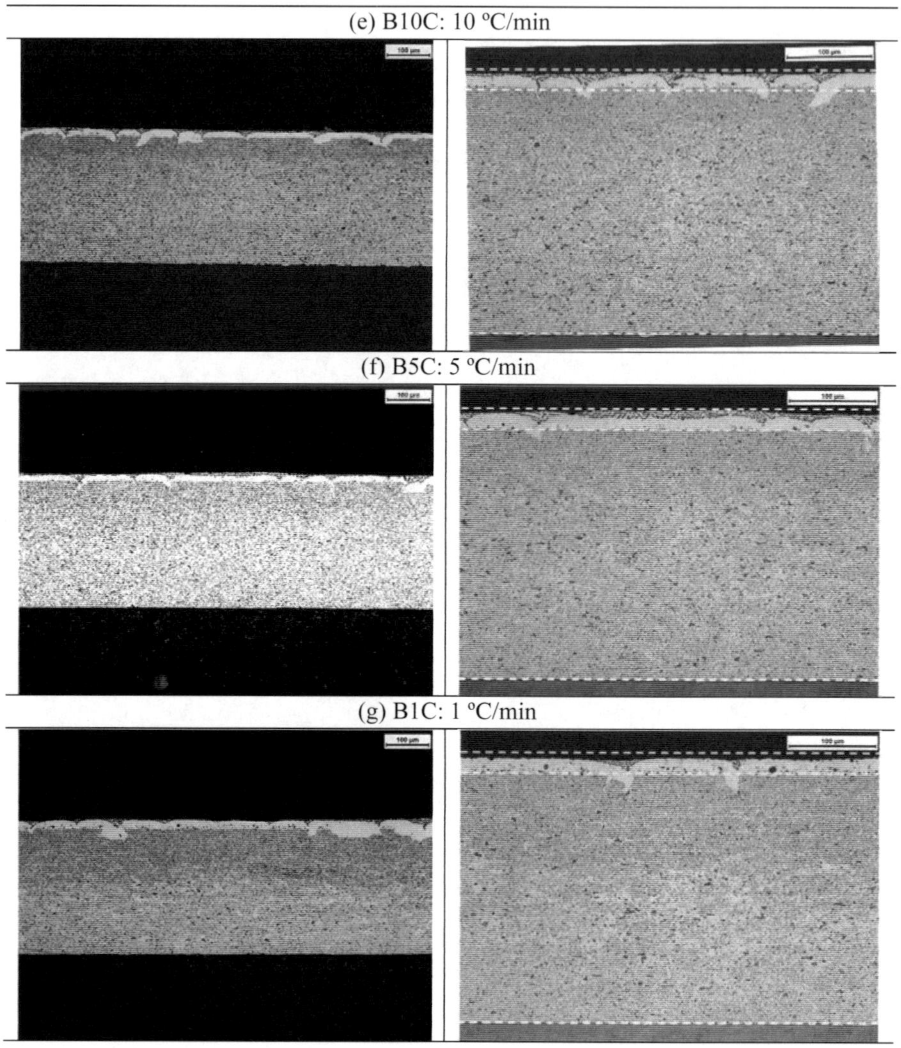

Trillium™ as the horizontal piece with AA3003 as the vertical piece. L_v length is used as the simplest representative parameter to evaluate the joint size (the vertical length of a triple line location) [16].

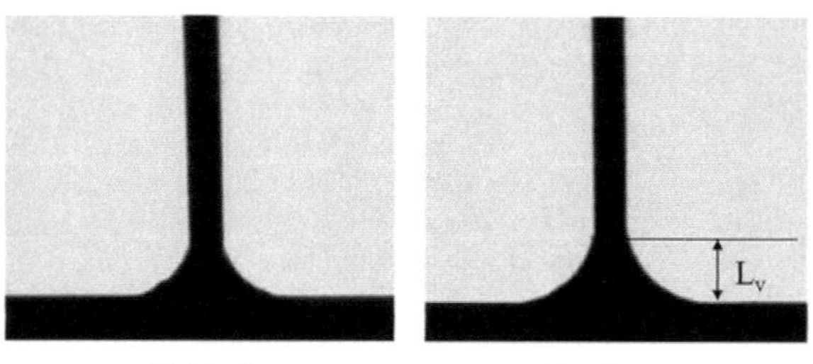

Fig. 6 Triple line location comparison at different heating rates

Table 6. Triple line location distance (L_v) at different heating rates

Heating rate (°C/min)	Triple line location (mm) (left side)	STD (mm)	Triple line location (mm) (right side)	STD (mm)
1	0.54	0.011	0.52	0.002
5	0.75	0.027	0.74	0.020
10	0.78	~ 0	0.80	0.008
20	0.78	0.023	0.78	0.013
40	0.80	0.008	0.81	0.015
60	0.79	0.028	0.80	0.009

One can see that the joint size is much larger when brazed at 10°C/min compared to that brazed at 1°C/min, as expected. Table 6 gives the values of L_v for joints brazed at different heating rates. The heating rate of 15°C/min is not included because the trend is obvious. The triple line location data were again illustrated in Fig. 7. One can see that for the lowest heating rate, say 1°C/min, joint size is the smallest. As heating rate increases, the joint size increases correspondingly, but after 10°C/min, the joint size would not change significantly. The Si depleted α-Al phase is almost the same for heating rates of 10°C/min, 20°C/min, 40°C/min. However, there is a dramatic increase of the Si depleted area for heating rate of 1°C/min, thus much less liquid metal is available to form the joint.

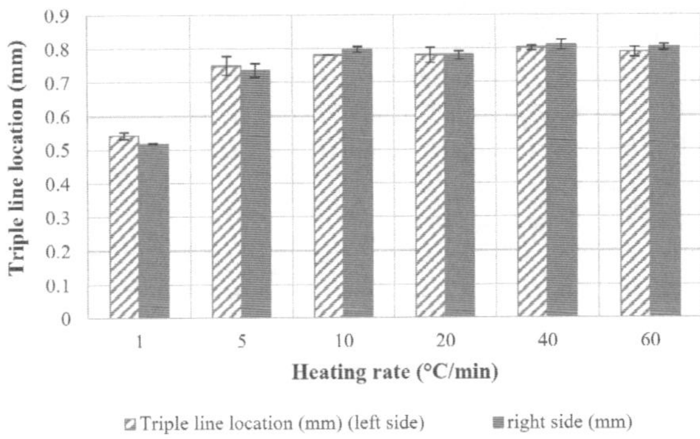

Fig. 7 Triple line location at different heating rates

Conclusions

In this paper, a series of tests have been performed by exposing Trillium™ brazing sheet to different heating rates: 1°C/min, 5°C/min, 10°C/min, 15°C/min, 20°C/min and 40°C/min. Quantitative data for the distribution of Si, K, Cu, Ti, Fe, Mn have been established by using Glow Discharge Optical Emission Spectroscopy (GDOES).

It is demonstrated that Si diffuses more from the clad to the core when heating rate decreases, leaving the concentration slope the steepest within the clad when heating rate decreases to 1°C/min.

Cu has the best mobility. Ti, Fe, Mn are less mobile. Potassium fluoro-aluminate flux may get entrapped for low heating rates, e.g. 1°C/min.

Si depleted α-Al zone is increasing and core is eroded more when heating rate decreases. However, the main clad/core interface does not appreciably move.

The triple line location distance in an equilibrium state at the onset of solidification is the smallest when heating rate is 1°C/min. As heating rate increases, the joint size increases correspondingly, but after 10°C/min, the joint size would not change prominently.

Acknowledgments

This work has been funded within NASA's Physical Sciences Research Program, Grant# NNX17AB52G. Materials and the early stage of this work were supported by Gränges AB, Finspång, Sweden. TRILLIUM™ Technology is protected by United States Patent No. 8871356 as well as by corresponding patents and pending patent applications in other major countries. TRILLIUM™ Technology is a registered trademark of Gränges AB.

References

[1] Brazing handbook [J]. 5th edition, AWS, Miami, 2007.

[2] Terrill, J. R. "Diffusion of silicon in aluminum brazing sheet." *Welding Journal* 45.5 (1966): S202.

[3] Gao, F., et al. "Solid state Si diffusion and joint formation involving aluminum brazing sheet." *Materials Science and Engineering: A* 337.1 (2002): 228-235.

[4] Gao, F., et al. "Formation of microlayers of clad residue on aluminium brazing sheet during melting and resolidification in brazing process." *Materials science and technology* 20.5 (2004): 577-584.

[5] Yu, C., et al. "Al brazing under severe alterations of the background atmosphere: A new vs. traditional brazing sheet." *Brazing and Soldering 2012: IBSC Proceedings of 5th International Conference*. ASM International, 2012.

[6] Zhao, H., and Sekulic, D. P. "Diffusion-controlled melting and re-solidification of metal micro layers on a reactive substrate." *Heat and mass transfer* 42.6 (2006): 464-469.

[7] Ogilvy, A. J. W., Hawksworth, D. K., & Abom, E. (2014). *U.S. Patent No. 8,871,356*.

[8] Hawksworth, D. K. et al, "A Mechanistic Study of Aluminium Controlled Atmosphere Brazing Processes," *12th Vehicle Thermal Management Systems Conference and Exhibition*, Nottingham, U.K., May 10-13, 2015.

[9] Sekulić, D. P., ed. Advances in brazing: science, technology and applications. Woodhead/Elsevier Science, Cambridge, 2013, pp. 566.

[10] Nelis, T., and Payling, R. *Glow discharge optical emission spectroscopy: a practical guide*. Vol. 7. Royal Society of Chemistry, 2003.

[11] Galindo, R. E., et al. "Towards nanometric resolution in multilayer depth profiling: a comparative study of RBS, SIMS, XPS and GDOES." *Analytical and bioanalytical chemistry* 396.8 (2010): 2725-2740.

[12] Shimizu, K., et al. "Radiofrequency GDOES: a powerful technique for depth profiling analysis of thin films." *Surface and interface analysis* 35.7 (2003): 564-574.

[13] Chuang, T. H., et al. "Development of a low-melting-point filler metal for brazing aluminum alloys." *Metallurgical and Materials Transactions A* 31.9 (2000): 2239-2245.

[14] Wang, S. S., et al. "Corrosion behavior of Al–Si–Cu–(Sn, Zn) brazing filler metals." *Materials characterization* 47.5 (2001): 401-409.

[15] Feng, J., Xue, S., and Dai, W. "Effects of Ti on the brazability of Zn-22Al-xTi filler metals as well as properties of Cu/Al brazing joints." *Rare Metal Materials and Engineering* 42.12 (2013): 2453-2457.

[16] Kawase, H., et al. "Study of a method for evaluating the brazeability of aluminum sheet." *Welding Journal* 68.10 (1989): 396s-403s.

Furnace Based FEA Model of the Diffusion Bonding Process

Gemse, Felix; Fey, Andreas; Dahms, Steffen; Jahn, Simon
ifw Jena - Guenter Koehler Institut fuer Fuegetechnik und Werkstoffpruefung GmbH, Jena, Germany

Abstract

Diffusion bonding as a solid-state joining technology provides outstanding bond quality. Due to the lack of filler material bond strengths comparable to the base material are achievable without a macroscopic deformation of the parts. One challenge in setting up the bonding process is always finding the right bonding parameters. To achieve a high quality bond a homogeneous load and temperature distribution throughout the bond area is necessary. Especially for the bonding of larger parts. Available recipes from publications have a limited validity because every part geometry and every diffusion bonding system has its own thermal behavior. Therefore, the experimental efforts to find the optimal set of parameters are comparatively high. One solution to enhance the parameter development and reduce the number of experimental steps could be the Simulation of the diffusion bonding process. Actual FEA models are based on the brazing process and often start their calculations on the part surface. Additionally, the press system, and therefore the load distribution and its thermal influence are not included.
In this contribution, a furnace based FEA-model to calculate the temperature and load distribution for diffusion bonding in vacuum will be introduced, which contains all important components of the diffusion bonding system (heating elements, isolation or press ram). The model will be presented, its boundary conditions as well as FEA results.

Introduction

The further development of technical solutions and their more sophisticated constructions, result in higher requirements for the joining technology. Therefore the use of the diffusion bonding increases, because it offers a large degree of design freedom and allows the joining of an enormous variety of materials. This joining process will be a commercial used technology, but currently it is not investigated as other manufacturing methods are. The main problem is the knowledge of the significant process parameters, which depend on the part material, the workpiece geometry and the batch alignment in the furnace. Furthermore they are furnace specific. To conclude, diffusion bonding is a complex furnace process. For these reasons further research of that joining technology need a large number of experimental furnace cycles to get more information about the important process parameters. That leads to a high energy, cost and time expenditure, which can be prevent by modern engineering. Numerical solution methods are an efficient alternative to the experimental determination of process parameters. Especially the FEM offers a wide range of possibilities, justified in its differential calculating functionality. Therefore it is possible to couple different physical exercises so that in consequence many advantages for the simulation of the diffusion bonding arise. Especially the coupling of thermal and structural numerical systems can lead to new findings of the significant process parameters, which include the set temperature, the time to hold this temperature and the process force. Hence, the main motivation for the present study is to generate new knowledge for the geometry and material depending design of diffusion bonding cycles. Furthermore it is currently necessary to design the bonding cycle of new workpieces, for which the process parameters are unknown, simply based on experience. Therefore the bond quality is assessed after the accomplishment of the cycle to adjust the process parameters by an iterative procedure of further attempts. That result in time and cost intensive experiments before the bonding of parts. The amount of such tests can be reduced by using FE-Analyses to determine the important process parameters of the diffusion bonding.

Diffusion bonding

The diffusion bonding is a joining technology that joins workpieces in their solid state. Because of the absence of a liquid phase, it is possible to bond a large variety of materials including the possibility of combining mixed compositions [1]. The workpieces are joined by using heat and force under a process atmosphere, such as vacuum or different protective gases, as it is schematically shown in **Fig. 1**.

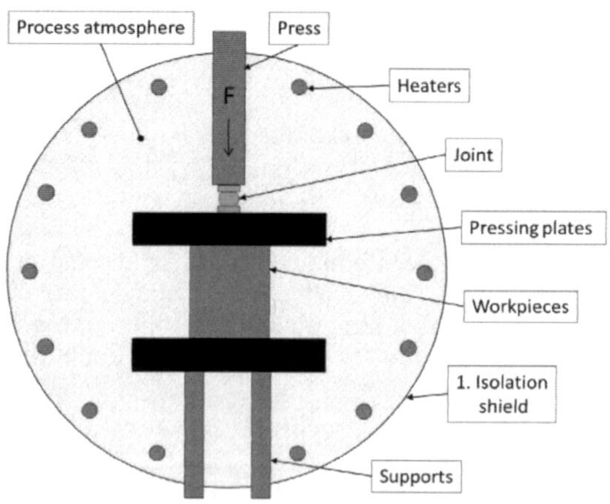

Fig. 1. Principle of the diffusion bonding

The generated firmly bonded joint is based on the effect of solid state diffusion. The parts have to be stacked and then heated to the elevated joining temperature. For the bonding process the temperature is typically set below the melting respectively solidus temperature (about 50%...90% Tm respectively Ts). An applied load is necessary to get the faying surfaces in close contact. The close contact allows atoms to "move" through the interface. Diffusion itself can be controlled by temperature, with the correlation that higher temperatures increase the diffusion rate. Temperature and stress driven, diffusion takes place across the faying surfaces. Due to the applied load, creep and plastic deformation of faying surface roughness occur, too [2]. At the end of the process, the bond line is not visible any more in the case of similar material joining.
Moreover the joined workpieces have a high dimensional accuracy, because there is no macroscopic plastically deformation [1]. For this reasons the diffusion bonding is mostly used in the aerospace technology, the power electronics and the optical industry. But the area of applications increases daily and the further research, for example using simulation technology, will promote the importance for the industrial use.

State of the art

The heat transfer during the simulation of a vacuum furnace includes mainly radiation. In contrast to conduction heat transfer, the numerical radiation calculation is highly non-linear, caused by the varying heat flow of the temperature differences between surfaces to the fourth power. This heat flow is calculated by **Equ. 1** [3].

$$\dot{Q} = A_i \cdot f_{ij} \cdot \varepsilon_i \cdot \sigma \cdot (T_i^4 - T_j^4) = C_{Str} \cdot (T_i^4 - T_j^4) \qquad (1)$$

(3.1)

$\dot{Q}$	Heat flow [W]
A_i	Area of element i [m^2]
f_{ij}	Shape factor between area i and j
ε_i	Emissivity of area i
σ	Stefan-Boltzmann-Constant: 5,67 10^{-8} [$\frac{W}{m^2\,K^4}$]
T_i	Temperature of area i [K]
T_j	Temperature of area j [K]
C_{Str}	Radiation exchange coefficient [$\frac{W}{K^4}$]

The heat conduction of the furnace simulation is limited to the heat dissipation by the cooled press and supports. They permanently extract heat from the batch, what result in a lower workpiece temperature.
There are some publications that are dealing with the simulation of furnace processes under vacuum. Thus, *Jiang* [4] explained the simulation of a soldering process, by which the temperature and stress distribution in a workpiece can be determined. He limited the thermal boundary conditions, as well as it did *Golanski* [5], to the direct workpiece heating by a defined temperature, which neglected the furnace and the radiation heat transfer. *Pissanetzky* [6] and *Hao* [7] created furnace simulations that can calculate the heating and cooling performance of the furnace and workpieces. Hence, deformation and stress distribution of workpieces cannot be determined. On the basis of these examples it can be seen, that there was no simulation of the diffusion bonding generated in the past, which includes every physical influence of the diffusion bonding system. Therefore the described investigation in the following presents a simulation model that includes temperature and load distribution of workpieces during the diffusion bonding process by taking the furnace isolation and the dissipated heat over the press and supports into account.

FEA-Model

The thermal part of the simulation was created by using a transient-thermal-analysis, which also calculates the heat storage capacity of every considered part. To get knowledge about the temperature dependent deformation of workpieces, the transient-thermal-analysis was indirectly coupled with a static-structural-analysis. So the thermal expansions of the parts affect the arising deformations, but not vice versa.
Different diffusion bonding furnaces were used for this work. All are high vacuum radiation shield furnaces. Therefore the isolation is constructed by several separated molybdenum shields. A CAD Model of one used system is shown in **Fig. 2.**

Fig. 2. CAD-Model of Furnace hotbox

All necessary material properties were determined by literature and the important ones were taken temperature dependent into account. **Fig. 3** shows exemplarily temperature dependent material properties of molybdenum and the bonded 1.4301 steel.
The numerical finite element model of the furnace was abstracted with a few simplifications. Therefore the thermocouple and the heaters were built as levitating bodies. There fastening in the real furnace consists of thermally insulating material. Thus, the heat, which is dissipated over these parts, can be neglected. The furnace isolation was included by integrating just the first radiation shield. The attempt to simulate the whole isolation with every shielding layer and their connections failed due to the fact that this model takes a very high amount of computing power and the thermal conductivity of the imperfect contacts between the isolation parts are unknown. For this reason the simulation was simplified and just the first radiation shield was included as it was mentioned above. For the workpiece temperature just the correct heat flow, which is emitted by the first radiation shield, is important. Hence, if the temperature of this part is known, the calculation deviation will be insignificant, because of the

correct radiation exchange between the shield and the workpieces. Furthermore the press and supports, the workpieces and supports by the several press plates as well as the joint were integrated in the numerical model without any simplification. By applying symmetry function, only a quarter of the structure needs to be calculated.

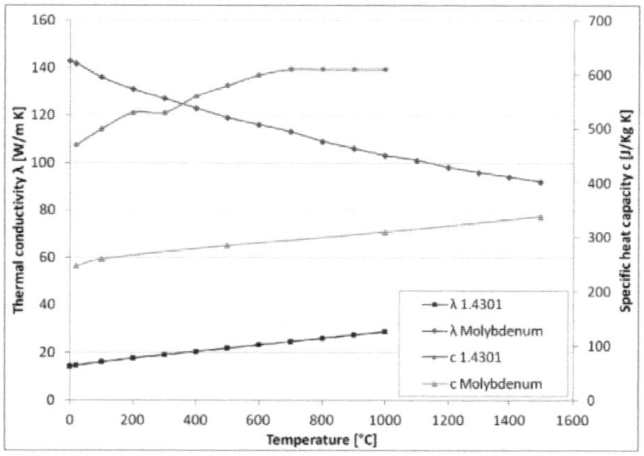

Fig. 3. Material properties of molybdenum and 1.4301

After creating the 3D geometry, the simulation model was discretized by a numerical grid. This mesh consists of several thousand elements, whereby 94 % of them are hexahedrons, which have a very good convergence behavior and a small numerical false [8]. The discretization accuracy of this mesh was controlled by a systematical mesh refining study. **Fig. 4** shows symmetrized geometry.

The numerical connections between the surfaces of the parts are all calculated as bonded, except the contact between the radiation shield and the press. This contact is frictionless to allow a movement of the press for the structural analysis possible.

After the complete discretization, the boundary conditions were defined. At the beginning a significant set temperature curve was determined for the simulation and the related experiment. Therefore a typical two stage heating cycle as was used. As already mentioned, the correct emitted heat flow of the first radiation shield, has to be integrated. Furthermore the cooling of the press and supports were included by setting the lower front faces of the supports and the upper shoulder faces of the press permanently to 23 °C (corresponds with cooling water temperature).

The last and the most important thermal boundary condition were to set the heaters temperature (transient). Therefore an intelligent PI temperature control was included by a user written subroutine, whereby the heaters temperature is automatically generated during the simulation. The functionality of this control is schematically shown in **Fig. 5**.

Thus, if a deviation between the set temperature and the current temperature of the thermocouple occur, the PI control will update the heaters temperature. This analysis loop is called "up in each" iteration, which is why the current heaters temperature depends on the furnace temperature and the user defined control parameters.

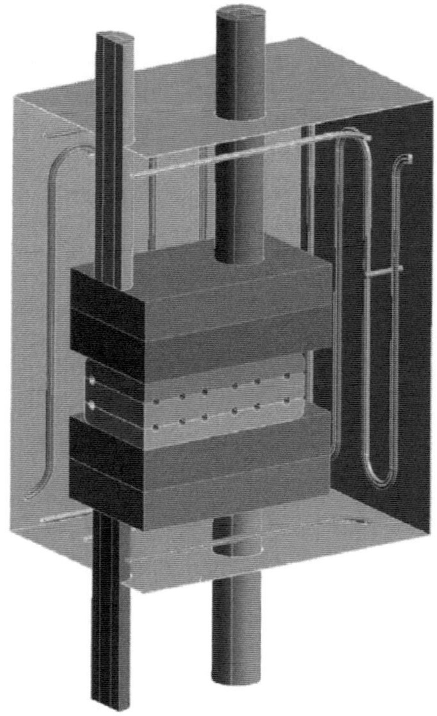

Fig. 4. Discretized and symmetrized geometry

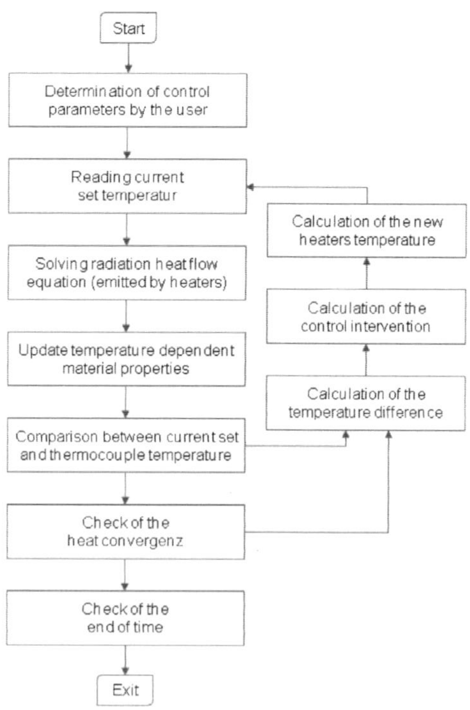

Fig. 5. Functionality of the PI temperature control

Hence, the thermal boundary conditions are completely determined, the thermal distribution can be calculated. These thermal results were subsequently imported to the already mentioned static-structural-analysis. Therefore it is possible to

calculate the deformation of the workpieces and the surface pressure in the joining zone at every time step and every temperature during the process. For this reason the definition of the structural boundary conditions were performed.
At the beginning, the degrees of freedom for all bodies that are not important for the structural analysis, like the heaters, were fixed for saving calculation time. The lower front faces of the supports were also fixed and the upper face of the press was applied with different forces, depending on faying surface size, e.g. 500 N for the quarter of a cube sample geometry with 15 x 15 mm². Thereby the surface pressure in the joining zone arises, which is calculated by **Equ. 2**.

$$p = \frac{F}{A} = \frac{500\,N}{15 \cdot 15\,mm^2} = 2{,}2\bar{2}\,\frac{N}{mm^2} \qquad (2)$$

(3.1)

p	Surface pressure in the bonding zone $[\frac{N}{mm^2}]$
F	Force (symmetrized) applied on press [N]
A	Symmetrized area of the bonding zone $[m^2]$

This over the whole process constant pressure is important as well as the workpiece temperature for the bond quality. Therefore a uniform surface pressure is desirable and should be achieved by supporting the workpieces with press plates.

Experiments

The experiment served primarily to verify the simulation results. Therefore a complete furnace cycle with the same setup as calculated by the simulation was performed. An example setup as it is shown in **Fig. 6**. Also the same set temperature profile was used for this thermal joining.
To measure the workpiece temperature during the process, holes with a diameter of 2 mm were drilled into each of the blocks for positioning additional thermocouples. Thus, there is one thermocouple for measuring the core temperature and one for the surface temperature in each block. This measured temperature curves were used to control the simulation results.

Fig. 6. Experimental setup

Results

Thermal analysis

A comparison between the simulated and the experimental temperature profile is given in **Fig. 7** for a small cube with edge length of 30 mm ("small part"). The "Set temperature" is the programed furnace temperature. For different locations the resulting temperature profile was analyzed. In **Fig. 7** the temperature profile of a single point of the assembly is given. The position of this point is equal to the core area thermocouple of the lower cuboidal block (calculated and measured temperature in **Fig. 7**).

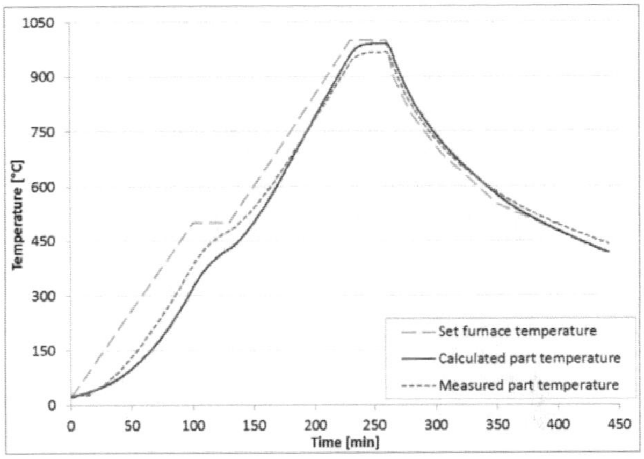

Fig. 7. Simulated and measured temperatures

The temperature profiles are qualitative similar, whereby especially the cooling of the workpieces shows a small difference
. The most significant deviations are at the end of the hold times. Thereby the calculated temperature is different at the end of the hold times. That indicates small differences between the prescribed and the real temperature dependent emission ratios of the involved surfaces. The literature does not offer a wide range of information about these complex surface properties with regard to radiation and the indicated values are imprecise in relation to the surface finish. However, at the for diffusion bonding important temperature range above 600° C, a maximum deviation between the simulated and the measured temperatures of 23 K could be achieved.
For determination of geometry influences of different workpiece sizes, another cube with the dimensions of 300 x 300 x 300 mm ("big part") was imported to the simulation model and temperatures calculated (**Fig. 8**). The setup is different to the chosen for the small part and the structural boundary conditions were adjusted. Therefore, the faying surfaces of the big part was a ribbed design to get once again the same surface pressure (p = 2 N/mm²) and to examine local deformations as well.
One calculation was done with the same thermal boundary conditions and cycle profile of the small part. The resulting thermal profiles for the big part are given as comparison between the simulated set- and actual temperature of the furnace control thermocouple in **Fig. 9**. If the PI temperature control is working accurate, there should be just small deviations between these temperatures. Furthermore the average cube temperature is integrated into the diagram. **Fig. 9** shows just a small deviation between the set- and actual temperature of the furnace thermocouple during the heating. Because of the higher mass and volume of the workpiece, the

temperature increase inside the part is delayed and the set temperature of 1000° C is not reached within the hold time.

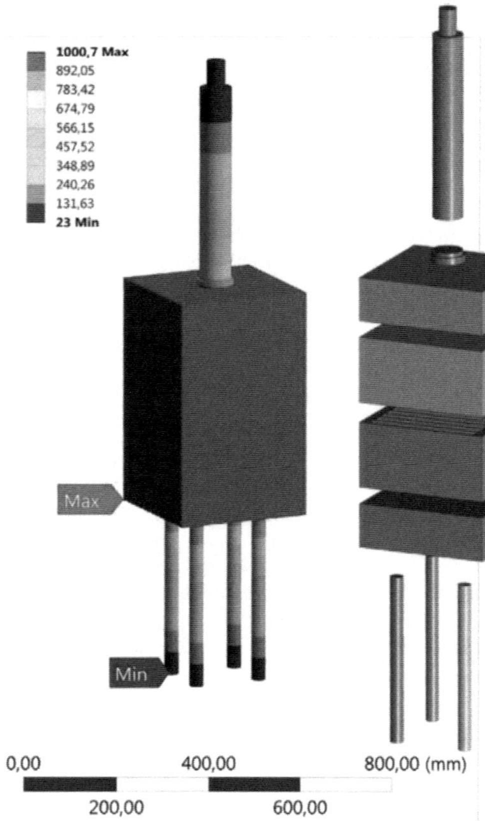

Fig. 8. Temperature distribution for big part at 1000° C after soak time (left) and setup (right)

During the cooling phase, the cooling is influenced by the stored heat (inside the part). Compared to the PI-controlled heating, the cooling is done in vacuum by the heat removal of the cooled vessel, press and supports only. Because of this, there is a temperature difference between set and measured furnace control TC temperature. With raising mass of part, the difference increase as well. To reduce the cycle time with regard to the cooling phase, gas can be introduced into the vessel or for even faster cooling, gas circulation combined with gas coolers are available.

The FEA model was verified for heating the parts to bond as well as for the influence of the part size. With increasing part mass, the thermal profile has to be adjusted to get enough heat into the part in order to reach the bonding temperature at the end of the heating phase. With the thermal FEA model it is now possible to estimate the thermal profile for the furnace cycle with regard to part size and mass without many experiments.

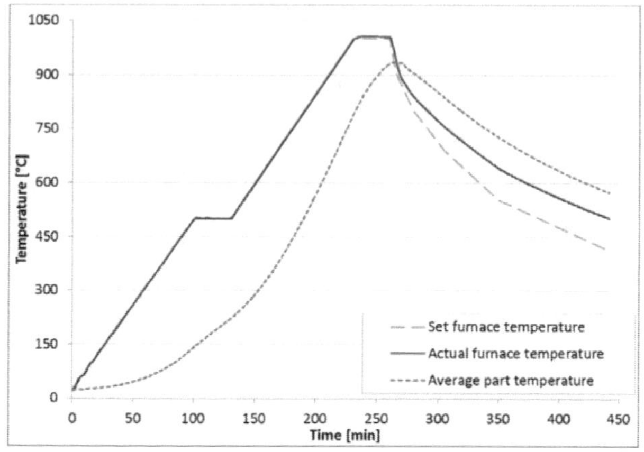

Fig. 9. Comparison between set- and measured temperature for furnace control TC and average big part temperature

Structural analysis

Beside the thermal profile, for a good bond a homogenous load distribution within the faying surfaces is necessary. To achieve this, pressing plates are used below and above the part to join. These plates deform beneath load as well. Depending on the pressing plate material Young´s modulus depends on temperature, too. Because of this and beside the thermal profile, another challenge of setting up a diffusion bonding process is to choose the right load train, especially for larger parts. In **Fig. 10** a scenario is shown. If the pressing plates / diameter of pressing ram exceed the size of the part to bond, the highest stresses are at the outer corners of the part due to the deformation (bending) of pressing plate (small part). In contrast, if the press ram diameter is small compared to part to join (big part), the pressure at the outer edges can be too low to get a good bond. If the bonding system is designed like one of the used systems ("single press ram design"), the typical solution to avoid this is to add additional or use thicker pressing plates.

With the FEA it is now possible, to estimate the load distribution within the part or study the effect of different load train setups without time and cost consuming experiments.

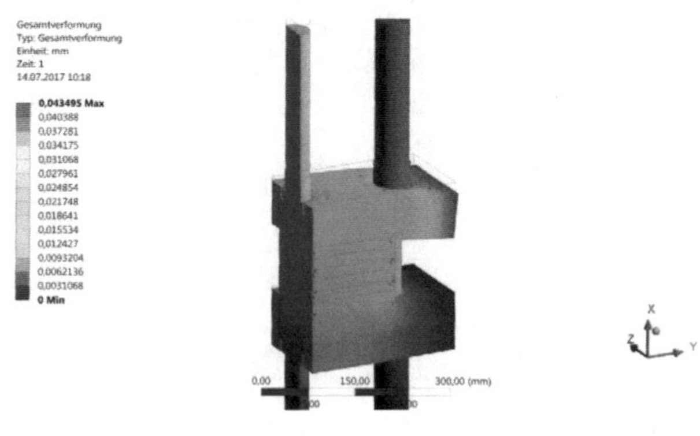

Fig. 10. Stress induced displacement distribution inside the load train (only quarter of setup is shown)

Conclusion

For the diffusion bonding process, a FEA model for calculating the thermal behavior as well as load distribution and resulting stresses was developed. Main reason for the development was, to forecast the thermal profile inside the parts to bond and set the thermal profile for the bonding system according to the part mass/volume. Further, the model helps to optimize the load train with regards to additional pressing plate size and thickness.

To achieve this, the heat is generated by the heating elements and not like before a surface temperature is applied to the outside of the part. Essential was the PI control development for the correlation between programmed thermal profile of furnace and heating element temperature. With this approach, it is possible to estimate the effect of bonding system in part (e.g. different part sizes).

The model was verified with experiments with high accuracy, e. g. maximum differences of 25° C for temperatures above 750° C. Further the resulting part deformation can be traced back to the stresses calculated with the FEA model.

For joining large areas without any additional filler material, diffusion bonding is the process of choice and becomes more popular. This can be recognized by the growing number of job-shopping companies as well as the increase of available diffusion bonding systems [10, 11, 12, 13, 14]. The FEA will help to promote further applications of diffusion bonding, hence the quantity of cost and time consuming experiments can be reduced significant.

Acknowledgements

The results were obtained in several research projects, in VF130047 and IGF project (IGF-Nr.: 18.020 B) of the Research Association on Welding and Allied Processes of the German Welding Society (DVS), supported via the AiF as part of the program to support Industrial Community Research and Development (IGF) by the Federal Ministry for Economic Affairs and Energy on basis of a decision by the German Bundestag.

References

[1] Günther, W.-D.; Mehlhorn, H.; Wiesner, P.: Betriebspraxis.- Diffusionsschweißen.; VEB Verlag Technik; Berlin, 1978.

[2] Kazakov, N.F. (Edt.): Diffusion Bonding of Materials, Mir Publishers, Moscow, 1985.

[3] Müller, G.; Groth, C.: FEM für Praktiker.- Band 1: Grundlagen.; 8. Auflage; Expert Verlag; Renningen, 2007.

[4] Jiang, W.; Gong, J.; Tu, S.; Chen, H.: Modelling of temperature field and residual stress of vacuum brazing for stainless steel plate-fin structure.; Journal of Materials Processing Technology (2009); Nr. 209, S. 1105-1110, 2009.

[5] Golanski, D.; Temperature distribution in a cylindrical Al_2O_3-steel joint during the vacuum brazing cycle.; Journal of Materials Processing Technology (1996); Nr. 56, S. 945-954, 1996.

[6 Pissanetzky, S.; Cingolani, H.; Volpi, R.; Almagro, J.C.: Numerical simulations of the transient temperature distribution within the cladding material during annealing in a high vacuum furnace.; Nuclear Engineering and Design (1981); Nr. 65, S. 141-150, 1981.

[7] Hao, X.; Gu, J.; Chen, N.; Zhang, W.; Zuo, X.: 3-D Numerical analysis on heating process of loads within vacuum heat treatment furnace.; Applied Thermal Engineering (2008); Nr. 28, S. 1925-1931, 2008.

[8] Lee, H.-H.; Finite Elemente Simulations with ANSYS Workbench 14.- Theory, Applications, Case Studies.; Schroff Development Corporation; Tainan, 2012.

[9] Jahn, S., Gemse, F., Broich, U., Saendig, S.: Efficient Diffusion Bonding of Large Scale Parts. In Materials Science Forum (Vol. 838, pp. 500-505). Trans Tech Publications, 2016

[10 N.n.: High-vacuum diffusion bonding – Introduction into industrial series application is being prepared, press release by PVA Tepla AG, Wettenberg, 27.10.2014.

[11] N.n.: Largest Vacuum Hot Press in the WORLD manufactured in the USA by AVS has been acquired by VPEI for the production of Diffusion Bonded Heat Exchangers and contract work, press release by Vacuum Process Engineering Inc., Sacramento, 15.04.2015.

[12] N.n.: www.refrac.com/, 25.01.2018.

[13] N.n.: http://www.vpei.com/, 25.01.2018.

[14] N.n.: www.kinzoku.co.jp/english/, 25.01.2018.

Aluminum Braze Triple Line Movement in a Wetting/Non-wetting Setting under Variable Surface Roughness and Inclinations

Yangyang Wu, Cheng-Nien Yu, Dusan P. Sekulic

Department of Mechanical Engineering, College of Engineering, University of Kentucky, Lexington, KY, 40506, USA
sekulicd@engr.uky.edu

Abstract

The triple line movement of a molten braze in an aluminum alloy/alumina wedge-tee configuration (a wetting/non-wetting assembly) was studied. One of the two components (the horizontal) was alumina, while the other, the vertical piece, was made of AA3003. Trillium™ composite in a liquid state (an Al-Si alloy with flux embedded), wets AA3003 only. When heated, molten Trillium™ climbs the vertical AA3003 piece driven by surface tension and retarded by gravity, thus forming a surface tension - gravity driven free surface configuration. The vertical AA3003 featured three different surface roughness topographies and five inclination angles. Two primary micro-grooves' directions (horizontal or vertical) were imposed. Samples were exposed to a CAB brazing cycle with a 40°C/min heating rate. The triple line movement was monitored in real time, *in situ*. Results show that wetting kinetics and the primary surface profile at the equilibrium state feature similar behaviors under variable surface roughness conditions. Both advancing and receding triple lines move the fastest within the first 2s upon the onset of spreading. However, the wicking distance, i.e. capillary rise front, gets longer as surface gets rougher. Moreover, vertical grooves are demonstrated to promote the wicking of liquid metal compared to horizontal grooves at the same roughness. Gravity does have an effect on wetting of the liquid metal on AA3003. As the inclination angle gets smaller, the surface profile gets leaner, the advancing contact angle gets smaller, and the wetted distance becomes longer.

Introduction

A number of theoretical models and quantitative experimental observations have been developed to give scientific understanding of the wetting phenomena [1]. Both non-reactive [2] and reactive wetting [3, 4] are widely discussed. For example, wetting behavior on substrates having different surface topographies (i.e. different roughness, grooves' profiles, etc.) [5, 6, 7, 8] and the effect of gravity on wetting are extensively studied [9]. Comprehensive research has been made for both wetting and non-wetting cases. However, the combination of wetting/non-wetting phenomena is less studied.

In this paper, the triple line movement and surface profile of Trillium™ composite, an Al-Si alloy with flux embedded, in an AA3003/Al_2O_3 wedge-tee configuration (a wetting/non-wetting assembly) at high temperatures were studied. In the first part of this paper, wetting features of the molten alloy on three different surface roughness, characterized with two primary micro-grooves' directions (horizontal or vertical), were discussed. In the second part, five different inclination angles of mating surfaces were tested to uncover the impact of the spreading plane orientation in the gravity field on the wetting behavior.

Materials and experimental procedures

Sample configuration. The configuration of the aluminum alloy/alumina wedge-tee assembly was illustrated in Fig. 1(a). The horizontal substrate is 96% alumina (Al_2O_3). It was cut into coupons with dimensions of 20 × 20 × 1.02 mm using a diamond saw. The vertical piece is AA3003 with dimensions of 15 × 10 × 0.39 mm. The third participant, the to be melted Al-Si alloy, is the so called Trillium™. This composite was prepared in form of a small bar, having a mass of 100 ± 0.4 mg and approximately a size of 10 × 2.5 × 1.6 mm. This alloy was pre-placed on the alumina substrate, in parallel with the vertical AA3003 piece. The location of the pre-placed alloy is at a small distance (of around 0.6 mm) from the vertical mating surface. Trillium™ composite is a metal matrix of Al-Si alloy with imbedded potassium-fluoro-aluminate flux [10, 11, 12]. The liquid metal formed upon heating wets the vertical AA3003 only. The small gap between the Trillium™ composite and AA3003 is to ensure melting of the composite before contacting and climbing over the vertical AA3003 piece. Two K-type thermocouples touch the upper side of the alumina substrate to control the furnace temperature history and to record the surface temperature. A stainless steel wire with a gauge of 0.38 mm was used to fix the sample assembly, positioning the vertical AA3003 mating surface onto a ceramic base.

Surface roughness. In order to study wetting of liquid Trillium™ composite (Al-Si alloy) on different AA3003 surface topographies, three different surface roughness were imposed by grinding AA3003 using SiC paper with different grits. Figure 2(a-c) shows the three different topographies of the AA3003 made visible by using Zygo New View™ 7300 3D optical surface profiler. A virgin AA3003 piece already has shallow grooves (micro-channels) formed during the rolling process. It has a roughness of 0.41 ± 0.03 μm. The AA3003 grinded by #220 and #50 SiC paper has a roughness of 1.74 ± 0.09 μm and 5.03 ± 3.18 μm respectively. Figure 2(d) shows the topography of the alumina substrate with a surface roughness of 2.00 ± 0.04 μm. Note that the surface roughness measurements were made at least at 3 spots with a scanning area of 0.27 mm × 0.27 mm (10 measurements for each spot). The present roughness features two primary micro-grooves' directions (horizontal or vertical). Figure 3 shows the schematic diagram of the vertical and horizontal grooves on AA3003.

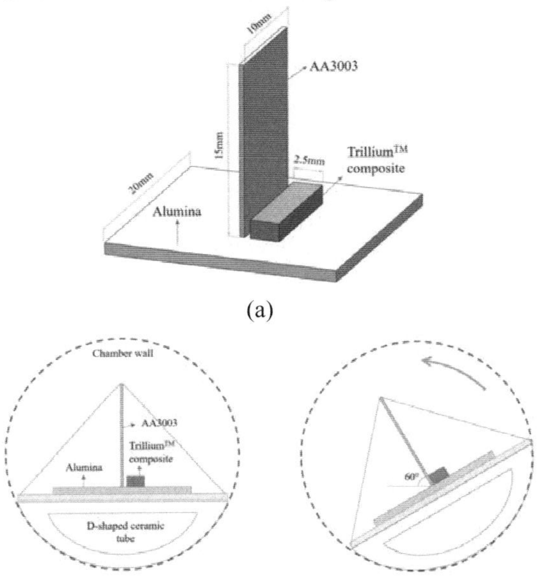

Fig. 1 Sample configuration: (a) 3-D view; (b) side view; (c) side view when sample is rotated counterclockwise, AA3003 having an inclination of 60° to the horizontal line.

Inclination. In order to study how orientation in a gravity field affects wetting of liquid Trillium™ composite (Al-Si alloy) on AA3003, variable inclinations were made for the sample assembly. Figure 1(b) shows that the sample assembly was positioned on a D-shaped ceramic hot zone inner tube-stand (in the cylindrical chamber furnace, refer also to the side view of the furnace in Fig. 4). This D-shaped ceramic tube can be rotated, hence allows for achieving variable inclinations of the assembly. Figure 1(c) illustrates the side view when sample is rotated counterclockwise, making AA3003 have an inclination of 60° to the horizontal line. We define the inclination angle as the angle between the AA3003 surface and the horizontal line. If one rotates the assembly counterclockwise further, 45°, 30°, 0° inclinations can be obtained. The 0° degree corresponds to the situation when the AA3003 is in the horizontal orientation (orthogonal to the gravity field). In this study, inclinations of 90°, 60°, 45°, 30°, and 0° were discussed.

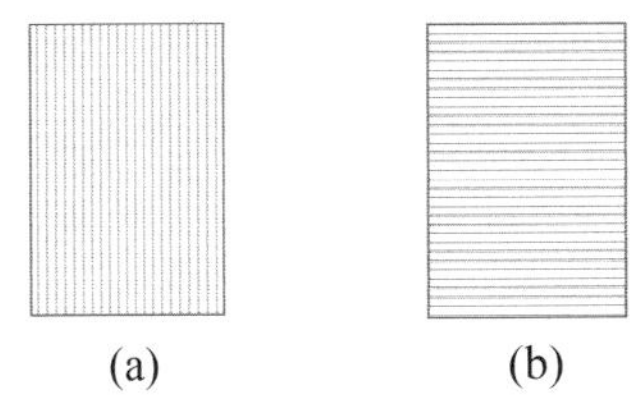

Fig. 3 Schematic diagram of the grooves on AA3003 surface: (a) vertical grooves; (b) horizontal grooves

Equipment configuration. Figure 4 shows the schematic of the high-temperature optical contact angle measuring system. This facility is a unique system manufactured by DATA PHYSICS based on a customized specification. It consists of OCA-15 LHT, HTFC 1200 and has a capability to control the background atmosphere. The CCD camera, furnace hot zone axis, and the halogen back-light source are in perfect horizontal alignment. Thus a real time, *in situ* recording can be made in the furnace cylindrical hot zone chamber. Two K-type thermocouples were in contact with the alumina substrate. One is used to provide a feedback to control the furnace hot zone, the other is connected to a Labview hardware to record temperature history. Note that the furnace is driven by Joule-heating, thus the hot zone chamber features a presence of a corresponding induced magnetic field. A ferromagnetic object

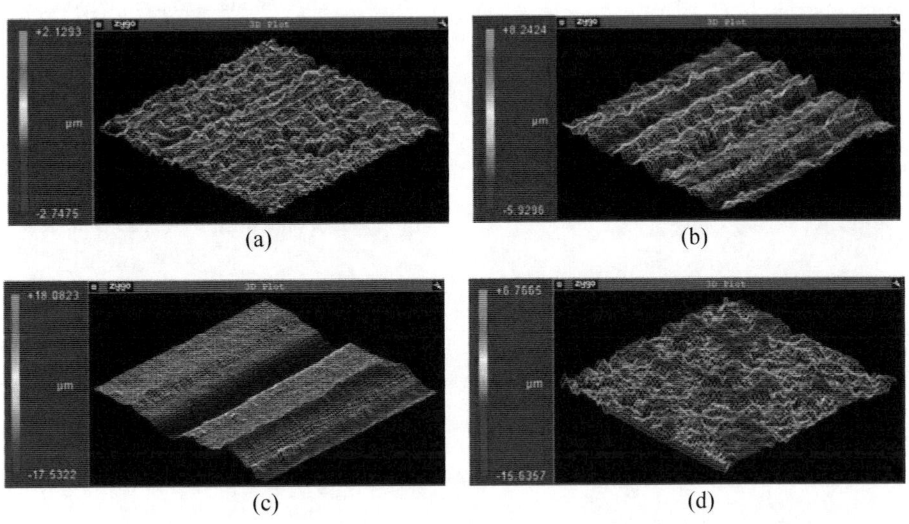

Fig. 2 3D scanning images of the topography of AA3003 and alumina substrates measured by Zygo New View™ 7300 3D optical surface profiler: (a) Virgin AA3003, Ra = 0.41 ± 0.03 µm; (b) AA3003 grinded by #220 SiC paper, Ra = 1.74 ± 0.09 µm; (c) AA3003 grinded by #50 SiC paper, Ra = 5.03 ± 3.18 µm; (d) Alumina substrate, Ra = 2.00 ± 0.04 µm. Scanning area is 0.27 mm × 0.27 mm. Ra is the arithmetic mean roughness value.

put into the chamber would sense more or less a vibration, thus a stainless steel base holder is not recommended.

Experiment procedures. Before testing, all the components of sample assemblies were ultrasonically cleaned in soapy water, water, and finally in 95% ethanol for at least two minutes to eliminate the oil deposits and any eventual other chemical residues. Subsequently, the assemblies were dehydrated in air. Ultra high purity nitrogen (99.999%) has been used to purge the furnace chamber at 1 scfh (472 ccm) for 2 hrs, resulting in the O_2 concentration level below 50 ppm before the tests execution. Nitrogen is continuously provided at the same rate during the test. Subsequently, the samples were exposed to a CAB brazing cycle starting from the room temperature (22°C) to 610°C at a heating rate of 40°C/min, followed by a dwell time of 3 mins, and then exposed to natural cooling. The sample temperature history is illustrated in Fig. 5.

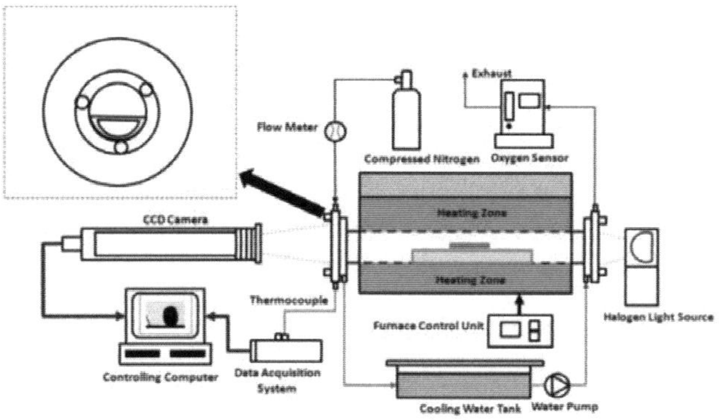

Fig. 4 Schematic of the high-temperature optical contact angle measuring system [13].

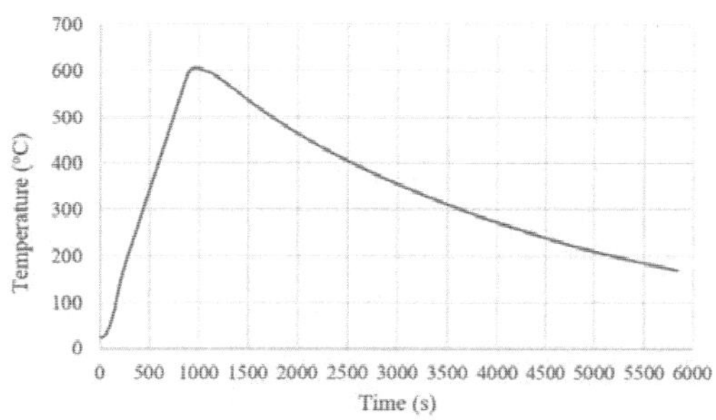

Fig. 5 Sample temperature history

Results and discussion

When heated, Trillium^{TM} composite will melt and climb the vertical AA3003 piece driven by surface tension, hence forming ultimately a surface tension - gravity driven equilibrium free surface. A CCD camera gives a real time, *in situ* recording of this process at a frame rate of 25 fps. Figure 6 illustrates the 5 stages of the process. As shown in Fig. 6(a), before heated, Trillium^{TM} composite was pre-placed at a small distance (of around 0.6 mm) from the vertical mating surface to ensure melting of the composite before contacting and climbing over the vertical AA3003 piece. Figure 6(b) shows that the Trillium^{TM} composite at the onset of melting forms first a cylinder-shaped bar. Due to non-wetting between the liquid metal and the alumina substrate, the liquid metal bar starts to "congregate", forming a ball (one can observe that the diameter of the liquid circular shape is larger in Fig. 6(c) than that in Fig. 6(b)). Because of the diameter expanding, the liquid metal would ultimately touch and subsequently "climb" the vertical AA3003 piece driven by surface tension. Figure 6(d) depicts a frame during the climbing process. Surface tension and inertia dominate at the early stage; gravity and viscosity retard the spreading at the later stages. Interfacial reaction between the AA3003 and molten Trillium^{TM} composite takes place along with the wetting process. Figure 6(e) indicates the ultimate equilibrium state of this wetting process.

Wetting of Trillium^{TM} composite on an AA3003 under variable surface roughness

The vertical AA3003 has three different surface roughness magnitudes (refer to Fig. 2(a-c)). Moreover, vertical AA3003 specimens feature two primary micro-grooves' directions (horizontal or vertical) as shown in Fig. 3. The triple line movement was monitored in real time, *in situ*. Figure 7(a) illustrates advancing and receding processes of the liquid metal. Figure 7(b) illustrates the advancing contact angle (θ_1), and the receding contact angle (the angle between the liquid metal and the alumina substrate (θ_2)) at the final state.

The advancing and receding triple line movement versus time for Trillium^{TM} composite on virgin AA3003 vertical grooves are provided in Fig. 8. Three repeated tests show identical trends, indicating good replicability of the experiments. During the first 2 seconds at the onset of spreading, the large amount of liquid metal forming a ball shape, sitting on the alumina

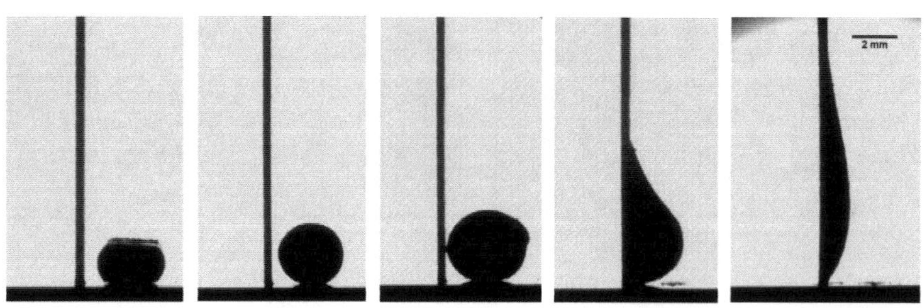

(a) Before heating (b) melting (c) congregating (d) climbing (e) equilibrium

Fig. 6 Five stages of the process: (a) before heating; (b) onset of melting; (c) congregating; (d) climbing; (e) equilibrium.

substrate, will leave the substrate and climb the vertical AA3003, resulting in the triple line movement being the fastest in this short time period. As time goes by, the liquid metal front will move slower due to gravity and viscosity impacts, as well as the host of interaction phenomena at the liquid/solid interface, and finally will approach the equilibrium state, asymptotically to a value of approximately 11 mm as shown in Fig. 8(a). The liquid metal, initially on the alumina substrate, will be depleted to zero at the equilibrium state. This is illustrated in Fig. 8(b). One can observe that the receding line location will stay at around 0.5 mm (or alternately ~1 mm) for a while and later reduce to 0 mm at the final stage.

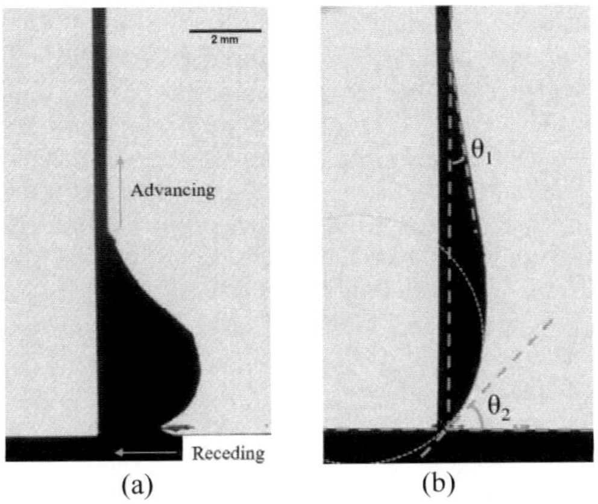

Fig. 7 (a) Illustration of advancing and receding spreading; (b) advancing and receding contact angles at the final stage.

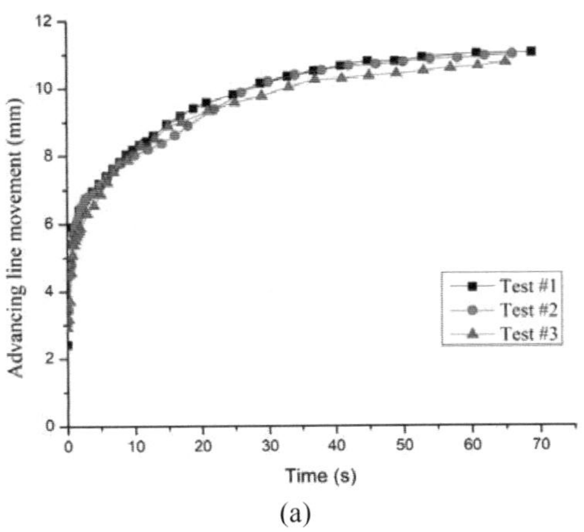

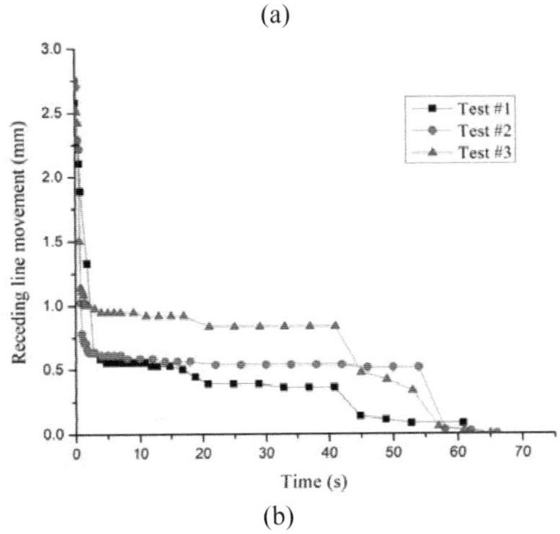

Fig. 8 (a) Advancing line location vs. time for Trillium^{TM} composite on the as is AA3003 vertical grooves; (b) Receding line movement vs. time for Trillium^{TM} composite on an as is AA3003 vertical grooves.

Using the OCA optical contact angle measuring device (DataPhysics Instruments) with the SCA 20 software, the advancing contact angle (θ_1) at the final stage is measured to be 7.3 ± 0.3° for the Trillium^{TM} composite on the virgin AA3003 surface with vertical grooves. The receding contact angle between the liquid metal and the alumina substrate (θ_2) at the final stage is 54.1 ± 1.9° for the same surface roughness and groove direction (note, the complementary angle is actually established on the AA3003 vertical surface).

It is observed that the wetting kinetics and the primary surface profile at the equilibrium state feature similar behaviors for all surfaces irrespective of the imposed roughness and groove directions. However, more apparent difference between samples can be noticed on AA3003 in the front view. As shown in Table 1, for the vertical grooves cases, a clear boundary line between the AA3003 and the molten liquid metal can be recognized when the surface is non-grinded. Nevertheless, the boundary gets fuzzy when the surface gets rougher, say in #220 and #50 SiC grinded cases. This is because liquid metal seems to be wicking along the grooves to the very end of the flow driven by capillary force, filling the grooves. In addition, the rougher the surface, the more liquid metal is driven forward. While, the volume of this liquid metal spread above the boundary line is quite small, hence the free surface configuration in the equilibrium state in the side view looks similar for all the three different roughness cases.

For the horizontal grooves situation, it can be concluded that roughness also promotes the wicking of the liquid metal on AA3003. This can be established after comparing the capillary rise front of the #220 and #50 SiC grinded samples with the non-grinded, original sample. However, when the surface gets much rougher, the surface topography may impede wicking of the horizontal grooves since a slightly lower and non-uniform wetting front is observed in the #50 SiC grinded sample, compared to that in the #220 SiC grinded sample.

Yet another comparison between a sample with vertical grooves and the one with horizontal grooves for the same roughness can be made. One can easily conclude that for the original, relatively smooth surfaces, the wetting distances are almost the same for both vertical grooves and horizontal grooves. Clear boundary lines exist in both cases. However, when the surface get rougher, say in #220, #50 SiC grinded cases, vertical grooves promote the wicking of liquid metal since the capillary rise fronts for vertical grooves samples are much higher than the ones in the horizontal grooves samples. Thus, the surface roughness does have an effect on the triple line movement in this configuration setting.

Table 1. Wetting distance of liquid Trillium™ composite on AA3003 at different surface roughness and groove directions.

	Original	#220 SiC grinded	#50 SiC grinded
Vertical groove			
Horizontal groove			

Wetting of Trillium™ composite on an AA3003 under variable inclinations

In order to uncover how gravity affects wetting, and the equilibrium state free surface configuration, a series of tests was conducted by rotating the whole assembly counterclockwise. This made the AA3003 to be with an inclination to the horizontal line of 60°, 45°, 30° and 0° respectively. Figure 1(c) shows the side view of the sample assembly when AA3003 has an inclination of 60° to the horizontal level. In this situation, liquid metal will climb on an inclined AA3003 surface instead of on a perpendicular one. If one rotates further the whole assembly, ultimately a 0° inclination of the AA3003 piece can be achieved. In this case, AA3003 is horizontal and the liquid metal will wet the surface without the localized (along the spread direction) change of the influence of gravity, although in a gravity field). Note that the AA3003 used here is non-grinded with vertical grooves.

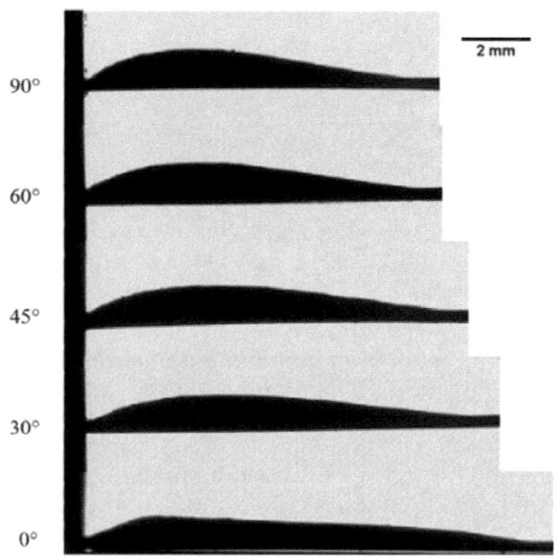

Fig. 9 Equilibrium state surface profiles of liquid Trillium™ composite on an AA3003 at different inclinations to the horizontal line: 90°, 60°, 45°, 30°, 0°. Note that these images were processed by rotating the original images to the horizontal orientation.

Figure 9 shows the equilibrium state surface profiles of liquid Trillium™ composite on AA3003, at inclinations of 90°, 60°, 45°, 30° and 0° respectively. These images were processed by rotating the original images vs. the horizontal plane (except for the inclination of 0°). It can be concluded that gravity does have an effect on wetting of the liquid metal on AA3003. As the inclination angle gets smaller, the surface profiles get "thinner," the advancing angle (θ_1) gets smaller, and the contact angle (θ_2) between the alumina substrate and the liquid metal becomes larger. Most importantly, the wetting distance gets longer when the inclination of AA3003 is decreasing. Figure 10 shows the sample images of the wetting distance of the liquid Trillium™ composite on AA3003 at inclinations of 90°, 60°, 45°, 30° and 0°. The values of the wetting distance for five different settings were given in Table 2. When AA3003 is vertical (90°), the wetting distance is the smallest with 10.91 mm. However, after rotating the sample assembly and making AA3003 horizontal (0°), the largest wetting distance can be obtained, which is 15.49 mm. The bar chart in Fig. 11 gives an additional illustration of the wetting distance under different inclinations, which in turn demonstrates the impact of gravity.

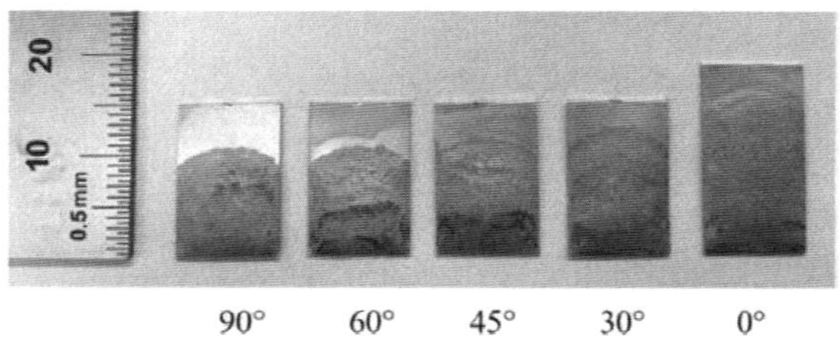

Fig. 10 Wetting distance of liquid Trillium™ composite on AA3003 at different inclinations: 90°, 60°, 45°, 30°, 0°.

Table 2. Wetting distance of liquid Trillium™ composite on AA3003 at different inclinations: 90°, 60°, 45°, 30°, 0°.

Inclination Angle (degree)	Distance (mm)
90	10.91 ± 0.28
60	11.27
45	11.89
30	12.80
0	15.49 ± 0.65

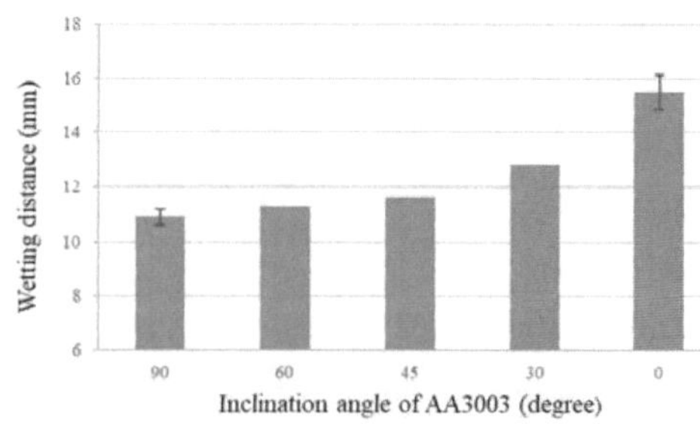

Fig. 11 Wetting distance of liquid Trillium™ composite on AA3003 at different inclinations: 90°, 60°, 45°, 30°, 0°.

Conclusions

The (i) triple line movement and (ii) surface profiles of the liquid Trillium™ composite, in an AA3003/Al_2O_3 wedge-tee configuration (a wetting/non-wetting assembly), have been studied in this paper. The impacts of surface roughness and gravity were considered.

Wetting kinetics and the equilibrium surface profiles feature similar behaviors under different surface roughness conditions. Both advancing and receding triple lines move the fastest within the first 2s. As time goes by, the liquid metal moves slower and the front location becomes asymptotic to the value of approximately 11 mm. The receding line location stays at around 0.5 mm (or alternately, ~1 mm) for a while, and ultimately reduces to 0 mm at the equilibrium state. The advancing contact angle (θ_1) at the final stage is measured to be $7.3 \pm 0.3°$ for the Trillium™ composite on the original AA3003 vertical grooves surface; the contact angle between the liquid metal and the alumina substrate (θ_2) is $54.1 \pm 1.9°$ for the same surface roughness and groove direction.

For the vertical grooves cases, a clear boundary (triple) line between the AA3003 and the molten liquid metal can be recognized when the surface is non-grinded. However, this boundary gets fuzzy when the surface gets rougher. In addition, the rougher the surface, the more liquid metal wicking beyond the boundary line along the grooves to the top end due to a capillary force.

In the horizontal grooves configuration, it can be concluded that roughness promotes wetting of the liquid metal on AA3003. However, when the surface gets much rougher, it may slightly impede the wetting.

For the samples with the same surface roughness but with different groove directions, the wetting distances are almost the same for the non-grinded, relatively smooth surface. However, when the surface get rougher, say in #220, #50 SiC grinded cases, vertical grooves promote more the wetting of liquid metal compared to horizontal grooves.

Gravity does have an effect on wetting of the liquid metal over AA3003. As the inclination angle gets smaller, the surface profile gets thinner, advancing angle (θ_1) gets smaller, and the contact angle (θ_2) between the alumina substrate and the liquid metal becomes larger. Most importantly, the wetting distance is longer when the inclination of AA3003 is decreasing.

Acknowledgments

This work has been funded within NASA's Physical Sciences Research Program, Grant# NNX17AB52G. The team thanks Dr. Doug Hawksworth, Diomedea Inc., Canada, and Gränges AB, Sweden, for assistance and providing the Trillium™ materials, respectively. The authors acknowledge the discussion on the subject with Dr. Fiqiri Hodaj, Grenoble Institute of Technology, France, as well as Dr. Paul Steen from Cornell University, USA, during their visits to the brazing laboratory at the University of Kentucky. The first author (Y. Wu) wants to thank the assistance she got from Dr. Hai Fu (Corning Inc., USA) and Dr. Chuanbao Jia (Shandong University, China) during their tenure as a graduate student and a visiting scholar at the University of Kentucky, respectively. Trillium™ Technology is protected by United States Patent No. 8871356 as well as by corresponding patents and pending patent applications in other countries. Trillium™ Technology is a registered trademark of Gränges AB.

References

[1] Eustathopoulos, N., Nicholas, M. G., and Drevet, B., eds. Wettability at high temperatures, Pergamon Press, (Oxford, UK, 1999).

[2] Tanner, L. H. "The spreading of silicone oil drops on horizontal surfaces." *Journal of Physics D: Applied Physics* 12.9 (1979): 1473.

[3] Yost, F. G., Rye, R. R., and Mann Jr., J. A. "Solder wetting kinetics in narrow V-grooves." *Acta Materialia* 45.12 (1997): 5337-5345.

[4] Zhao, H., and Sekulic, D. P. "Wetting kinetics of a hypo-eutectic Al–Si system." *Materials Letters* 62.15 (2008): 2241-2244.

[5] Quéré, D. "Wetting and roughness." *Annu. Rev. Mater. Res.* 38 (2008): 71-99.

[6] Rye, R. R., Mann, J. A., and Yost, F. G. "The flow of liquids in surface grooves." *Langmuir* 12.2 (1996): 555-565.

[7] Sekulic, D. P. "Wetting and Spreading of Liquid Metals Through Open Micro Grooves and Surface Alterations." *ASME 2009 7th International Conference on Nanochannels, Microchannels, and Minichannels*. Pohang, South Korea, 2009.

[8] Liu, W., and Sekulic, D. P. "Capillary driven molten metal flow over topographically complex substrates." *Langmuir* 27.11 (2011): 6720-6730.

[9] De Gennes, P. G., Brochard-Wyart, F., and Quéré, D. "*Capillarity and Gravity*." Capillarity and Wetting Phenomena. Springer (New York, 2004), pp. 33-67.

[10] Hawksworth, D. K., Westergard, R. G. J., and Ogilvy, A. J. W. "Trillium technology-aluminium brazing with a composite liner." *7th International Congress of Aluminium Brazing*. May 22-24, 2012.

[11] Hawksworth, D. K., Westergård, R. G. J., Nilsson, P. "Brazing properties of TRILLIUM™ composite." *8 Int. Congress Aluminium Brazing*, June 03-05, 2014.

[12] Ogilvy, A. J. W., Hawksworth, D. K., and Abom, E., (2014). *U.S. Patent No. 8,871,356.*

[13] Fu H, et al. "Kinetics of the molten Al–Si triple line movement during a brazed joint formation." *Journal of materials science* 51.4 (2016): 1798-1812.

Eutectic phase formation during friction stir spot welding of aluminum and copper dissimilar joints

Tobias Köhler, Anna Regensburg, Franziska Petzoldt, Jean Pierre Bergmann
Production Technology Group, Technische Universität Ilmenau, Germany
info.fertigungstechnik@tu-ilmenau.de, +493677/692980

Abstract

Several approaches have shown, that there is a good opportunity in the automotive on board power system, to reduce the weight and the costs of a vehicle. One possibility to accomplish this is the substitution of Cu by Al. However dissimilar material related factors like oxide layers and IMC (intermetallic compounds) formation represent challenges for joining. In this work, a dissimilar EN AW 1070/ EN CW 004A joint was connected by a friction stir spot welding (FSSW) process at which the copper element is placed on the end face of the aluminum strand. The tool penetrates into the copper element to generate a friction based heat input and a plastic deformation. The objective of this work is the generation of a local melt formation at the interface with a eutectic composition by a fast and targeted heat input. Hence, the wetting of the aluminum wires leads to an increase of the cross-conductance. By considering the eutectic point in the Al/Cu phase diagram, a diffusion based material locking can be generated without reaching the melting temperatures of the base materials. The growth of brittle intermetallic phases in the contact area is reduced and therefore a ductile connection is produced.

Introduction

The efforts of the automotive industry to promote lightweight engineering on the basis of political, ecological and economic reasons, demonstrate the importance of this subject. The development towards lightweight design is based on the need for cost savings and an improved carbon dioxide balance. In the automotive industry, lightweight design is not only used in the vehicle body, but also in the on board power system. Costs and weight can be reduced by substituting copper with aluminum in the vehicle's electrical system [1]. With a consequential increase of the wire cross-section of the aluminum wire about 60% [2], there is still a possible reduction of weight around 40% to 50% [2] and a possible reduction of material costs around 85 % by means of the named substitution.
However, the natural oxide layer and creep behavior of aluminum is a drawback for force locked connections and the sole use of aluminum is not suitable for applications with high requirement on mechanical and electrical stability under elevated temperature. Hence, the use of copper elements and respective dissimilar joining technologies is needed. In this case, there are different challenges to fulfill. The oxide layer of the aluminum needs to be disrupted, brittle intermetallic compounds shall be avoided as far as possible and the connection must be protected against galvanic corrosion. Because of the high affinity to each other at temperatures higher than 120 °C, the Al-Cu-diffusion leads to several kinds of intermetallic phases, depending on the temperature-time-regime [3]. These intermetallic phases like Cu_9Al_4, CuAl and $CuAl_2$ are up to ten times harder than copper and the specific resistance of these phases are up to 4 to 5 times higher than the specific resistance of aluminum [4]. The challenge is to produce a material locking which provides a ductile bond with a comparatively low specific resistance. By using the eutectic point in the aluminum copper phase diagram, a ductile connection with a comparatively low specific resistance of around 6,6 $\Omega mm^2/m$ [5] can be produced.
A eutectic is defined as the balance of two phases with different compositions, which occurs at a certain concentration and a certain temperature in phase diagrams [6, 7]. The characteristic of a eutectic is the direct transition of the solid phase to the liquid phase at a certain temperature. In the aluminum-copper-phase-diagram, the eutectic concentration amounts to 67 wt.-% Al and 33 wt.-% Cu, the eutectic temperature amounts to 548 °C. The composition of the Al/Cu eutectic consist of a lamellar structure of α-aluminum matrix and Al_2Cu (θ)-phase. [6]
At the moment, the state of art shows different ways for joining copper and aluminum cables in electronic applications. The most common processes are crimping for smaller cross sections and ultrasonic welding for lager cross sections. Other processes are plasma soldering [8] and ultrasonic tinning [9]. In this work, the friction stir spot welding (FSSW) process was used to generate a eutectic phase in the interface. Friction stir spot welding is a friction based solid state welding technology, where a rotating symmetrical tool plunges into the material, which generates a friction based heat input and a plastic deformation. Figure 1 a) shows a lap joint of an aluminum and a copper sheet and a rotating FSSW tool.
In the preliminary study the influence of the different welding parameters was investigated on sheet metal lap joints. After that the process was adapted to the cable-arrester connection where the copper element was formed like a lid and positioned on the end face of the aluminum cable. After the lid was crimped on the aluminum cable, the connection was welded

from the end face of the lid. Figure 1 b) shows the copper lid placed on the aluminum cable.
The main advantages of a copper lid placed on the end face of the aluminum strand are the increased cross section in the interface and a direct contact of each wire to the copper element.

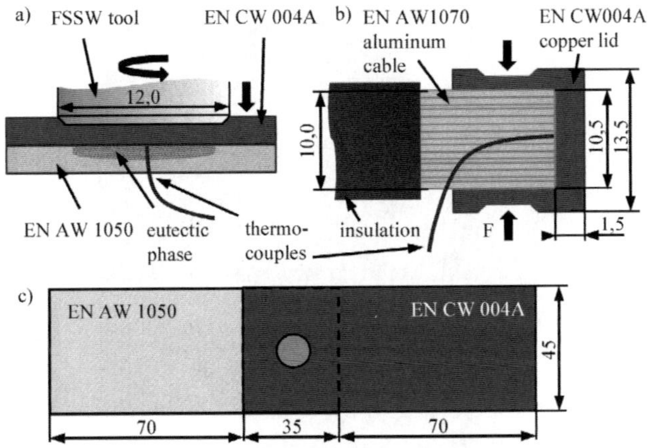

Figure 1: FSSW process for joining aluminum cables and copper conductors

Figure 2 shows the three steps of the FSSW process for joining cables and conductors. The tool plunges into the copper with a certain rotational speed and plunging speed. The friction and the plastic deformation generate thermal energy, which leads to the diffusion of copper atoms into aluminum and vice versa. In the second step, when the plunging depth is reached, the dwell time starts. The rotating tool produces more friction-based energy. Hence, the diffusion continues, and therefore the growing of the eutectic melting layer begins or continues. After that, in the third step, the tool is removed, the cooling and solidification starts. Because of the comparatively fast heat transfer into the clamping jaws and the aluminum cable, fast cooling rates can be achieved. Therefore, a structure with a eutectic composition is produced.
These steps result into the adjustable parameters of this process: rotational speed, feed speed, dwell time, plunging depth, tool geometry and backing structure.

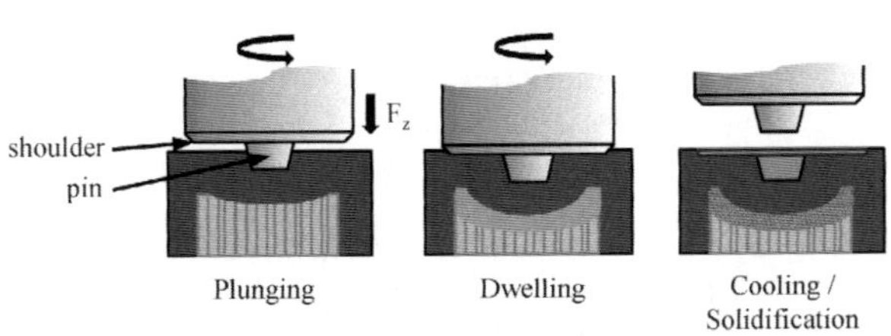

Figure 2: FSSW process for joining an aluminum cable and a copper lid

Experimental Procedure

For the development of the FSSW process for electrical applications, EN AW 1070 aluminum cables with a wire cross section of 59 mm² and EN CW 004A copper conductors (as shown in Figure 1 b)) with a wall thickness of 1,5 mm were welded to obtain a eutectic structure in the interface with a good wetting of the aluminum strand. In the preliminary investigations 2 mm EN CW 004A sheets were welded onto 2 mm EN AW 1050 sheets in order to evaluate the basic conditions for the melt formation. The examinations were performed on a Hermle UWF 1001, a position controlled milling machine with a maximum rotational speed of 4200 rpm and a high rigidity. Figure 3 shows the device for clamping the dissimilar joints. The cable and the arrester are clamped by two clamping jaws, one to chuck the cables and one to chuck the arrester. The spindle torque and the axial force were measured by a strain gauge based measurement system by Artis MARPOSS. The temperature was measured by a DAQ system data logger and type K (Ø 1 mm) thermocouples, which were positioned in the middle of the aluminum strand in the interface. The characterization of the specimen was performed by tensile tests and metallographic preparation.

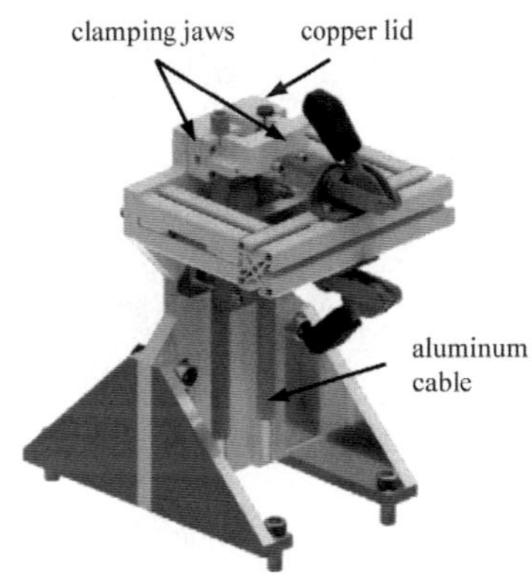

Figure 3: Clamping device

Results and Discussion

The objective of this work is the development of a dissimilar joint of an aluminum cable with a copper arrester.
In the preliminary investigations, 2 mm thick EN CW 004A sheets were welded onto 2 mm EN AW 1050 sheets by the FSSW process, to have a more precise view on the dependency of the rotational speed and the feed speed on the wetting effect and the generation of the eutectic phase. Figure 4 and Figure 5 show the results of this investigation. The rotational speed was varied between 2400 rpm and 2800 rpm. The feed speed was varied between 30 mm/min and 50 mm/min and the dwell time was set to 0 s. The plunging depth was set to 0,7 mm and a tool without pin was used. Due to the axial force, the liquid phase is pushed out of the interface during the welding process. Hence, the eutectic phase is located in the outer area of the interface.
Regarding the variation of the rotational speed at the respective feed speed, a clear tendency of an increasing melt formation at higher rotational speeds can be deduced. In addition, a higher feed speed means a lower temperature; hence, a lower generation of liquid phase is shown in Figure 4. The influence of the feed speed at the respective rotational speed shows, that a higher feed speed results in less phase formation. However, regarding the phase formation, the rotational speed is the most significant parameter.

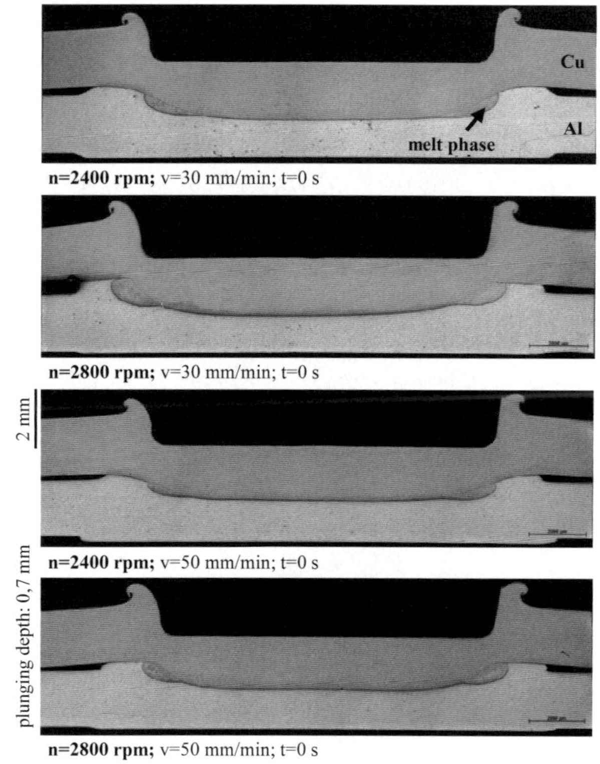

Figure 4: melt formation depending on rotational speed and feed speed

Figure 5 shows the axial force, the torque and the temperature during the welding process. The parameter setting was n=2400 rpm, v=30 mm/min, t=0 s and x=0,7 mm. Considering the axial force and the torque, an unsteady trend in the first second can be observed, which is caused by the first contact of the tool with the copper. After that, a steady raise of the force, the torque and the temperature is obtained. The temperature reaches its maximum of 554,5 °C after the steady increase and stays then at around 525±25 °C until the end of the process.

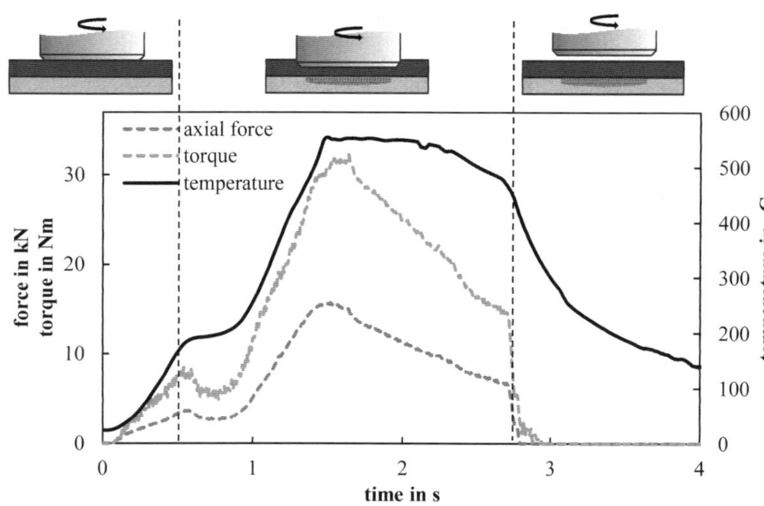

Figure 5: torque, axial force and temperature during the process (n=2400 rpm; v=30 mm/min; t=0 s; x=0,7 mm)

The axial force and the torque also reach their maximum at the point of the highest temperature after the steady increase (F=15,69 kN; M=31,76 Nm). During temperature around the eutectic temperature, the liquid phase is generated. The force and torque decrease while the liquid is pushed out of the interface and the yield stress is decreased.

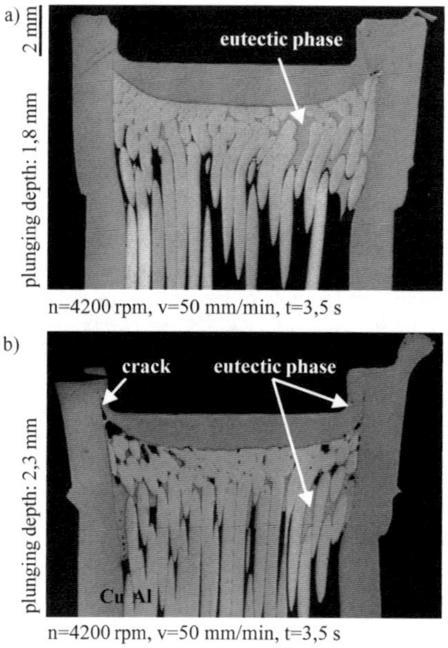

Figure 6: development of the tool geometry

The results of the sheet/sheet connections were adapted on the FSSW-process for joining an aluminum cable and a copper arrester. This connection should feature a maximum binding of the wires by the use of a eutectic melt and without damaging the cable. In this investigation the rotational speed, the feed speed and the dwell time were varied. The immersion depth is obtained by the developed tool. The rotational speed range was set from 3200 rpm to 4200 rpm and was varied in steps of 500 rpm. Due to the limitation of the used machine, the rotational speed of 4200 rpm was set as the maximum rotational speed. For the feed speed the values, 25 mm/min and 50 mm/min were selected. In order to reduce the influencing factors, the dwell time was set to 1 s in the main investigations and was varied between 0 s and 5 s in the preliminary investigations. Shen [10] showed that the temperature of the eutectic point can be reduced by a large plastic deformation. Therefore, the tools geometry was designed in order to obtain a high degree of plastic deformation.

The tool was carried out with two shoulders in the preliminary trials, the first shoulder for generating the friction-based heat input and the second shoulder for reducing the occurring flash. Figure 6 shows the results of these specimen.

A eutectic melt was generated with a comparatively high dwell time of 3,5 s.

The failure load of this specimen amounts around 2600 N. To ensure a higher failure load, plunging depth was increased in order to increase the deformation-based energy input. The plunging depth of the first generation tool was limited due to the damage of the copper lid. The high deformation leads to a defective weld, because the liquid phase could flow through the crack (see Figure 6 b)). For a higher degree of plastic deformation, a different tool geometry was used in the main investigations. In these examinations, the tool was designed with an additional pin and two shoulders. The pin was used to ensure a higher plunging depth, hence, a higher surface was generated and more energy was produced. Sound welds with failure loads up to 3572,3±109,9 N were produced.

Figure 7 shows the failure load and the temperature over the welding parameters rotational speed and feed speed. The diagram in Figure 7 shows the failure load of a crimped connection before welding. The failure load of this sample amounts to 884,7±113,0 N. Moreover, the graph shows a clear tendency to higher failure loads at higher rotational speed rates at different feed speeds. Hence the friction-based heat input is increased by the increased relative movement, which can be seen by considering the temperature at the different rotational speeds and feed speeds. The arithmetically averaged temperature of 545,7 °C at n=3200 rpm and v=25 mm/min results in a failure load of 3203,7±202,6 N and the temperature at the setting n=4200 rpm and v=25 mm/min of 578,7 °C results in an average failure load of 3572,3±110,0 N. A higher temperature signifies a higher failure load. However, the difference in failure load from 3200 rpm to 3700 rpm is higher than the difference in failure load from 3700 rpm to 4200 rpm. This effect is likely according to a saturation effect at a raising rotational speed, because of the high dependency of the coefficient of friction on the temperature [11]. Similar temperatures at different parameter settings lead to similar failure loads. Comparing the failure load of specimen with similar temperatures at different rotational speeds shows that the results stay in a similar range.

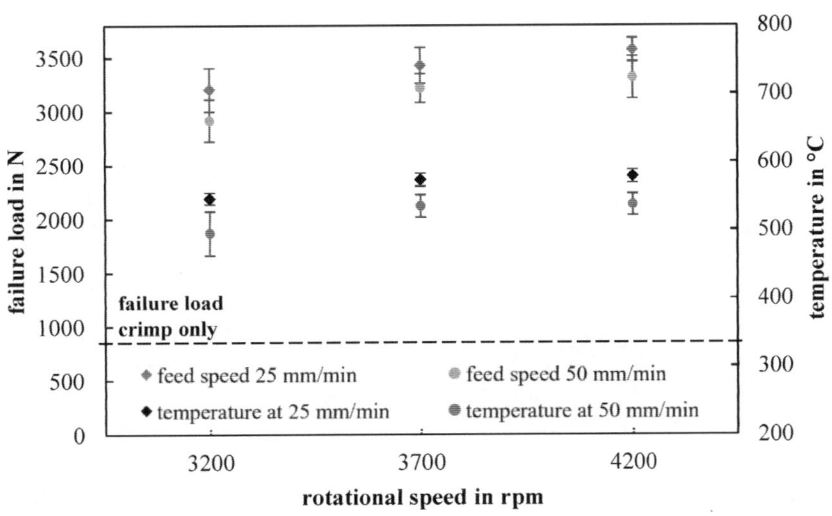

Figure 7: failure load and temperature over the process parameters rotational speed and fees speed

The temperature of connection with n=3200 rpm and v=25 mm/min amounts to 545,7±8,6 °C which can be compared to a joint with n=3700 rpm and v=50 mm/min where the temperature amounts to 534,8±16,2 °C. The first parameter set leads to a failure load of 3203,7±202,6 N and the second parameter set to a failure load of 3215,3±133,8 N. The conclusion is that an increase of rotational speed of around 16% leads to a decrease of process time of around 50%. Considering the feed speed a constant rotational speed of 4200 rpm for instance, an increase of the temperature and an increase of the failure load of around 7,8% is reported at a simultaneous decrease of the feed speed.

Which is caused by the total welding time, which is 100% higher at the lower feed speed of 25 mm/min compared to the feed speed of 50 mm/min.

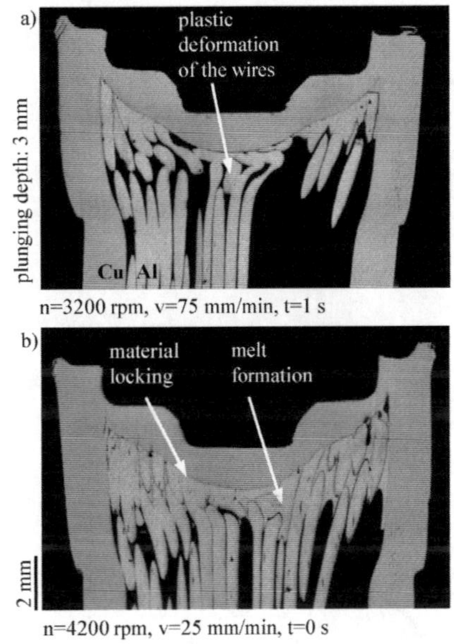

Figure 8: comparison of different parameter settings in cross section. a) n=3200 rpm; v=75 mm/min; t=1 s; x=3 mm; b) n=4200 rpm; v=25 mm/min; t=0 s; x=3 mm

Figure 8 shows two cross sections of the welded specimen. In Figure 8 a) a relatively cold and fast regime (n=3200 rpm, v=75 mm/min, t=1 s, x=3 mm) leads to an insufficient bonding and a plastic deformation of the wires. As the generated heat input is too low for the fast feed speed, the wires were pushed out of the welding zone. For a better binding and a high strength, the liquid phase should flow between the wires, to ensure a good wetting of the strand. The picture of Figure 8 b) shows a sample welded with another parameter setting (n=4200 rpm, v=25 mm/min, t=0 s, x=3 mm). Due to the higher rotational speed and the longer process time, the energy input is increased in comparison to Figure 8 a). Hence, a eutectic phase was generated and a good wetting of the aluminum strand was reached.

The behavior on the temperature during the FSSW process is shown in Figure 9. Here, the process is separated in three parts. The penetration of the pin starts at approx. 0,5 s with a rise of the temperature. Regarding the temperature during the plunging of the first shoulder, an enhancement up to approx. 530 °C can be registered. When the second shoulder starts penetrating the copper, the temperature stays at a temperature of around 510 °C before it raises to its maximum at the end of the process. This constant temperature is likely according to the diffusion process of the eutectic.

Figure 10 shows the influence of the rotational speed on the liquid phase formation and on the wetting effect. That the thickness of the amount of melt in the interface and between the wires is increased by a higher rotational speed.

With increasing rotational speed and thus increasing temperature, there is a wetting of the strand by the eutectic melt, which leads to an enlargement of the cross-conductance. At n=3200 rpm, the melt layer in the interface is approximately 20 µm thick, at 3700 rpm approximately increased by 100%.

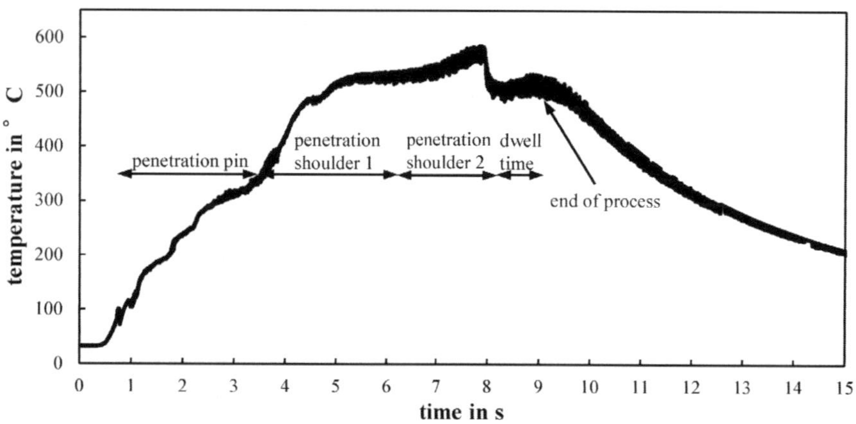

Figure 9: temperature in the contact area during the process (n=4200 rpm; v=25 mm/min; t=1 s; x=3 mm)

At 4200 rpm, the melting layer does not become much thicker as the eutectic melt flows between the strands and the saturation effect of the temperature takes place (see Figure 7).

Figure 11 a) and c) show the failure behavior after the tensile test. A partial failure in the contact area, in the melting zone and a partial failure in the aluminum base material was discovered. Figure 11 b) shows the wetting effect of the liquid phase on the aluminum wires. Figure 12 b) shows the composition of the diffusion layer of a sheet/sheet connection. The melting zone consists of the intermetallic phases γ_2 (Cu_2Al) and η_2 ($CuAl$) next to the copper, the θ-phase ($CuAl_2$), the eutectic layer and α-aluminum participation.

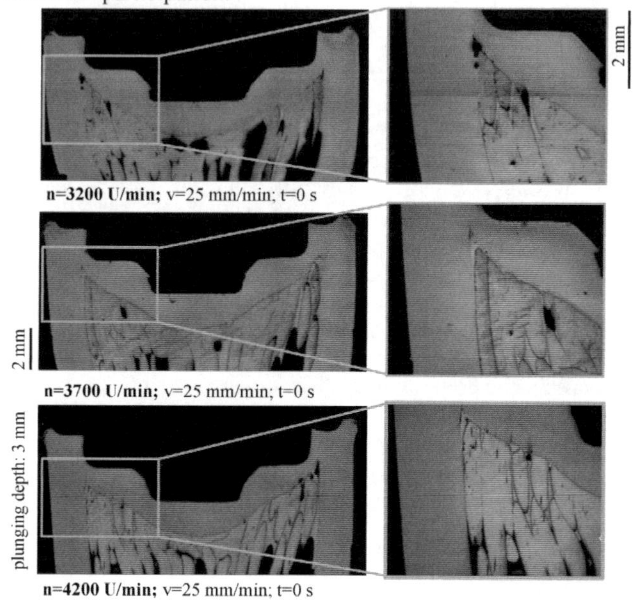

Figure 10: wetting effect and eutectic phase volume depending on the rotational speed

The joint failed in the contact area in the range of the brittle intermetallic phases γ_2 and η_2. The eutectic adhesive bond is stronger than the wires in the edge section. Hence, the joint failed in the aluminum base material. In this area a eutectic bond was generated, which can be seen in the magnification b) in Figure 11. Due to the higher angular speed in the outer areas, more friction-based heat was generated. Hence a better wetting of the aluminum wires was reached.

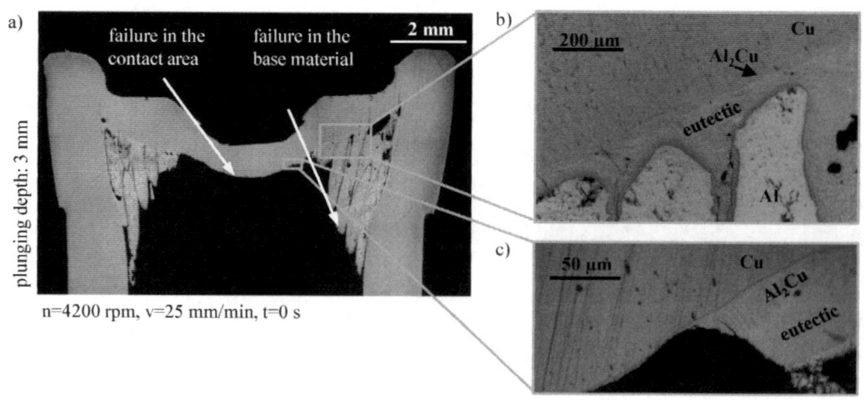

Figure 11: failure behavior of the aluminum/copper connection

a)

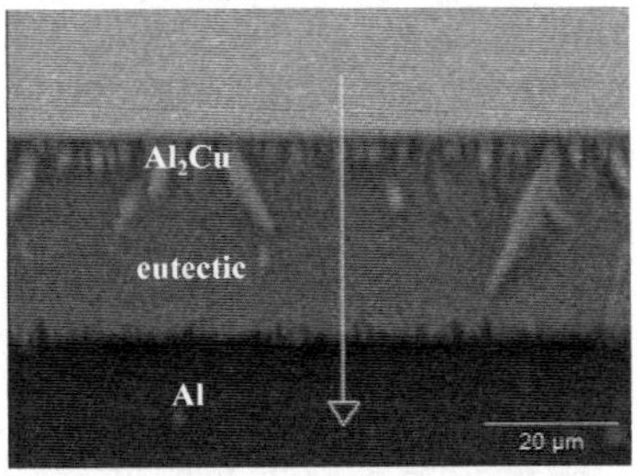

b)

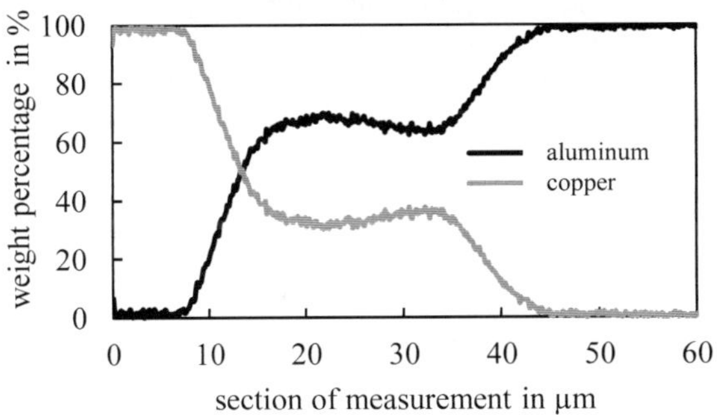

Figure 12: EDX-analysis of the melting layer of a sheet/sheet connection (n=2200 rpm; v=25 mm/min; t=3 s; x=0,3 mm)

Conclusions

In this work, a friction-based process for joining an aluminum cable with a copper conductor was developed and optimized. This friction-based process is referred to as friction stir spot welding. The tool was designed with a pin and a shoulder. These were designed so that the largest possible plastic deformation is initiated without destroying the arrester, to generate a material locking between aluminum and copper. A eutectic phase formation was generated in the contact area with a simultaneous reduction of the intermetallic phases.

The variables influencing the welding during this process are the rotational speed, the feed speed, the immersion depth and the dwell time of the tool. The experiments have shown that the rotational speed has the greatest influence on the formation of a eutectic melt layer. For the tested cables with a wire cross-section of 59 mm², a mean separation force of 3572,3 ± 110,0 N was achieved with the parameters rotational speed 4200 rpm, feed rate v=25 mm/min, plunging depth x=3 mm and dwell time t=1 s. It was shown, that higher rotational speed results in higher temperatures and hence in an increase of the failure load until a certain point.

The friction stir spot welding process provides numerous advantages for joining dissimilar cable conductor joints. It was also shown, that the wetting effect of the phase layer has a positive influence on the strength of a weld. In addition, a reduction of the process sensitivity was reached by a high deformation and relative motion in the interface. Hence, oxide layers and particles were displaced by the deformation and the fast and targeted head input.

The intermetallic γ_2 and η_2, as well as the θ-phase were detected, as well as α-aluminum precipitations and eutectic phase. By generating a liquid phase without reaching the melting point of the base materials, a reduction of the brittle intermetallic phases was reached. Hence, a ductile failure behavior of the joint was achieved.

Acknowledgments

The IGF Project No. 19036B of the research association "Schweißen und verwandte Verfahren e.V." of the DVS, Aachener Straße 172, 40223 Düsseldorf was, on the basis of a resolution of the German Bundestag, promoted by the Federal Ministry of Economic Affairs and Energy via AiF within the framework of the program fort the promotion of joint industrial research and development (IGF). The authors thank all the participants for the funding and the support.

References

[1] Bergmann J P, Regensburg A, Petzoldt F, Schürer R, Hellwig P and Steinberg H 2016 *Prozess- und werkstofftechnische Aspekte beim Fügen von Aluminium-Kupfer- Verbindungen für Bordnetzanwendungen* (Ilmenauer Fertigungstechnik)

[2] Scheidhammer G and Himmel J 2014 *ATZ Automobiltech Z* **116** 58–63

[3] Abbasi M, Karimi Taheri A and Salehi M T 2001 *Journal of Alloys and Compounds* **319** 233–41

[4] Braunovic M, Myshkin N K and Konchit Łs V V £ 2007 *Electrical contacts: Fundamentals, applications and technology (Electrical and computer engineering* vol 132*)* (Boca Raton, Fla, London: Taylor & Francis distributor)

[5] Aksöz S, Ocak Y, Maraşlı N, Çadirli E, Kaya H and Böyük U 2010 *Experimental Thermal and Fluid Science* **34** 1507–16

[6] Ilschner B and Singer R F 2010 *Werkstoffwissenschaften und Fertigungstechnik: Eigenschaften, Vorgänge, Technologien (Springer-Lehrbuch)* 5th edn (Berlin Heidelberg: Springer-Verlag Berlin Heidelberg)

[7] Sahm P R, Egry I and Volkmann T 1999 *Schmelze, Erstarrung, Grenzflächen: Eine Einführung in die Physik und Technologie flüssiger und fester Metalle* (Berlin, Heidelberg, s.l.: Springer Berlin Heidelberg)

[8] Karl Froschl 2006 Process for connecting terminal elements to an electrical conductor consisting of aluminum, and electrical conductor produced by the process *US20060292922A1*

[9] Wolfgang Hentschel 2011 Anschlussteil und Verbindung für elektrische Anlagen *DE102011011409 A1*

[10] Shen J, Suhuddin U F H, Cardillo M E B and dos Santos J F 2014 *Appl. Phys. Lett.* **104** 191901

[11] Bouarroudj E-o, Chikh S, Abdi S and Miroud D 2017 *Applied Thermal Engineering* **110** 1543–53

Improved binder degassing for defect reduction in large area paste brazing

K. Bobzin, M. Öte, S. Wiesner, J. Hebing
Surface Engineering Institute, RWTH Aachen University, Germany
hebing@iot.rwth-aachen.de

Abstract

Brazing pastes are essential in modern vacuum and inert gas brazing. Consisting of a metal powder and a binder, brazing pastes can easily be applied especially on complex geometries. The simple dosing and the uncomplicated production by mixing different elements in powder form make this method a very efficient and modern application technique. However, increased defect density can be observed when brazing pastes are used to braze large surfaces areas or complex geometries. For this reason, two approaches have been introduced with which the formation of defects is to be reduced. A process has been developed for improved degassing of the brazing paste during heating. For this purpose, the brazing paste is applied as a defined structure on the brazing surfaces by means of a template. The structure includes degassing channels which are intended to ensure an improved escape of the decomposition products of the binder out of the brazing gap. In addition, it was investigated whether carbon-containing binder residues can be avoided by the addition of titanium and thus further defects are forged. The results show that the methods can notably reduce the defect density of large-area compounds.

Introduction

Joining using brazing pastes is a widely used process variant. Typical application examples are the brazing of charge air coolers and heat exchangers. Brazing with paste has also established itself in aviation and energy technology e.g. in the repair brazing process of Ni- and Co-based superalloys. Brazing pastes offer the advantage of being easily automatably applicable, even in case of complex component geometries. Furthermore, very brittle filler metals, e.g. based on Ni or Fe, can be processed and applied almost effortlessly. Brazing pastes consist of a homogeneous mixture of metallic brazing powder and a proportion of 10 wt.-% $< m_{binder} <$ 16 wt.-% binder [1]. The binder usually consists of at least two components: a low-boiling solvent and a high-boiling binder component. A distinction is made between binders whose high-boiling constituents are water-soluble and those whose high-boiling constituents are dissolved in organic solvents [2]. The aqueous systems use dextrins, glycol and other substances as adhesives. For binders based on an organic solvent, polymers, vegetable oils and mineral oils are added as adhesives [3].

Currently, brazing pastes are applied outside of the brazing gap. Upon reaching the working temperature, the molten filler metal is pulled into the brazing gap due to the capillary effect and generates a cohesive bonding. Through the technical and economic advantages of brazing pastes, an extension of the field of application to large-area joints is of interest. However, the limited degassing capabilities between two substrates, lead to defects in the form of pores and precipitations. Previous studies have shown that binders can be distinguished based on at least two types of decomposition [4]. In the case of type A (Figure 1, B-I), the solvents evaporate at temperatures below T < 300 °C. Due to the strong increase in volume, pores are formed in the brazing paste and the paste can be partially forced out of the brazing gap. At higher temperatures of T < 500 °C, the high-boiling binder components dissociate. In contrast, type B (Figure 1, B-II), is characterized by one-stage decomposition and larger amounts of binder residues. Extrusion of brazing paste out of the gap was not observed with B-type binders. Furthermore, it is possible that high-carbon residues of these binders on the faying surface can lead to defects in the form of wetting defects or undesired precipitations.

In the context of the presented work, a methodology has been developed which improves the degassing of the brazing paste and consequently improves the brazing gap quality. In addition, it was examined whether carbon-containing binder residues can be bonded by modifying the brazing paste with carbide-forming elements.

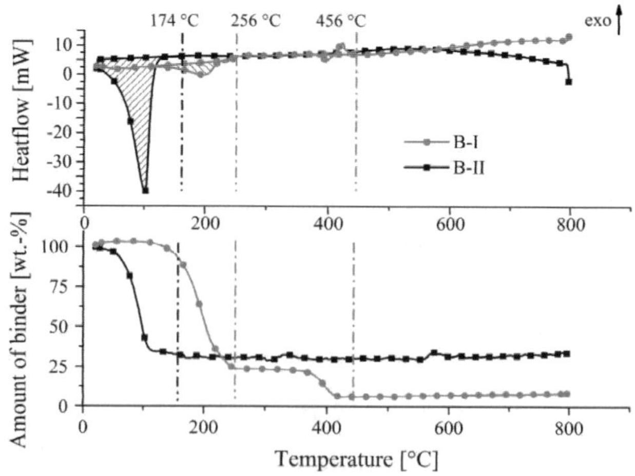

Figure 1: DSC/TG analyses of two different binders

Experimental Procedure

In order to reduce binder defects due to the brazing paste, two approaches were investigated. The first approach is the targeted improvement of the degassing behaviour of the paste by introducing structures into the brazing paste. Another way to reduce defects is to add active elements such as titanium, with the intention to turn carbon containing binder residues into carbides. Both approaches were examined in the presented work. In addition, a series of tests was carried out to investigate the influence of titanium and the binder on the mechanical properties of the brazed joints.

Degassing Structure

In order to avoid the formation of defects within the brazing seam, it is important to ensure adequate degassing of the binder components from the brazing paste. For this reason, a template for structuring the brazing paste was developed. Figure 2 shows the design drawing (A) and the 3D-printed plastic template (B). Paste fields with dimensions of $A = 3 \times 3$ mm^2 and a thickness of $d = 400$ µm can be adjusted by means of the template. The channels for discharging the degassing products have a width of $w = 1.5$ mm. The structure is intended to remove the gaseous decomposition products and thus suppress the formation of defects.

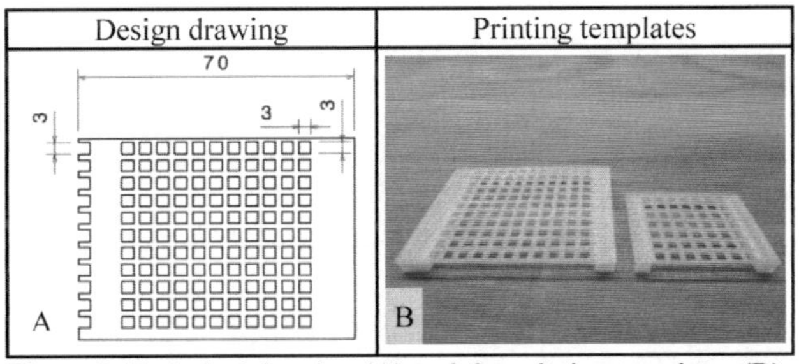

Figure 2: Technical drawing (A) and the printing templates (B) for structuring the brazing paste

Materials and Brazing Process

In the presented investigations, the brazing powder BNi-5 was used. This Ni-based filler metal has a melting interval between 1,080 C° < T < 1,135 °C [5]. In addition, a second filler metal was used, which is also based on the BNi-5 filler metal. However, $m = 2$ wt.-% titanium was added as an active element in the brazing powder in order to bind possible carbon-containing precipitations and residues of the binder in the form of carbides. Both filler metals were mixed with two different binders B-I and B-II to obtain four homogeneous brazing pastes. The used binders differ significantly in their decomposition and evaporation behaviour (Figure 1) and are representative for available industrial binders. In order to ensure that the structures in the pastes are sufficiently stable and do not deliquesce, the pastes were made with a relatively low binder content. The pastes consisted of $m_{binder} = 7.5$ wt.-% binder and $m_{braze} = 92.5$ wt.-% filler metals.

The substrate used for the investigations was the austenitic stainless steel AISI 304 with the dimensions of $A = 30 \times 30$ mm^2 and a thickness of $d = 8$ mm. In order to be able to determine the effect of structuring, comparative samples without structuring were prepared. Cross-sections of the samples were analyzed by light microscopy, scanning electron microscopy and energy-dispersive X-ray spectroscopy (EDS). Table 1 shows an overview of the samples.

Table 1: Overview of the samples

Sample	Binder	Comment
A1	B-I	Without structure
A2	B-I	Structured
A3	B-I	Modified with Ti
A4	B-I	With Ti and structuring
B1	B-II	Without structure
B2	B-II	Structured
B3	B-II	Modified with Ti
B4	B-II	With Ti and structuring

The brazing processes were carried out in a PVA MOV 553T vacuum furnace from PVA TePla AG (Wittenberg, Germany). A brazing temperature of $T = 1{,}190$ °C was chosen in high vacuum. The holding time at brazing temperature was $t = 10$ min with a subsequent rapid cooling at $p = 500$ mbar in an Ar atmosphere.

Shear Testing

In order to investigate the influence of the various binders as well as titanium on the mechanical properties of the brazing joints, shear tests were carried out based on AWS C3.2/C3.2: 2008 [6]. The substrate consisted of sheets of AISI 304 with the dimensions of $A = 60 \times 190$ mm^2 and a thickness of $d = 3$ mm. The overlap with which the sheets were brazed was $O = 3$ mm (Figure 3). The sheets were ground in the area of the overlap, cleaned and then the brazing paste was applied. The thickness of the brazing paste was adjusted to $d = 400$ µm by means of a blade. The brazing paste was not structured in this test, since the paste can still degas without structuring with an overlap of $O = 3$ mm. In order to avoid an offset of the sheets during the brazing process and to ensure a constant overlap, the sheets were fixed by means of spacers. After the brazing process, the samples were separated and milled to the nominal width. The testing speed was $v_{test} = 5$ mm/min.

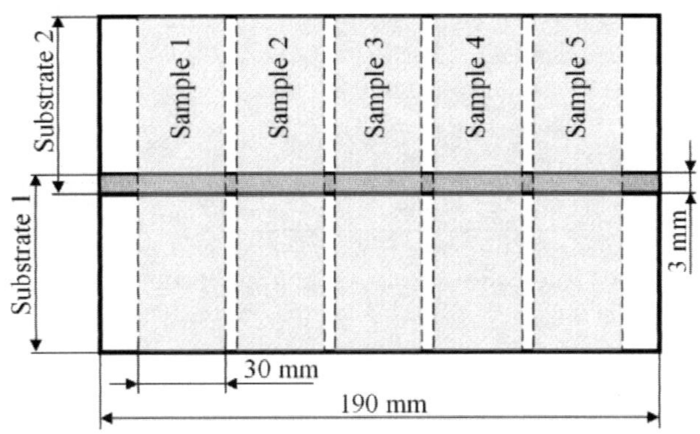

Figure 3: Sketched sample preparation of the shear samples

Results and Discussion

Structured Brazing Paste

Cross-sections of four samples are shown in Figure 4. Samples A1 and B1 were brazed without a previously structured brazing paste. In the preparation of samples A2 and B2, the brazing paste was applied in a structured form by means of the template (Figure 2). The final width of the brazing gaps are all between 70 µm < w_{gap} < 90 µm. When comparing sample A1 with sample A2, it can be observed that structuring in combination with Binder B-I results in a significant reduction in defect number and defect size. The effect is less pronounced in the samples which use binder B-II. Here it must be noted that sample B1 made with binder B-II already exhibits fewer defects compared to sample A1 with binder B-I in the initial state. The reason for this difference is the significantly different evaporation and decomposition behaviour of the binders. As has already been shown in Figure 1, the binders have distinctly different thermal properties due to their different chemical compositions. For this reason, a division into the two decomposition types A and B was made. Compared with the brazing pastes with the binder B-II (type B), brazing pastes containing the binder B-I (type A) tend to squeeze the brazing paste out of the gap to a greater extent. This leads to the formation of larger areas in the brazing gap that are depleted of brazing paste. These areas cannot be completely filled by the molten filler metal during the brazing process. Binder B-II also produces gases, but the paste is not forced out of the gap so strongly and the molten filler metal can fill the brazing gap better.

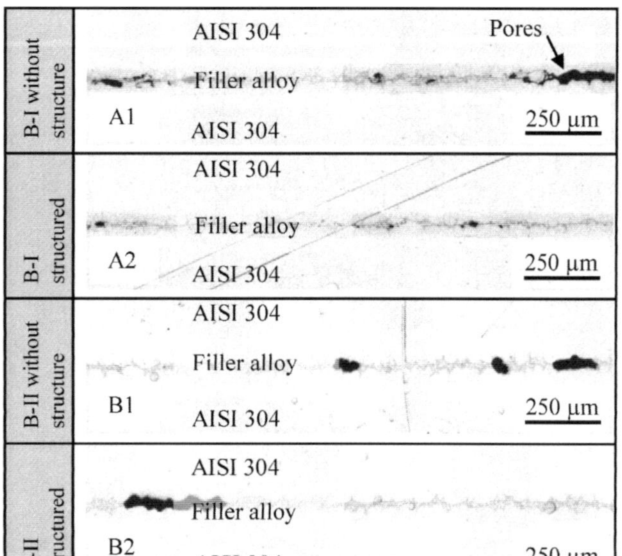

Figure 4: Cross-section of brazing samples consisting of AISI 304 substrate and structured and non-structured brazing paste made of BNi-5 and two different binders

Influence of Titanium Modification

Figure 5 shows samples made with non-structured titanium-containing brazing paste. Compared to the initial state in Figure 4 A1 and B1, the samples have fewer and smaller pores. The final width of the brazing gaps is between 70 µm < w_{gap} < 90 µm for the samples comparable to the initial state. Both samples show smaller precipitations. The precipitations in the sample A3 are distributed relatively homogeneously in the brazing seams. In comparison, sample B3 shows pronounced precipitations in the interface between the seam and the steel substrate.

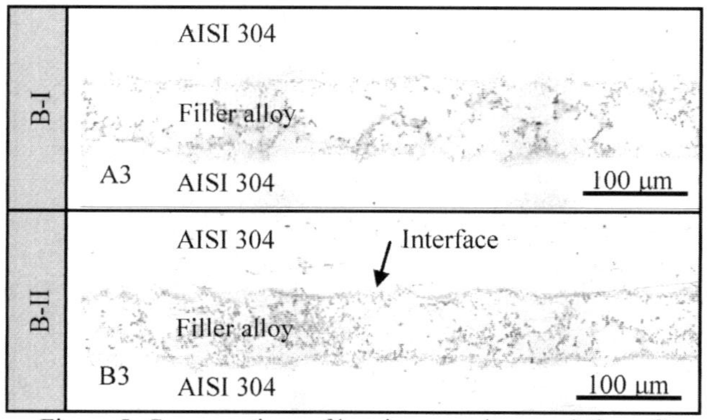

Figure 5: Cross-sections of brazing samples consisting of AISI 304 substrate and non-structured brazing paste made of BNi-5+Ti and two different binders

Figure 6 shows a detailed view of sample A3 from Figure 5. Precipitations can be seen, which partially have a rectangular shape. The precipitates accumulate in two areas of the brazing seam. In finely dispersed form, the precipitates are visible at the interface near the steel substrate. This may indicate a reaction with the substrate. In addition, excretions are visible within the brazing seam, but most of the precipitates are embedded in a darker phase.

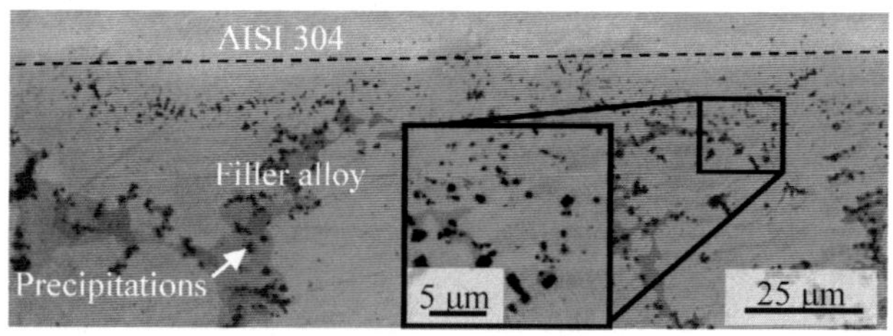

Figure 6: Detailed view of Figure 5A3, brazing samples consisting of AISI 304 substrate and non-structured brazing paste made of BNi-5+Ti and binder B-I

Structured and Modified Brazing Paste

After the investigation of the structured brazing paste as well as the brazing paste modification with titanium, both methods were brought together. Figure 7 shows two SEM images of samples prepared using structured and modified BNi-5 brazing paste. Defects in the form of pores can only be found sporadically. As was the case with the light-microscopic images, it can be observed that the precipitations increasingly appear at the interface to the substrate. In sample A4, it can be seen that the precipitations are mainly embedded in a lighter phase. By comparison, the precipitations in sample B4 are found mainly at the interface. Furthermore, a pronounced central phase band can be seen in sample B4, which is typical for Ni based filler metals in a large brazing gap [7, 8]. Typically, the phase bands are formed at gap widths of approximately $w_{gap} > 50$ µm. The sample A4, however, has a more homogeneous seam without a continuous phase band.

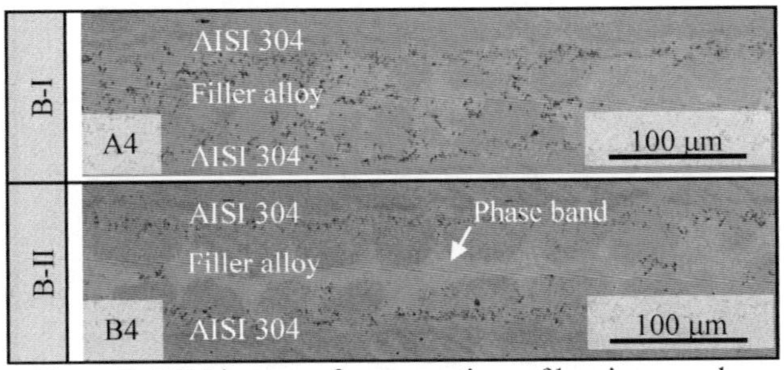

Figure 7: SEM images of cross-sections of brazing samples consisting of AISI 304 substrate and structured brazing paste made of BNi-5+Ti and two different binders

Figure 8 shows the SEM image of sample A4 in Figure 7 brazed using the binder B-I. The brazing seam has a second phase within the matrix phase. Clustered in this second phase, a variety of precipitations can be seen. The chemical composition at several points was analyzed by EDS. The EDS results can be found in Table 2. At the points P1 and P2, high values for titanium were measured which is an indicator for titanium carbide. On the other hand, carbon could not be clearly identified. The second phase in P3 has a lower concentration of nickel compared to the matrix phase P5 with $m_{P3} = 66.8$ wt.-% Ni versus $m_{P5} = 52.9$ wt.-% Ni. Furthermore, the concentration of Si in P3 is higher with $m_{P3} = 10.5$ wt.-% compared to P5 with $m_{P5} = 5.1$ wt.-%. Due to the increased silicon concentration in the second phase, this phase is likely to be a relatively low melting phase. The EDS results of sample B4 are comparable to the results shown. For this reason, their presentation is omitted.

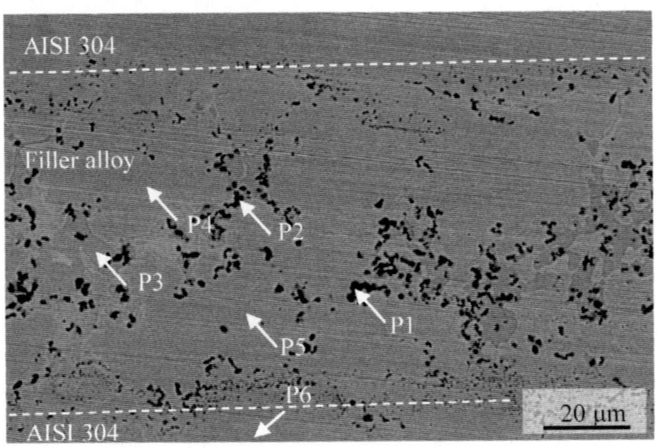

Figure 8: Detailed view of Figure 7A4, brazing sample consisting of AISI 304 substrate with structured brazing paste made of BNi5+Ti and the binder B-I

Table 2: EDS results for Figure 8 in wt.-%

Pos.	O*	Si	Ti	Cr	Fe	Ni
P1	3.5	1.1	83.5	2.5	1.7	7.7
P2	10.5	3.2	57.6	4.9	2.4	21.4
P3	0.0	10.5	2.9	12.3	7.5	66.8
P4	0.2	5.2	0.2	19.4	22.0	52.9
P5	0.0	5.1	0.4	19.9	21.7	52.9
P6	0.3	1.2	0.1	17.1	72.2	9.1

*The measurement of oxygen and other low-density elements is limited by EDS methods. Accordingly, the measured values can only be assessed qualitatively.

Figure 9 shows the results of the shear tests. Strengths of the samples are in the range of 110 MPa $< \tau_{shear} <$ 165 MPa. These are relatively low shear strength values for the filler metal used. The cause is assumed to be the wide brazing gap of 70 µm $< w_{gap} <$ 90 µm. It is known that with broader brazing seams a loss of strength is associated [1]. It must be further noted that the samples brazed using binder B-II (B1, B3) exhibit lower shear strength compared to the samples brazed using binder B-I (A1, A3). The reason for the lower strength is probably the central phase band, which forms especially when using the binder B-II (Figure 7). The phase bands, which occur especially at gap widths of $w_{gap} > 50$ µm, represent metallurgical notches and consequently have negative effects on the mechanical bonds of a brazing joint. Furthermore, it can be seen in the graph that the addition of Ti resulted in an increase in the shear strength of 20 MPa $< \Delta\tau <$ 35 MPa (A3, B3). Despite the likely brittle inclusions, no impairment of toughness was noted during the shear tests. The improved mechanical behaviour of the brazed joints is attributed to a lower number of defects as well as the formation of more homogeneous microstructure of the brazed joints as a result of the Ti modification.

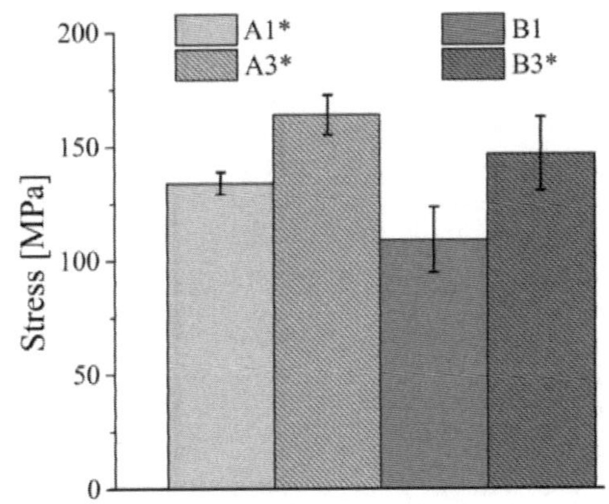

* Reduction to 4 samples due to cracks in the brazing gap

Figure 9: Results of the shear tests according to AWS C3.2/C3.2:2008 of samples consisting of AISI 304, brazed with BNi-5 each with and without $m_{Ti} = 2$ wt.-% titanium and different binders

Conclusions

Within the scope of this work, it could be shown that the used binder has a considerable influence on the brazed joint quality. The evaporation and decomposition of binder components causes the extrusion of brazing pastes and the formation of defects. In particular, the extrusion of brazing paste can lead to large defects within the brazing seam. The evaporation and decomposition behaviour of the binder also leads to different microstructures of the joints. Consequently, the binders have an influence on the strength and quality of the brazed joint. Brazing tests have shown that the degassing channels in the brazing paste can reduce the defects and thus positively influence the brazing seam quality. Thus, evaporation and decomposition products of the binder can be guided out of the gap, resulting in an improved degassing of the brazing paste. Furthermore, the results show that adding of $m_{Ti} = 2$ wt.-% titanium to the BNi-5 filler metal also has a positive effect on the brazing seam quality and the mechanical properties of the brazing joint. Further investigations must show whether the effect of the structuring of the brazing paste and the addition of active elements can also be transferred to other filler metals.

Acknowledgments

The IGF Project No. 17907 N of the DVS (DVS-No. 07.075) was, on the basis of a resolution of the Lower House of the German Parliament, promoted by the German Ministry of Economic Affairs and Technology via AiF within the framework of the program for the promotion of joint industrial research and development (IGF). The authors thank all the participants for the funding and the support. Many thanks also go to the project-accompanying committee for their constructive co-operation and the involved colleagues.

References

[1] AWS Brazing Handbook, American Welding Society (Miami, 1991), pp. 79/80, pp. 9/10.
[2] DVS Merkblatt 2605 (2003), “Bindemittel für pulverförmige Lote zum Löten unter Vakuum oder Schutzgas“, Düsseldorf.
[3] Günther Linke, EP1637275 A2, “Lötpasten zum Hartlöten“.
[4] K. Bobzin (2017) “Vermeidung binderbedingter Fehlstellen durch prozesssichere Verarbeitung von Lotpasten bei flächigen Bauteillötungen“, Abschlussbericht: DVS-Nr.: 07.075/IGF-Nr.: 17.907 N, Aachen, Germany.
[5] Technisches Datenblatt Ni 650 (2011), Umicore Technical Materials AG & Co. KG, Hanau.
[6] AWS C3.2M/C3.2:2008 (2008), “Standard Method for Evaluating the Strength of Brazed Joints”, American Welding Society.
[7] E. Lugscheider, K.-D. Partz (1983), “High Temperature Brazing of Stainless Steel with Nickel-Base Filler Metals BNi-2, BNi-5 and BNi-7”, American Welding Society.
[8] F. W. Bach, K. Möhwald, U. Holländer, J. Schaup, A. Langohr, C. Roxlau (2011), “Boron and phosphorous free nickel-based filler metals for brazing stainless steel in shielding gas furnace”, International Journal of Materials Research Vol. 102, No. 8, pp. 964-971.

Development of laser brazing system with precise temperature control

Kotaro Matsu and Kangdao Shi from Tokyo Braze Co., Ltd.
3-23-10 Minamikarasuyama Setagaya-ku Tokyo 157-0062 JAPAN
Phone 03-3300-1141 Fax 03-3300-1145
email: kotaro@tokyobraze.co.jp and ishi@tokyobraze.co.jp

Yoshihisa Sechi from Kagoshima Prefectural Institute of Industrial Technology
1445-1 Oda Hayato-cho Kirishima-city 899-5105 Kagoshima JAPAN
Phone 0995-43-5111 Fax 0995-64-2111
email: sechi@kagoshima-it.go.jp

Abstract

Among the heating process of brazing, laser heating has excellent characteristics such as high-speed heating in local area. The laser brazing technology has been successfully applied to aluminum to steel joint of car body. However, this laser brazing method controls only power output and scanning velocity but not brazing temperature. If brazing temperature can be controlled by measuring actual temperature during heating, it can be control forming intermetallic compounds at the joint and this will be big benefit for brazing dissimilar metals. In this study, laser brazing system with precise temperature control was developed. Actual temperature was measured during heating by infrared thermometer and feedback to the pulse generation system to control heating temperature. In tool industry, market demand of high performance cutting tools which are combination of diamond to carbides has been increased and they are all brazed. Current brazing method is by induction heating or in the furnace. These heating methods have tendency to graphitization of diamond in a long brazing time at high temperature. Laser brazing method is suitable for brazing of diamond to carbides in short brazing time with controlled temperature. PCD and WC was brazed by developed laser brazing system in this study. Mechanical properties were evaluated by shear strength test and compared with furnace brazing parts. Cross section of the brazed joint was observed by optical microscope and SEM. Elemental distribution was analyzed by EPMA. Sound brazing joints were obtained with higher joint strength than furnace brazing parts. Minimum forming of intermetallic compound at the joint was observed in short brazing time.

Introduction

Brazing is known as joining process with heating so that it is impossible to avoid from heat affecting for mechanical properties of joint. Therefore, controlling brazing temperature and holding time is very important but it was not done in any of current brazing processes. Particularly, short brazing time and precise brazing temperature is very difficult to control. To solve this difficult problem, it is understood that laser heating can be the one of good solutions.

There are many types of lasers; YAG laser as solid-state laser, CO_2 laser as gas laser, liquid laser, semiconductor laser, fiber laser and others. Those lasers are used in various fields and applications. With the spread of its applications, prices of laser systems have gradually declined, so further expansion of its application is expected.

As well as in brazing field, laser brazing technology has already been put to practical use in the automobile industry. They have successfully brazed dissimilar metals with less distortion by controlling the heat input to the joint by locally heating in short brazing time. However, the application of laser brazing is hardly found other than the automobile industry. Furthermore, heating time and temperature control of laser brazing are only adjusted by the laser output and the moving speed of robot. To perform the best characteristics of laser brazing, it is necessary to control the heating temperature and time more accurately to make brazing joint properly. Then, we have developed a laser brazing system which is capable with precise temperature control by feedback from measured temperature value of base metal.

Developed laser brazing system

Appearance of the developed laser brazing system is shown in Photo. 1. The developed laser brazing system is equipped a radiation thermometer which is a non-contact thermometer, and the temperature measurement data of the base metal is input to the main unit control unit. The brazing temperature is controlled by controlling the pulse output of the laser with the function generator (pulse generator) via the control PC. In order to enable stable temperature control in this system, it is necessary to get fast enough response speed of the radiation thermometer and the laser oscillator, and sufficient speed of the sequencer calculation of the main control unit. The specifications of this system allow operation at all speeds within few msec.

Vacuum chamber is installed for making different condition of brazing atmosphere as vacuum, vacuum + Ar gas or Ar gas only. Appearance of the vacuum chamber is shown in Photo.

2. The laser is irradiated from directly above the sample. Irradiation of any figures is possible by Galvanometer scanner.

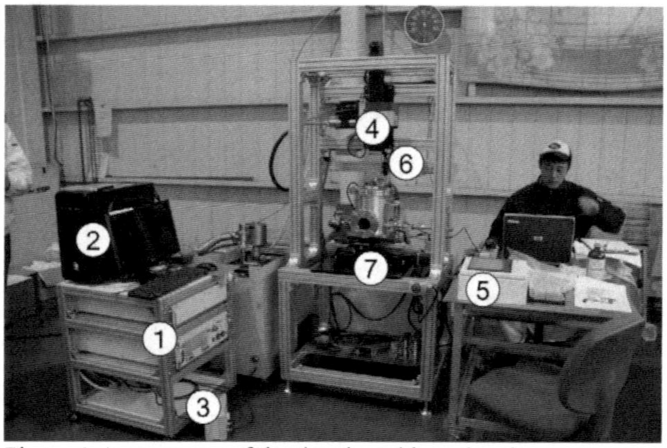

Photo. 1 Appearance of the developed laser brazing system

1. Fiber laser oscillator: Power and optical unit
2. Control PC
3. Function generator (Pulse generator)
4. Galvanometer scanner
5. Main control unit
6. Radiation thermometer
7. X-Y stage

Photo. 2 Appearance of the vacuum chamber

Experimental procedure and materials

Cemented carbide (ISO class K10) and PCD (polycrystalline diamond) were selected as the specimens. This material combination is applied to high performance cutting tools. Actual products are brazed with active brazing alloy in vacuum or Ar gas. However, the diamond can be carbonized depending on the brazing conditions so that there is a problem that not only the joint quality is affected but also the original performance of diamond becomes insufficient as high performance cutting tool. In evaluating this developed laser brazing system, brazing of the cemented carbide and PCD was done in varied brazing conditions and brazing joints were investigated.

Schematic diagram of specimen is shown in Fig.1 The specimen is used the actual product, so it is a near net shape of the cutting tool. Cemented carbide backed PCD chips are commonly used and brazing surface is always carbide to carbide. But, PCD chip surface was brazing surface to make the specimens. The specimens were degreased and cleaned with acetone before brazing. Appearance of brazed specimens are shown in Photo. 3.

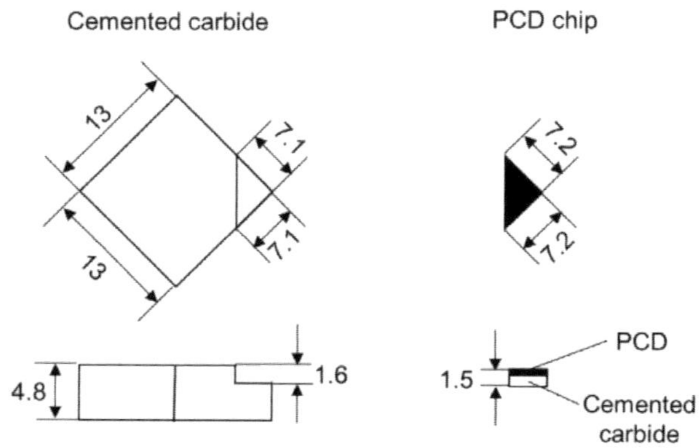

Fig. 1 Schematic diagram of specimen

Photo. 3. Appearance of brazed specimens
Left: For shear test and cross-section
Right: Actual product shape

Paste form of TB-629T was used as brazing alloy. Chemical compositions and melting range of TB-629T is shown in Table 1. A certain amount of brazing alloy was applied to the whole surface of the chip side, then set to carbide. Brazing paste was dried sufficiently at 150 degrees C.

Table 1 Chemical compositions and melting range of TB-629T

Type	Chemical compositions (mass%)				Melting range (°C)
	Ag	Cu	In	Ti	
TB-629T	60	24	14	2	620 ~ 720

Brazing was carried out after inside of the chamber was vacuumed to 1.5 Pa, then Ar gas was introduced while vacuum was continued. Laser irradiation was started with the pressure inside the chamber controlled at 1.5 x 10^3 Pa. Brazing temperature was fixed at 780 degrees C and the brazing time

was varied as 5, 30 and 60 sec. Accuracy of temperature control at brazing temperature was within ±10 degrees C.
Laser irradiation position and measurement point of the radiation thermometer by the guide laser, and PCD chip position was shown in Photo. 4. Laser irradiation position was circular by Galvanometer scanner and it was moved at a speed of 500 mm/sec.

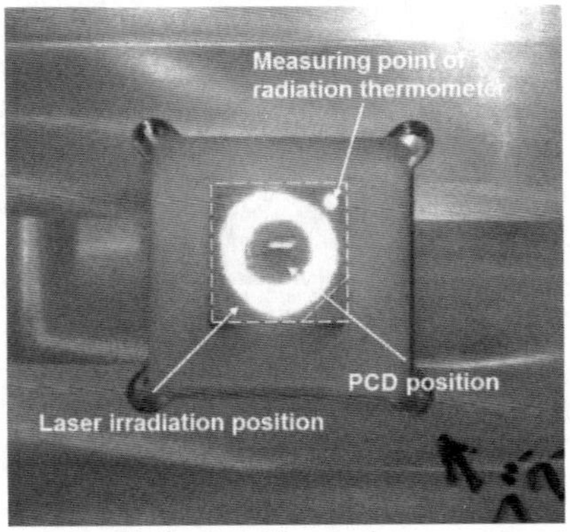

Photo. 4 Laser irradiation position and measurement point of the radiation thermometer by the guide laser, and PCD chip position

For comparison with making specimens by laser brazing, specimens by vacuum furnace was prepared. Vacuum brazing was carried out that vacuum pressure was reached to 1.0 x 10^{-1} Pa, then Ar gas was introduced while vacuum continued, and vacuum pressure was controlled at 50 Pa. Brazing temperature was fixed at 780 degrees C and the brazing time was fixed at 600 sec. Accuracy of temperature control at brazing temperature was within ±10 degrees C.

Results and discussion

Sound brazing joints without discoloration, oxidation or lack of brazing alloy flow were obtained in any brazing conditions. SEM images of cross-section of specimens were shown in Fig. 2. Thickness of the brazing alloy layer did not become constant regardless of brazing time. It is assumed that there was variation as specimen was prepared even though the amount of brazing alloy was controlled to some amount. No gap control of brazing alloy thickness was done in both laser brazing and vacuum brazing. It is understood that these factors were influenced the thickness of the brazing alloy layer. The microstructure of the brazing alloy layer tended to become larger as the brazing time increased.

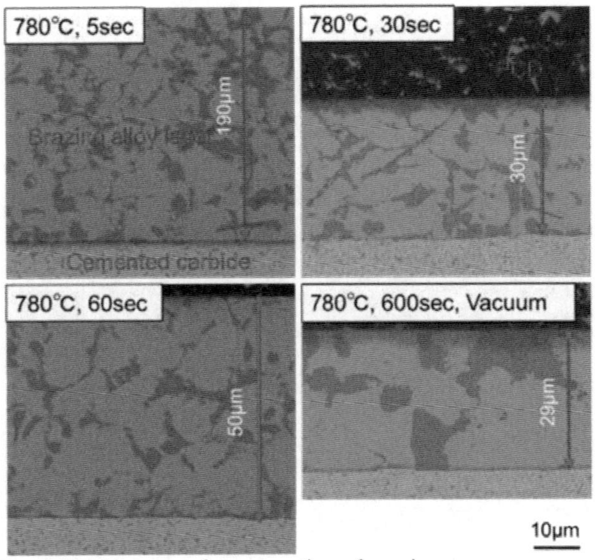

Fig. 2 SEM images of cross-section of specimens

Relationship between brazing time and shear strength is shown in Fig. 3. In the case of laser brazing, the maximum strength showed 184MPa at brazing time of 5 sec and the joint strength was decreased as brazing time increased. In the case of vacuum brazing, joint strength decreased to 107MPa as brazing time increased to 600 sec.

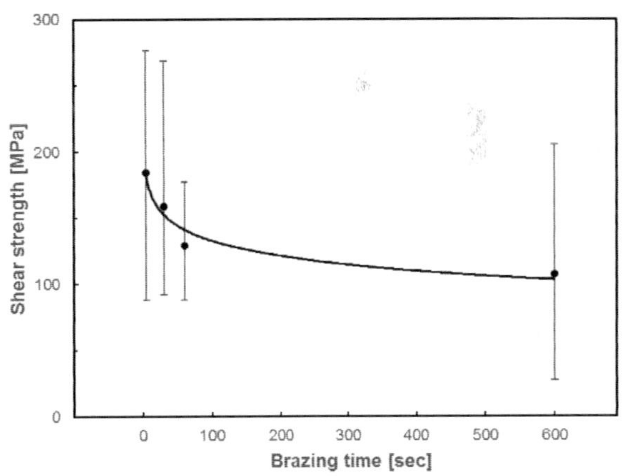

Fig. 3 Relationship between brazing time and shear strength

Typical fracture appearance of the specimens after the shear test on each brazing condition is shown in Fig. 4. As a result of visual observation, all the fracture was occurred at the interface at the cemented carbide side. The fracture occurred in the gray layer as brazing time increased in laser brazing. This gray layer is understood to be intermetallic compound layer of TiC and is considered to be an interfacial reaction layer formed by the reaction between Ti in the brazing material and C in the cemented carbide.

On the other hand, the fracture surface of the specimens brazed in a vacuum furnace showed complicated shape, and metallic surface and gray layer of TiC were mixed. From the appearance observation on the PCD chip side, there were cracks in the chip layer. In the case of vacuum brazing, it is understood that the brazing time is remarkably long as compared to laser brazing so that the carbonization of PCD was progressed then joint strength was decreased.

Fig.4 Typical fracture appearance of the specimens after the shear test

In order to understand the behavior of Ti at the PCD side, cross-section of brazed joint was observed by SEM image and elemental distribution was analyzed by EPMA. SEM image and elemental distribution map of Ti and C was shown in Fig. 5. In previous paper, it was reported that when diamond and cemented carbide are brazed with an active brazing alloy, submicron size of intermetallic compound layer of TiC is formed in the brazing alloy layer of diamond side. In laser brazing, the thickness of the intermetallic compound layer at the brazing time of 5 sec was very thin with weak detection. However, Ti was gradually detected clearly as brazing time increased to 30 and 60 sec. In consideration of the distribution condition of C, formation of TiC at the interface is presumed at any brazing time. Since all fracture was occurred on the cemented carbide side in any of the laser brazed specimens, the time for forming the TiC at the interface between the brazing filler metal and the diamond was very short. And it was found that the TiC formed in a very short time is very strong.

On the other hand, the TiC did not grow remarkably in the case of vacuum brazing even though Ti is clearly detected at the interface. It is considered that the Cu-Ti layer was formed on the brazing alloy side of TiC after forming the intermetallic compound layer which was formed by originally contained Ti in the brazing alloy reacted sufficiently to C in the PCD. Regarding the decrease of joint strength, it is considered that not only the growth of the TiC layers but also the long-time heating at elevated temperature to PCD.

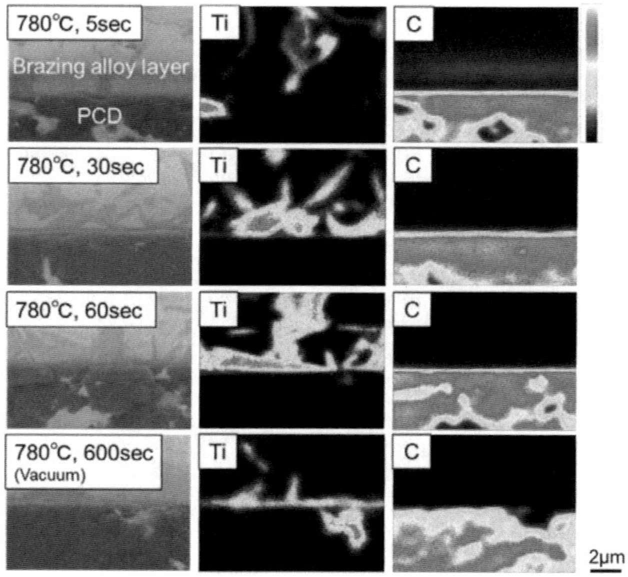

We successfully observed the influence of the joint strength with very short brazing time while precisely controlling the temperature. It was found that sufficient joint strength was obtained at the brazing time of 5 sec, and the interfacial reaction of active brazing alloy was extremely faster than we expected.

Conclusions

We have developed the laser brazing system capable with precise temperature control and it was confirmed that brazing can be performed with excellent interfacial reaction control. Furthermore, this laser brazing system is capable with precise temperature control rather than conventional laser systems, so it is possible to suggest precise heat input control to the base metals which are sensitive to heat affect. For further investigation, we will braze various base metals by this laser brazing system to obtain basic data.

Acknowledgments

The authors would like to extend a special thanks to Saitama prefecture, Japan who supported our study. The budget of this study was sponsored by "Saitama prefecture new technology/product development grant subsidy".

埼玉県 Powered by Saitama Prefecture
先端産業創造プロジェクト

References

[1] Y. Sechi and K. Nakata: Transactions of JWRI, Vol.39, No.2, (2010), pp340-342.

Influence of the bulk material surface condition on AISI H13 / BNi-2 brazed joints

Tillmann, W.; Bobzin, K.; Wojarski, L.; Öte, M.; Manka, M.; Wiesner, S.; Eilers, A.
Institute of Materials Engineering, Faculty of Mechanical Engineering, TU Dortmund University (TU Dortmund), Dortmund, Germany
wolfgang.tillmann@tu-dortmund.de, 0049-231-7552581

Surface Engineering Institute, Faculty of Mechanical Engineering, RWTH Aachen University, Aachen, Germany
info@iot.rwth-aachen.de, 0049-241-8095327

Abstract

Usually, faying surfaces of brazing couples are prepared by means of machining, using miscellaneous lubricants. This can lead to significantly different surface conditions compared to the bulk materials. Residues of the lubricants can prevent a wetting of the base material with the molten filler during the brazing process, which can lead to the formation of voids.
This paper investigates the influence of purposely contaminated surfaces of the hot work steel AISI H13 by applying different lubricants and subsequent cleaning procedures on the joint quality. The joints were brazed at T = 1.090 °C in high vacuum, utilizing a BNi-2 filler alloy. In order to ensure a reproducible contamination of the surfaces, the samples were coated with a water-soluble and synthetic lubricant, respectively. For the subsequent cleaning, a hydrocarbon based cleaner or acetone were used. The effect of the residues on the wetting behavior, the appearance as well as on the strength of the joints was analyzed by means of light and electron microscopy, energy dispersive X-ray spectroscopy, and tensile testing. The results showed various influences of the surface condition on the microstructure and joint strength.

Introduction

In order to achieve high quality brazing seams, a good wetting behavior of the filler metal on the base material is required. The wettability depends on multiple parameters such as the temperature, alloy formation, and surface conditions. A technical surface consists of a (cold) deformed layer, an oxidic reaction layer, as well as an adsorption layer, on which contaminations deposit (Fig. 1a).

In order to achieve a good wettability, the contaminations as well as the reaction and adsorption layers have to be removed prior to brazing by using mechanical, (electro)chemical, or thermal cleaning processes. The technical data sheet "DVS-Merkblatt 2606" of the German Welding Society (DVS) shows some eligible cleaning processes for brazing and soldering applications [1, 2].

During the manufacturing chain of brazed joints, almost every surface runs through a machining process in which lubricants are used. The influence of lubricant residues on the microstructure and the mechanical behavior of brazed joints is hardly known [3-5]. FÜSSEL et al. investigated i.e. the influence of a surface contamination with different oils and cleaners on brazing joints made of aluminum alloys. No influence of the contaminations on the quality of the brazed joints could be observed [6, 7].

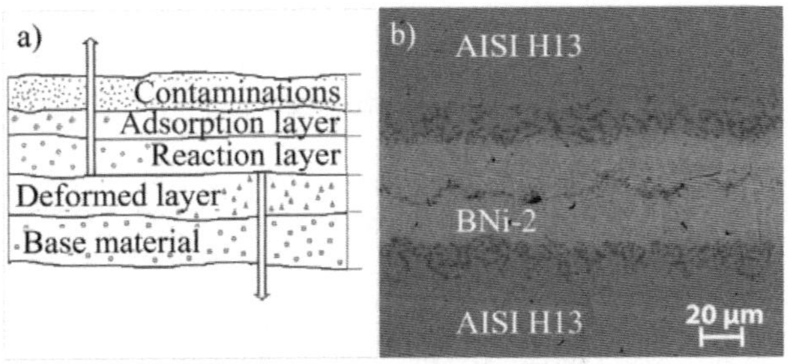

Figure 1: a) Schematic image of a technical surface with different layers [1]; b) Microstructure of the brazed AISI H13 / BNi-2 joint [8]

A typical microstructure of AISI H13 / BNi-2 is shown in Fig. 1b. Dark boron-rich brittle phases are formed in the diffusion zone and in the center of the brazing seam [8]. Regarding the mechanical behavior of the AISI H13/ BNi-2 joint, there exist only few research results. Therefore, the tensile strength of brazed joints of austenitic chromium nickel steel with the filler metal BNi-2 are used as a reference. They show tensile strength values of approx. $R_m = 350 - 513$ MPa [9-11]. The fracture analyses show ductile as well as brittle fracture, depending on the location of the failure. When the failure occurs in the brazed seam, a brittle fracture can be observed, while a base material failure leads to a ductile fracture [10].

In most cases, the failure of brazed joints occurs in the brazing seam or in the diffusion zone. Especially imperfections of the brazed seam, such as invaded fillets, notches, pores or offsets, result in higher stresses leading to a premature failure of the joint. Fig. 2 shows a simulative comparison of the strain distribution of an ideal brazed seam and brazed seams with imperfections. The highest strain in the brazed seam appears when the notch effect occurs i.e. due to an invaded fillet or a pore inside the seam [8, 12].

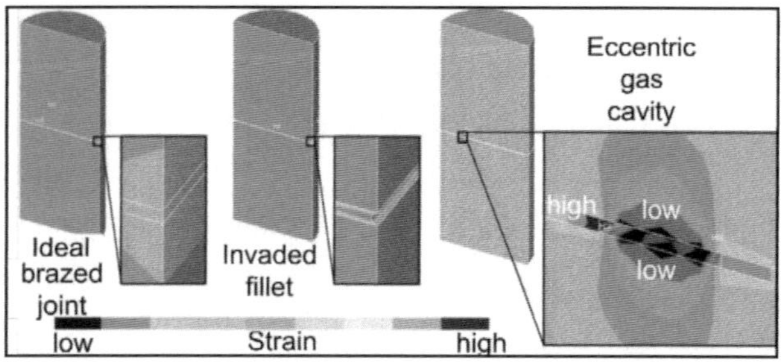

Figure 2: Simulation of the mechanical strain distribution of a brazed joint during tensile load [8]

The aim of this study is to analyze the influence of contaminations and cleaning processes on the microstructure development and mechanical properties of AISI H13 / BNi-2 brazed joints. For this purpose, wetting and joining tests are performed with substrates with different surface conditions. Additionally, the fracture surface of the tensile test samples was analyzed to correlate the residues of the contaminations with the mechanical properties and the type of fracture of the joints.

Experimental Procedures

Materials

For the experiments, the hot work tool steel AISI H13 was used as base material. The specimen dimensions for the microstructural investigation were 25 mm x 15 mm x 8 mm. The specimens for the mechanical testing had a diameter of d = 18 mm and a length of l = 50 mm. As filler metal the nickel-based BNi-2 was used. The chemical composition of base and filler metal is listed in Tab. 1.

Table 1: Composition of the AISI H13 steel and the filler metal BNi-2

[wt.%]	Fe	Cr	Ni	Si	B	Mo	V	C
AISI H13	90.9	5.3		1.0		1.4	1.0	0.4
BNi-2	3.0	7.0	82.4	4.5	3.1			

In order to investigate the influence of the surface condition on the quality of the brazed seam, two lubricants were selected: the water-soluble lubricant "Avilub V 3352" (hereafter called "lubricant 1") and the synthetic lubricant "Avia Metacool Syntogrind" (hereafter called "lubricant 2"). Either acetone or "IBS-Spezialreiniger EL/Extra", a hydrocarbon-based cleaner, consisting of more than 90 % isoparaffin hydrocarbon ($C_{10\text{-}13}$), was used to clean the surfaces after the contamination with the lubricants.

Surface preparation

The surface of the substrates used for the wetting and joining tests was milled to a roughness of approx. Sz = 5.3 ± 1.1 µm as determined by confocal laser scanning microscopy. The joining surfaces of the tensile test samples were machined to obtain planar surfaces. The roughness after machining was comparable to the roughness of the flat brazing specimens.

In order to create different homogeneous surface states, a continuous fluid layer made of 10 vol.-% lubricant and demineralized water was applied onto the substrate surfaces. Afterwards, the samples were dried in air for several hours until the water had evaporated. Some samples were cleaned subsequently in an ultrasonic acetone bath or IBS cleaning agent for $t_{cleaning}$ = 15 min at $T_{cleaning}$ = 30 °C, followed by a rinsing process with demineralized water.

Brazing

As filler metal, a BNi-2 foil with a thickness of t = 50 µm was employed. For the wetting tests, circular foil segments with a diameter of approx. d = 10 mm were used. For the joining and tensile test samples, the foil was cut to fit the dimensions of the brazing surfaces. The parts of the cylindrical tensile test samples were temporarily fixed prior to the brazing process in order to avoid an offset of the parts. The brazing process was performed in the high vacuum furnace MOV 553 T (PVA TePla AG, Wettenberg, Germany) with a holding time of t = 10 min at a brazing temperature of T = 1,090 °C.

Microstructure and Mechanical Testing

The tensile strength of the brazed joints with different surface conditions was determined using a hydraulic tensile testing machine (Mohr und Federhoff) at room temperature and with a speed of v = 0.04 mm/s. Images of the fractured surface were taken after tensile testing. In some cases, the filler metal did not wet the whole joining surface (Fig. 3a, marked area). Therefore, the percentage of the wetted area A was determined with the help of an image analyzing software and used for the calculation of the tensile strength (Fig. 3).

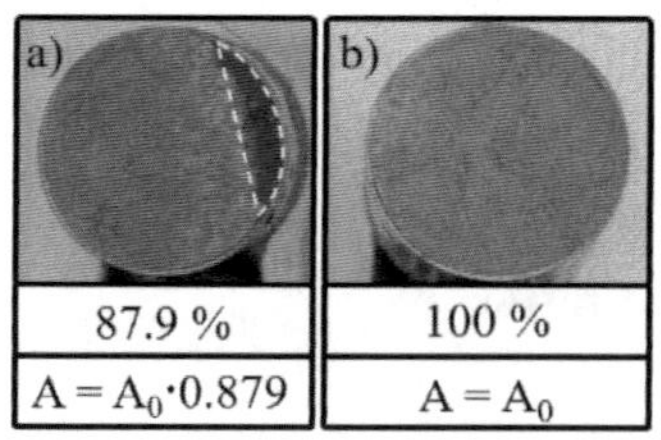

($A_0 = \pi \cdot r^2$
with r = 9 mm)

Figure 3: Images of two fractured surface and the amount of wetted area; a) a wetted area smaller than the diameter ($A < A_0$), b) completely wetted diameter ($A = A_0$)

Subsequently, the fractured surfaces were investigated by using the scanning electron microscope (SEM) JSM 7001F (Joel) and energy-dispersive X-ray spectroscopy (EDS) for element analyses. Additionally, cross sections of the fractured surfaces were prepared and examined by SEM and EDS for selected samples. Furthermore, cross-sections of the joining specimens with different surface conditions were analyzed to obtain information concerning the undamaged brazed seam.

Results

Wetting Behavior

Fig. 4 shows the top view of AISI H13 / BNi-2 wetting samples. Prior to brazing, different surface conditions were generated by contaminating them with different lubricants and with a subsequent cleaning process. In Fig. 4, no notable influence of these surface conditions on the wetting behavior could be observed by optical inspection. The molten filler metal wets the steel surface exhibiting no contraction behavior which would imply a negative influence of the contaminations on the wetting behavior.

However, differences concerning the surface roughness of the solidified filler metal can be observed. SEM analyses and EDS measurements were conducted to analyze the surface in detail.

The observed microstructure is typical for the Ni-based filler metal with rectangularly shaped Cr-rich phases in a Ni-based matrix. These phase typically contains boron [11], which, however, could not be detected in this case due to the detection limit of the EDS technique for light elements. The morphology of the Cr-rich phases seems to differ slightly depending on the cleaning agent. However, no influence of the lubricant can be detected. It has to be noted that cracks occur in the Ni matrix of the filler metal, especially at the surface of the samples cleaned with the IBS cleaning agent. A change in the chemical composition could be excluded as a possible cause for the crack formation.

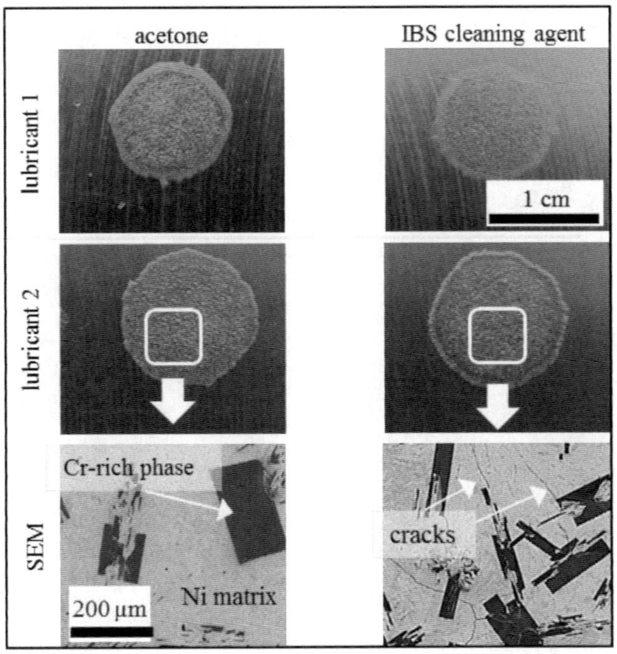

Figure 4: Top view of AISI H13/BNi-2 wetting samples with different surface states prior to brazing and SEM images of the surface of the solidified filler metal

Microhardness measurements also showed differences between the samples cleaned with different liquids. The hardness of the Ni matrix corresponds to $H_{Ni\ matrix;\ acetone} < 400$ HV 0.1 for the crack-free sample cleaned with acetone. For the sample cleaned with the IBS cleaner, it amounts to $H_{IBS} = 650$ HV 0.1. While the actual reason for the differences has yet to be investigated, the observation already shows that the mechanical properties of the brazed joints can be influenced by the surface condition prior to brazing.

Microstructure

Cross-sections of the brazed AISI H13 / BNi-2 joints with different pre-brazing surface conditions are shown in Fig. 5. Both cross-sections with contaminated surfaces show a high number of voids, especially at the interface of the filler metal and the base material (Fig. 5a/b). A contamination with lubricant 1 leads to fewer yet larger voids (Fig. 5a) while contamination with lubricant 2 results in a higher amount but smaller voids (Fig. 5b). If the surfaces were cleaned with the respective cleaner subsequent to contamination, a significant reduction of the voids can be achieved (Fig. 5c/d). Nevertheless, the typical occurrence of borid-containing phases in the center of the brazing seam and in the diffusion zone is not affected by a pre-brazing cleaning. Regarding the mechanical properties of the joints, a greater number of voids should lead to lower tensile strength, i.e. due to the higher strains (Fig. 2, right).

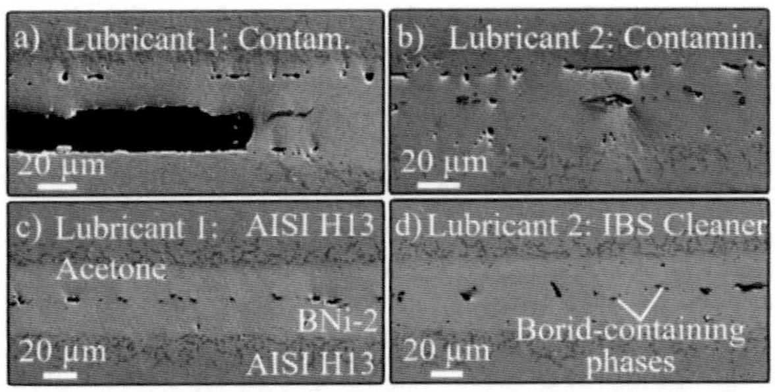

Figure 5: SEM images of the cross-sections of an AISI H13/ BNi-2 brazed joint with different pre-brazing surface conditions: a) contaminated with lubricant 1; b) contaminated with lubricant 2; c) contaminated with lubricant 1 and cleaned with acetone; d) contaminated with lubricant 2 and cleaned with IBS cleaner.

Mechanical Properties and Fracture Surface Analysis

In Fig. 6, the tensile strength mean values as well as the maximum and the minimum strength of the brazed joints with different pre-brazing surface conditions of the base material are shown. For each surface condition, three samples were tested. As expected, by far the highest values are achieved with cleaned surfaces, even if not all samples with cleaned surfaces have such high values. As most of the values are below $R_m = 300$ MPa, only three samples have "normal/standard" tensile strength values within the range of values in other publications that investigate brazed joints of austenitic chromium nickel steel with BNi-2 [10, 11]. One of the three samples was contaminated with lubricant 1 and cleaned with acetone, the other two were contaminated with lubricant 2 and cleaned with the IBS cleaner. Nevertheless, there are samples with the same surface treatment, which have much lower tensile strength values.

All samples with contaminated surfaces show lower tensile strength values. This corresponds with the findings of the examined cross-sections where a greater number of voids occur (Fig. 5a/b).

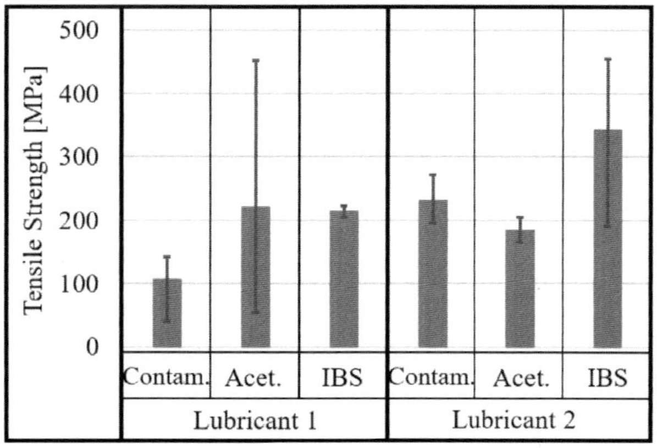

Figure 6: Tensile strength mean values with maximum and minimum values of the brazed AISI H13 / BNi-2 joints with different surface conditions before brazing.

In order to understand the variation of the tensile strength (Fig. 6, lubricant 1/ acetone and lubricant 2/ IBS), Fig. 7 shows the fractured surfaces of each sample with the above mentioned surface treatments in more detail. Even after adjusting the tensile strength values to the wetted area (Fig. 3), the samples with a smaller wetted area show much lower tensile strengths (Fig. 7a, samples 1 and 3; Fig. 7b, sample 1). Two possible influencing factors for this behavior are the eccentrical load and the notch effect. In all cases of low tensile strength values, the wetted area of the samples is asymmetric. Therefore, the applied load leads to excessive shear stresses in the brazed seam during the tensile test and thus to lower values. Furthermore, if the edge of the sample is not wetted (especially in asymmetric cases), the notch effect occurs and further lowers the tensile strength values (cf. Fig. 2, middle).

So far, it could not be clarified when smaller wetted areas occur. There are many influences that can affect the wetting so that further investigations regarding this subject are neccessary.

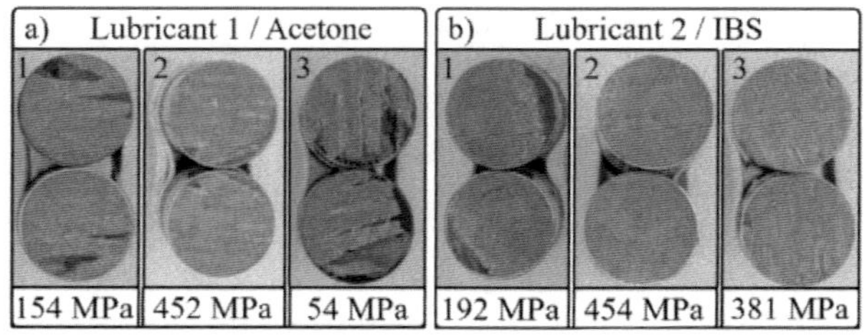

Figure 7: Images of the fractured surfaces and corresponding tensile strength values of a) three samples contaminated with lubricant 1 and cleaned with acetone before brazing, b) three samples contaminated with lubricant 2 and cleaned with IBS cleaner before brazing.

After analyzing the tensile strength and the fractured surfaces on the macroscopic level, the SEM was used for microcopic analyses in both backscattered electron mode (Fig. 8b) and secondary electron mode (Fig. 8c/d). In Fig. 8a, the fractured surface of an asymmetrically wetted sample is shown. Besides the dark unwetted zone, two different failure areas can be observed as well. On the SEM image in Fig. 8b, these areas are investigated using EDS analysis. At position 1 elements of the filler metal BNi-2 as well as elements of the base material AISI H13 are detected. Thus, it can be assumed that the failure of this fractured surface area (Fig. 8b, top left) takes place at the interface between base and filler metal. Another area (Fig. 8b, diagonally from top right to bottom left) shows two different phases. Their compositions are represented with positions 2 and 3. The brighter matrix (position 2) has almost the same composition as the filler metal while the darker phases (position 3) show high contents of chromium and boron, which leads to the assumption that chromium borides are formed [1, 11, 13]. Considering that brittle chromium borides exist in the center of the brazing seam (cf. Fig. 1b/5), a failure within that area is likely. However, due to the detection limits of EDS for light elements like boron the results should not be evaluated quantitatively.

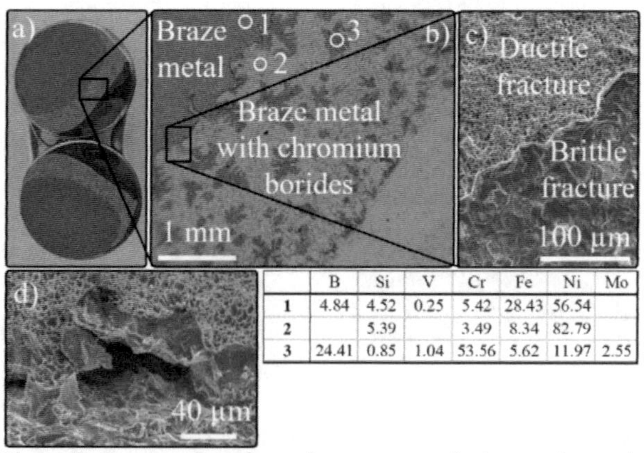

	B	Si	V	Cr	Fe	Ni	Mo
1	4.84	4.52	0.25	5.42	28.43	56.54	
2		5.39		3.49	8.34	82.79	
3	24.41	0.85	1.04	53.56	5.62	11.97	2.55

Figure 8: Fractured surface of an asymmetrical wetted sample (lubricant 1, IBS cleaner): a) macroscopic image b) SEM image in backscattered electron mode, c) SEM image in secondary electron mode, d) SEM image at an angle ($\alpha = 60°$) between brittle and ductile fractured surface.

In order to evaluate the fracture type, a higher resolution as well as the secondary electron mode were chosen for the SEM image in Fig. 8c. Regarding the surface structure, a ductile fracture can be observed for the failure at the interface and a brittle fracture for the failure in the brazing seam.

In Fig. 9, the fractured surface of a sample with a pre-brazing contamination with lubricant 1 is shown. Dark areas are already visible on the fracture surface on the macroscopic image in Fig. 9a which were analyzed at the SEM using EDS. These areas are oxygen- and carbon-rich phases that are presumably residuals of the lubricant. Despite showing a large wetted area, the tensile strength values of the contaminated samples are low (Fig. 6). Therefore, it can be assumed that the residues decrease the tensile strength.

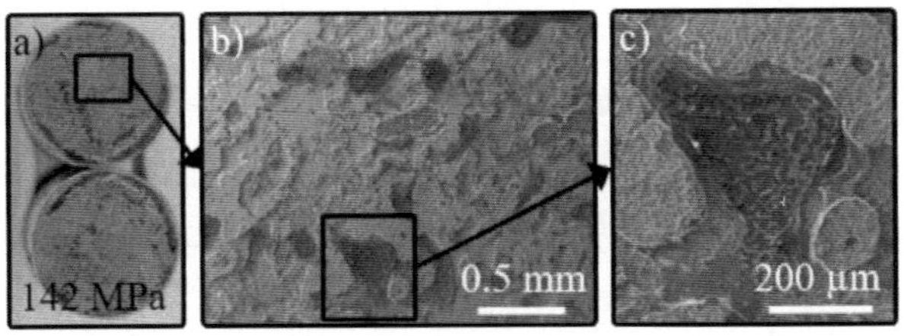

Figure 9: Fractured surface of an sample with pre-brazing contamination with lubricant 1: a) macroscopic image, b/c) SEM images

Cross-sections of the fractured surfaces were investigated in order to confirm the assumptions regarding the location of the failure position. Basically three different types of failures can be observed: Failures at the interface between filler and base metal, in the diffusion zone, and in the brazed seam. The failure at the interface mainly propagates along the pores, which are probably Kirkendall pores [14], where higher strains occur during the load (Fig. 2, right). The side on which the failure propagates can sometimes change abruptly as shown in Fig. 10a. The failure in the diffusion zone occurs probably in the brittle ‚boron-rich phase areas (Fig. 10b). The boron-rich phases do not only occur in the diffusion zone but also in the center of the brazed seam. Thus, in some cases, the failure occurs along these brittle boron-rich phases in the center of the brazed seam (Fig. 10c). These findings can be related to the findings in Fig. 8.

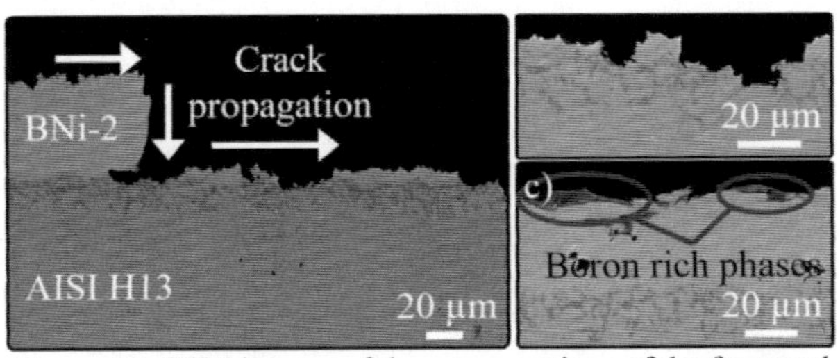

Figure 10: SEM images of the cross-sections of the fractured surfaces (lubricant 2, IBS cleaner): a) failure at the interface, b) failure in the diffusion zone c) failure in the center of the brazed seam

Conclusions

For the brazed AISI H13 / BNi-2 joints with different pre-brazing surface conditions the following results can be summarized:

1. All wetting samples show chromium rich phases, while the Ni matrix of the samples cleaned with the IBS cleaner exhibit higher hardness values than that of the samples cleaned with acetone.
2. The highest tensile strength values are achieved by acetone cleaning for the water-soluble lubricant and by the hydrocarbon based cleaner for the synthetic lubricant.
3. An effectively wetted area is crucial to achieve high tensile strengths.
4. The contaminated samples show residues, probably due to the lubricant, which decrease the tensile strength.
5. Three different types of failure occur on the fractured surfaces: Failure at the interface, in the diffusion zone, and in the center of the brazing seam. Thereby, the brittle boride phase has a significant influence on the failure position.

Acknowledgments

The project IGF 18387 N of the German Welding Society (DVS) research association is funded by the AiF within the framework of the program to promotion of the IGF (Industrielle Gemeinschaftsforschung) by the German Federal Ministry of Economic Affairs and Energy according to a resolution of the German Bundestag. The authors are thankful for this support.

References

[1] Feldmann, K., Schöppner, V., Spur, G., "Handbuch Fügen, Handhaben und Montieren", (2013), pp. 103–106.

[2] DVS-Merkblatt 2606, "Hinweise auf mögliche Oberflächenvorbereitungen für das flussmittelfreie Hart- und Hochtemperaturlöten", (2000).

[3] Bobzin, K., Öte, M., Wiesner, S., "Investigation of the effect of contaminations and cleaning processes on the surface properties of brazing surfaces" IOP Conf. Series: Materials Science and Engineering, 181, (2017), pp. 526–529.

[4] Bobzin, K., Öte, M., Wiesner, S., Tillmann, W., Wojarski, L., Manka, M., Eilers, A., "Investigation of the effect of surface contamination on the wetting behavior of Cu-based filler alloys", DVS-Berichte, 325, (2016), pp. 128–132.

[5] Battenbough, A.J., Osmanda, A.M., Staines, A.M., "Surface Preparation for High Vacuum Brazing", Nicrobrazenews, (2011)

[6] Füssel, U., "Korrelation zwischen der Oberflächenhistorie, den Prozessbedingungen und der Löteignung von Aluminiumwerkstoffen", final report of IGF project 17.748 BR, (2015)

[7] Hofmann, E., Zschetzsche, J., Füssel, U., „Lötbarkeit von Aluminiumwerkstoffen nach Durchlaufen des

Fertigungsprozesses“, Schweißen und Schneiden 69, Heft 9, (2017), pp. 640–648
[8] Bobzin, K., Tillmann, W., “Systematische Untersuchung der Eigenschaften gelöteter Fügeverbunde mit anwendungsrelevanten Prüfverfahren II”, final report of IGF project 16.558 N, (2012)
[9] Hartmann, T., Nuetzel, D., „Nickel-chromium-based amorphous brazing foils for continuous furnace brazing of stainless steel“, DVS-Berichte 263, (2010), pp. 42–47
[10] Lee, Y.-W., Kim, J.-H., Song, Y.S., Seok, C.-S., “Mechanical Strength and Microstructural Characterization of SUS304 Brazed Joints with BNi-2 Filler Metal: Influence of Brazing Temperature and Time”, Solid State Phenomena Vols. 124-126, (2007) pp. 1673–1676
[11] Lugscheider, E., Partz, K.-D., “High Temperature Brazing of Stainless Steel with Nickel-Base Filler Metals BNi-2, BNi-5 and BNi-7”, Welding research supplement, (1983), pp. 160–164
[12] Tillmann, W., Wojarski, L, Manka, M., Lehmert, B., Bobzin, K., Öte, M., Wiesner, S., „Rechnergestützte Berechnung von Spannungen in Lötverbindungen“, DVS Congress, DVS-Band 315, (2015), pp. 573–577
[13] Ruiz-Vargasa, J., Siredey-Schwallera, N., Geya, N., Bocherb, P., Hazottea, A., “Microstructure development during isothermal brazing of Ni/BNi-2 couples„ Journal of Materials Processing Technology 213, (2013), pp. 20–29
[14] Shiue, R.K., “Brazing of 422 stainless steel using the AWS classification BNi-2 Braze alloy”, Journal of Materials Science 41, (2006), pp. 6353–6361

A study on bonding method for nickel foam heat exchangers

Hoejun Heo[1], Yongchan Park[1], Chungyun Kang[1]

[1] *Division of Material Science and Engineering, Pusan National University, Busan, South Korea*
hjheo@pusan.ac.kr, +82-10-8533-2460

Abstract

Effect of brazing condition on the microstructure and mechanical properties of Ni foam/stainless steel joint was investigated in this study. Open-cell Ni metal foams, which have several thermal conductivity due to its large surface area to ratio and a tortuous flow path, were brazed using Ni based filler metal. It turns out that microstructure and tensile properties of the brazed joint were dependent to bonding time. 0~10 min bonding time provided the best results in terms of the tensile load and elongation, but 30~60 min did not. Ni foam were partially reacted and altered form pure Ni to Ni solid solution (Ni-Si-Cr) and Ni3B intermetallic compound until 10 min, but, after 30 min, the entire Ni foam was changed into Ni solid solution and interconnected Ni_3B phases in the horizontal direction of the Ni foam, which is a main culprit to decrease the mechanical property.

Introduction

Metal foam is fabricated as cellular structure with large pores so that it has a number of advantages such as a light density and a larger surface per volume than the bulk material [1, 2]. In particular, open-cell structured metal foam, which has interconnected pores, can let fluids move through the pores and have a high surface ratio per unit volume. In addition, since the turbulence generated by its complicated flow path between the solid media has a large effect of thermal convention, the metal foam can be utilized in heat exchangers to increase a heat transfer [3-8].

There are two main types of the heat exchanger using the metal foams [3,4]. One is a plate-type in which the plate material and metal foam has a sandwich construction, and the other is a shell and tube type which surrounds the metal foam material outside the tube. Although the mechanical fastening method has been applied to join metal foams and base substrates such as plates or tubes, the strength of the mechanical fastening part is low, and a thermal conductivity is low due to high thermal resistance generated on surface of base substrates [7]. In order to fabricate a high thermal transfer heat exchanger using the metal foams, a proper joining process between them is important.

Bonding methods for metal foams include adhesive bonding, soldering, welding and brazing [8-11]. Adhesive bonding and soldering are limited in high temperature environments due to the low melting point of the adhesive and solder [8]. Mrityunjay Singh et al. applied laser welding process without filler metals, and the metal foam and base materials can be welded in a narrow area by locally focused energy input. However, shrinkage deformation of metal foam occurred in the laser-heated area, the pores became blocked with molten liquids, which interrupts the flow through metal foam [9]. Brazing, which the filler metal was inserted between work pieces and melted at bonding temperature, would be a proper method because the molten filler metal fills the gap via a capillary phenomenon with maintaining the original shape of adjoining materials. Unfortunately, few studies for brazing metal foam has been studied [10,11]. The purpose of this study is to develop a brazing process between a Ni metal foam and SS 304 plate material for a micro-channel heat exchanger.

Experiment

25PPI (pore per inch) open-cell nickel foam (Ni 99.99%, 10*10*5), which were fabricated by electro-deposition method, and 10mm diameter of SS 304 (Fe-Cr-Ni) rod were used in this study. In the Figure 1, the nickel foam is composed of a cellular shape with an average pore size of ≈ 600μm porosity, and struts, which are cylindrical structure, are connected at a central structure node forming a regular shape to a hexagonal dodecahedron. As a filler metal used for brazing MBF20 amorphous sheet (t =50 μm) was applied, and chemical composition of the filler metal and SS304 was shown in Table 1.

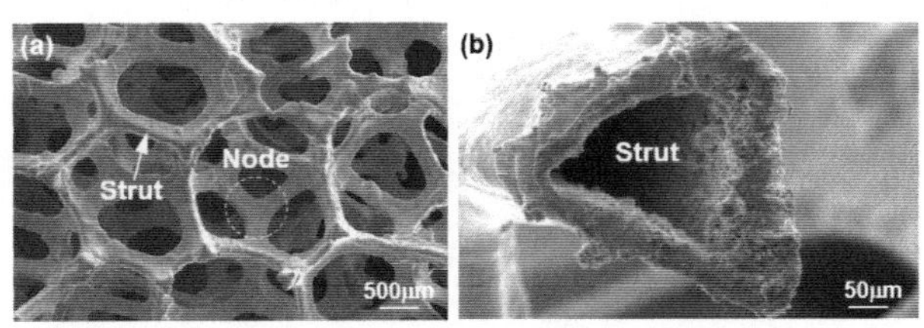

Figure 1 SEM image of 25PPI (pore per inch) Ni foam and its cross-sectional magnified shot of cylindrical structure.

Table 1 Chemical composition of SS 304 and MBF 20 filler alloy

	B	Si	Cr	Fe	Ni
SS 304	-	-	22.4	Bal.	10
MBF 20	2.8	4.2	7.4	3.0	Bal.

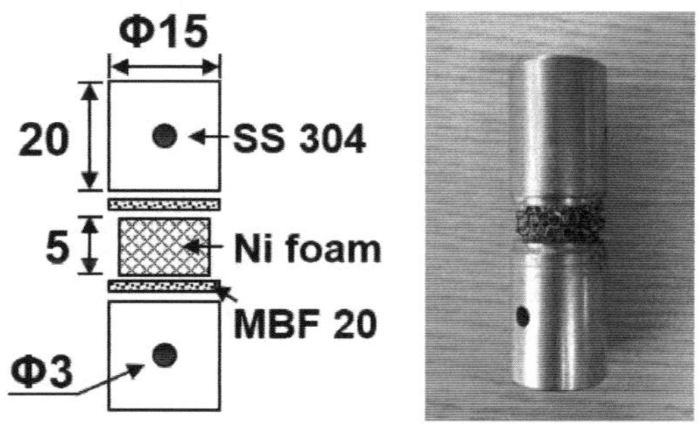

Figure 2 The specimen used for mechanical shear test

The filler metal was inserted between Ni foam and SS 304 in a sandwich structure. Prior to brazing, the surfaces of the nickel foams and SS 304 rod were grounded with 600 grit (P1200) SiC abrasive papers and ultrasonically cleaned in ethanol for 5 min. The specimens were heated to 1050 °C with 20 °C /min in a high vacuum (up to 5.0×10^{-5} torr) furnace. Holding time were varied 0, 10, 20, 60 minutes and then cooled. The $0.02 g/mm^2$ pressure was applied perpendicular to the bonding area during heating cycle. Tensile test specimens were shown in Fig. 2, and it was pulled with 1 mm/min crosshead speed.

The cross-sectional specimen of as-brazed joints was polished up to 1µm. The specimens were electrically etched with solution 0.8wt. % oxalic acid. The cross-sectional microstructure of as-brazed joints was characterized via optical microscopy (OM), secondary electron microscopy (SEM) and field-emission electron probe microanalysis (FE-EPMA).

Results and Discussion

Microstructure of the joint

Figure 3 shows close images of a Ni foam/SS 304 specimen before(a) and after(b) brazing at 1050°C for 10 minute. The specimen before brazing has a gap between the strut of Ni foam and base metal(SS 304) at the interface, but the specimen after joining were succesfully bonded showing that the gap is filled without defect such as void and non-bonded zone. It can be seen that the filler metal was liquefied to bridge the gap between the Ni foam and the SS 304 making a joint and finally sound joint was obtained.

Figure 4 shows an optical micrograph of a Ni foam/ SS304 joint brazed at 1050°C for 30 minute using MBF 20 filler metal, and Ni, B, Si, Cr and Fe element mapping using FE-EPMA. It is shown that B, Si, Cr and Fe are distributed from

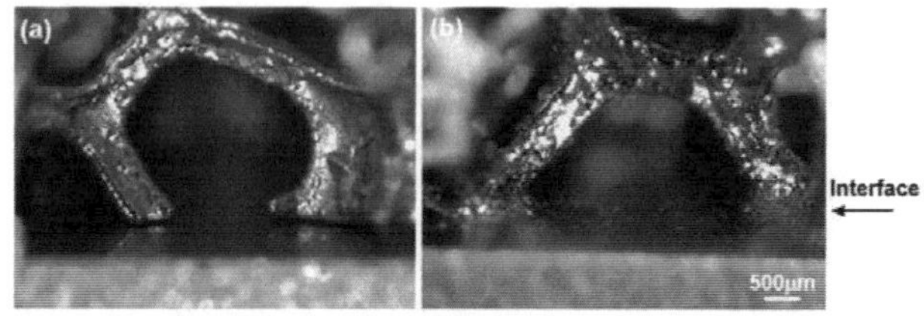

Figure 3 Close images of (a) non-brazed and (b) brazed joints between Ni foam and SS 430.

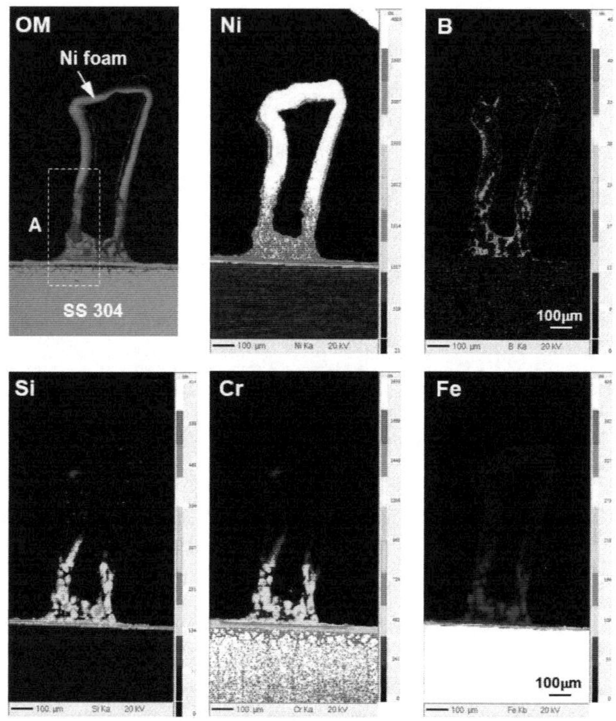

Figure 4 An optical micrograph of a cross-sectional Ni foam/SS 304 brazed joint along with FE-EPMA elemental mapping for Ni, B, Si, Cr and Fe.

the SS304 interface to the center of the sturt of the Ni foam. In other words, a part of the Ni foam as base metal reacted with the filler metal and new microstructures were fabricated. In addtion, Cr-depleted zone appears inside SS 304 base metal.

In order to observe the microstructure characateristics of the brazed joint, the magnified SEM images of dotted area in Figure 4 was shown in Fig. 5., and FE-EPMA spot quantity analysis of numbered phases is provided in Table 2. In upper part of the Ni foam (X region), Ni foam (numbered as 1) is existed as original base metal, and small piece of Ni_3B (numbered as 2) phases were distributed along the interface. The middle(Y region) of the Ni foam pillar contains Ni solid solution (numbered as 3) containing 4.8wt.%Cr and 6.3wt.%Si, and Ni_3B (numbered as 4) phases were existed along grain boundary of the Ni solid solution. At the interface of the SS 304(Z region), Ni_3B (numbered as 5) and the Ni solid solution (numbered as 6) phases were existed above a

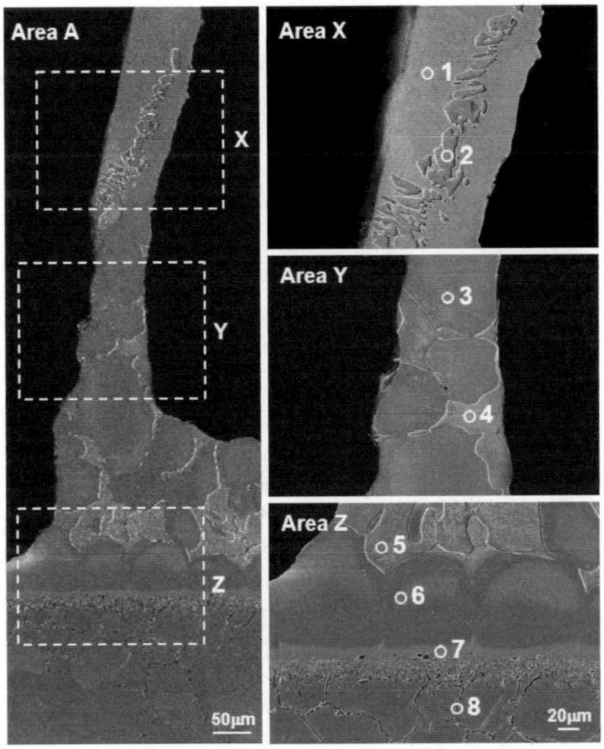

Figure 5 Magnified SEM micrographs of the Ni foam/SS 304 joint in Fig. 4.

Table 2 FE-EPMA quantity analysis (at. %) of the point in Fig. 4.

No.	B	Si	Cr	Fe	Ni	Phase
1	0.0	0.0	0.0	0.0	100.0	Ni
2	29.1	0.1	1.2	0.5	69.2	Ni_3B
3	3.3	6.3	4.8	2.4	83.3	α-Ni
4	27.2	0.0	1.2	0.4	71.3	Ni_3B
5	27.2	0.0	1.2	0.4	71.3	Ni_3B
6	4.9	5.9	4.5	2.0	82.7	α-Ni
7	0.1	1.68	11.6	42.3	42.3	Fe(Ni)
8	28.1	0.3	29.1	40.1	2.4	Cr_xB_y

Fe-Ni solid solution (numbered as 7), and Cr_xB_y phases(numbered as 8) formed along the grain boundaries were present inside SS 304 base material. This is considered to be the phase generated by the reaction of B existing in the filler metal and Cr in the SS 304 base material. In addtion, it is seemed that existance of CrB phases along the grain boundary means boron(B) element diffuses preferentially along the grain boundary.

To elucidate the microstructural changes of the Ni foam /SS 304 joints depending on holding times, the optical microscopics of the joints brazed at 1050°C for 0~60 minutes are shown in Fig. 6. For the specimens holding 0~10 minutes, the Ni foam originally remained and the liquified filler metals partially reacted with the Ni foam with forming a fillet. However, at 30 miniut holding time, the reacted Ni foam was composed of Ni solid solution and Ni_3B phases, and especially, the Ni_3B phases are horizontally connected each other penetrating the entire Ni foam. At 60 minute, it can be seen that thickness of the Ni foam was lessened by composing larger Ni solid solution and Ni_3B phases.

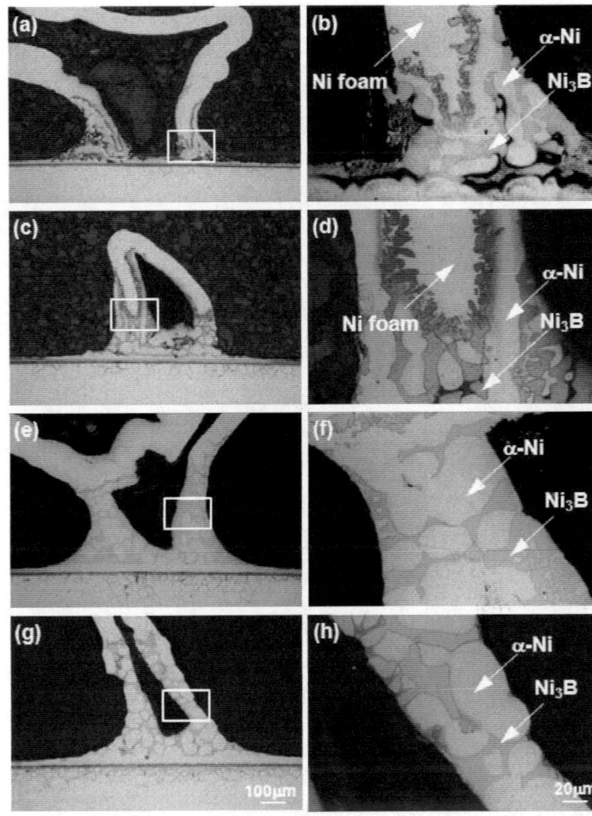

Figure 6 Cross-sectional optical micrographs of the 25PPI Ni foam/SS 304 joints brazed at 1050°C for (a,b) 0 minute, (c,d) 10 minute, (e,f) 30 minute, and (g,h) 60 minute.

Mechanical property of the joint

Fig. 6 represents the average tensile load and elongation at 1050°C for the different holding time. The tensile load and elongation results were divided into two grpous between 10 and 30 minute. In detail, the tensile load has a value of 470 ~ 600N from 0 to 10 minutes, but after 30 minutes, it became relatively low obtaining a 300 ~ 400N value. From 0 to 10 minutes the elongation of the joint is 25 to 28%, but it is 15 to 20% after 30 minutes, which shows a relatively low elongation.

In order to understand why the mechanical properties were separated into two groups according to the bonding time, fracture surface of 10 and 60 minute specimens were investigated. Figure 7 shows SEM images of the fractured surface and its crossc section optical images of fractured specimens for 10 min (a,c,e,g) and 60 min.(b,d,f,h), respectively. For 10 minute condition, fracture path occurred at the middle of the Ni foam , and dimple fracture

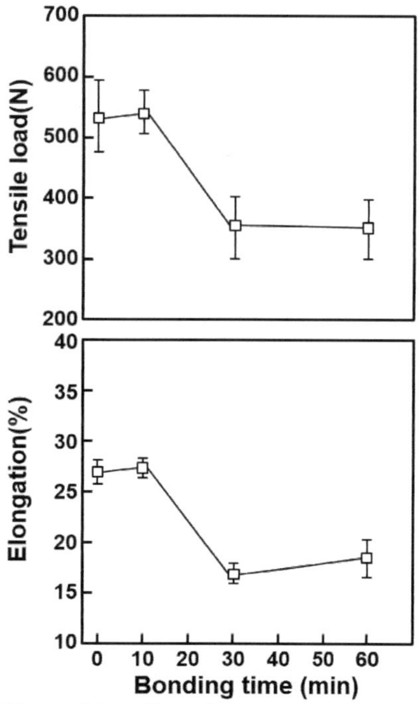

Figure 7 Effect of bonding time on the tensile load and elongation of Ni foam/SS 304 plate joint brazed at 1050°C for 0~60 min.

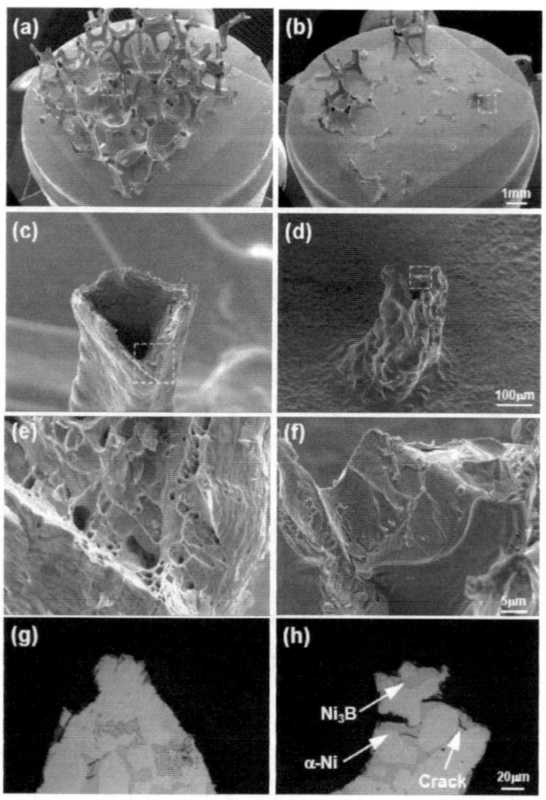

Figure 8 Fracture surface of the joints brazed at 1050°C for (a,c,e) 10 min. and (b,d,f) 60 min. and (g,h) optical micrographs of its cross-sectional fracture surfaces.

morphology and tearing regions were observed ,and necking phenomenon can be seen at cross section image inferring that plastic deformation of Ni foam took place during tensile loading. It is asuumed that since, at 0~10 minute, the bottom part of the Ni foam was not completely eroded by liquid filler metal and Ni_3B phases were not grown toward Ni foam, the Ni foam woud have a plastic deformation resulting higher tenislie load and elongation.

At a specimen of 60 minute condition, most of the fracture occurred in the Ni foam near SS 304 interface, and cleavage fracure can be seen on the fracture surface, and fracture path propagated mainly through the Ni_3B phases surrounding Ni solid solution. It is seemed that these results reflected low tensiled load and elongation.

It can be seen that when the heat input increases as the holding time increases, Ni solid solution and Ni3B phases tends to become united each other. It is judged that crack propagated thorugh intermetallic Ni_3B phases in the middle of Ni foam near SS 304 interface resulting in low mechanical properties.

Hardness of the phases

The micro vickers hardness of the phases were measured to characterize the fracture phenomenon. Figure 9 shows an optical images of hardness evaluation for the specimens brazed at 1050°C for 0 minute and its average results are displayed in Table 3. The microstructure has 100Hv for Ni foam as base material, 307Hv for Ni solid solution, and 1018Hv for Ni_3B intermetallic compound. This result inferred that brittle intermetallic compound Ni_3B gave rise to cracks under tensile loading and play a role to have a crack initiation causing low tensile load and elongation.

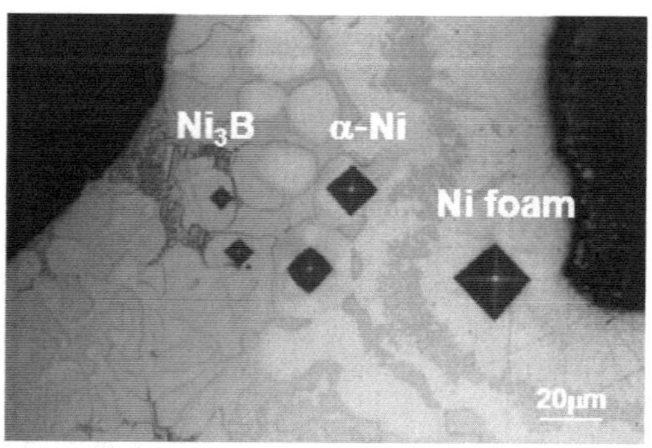

Figure 9 Optical images of Vickers hardness for the phases in the brazed joints

Table 3 Hardness evaluation of the phases in Figure 9.

Phase	Hardness (H_v)
Ni(foam)	**107**
α-Ni	**307**
Ni_3B	**1018**

Conclusions

A set of Ni foam for an advanced heat exchanger has been successfully brazed at 1050°C with SS 304 plate using MBF 20 filler metal in a high-vacuum furnace. The microstructural evolution and mechanical properties were investigated in this study and summarized as follows:

1. Microstructure and tensile properties of the brazed joint were dependent to bonding time. Bonding time of 0~10 minutes provided the best results in terms of the tensile load and elongation better than specimens conducted at 30~60 minutes.

2. In the case of 0~10 minutes, the liquid metal and the Ni foam were partially reacted, and it is observed that Ni foam as base material was composed of Ni solid solution and small morphology of intermetallic compound Ni_3B. However, at 30~60 minutes condition, the entire Ni foam was changed into Ni solid solution (Ni-Si-Cr) and Ni_3B phases, and the brittle intermetallic compound Ni_3B were connected in the horizontal direction of the Ni foam.

3. Tensile test results showed that the tensile load and elongation were high in 0 ~ 10 minutes condition and most of them were fractured in middle of the Ni foam base material by revealing plastic deformation. However, in 30~60 minutes, mechanical properties were lowered because fracture propagated through the brittle intermetallic compound Ni_3B phases existed along the grains of the Ni solid solution.

References

[1] J. Banhart, "Manufacture, characterisation and application of cellular metals and metal foams", Prog. Mater. Sci. 46 (2001), 559-632.

[2] J. M. Jang, W. S. Lee, S. H. Ko, "Energy Absorbing Efficiency of Various Aluminum Foam Filled Tube", J. Kor. Inst. Met. & Mater., 43 (2005), pp. 292-297.

[3] A. Bhattacharya, V. V. Calmidi, R. L. Mahajan, "Thermophysical properties of high porosity metal foams", Int. J. Heat Mass Transf. 45 (2002), pp. 1017-1031

[4] 7. W. Lu, C. Y. Zhao, S. A. Tassou, "Thermal analysis on metal-foam filled heat exchangers. Part I: Metal-foam filled pipes", Int. J. Heat Mass Transf. 49 (2006), pp. 2751-2761

[5] 9. C. Y. Zhao, "Review on thermal transport in high porosity cellular metal foams with open cells", Int. J. Heat Mass Transf. 55 (2012), pp. 3618-3632.

[6] 10. Z. Liu, Y. Yao, H. Wu, "Numerical modeling for solid–liquid phase change phenomena in porous media: Shell-and-tube type latent heat thermal energy storage", Appl. Energy. 112 (2013), pp. 1222-1232

[7] 13. P. De Jaeger, C. T'Joen, H. Huisseune, B. Ameel, S. De Schampheleire, M. De Paepe, "Assessing the influence of four bonding methods on the thermal contact resistance of open-cell aluminum foam", Int. J. Heat Mass Transf. 55 (2012), pp. 6200-6210.

[8] 15. Y. Huang, J. Gong, S. Lv, J. Leng, Y. Li, "Fluxless soldering with surface abrasion for joining metal foams", Mater. Sci. Eng. A-Struct. Mater. Prop. Microstruct. process. 552 (2012), pp. 283-287

[9] 16. M. Singh, R. Asthana, A. L. Gyekenyesi, C. E. Smith, "Bonding and integration of titanium to graphitic foams for thermal management applications", Int. J. Appl. Ceram. Technol. 9 (2012), pp. 657-665

[10] 18. A. A. Shirzadi, M. Kocak, E.R. Wallach, "Joining stainless steel metal foams", Sci. Technol. Weld. Join. 9 (2004), pp. 277

[11] 14. T. Jarvis, W. Voice, R. Goodall, "The bonding of nickel foam to Ti–6Al–4V using Ti–Cu–Ni braze alloy", Mater. Sci. Eng. A-Struct. 528 (2011) , pp. 2592-2601

Influence of Nitrogen in Process Atmospheres on the Fatigue Behavior of Brazed Stainless Steel Joints

V. Fedorov, T. Uhlig and G. Wagner
Chemnitz University of Technology, Institute of Materials Science and Engineering, Chemnitz, Germany
vasilii.fedorov@mb.tu-chemnitz.de

A. Langohr and U. Holländer
Leibniz University Hannover, Institute of Materials Science, Hannover, Germany
langohr@iw.uni-hannover.de

Abstract

Stainless steel components, such as heat exchangers for energy and air-conditioning technologies are brazed using nickel-based fillers, which provide the highest corrosion and oxidation resistance of the resulting joints. The brazing process was carried out for the temperatures 1000 °C, 1125 °C and 1150 °C in vacuum furnaces or continuous furnaces. In both cases, the parts interact with process gases like nitrogen within the brazing process, especially during cooling. The nitrogen enrichment within the braze metal and the base material influences the corrosion behavior of the brazed joints. Moreover, the mechanical properties of the joints are affected. Therefore, the monotonic tensile properties and the fatigue behavior of the joints were investigated with regard to the process conditions. The amount of nitrogen in the braze metal as well as in the base material was determined using a carrier gas hot extraction technique. Fatigue tests were carried out at ambient temperature on a RUMUL resonance pulsator under load controlled conditions with a load ratio of R = 0.1. The microstructure of the brazed joints and the fracture surfaces of the tested samples were investigated by SEM.

Introduction

Brazed stainless steel components such as heat exchangers for energy and air conditioning technology are manufactured millions of times in a wide range of sizes and types. A significant part of these components is brazed with nickel-based fillers, which offer the highest corrosion and oxidation resistance of the resulting joints [1-3]. By alloying with the elements silicon, boron and phosphorus, the melting temperature of the nickel or the nickel-chromium matrix is reduced and thus brazing temperatures between 1000 °C to 1200 °C are possible. Nickel-based fillers like (NiCoCr)-B-Si amorphous brazing filler metals, Ni-Cr-Si-B-P amorphous brazing foils and new nickel-based brazing alloy Ni-29Cr-8Mo-xFe-ySi offer a great potential for joining of stainless steels. The processing, structure and properties of the resulting brazed joints have been extensively reported [4-6]. The alloying elements provide excellent flowing and wetting properties, because they are able to eliminate the oxide film on the surface of the base materials. However, the alloying elements adversely affect the microstructure and the mechanical properties of the joints when the gap size exceeds 50 µm [7]. The joints are commonly brazed in vacuum furnaces or continuous furnaces. Nitrogen (N_2) or mixtures of nitrogen and hydrogen (N_2+H_2) are used as process gases in continuous furnaces. In vacuum furnaces, N_2 is often added during the cooling process. For this reason, a nitrogen enrichment takes place in the braze metal and in the near-surface area of the base material. However, the influence of the nitrogen enrichment on the properties of the brazed joints has not been investigated yet. In this context, it was supposed that nitrogen is able to influence the fatigue behavior of stainless steel joints, brazed with nickel-based fillers [8]. These joints often need to withstand a high number of high cyclic loads during application. Therefore, in addition to the monotonic mechanical properties, the fatigue behavior must be considered. In this work, the influence of nitrogen in process atmospheres on the fatigue behavior of brazed stainless steel joints was investigated. The nitrogen content in the braze metal as well as in the base material was determined using the carrier gas hot extraction technique (CGHE). Fatigue tests were conducted at ambient temperature on a RUMUL resonance pulsator under load controlled conditions with a load ratio of R = 0.1. The properties of samples, brazed with and without the influence of nitrogen, were compared. In order to determine the damage causes, the fracture surfaces of the fatigue tested samples were investigated by SEM.

Experimental Procedure

Austenitic stainless steel (AISI 304) was used as base material. The components to be brazed (circular and semi-circular blanks with a diameter of 30 mm) were made of finely rolled sheets by laser cutting, which leads to burr-free and distortion-free samples. The fillers Ni710, Ni60CrPSi and Ni650 were applied on the base material as powder. The thickness of the produced brazed joint was adjusted at 50 µm by means of inserted spacer foils. The quantity of applied filler was the same for all samples and was dimensioned, that a joint thickness of at least 50 µm was ensured on average. Binder for a brazing powder was avoided to eliminate carbon residues and their consequences, for example, the formation of chromium carbide in the braze metal.

The brazed single lap samples, used for the analysis of the nitrogen enrichment and the corrosion measurements, were produced at temperatures of 1000°C, 1125°C and 1150°C in a vacuum furnace and a continuous furnace in process gas, Fig. 1. The heating rate and cooling rate was adjusted at 50 K/min in both furnaces. The brazing process in the continuous furnace was carried out using different holding times of 5 min and 10 min and different process gases, 100% N_2 and 50% N_2/50% H_2. The brazing temperature was recorded as sample temperature by a trailing thermocouple attached just below the sample surface. In the vacuum furnace, a holding time of 10 min and a cooling process in an argon atmosphere were performed. The brazing temperature was measured by a thermocouple, attached in the base material below the brazed surface. For the detailed evaluation of the monotonic mechanical properties and the fatigue behavior, joints brazed with Ni60CrPSi in vacuum and process gas are chosen.

Brazing filler	**Continuous furnace**				**Vacuum furnace**
	Holding time				
	5 min		10 min		10 min
	Process gas				
	100% N_2	50% N_2/50% H_2	100% N_2	50% N_2/50% H_2	0% N_2
NiCrP (Ni710) $T_{Brazing}$ = 1000°C		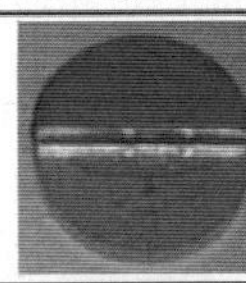		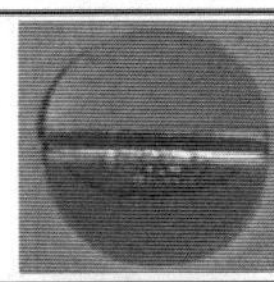	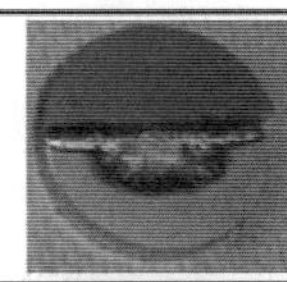
Ni60CrPSi $T_{Brazing}$ = 1125°C	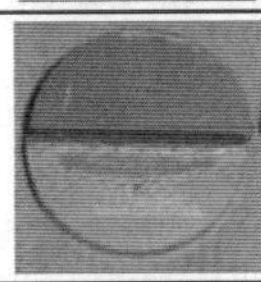		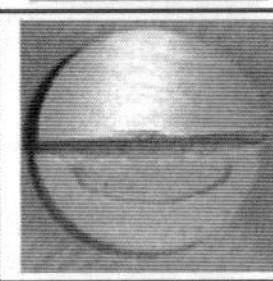		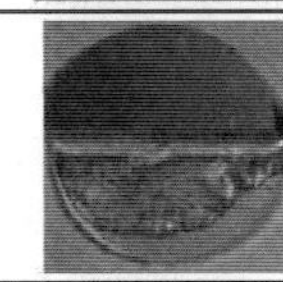
NiCrSi (Ni650) $T_{Brazing}$ = 1150°C	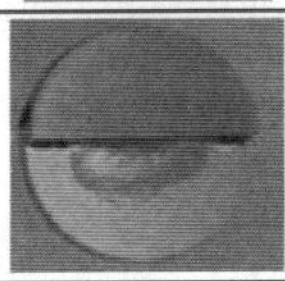	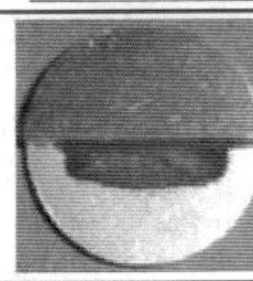	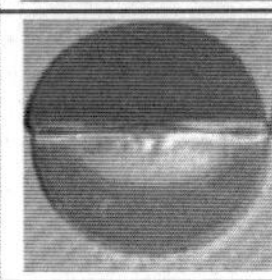	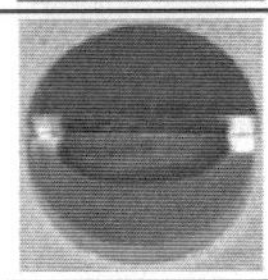	

Figure 1: Overview of the brazed samples

The quantitative determination of the nitrogen enrichment in the braze metal as well as in the base material are possible by means of CGHE. For this purpose, sample masses of about 100 mg are required. The process principle is described by Bagel [9]. Specific areas of the overlap samples are extracted by milling using a cutter with an associated cemented carbide indexable insert. The sample preparation for the analysis of the nitrogen enrichment is schematically presented in Fig. 2. The surface of the base material is cut off on the surface of the upper part of the sample in a depth of 100 µm. The braze metal is cut out of the braze fillet and the wetted surface without contact to the base material. The emerging material chips are used for CGHE.

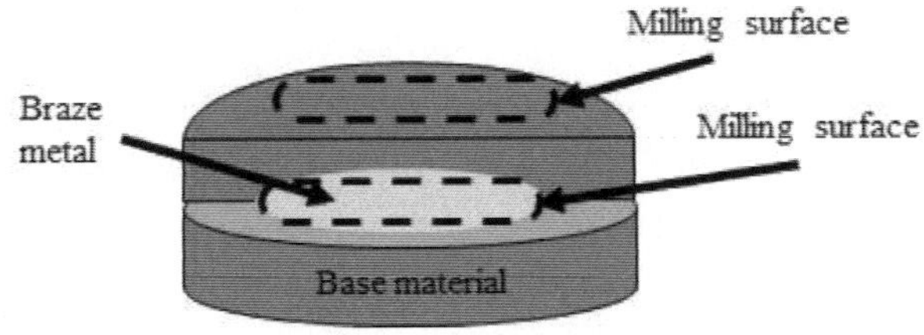

Figure 2: Scheme of sample preparation for the analysis of nitrogen enrichment

Results

In Fig. 3, the microstructure of the joint, brazed in vacuum using the Ni60CrPSi filler, is presented. Due to diffusion of Fe into the braze metal, Fe-enriched Ni-Cr-Si solid solutions (1) are formed at the interface to the stainless steel and inside the braze metal. Additionally, the resulting braze metal consists of Ni-Cr-Si solid solutions (2), and the Ni-Cr-P intermetallic (3).

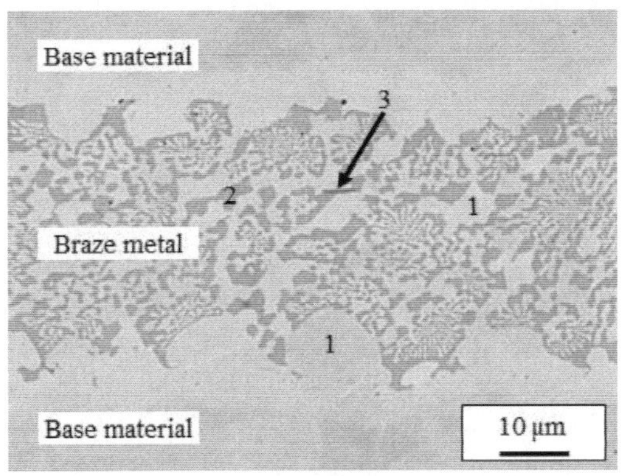

Figure 3: Microstructure of the vacuum brazed joint (Ni60CrPSi filler): 1 - Fe-enriched Ni-Cr-Si solid solutions, 2 - Ni-Cr-Si solid solutions, 3 - Ni-Cr-P intermetallic

At first, the contents of N_2 in the brazing fillers and the base material were measured in the as-delivered condition. The values are 0.02 wt.% for Ni710, 0.04 wt.% for Ni60CrPSi and 0.05 wt.% for Ni650. The difference in the contents can be explained by the different chemical compositions of the used fillers. The N_2 content of the base material in the as-delivered condition is 0.07 wt.%. After that, the N_2 contents of vacuum brazed samples were measured. The values are 0.06 to 0.07 wt.% in the base material and 0.03 to 0.06 wt.% in the braze metal. As expected, the vacuum brazing process does not influence the N_2 contents of base material and braze metal in comparison to the as-delivered condition. In case of samples that were brazed in the continuous furnace, the brazing process influences the N_2 contents of base material and braze metal significantly. In Fig. 4, the measured values of N_2 in the braze metal and in the base material are shown as a function of the brazing temperature resp. brazing filler.

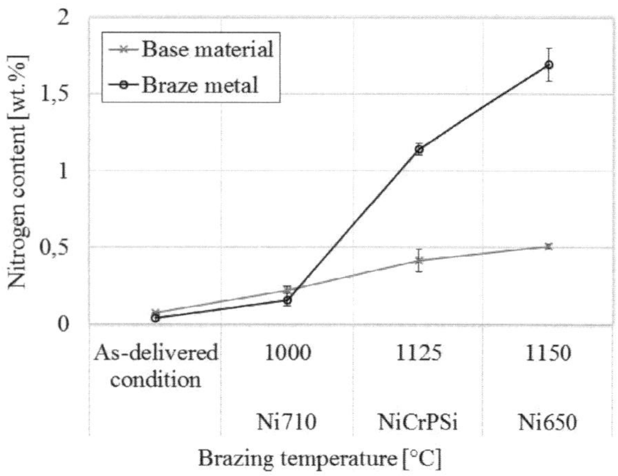

Figure 4: Nitrogen contents in the braze metal and in the base material as a function of the brazing temperature resp. brazing fillers, determined by CGHE

With an increase of the brazing temperature and the use of a higher melting brazing filler, the N_2 contents in the braze metal as well as in the base material increase. This can be explained by an enhanced diffusion of nitrogen atoms into the braze metal and the base material, as it is known from nitriding processes of steels. The higher the temperature, the faster the diffusion of N_2 atoms takes place [10]. Additionally, it was found out that the process parameters holding time and N_2 content in the process gas do not have a significant influence on the N_2 contents of base material and braze metal. Hence, the scatter of the values is shown as standard deviation in the diagram. The braze metal is more intensely enriched with N_2 in comparison to the base material. The difference can be explained by the fact that liquid metals can absorb a higher amount of N_2 in comparison to solid metals, which is dissolved in the braze metal after solidification. During the brazing process, the braze metal is liquid while the base material is in the solid state.

To evaluate the mechanical properties and fatigue behavior of the joints, brazed with Ni60CrPSi, monotonic tensile tests as well as fatigue tests at ambient temperature were carried out. Single lap samples with a joining area of 6 x 5 mm^2 were used for mechanical testing. The tensile tests were conducted in a material testing machine Zwick Allround-Line 20 kN. A tensile shear strength of 185 MPa was determined for all joints, brazed in vacuum and process gas. The fatigue tests were carried out on a RUMUL resonance pulsator under load controlled condition with a load ratio of $R = 0.1$. During the tests, a resonance frequency of 100 Hz arises. Three samples per stress level were tested for the joints, brazed in vacuum and process gas. The fatigue tests were carried out in the high cycle fatigue (HCF) range up to a limited number of cycles of $N = 2\times10^6$. Therefore, three stress levels (70, 50 and 30%) are set based on the measured tensile shear strength of the brazed joints. In Fig. 5, the results of the fatigue tests of the investigated brazed joints are presented. The comparison of the S-N data shows that similar numbers of cycles were reached at the highest as well as lowest stress levels for the joints, brazed in vacuum and process gas. At the highest stress level (58 MPa), which corresponds to 70% of the monotonic tensile shear strength, a number of 5×10^4 cycles was determined. At a lower stress amplitude of 37.5 MPa, which corresponds to 50% of the monotonic tensile shear strength, the number of cycles increases to 5×10^5. At the lowest stress level, the defined limited number of cycles was reached at a stress amplitude of 25 MPa for both cases. Consequently, it was found out that the investigated joints show the same fatigue behavior at all defined stress levels. In order to determine the damage causes of the joints after fatigue testing, the fracture surfaces were investigated by SEM.

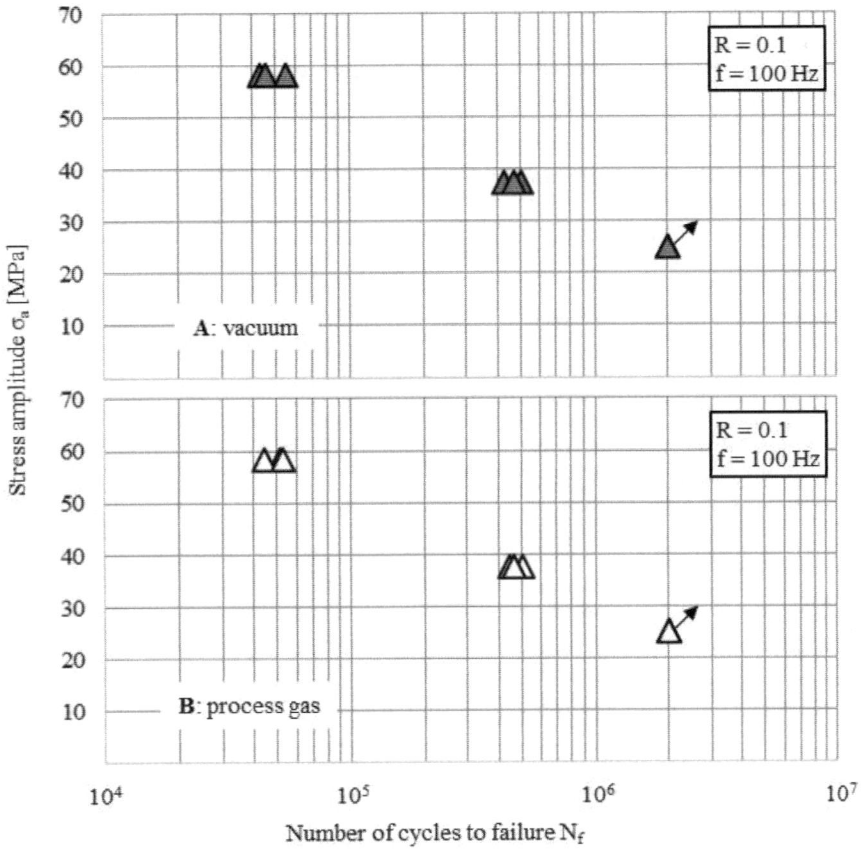

Figure 5: S-N data of the stainless steel joints, brazed: a) in vacuum, b) in process gas

In Fig. 6, a cross section of the fracture surfaces of the fatigue tested sample is presented. It was found out that the joints, tested at all defined stress levels, show the same fracture mechanism. It can be seen that the fracture takes place in the ductile braze metal near the interface to the stainless steel, where the thermal induced stresses are concentrated. The cracks occur in the phosphorus-containing intermetallics. This can be the cause of the failure of the brazed joints.

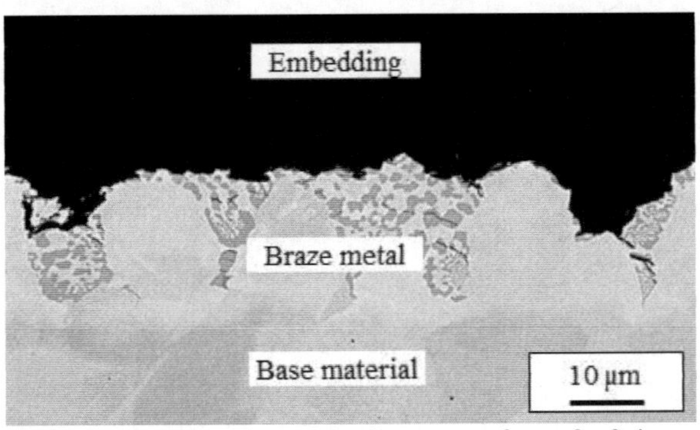

Figure 6: Cross section of the fracture surface of a fatigue tested sample, S°=°58°MPa, N =°5×10^4

The fracture surface shows a honeycomb structure, Fig. 7. Consequently, the investigation of the fracture surfaces confirms the fact that the joints have the same fatigue behavior at all defined stress levels. It can be summarized that the N_2 content in process gas does not influence the tensile shear strength and the fatigue behavior of the joints.

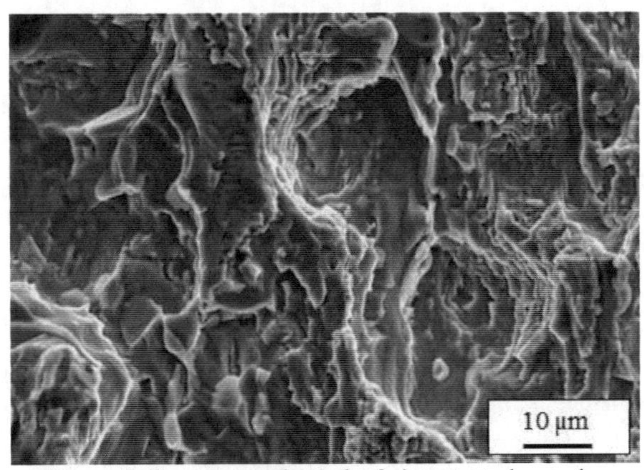

Figure 7: Fracture surface of a fatigue tested sample, S°=°58°MPa, N =°5×10^4

Conclusions

The joints were produced using nickel-based brazing fillers at temperatures of 1000 °C, 1125 °C and 1150 °C in a vacuum furnace and a continuous furnace. The N_2 content in the braze metal as well as in the base material was determined by CGHE. It was found out that the different process parameters (holding time and N_2 content in the process gas) during continuous furnace brazing do not have a significant influence on the N_2 contents of the base material and the braze metal. In this turn, the samples, brazed in a continuous furnace, are more intensely enriched with nitrogen in comparison to samples in the as-delivered condition and vacuum brazed samples. With an increase of the brazing temperature respectively higher melting brazing filler, the N_2 content in the braze metal as well as in the base material increases for the joints, brazed in a continuous furnace. Additionally, it was determined that the braze metal is more intensely enriched with nitrogen in comparison to the base material. For the joints, brazed in vacuum and process gas, a tensile shear strength of 185 MPa was determined. Additionally, it was found out that the investigated joints show the same fatigue behavior at all defined stress levels for both cases. A fatigue endurance limit of 2×10^6 cycles was reached at a stress amplitude of 25 MPa. Consequently, the N_2 content in process gas does not influence the tensile shear strength and the fatigue behavior of the joints. On the other hand, the heat exchangers often need to endure high pressures in a corrosive environment during operation. The investigation results from the work [12] show that the nitrogen enrichment has a significant influence on the corrosion behavior of the brazed joints. Therefore, the fatigue behavior of corrosively affected brazed joints will be observed in further investigations.

Acknowledgments

Part of the research was carried out within the scope of the project # 19056 BR funded by the Bundesministerium für Wirtschaft und Energie (BMWi) via the Arbeitsgemeinschaft industrieller Forschungsvereinigungen "Otto von Guericke" e.V. The authors thank the BMWi for financing this project.

References

[1] Hoffmann, E. E. *et al*, "An Evaluation of the Corrosion and Oxidation Resistance of High-Temperature Brazing Alloys," *Welding Journal* (1956)

[2] Achete, M. *et al*, "Korrosionsbeständigkeit von Hochtemperaturlötverbindungen in organischen und wässrigen Lösungen," *Schweißen & Schneiden*, Vol. 49 (1987), pp. 280-284

[3] Brandl, W. *et al*, "Korrosionsverhalten von gelöteten CrNi-Stahlverbindungen in Trinkwässern," *DVS-Berichte,* Vol. 132 (1990)

[4] Rabinkin, A., "Brazing with (NiCoCr)–B–Si amorphous brazing filler metals: alloys, processing, joint structure, properties, applications," *Science and Technology of Welding and Joining*, Vol. 9 (2004), pp. 181-199

[5] Hartmann, T. and Nuetzel, D., "New amorphous brazing foils for exhaust gas applications," *Proc 4th IBSC*, 2009, pp. 110-117

[6] Persson, U. and Bornegård, N., "New nickel based brazing alloy suitable for very corrosive environments," *DVS-Berichte,* Vol. 325 (2016), pp. 224-226

[7] Lugscheider, E. *et al*, "Metallurgical Aspects of Additive-Aided Wide-Clearance Brazing with Nickel-Based Filler Metals," *Welding Research Supplement* (1989), pp. 9-13

[8] Holländer, U. *et al*, "Ermittlung von Versagenskriterien mechanisch-korrosiv belasteter, hartgelöteter Edelstahlblechverbindungen unter Berücksichtigung der Nickellotmetallurgie und der Fertigungsbedingungen," *Schweißen & Schneiden*, Vol. 66 (2015), pp. 174-182

[9] Bagel, A., Die Ermittlung des Gesamtstickstoffanteils Handbuch für das Eisenhüttenlaboratorium, Verlag Stahleisen, (Düsseldorf, 1985)

[10] Liedtke, D., Wärmebehandlung von Eisenwerkstoffen, II: Nitrieren und Nitrocarburieren, Expert Verlag, (Düsseldorf 2006)

[11] VDA 230-214 2010, "Beständigkeit metallischer Werkstoffe gegen Kondensat-Korrosion in abgasführenden Bauteilen," *Verband der Automobilindustrie*

[12] Fedorov, V. *et al*, "Investigation on the influence of nitrogen in process atmospheres on the corrosion behavior of brazed stainless steel joints," *IOP Conference Series: Materials Science and Engineering* (2018)

Specialized C/C Fixtures for Today's Brazing Needs

Lloyd Nagamine
Across USA, Inc., 1480 Beachey Place, Carson, CA 90746
E-mail: nagamine@acrosscc.com Phone: 310-635-3555

Dan Kay
Kay & Associates, 4 Lawton Drive, Simsbury, CT 06070
E-mail: dan.kay@kaybrazing.com Phone: 860-651-5595

Abstract

Brazing furnaces today often contain a lot of very heavy materials in their brazing chambers (such as grates, fixtures, baskets, etc.), all of which must be heated up to brazing temperature and then brought safely back down to room temperature, hopefully distortion-free. All this extra weight in the chamber results in a significant percentage of the brazing "load" actually being "non-productive", meaning that it costs a lot of money to heat all that heavy weight up to brazing temp, the cost of which does not help the brazing shop's bottom-line profitability. This has resulted in some brazing shops suggesting that "we're in the business of heat-treating fixtures" rather than their being in the brazing-business. Carbon-fiber reinforced carbon (C/C) is a material being used more and more for making brazing fixtures today, knowing that such C/C fixtures not only greatly reduce the weight of fixturing going into the furnace, but have the added benefit of excellent thermal stability over the long term. However, there are some strong precautions that should be known about the use of C/C fixtures in brazing, and these concerns and precautions will also be reviewed and discussed during this presentation.

INTRODUCTION

When we hear the word "carbon" used in today's news media, much of what we hear is about "carbon emissions" into the atmosphere, and the negative impact on global warming through the greenhouse-gas effect. Although this is important, it is a result of carbon in the atmosphere in the form of carbon-dioxide gas. But carbon also takes on other forms, and this paper will discuss a very interesting solid form of carbon known as "carbon-fiber reinforced carbon" or "C/C" for short, which is quietly gaining acceptance in the brazing world as a fixturing material that is able to significantly reduce energy consumption in today's furnace brazing industry.

BASICS

The inner structure of a "C/C" fixture consists of carbon-fibers embedded in a matrix of carbon to form a composite structure that is very strong. "Carbon-fiber reinforced carbon" material (or C/C-material) can be tailor-made into a wide variety of products by controlling the length of carbon-fibers used, and the orientation of those fibers in the carbon matrix. C/C composites are among the strongest and lightest high-temperature engineered material in the world. Compared to other fixture materials such as compressed pure graphite, or ceramics, or metal, C/C material is lightweight, very strong, can withstand brazing temperatures over 2000°C without any loss in performance, and is more dimensionally stable, over a longer time, than metal fixtures.

C/C fixtures combine the desirable properties of both the carbon-fibers (high-strength, high elastic modulus) and the carbon matrix material (heat resistance, chemical resistance, low thermal expansion coefficient, high-thermal conductivity, low electric resistance, low specific gravity). When the carbon fibers are blended together in the solid carbon matrix material, a much better performance material is created. The reinforcing fibers themselves can be either long or short, depending on the size needed for a given C/C application. Although C/C materials will be considered in this paper primarily for its extreme high temperature use, we are very familiar with its use in everyday applications such as in sporting equipment (golf-clubs, and tennis rackets, as two examples) because of its light weight, high strength, and flexibility.

As one of the most advanced and promising engineering materials, the development of carbon-carbon composites started during the 1950's with US Air Force sponsored work, and which later received a massive boost with the onset of the Space Shuttle program. Consequently, the predominant uses of C/C composite material have been mainly in aerospace applications such as aircraft brakes, re-entry vehicles and rocket nozzles.

In the late 20th century and now in the 21st century, manufacturing technology and industrial research-and-development (R&D) has accelerated the penetration of C/C materials into many other industries and end-use applications. The heat-treating and furnace-brazing industry have especially benefited from increased use of C/C material in furnace heating elements, trays and supports, fixtures, insulating sheets, convection fans, mechanical fasteners, etc. All of these components used to be made from metals, but now, with the advancement of C/C composite materials, these components can now have many of the advantages of graphite (such as chemical resistance, etc.) combined with the mechanical strengths of metal. Indeed, C/C parts are slowly seeing more and more use in wider markets.

C/C manufacturing and the method of processing the Carbon-Carbon matrix can be quite complex, and we will not cover that in detail during this presentation. Instead, we will just give you a quick overview here instead. C/C manufacturing process starts from the selection and procurement of raw materials, chiefly the type of carbon-fiber and the type of matrix. The types of carbon fiber are usually PAN (polyacrylonitrile) based, pitch based, or cellulose based for special applications. These raw-materials are typically processed through the following stages: blending of carbon powder / carbon binder / resin / carbon-fiber sheet and lamination / hot press molding / carbonization / graphitization / primary machining / surfacing / thickness adjustment / sizing / final cleaning. Most commercial applications for C/C material are 2-dimensional woven structures, but there can be 3-dimensional and multi-dimensional structures for more advanced C/C applications. The majority of the C/C shapes and forms are available as plates, bars, or sheets, which are later machined and fabricated into individual components. Other unique shapes that can be made include coil-springs, channels, angles and even tubes/pipes can be made via filament winding.

PROPERTIES OF C/C

In order to understand the role C/C fixtures play, the attributes of C/C composite material must first be considered.

Advantages are:

(1) Available in many grades and forms with wide-ranging properties
(2) High modulus, good strength, low density
(3) Good thermal stability (in the absence of O2)
(4) High thermal conductivity
(5) Low thermal expansion coefficient
(6) Excellent creep resistance
(7) Good chemical resistance
(8) Low electrical resistivity
(9) Biocompatibility

Disadvantages are:

(1) Relatively high cost
(2) Low/poor impact strength
(3) Care required during handling
(4) Oxidizes in air at temperatures above 450degC
(5) Exhibits anisotropy in the axial and transverse directions
(6) Cannot weld together or fabricate curved/thin walls
(7) Only gaseous processes are recommended (not oil quench)
(8) Carbon dust/residues
(9) Eutectic melting with Iron at temps above 2100F

The end-use conditions in which C/C material will be used must be carefully analyzed. Because this material is carbon, it can readily oxidize in air at temperatures above 450 C (840F). Therefore, they should only be used in either a vacuum furnace or in a furnace utilizing a non-oxidizing atmosphere.

Furnace braze applications. Proper handling of C/C fixtures by shop personnel is very important, and requires instruction and training. Great care is needed when using C/C fixturing material during loading, or when transporting parts through the shop, etc. Carelessly throwing C/C fixtures across the ground, or dropping them on the floor, or loading them too aggressively, can damage the C/C material, and greatly shorten the life of such fixtures. They MUST be treated carefully, since they do not absorb shock and mishandling as current metal fixturing might do.

As with all materials, C/C fixtures will expand and contract during heating, but should be able, over time, due to its ability to resist distortion better than metals can, be able to maintain its dimensional flatness better than metal fixtures can. Please bear in mind, too, that metal components that are being heat-treated or brazed while sitting on top of those C/C fixtures, those metals will expand and contract at a different rate than the C/C fixture, and must be allowed to "move" on the C/C surface without undue restriction.

This brings up another VERY important factor when using C/C material in your fixtures: will the metals you are brazing react with the carbon-based C/C fixtures on which they are sitting? Therefore, will you need to provide a "separating layer" between the metal parts and the C/C fixture? We strongly advise you to always use some form of separating barrier between the metal and the C/C, by using such materials as a layer of brazing stop-off on the surface of the C/C fixture, or a thin sheet of ceramic, or a ceramic-fiber cloth, etc. Once again, because C/C is more than 99% pure carbon, and pure carbon likes to react both chemically and metallurgically with many metals --- particularly those that contain iron (such as stainless steel) --- that reaction between carbon and metal may form low-melting eutectic compositions at temperatures under 2100°F/1150°C. You need to know this, and take precautions to insure that metal does NOT come into direct contact with the C/C fixtures during heating in the furnace.

REMINDER: It is very important that stainless components are kept from directly contacting C/C fixtures during brazing cycles. And once again, this can be easily accomplished by a layer of solid ceramic, or a ceramic-fiber paper/cloth, or by an adequate layer of an appropriate brazing stop-off painted or coated onto the surface of the graphite.

C/C BRAZE FIXTURE DESIGN

Carbon-carbon (C/C) material brazing fixtures and Carbon springs are used for processing of aluminum and copper brazing of automotive heat-exchanger components such as for car radiators and oil coolers. The material properties of C/C when compared to metal include a high heat resistance over 2000°C (3500°F); low thermal expansion; light weight (20% the weight of metal); and excellent resistance to corrosion and radiation. Compared to ceramics, C/C has a high resistance against fracture, it does not readily bond to itself and it can be made into complex shapes. When compared to pure-graphite, C/C structures have high strength and rigidity and a high resistance to fracture.

Structural Design of Fixtures

Let's now look at how C/C material can be used in constructing braze-fixtures, and what their advantages might be over similar types of metal fixtures.

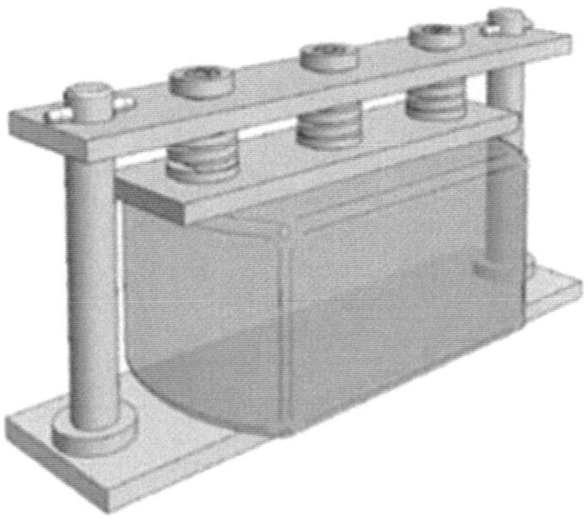

Fig. 1 Fixture using C/C plates, rods, and springs

Figures 1, 2, and 3 show examples of C/C brazing fixtures. The fixtures are designed to hold the braze product between C/C top and bottom contact plates. C/C coil springs are used to apply equal pressure during the brazing process. C/C springs are springs made of unidirectional carbon fiber composites. C/C coil springs are very light in weight, and have high strength and high modulus. Excellent spring characteristics are maintained even at temperatures higher than 1000°C (1832°F) with high toughness, excellent wear resistance and corrosion resistance.

A usage example would be brazing heat exchangers during the production process. After the parts are set up into the C/C jig (as shown), a constant force will be applied by the spring. The C/C spring maintains the same force as temperature increases, and brazing will start when the temperature reaches the melting point of the filler metal.

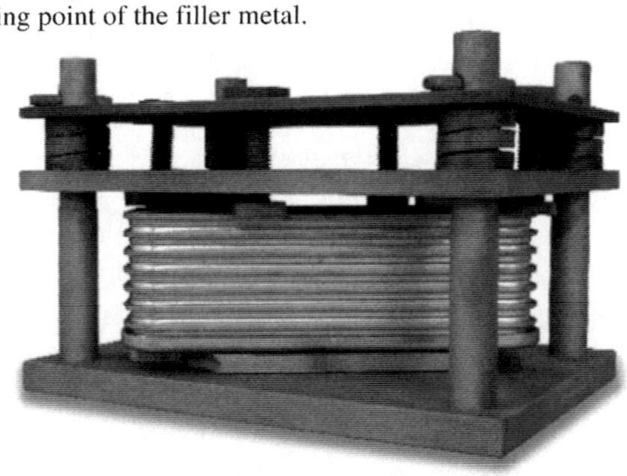

Fig. 2 Careful design of the fixturing is required so that very even force will be applied to the heat-exchanger as the temperature rises, the metal heat-exchanger expands, and the C/C springs compress.

In both heat-treating and brazing, the furnace's heating capability and maximum load capacity will help you determine the total amount of weight (fixture weight + parts weight) you can process in any given furnace cycle. Figure 3 illustrates one particular case in which C/C composite brazing

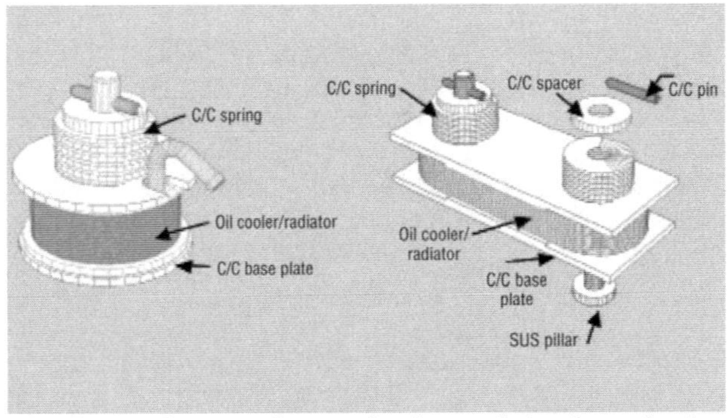

Fig. 3 Another actual example of parts being brazed using lightweight C/C fixturing with C/C springs.

fixtures were used to increase production volume. C/C braze fixture jigs of 60 kg (132 pounds) replaced stainless jigs weighing 200 kg (440 pounds) on a heat-treat furnace with total load capacity of 350 kg (772 pounds). The ratio of total jig weight to total load capacity was drastically reduced by 70%, which made it possible to carry more parts. Using C/C jigs, a significant increase is achieved in the quantity of parts that could be processed per cycle.

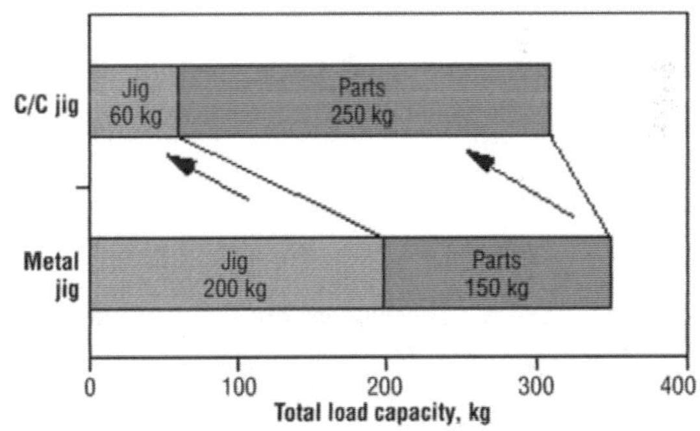

Fig. 4 This chart clearly shows that many more parts can be processed in each furnace load due to the significant reduction in fixturing weight when using C/C fixtures

C/C Springs vs. Dead weights for compression.

As many of you know, it is fairly common for braze shops to place a number of metal blocks (dead-weights) on top of parts so that the weight of the metal blocks can help to compress the load beneath them, thus keeping those metal parts in close contact with each other throughout the brazing cycle. This can often add hundreds of pounds/kilograms to the furnace load being processed.

VERY IMPORTANT: There is NO free ride, so to speak! Every pound/kilogram put into the furnace absorbs heat, and must continue to do so more and more as the furnace comes up to brazing temp. Every BTU/calorie absorbed by those dead-weights costs you money! Thus, a significant percentage of the cost of each furnace run can be used-up just in heating the fixturing! I have personally seen brazing-furnace loads in which more than 90% of the energy for that entire braze-run was used just for heating and cooling the fixtures. This is very NON-productive!

With C/C braze fixtures, the dead-weights could be replaced with C/C coil springs to apply equivalent constant load, as shown in Fig.5. Here you see an example of a furnace load in which a 20-kg (44-pounds) load is needed to compress a part. But with a C/C brazing fixture setup, only two C/C coil springs of 70 grams each are needed for an equivalent amount of pressure on that load. An option might be to use a C/C leaf-spring setup instead, if the conditions allow.

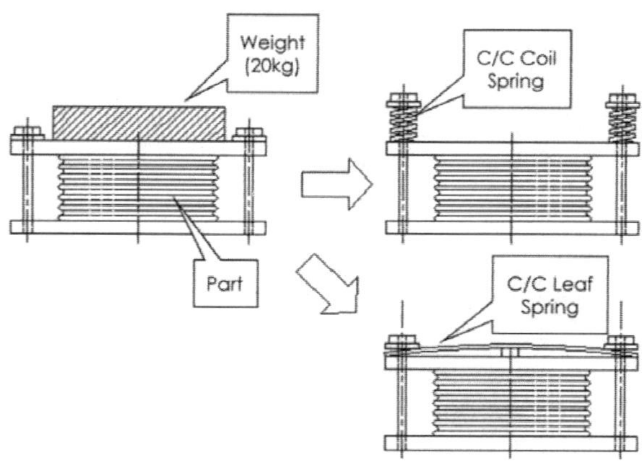

Fig. 5 C/C coil-springs or leaf-springs can replace a lot of dead-weight fixturing used previously, to achieve the same amount of pressure on the load.

As can be seen, there is huge difference between using 20-kg dead-weights or in using 70-gram (2.5-ounce) C/C springs as far as load-efficiency is concerned. Which method do you think will yield greater productivity in that braze shop?

Shadowing. Another consideration in furnace brazing is the "shadowing" effect that massive fixturing can have on loads containing many parts that are to be brazed. With large metal-fixtures, perhaps loaded with many pounds/kilograms of dead weight, spacing between neighboring stacks of parts becomes critical because of the heat-absorption by the heavy metal mass of the furnace's radiant heat, perhaps causing neighboring parts to become a bit "starved" of heat if they are too close to their neighbor, thus causing them to take much longer to come up to brazing temperature. Because of this neighboring stacks might have to be placed much farther apart, to avoid this "shadowing" effect. This in turn then results in less parts being able to be brazed in each furnace load, and this can negatively affect productivity. By contrast, when C/C brazing-fixtures are used, as illustrated in Fig. 5, then the fixture itself can be very compact and does not encounter these stack-positioning problems nearly as often.

Deformation Resistance. C/C fixtures have a very high deformation resistance, which leads to longer service life. After long-term service, metal fixtures tend to deform due to high-temperature cycling, which can also then have a negative effect on the distortion of the parts being brazed. On the other hand, C/C fixtures maintain high strength and deformation-resistance (resistance to distortion) from room temperature up to 2000°C (3500°F), thereby assuring accurate brazing with little, if any, distortion concerns.
In addition to C/C braze fixtures, the base-metal trays and furnace-grids used as flat surfaces on which to set the brazing-fixtures could also be replaced with C/C material for further improvements in production efficiency. There have been reports that production volume has doubled merely from the optimized use of C/C jigs and fixtures, grates, etc., replacing many of the metal fixtures, grates, etc., thus allowing for much greater load efficiency in a furnace, without having to modify anything else in the furnace.

SUMMARY

In summary, there are many advantages to C/C fixturing when compared to metal fixturing being used for the same purpose. Among these are:

- Less gross weight of fixturing means added productivity and drastic reduction in cost of energy per part brazed. Each carbon spring can generate force up to 24.5 kg (53.9 pounds), but weighs only 26-84 grams. When this is compared to the inefficiency of trying to use heavy dead-weights, the improvement in productivity potential becomes quite clear.
- Speed up in processing time. Small heat capacity will allow faster heating time.
- Less need for maintenance/replacement due to fatigue. A C/C fixture does not wear-out or deform (distort) when used in atmospheres without oxygen. Its spring constant is maintained under repeated heat cycles.

As it was initially mentioned, carbon-fiber reinforced carbon (C/C) composite fixtures were originally introduced as an advanced composite-materials for the aerospace market. But as C/C production technology has evolved and innovations continue to develop, it is expected that more and more people will see C/C material as a fairly simple way to increase their productivity and profitability in the brazing world. Carbon composite technology will indeed continue to play an important and even wider role not only in the brazing world, but in many other industrial fields too.

Development of thin film Active-Brazing Alloy

Takaomi Kishimoto
Tanaka Precious Metals, Tomioka, Gunma, Japan
kishimoto-t@ml.tanaka.co.jp, +81-274-62-5670

Abstract

Active-brazing does not require a metalized layer because it is a method that enables direct joining to ceramics by improving the wettability to ceramics by an adding active metal such as Ti, etc. Ag-Cu-Ti alloy is known as an active-brazing alloy. But Conventional Ag-Cu-Ti active-brazing alloy has a problem. The Cu-Ti intermetallic compounds with a diameter of 500 μm are deposited in the material. So, the deposited Cu-Ti intermetallic compound is extremely hard and cannot be molded in plastic forming, and only Ag-Cu undergoes plastic forming. The Cu-Ti intermetallic compound falls out when finely processed, and this can cause brakes if processed in thin films. In order to solve this problem, we investigated to the influence of various additives.
And, we found that this problem is solved by adding a Tin (Sn). In the newly developed Ag-Cu-Sn-Ti alloy, there is not Cu-Ti intermetallic compound but Sn-Ti intermetallic compound. And, Sn-Ti intermetallic compound is finely dispersed in the material. So, we are able to supply thin film Active-Brazing Alloy.

Introduction

Ceramic is material having various excellent properties such as heat resistance, corrosion resistant, electrical insulation, etc. [1]. Further, by joining ceramic-to-ceramic or, ceramic-to-metal, it is possible to produce higher-performance materials combining the respective material properties. Such materials are expected to be applied to electronic parts, cutting tools [2], and jewelry articles.

In order to answer those needs, the technique of joining ceramic is an important factor. As a method of joining ceramic, a Mo-Mn process is common and has been adopted industrially for a long time [3]. Among them, the Mo-Mn process is widely used for joining alumina. However, there is a problem that the Mo-Mn process cannot be applied to ceramics other than alumina. So, there is a demand for development of bonding technology applicable to various ceramic other than alumina. One method of solving this problem is an active-brazing. The active-brazing is a method in which ceramics is directly brazed without metallization by using an active-brazing alloy. An active-brazing metal is containing an active metal such as titanium, zirconium, and hafnium in a brazing material. As an active-brazing, Ag-Cu-Ti alloy in which Ti, which is an active metal, is added to an Ag-Cu alloy is widely known [4]. But, Ag-Cu-Ti alloy has large problems. Generally, Ag-Cu-Ti alloy is produced by melt casting method. The cast ingot is plastically worked to a predetermined plate thickness by rolling processing. However, in the Ag-Cu-Ti alloy, a coarse Cu-Ti intermetallic compound of 500 μm or more is likely to precipitate during casting. The intermetallic compound remarkably deteriorates workability and it is difficult to process into a minute shape corresponding to brazing of electronic parts or the like which has been made lighter, thinner and smaller. (Fig. 1) In addition, it was also a problem that coarse Cu-Ti intermetallic compound falls off during processing, thereby locally decreasing the amount of Ti and lowering the reliability of brazing.

Figure 1. Cross section view of 71Ag-27.5Cu-1.5Ti alloy

Experimental

In this study, we attempted to refine the Cu-Ti intermetallic compound by adding the fourth metal element. Several additional elements were investigated. As a result, improvement effect was observed by adding Sn, so detailed investigation was carried out. Ag having a purity of 4 N, Cu, Ti and Sn having a purity of 3 N was weighed so as to have the composition shown in Table 1. Each sample was melted and cast to prepare an ingot having a width of 100 mm, a length of 150 mm, and a thickness of 20 mm. A sample was cut out from the manufactured ingot and the metallographic structure was observed with a metallurgical microscope. And, the detail of the intermetallic compound was inspected by scanning electron microscopy (SEM) and energy dispersive x-ray spectroscopy (EDS).

Table 1. Compositions of Ag-Cu-Sn-Ti alloy

Sample No.	Alloy (weight %)			
	Ag	Cu	Ti	Sn
a	69.0	28.0	2.0	1.0
b	68.0	28.0	2.0	2.0
c	68.4	26.6	2.0	3.0
d	67.67	26.33	2.0	4.0
e	65.0	28.0	2.0	5.0
f	65.53	25.47	2.0	7.0

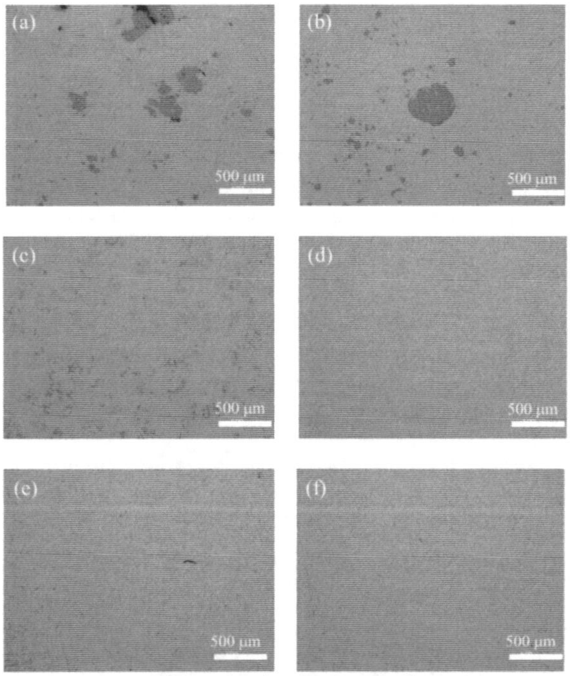

Figure 2. Ag-Cu-Sn-Ti alloy
(a)Sn: 1wt%. (b) Sn: 2wt%. (c) Sn: 3wt%.
(d) Sn: 4wt%. (e) Sn: 5wt%. (f) Sn: 7wt%.

Figure 2 shows the metallographic structure of Ag-Cu-Sn-Ti alloy. At first, in the case of adding 1 or 2 wt% Sn, as in Ag-Cu-Ti alloy, an intermetallic compound exceeding 500μm was observed. At 1 wt% of Sn added, the intermetallic compound was composed of Cu and Ti. On the other hand, 2 wt% of Sn added, the intermetallic compound was Cu, Sn and Ti. Next, when the amount of Sn added reached 3wt%, the size of the intermetallic compound began to decrease. At this time, the constituent element of the compound is Cu, Sn and Ti. Furthermore the addition amount of Sn was 4, 5 or 7wt%, no coarse intermetallic compound was observed, and the intermetallic compound was finely dispersed. Fig. 3 shows SEM-EDS images of sample (e) (65Ag-28Cu-5Sn-2Ti).

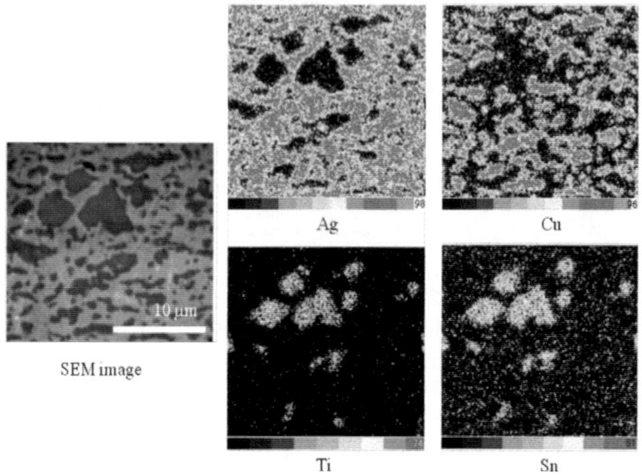

Figure 3. SEM-EDS images of 65Ag-28Cu-5Sn-2Ti

The intermetallic compound is not Cu-Ti but Sn-Ti. It was confirmed that the addition of Sn changed the component of the intermetallic compound from Cu-Ti to Sn-Ti and the size of the intermetallic compound became fine.

To research this reason, we check the detail of the intermetallic compound by XRD. In the case of not adding Sn, the compound was Cu_3Ti. On the other hand, when Sn was added, the compound was Sn_3Ti_5. According to the state diagram, the melting point of each compound was about 880 ° C. for Cu_3Ti and 1500 ° C. for Sn_3Ti_5. From this result, it is presumed that the Sn_3Ti_5 was produced because the melting point of the Sn_3Ti_5 is higher than that of the Cu_3Ti, so that the Sn_3Ti_5 preferentially precipitated during melting and casting. Also, in the case of precipitating with Sn_3Ti_5 and the case of precipitating with Cu_3Ti, the amount of compound generated for the same amount of Ti is different. Since the amount of refining of the Sn_3Ti_5 is smaller than that of the Cu_3Ti, it is presumed that a metallic structure in which the Sn_3Ti_5 compound is finely dispersed was obtained. Although it is the optimum amount of Sn addition amount, when the amount of Sn added is 7%, a fine structure can be obtained, but since the amount of Sn added is large, the material becomes brittle and the workability of the material cracks during rolling process. Therefore, excessive addition of Sn is not preferable, and it is considered that adding Sn which is about twice weight % of Ti to be added is a preferable addition amount compatible with fine structure and processability. Figure 4 shows the cross section image of 65Ag-28Cu-5Sn-2Ti alloy at thickness of 50 μm. There was no coarse intermetallic compound, even when rolling worked to 50 μm, the structure was fine and could be processed stably. This made it possible to stably produce thin film active-brazing alloy, which was difficult to produce so far.

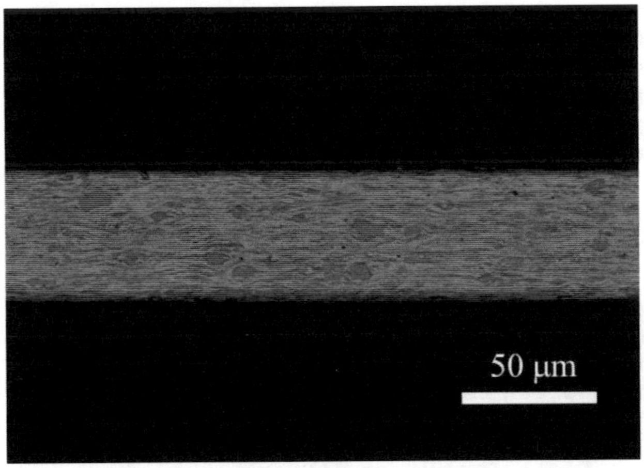

Figure 4. 65Ag-28Cu-5Sn-2Ti alloy at thickness of 50 µm

Brazability of Ag-Cu-Sn-Ti active-brazing alloy

In order to evaluate the brazability of the Ag-Cu-Sn-Ti alloy, Al_2O_3 (alumina) was brazed to each other and the strength was measured by a 4-point bending test. Brazing was carried out under a condition of 830 ° C. in a vacuum atmosphere. For comparison, 66 Ag - 29.5 Cu - 3 Sn - 1.5 Ti and 72 Ag - 28 Cu (Al_2O_3 waste metalized by Mo - Mn process) were carried out. A plate material of 0.1 mm × 10 mm × 20 mm was used as a brazing material, and alumina blocks of 10 mm × 20 mm × 20 mm were brazed to each other. From the brazed sample, a test piece of 3 mm × 4 mm × 40 mm was cut out, and the breaking strength was measured by 4 point bending test for each sample, and the average strength of 5 times was calculated.

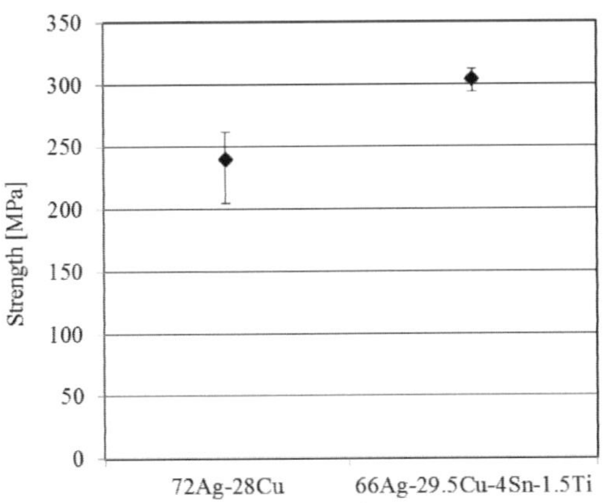

Figure 5. 4-point bending strength

In the case of brazing with 72Ag-28Cu, fracture at the interface between the metallized layer and the ceramic was confirmed, and the strength was 240 MPa on average. On the other hand, when an active-brazing alloy was used, breakage at the Al_2O_3 was confirmed, and strength close to the Al_2O_3 strength of 300 MPa or more was confirmed. Therefore, when Al_2O_3 is joined using an active-brazing alloy, higher strength than the conventional metallizing method is obtained and there is no influence on strength due to the addition of Sn, and by using Ag-Cu-Sn-Ti alloy. Also, the results of observing the bonded cross section is shown Fig. 6. Ti forms a layer between the ceramics and the brazing filler metal and shows the bonding behavior of the active-brazing alloy, and it is considered that the behavior of Ti is not inhibited by Sn.

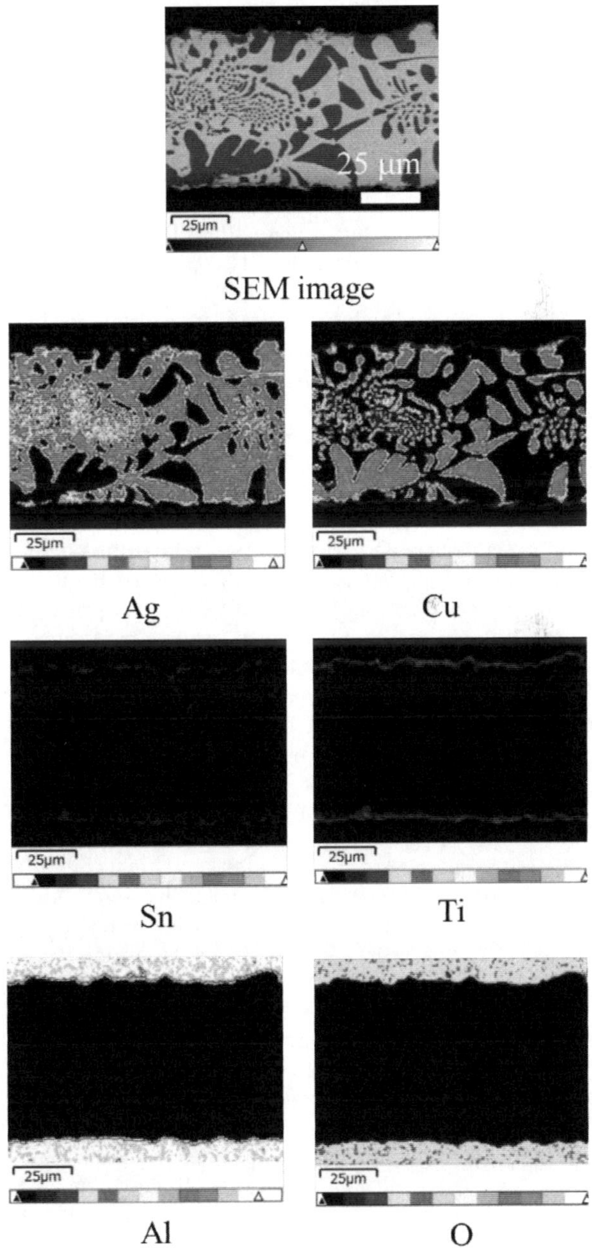

Figure. 6 SEM-EDX images of bonded cross section

Application

The active-brazing filler metal of Ag-Cu-Sn-Ti alloy developed this time has the feature that thin film can be made by rolling. We applied this feature and developed active-brazing alloy and Cu cladded material. The cross sectional view of the Ag-Cu-Sn-Ti active-brazing alloy and Cu (OFC) clad material is shown in the Figure 7. Copper thickness is 0.8 mm. The thickness of the brazing filler metal is 0.02 mm, and the component is 66Ag-29.5Cu-Sn3-Ti1.5. It is also possible to perform rolling after composing the brazing filler metal into Cu. Further, the thickness of the brazing filler metal can be processed to 20 μm, which is thinner than 50 μm of the brazing filler metal itself.

Application of this clad material is supposed to be a ceramic substrate for power modules. Generally, a ceramic substrate is made using paste type active-brazing alloy [5]. Paste brazing filler metal is coated on ceramics, and copper plate is placed on it and brazed. After brazing, masking is done on the surface of the copper, and the copper plate and the brazing material layer are etched to make a circuit pattern. In recent years, the thickness of copper has become thick in order to improve the heat radiation performance of the ceramic substrate. Therefore, it is necessary to carry out etching of thick copper, which is a technical problem, and a simple manufacturing method is desired.

By using active-brazing alloy with Cu clad material, the following method is proposed. As a usage method, a circuit pattern is punched out and brazed onto a ceramic plate. At this time, it is contemplated to directly form a circuit pattern without etching process. FIG. 8 shows an example of a pressed piece of a clad material and an appearance of a sample obtained by brazing it. The press shape is 0.8 mm × 30 mm × 30 mm. One piece of this pressed piece was placed on each side of SiN ceramics (0.32 mm × 31 mm × 31 mm) and heated at a temperature of 830 °C. in a vacuum and brazed (FIG.9).

The results of Scanning Acoustic Tomography (SAT) after brazing and after heat cycle test are shown in the figure 10. Regarding the state after brazing, it was confirmed that brazing could be performed without voids. This is considered to be the effect that the brazing filler metal is clad with copper and the structure in which the brazing filler metal is uniformly spread over the brazed surface. Also, a heat cycle test was carried out under the condition of holding at -50 to 175 °C. for 30 minutes each. Even when 1,500 cycles or more was performed, peeling of material, cracking of ceramics and the like were not observed. From this, we think that the ceramic substrate fabricated using the Ag-Cu-Sn-Ti active-brazing alloy with Cu clad material is a new method that can easily produce a highly reliable ceramic substrate.

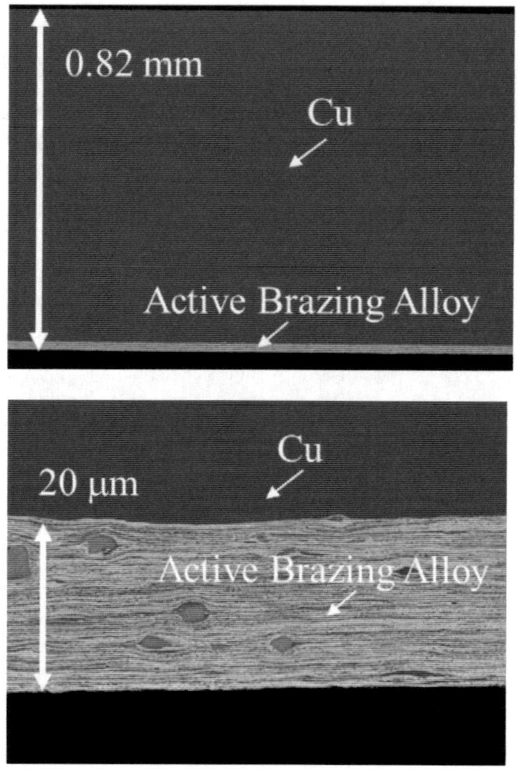

Figure 7. Cross Section View (Active Brazing Alloy with Cu)

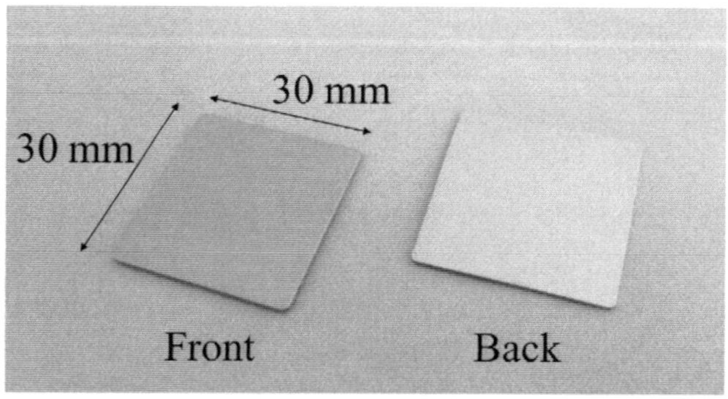

Figure 8. Active Brazing Alloy with Cu Clad material

Figure 9. Brazing sample of a ceramic substrate

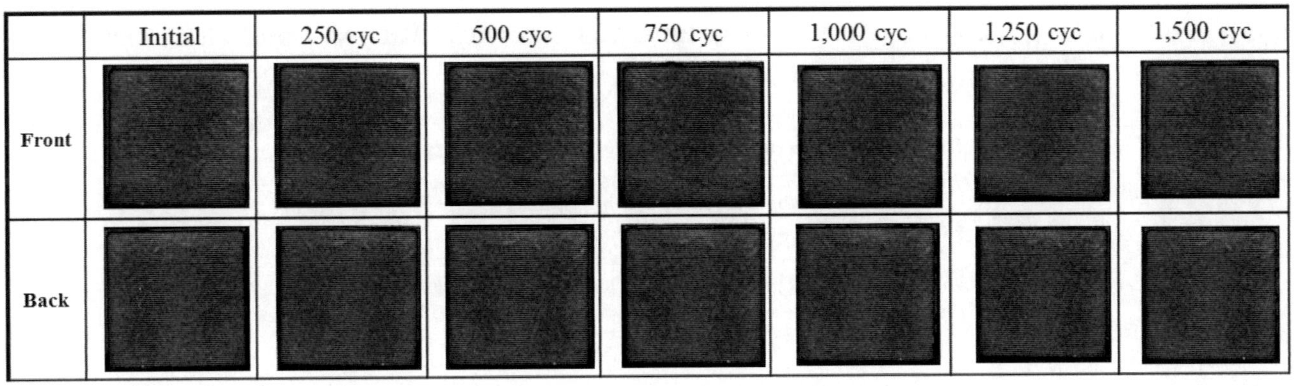

Figure 10. SAT images of heat cycle test

Conclusions

The active-brazing using active-brazing alloy is expected as a joining technology of ceramics instead of metallizing method. In the active metal brazing material of Ag-Cu-Ti alloy known so far, coarse Cu-Ti intermetallic compound precipitates, so there was concern about the reliability of brazing. In this time, the newly developed active metal brazing material of Ag-Cu-Sn-Ti alloy suppresses coarse precipitation of Cu-Ti intermetallic compound by adding Sn as an additional element, and finely disperses Sn-Ti intermetallic compound. As a result, workability was improved and it was possible to provide thin film materials that could not be dealt with so far, or composite materials clad with other metals.

References

[1] O. M. Akselsen., "Review Advances in brazing of ceramics." *Journal of Materials Science*, April 1992, Vol. 27, Issue 8, pp. 1989–2000.

[2] Yoshihisa S. et al, "Effect of composition of titanium in silver–copper–titanium braze alloy on dissimilar laser brazing of binder-less cubic boron nitride and tungsten carbide." *2014 IOP Conf.* Ser.: Mater. Sci. Eng. **61** 012019

[3] Haruo T., "Alumina Ceramic-to-Metal Seals by the Mo Process," *Journal of Ceramic Association Japan*, Vol. 81(1973), Issue 931, pp. 108-114.

[4] Chen B. et al, "Joining of SiO_{2f}/SiO_2 Composite to Metals Using AgCuTi Brazing Fille, the Joint Strengths and Microstructures." *IBSC 2012*, April, pp. 109-113.

[5] William W. Sheng、Ronald P. Colino, Power Electronic Modules: Design and Manufacture.

A novel design of a multiple transition metals interlayer to diffusion bond ZrC_x ceramics

Rui Pan
State Key Laboratory of Advanced Welding and Joining, Harbin Institute of Technology, Harbin, 150001, China
College of Engineering, University of Kentucky, Lexington, KY 40506, United Stated
Tiesong Lin, Zhao Huang, Peng He
State Key Laboratory of Advanced Welding and Joining, Harbin Institute of Technology, Harbin, 150001, China
Dusan Sekulic
State Key Laboratory of Advanced Welding and Joining, Harbin Institute of Technology, Harbin, 150001, China
College of Engineering, University of Kentucky, Lexington, KY 40506, United Stated

Abstract

To meet stringent high temperature application's requirements of ZrC_x ceramics, a novel joint design approach has been proposed. It involves a use of multiple transition metals, e.g., Ti/M/Ti (M=Zr, Ta, Hf and Nb, respectively) as the interlayer to form a homogenous joint by diffusion bonding. We have verified that when using a single Ti foil as an interlayer, ZrC (stoichiometric, nominally) ceramics would not form homogenous joints. TiC formed at the interface is difficult to dissolve into the base ceramics under the considered bonding conditions. Hence, a new design of the interlayer has been proposed. It is verified that a seamless and homogenous joint can be obtained by using a Ti/Zr/Ti interlayer at 1400°C/1h/20MPa. This design also applies to ZrC_x with a larger population of carbon vacancies, like $ZrC_{0.85}$. It is established that both the room and the high temperatures bending strengths of the joints have been dramatically improved by the induced homogenization of the joint. The strength is comparable to that of the base ceramics. The homogeneous microstructures and the outstanding mechanical properties demonstrate the significant application potential of this multiple interlayer design.

Introduction

Transition Metal Carbides (TMCs) (such as ZrC_x, TiC_x, TaC_x VC_x, etc.) display a wide range of nonstoichiometric characteristics. This is due to the presence of vacancies in the carbon sub-lattices. The carbon-to-metal ratio x ranges from 0.47 to 1.0 [1, 2]. These carbides have a combination of remarkable properties, such as a high melting point (>3000°C), high hardness, excellent wear and corrosion resistances and good electrical conductivities [1-3]. Such superior materials' features make these ceramics to be the prime candidates for various ultra-high-temperature applications. These applications include the next-generation rocket engines, sharp leading edges of future-generations of reentry vehicles, gas-cooled fast nuclear reactors, to mention just a few [1-5]. Stringent high temperature working environments of TMCs require the material's joints to have as high quality as possible. The requirement is not only to obtain a joint with as good mechanical properties as possible, but also an ability to sustain a highly-enhanced temperature resistance. To meet the above requirement, a novel joining method of an active diffusion bonding was suggested, [6, 7]. The objective was to bond ZrC_x ceramics by forming homogenous joints using single transition metal, Ti [6] and Zr [7], as the interlayer. These studies promoted the method based on the presence of carbon vacancies and the mutual solubility between TMCs. But for the ZrC_x ceramics with a small population of carbon vacancies, like ZrC (stoichiometric, nominally), it is more difficult to form homogenous joints due to a weak promoting effect of carbon vacancies. To form homogenous joints for ZrC ceramics, in the present study, a novel approach is used. Multiple transition metals, e.g., Ti/M/Ti (M=Zr, Ta, Hf and Nb, respectively), are proposed as the interlayer. This paper offers the first results of the study of the microstructures and mechanical properties of different ZrC_x joints.

Experimental procedure

The ZrC_x ceramics employed in this study are ZrC and $ZrC_{0.85}$, prepared by the reaction hot-pressed sintering using graphite and ZrH_2 powders as the raw materials. The value of x indicates the concentration of carbon vacancies (e.g. $1\text{-}x$), which was controlled when weighting the two powders proportionally. Each of the resultant ZrC_x ceramics were cut into (i) 5 mm × 4 mm × 3 mm pieces for a microstructure analysis, and (ii) 18 mm × 4mm × 3 mm specimens for mechanical three-point bending testing. Prior to bonding, the mating specimen surfaces were polished using diamond millstones, from 500-grit to 2000-grit, successively. The transition metal's foils, e.g. Ti (10 μm) and Zr (30 μm), used in this study, are commercially available and have a high purity (more than 99.7%). The foils were cut into 6 mm × 6 mm squares. Prior to the assembly, all pieces to be joined were ultrasonically cleaned for 20 minutes in acetone. The transition metal interlayer (single Ti or multiple Ti/Zr/Ti) was sandwiched between two ZrC_x pieces, followed by vacuum diffusion-bonding. The schematic diagram of the assemblies' design is shown in Fig. 1. The assemblies of ZrC_x ceramics were subsequently heated to 700 °C at 20 °C/min and held at that temperature for 10 min. Then, the assemblies were heated to the peak bonding temperatures (ranging from 900 °C to 1500 °C) at a rate of 10 °C/min and were held at the peak temperature between 10 min and 5 h. The specimens were slowly cooled to 400 °C at a rate of 10° C/min, and ultimately to the room temperature. The setup temperature history of the hot zone is presented in Fig. 2. During the process, 20 MPa pressure was applied to the assemblies. After bonding, the three-point bending strength of the joints was measured by Instron-1186 universal mechanical testing machine, at both room temperature and 1000 °C high temperature. The displacement rate of 0.5 mm/min was imposed. The configuration of the bending testing of the samples is shown in Fig. 3. The average measured value for each experimental condition was calculated from at least 3 measurements. The margins of error are presented in Fig. 10.

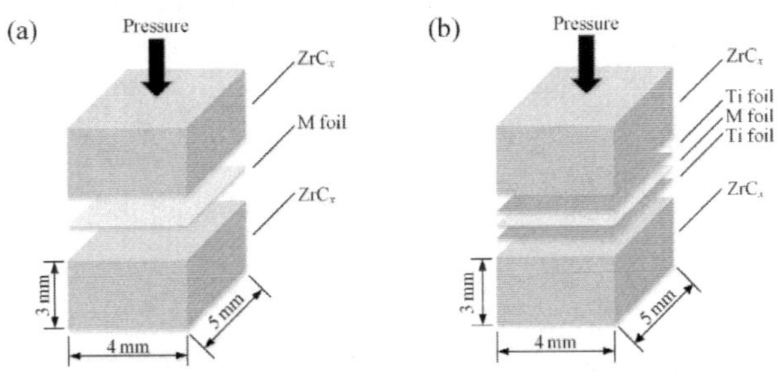

Fig. 1 The schematic diagram of the assemblies of ZrC ceramics using (a) a single Ti as the interlayer, and (b) the multiple Ti/Zr/Ti as the interlayer.

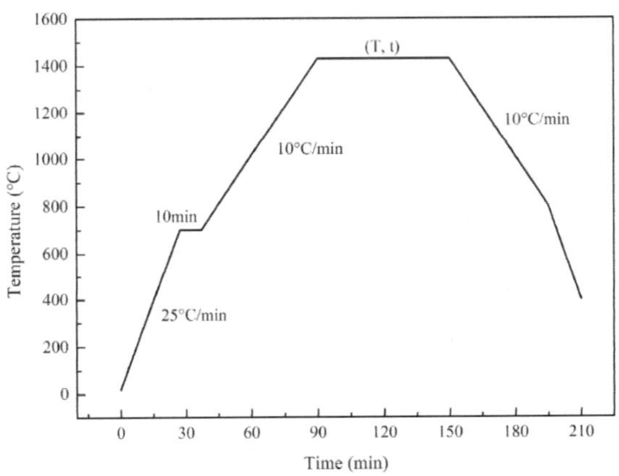

Fig. 2 The setup temperature history of the hot zone

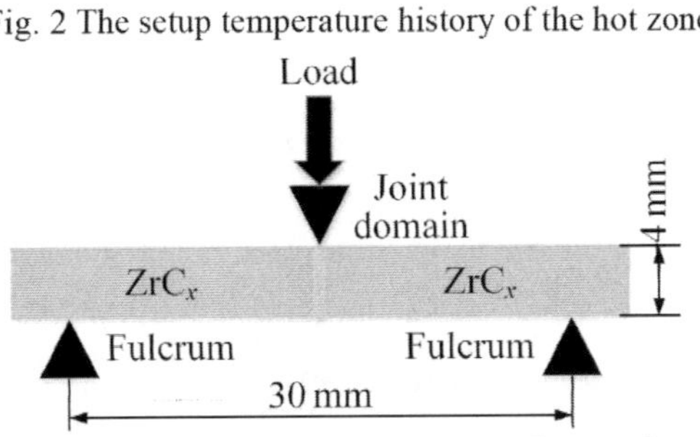

Fig. 3 Configuration of the three-point bending testing; not to scale

Results and Discussion

Microstructures of ZrC joints formed by using single Ti as the interlayer

As a benchmark case, the microstructures of ZrC ceramic joints using 10 μm Ti foil as the interlayer are investigated first. The typical microstructure of ZrC joints is shown in Fig. 4(a). The composition analysis results are shown in Table 1. The phases of each zone are identified as TiC layer and a residual (α+β)-Ti layer. Figures 4 and 5 show the microstructures of ZrC joints diffusion bonded at different temperatures and for different times, respectively. It can be seen from Fig. 4 that, when diffusion bonded at 1100 °C, TiC is formed at the interface and part of β-Ti remains in the α-Ti matrix due to the inter-diffusion of C, Ti and Zr. When bonding temperature increases to 1200 °C, almost all the Ti interlayer is transformed into TiC (with a thickness of 7 ± 2 μm) and a diffusion layer of ZrC(Ti) with a thickness of 15 ± 2 μm are formed at the interface. However, when the temperature is further increased to 1400 °C, and even to 1500 °C, the microstructures of the joint do not change significantly compared with that obtained at 1200 °C. The final structure of a joint is always composed of TiC and a ZrC(Ti) diffusion layer, as shown in Fig. 4(c) and (d). This indicates that TiC and ZrC do not dissolve with each other when the bonding temperature is increased. Thus, a homogenous joint cannot be formed at the interface.

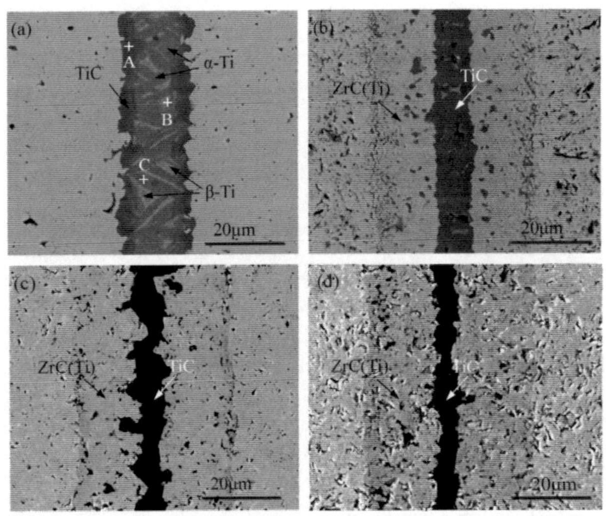

Fig. 4 Microstructures of ZrC joints diffusion bonded for 1h under different temperatures using 10 μm thick Ti as the interlayer:
(a) 1100 °C; (b) 1200 °C; (c) 1400 °C; (d) 1500 °C

Table 1 EDS analysis of different locations marked in Fig. 4(a)

Point	Elements (at. %)			Possible phase
	Zr	C	Ti	
A	10.16	29.13	60.71	TiC(Zr)
B	9.65	14.21	76.14	α-(Ti, Zr)
C	24.66	30.11	45.23	β-(Ti, Zr)

The effect of the holding time on the microstructure of ZrC joints (Fig. 5) follows the same trend. After all the Ti interlayer was converted into TiC, there is no obvious further change of the joint microstructure. TiC and ZrC do not dissolve with each other when extending the holding time at a bonding temperature.

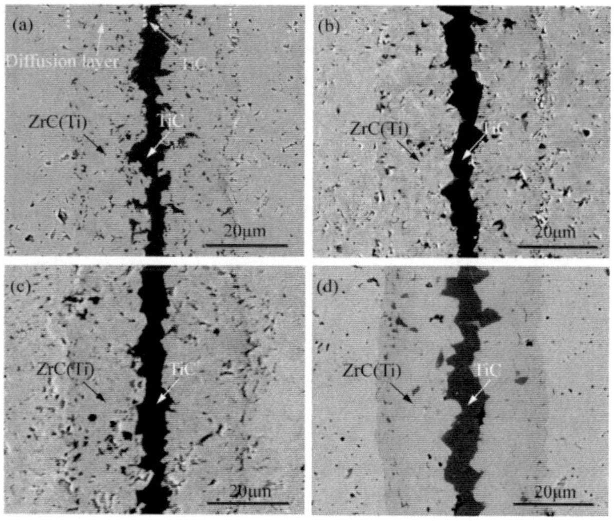

Fig. 5 Microstructures of the ZrC joints diffusion bonded at 1400°C for different times using 10μm thick Ti as the interlayer
(a) 10min; (b) 0.5h; (c) 2h; (d) 5h

Microstructures of ZrC joints formed by using Ti/Zr/Ti as the interlayer

To obtain a homogenous joint domain in a ZrC joint, a new approach involving multiple transition metals, e.g., Ti/M/Ti (M=Zr, Ta, Hf and Nb, respectively), as a "compounded" interlayer is proposed. In present study, we take Ti/Zr/Ti interlayer as an example to verify our main research hypothesis.

The typical microstructure of the ZrC joint, facilitated by using Ti/Zr/Ti as the interlayer, is shown in Fig. 6 (diffusion bonded at 1300 °C for 1 h under the 20 MPa pressure). It can be seen from Fig. 6 that the joint consists of three layers, marked as the layers Ⅰ, Ⅱ and Ⅲ, respectively. According to the EDS analysis summarized in Table 2, the possible phases of the corresponding layers Ⅰ, Ⅱ and Ⅲ, are identified as (i) ZrC_x diffusion layer, (ii) pure α-(Ti, Zr) layer, and (iii) α-(Ti, Zr) with the newly formed ZrC_x(Ti) particles layer, respectively.

Fig. 7 shows the microstructures of ZrC joints diffusion bonded for 1 h at different temperatures using Ti/Zr/Ti as the interlayer. As can be seen in Fig. 7(a), when diffusion bonded at 900 °C, the original Ti/Zr/Ti interlayer has become (Ti, Zr) solid solution due to the inter-diffusion between Ti and Zr. With the increase of bonding temperature, due to the diffusion of Ti and Zr in (Ti, Zr) solid solution into the base ceramics, the thickness of (Ti, Zr) solid solution decreases and a ZrC_x(Ti) diffusion layer gradually forms at the interface. At the same time, the newly formed ZrC_x particles also grow bigger. When bonding temperature increases up to 1350 °C, the newly formed ZrC_x particles get connected and merged into ZrC_x(Ti) diffusion layer. Only a small amount of (Ti, Zr) solid solution is left in the joint. When bonding temperature further increases up to 1400 °C, all of the (Ti, Zr) solid solution is consumed up and a homogenous joint composed only of a ZrC_x(Ti) diffusion layer, is formed at the interface. This verifies feasibility of our hypothesis involving multiple Ti/Zr/Ti as the interlayer to obtain homogenous joints of ZrC_x ceramics. In addition, the uniform distribution of Ti shown by the line scan results (along the black line in Fig. 7 (e)) in Fig. 8 indicates an obvious homogenization of the joint composition.

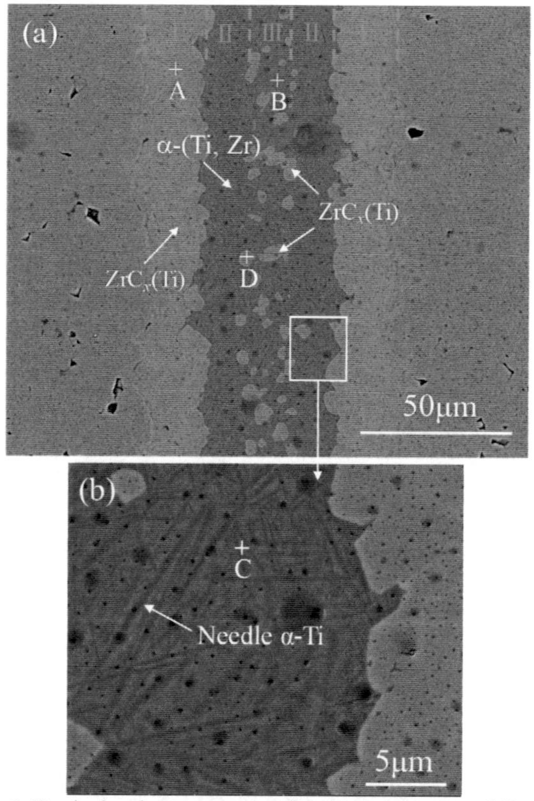

Fig. 6 Typical microstructure of a ZrC joint (a) diffusion bonded at 1300 °C for 1 h under 20 MPa pressure using 50μm thick Ti/Zr/Ti as the interlayer; (b) the enlarged image of the zone marked in (a)

Table 2 EDS analysis of different locations marked in Figs. 6

Point	Elements (at. %)			Possible phase
	Zr	C	Ti	
A	39.99	57.29	2.71	ZrC_x(Ti)
B	30.82	40.03	29.16	Ti(Zr,C)
C	32.20	34.93	32.87	Ti(Zr,C)
D	41.52	53.59	4.89	ZrC_x(Ti)

Compared with the case of a single Ti as the interlayer, the main reason for the formation of a homogenous joint by using Ti/Zr/Ti as the interlayer is hypothesized as follows. When Ti/Zr/Ti is used as the interlayer, Ti forms (Ti, Zr) solid solution with Zr interlayer, instead of an intense reaction with the base ceramics, observed when using Ti as the interlayer. This prevents a heavy formation of TiC_x which is in turn difficult to get dissolved into the base ceramics at the interface.

The proposed method is applicable to other ZrC_x ceramics as well. Figure 9 shows the microstructures of $ZrC_{0.85}$ joints diffusion bonded for 1h at different temperatures using Ti/Zr/Ti as the interlayer. As can be seen from Fig. 9, a homogenous joint is formed at the interface when the bonding temperature is increased to 1400 °C. Comparison of ZrC and $ZrC_{0.85}$ ceramics' homogeneous joints indicates that the thickness of $ZrC_{0.85}$ joint becomes 13 μm thicker than that of ZrC joint. This is because of the promoting effect of carbon vacancies on the diffusion of interfacial elements.

In addition, it is also verified in this study that using other Ti/M/Ti interlayers leads to homogenous joints with ZrC_x ceramics. These results will be discussed elsewhere.

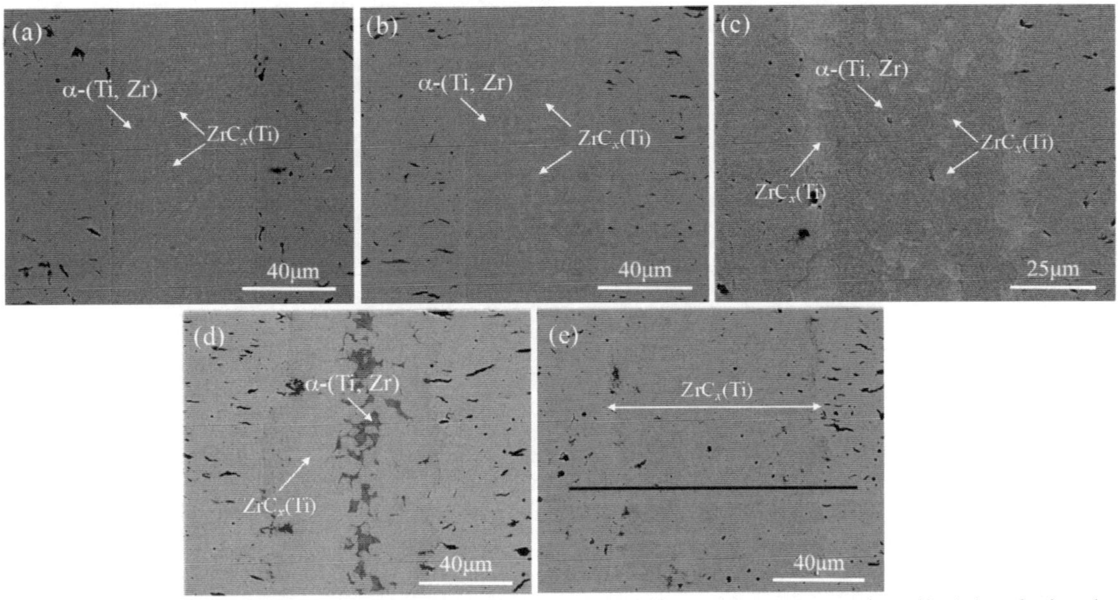

Fig. 7 Microstructures of the ZrC joints diffusion bonded for 1 h at different temperatures using Ti/Zr/Ti as the interlayer (a) 900 °C; (b) 1150 °C; (c) 1250 °C; (d) 1350 °C, (e) 1400 °C

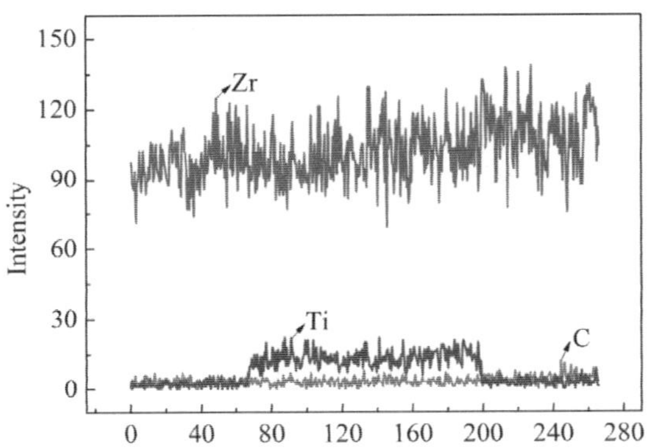

Fig. 8 Line scan along the black line marked in Fig. 7(f)

Mechanical properties of ZrC_x joints

The three-point bending testing at both room temperature and at an elevated temperature of 1000°C were performed to establish the mechanical properties of ZrC_x joints. In addition, as a reference, the same testing was carried out for the base ceramics. Figure 10 offers data for (i) the room temperature and (ii) a high temperature, three-point bending strengths of different $ZrC_{0.85}$ joints. Ti/Zr/Ti is used as the interlayer (diffusion bonding for 1h with a load of 20MPa). It can be seen from Fig. 10 that with an increase of the bonding temperature, both the room and high temperature bending strength increase. When the bonding temperature increases up to 1400 °C, both the room and high temperature bending strength reach to the maxium, which are 192.6 MPa and 132.5 MPa, respectively, comparable to these of the base ceramics. This is clearly due to the formation of homogenous joints. These data verify that the homogenization of the joint domain improves both the strength and the high temperature resistance of ZrC_x joints.

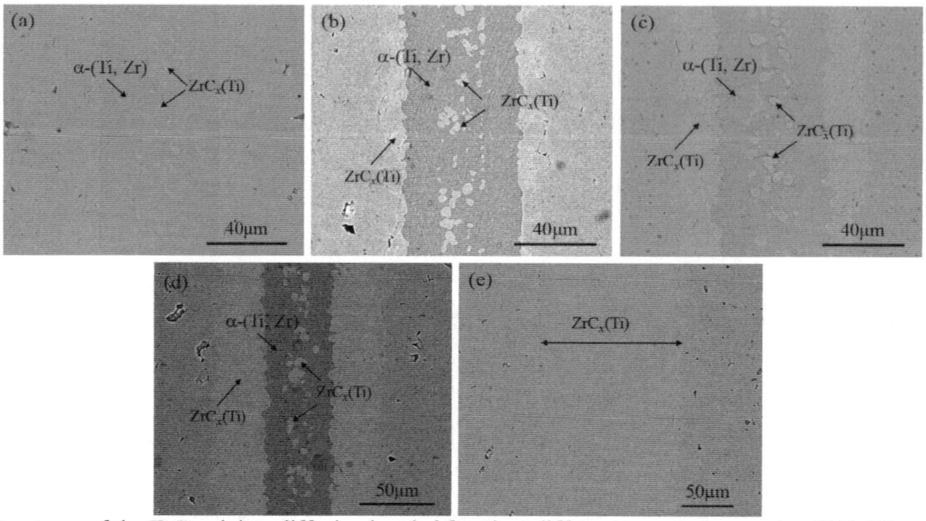

Fig. 9 Microstructures of the $ZrC_{0.85}$ joints diffusion bonded for 1h at different temperatures using Ti/Zr/Ti as the interlayer (a) 1000 °C; (b) 1100 °C; (c) 1200 °C; (d) 1250 °C; (e) 1300 °C, (f) 1400 °C

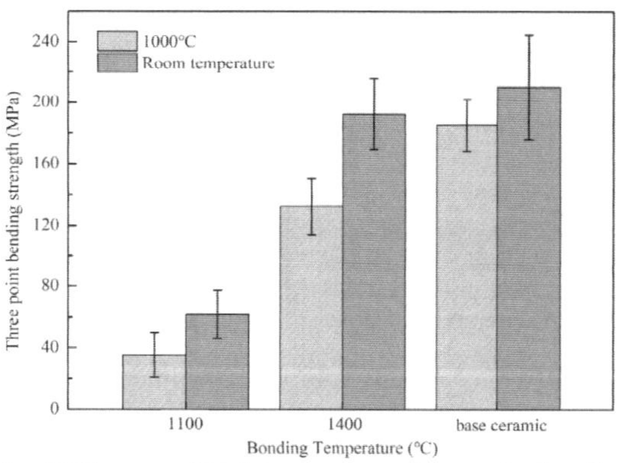

Fig. 10 Room and high temperature three-point bending strengths of different $ZrC_{0.85}$ joints using Ti/Zr/Ti as the interlayer, diffusion bonded at different temperatures for 1h with a load of 20MPa

Conclusions

To meet a high temperature working requirement of ZrC_x ceramics bonds, both single Ti and multiple transition metals, e.g., Ti/Zr/Ti, were used as the interlayer to obtain homogenous joint by diffusion bonding of ZrC ceramics. The primary findings are as follows.

(1) It is established that when using single Ti foil as an interlayer, ZrC (stoichiometric) ceramics would not form a homogenous joint. This is because TiC formed at the interface is difficult to dissolve into the base ceramics under the considered bonding conditions.

(2) A new design of the interlayer consisting of transition metals Ti/M/Ti is proposed. It is verified that a seamless and homogenous joint would be obtained for ZrC by using a Ti/Zr/Ti interlayer at 1400°C/1h/20MPa. This design also applies to ZrC_x with a larger population of carbon vacancies, like $ZrC_{0.85}$.

(3) It is established that both the room and high temperature bending strengths of a joint may be dramatically improved by the homogenization of the joint domain.

Acknowledgments

This work is supported by the National Natural Science Foundation of China (NSFC, grant number 51305102 and 51475103) and the Funds for Distinguished Young Scientists of Heilongjiang Province (grant number JC2015011). One of the authors (DPS) acknowledges a support through the Distinguished 1000 Foreign Expert Professor Plan. The first author acknowledges a support through CSC for research done as a visiting PhD student at the University of Kentucky.

References

[1] Nachiappan, C. *et al*, "Synthesis and densification of monolithic zirconium carbide by reactive hot pressing," *J. Am. Ceram. Soc.*, Vol. 93, No. 5 (2010), pp. 1341-1346.

[2] Wang, X. G. *et al*, "Effect of solid solution formation on densification of hot-pressed ZrC ceramics with MC (M= V, Nb, and Ta) additions," *J. Eur. Ceram. Soc.*, Vol. 32, No. 8 (2012), pp. 1795-1802.

[3] Vasudevamurthy, G. *et al*, "Laboratory production of zirconium carbide compacts for use in inert matrix fuels," *J. Nucl. Mater.*, Vol. 374, No. 1 (2008), pp. 241-247.

[4] Dridi, Z. *et al*, "First-principles calculations of vacancy effects on structural and electronic properties of TiC_x and TiN_x," *J. Phys.: Condens. Matter.*, Vol. 14, No. 43 (2002), pp. 10237.

[5] Li, Q. *et al*, "Fabrication and comparison of 3D C_f/ZrC-SiC composites using ZrC particles/polycarbosilane and ZrC precursor/polycarbosilane," *Ceram. Int.*, Vol. 38, No. 6 (2012), pp. 5271-5275.

[6] Pan, R. *et al*, "Homogenization of the zirconium carbide-titanium interface domain," *Scripta Materialia*, Vol. 112 (2016), pp. 42-45.

[7] Pan, R. *et al*, "Cross-diffusion phenomena within a ZrC_x – Zr – ZrC_x joint, *J. Eur. Ceram. Soc.*, Vol. 37, No. 8 (2017), pp. 2779-2786.

A Novel Method for Testing Mechanical Bonding of Diamond Single Crystals to a Braze Alloy Coating

Jussi Laurila and Petri Vuoristo

Tampere University of Technology, Laboratory of Materials Science

Korkeakoulunkatu 6, 33720 Tampere, Finland

jussi.laurila@tut.fi

Olli Tuovinen

Valmet Technologies Inc., Yrittäjänkatu 21, 33101 Tampere, Finland

Abstract

Brazed diamond crystals are used in applications where objects are stressed and carrying high mechanical loads, such as in machining tools. For example, when tooling or grinding hard ceramics it is important that the diamond is firmly fixed on the tool surface and that the individual crystals do not fracture when subjected to stresses. In order to get quantitative information about the mechanical strength of the interface bonding between the diamond and the brazing filler metal, a novel mechanical testing method suitable for the diamond single crystal testing with large number of diamond crystals was developed and used. For mechanical testing, diamonds were brazed in rows on a flat stainless steel plate surface with nickel based chromium alloyed braze filler metal (Ni-Cr-P). The distances between the individual diamonds was set to 1 mm with smaller diamonds or to 2 mm with the larger ones. The diameters of the smaller diamond single crystal grits tested were approximately 300 μm and the larger ones 410, 490 and 580 μm. The results from the mechanical testing can give basic information about the breaking load of stress-loaded individual diamond crystals. Development of the novel method included development of the testing system mechanics and testing of the sensitivity. Repeatability of the method, methods sensitivity is in range +-2 N when 50 kN load cell is used. According to the results measured from the smaller diamonds 80 % of the tested diamond single crystals are fractured between 20 N and 40 N loads. For the larger diamonds, the similar load range was measured to be 48.5 N and 124 N. The variation of the braze filler material layer thickness seems to have only a minor effect to the load bearing ability of the diamond single crystals.

Introduction

Mechanical strength of braze-diamond construction commonly defines the life span of the tool. This makes the mechanical testing of the tool a very important issue. The strength of diamond to metal-brazed joints cannot be measured by using standard testing methods for mechanical properties due to the small size of the brazed diamond particles or inserts. The average diameter of the synthetic diamonds is typically under 0.5 mm when they are brazed on a tool base. Therefore, indirect methods are developed and used for the mechanical characterization and quality assessment of brazed joints.

Leinenbach et al. [1 - 4] and also Bojiang et al. [5] studied the mechanical strength of brazed single crystal diamonds. In their studies, they used Cu-Sn [1] and Ag–35 wt.-% Cu–1.75 wt.-% Ti brazing filler materials, but not any nickel-based alloys, to join the single crystal diamonds on their substrates [4]. The test was used to detect differences in the mechanical strength when brazing parameters such as temperature and holding time are adjusted [1]. Only a few studies have been focusing on the mechanical strength of diamond grits brazed using Ni-Cr-P brazing filler material [6].

In order to study the stress-bearing characteristics of the diamond–braze composites it was necessary to test the breaking loads of individual single crystal diamonds. For this purpose a mechanical test method suitable for testing statistically sufficient amount of diamond grits was developed. Because there are several factors, which are affecting to the properties of base material - brazing filler - diamond grit construction it is important to achieve statistical information about the load carrying capability of it. On this basis, the mechanical tests presented in this paper were carried out.

Experimental

All the studied specimens were manufactured and brazed by Valmet Technologies Inc. locating in Tampere, Finland. Before the brazing process, the surfaces of the base materials were cleaned of any oil remainders by wiping them with pure acetone. The diamond single crystals were placed on the base material surface and the brazing filler metal was sprayed as a layer on the surface. The studied materials were brazed in the Schmetz vacuum furnace (Menden, Germany) with a load dimensions of 90 cm x 120 cm x 70 cm. The raising speed of the temperature of the furnace was set to 3°C/min and the holding time at the brazing temperature from five to eight minutes. Cooling rate after brazing was 2°C/min and the specimens were cooled down with the furnace in vacuum. Vacuum level during the brazing was between $2*10^{-6}$ and $6*10^{-6}$ mbar. The brazing temperatures were between 950°C and 980°C.

For the mechanical testing diamond grits of different diameters were brazed in rows on flat stainless-steel plate surface by using a nickel-based chromium alloyed phosphorus containing braze material (BNi-7 [7]). The diameter of the smaller grits was 300 µm and larger grits 410, 490 and 580 µm. The distance between individual diamonds was set to 1 mm with the smaller diamonds and to 2 mm for the larger ones. The diameters of the diamond grits and the braze filler material thicknesses are given in Table 1.

Table 1. The diameters of the diamond grits and the layer thickness the braze filler material.

	Braze layer thickness/µm	Diameter of the diamond grits/µm
1	70	300
2	99	410
3	116	490
4	141	580

The phosphorous content of the braze filler material is near to the eutectic point of the solution, being about 10 wt.-%. This is to have a low melting point, a narrow melting temperature range and to improve the fluxing and wetting behaviour of the braze filler material on the surfaces involved. Before brazing the braze material was spread on the substrate over the diamond single crystals. The substrate material was molybdenum alloyed stainless steel EN1.4404.

The tests were carried out with a hydraulic Instron 8800 heavy mechanical testing machine (Norwood, MA, USA). A simplified graph of the testing set-up is presented in Figure 1. Such a heavy instrument for mechanical testing was used to avoid the yielding of the test set up during testing. Samples were tested using 50 kN load cell.

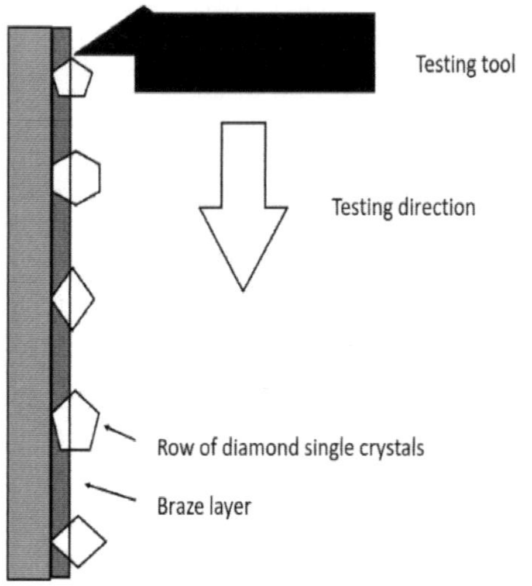

Figure 1. Configuration of the set-up for mechanical testing. An optical scope was used to position the diamond rows under the testing tool. [6]

Testing speed for the tests was selected to be 2 or 3 mm/min and testing range per test was 20 mm, so with one test performed the breaking load of 10 - 20 diamonds was measured. The position of the testing tool and the level of the load was measured 100 times per second. Because of noise detected both in position and load signals, the signals were filtered by using averages of 5 measurements. For each diamond grit tested the peak height of the force to break up the diamond single crystal was measured and the background was subtracted from the measured height of the peak.

Results and discussion

To gain an understanding about the stress-bearing characteristics of the diamonds and tools, tests on the breaking loads of individual diamonds were performed. For these tests, a mechanical testing method suitable for single diamond crystal testing was utilized. A set of diamonds were brazed in rows on a flat stainless-steel plate with nickel-based chromium alloyed braze filler material. An example of the test sample is presented in Figure 2, where the same specimen area is presented before and after the mechanical test documented with Alicona optical profile meter. A diamond single crystal of an untested row of diamonds found from both images is marked with a white ring on both images.

a)

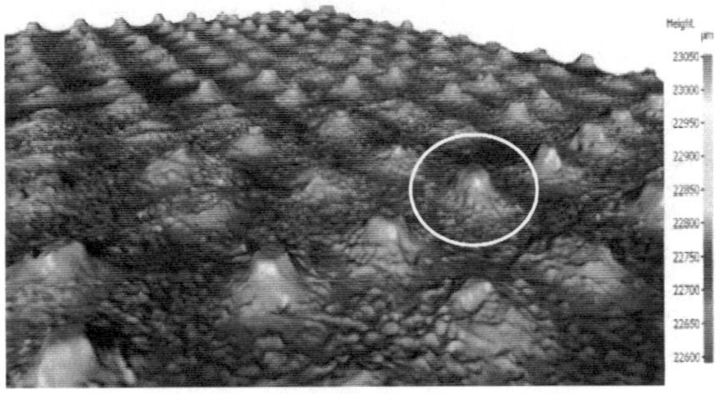

b)

Figure 2. The mechanically tested specimen area before (a) and after (b) the test measured using the optical profilometer. A diamond single crystal found from both images is marked with a white ring on both images. [6]

It should be noted that the tested diamonds were covered with the braze filler material and carbide layers as presented in Figure 3 a). Figure 3 b) presents the fracture surface of the diamond after mechanical testing. Tested diamond single crystals were typically fractured this way, and pull offs of the crystals were not observed. This result means that the bonding between the diamond and the braze filler material is strong enough to hold the stressed diamond until the diamond breaks by fracturing. Some marks of mechanical deformation are found on the braze filler material surface in Figure 3 b) around the diamond indicating that the testing tool touched the braze surface during the test.

Because the diamonds were not in perfect rows, it was not possible to test every one of them, as presented in Figure 2, where the same test area is presented before and after mechanical testing. Some of the diamonds at the tested row has remained untested.

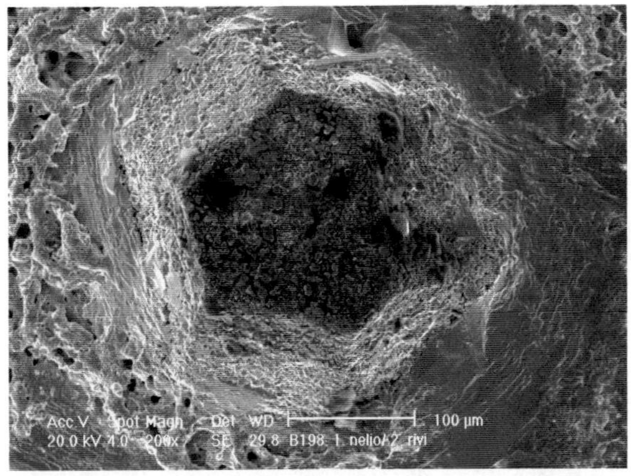

a)

b)

Figure 3. Top view of mechanically tested diamond before (a) and after (b) the test. Diamond is brazed on the stainless steel surface, the diamond is covered with braze/carbide layer. Note the different magnification when comparing the details of the images. [6]

Results of one of the testing cycles is presented in Figure 4. When analysing the results, the peak height of the force to break up the single diamond crystal was measured for each diamond single crystal tested and the background was subtracted from the measured peak. There is some scatter in the diamonds' breaking loads. A reason for this could have been the variations in size and orientation of the single crystals, but also due to differences in the height of the braze layer. This would mean that, for some diamonds, the mechanical testing tool touches more of the braze layer than with others. This error can be minimized by subtracting the background level from the measured peak height, and this was done when analysing the test results.

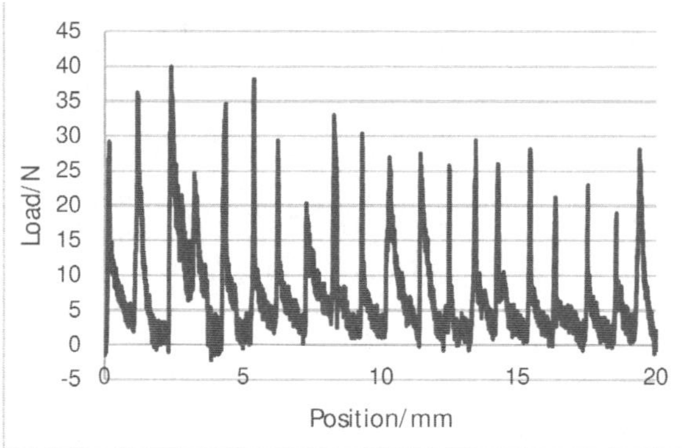

Peak	29.5	36.9	39.9	35.3	39.9	29.4	22.3	33	31.8	27	28	26.5	31	27	29.4	21.9	23.5	18.9	28.4
Background	0.5	3.8	1.2	0.4	1.8	3.6	2.8	5.4	2.7	2.8	2	3.1	1.6	2.8	1.4	1.9	1.6	1.6	2.1
Remainder	29	33.1	38.7	34.9	38.1	25.8	19.5	27.6	29.1	24.2	26	23.4	29.4	24.2	28	20	21.9	17.3	26.3

Figure 4. The results of one 20 mm test cycle with measured peaks, the background level and calculated force values (remainder) needed to break up the test diamonds. [6]

Figure 5 presents a measured peak of when one diamond is tested and broken. Before hitting the diamond there is a base level of background force and when the diamond is hit the force value rapidly increases. The distance moved by the testing tool between hitting the diamond and breaking was measured to be 0.038 mm. The slope of the curve was calculated as y = 680.65x. Behind the measured load peak presented in Figure 5 the load is still seen at higher level compared to the load level before hitting the diamond. This indicates that the tool is touching the diamond fracture surface.

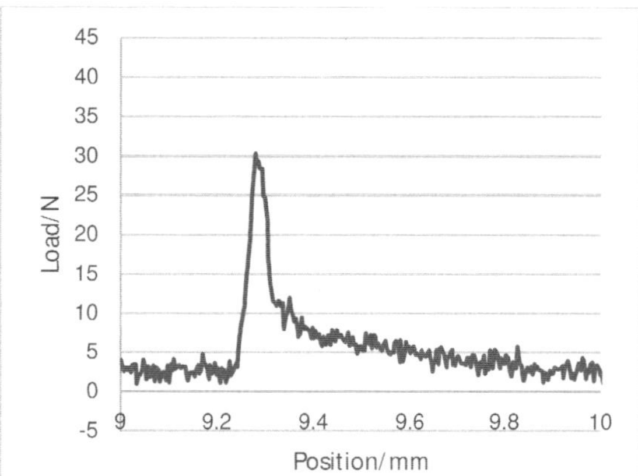

Figure 5. One of the measured load peaks to break up diamond single crystal. The slope of the linear section of the curve when breaking the diamond is calculated to be y = 680.65x.

All together 296 diamond grits with smaller grit size were mechanically tested. The thickness of the braze layer on the substrate surface was 70 μm when measured in the specimen cross-section. All the results performed for the smaller grits are collected in Figure 6 where the breaking load is presented as a function percent of all specimens tested. Figure 6 shows that 80 % of the diamonds broke at load levels between 20 N and 40 N and 50 % of the specimens broke at load levels over 28 N. A minor portion, about 10 %, of the diamonds can hold over 40 N loads.

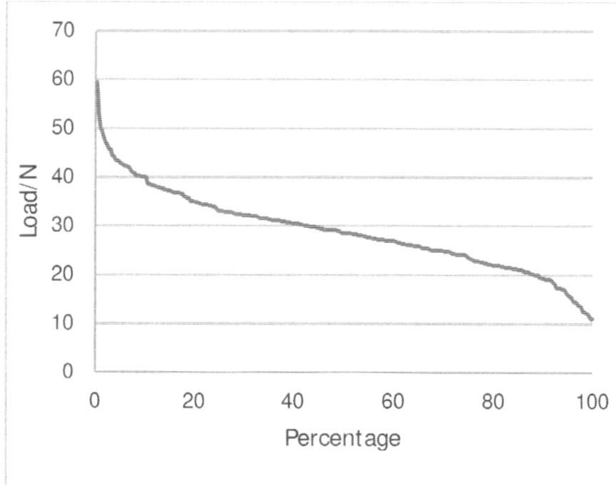

Figure 6. Mechanical testing of smaller diamond grits, size 300 μm: The load for breaking a diamond single crystal as a percentage of tested specimens.

All the results performed for the larger grits (diameter 410, 490 and 580 μm) are collected in Figure 7. For the test for larger diamond grits, three specimens were prepared with brazing filler material with differing thickness. The thickness of the brazing filler were 99, 116 and 141 μm, as presented earlier in Table 1.

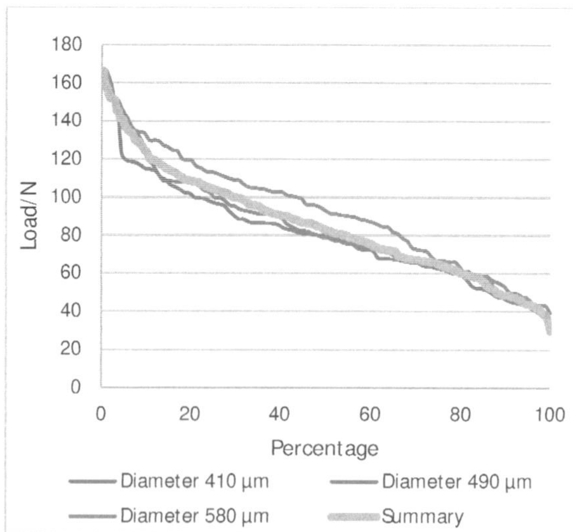

Figure 7. Mechanical testing of larger diamond grits: The load for breaking a diamond single crystal as a percentage of tested specimens.

The breaking loads on individual diamond grits are presented in Figure 7 as a function of percent of all specimens tested. Total number of tested diamonds in this case was 272 grits. Summarized results in Figure 7 shows that 80 % of the diamonds broke at load levels between 48.5 N and 124 N and 50 % of the specimens broke at load levels over 83 N. A minor portion, about 10 %, of the diamonds can hold over 124 N loads. The variation in the grit size from 410 μm to 580 μm is not giving significant difference to the measured breaking loads of diamond crystals. According to these results it seems, that the thickness of the brazing filler layer is not effecting significantly to the measured braking load value.

Conclusions

A mechanical testing method was developed to have statistical information about the mechanical diamond/braze alloy interface properties of brazed diamond single crystals. Achieved results give basic information of the breaking load of the stressed individual diamond crystals. Development of novel method should include in the future development of testing system mechanics, testing suitability of different kinds of loading intenders and load cells and testing of sensitivity and repeatability of the method. The developed testing method can be used for testing the effect of different braze materials and brazing parameters on the load bearing ability of the diamond single crystals.

According to the results measured from the smaller diamonds 80 % of the tested diamond single crystals are fractured between 20 N and 40 N loads. For the larger diamonds, the similar load range was measured to be 48.5 N and 124 N. The variation of the braze filler material layer thickness seems to have only a minor effect to the load bearing ability of the diamond single crystals

Acknowledgements

The work was carried out under the HYBRIDS program in DIMECC Ltd. The researchers gratefully acknowledge the financial support of Valmet Technologies Inc. and Business Finland.

References

[1] S. Buhl, C. Leinenbach, R. Spolenak, K. Wegener, Microstructure, residual stresses and shear strength of diamond–steel joints brazed with a Cu Sn-based active filler alloy, Int. Journal of Refractory Metals and Hard Materials 30, pp. 16–24, 2012

[2] M. Koster, A. Lis, W. J. Lee, C. Kenel, C. Leinenbach, Influence of elastic–plastic base material properties on the fatigue and cyclic deformation behavior of brazed steel joints, International Journal of Fatigue, in press, 2015

[3] C. Leinenbach, R. Transchel, K. Gorgievski, F. Kuster, H. R. Elsener, and K. Wegener, Microstructure and Mechanical Performance of Cu-Sn-Ti-Based Active Braze Alloy Containing In Situ Formed Nano-Sized TiC Particles, Journal of Materials Engineering and Performance, Vol. 24(5), May 2015

[4] S. Buhl, C. Leinenbach, R. Spolenak, K. Wegener, Influence of the brazing parameters on microstructure, residual stresses and shear strength of diamond–metal joints, J. Mater. Sci. 45: pp. 4358–4368, 2010

[5] Bojiang Ma, Guanglei Yang, Fanning Bu, Study of brazed diamond micro-powder burs fabricated using induction brazing with either an amorphous or a crystalline Ni-based filler alloy, Int. Journal of Refractory Metals and Hard Materials 62, pp. 58–63, 2017

[6] V-J. Laurila, Structure and properties of nickel based braze surface with diamond grits – interaction phenomena between brazing alloy and diamond grits, thesis, Tampere University of Technology, 2017

[7] Metals Handbook vol. 6, ASM, asminternational.org, viewed 2016

Joining of SiBCN ceramic with CuPd-V filler alloy and the interfacial reactions

Wen-Wen Li, Hua-Ping Xiong, Hai-Shui Ren, Bo Chen, Xin Wu, Yao-Yong Cheng*

Welding and Plastic Forming Division, Beijing Institute of Aeronautical Materials, Beijing 100095, China

**Corresponding author, Tel: +86-010-62496680.*

E-mail address: xionghp69@163.com

Abstract

The joining of SiBCN ceramic was carried out using CuPd-V filler alloy. The wettability of CuPd-V filler alloy on SiBCN ceramic was studied with the sessile drop method. The interfacial reactions were analyzed by SEM, EDS and XRD. The results showed that after heated at 1200°C for 10min, CuPd-V filler alloy exhibited very good wettability on SiBCN ceramic with an average contact angle of 5.2°. Through the analysis of the joint microstructure, it can be indicated that V played active role in the interfacial reactions, with the formation of two reaction layers V-(C,N) and V_5Si_3. The central part of the joint consisted of Cu(Si) solid solutions and Pd_2Si refractory compound. Furthermore, the filler alloy diffused into the based SiBCN ceramic, and formed two (Pd,Cu)-Si diffusion zones with a width of 70μm.

Introduction

Quaternary precursor-derived Si-B-C-N ceramic has gained large interstates in field of the ultra-high temperature materials. This amorphous silicoboron carbonitride ceramic frequently resists thermal decomposition even at temperatures exceeding 1800°C[1,2], and the oxidation resistance at 1700°C is superior to SiC or Si_3N_4 ceramic [3]. As a result, SiBCN ceramics are considered to be the most potential materials in aerospace and anti-ablation structures (nose caps, satellite nozzles, rocket engines, etc.) in future [4,5].

Nevertheless, the research on the SiBCN ceramic has just started from the last two decades, and the theoretical study on the microstructure of SiBCN ceramic is still insufficient. The HRTEM image indicated that nano SiC crystallites and BCN turbostratic layers existed in as-pyrolysed amorphous $Si_{2.4}B_{0.8}C_{4.5}N_{2.3}$ ($O_{0.1}$) ceramic [6]. After heat treatment at 1800 °C in nitrogen, the crystallization products of $Si_3B_1C_{5.5}N_3$ ceramic were composed of SiC and Si_3N_4 [7,8]. Both the precursor type and synthesis method would determine the decomposition temperature, constituent and microstructure of the fabricated Si-B-C-N ceramic [9]. Compared with the other ceramics such as SiC or Si_3N_4, the brazing of SiBCN ceramic has not been investigated systemically. Up to now, the reports on the SiBCN ceramic joining are extremely lacking.

Sun et al. fabricated a functional gradient SiBCN-Nb composite by hot pressing[10], which can be used to the joining of dissimilar materials. Considering the direct active brazing, it has been widely used for ceramics or CMCs among the various ceramic joining techniques due to its simplicity, lower cost investment and potential as a mass production process[11]. AgCuTi filler alloy has been used to braze the 2Si-B-3C-N ceramic and obtained the maximum shear strength of 138 MPa at room temperature [12]. However, obviously, the service temperature of AgCuTi system brazing alloy is still a problem. In the previous work, C_f/SiBCN composite was brazed with Ni-based filler alloys with the bend strength of 62.9MPa[13]. But regretfully, no further study on the high-temperature strength was conducted.

According to the previous references[7,8], in amorphous Si-B-C-N at high temperature the phases of SiC and Si_3N_4 easily crystallized to precipitate. Therefore, it is likely possible to join SiBCN ceramic with the filler metals which can be used for brazing SiC and Si_3N_4 ceramics or C_f/SiC composite. For the joining of C_f/SiC composite, CuPd-V system filler alloy was used. V played the interfacial activity to form the $VC_{0.75}$ reaction layer and the room-temperature strength of 135MPa can be maintained up to 800°C[14]. For brazing C_f/SiBCN, a new composition of CuPd-V filler alloy was designed[15], and the results verified that CuPd-V alloy after modifying the concentration of vanadium can be successfully used for brazing C_f/SiBCN composite. In this work, new CuPd-V filler alloy was applied into the joining of SiBCN ceramic, the wettability and the interfacial reactions were investigated.

Experimental procedure

The nominal composition of CuPd-V filler alloy can be expressed as Cu-(35~45)Pd-(10~15)V, which can be fabricated into brazing foil with the thickness of 100μm by alternating room-temperature rolling process and vacuum annealing treatment, attributing to the good plasticity. The liquidus temperature is 1155.6℃ measured by differential thermal analysis (DTA) method. The wettability of CuPd-V filler alloys on SiBCN ceramic was studied with the sessile drop method. Before the wetting experiment, five CuPd-V foils with a diameter of 3mm were prepared, and then 5-layer foils were laminated by low-electric-current spot welding. Afterwards the laminated filler plate was placed on the SiBCN substrate $(10\times10\times3)mm^3$. Next, the assembly was put into a vacuum furnace for the wettability study. The wetting sample was firstly heated to 1200°C at a heating rate of 10°C /min. After being held at 1200°C for 10 min, the samples were cooled to 500°C at a cooling rate of about 5°C /min, followed by furnace cooling to room temperature. During the heat process, the vacuum of furnace was kept between 3.0×10^{-3}Pa and 7.0×10^{-3}Pa.

SiBCN ceramic used as the joined material was fabricated by hot pressed sintering. For microstructure analysis, the SiBCN ceramic samples were cut into $(5\times5\times3)mm^3$, whereas, the CuPd-V filler foil should be cut into a little larger size than the joined material. One layer CuPd-V filler foil was placed between two SiBCN substrates to form a butt joint. A little pressure perpendicular to the joining surface was applied to ensure good contact between the interlayer and the composite.

The parameters of the joining experiment were identical to the wetting experiment.

The joint microstructures were analyzed by a scanning electron microscope (SEM) equipped with an x-ray wave-length dispersion spectrometer (XWDS) and an X-ray energy-dispersive spectrometer (EDS). For determination of the interfacial reaction products, X-ray diffraction (XRD) analysis was performed on a simulated specimen of the SiBCN joint brazed at 1200°C for 10min. The simulated specimens were prepared by a wetting experiment, and then gradual polishing was conducted through the reaction zone on the surface of the wetting specimens. XRD analysis was carried out successively on the surface containing the reaction layer.

Results and discussions

After being held at 1200°C for 10min, the CuPd-V filler alloy melted and spread on the SiBCN ceramic. The morphologies of molten droplet were shown in Fig. 1. From the over-view picture, the spreading area was nearly twice than the original wetting sample. According to the cross-section image, the average contact angle can be measured as 5.2°. It can be indicated that the CuPd-V filler alloy exhibited good wettability on the SiBCN ceramic. And it was a precondition for the brazing of SiBCN ceramic.

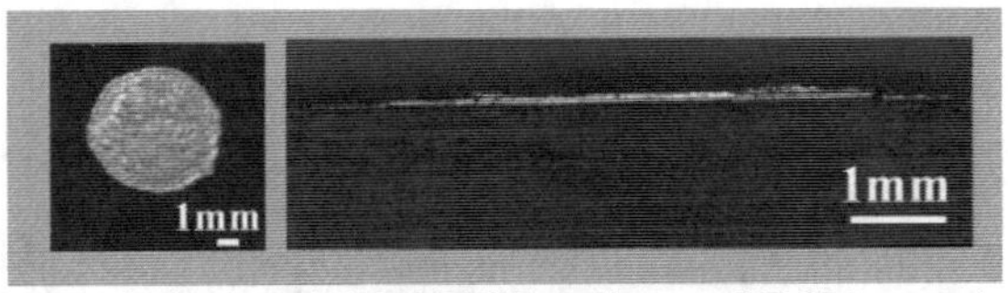

Fig. 1 Morphologies of CuPd-V filler alloy on SiBCN ceramic after wetting experiment.

Similarly, CuPd-V system filler alloy showed a contact angle of 6°after heating at 1200 °C for 10min[14]. However, for Si_3N_4 ceramic, the CuPd-V filler alloy reached an equilibrium contact angle of 71° when heated to 1200 °C, and the contact angle no longer changed with the prolonging of the holding time [16]. As a consequence, in terms of the wetting behavior, CuPd-V system filler alloy was close to the C_f/SiC composite to some extent.

The microstructures of SiBCN-SiBCN joint brazed with CuPd-V filler alloy was presented in Fig. 2. From the picture with a low magnification (Fig. 2(a)), it can be seen that the braze seam has good quality and the filler alloy has reacted with SiBCN ceramic sufficiently at the interface. During the brazing procedure, the filler alloy melted and infiltrated into the SiBCN base materials, then a diffusion reaction zone with a width of 70μm formed.

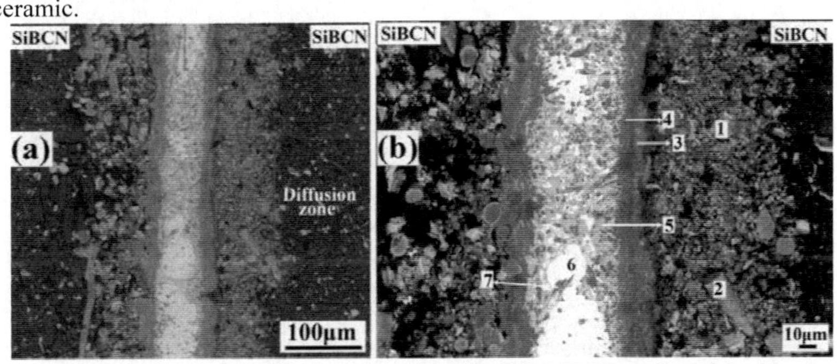

Fig. 2 Back-scattered electron image of SiBCN-SiBCN joint brazed with CuPd-V filler alloy: (a) with a low magnification; (b) with a high magnification.

The detailed microstructure of SiBCN-SiBCN joint was showed in Fig. 2(b), and the compositions for the typical microzones were listed in Table 1. At first, in the diffusion reaction zone, the microzone "1" and "2" were enriched with Pd, Cu and Si. This phenomenon was good agreement with the element distribution maps (as shown in Fig. 3(b), (c) and (e)). It can be deduced that when the filler alloy melted, elements Pd and Cu diffused into SiBCN ceramic and reacted with element Si. According to the XRD analysis (Fig. 4), Pd_2Si compound was detected at the wetting interface between SiBCN and CuPd-V filler alloy. As well known, the enthalpy of Si in melted Pd was -145 kJ/mol [17], that means the element Pd has a large affinity with Si[18~20]. Therefore, it was not strange for the formation of Pd_2Si compound through the reaction between the SiBCN ceramic and the CuPd-V filler alloy. In addition, Cu and Pd can form an infinite solid solution according to the Cu-Pd binary phase diagram. As a result, the diffusion reaction zone should be (Pd,Cu)-Si silicides.

Table 1. EDS analysis results for the microzones in Fig. 2(b).

Micro-zone	Composition (at.%)							Deduced phase
	Cu	Pd	V	Si	B	C	N	
1	29.35	34.19	1.93	34.53	/	/	/	(Pd,Cu)-Si
2	22.53	41.21	0.58	35.69	/	/	/	(Pd,Cu)-Si
3	2.36	1.12	46.71	1.54	3.38	26.32	18.58	V_4C_3, V-N
4	0.86	0.17	52.43	28.79	/	6.78	10.97	V_5Si_3, V-N
5	60.04	9.87	0.29	15.33	5.32	7.55	1.60	Cu(Si)ss
6	6.99	58.71	0.44	29.05	/	3.55	1.27	Pd_2Si
7	5.19	1.16	54.06	33.90	/	1.48	4.20	V_5Si_3

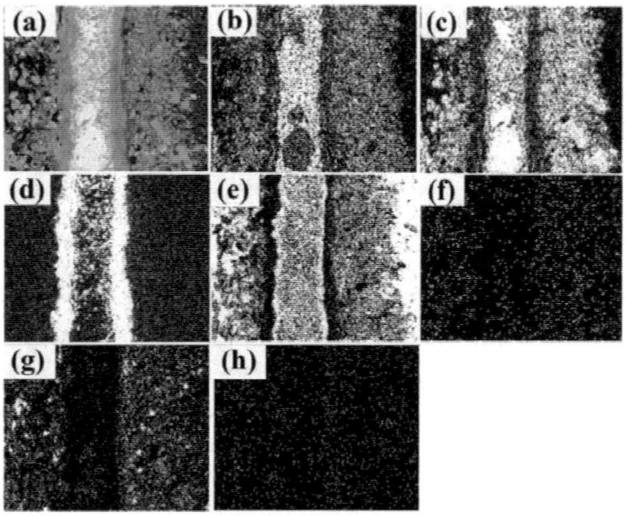

Fig. 3 SEI of the SiBCN-SiBCN joint brazed with CuPd-V filler alloy (a) and the area distribution maps of elements Cu(b), Pd(c), V(d), Si(e), B(f), C(g), N(h).

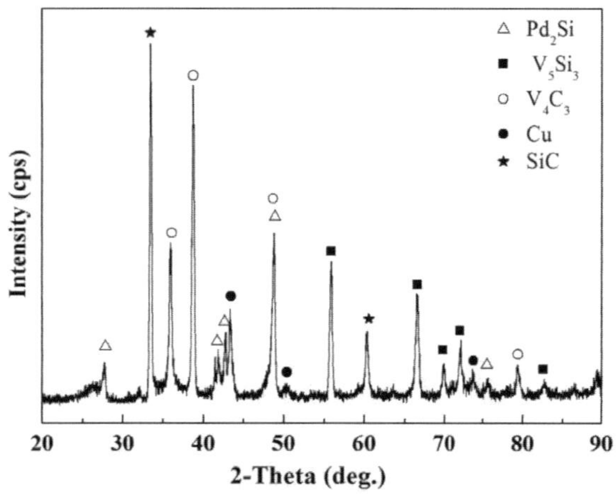

Fig. 4 X-ray diffraction pattern of the interface between SiBCN ceramic and CuPd-V filler alloy

At the interface, there formed continuous reaction layers where element V segregated according to the distribution map (Fig. 3 (d)). It was indicated that V played obvious active role in the interfacial reactions. In actual, the reaction layer consisted of two sub-layers marked as "3" and "4". For these two layers, the same is the enrichment of the active element V. However, the microzone "3" was enriched with 26.32 at.% C and 18.58 at.% N, whereas 28.79 at.% Si and 10.97 at.% N existed in microzone "4" according to the EDS results in Table 1. With the XRD analysis for the reaction products at the interface as shown in Fig. 4, V_4C_3 and V_5Si_3 compounds were found. It should be noticed that the XRD peaks associated with SiC come from the joined SiBCN substrate. That is, in the SiBCN ceramic used in our work, SiC crystal particles should be present. It is reasonable to deduce that after the interfacial reaction, the main product in layer "3" was V_4C_3, and the main product in microzone "4" was V_5Si_3.

It should be noticed that the thickness of the V_4C_3/V_5Si_3 was about 15μm. But when brazed C_f/SiC composite used CuPd-V filler alloy, the $VC_{0.75}$ reaction layer was less than 2μm[14]. Compared with C_f/SiC composite, the filler alloy reacted with SiBCN ceramic more intensively. The over-reaction would weaken the bonding strength, and the composition of filler metal would be modified in the future.

Previous studies reported different reaction products of Si_3N_4/V system. For example, Xiong[21] studied the wettability of V-containing filler alloy on Si_3N_4 ceramic, and V_6Si_5 compound was verified. However, Y. Ito *et al.* [22] reported the formation of V_3Si and V_5Si_3 at 1523K at the diffusion banded interface of Si_3N_4/V foil. For the SiBCN/V system, according to the XRD pattern, only V_5Si_3 compound can be verified.

V-containing filler alloys were commonly used to brazed SiC or SiC ceramic matrix composite. Fukai[23] studied the reaction mechanism of SiC/V joint and detected V_2C in the central part of the joint bonded at 1573K. Xiong *et al.*[24] designed Cu-Au-Pd-V brazing filler for the C_f/SiC composite joining, and $VC_{0.75}$ reaction band formed by the expression: $C/SiC+V+Pd \rightarrow VC_{0.75}+Pd_2Si$. It was consistent with the formation of V_4C_3 and Pd_2Si in this paper.

Furthermore, some content of V-N compound was coexisted in the two layers. In general, when brazed Si_3N_4 ceramic used V-containing filler metals, VN or V_2N compound should be the main products at the Si_3N_4 interface [15,25,26]. However, in the SiBCN joint, there was no evident diffraction peak of vanadium nitride in Fig. 4. Maybe, in SiBCN ceramic, the Si-N covalent bond was stable compared with Si-C bond. That is, intensive reactions of filler metal with Si-C were easy to generate.

In the central part of the joint, the matrix "5" was enriched with 60.04 at.% Cu and 15.33 at.% Si, and it can be deduced as a Cu solid solution with some Si dissolved. The constitute of the light white phase "6" was 58.71 at.% Pd and 29.05 at.% Si, the atomic ratio of Pd and Si was 2:1, so that Pd_2Si phase also formed in the central part of the joint. It should be noticed that, some small dark phases "7" were scatteringly distributed in the joint matrix, where element V and Si had a high concentration. According to the EDS result, the formation of the V_5Si_3 compound can be deduced.

Conclusions

(1) When heated at 1200°C for 10min, CuPd-V filler alloy exhibited very good wettability on SiBCN ceramic with an average contact angle of 5.2°.

(2) Under the same condition, CuPd-V filler alloy was used to braze the SiBCN to itself. A (Pd, Cu)-Si diffusion zone with a width of 70μm formed inside of SiBCN base material.

(3) Near the SiBCN ceramic surface, two V-containing reaction layer V_4C_3/V_5Si_3 were formed. In the central part of the joint, the microstructure was composed of Cu solid solution with some Si dissolved and Pd_2Si refractory compound, as well as V_5Si_3 scatteringly distributed.

Acknowledgments

This research work was sponsored by National Natural Science Foundation of China (Contract Nos. 59905022, 50475160 and 51275497). We also would like to thank Aeronautical Science Foundation of China (Grant 2008 ZE21005).

References

[1] J. Haug, P. Lamparter, M. Weinmann, F. Aldinger. Diffraction study on the atomic structure and phase separation of amorphous ceramics in the Si-(B)-C-N system. 2. Si-B-C-N ceramics. Chem. Mater. 16 (2004) 83-92.

[2] R. Riedel, A. Kienzle, W. Dressler, L. Ruwisch, J. Bill, F. Aldinger. A silicoboron carbonitride ceramic stable to 2000°C. Nature. 382 (1996) 796-798.

[3] M. Weinmann, J. Schuhmacher, H. Kummer, S. Prinz, J.Q. Peng, H.J. Seifert, M. Christ, K. Muller, J. Bill, F. Aldinger. Synthesis and thermal behavior of novel Si-B-C-N ceramic precursors. Chem. Mater. 12 (2000) 623-632.

[4] P. Baldus, M. Jansen, D. Sporn. Ceramic fibers for matrix composites in high-temperature engine applications. Science. 285 (1999) 699-703.

[5] J.Y. Wang, X.M. Duan, Z.H. Yang, D.C. Jia, Y. Zhou. Ablation mechanism and properties of SiC_f/SiBCN ceramic composites under an oxyacetylene torch environment. Corros. Sci. 82 (2014) 101-107.

[6] A. Muller, A. Zern, P. Gerstel, J. Bill, F. Aldinger. Boron-modified poly (propenylsilazane)-derived Si-B-C-N ceramics: preparation and high temperature properties. J. Eur. Ceram. Soc. 22 (2002) 1631-1643.

[7] J. Bill, T.W. Kamphowe, A. Muller, T. Wichmann, A. Zern, A. Jalowieki, J. Mayer, M. Weinmann, J. Schuhmacher, K. Muller, J.Q. Peng, H. J. Seifert, F. Aldinger. Precursor-derived Si-(B-)C-N ceramics: thermolysis, amorphous state and crystallization. Appl. Organomet. Chem. 15 (2001) 777-793.

[8] S. Bernard, M. Weinmann, P. Gerstel, P. Miele, F. Aldinger. Boron-modified polysilazane as novel single-source precursor for SiBCN ceramic fibers: synthesis, melt-spinning, curing and ceramic conversion. J. Mater. Chem. 15 (2005) 289-299.

[9] D.C. Jia, P.F. Zhang, Z.H. Yang, D. Ye, Z.L. Sun, Y. Zhou. Progress of amorphous and nanostructured Si-B(Al)-C-N ceramics. Mater. China. 30 (2011) 5-11. (in Chinese)

[10] M. Sun, R.Y. Fu, J. Chen, X.F. Mao, J. Zhang, Z.H. Yang, B. Liang. Fabrication and microstructures of functional gradient SiBCN-Nb composite by hot pressing. Mater. Charact. 114 (2016) 115-121.

[11] J.M. Fernandez, R. Asthana, M. Singh, F.M. Valera. Active metal brazing of silicon nitride ceramics using a Cu-based alloy and refractory metal interlayers. Ceram. Int. 42 (2016) 5447-5454.

[12] R. Pan, T.S. Lin, P. He, H.M. Wei, Z.H. Yang, Y.X .Shen. Micro-structure and mechanical property of 2Si-B-3C-N ceramic joints brazed by AgCuTi filler metal. J. Chin. Ceram. Soc. 43 (2015) 1719–1724. (in Chinese)

[13] W.W. Li, B. Chen, Y. Xiong, H.P. Xiong, Y.Y. Cheng, W.J. Zou. Joining of C_f/SiBCN composite with two Ni-based brazing fillers and interfacial reactions. J. Mater. Sci. Technol. 33 (2017) 487-491.

[14] H.P. Xiong, B. Chen, Y. Pan, W. Mao, Y.Y. Cheng. Interfacial reactions and joining characteristics of a Cu-Pd-V system filler alloy with C_f/SiC composite. Ceram. Int. 40 (2014) 7857-7863.

[15] W.W. Li, B, Chen, L.M. Cao, W. Liu, H.P. Xiong, Y.Y. Cheng. Joining of C_f/SiBCN composite with CuPd-V filler alloy. J. Mater. Sci. Technol. Doi: 10.1016/j.jmst.2018.01.001.

[16] H.P. Xiong, B. Chen, Y. Pan, H.S. Zhao, W. Mao, Y.Y. Cheng. A Cu-Pd-V system filler alloy for silicon nitride ceramic joining and the interfacial reactions. J. Am. Ceram. Soc. 97 (2014): 2447-2454.

[17] A.R. Miedema, F.R. de Boer, R. Boom. Model predictions for the enthalpy of formation of transition metal alloys. CALPHAD 1(4) (1977) 341-359.

[18] J.S. Park, K. Landry, J.H. Perepezko. Kinetic control of silicon carbide/metal reactions. Mater. Sci. Eng. A. 259 (1999) 279-286.

[19] K. Bhanumurthy, J.C. Schuster. Interface reactions between silicon carbide and metals (Ni, Cr, Pd, Zr). Compos. Part A-Appl. S. 32 (2001) 569-74.

[20] M. Tanaka, M. Takeguchi, H. Yasuda, K. Furuya. In-situ observation of deposition process of Pd on clean Si surfaces by ultrahigh vacuum-transmission electron microscopy scanning tunneling microscopy. Thin Solid Films. 398-399 (2001) 374-378.

[21] H.P. Xiong, W. Dong, B. Chen, Y.S. Kang, A. Kawasaki, H. Okamura, R. Watanabe. Wettability of Ni-V, Co-V, and Ni-Cr-V system brazing alloys on Si_3N_4 ceramic and interfacial reactions. Mater. Sci. Eng. A 474 (2008): 376-381.

[22] Y. Ito, K. Kitamura, M. Kanno. Joint of silicon nitride and molybdenum with vanadium foil, and its high-temperature strength. J. Mater. Sci. 28 (1993): 5014-5018.

[23] T. Fukai, M. Naka, J.C. Schuster. Interfacial structure and reaction mechanism of SiC/V joints (materials, metallurgy & weldability). Trans. JWRI. 26 (1997): 93-98.

[24] H.P. Xiong, B. Chen, Y. Pan, H.S. Zhao, L. Ye. Joining of C_f/SiC composite with a Cu-Au-Pd-V brazing filler and interfacial reactions. J. Eur. Ceram. Soc. 34 (2014) 1481-1486.

[25] Y. Sun, Y.P. Geng, K. Ikeuchi, T. Shibayanagi. Microstructure and mechanical properties of an Si_3N_4/Si_3N_4 joint brazed with Au-Ni-Pd-V filler alloy. Script. Mater. 64 (2011) 414-417.

[26] H.P. Xiong, B. Chen, Y. Pan, H.S. Zhao, W.L. Guo. Joining of Si_3N_4 to Si_3N_4 using a Au Pd(Co,Ni)-V filler alloy and the interfacial reactions. Ceram. Int. 40 (2014): 4141-4148.

Effect of Different Brazing Temperatures and Holding Times on the Joint Microstructure and Bond Strength of SiC/SiAlTi/SiC Braze Joints

J. Wei, S. Liu*, J.C. Madeni*, S. Nelson** and R. Xu***

**Center for Welding, Joining, and Coatings Research*
George Ansel Metallurgical and Materials Engineering Department
Colorado School of Mines, Golden, CO, United States

***Rolls-Royce Corporation*
Indianapolis, IN, United States

Abstract

Reaction bonded silicon carbide (RB SiC) braze joints were produced utilizing a SiAlTi filler material (FM). Joints were brazed under high vacuum using five different brazing cycles, varying both hold temperature and time. The FM exhibited good wettability onto the SiC surface, as evidenced by full penetration of the FM into joint faying surfaces. The joint microstructure, interfacial reactions, and the variation of these items as a function of the braze cycle were analyzed using Scanning Electron Microscopy (SEM) and Electron Dispersive Spectroscopy (EDS). Phase identification revealed that the active element titanium diffused into the base substrate (BS) forming a silicide within a Diffusion Affected Zone (DAZ). Additionally, shear testing of the joints brazed under the different cycles show that the shear strength of the joints increased with longer brazing hold times and reached a peak strength before degrading as the holding time further increased. This testing has indicated that a properly selected braze cycle has a large effect on the joint strength when joining SiC substrates using the SiAlTi filler material.

Introduction

Reaction Bonded Silicon Carbide (RBSiC), also known as Siliconized Silicon Carbide (SiSiC), is well known for its chemical resistance, heat resistance, abrasion resistance, and high-temperature mechanical properties; thus, it is widely used in high-temperature structural applications [1,2]. Practical industrial use of SiC requires appropriate joining techniques because of the processing difficulties and expense associated with making large or complex ceramic components.

Of the many different methods of joining ceramics, brazing is one of the most widely used and most effective methods owing to its simplicity, small capital investment, and potential in mass-production [3].

There are two main problems in ceramic brazing. The first is the poor wettability of ceramics by most traditional filler alloys; This problem can be approached using FM containing active elements such as Ti, Zr, Cr, V, and Nb that have the ability to react with the ceramic substrate [4-12]. Utilization of these elements is the basis of active metal brazing, a method without the need to apply any metallization to ceramic substrates. The second problem is the thermal residual stress in the braze joints resulting from the differences in coefficient of thermal expansion (CTE) and modulus of elasticity between dissimilar ceramics and metals [13]. This problem can be overcome by inserting or forming ductile interlayers between two dissimilar materials to relieve residual stresses through plastic deformation during brazing [14,15].

The most widely used active metal brazing filler metals to join SiC have been Cu-based and AgCu-based systems with additions of Ti. Generally, Ag-Cu-Ti alloys are used for brazing at temperatures below 600 °C. There are, however, higher melting active brazing metals. For example, copper ABA (92 wt% Cu with the addition of Ti, Al, Si), which is the most important Cu-based braze alloy for the certain applications; it has a liquidus temperature of 1025 °C. Ormston [16] reported the presence of Ti_5Si_3 and Cu_5Si in the braze joint when brazing SiC using Cusil ABA (63wt% Ag, 1.75Wt% Ti, 35.25wt% Cu) foil, and Ticusil (68.8wt% Ag, 26.7wt% Cu, 4/5wt% Ti) foil at brazing temperatures between 900 °C and 950 °C. Four-point bend tests performed on these brazed joints showed strengths ranging from 300 to 400 MPa. Lee et. al. [17] reported a flexural strength of 178 MPa achieved in samples brazed at 985 °C for 10 minutes using Ag–5Ti (at%) filler metal. In Liu's study [18], a higher flexural strength (342 MPa) was achieved in samples brazed at 900 °C for 30 minutes using Ag–35.25Cu–1.75Ti (wt%) traditional brazing filler metal.

Other brazing filler metals can also be used in ceramic brazing. It was reported that 400 MPa strength was obtained in Si_3N_4 braze joints produced using Au-15.5Ni-0.75Mo-1.75V brazing FM at 1020 °C, resulted in an interfacial reaction product VN [19]. Xiong et. al. [3, 20-21] have performed work using CoFeNi(Si,B)-Cr-Ti braze to join SiC to itself or to Ni-based alloys at high temperatures. They reported room temperature joint strength of 160 MPa for brazed joints processed at 1150 °C. The behavior of Ti-Si braze alloy with high silicon content at brazing temperatures over 1300 °C was studied by Riccardi et. al. [22] and Li et. al. [23,24]. In Riccardi's work, a shear strength of 71 MPa was achieved in SiC/SiC joints processed using Ti–84Si (at%) braze alloy at 1330 °C for 10 minutes. A higher shear strength of 125 MPa was also reported by Li et. al. using Ti–78Si (wt%) braze alloy at 1400 °C for 10 minutes. While at an even higher brazing temperature of 1600 °C, a shear strength of 154 MPa was achieved in SiC joints using Ni–50Ti (at%) alloy after holding for 30 minutes [25].

In this study, (Si, Al)-based brazing filler metal with active element Ti addition was used to braze SiC to itself. The wettability of these filler alloys on SiC substrate and the braze joint microstructure were investigated. The bond strength of braze joints was measured and the fracture surfaces were characterized.

Experimental Procedures

In this study, an active Si-Al-Ti filler metal with a liquidus temperature of 1163 °C was used to join RBSiC to itself. Cylindrical filler metal pellets were produced by arc melting of pure metals and then sliced into thinner pieces, each of 5/16 × 5/16 × 1/16 inch (8 × 8 × 1.5 mm), respectively, using a slow diamond cutting saw.

Two different sizes of SiC coupons were used. The first type of coupons were those with the as-received geometry of 1 × 1 × 1/4 inch (25.4 × 25.4 × 6.4 mm); the second coupons were those cut into a geometry of 7/16 × 7/16 × 1/4 inch (11.1 × 11.1 × 6.4 mm). All SiC coupons were ultrasonic cleaned in acetone for 30 minutes to remove impurities from the surface before brazing.

All brazing tests were conducted under a vacuum furnace with the ability to provide 10^{-5} Torr maximum vacuum. This pressure (10^{-4} -10^{-5} Torr) was achieved, then isothermal holding started. The five thermal cycles used during brazing are shown in Figure 1; these included a continuous heating with a heating rate of 15 °C/min, which took about 82 minute heating time, and three-hour furnace cooling. For the joints produced for microstructure analysis, the brazing temperature was 1300 °C and the holding time was 30 minutes; for the joints produced for shear testing, the brazing temperature ranged from

1200 to 1300 °C, and the isothermal holding time varied from 10 minutes to 1 hour. During brazing, the smaller coupon of SiC with a geometry of 7/16 × 7/16 × 1/4 in (11.1 × 11.1 × 6.4 mm) was placed at the bottom, while the larger as-received SiC coupon was placed on the top, and a thin piece of filler metal with a geometry of 5/16 × 5/16 × 1/16 in (8 × 8 × 1.5 mm) was sandwiched between the two SiC coupons. In addition, a specially designed open-faced refractory holder produced by investment casting was used in the brazing in order to assure proper alignment of the sandwiched specimens.

An MTS servohydraulic testing machine was used in the bond strength measurement of the brazed joints. A 1018 steel fixture was used under slight compression to hold the braze joints in place, a schematic drawing of the fixture is shown in Figure 2. The test was conducted with a cross-head speed of 0.5 mm/min^{-1} until the braze joint fractures.

The microstructure of braze joint interface and the morphologies of the fracture surfaces were analyzed using SEM and EDS. Sample preparation started by obtaining a cross-section of the braze joint; then, the samples were mounted, grinded and polished to 1 µm. Then, the polished braze joint samples were ultrasonically cleaned in acetone for 30 minutes before use.

Results

In this study, brazing of SiC using the Si-Al-Ti filler metal was successfully conducted; no deformation and cracking was detected after brazing, as shown in Figure 3(a)-(e). The braze alloy showed good wettability and spreadability on the RB SiC base substrate; and the amount of FM used was able to fill the overlapping gap between the top and the bottom SiC coupons, leaving no cavity in the brazing gap.

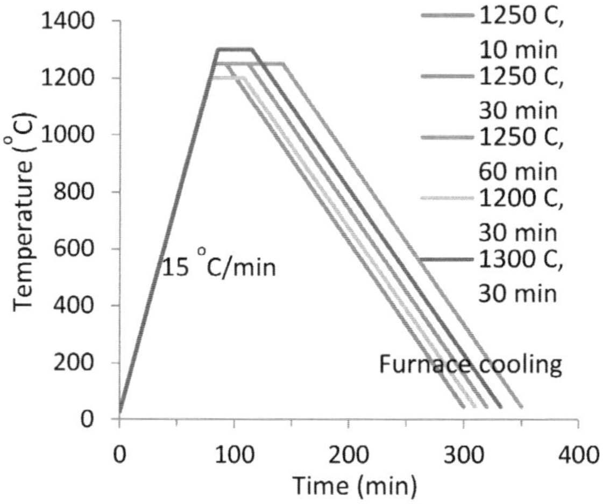

Figure 1: Thermal cycles of SiC brazing using Ti-Al-Si FM for developing shear testing joints.

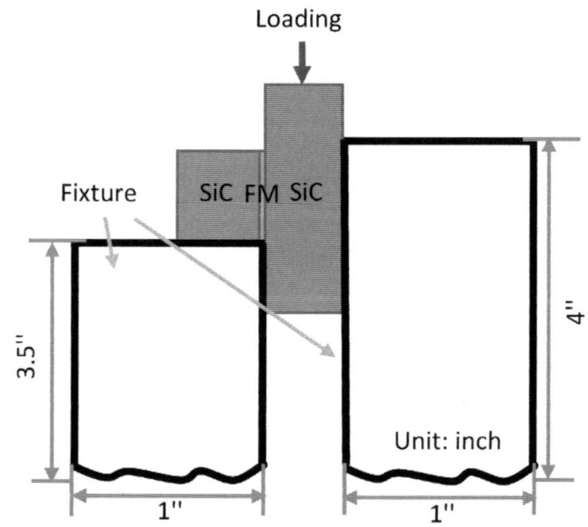

Figure 2: Schematic drawing of shear testing and fixture

Interfacial microstructure characterization

The SEM images in Figure 4(a) showed a smooth and uniform brazing joint with a thickness of about 100 microns obtained under tested parameters of isothermal holding at 1300 °C for 30 minutes. Good wetting and spreadability of the braze alloy were obtained in all joints.

(a)1200 °C-30 min (b)1300 °C-30 min (c)1250 °C-10 min

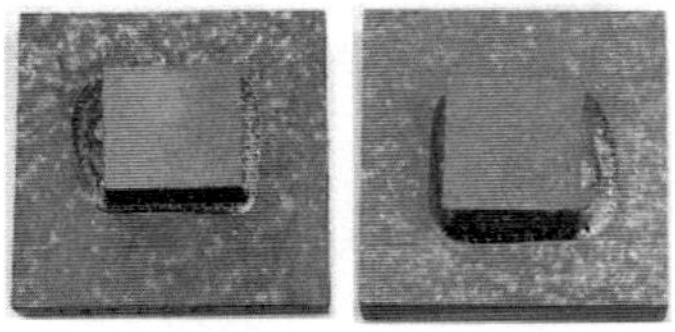

(d)1250 °C-30 min (e)1250 °C-60 min

Figure 3: As-brazed five specimens from different brazing cycles

Higher magnification of SEM images were taken to analyze the phases in the brazing joint and in the interaction zone, as shown in Figure 4(b)-(d). Selected areas marked by the squares in the micrographs were enlarged in order to better visualize the phases that were present. The relevant EDS results of the different phases detected in the joints in Figure 4 are listed in Table I. According to Figure 4(a), over 90 % of the interface area was uniformly occupied by the grey colored phase, which was identified to be Si (spot 1). Other than Si, light grey-colored polygons were identified as $TiSi_2$ (spot 2), and the Al-Si eutectic (squares I and III) was identified to be composed by small Si flakes (spot 3) and the alpha phase matrix (spot 4).

Cracks and voids were observed in the brazing joint, mainly near or inside the eutectic microstructure, resulting from a significant difference between these eutectic areas and the rest of the solidified braze alloy (mainly Si). Voids also formed at the interface between alpha-Al and Si flakes inside the eutectic area marked as a yellow circle in the micrograph. Two large voids and cracks were found in the micrographs in Figures 4 (c) and (d).

Fine pores were detected in the SiC base substrate shown in Figure 6(e), which likely were formed during the production of the reaction bonded SiC. These micropores inside the SiC base substrate might lead to cracking, which will be discussed later.

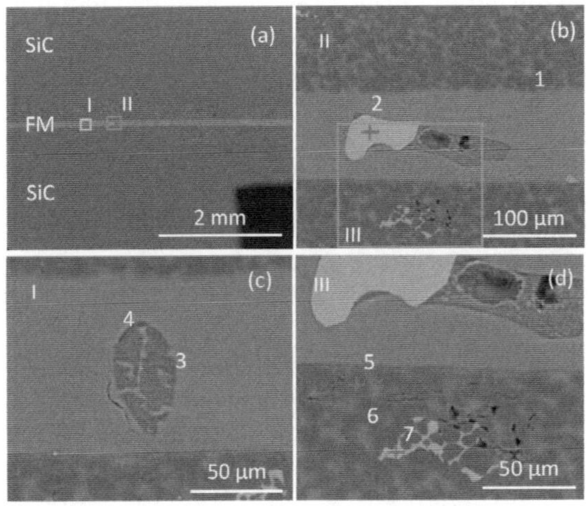

Figure 4: Braze joint: (a) Interface with magnification 25x; (b) Phase identification in area II; (c) Phase identification in area I; (d) Phase identification in area III.

Table I: EDS results and phase identification at braze joint interface.

At%	Spot 1	Spot 2	Spot 3	Spot 4	Spot 5	Spot 6	Spot 7
Si	98.65	66.15	89.03	1.50	67.72	98.98	66.16
C	0.63	0.70	0.00	0.00	31.67	0.00	4.85
Al	0.60	1.36	10.78	98.01	0.47	0.78	4.07
Ti	0.18	28.78	0.20	0.49	0.14	0.25	27.91
Phase	Si	$TiSi_2$	Si flake	α-Al	SiC	Si network	$TiSi_2$

Furthermore, the phases inside the SiC substrates were also analyzed; reaction bonded SiC is mainly formed by cell-like SiC grains embedded in Si network. SiC grains (spot 5) and Si network (spot 6) were found as expected, and $TiSi_2$ (spot 7) "network" was also observed. Decomposition of SiC and formation of carbides in the SiC base substrate were not observed;

therefore, the formation of $TiSi_2$ probably have originated from the reaction between the Si network and Ti. A reaction $Ti+Si=TiSi_2$ should have occurred when the parts were exposed to the brazing cycle, and Ti diffusion into the SiC substrate must have occurred at the brazing temperature. Also, it can be presumed that the $TiSi_2$ network shape resulted from the replacement of Si network by $TiSi_2$ after the metallurgical reaction between Ti and Si was completed as Ti diffusion occurred. Since the CTE mismatch between $TiSi_2$ and Si is not as significant as in the case of Al and Si, cracking was not formed at the interface between $TiSi_2$ and Si.

Shear testing results and discussion

The shear strengths of the five sets of brazed joints were measured by performing compression test at room temperature. A sudden fracture occurred at the peak load, indicating a brittle nature of the failure with very little plastic information; the relationship between phase distribution at the fracture surface and the fractographic characteristics of the sheared surface will be discussed later.

Average shear strengths for the braze joints under each brazing condition were given in Table II, Joints processed under isothermal holding at temperature of 1250 °C for 30 minutes had the highest shear strength of 88 MPa, and those joints processed at 1200 °C for 30 minutes and at 1250 °C for 60 minutes resulted in a shear strength of about 70 MPa, which was a little bit lower than those joints produced at 1250 °C for 30 minutes. Braze joints processed at of 1250 °C for 10 minutes and at 1300 °C for 30 minutes showed the lowest strength of around 43 MPa.

The relationships between shear strength and brazing conditions (temperature and holding time) were plotted in Figures 5 (a)-(b). The shear strength of the brazed joints reached a maximum in samples processed at 1250 °C held for 30 minutes. Both temperature and time affect significantly on the diffusion of certain elements during brazing, which affects the strength of the joint.

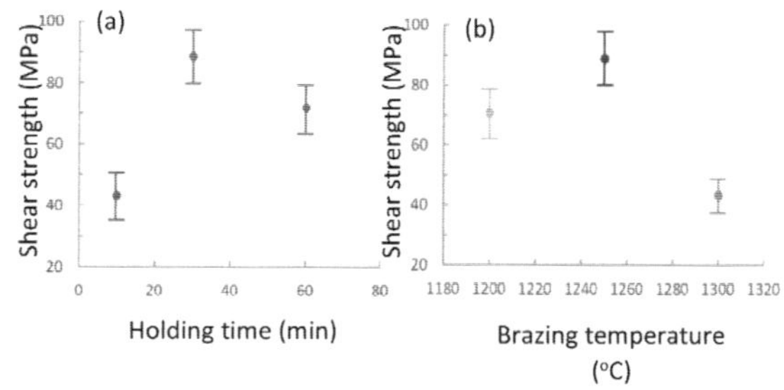

Figure 5: The relationship between (a) shear strength and holding time at a brazing temperature of 1250 °C; (b) shear strength and brazing temperature with 30 min holding time.

Table II: Shear strength of braze joints.

Brazing cycles	1200 °C 30 min	1300 °C 30 min	1250 °C 10 min	1250 °C 30 min	1250 °C 60 min
Avg. shear strength (MPa)	70.5	42.8	43.0	88.4	71.9

According to the shear testing data, isothermal holding time should be longer than 10 minutes. It was observed that increasing holding time from 10 minutes to 30 minutes at 1250 °C increased the shear strength of the joints from 43 MPa to 88 MPa; however, increasing holding time even further, the strength of the joints decreased slightly to 71.9 MPa, this result was still much higher than the strength obtained in samples processed with insufficient holding time. Also, as brazing temperature increased from 1200 °C to 1250 °C, the shear strength increased from 70.5 MPa to 88.4 MPa; however, increasing the temperature to 1300 °C resulted in lower shear strength, 42.8 MPa. The reason for the effect of temperature differed from that of holding time on brazing was that all selected brazing temperatures were high enough to ensure diffusion of elements into the base substrate during brazing. A large drop in shear strength occurred in samples processed with 30 minutes isothermal holding at 1300 °C, which probably resulted from excessive diffusion of braze elements into the substrate.

Low magnification SEM micrographs of fracture surfaces were analyzed in Figures 6, which illustrate the details of the fracture surfaces, including overlapping areas, spreading areas, brittle flat fracture ratio, and rougher fibrous fracture ratio. The total spreading area of filler metal was marked as the region inside the blue dot-dash line, the overlapping area was the area inside the yellow dash line, and the area between the two lines was the area of over-spreading of filler metal out of the overlapping region. The region marked by the green dashed line inside the yellow square was the brittle flat fracture zone and rest of the region inside the yellow square was the rougher fibrous fracture zone. The measured over-spreading area and brittle fracture fraction were listed in Table III. From the summary in Table III, joints with higher shear strength had a lower fraction of brittle flat-looking morphology at the fracture surface and better spreadability than those joints with lower strength.

Higher magnification micrographs of the fracture surfaces are shown in Figure 7 (a)-(b). The difference in fracture surface morphology between those samples with lower strength (Figure 7(a)) and higher strength (Figure 7(b)) was significant; the one with lower strength displayed a flat cleavage fracture surface, and the one with higher strength displayed a fibrous and rough fracture surface. The more fibrous in the fracture surface, the higher the joint shear strength was.

SEM and EDS were also used in the phase identification and microstructure characterization in the fracture analysis. The lighter grey color phases in all SEM micrographs were identified as $TiSi_2$, and the coarse grey feature with cracking was identified as Si. Al was mainly located along or close to cracks. Over 80% of the fracture surface was coarse Si phase. $TiSi_2$ and Al-rich zones were also found at the joint interface.

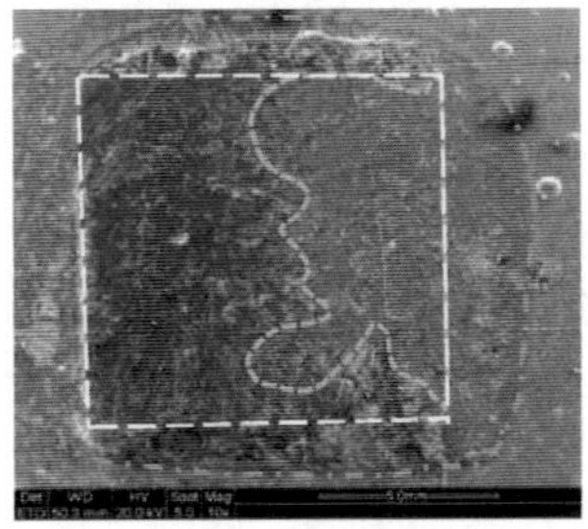

Figure 6: SEM images of the fracture surface of joint obtained at 1200 °C for 30 min.

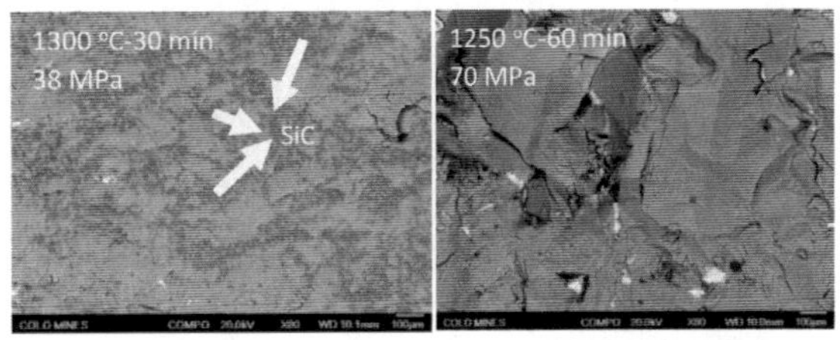

(a) (b)

Figure 7: Higher magnification micrographs of the fracture surface of joints obtained (a) at 1300°C for 30 min; (b) at 1250 °C for 60 min.

Table III: Measurements of spreading areas and ratio of flat brittle fracture.

Brazing cycles	1200 °C 30 min	1300 °C 30 min	1250 °C 10 min	1250 °C 30 min	1250 °C 60 min
Spreading area (mm^2)	64	42	47	87	82
Brittle fracture ratio (%)	37	95	51	12	43

The volume fractions of the different phases at the fracture surface were determined and the results are shown in Table IV. Braze joints with lower shear strength were found to fracture near the braze alloy/SiC interface. Braze joints processed at 1300 °C for 30 minutes and at 1250 °C for 10 minutes had 21 vol. % and 2 vol. % of SiC particles at the fracture surface, respectively.

Other braze joints fractured in locations inside the brazing gap, so no SiC was found at the fracture surfaces of these joints. The Al content in the joints with lower strength was higher than in the rest of the samples, this result was in accord with the previous analysis indicating that Al-rich zones were related to fracture sensitivity and crack initiation.

Table IV: Volume fraction of phases at the fracture surfaces of braze joints.

Vol. % of phases	1200 °C 30 min	1300 °C 30 min	1250 °C 10 min	1250 °C 30 min	1250 °C 60 min
Si	84	69	85	85	93
Alpha-Al	5	8	9	6	4
$TiSi_2$	11	2	4	9	3
SiC	-	21	2	-	-

Table V: Diffusion distance of FM into DAZ.

Brazing cycle	1200 °C 30 min	1300 °C 30 min	1250 °C 10 min	1250 °C 30 min	1250 °C 60 min
Diffusion distance (μm)	286	1190	246	295	358

Braze joints after shear testing were cross sectioned, grinded and polished to analyze the interaction zone, which showed any existing transverse cracking and the diffusion zone of braze alloy elements into the base substrate. As in previous microstructure analysis, the light grey color feature was identified as TiS_2, and the main grey feature was Si. A Diffusion Affected Zone (DAZ) was found in the SiC substrate in all braze joints. Table V shows the diffusion distance of FM into the substrate in braze samples processed under different brazing conditions.

It should be noted that the amount of braze alloy attached to the SiC base substrate varied in different braze joint samples. In joints with low shear strengths, which were brazed at 1300 °C for 30 minutes or at 1250 °C for 10 minutes, the amount of solidified braze alloy on the fracture surface was much less compared to that of the joints with shear strength higher than 70 MPa. Insufficient braze alloy on fracture surface indicated an insufficient interaction between the braze alloy and SiC substrate; in these samples, the fracture was initiated at the braze alloy/SiC interface. Higher amount of filler metal attached to the SiC substrate indicated a significant reaction between the filler metal and the substrate; in this case, the crack initiation was located inside the braze gap in the braze filler.

DAZ size was found to increase with increasing temperature and isothermal holding time; the DAZ was a result of the diffusion of Ti into the SiC substrate, this was also an indication of more interaction between the braze alloy and SiC substrate. Neither an over-interaction process nor an insufficient-interaction process between brazing alloy and base substrate can produce joints with maximum shear strength. A moderate DAZ size about 290-350 microns can produce braze joints with good bond strength higher than 70 MPa.

Comparing these chemical analysis results with the previous brazed joint interfacial microstructural analysis may suggest that the fracture of the joints should start in the solidified braze alloy instead of the SiC substrate. The previous microstructural analysis also showed the existence of cracks around the eutectic regions, and these cracks would be the initiation sites of failure during mechanical testing. Since eutectic microstructure contained a high content of ductile Al, crack propagation is expected to be avoided in this region; for that reason, low Al content (less than 5 at%) was observed on the fracture surface.

Transverse Cracking

Two types of transverse cracking were found in the cross-section of the braze joints, as shown in Figure 8. The first type of transverse cracking, about 130 microns, was found in the DAZ of the brazed joint produced at 1200 °C for 30 minutes; this type of cracking occurred inside the DAZ zone but it did not propagate through the braze alloy on the fracture surface. The second type of transverse cracking, of about 280 microns, went through both the braze alloy (Coarse Si phase) and the SiC substrate. Also, in both cases, the cracking propagated through the Si network instead of through SiC grains in the substrate, indicating that the Si network is more brittle than the cell-like SiC grains.

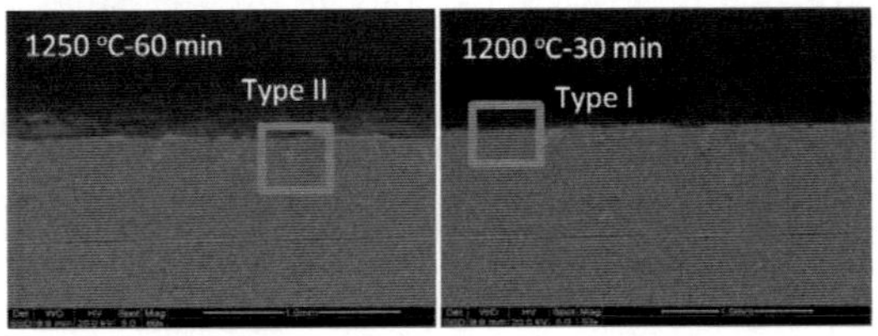

Figure 8: Transverse cracking in DAZ zone: (a) type II cracking; (b) type I cracking.

In the previous microstructure analysis, fine pores were found in the SiC base substrate resulting from RBSiC production, under loading condition, these pores can propagate and form transverse cracking in the base substrate. While for the second type of cracking, the initiation of cracking might initiate in the base alloy and propagated all the way through the Si work in the base substrate.

Diffusion of Ti during brazing

The schematic representation of Ti diffusion in the system is shown in Figure 9. Since no carbides or SiC decomposition were observed in this study, the reaction between the braze alloy and base SiC substrate can only be that between the active element, Ti, and the Si network in the RBSiC to form $TiSi_2$ through the chemical reaction $Ti+Si=TiSi_2$. Thus, the diffusion of Ti is suggested to follow the pattern illustrated in Figure 9 of titanium atoms moving through the Si network between SiC grains. If there happen to be fine pores in regions close to the interface in the base substrate, type I or type II transverse cracking can occur in the DAZ under loading conditions. If transverse cracking occurs in the braze alloy it might also propagate though Si network under loading conditions.

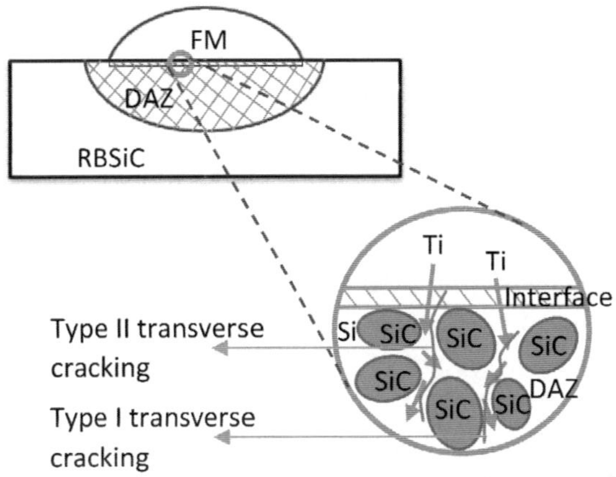

Figure 9: Schematic drawing suggesting the diffusion path of Ti into SiC substrate

Conclusions

In this study, the following conclusions can be drawn:

1). Consistent brazed joints were produced using SiAlTi filler metal to join RBSiC coupons at selected brazing cycles. The braze alloy used in the brazing process exhibited good wettability and spreadability on RBSiC.

2). Interfacial microstructure characterization showed that over 90 at% of coarse Si, some $TiSi_2$, and Al-rich eutectic zones existed at the brazing joints. Only $TiSi_2$ was found in the DAZ, indicating diffusion of Ti from the FM into SiC.

3). Voids and cracks were found at the phase boundaries between Al-rich zone and Si, or between the Al-rich zone and Si flakes inside the eutectic region after cross-sectioning. Cracks were mainly found along or close to the Al-rich eutectic zones. Formation of voids at the phase boundaries between Al/Si resulted from the significant CTE difference between these two; and the crack initiated at void alignment when the sample was under loading conditions are avoided by the ductile Al-rich zones.

4). Consistent braze joints processed at five different brazing cycles were produced for shear strength testing using brazing fixturing produced with a homemade shell investment casting mold and a compression steel fixture.

5). Among all five brazing processes, the joints produced at 1200 °C/30 min, 1250 °C/30 min, and 1250 °C/60 min were found to have a good bond strengths of over 70 MPa; and the highest bond strength of 88 MPa was obtained in joints produced at 1200 °C/30 min. The higher strength joints had lower brittle fracture ratio and more ductile fibrous morphology compared to those with lower bond strengths. The braze alloy spreadability of the higher strength joints was also better than that of the joints with lower strength.

6). Increasing both temperature and time promoted diffusion and increased interaction between the braze alloy and SiC substrate. To get braze joints with good strength, the holding time at brazing temperature should not be less than 10 minutes.

7). Over 80 vol. % of Si was found at the fracture surfaces, low Al and $TiSi_2$ were also found. Joints with a low shear strength tend to have higher Al content at the fracture surface since Al was related to crack initiation; Al was found to accumulate mostly in regions close to cracks and voids.

8). Joints with a moderate DAZ size of about 300 microns and a rough fracture surface morphology tend to have better strength than joints with insufficient braze/substrate interaction and with small DAZ, or excessive braze/substrate interaction resulting in large DAZ.

9). Two types of transverse cracking were identified. One was located inside the RBSiC, and the other propagated through both SiC and braze alloy. The initiation of transverse cracking might be related to intrinsic voids formed during SiC production, or it could form by the propagation of existing cracking through the Si network in the DAZ and inside the braze alloy.

10). Diffusion of Ti occurred along the Si network since no carbides and no decomposition of SiC was found in the DAZ.

Acknowledgements

This work was supported by Rolls-Royce Corporation. One of the authors (J. Wei) also received the Li Foundation scholarship. Additional support from the CSM Center for Welding, Joining and Coatings Research (CWJCR) is also acknowledged.

Reference

[1] Colombo, P., Sglavo, V., Pippel, E., et al. (1998). Joining of reaction-bonded silicon carbide using a preceramic polymer. Journal of Materials Science, 33, pp. 2405-2412.

[2] Rado, C., Kalogeropoulou, S., and N. Eustathopoulos, N. (1999). Wetting and bonding of Ni–Si alloys on silicon carbide. Acta Mater, 47(2), pp. 461-473.

[3] Xiong, H., Mao, W., Xie, Y., Guo, W., Li, X. and Cheng, Y. (2007). Brazing of SiC to a wrought nickel-based superalloy using CoFeNi(Si, B)CrTi filler metal. Materials Letters, 61, pp. 4662-4665.

[4] Hernandez, X., Jimenez, C., Mergia, K. et al. (2014). An Innovative Joint Structure for Brazing Cf/SiC Composite to Titanium Alloy. Journal of Materials Engineering and Performance, 23(8), pp. 3069-3076.

[5] Blugan, G., Janczak-Rusch, J. and Kuebler, J. (2004). Properties and fractography of Si_3N_4/TiN ceramic joined to steel with active single layer and double layer braze filler alloys. Acta Materialia, 52, pp. 4579-4588.

[6] Singh, M., Martinez Fernandez, J., Asthana, R. et al. (2012). Interfacial characterization of silicon nitride/silicon nitride joints brazed using Cu-base active metal interlayers. Ceramics International, 38, pp. 2793-2802.

[7] Matteo Galli, John Botsis, Jolanta Janczak-Rusch, et al. (2009). Characterization of the Residual Stresses and Strength of Ceramic-Metal Braze Joints. Journal of Engineering Materials and Technology, 131, pp. 021004–1-021004–8.

[8] Kar, A. and Ray. A. (2007). Characterization of Al_2O_3–304 stainless steel braze joint interface. Materials Letter, 61, pp. 2982-2985.

[9] Lee, H. and Lee, J. (1996). A study of the wetting, microstructure and bond strength in brazing SiC by Cu-X (X = Ti, V, Nb, Cr) alloys. Journal of Materials Science, 31(15), pp. 4133-4140.

[10] Liu, Y., Zhang, L., Liu, C., Yang, Z., Li, H. and Feng, J. (2011). Brazing C/SiC composites and Nb with TiNiNb active filler metal. Science and Technology of Welding and Joining, 16 (2), pp. 193-198.

[11] Xiong, J., Huang, J., Zhang, H. and Zhao, X. (2010). Brazing of carbon fiber reinforced SiC composite and TC_4 using Ag–Cu–Ti active brazing alloy. Materials Science and Engineering A, 527, pp. 1096-1101.

[12] Prakash, P., Mohandas, T. and Dharma Raju, P. (2005). Microstructural characterization of SiC ceramic and SiC–metal active metal brazed joints. Scripta Materialia, 52, pp. 1169-1173.

[13] He, Y., Zhang, J., Liu, C. and Sun, Y. (2010). Microstructure and mechanical properties of Si_3N_4/Si_3N_4 joint brazed with Ag–Cu–Ti + SiC_p composite filler. Materials Science and Engineering A, 527(12), pp. 2819-2825.

[14] Galli, M., Botsis, J. and Janczak–Rusch, J. (2006). Relief of the Residual Stresses in Ceramic-Metal Joints by a Layered Braze Structure. Advanced Engineering Materials, 8(3), pp. 197-201.

[15] Zhong, Z., Zhou, Z. and Ge, C. (2009). Brazing of Doped Graphite to Cu Using Stress Relief Interlayer. Journal of Materials Processing Technology, 209, pp. 2662-2670.

[16] Ormston, D. (2001). The particulate reinforcement of active braze alloys for joining silicon carbide. University of Cambridge, Cambridge, UK.

[17] Lee, H., Hwang, S. and Lee, J. (1993). Effects of the relative contents of silver and copper on the interfacial reactions and bond strength in the active brazing of SiC. Journal of Materials Science, 28, pp. 1765-1774.

[18] Liu, Y., Huang, Z. and Liu, X. (2009). Joining of sintered silicon carbide using ternary Ag–Cu–Ti active brazing alloy. Ceramics International. 35, pp. 3479-3484.

[19] Paulasto, M., Ceccone, G., Peteves, S., Voitovich, R. and Eustathopoulos, N. (1997). Brazing of Si_3N_4 with Au–Ni–V–Mo filler alloy. Ceramic Transactions, 77, pp. 91-98.

[20] Xiong, H., Li, X., Mao, W. and Cheng, Y. (2003). Wetting behaviour of Co-based active brazing alloys on SiC and the interfacial reactions. Materials Letters, 57, pp. 3417-3421.

[21] Xiong, H., Mao, W., Xie, Y., Chen, B., Guo, W., Li, X. and Cheng, Y. (2007). Control of interfacial reactions and strength of the SiC/SiC joints brazed with newly-developed Co-based brazing Alloy. International Journal of Materials Research, 22, pp. 2727-2736.

[22] Riccardi, B. Nannetti, C., Woltersdorf, J., Pippel, E. and Petrisor, T. (2002). Brazing of SiC and SiCf/SiC composites performed with 84Si–16Ti eutectic alloy: microstructure and strength. Journal of Materials Science, 37, pp. 5029-5039.

[23] Li, J., Liu, L., Wu, Y., Zhang, W. and Hu, W. (2008). A high temperature Ti–Si eutectic braze for joining SiC. Materials Letters, 62, pp. 3135-3138.

[24] Li, J., Liu, L., Wu, Y., Li, Z., Zhang, W. and Hu. W. (2009). Microstructure of high temperature Ti-based brazing alloys and wettability on SiC ceramic. Materials Design, 30, pp. 275-279.

[25] Naka, M., Taniguchi, H. and Okamoto, I. (1990). Heat-resistant brazing of ceramics (report I): brazing of SiC ceramic using Ni-Ti filler metals. Transactions of Japanese Welding Research Institute, 19, pp. 25-31.

Accelerated and Directed Diffusion during Electric Current-Assisted Brazing

K. Bobzin, M. Öte, S. Wiesner, A. Schmidt
Surface Engineering Institute, RWTH Aachen
schmidt@iot.rwth-aachen.de

Abstract

Transient liquid phase (TLP) bonding, a diffusion controlled joining process, results in remelting temperatures of the brazed joints which are higher than the brazing temperature itself. One of the major disadvantages of this process is the long brazing durations, which lead to an undesired heat treatment of the base material and thus -depending on the base material- eventually to undesired grain growth and phase transformation. In cases of shorter brazing time, intermetallic compounds may form in the filler metal or in the reaction zone. Thus, it can be stated that process requirements regarding base material and the formation of the brazed seam are partly in contradiction.

Inspired by research from the electronics industry that describes direct current induced diffusion in soldered joints, the focus of this work is set on an innovative in-line approach to influence the diffusion process in brazing by applying an external electric current. In this work, possibilities of accelerated and directed diffusion by applying electric current on the joint are analyzed. Furthermore, a concept for implementation of the current-assisted diffusion in a brazing process and a possible multistage electric current-assisted TLP bonding process are introduced.

Introduction

The influence of electric current on soldered joints is well known in the electronics industry since many decades. The latest results demonstrate that the morphology of joints can be significantly influenced via direct current. In their respective work, Chen et al. show the possibility to enhance or retard the growth of intermetallic compounds Ag_3Sn and Ni_3Sn_4 in the interface of soldered joints depending on the current direction with the material combinations of Sn/Ag and Sn/Ni [1]. Among other applications, it is thus possible to support or suppress the formation of intermetallic phases. The electric current also has an influence on the type and size as well as the distribution of the intermetallic phases [1-6].

Recent studies have verified that a similar effect can also be achieved in case of brazing process as a result of application of an external current. In contrast to the electrically induced diffusion in the electronics industry, which takes place at high current densities of 10^2 A/cm^2 $\leq J \leq 10^6$ A/cm^2 as well as at relatively low temperatures, similar effects could be identified in the case of brazing process at significantly lower current densities and in the liquid state of the braze [7].

Effect of Current on Diffusion Processes

In the electronics industry, diffusion processes supported with an electric current are known as an unwanted result from high current density in circuit paths, leading to material transport. This can result in an accumulation of material, the creation of vacancies and the formation of new phases. The microstructure of a material plays an important role in this process since material transport preferably takes place at outer or inner surfaces, interfaces between two different materials, along grain boundaries and at lattice defects, e.g. vacancies [8]. In all cases, the material is transported along the direction of the electric current. Consequently vacancies wander against the current flow direction.

The driving force of electrically induced diffusion processes is described as being composed of two different force components; an electrostatic force and an "electron wind". In this context, the electrostatic force is caused by the influence of an external electric field on the metal and the electron wind is a force from the exchange of momentum with other charge carriers, toward the flow of electrons. The electrostatic force can take effect in any direction, depending on the external electric field. By contrast, the electron wind always is in the direction of current flow and usually is the dominating force of the named two force components [8]. Therefore, as stated previously, the material transport is also predominantly in direction of the current. The extent of electrically induced diffusion depends on a number of physical, geometric and metallurgical factors, such as temperature and resistivity. The extent of the diffusion processes increases with the increase of temperature and electrical resistivity. Further, the chemical composition of alloys has an influence on the extent of diffusion of the respective elements. Finally, the geometric size, i.e. the distance from anode to cathode, is also an influencing factor, where a larger distance mainly causes a decrease in electrically induced diffusion [9].

State of the Art

A recent study conducted at the Institute of Materials Engineering at the Technical University Dortmund, aiming on in situ observation of the brazing process via electric resistance measurements along the joining gap, shows that a similar effect to current-driven diffusion in the electronics industry, might be also taking place in the case of brazing process. During the measurements of electrical resistance, it has been realized that the measurement current influences the diffusion process and phase formation in the brazed joint. The diffusion processes were significantly influenced by the electric current during the in situ observation of the electric resistance of the joint. Furthermore, it has been pointed out that the diffusion processes were influenced by the measurement current direction. A significant effect of the electric current on the phase formation was reported in [7]. The described effect was observed while brazing pure iron and cemented carbides. As filler metal copper was used. The experiments were conducted in a vacuum furnace at $T = 1{,}150$ °C in an Argon atmosphere with a pressure of $p = 10^{-2}$ mbar.

Figure 1 shows details of the interface between the cemented carbide substrate and the filler metal of samples brazed without and with electric current. Without electric current, iron diffuses through the copper filler metal to the cemented carbide. Consequently, brittle η-carbides were formed at the interface between filler metal and cemented carbide.

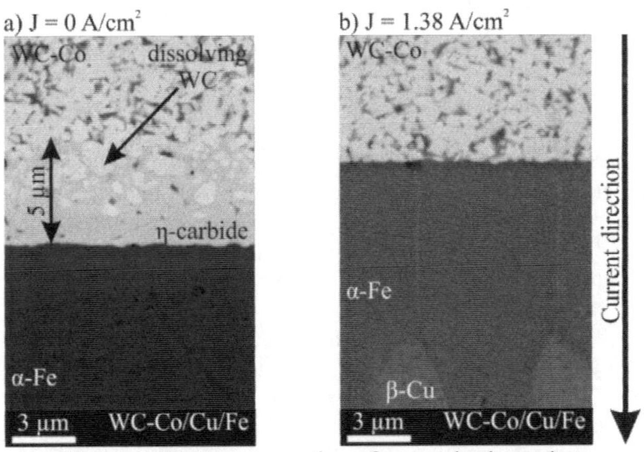

Figure 1: SEM cross section of a samples brazed at $T = 1{,}150$ °C for $t = 5$ min; a) without an application of electric current, formation of brittle η-carbides; b) with a direct current density of $J = 1.38$ A/cm² with a current direction from cemented carbide to iron [7]

When a direct current was applied with a current density of $J = 1.38$ A/cm² with a current direction from cemented carbide to iron, no formation of η-carbides could be observed, Fig. 1 b). Based on the microstructure of the whole brazed joint, it can be assumed that the diffusion of the copper was enhanced and the diffusion of the iron towards the cemented carbide was slowed down by the influence of electric current. Thus, the formation of the η-carbides was prevented.

Electric Current-Assisted Brazing (ECAB) in Transient Liquid Phase (TLP) Bonding

It has been already reported in [7] that for the material combination cemented carbide/copper/iron, the diffusion and consequently the microstructure of the brazed joint can be influenced by electric current. In particular, the phase formation as well as the brazing seam thickness of the brazed joint has been significantly affected.

A nickel-based filler metal is used commonly for TLP bonding. In order to achieve better material properties, such as strength, hardness, corrosion and oxidation resistance, other elements such as iron, copper, manganese and chromium are often alloyed. Additional alloying elements such as boron, silicon or phosphorus are added to the nickel base to lower the melting point of the filler metal [10-11]. Due to the interactions of the elements, undesirable intermetallic compounds, in particular Cr-borides, are formed in the middle of the brazed joint or at the interface between filler metal and base material [11-12]. In order to prevent the formation of undesired phases, a long brazing time can be used. This results in grain growth and phase transformation of the base material. Thus, process requirements regarding base material and the formation of the brazed seam are usually contradictory. A way to reconcile these contradictions is the use of the described effect of current-assisted diffusion. Due to controlled diffusion, most of the elements forming brittle phases can be guided into the base material. Thus, the formation of deleterious intermetallic phases in the joint can be suppressed. Additionally, a significant reduction in brazing duration can be achieved as a result of accelerated diffusion of some of the elements. Thus, the holding time can be reduced and the effect of the undesired heat treatment of sensitive base materials during brazing process can be minimized.

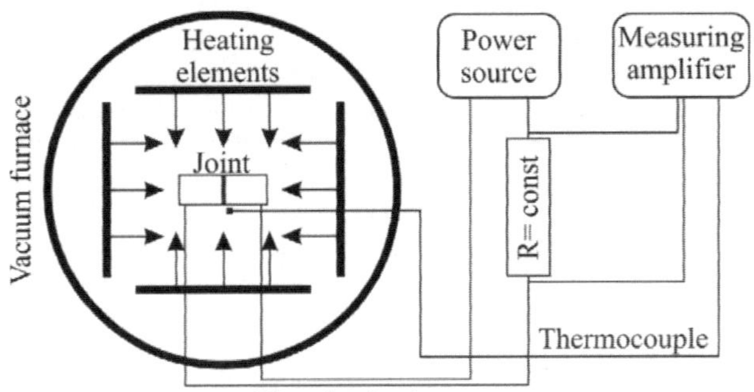

Figure 2: Schematic sketch of the experimental setup for current-assisted brazing

Since this approach shows high potential but has not yet been studied or proven to be effective in brazing technology, an experimental setup was realized to investigate electric influences on the diffusion. Figure 2 shows a schematic sketch of the experimental setup. Thus, it is possible to carry out current-assisted brazing experiments in a vacuum furnace. The AC/DC power source is programmable. In AC mode, different current frequencies can be adjusted. The measuring amplifier can be used to measure temperature and electric current simultaneously. With this setup, the influence of a combination of brazing parameters with different electric factors on the microstructure of the joint can be determined by means of subsequent SEM investigations. To prevent the influence on the temperature measurement, the thermocouple needs to be isolated electrically. Due to the electrical isolation properties together with sufficient thermal conductivity of the material, an alumina shell is selected for this purpose.

Since pure nickel has both sufficient electrical conductivity even at high temperatures, and a high melting point at the same time, it is used as the electrical conductor. In order to avoid an electrical short circuit in the vacuum furnace, the conductors are electrically isolated with alumina ceramic. A device to fix the samples and to ensure a secure and stable contact of the joint was developed, see Fig. 3. The base is electrically conductive. The use of alumina ceramic linear guides ensures that the current passes through the joint and allows the use of different sample geometries at the same time.

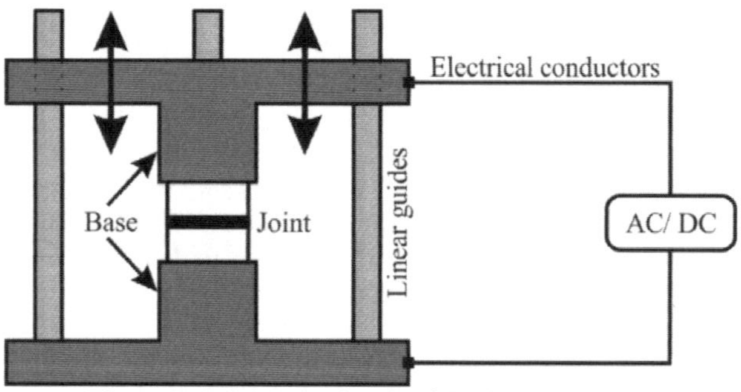

Figure 3: Schematic sketch of the holding device

In the next step, the influence of possible electric influencing factors in combination with brazing parameters on the formation of joints made of different material combinations will be examined. In detail, the influence of DC and AC current density on the microstructure of the joint will be determined. Furthermore, the influence of current frequency on the diffusion processes will be examined in the case of alternating current.

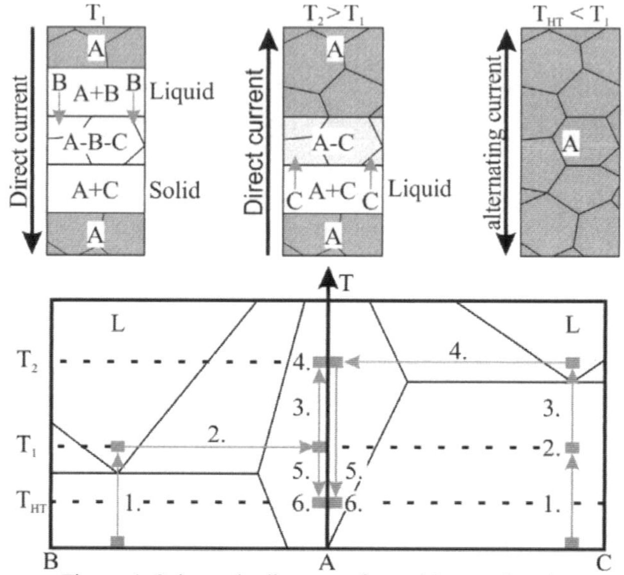

Figure 4: Schematic diagram of a multistage electric current-assisted TLP bonding process

Based on the results, a multistage electric current-assisted TLP bonding process is developed. Figure 6 introduces the conceptual design of such a process. Three types of filler metals with similar chemical composition are to be applied between joining partners. The melting points of the filler metals shall be different according to the chemical composition of the filler metals. In the first step, the upper filler metal A + B melts at T_1. During the brazing time carried out in the second step, the alloying element B diffuses -supported by external direct electric current- from the upper into the middle filler metal. Due to the diffusion, the chemical composition of the upper filler metal approaches that of the base material A and the filler metal solidifies isothermally. Since the lower filler metal is in a solid state, it has not yet wetted the base material or the middle filler metal. Thus, there are no interactions between these joining partners and the composition of this filler metal has not changed yet.

In the third step, the system is heated to the temperature T_2. The upper filler metal remains solid and the lower filler metal A + C melts. During the subsequent brazing time at T_2, the lower filler can wet the base material and the middle filler. By applying a direct current, this time in the opposite direction, C diffuses into the middle filler metal, whereby the chemical composition of the lower filler metals is adapted to that of the base material, leading to isothermal solidification. After the fifth step, the cooling of the joint to T_{HT}, the joint is heat-treated in the sixth step and by applying an alternating current, an accelerated homogenization step is carried out.

Summary and Conclusion

In this work the influence of electric current density in circuit paths on diffusion and phase formation was analyzed. Similar effect was shown for the case of measurement current during in situ resistance measurement of joints in a brazing processes.

Based on this, a concept for the implementation of the electric current assisted diffusion in a brazing process and conceptual design of a multistage electric current-assisted TLP bonding process was introduced. The presented experimental setup has already been realized, Fig. 5 and 6.

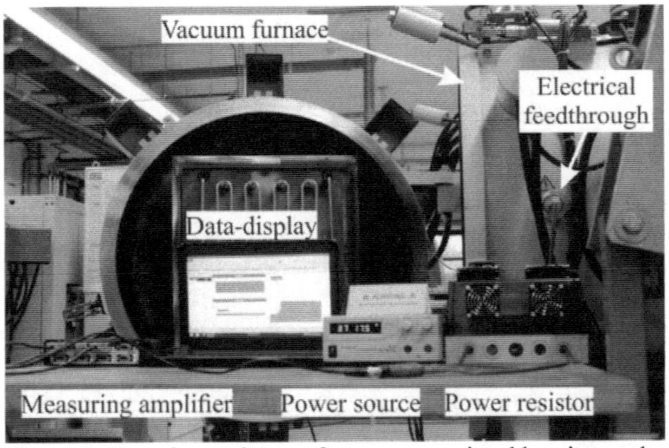

Figure 5: Experimental setup for current-assisted brazing at the Surface Engineering Institute

The above-mentioned experimental setup opens up new possibilities to examine the influence of the electrical factors on diffusion processes and the potential of ECAB. Based on this, the introduced multistage current-assisted brazing process will be developed in order to improve the quality of joints in TLP bonding. The methodologies developed for (a) accelerated (b) directed and (c) selective diffusion will be used to avoid the formation of intermetallic compounds as well as to suppress the grain growth in base materials, i.e. hot work tool steels.

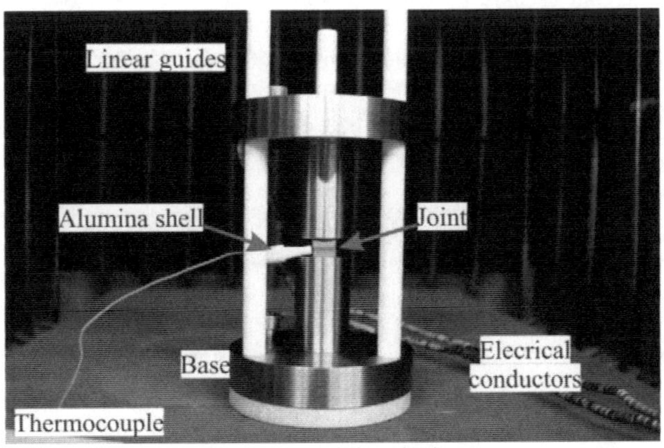

Figure 6: Holding device

Acknowledgments

All presented investigations were conducted in the context of the Collaborative Research Centre SFB1120 "Precision Melt Engineering" at RWTH Aachen University and funded by the German Research Foundation (DFG). For the sponsorship and the support we wish to express our sincere gratitude.

References

[1] Chen, C. M. and Chen, S.-W., "Electromigration effect upon the Sn/Ag and Sn/Ni interfacial reactions at various temperatures," *Acta Materialia,* Vol. 50, No. 9 (2002), pp. 2461-2469. ISSN: 1359-6454

[2] Chiu, T. C. and Lin, K. L., "The difference in the types of intermetallic compound formed between the cathode and anode of an Sn-Ag-Cu solder joint under current stressing," *Intermetallics,* Vol. 17, No. 12 (2009), pp. 1105-1114. ISSN: 0966-9795

[3] Wang, C. H., Kuo, C.Y., Chen, H. H. and Chen S. W., "Effects of current density and temperature on Sn/Ni interfacial reactions under current stressing," *Intermetallics,* Vol. 19, No. 1 (2011), pp. 75-80. ISSN: 0966-9795

[4] Yamanaka, K., Tsukada, Y. and Suganuma K., "Electromigration effect on solder bump in Cu/Sn-3Ag-0.5Cu/Cu system," *Scripta Materialia,* Vol. 55, No. 10 (2006), pp. 867-870. ISSN 1359-6462

[5] Zhang, L., Wang, Z. G. and Shang, J.K., "Current-induced weakening of Sn3.5Ag0.7Cu Pb-free solder joints," *Scripta Materialia,* Vol. 56, No. 5 (2007), pp. 381-384. ISSN: 1359-6462

[6] Zhang, X. F., Guo, J. D. and Shang, J. K., "Abnormal polarity effect of electromigration on intermetallic compound formation in Sn-9Zn solder interconnect," *Scripta Materialia,* Vol. 57, No. 6 (2007) pp. 513–516. ISSN: 1359-6462

[7] Sievers, N., Die elektrische Widerstandsmessung zur zerstörungsfreien Prüfung von Lötnähten am Beispiel von Hartmetall-Stahl-Verbunden, Vulkan Verlag (Essen, 2017), pp. 154-166. ISBN: 978-3-8027-8822-2

[8] Reiss, G. and Schürer, H., "Elektromigrationsvorgänge," *Physik in unserer Zeit,* Vol. 2 (1997), pp. 82-87.

[9] Hsieh, M. Y., Huntington, H. B., "Electromigration of copper in lead," *Journal of Physics and Chemistry of Solids,* Vol. 8 (1978), pp. 867-871.

[10] Cook, G. O. and Sorensen, C. D., "Overview of transient liquid phase and partial transient liquid phase bonding," *Journal of Materials Science,* Vol. 46, No. 16 (2011), pp. 5305-5323. ISSN: 0022-2461

[11] Hoyer I. and Wielage B., "Brazing of nickel-based filler metals for pipes and other components in contact with drinking water," *Advances in brazing: Science, Technology and Applications,* (2013), pp. 545-565. ISBN: 978-0-85709-423-0

[12] Bach, F. W., Möhwald, K., Holländer, U., Schaup, J., Langohr, A. and Roxlau, C., "Boron and phosphorous free nickel-based filler metals for brazing stainless steel in shielding gas furnaces," *International Journal of Materials Research,* Vol. 102, No. 8 (2011), pp. 964-971. ISSN: 1862-5282

Repairing oxidized surface of heat resistant steels by micro-plasma arc welding with transition metals

Yosuke Tajiri, Takahisa Yamazaki

Tokyo Institute of Technology, 2-12-1, O-okayama, Meguro ward, Tokyo, 152-8550, Japan

E-mail: tajiri.y.aa@m.titech.ac.jp, Tel: +81-3-5734-3236

Abstract

For repairing the oxidized surface without removing the crack and damaged substrate, active filler was applied for infiltration to heat resistant steels in AA-TIG welding. The reason why the process of gouging was lacked is due to application for vacuum container and nuclear container in situ repairing at low temperature. ANSI 316L steel is superior to corrosion resistance, mechanical properties and weldability. When Si flux was applied to the surface of ANSI 316L steel in AA-TIG welding as submerged arc welding, Si solved into the molten pool by Marangoni convection and the formation of MnO was depressed. On the other hand, SiO_2 flux was applied, Marangoni convection reversed and SiO_2 spread over the oxides of the surface and covered HAZ then the formation of MnO was depressed. In this study, micro-plasma arc welding was conducted at low temperature to ANSI 316L steel on crack forming surface using Zr-Ti based powder filler containing copper as active metal. EPMA analyses would be carried out for the examine the correlation with copper and oxygen.

Introduction

ANSI 316L steel and zircaloy-4 are used for vacuum container and nuclear container from the viewpoint of high temperature resistance, corrosion resistance and mechanical properties. Once broken, however, it is difficult to repair because it is often used at low temperature and the process of gouging is lacked. Therefore, there is a need for a technique that can repair damaged surface without thermal approach to damaged parts.

Figure 1 shows temperature characteristics of Marangoni convection. Marangoni convection is flowing from lower surface tension to higher surface tension. For materials with negative temperature gradient in surface tension, Marangoni convection flows from inside of a molten pool at high temperature to the outside at low temperature. However, when elements such as oxygen and sulfur dissolve in a molten pool, these elements act as surface active elements and reverse the temperature gradient of surface tension. The AA-TIG welding is a welding method that utilizes this phenomenon to obtain a deep narrow molten pool shape.

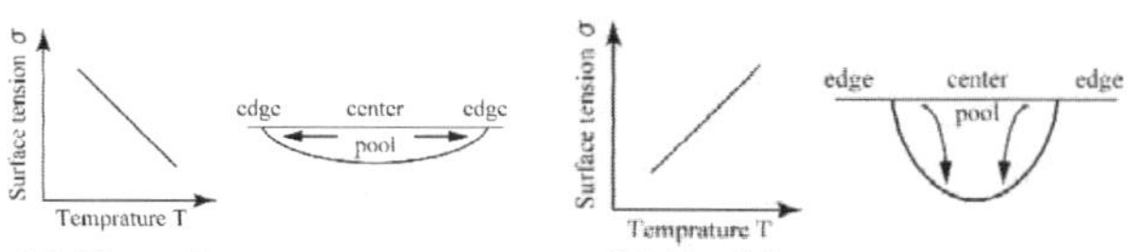

(a) Negative temperature gradient in surface tension **(b) Positive temperature gradient in surface tension**

Fig.1 Temperature characteristics of Marangoni convection

It is known that when AA - TIG welding is applied to ANSI 316L steel, oxygen contained in the shielding gas reacts with manganese in ANSI 316L steel to form manganese oxide. Corrosion resistance is impaired by the formation of manganese oxide and the quality of the product deteriorates. Therefore, it was found that formation of manganese oxide can be suppressed by coating Si flux on the surface of ANSI 316L steel, but Si solved into the base metal due to Marangoni convection and changed the mechanical properties and corrosion resistance of the welded part to get away. On the other hand, when SiO_2 flux was applied, manganese oxide could be formed in the same way, furthermore it was found that there was no invasion into the base material of Si and spread over the surface[1].

In this study, as shown in Table 1, using the active metal filler containing titanium and zirconium which are higher in surface energy than silicon, utilizing the phenomenon that the brazing filler metal spreads on the surface of the base material due to Marangoni convection, the damaged heat resistance Propose steel repair technology. For the base material, ANSI 316 L steel and zircaloy-4 were used. Plasma arc was applied to the base material which simulated cracks on the surface. The behavior of the melted brazing material was confirmed with an optical microscope and the intermetallic compound formed between the brazing material and the base material was investigated by EPMA analysis.

Table 1: Surface Energy of Titanium and Zirconium

	Surface Energy (mJ/m^2)	Reference
Si	865	[2]
Ti	1402 ~ 1675	[3]
Zr	1400 ~ 1543	[4]

Experiment

Sample materials were ANSI 316L steel, zircaloy-4 and zirconium-titanium based powder filler. Table 2 shows the composition of ANSI 316L steel, Zircaloy-4 and Zirconium-Titanium based powder filler.

Figure 2 shows schematic diagram of ANSI 316L steel and zircaloy-4 used in experiment as based material. ANSI 316L steel and Zircaloy-4 were disc with diameter of 25 mm and thickness of 1 mm. In order to simulate cracked surface, grooves having width of 0.3 mm and depth of 0.5 mm were dug at equal intervals on surface of ANSI 316L steel and zircaloy-4.

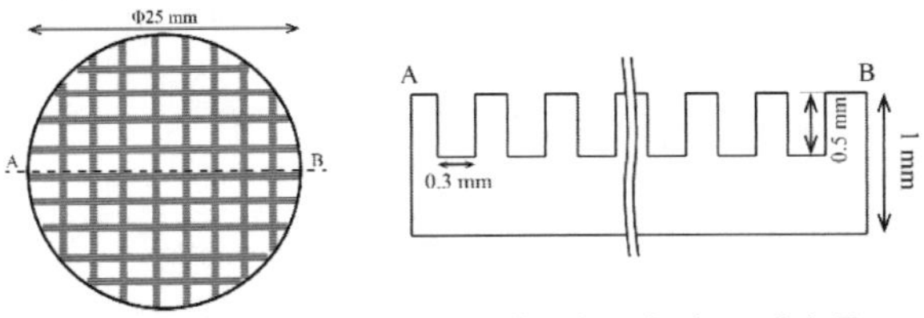

(a)Disk top **(b)Sectional view of A-B**

Fig.2 Schematic diagram of based materials

Zirconium-titanium based powder filler has particle size of 10 μm to 100 μm and melting point is 842 °C. Table 2 shows the composition of ANSI 316L steel, zircaloy-4 and Zirconium-titanium based powder filler.

Figure 3 shows schematic diagram of arc melting furnace. The arc melting furnace consists of vacuum chamber, water-cooled electrode, handle and water-cooled crucible. The water cooling electrode is tungsten and the water cooling crucible is copper. The inside of the chamber could be observed from the observation window. Condition of current was 50 A and condition of voltage was 100 V. After setting the base material in the water-cooled crucible, sprinkle powder filler on the base material as shown in figure4. After completing the setting of the base material and the brazing filler metal, close the chamber and start evacuation. Vacuuming inside the chamber was initiated with a rotary pump, and when the inside of the chamber became 5 Pa or less, it switched to a diffusion pump. After turning on the power supply, bring the electrode into contact with the water-cooled crucible to generate arc plasma. At this time, in order to stabilize the arc, the distance between the tip of the tungsten electrode and the water-cooled crucible should be about 5 mm.

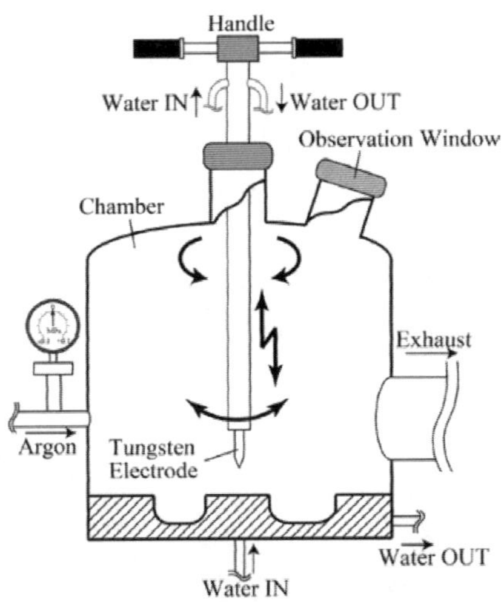

Fig.3 Schematic diagram of arc melting furnace

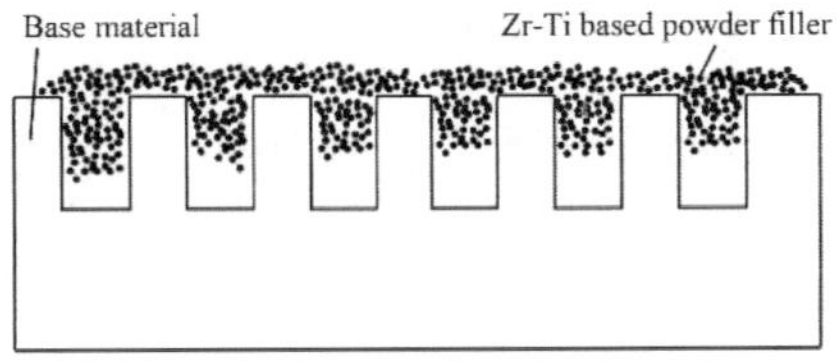

Fig.4 Set of powder filler

Experiment was conducted in argon atmosphere. In order to lower oxygen partial pressure, repeat the following procedure three times. First, the chamber was evacuated until it achieved 10^{-3} Pa. Second, it was filled with argon gas until it achieved 0.15 MPa.
After it was filled with argon gas, arc discharge was started.

Specimen was cut and specular polishing was carried out in order to observe the cross section. Behavior of the brazing filler metal was confirmed with an optical microscope and then element analysis was carried out by EPMA analysis.

Results and Discussions

Figure 5 shows groove cross section of ANSI 316L steel taken with optical microscope. Although it could be confirmed that the brazing material adhered to surface of ANSI 316L steel but dipping wetting could not be confirmed. Titanium and zirconium were expected to adhere more to ANSI 316L steel surface than silicon because of higher surface energy than silicon, but spreading like silicon was not confirmed. It was considered that titanium and zirconium were not actively acting on the oxide on the surface.
When repairing ANSI 316L steel, it was thought that silicon is better than titanium or zirconium. However, if the amount of silicon was too large, silicon will melt into the inside of molten pool and the quality of the bond part lowered, so it was necessary to consider the amount of silicon.

Fig.5 Cross section of ANSI 316L steel

Table 2: Chemical composition of base material and active metal filler (weight percent)

	Cr	Ni	Mn	Si	C	Mo	Fe	Zr	Cu	Ti	Sn	Hf
ANSI 316L steel	17.18	12.65	1.58	0.54	0.009	2.06	Bal.	-	-	-	-	-
Zircaloy-4	0.1	-	-	-	-	-	0.21	Bal.	-	-	1.45	0.01
Zr-Ti Based powder filler	-	10	-	-	-	-	-	37.5	15	37.5	-	-

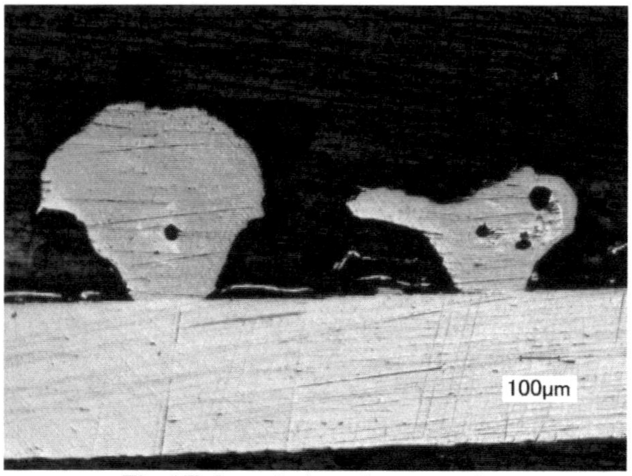

Fig.6 Enlarged view of cross section

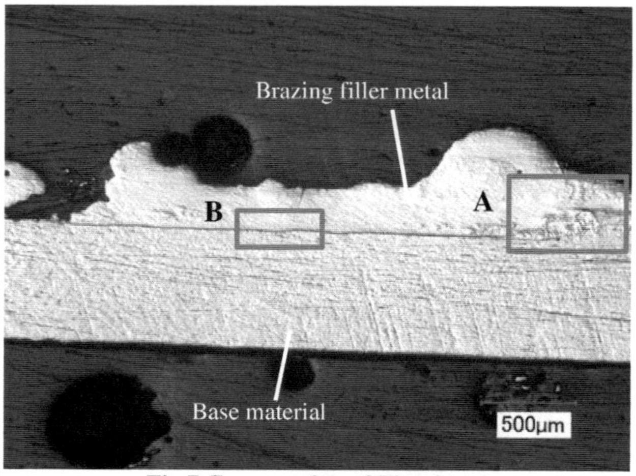

Fig.7 Cross section of Zircaloy-4

Figure 7 shows groove cross section of Zircaloy-4 taken with optical microscope. Since the contact angle of the brazing filler metal is $0 < \theta < 90\,°$, it was considered that both adhesion wetting and dipping wetting of the brazing filler metal occurred. Intermetallic compound was formed between base material and filler material. In area-A, base metal melted and metallic structure grew. In area-B, intermetallic compound was formed but metallic structure growth was not observed.

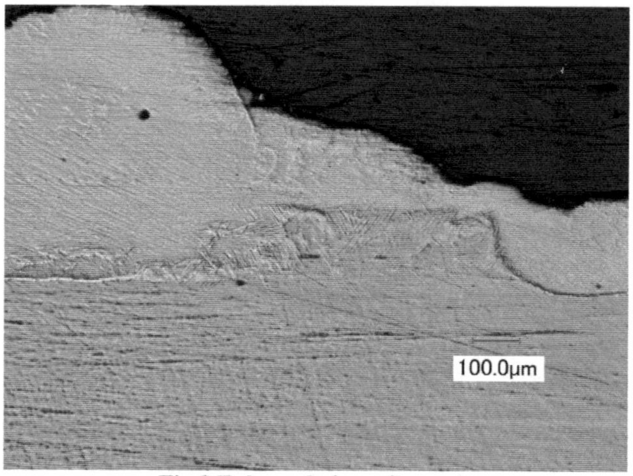

Fig.8 Enlarged view of area A

Figure 8 shows enlarged view of area A. In the area A, acicular structure was observed. The binary phase diagram of titanium and zirconium was classified as complete solid solubility type. This acicular structure was considered to be acicular α structure which appeared when heat treatment was performed in β phase. In this micro-structure, the formation tendency of micro-cracks was strong, stress relaxation was caused, and toughness and fatigue crack propagation resistance increase[5]. Even in zirconium alloys, acicular α-structure appeared when heat treatment was performed in β phase like titanium alloy. The acicular structure on the base material side was thought to be acicular α structure when heat was transferred from the arc, the temperature of the base material reached the β phase, and then appeared as it was cooled down.

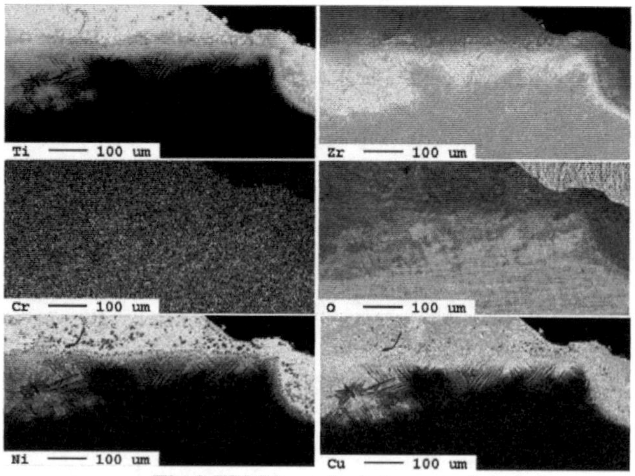

Fig.9 EPMA analysis in area-A

Figure 9 and 10 show result of EPMA analysis. Elements of copper and nickel were confirmed in the acicular structure on the brazing material side. Figure 11 shows binary phase diagram of zirconium and copper. Figure 12 shows binary phase diagram of titanium and copper. It was considered ZrCu and $TiCu_2$ were formed. Also in Cu and Ni, since the binary phase diagram is classified as complete solid solubility type, intermetallic compound did not form.

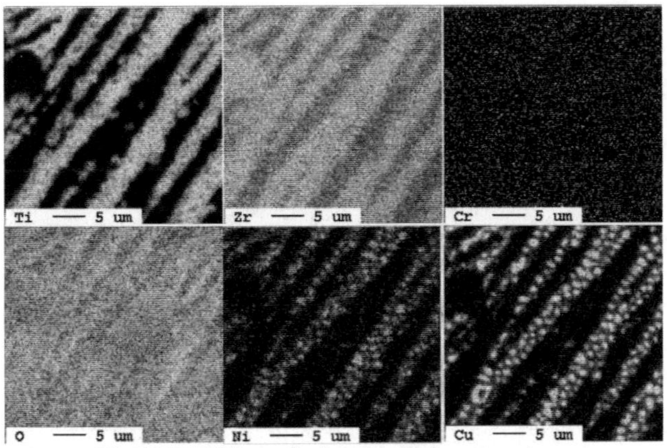

Fig.10 EPMA analysis of acicular structure

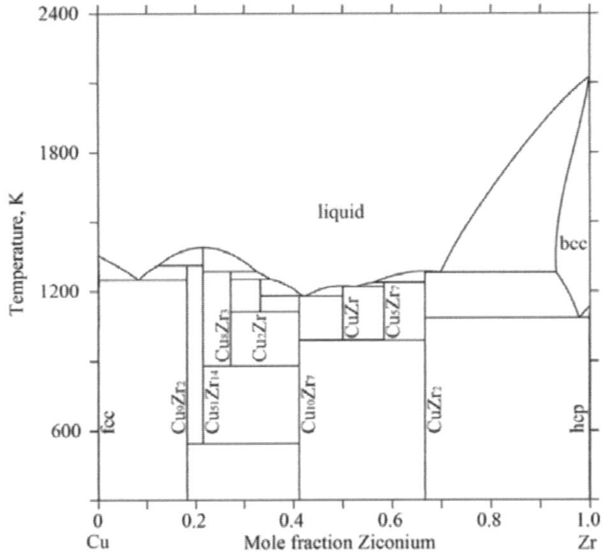

Fig.11 Binary phase diagram of Cu-Zr [6]

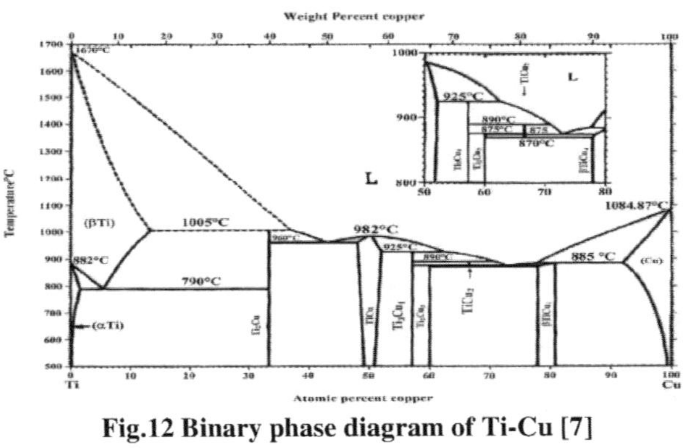

Fig.12 Binary phase diagram of Ti-Cu [7]

Figure 13 shows enlarged view of area B. Area B was thought to be the part into which molten brazing material flew from area A by Marangoni convection. However, almost no acicular structure was confirmed in Area B. It was thought that acicular α structure was formed in area A because heat from the arc was sufficiently transmitted from the surface of base material in Area A, but in area B, only heat from the molten brazing material was transmitted to surface of the base material. It was considered that temperature did not reach the β phase, so acicular α structure was not formed.

Fig.13 Enlarged view of area B

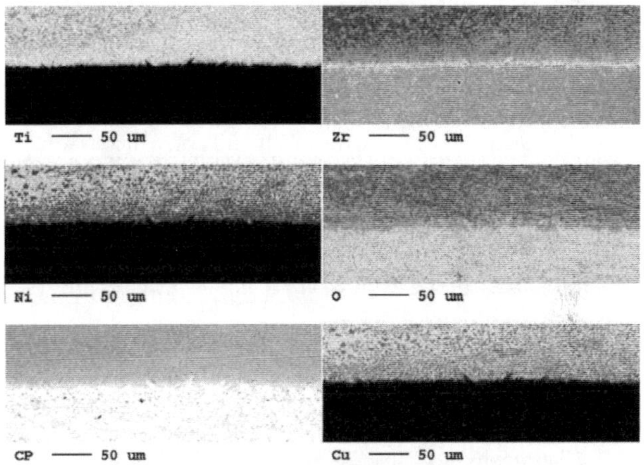

Fig.14 EPMA analysis in area-B

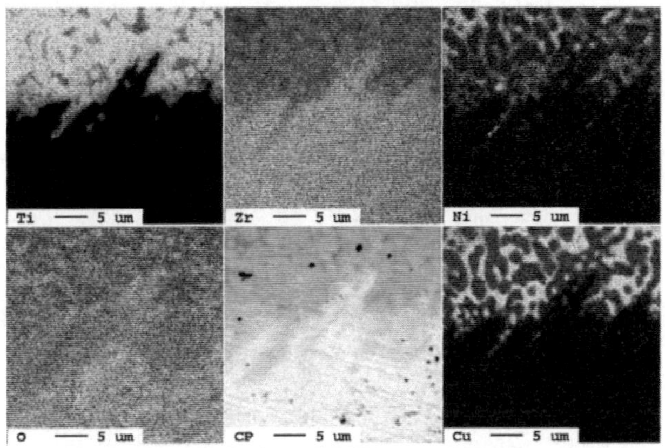

Fig.15 Enlarged view of EPMA analysis in area-B

When repairing Zircaloy-4, it was considered raising the temperature to the β or α + β phase and forming acicular α structure at the joining part were effective.

Conclusions

Zirconium-Titanium based powder filler was melted by arc plasma and the wetting behavior of filler meal on surface of ANSI 316L steel and zircaloy-4 was confirmed.

When zirconium-titanium based powder filler was used, adhesion wetting of the brazing filler metal could be confirmed on surface of ANSI 316L steel but dipping wetting could not be confirmed. It was considered difficult to repair cracks of ANSI 316L steel using zirconium-titanium based powder filler.

For Zircaloy-4, adhesion and dipping wetting of the brazing filler were confirmed. Where heat was sufficiently transmitted to the base material, acicular α-structure was confirmed between the brazing filler metal and the base metal, and growth of the microstructure could not be confirmed where dipping wetting occurred. When repairing zircaloy-4, zirconium-titanium based powder filler was effective in zircaloy-4 and formation of acicular α-structure in joint part was important.

References

[1] KUBO Toshiki, *et al.* "Effect of Si or SiO_2 on bead shape of 316L stainless steel during AA-TIG welding" Japan welding society, 2016

[2] Djamila Bahloul-Hourlier, Pierre Perrot "Thermodynamics of the Au-Si-O System: Application to the Synthesis and Growth of Silicon-Silicon dioxide Nanowires" Journal of Phase Equilibria and Diffusion, Vol.28, No.2, pp. 150-157, 2007

[3] K. Zhou, H.P. Wang, J. Chang, B. Wei "Experimental study of surface tension, specific heat and thermal diffusivity of liquid and solid titanium" Chemical Physics Letters, 639, pp.105-108, 2015

[4] P.-F. Paradis, T. Ishikawa and S. Yoda1 "Non-Contact Measurements of Surface Tension and Viscosity of Niobium, Zirconium, and Titanium Using an Electrostatic Levitation Furnace" International Journal of Thermophysics, Vol. 23, No. 3, pp.825-842, 2002

[5] Mitsuo Niinomi *et al.* "Fracture Characteristics and Microstructures of Intermetallic Compound Ti-24Al-11Nb (at%)" Journal of Japan Institute of Metals and Materials, Vol. 59, No. 7 (1995), pp. 708-716

[6] Chenyang Zhou *et al.* "Thermodynamic assessment of the phase equilibria and prediction of glass-forming ability of the Al-Cu-Zr system" Journal of Non-Crystalline Solids, 461, pp.47-60, 2017

[7] M. Reza Bateni *et al.* "THE EFFECT OF NOVEL Ti -Cu INTERMETALLIC COMPOUND COATINGS ON TRIBOLOGICAL PROPERTIES OF COPPER" NATIONAL TRIBOLOGY CONFERENCE 24-26, September, 2003, pp.55-62

Eutectic high entropy alloys – A novel class of materials for brazing applications

W. Tillmann[1, a], L. Wojarski[1, b], M. Manka[1, c], C. Timmer[1, d]

Institute of Materials Engineering, Faculty of Mechanical Engineering, TU Dortmund University (TU Dortmund), Dortmund, Germany

[a] Wolfgang.Tillmann@tu-dortmund.de, [b] Lukas.Wojarski@tu-dortmund.de, [c] Matthias.Manka@tu-dortmund.de, [d] Christian.Timmer@tu-dortmund.de

Abstract

Most recently, due to their unique properties, high entropy alloys attract the interest of a steadily increasing number of researchers. The high levels of resistance against oxidation, corrosion, and heat impact as well as their outstanding mechanical properties, which combine a distinct ductility with high ultimate strength, make these materials promising candidates for brazing applications, such as the sealing of SOFCs.

In this paper, the eutectic filler alloy Nb0.73CoCrFeNi2.1 was used to braze Hf-metallized YSZ ceramic to stainless steel X1CrTiLa22. For the application, Nb0.73CoCrFeNi2.1 was smelted from pure elements and subsequently cut into 300 µm thick foils. The samples were heated with 50K/min up to 1200°C, brazed for 5 min in a high vacuum of $4.3 \cdot 10^{-4}$ mbar, and slowly cooled down to room temperature.

Sound joints with a distinct HfO_2 reaction layer at the ceramic interface and a slight erosion of the steel counterpart were obtained. The eutectic structure in the center of the joint did not change compared to the smelted foil and featured a hardness of 487 HV0.01. The bonds reached a maximum shear strength of ~103MPa, which was limited by the reaction layer, and is 2.4 times higher when compared to YSZ-X1CrTiLa22-joints brazed at 920°C, using AgCuTi3.

Introduction

Since the revolutionary discovery of metallic materials and their successive extraction, tremendous efforts have been put into the improvement of their properties by alloying. In this regard, contemporary alloys still are based on one principal element or compound that is doped with various additional elements in order to achieve certain material properties. The reason for this are the experiences gained from the well-known binary and ternary material systems, where larger amounts of the alloying elements lead to a severe formation of various intermetallic compounds, which can dramatically decrease e.g. the strength, corrosion resistance and the electrical conductivity of the alloy [1].

Therefore, no phase diagrams of multicomponent materials exist, which are not based on CALPHAD and are not calculated solely based on binary and ternary systems. Especially in multicomponent material systems only the exterior regions of the phase diagrams are known, whereas the central region is largely unexplored. In order to remedy this perceived flaw Jien-Wei Yeh and Brian Cantor conducted intense research on multicomponent alloys and published their results independently in 2004 [2,3]. Cantor smelted alloys containing 20 different elements with an equiatomic composition and revealed that the microstructures mainly consisted of a solid solution. The main alloying elements of the FCC structured solid solution were the transition metals Cr, Mn, Co, Fe, and Ni. Based on this finding, it was determined that when using these elements in an equiatomic composition, only one primary phase with a dendritic microstructure is formed [2]. At the same time, Yeh also analyzed 40 different alloys, using a total of 59 different elements and found out that a high mixing entropy can lead to a decrease of the number of phases in the microstructure [3].

Whether a certain phase forming depends on the Gibbs energy, which is composed of the forming enthalpy and the product of temperature and forming entropy. Intermetallic phases have a low entropy due to their stoichiometric composition and, at the same time, a high forming enthalpy. In contrast, solid solutions have a low forming enthalpy and the entropy is dependent on the number of alloying elements. The forming entropy of solid solutions is dominated by the configurational entropy. Hence, the forming entropy (S) in multicomponent systems can be calculated with the following simplified rule [4]:

$$S = -R \cdot \Sigma x_i \cdot \ln(x_i)$$

where x_i is the molar fraction of the element i and R being the gas constant. This indicates that an increasing number of alloying elements in multicomponent systems increases the entropy as well. Thus, the solid solution with more than four alloying elements, especially in equiatomic configurations, have a high Gibbs free energy due to the high entropy, which can be higher than the Gibbs free energy of the possible intermetallic phases in the according material system. This fact leads to an impediment of the formation of intermetallic phases in multicomponent alloys. Because this effect is strongly driven by the entropy, the alloys are named high entropy alloys (HEA) [5].

Hence, the term HEA encompasses, according to the compositional interpretation, multicomponent alloys with at least 5 alloying components, whereby the amount of the respective elements has to range between 5 and 35 at-%. Besides, a definition of the term related to the entropy is possible as well. According to the latter definition, an alloy is considered a HEA when the entropy is higher than 1.61·R. These definitions cover the majority of existing alloys. Yet, there exist alloys with 5 elements with an entropy smaller than 1.61·R, which are still considered to be HEAs in accordance with the first definition. Thus, alloys which match either of the two definitions are considered to be HEAs [6]. The strong influence of the solid solutions on the microstructure of the HEA leads to interesting technological properties, which aroused the interest of many research groups, leading to a significantly increasing number of publications in this field since their discovery (see Fig.1)

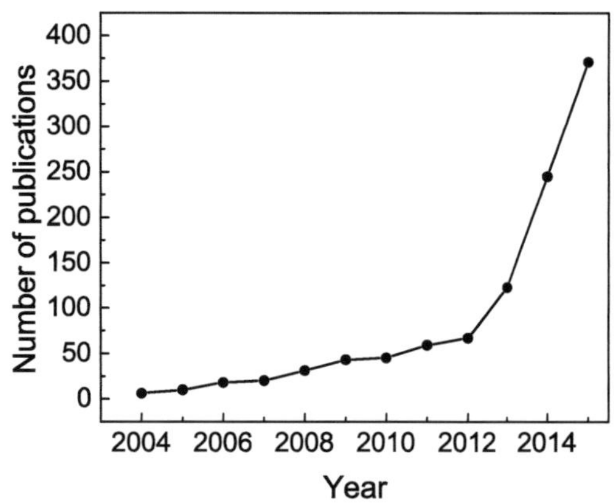

Figure 1: HEA related papers released until 2015 [7]

In general, these alloys distinguish themselves with either high ductility in the FCC structure or comparatively high hardness and strength in case of bcc structured alloys (see Fig. 2). The divergent characteristics of the HEA in dependence of the structure of the solid solution encouraged various research groups to develop eutectic high entropy alloys (EHEA). The eutectic alloys provide a high ductility and hardness. The benefits of these eutectic alloys are the improved castability and the overall fine grained structure [8,9,10,11].

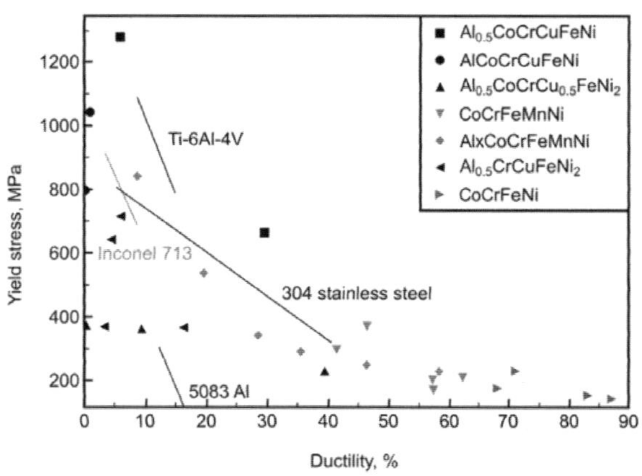

Figure 2: Mechanical properties of selected HEAs compared to conventional alloys at room temperature [7]

The excellent mechanical properties of HEAs in combination with their high resistance against corrosion and oxidation, even at high working temperatures, predestine HEAs for high temperature applications such as sealants in solid oxygen fuel cells (SOFC) [12,13].

SOFCs are comprised of a solid electrolyte which has to be bonded to a metallic counterpart. In this regard, yttrium-stabilized zirconia is frequently used as an electrolyte that is joined to the standard oxidation resistant interconnect Crofer 22 APU steel [14,15]. The elevated operation temperatures in ambient atmosphere necessitates the application of sealants with a high resistance against high temperature oxidation and static thermal stress. Typically, glass or glass-ceramic sealants are used for this purpose. Even though the properties of glass sealants can be widely adjusted according to the requirements of the specific application, they feature a brittle behavior that limits the possible applicable heating and cooling rates for SOFCs [16]. Within this context, the newly developed EHEA Nb0.73CoCrFeNi2.1 is a promising sealing material due to its high oxidation resistance in combination with an exceeding ductility when compared to glass solders [8].

Consequently, this publication focuses on the feasible application of Nb0.73CoCrFeNi2.1 as a brazing filler with respect to a subsequent application as a sealant in SOFCs. The investigations involve the manufacture of the filler alloy by smelting, the characterization of the microstructure of both the EHEA as well as of the YSZ-Crofer-joints and, finally, the determination mechanical properties of the joint.

Experimental procedures

Materials

The employed EHEA Nb0.73CoCrFeNi2.1 was created from pure elements (Co, Cr, Fe, Ni as powder with purity > 99.5 wt.-%, and Nb foil with purity 99.9 wt.-%). The powder materials were pressed into a pellet, enclosed by a Nb foil and were smelted in an arc furnace in an argon gas atmosphere. The samples where molten and afterwards turned and remolten three times in order to ensure a homogeneous element distribution within the alloy. The chemical composition of the EHEA is given in Table 1.

Table 1: Chemical composition of $Nb_{0.73}CoCrFeNi_{2.1}$ [wt.-%]

Nb	Co	Cr	Fe	Ni
18.95	16.47	14.53	15.61	34.44

As mentioned above, the metallic interconnect steel utilized in this paper is VDM® Crofer 22 APU (UNS S4 4535) [17]. Crofer 22 APU is a ferritic steel developed by ThyssenKrupp VDM GmbH especially for the use in SOFCs. The thermal expansion was adjusted to fit the thermal expansion coefficients of the ceramics used in SOFCs. Table 2 summarizes the chemical composition of the steel substrate.

Table 2: Chemical composition of Crofer 22 APU [wt.-%] [17]

Fe	C	Cr	Ti	La
Bal	0.015	22	0.11	0.11

The steel samples were cut into 20 mm x 20 mm x 2 mm coupons and machined with a 1200 grit grinding paper. Afterwards, every sample was cleaned with ethanol immediately prior to the brazing process.
The ceramic part consists of yttrium-stabilized zirconia with an yttrium content of 3mol-%. The dimensions of the samples were 8 mm x 8 mm x 3 mm. In order to ensure the wettability of the ceramic during the subsequent brazing process, a hafnium interlayer with a thickness of 6 µm was applied by means of arc PVD coating. Prior to the coating process, the surface of the YSZ was polished and cleaned with ethanol. Within the PVD deposition, the substrates were initially plasma-etched. During the plasma-etching process, a high bias voltage was applied to accelerate highly energized ions in order to remove impurities on the surface.

Brazing condition

After smelting the Nb0.73CoCrFeNi2.1 alloy was cut into foils and ground in order to adjust a constant thickness of 0.3 mm. The foil was placed between the joining couple while the Crofer coupons were placed at the bottom. The brazing experiments were conducted in a vacuum furnace, whereby the brazing temperature is limited to 1200°C due to the temperature limit of the vacuum furnace. Hence, the melting range of the EHEA has to be determined by means of differential thermal analysis (Linseis STA PT 1400). The brazing process was conducted with a heating rate of 50 K/min up to the brazing temperature. The samples were held at the brazing temperature for 5 min in a $4.3 \cdot 10^{-4}$ mbar vacuum. Subsequently, the samples were cooled down to 900°C with a cooling rate of approximately 50 K/min to prevent an excessive formation of a reaction layer. Afterwards, the samples were cooled with 5 K/min to 500°C to avoid high induced stresses. Finally, the samples were freely cooled down to room temperature.
For benchmarking, additionally brazed joints with a conventional AgCuTi3 filler alloy, provided by Umicore as CB4, were conducted. AgCuTi3 was used as a foil with a thickness of 50 µm. The samples were vacuum brazed at 920°C. In accordance with the EHEA brazements, a heating rate of 50 K/min was applied. After a holding time of 5 min at brazing temperature, the samples were cooled down to 900°C with a cooling rate of 5 K/min and were subsequently freely cooled down to room temperature.

Microstructural characterization

For the microstructural characterization of both the EHEA in the as-cast state as well as of the brazed joints, cross-sections were prepared by means of metallographic preparation. The samples were analyzed using a scanning electron microscope (SEM), equipped with an energy dispersive X-ray spectroscope (EDS). Furthermore, X-ray diffraction measurements of the samples were conducted in order to verify the phases predicted from their chemical composition. The micro hardness of the joint area was measured using a Vicker's indenter with an impact force of 0.01 N.

Joint fractography

Shear test were conducted to determine the mechanical properties of the joints. The brazed samples were sheared using the apparatus shown in Figure 3. Whereas the position of the YSZ was fixed by the shearing edge of the apparatus, the steel counterpart was pressed through the slot in the apparatus, which was adjusted according to the dimensions of the Crofer coupons. The ultimate shear strength of the joints was calculated as the mean value of the quotient of the maximum force and the joint area of three samples in total. In order to characterize the fracture of the joints, the fractured samples were rearranged and cross-sections were prepared. Finally, the cross-sections were analyzed by SEM.

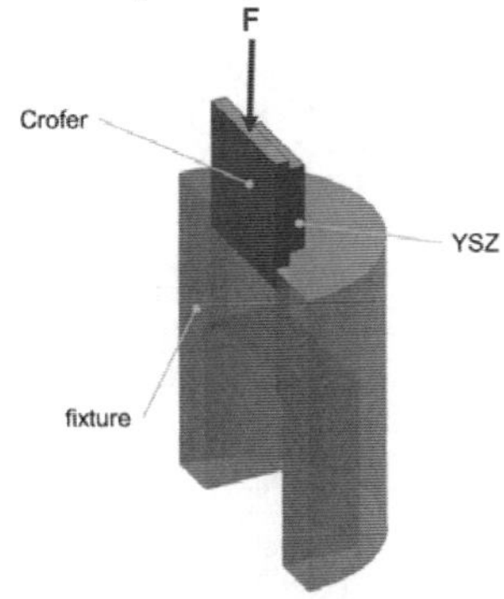

Figure 3: Schematic illustration of the shear strength test device in profile

Results and discussion

Based on the findings of Lu et al. [7], Calphad simulations with ThermoCalc of the Nb0.73CoCrFeNi2.1 EHEA were conducted in order to verify the melting range and the formed phases. The phase fraction diagram of the Nb0.73CoCrFeNi2.1 in Figure 4 shows the high temperature configuration of the EHEA. According to the calculation two phases occur before the alloy starts to melt. The solid microstructure consists of a FCC structured phase with high amounts of nickel and a Laves phase with high amounts of niobium. The calculation predicts the initiation of the melting at 1245°C and the complete liquid state at 1274°C. Hence, Nb0.73CoCrFeNi2.1 features a near eutectic composition.

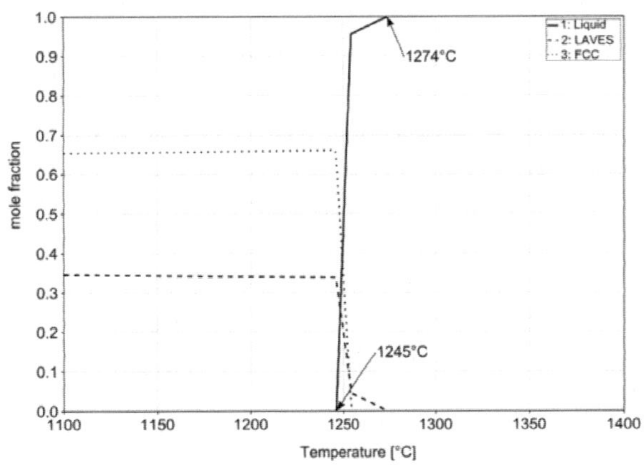

Figure 4: Phase fraction diagram of Nb0.73CoCrFeNi2.1

Microstructure of the filler metal

In well accordance with the phase fraction diagram, the microstructure of the smelted Nb0.73CoCrFeNi2.1 is comprised of only two finely dispersed phases. The backscattered electron (BSE) image of the microstructure is shown in Figure 5. Taking into consideration that the element with the highest fraction of the Laves phase is niobium, which is by far the heaviest element in the EHEA, and given the fact that lighter elements appear to be dark in the BSE mode, it is obvious that the brighter phase is the Laves phase. Hence, the darker phase has to be the FCC structured solid solution, which is in well agreement with the observations made in the study of Lu et al. [2].

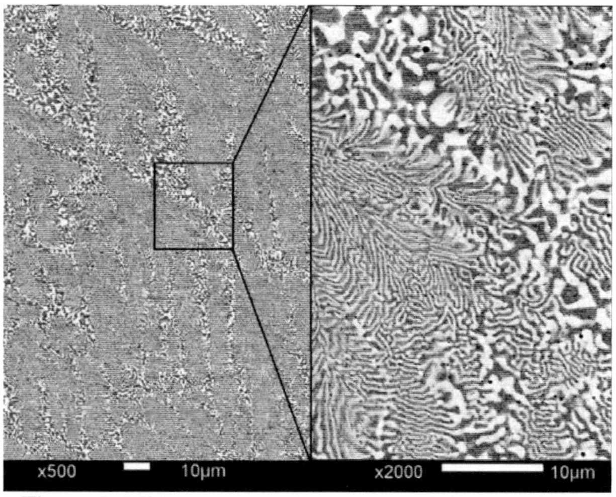

Figure 5: Microstructure of a $Nb_{0.73}CoCrFeNi_{2.1}$ sample

The average chemical composition of the FCC structured solid solution as well as of the Laves phase are summarized in Table 3. In order to avoid measurement errors, three different samples were analyzed. For each phase, ten measurements in each sample were conducted to determine the mean value. Both phases comprise of all five elements and mainly differ in their amount of niobium, as already mentioned above.

Table 3: Chemical analysis of the Nb0.73CoCrFeNi2.1 EHEA [wt.-%]

Phase	Co	Cr	Fe	Ni	Nb
FCC	17.33	19.32	15.96	35	12.39
Laves	11.38	13.04	14.34	27.58	33.66

The average micro hardness of the as-cast state was derived from 10 measurements in all three specimens as well. The micro hardness is 525.3 HV0.01. This high value can be attributed to the high cooling rates after smelting in combination with the tension of the atomic lattice.

Whereas the microstructure analyses of the smelted alloy verified the Calphad-based calculation regarding the existence of only two phases, the measured melting range differed significantly from the calculated melting behavior. The DTA measurement reveals a melting range from 1170°C to 1222°C, as shown in Figure 6. Hence, the measured values are significantly lower than the calculation had predicted. Unfortunately, the liquids temperature of the alloy is still higher than the desired maximum brazing temperature. Since most of the alloy is molten at 1200°C, this temperature was set as brazing temperature.

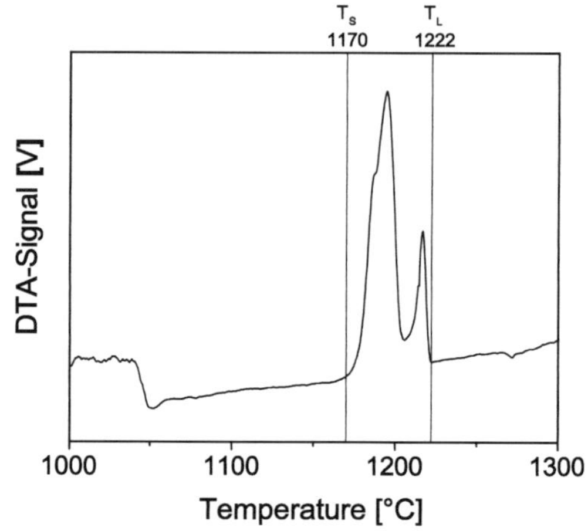

Figure 6: DTA curve of the filler alloy

Microstructure of the Nb0.73CoCrFeNi2.1 joints

Sound joint without significant pores or voids could be obtained using Nb0.73CoCrFeNi2.1 foils as a brazing alloy (see Figure 7). A white dense reaction layer has formed at the YSZ/braze interface, which is comprised of HfO_2 according to its chemical composition (see Table 4, point 2). Below the reaction layer, a fine eutectic microstructure is visible, which is coarser compared to the microstructure of the Nb0.73CoCrFeNi2.1 EHEA in the as-cast condition. The chemical analysis of the eutectic phases formed in the joint area reveals that even though the microstructure is highly similar to the as-cast microstructure, the composition of the occurring phases differs significantly. Furthermore, with increasing distance from the reaction layer, the microstructure turns coarser and the fraction of the solid solutions increases.

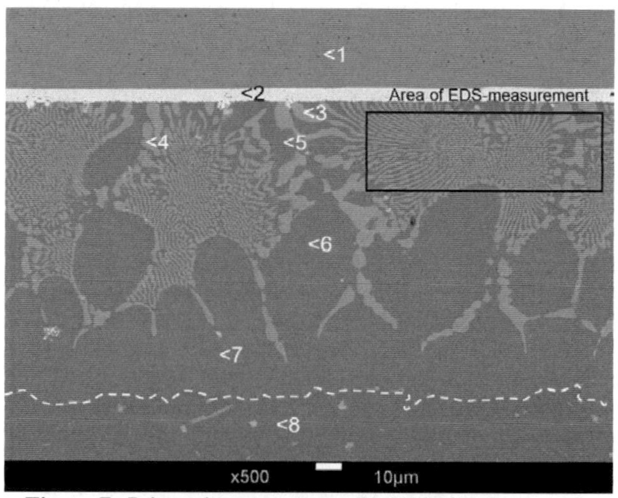

Figure 7: Joint microstructure of a YSZ-Crofer sample using a $Nb_{0.73}CoCrFeNi_{2.1}$ filler alloy

The bright niobium-rich phase in the joint area (points 3 and 4) features a significantly higher niobium and iron content when compared to the Laves phase in the as-cast condition. Furthermore, it contains a small amount of the sixth alloying element hafnium. The dark phase in the brazing area (points 5 and 6) is composed of only four alloying elements and does not contain any niobium at all. The quaternary phase comprises mostly of iron. Furthermore, this phase contains almost equal amounts of chromium and nickel, whereas it features only a small amount of cobalt that recedes with increasing distance from the reaction layer. The mean value of the micro hardness is 369 HV0.01 and hence significantly lower when compared to the Nb0.73CoCrFeNi2.1 EHEA in the as-cast condition. The lower hardness can be attributed to the slower cooling.

Table 4: Chemical analysis of the joint microstructure of the YSZ-EHEA-Crofer sample in wt.-%

No	Co	Cr	Fe	Ni	Nb	Zr	Hf	O
1	-	-	-	-	-	68.3	2.1	29.6
2	-	-	0.5	-	-	-	79.2	20.3
3	5.9	10.0	29.8	11.3	40.0	-	3	-
4	5.8	10.3	29.7	11.1	40.8	-	2.3	-
5	7.1	21.0	52.6	19.3	-	-	-	-
6	5.2	20.2	61.9	12.7	-	-	-	-
7	-	20.5	68.9	10.6	-	-	-	-
8	-	26.3	67.7	6	-	-	-	-

No evident diffusion of any elements from the filler alloy into the ceramic could be detected (see point 8) and the interface of the ceramic is still clearly visible, whereas at the steel side, no distinct interface line between the brazing seam and the base material is noticeable. The melted filler alloy partially dissolved the base material and, due to diffusion a small amount of nickel can still be detected in the base material close to the brazing seam (point 8).
In order to obtain the chemical composition of the alloy, which has formed in the brazing seam, an area scan by means of EDS was conducted of the cross-section of the joint. The chemical composition is summarized in Table 5. It is obvious, that the resulting alloy still fits the entropy-based definition for HEA.

Table 5: Measured average composition of the eutectic alloy formed in the brazing seam in wt.-%

Co	Cr	Fe	Ni	Nb	Hf
6.1	16.0	42.4	14.7	18.5	2.3

The phase composition in the brazing seam was different when compared to the phase composition in the as-cast state of the filler alloy. Therefore, Calphad-based calculations were conducted in order to determine the melting range of the resulting alloy in the brazing seam and to characterize its phases. The phase fraction diagram of the formed alloy in the brazing area is shown in Figure 8. According to the calculations, the obtained microstructure still consists of the same FCC structured solid solution and the Laves phase. Due to the fraction change of the two phases, compared to the as-cast state, a shift of the melting range occurred. Because of chromium and iron diffusing into the brazing area, the solidus temperature of the formed alloy was increased by ~50°C, whereas the liquidus temperature was even increased by ~100°C, leading to a comparably wide melting range.

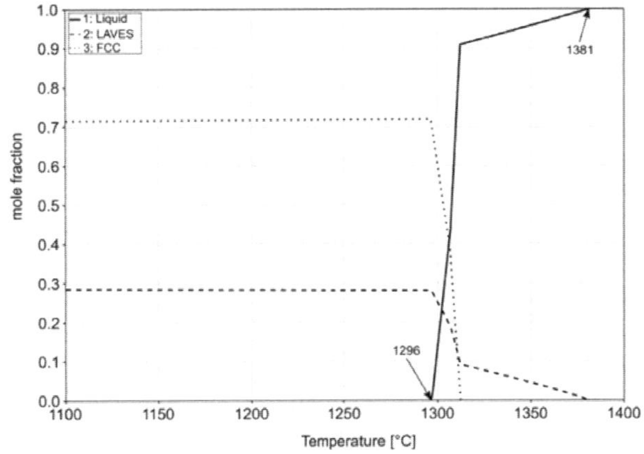

Figure 8: Phase fraction diagram of the alloy formed in the brazing seam

Microstructure of the AgCuTi3-joints

Brazing YSZ ceramic to Crofer using the conventional AgCuTi3 filler alloy was successful as well, as visible in Figure 8. The filler alloy shows a good connection to the ceramic base material and a slight dissolution of the Crofer substrate. In contrast to the EHEA joints, no distinct continuous reaction layer at the YSZ / braze interface has formed and the width of the brazing seam is significantly thinner, which can be attributed to the smaller thickness of the AgCuTi3 foil. Furthermore, the microstructure of the brazing area features a few scattered pores in the center of the joint.

Figure 9: Joint microstructure of a YSZ-Crofer sample using AgCuTi3 filler alloy

The microstructure of the joint is comprised of three different phases. An irregularly distributed bright silver-rich phase (see Table 6; points 2 and 3), a medium-grey copper-rich phase (points 4 and 5), and a dark-grey iron-rich phase, which contains high amounts of chromium and titanium. The copper-rich phases are agglomerated at the ceramic side and are partially framed by the silver-rich phase fragments. In contrast, the iron-rich phase is mostly located at the Crofer side. Whereas distinct amounts of titanium and a small amount of copper diffused into the Crofer substrate (point 8), no elements of the braze alloy or of the base material could be detected in the ceramic counterpart (point 9).

Table 6: Chemical analysis of the joint microstructure of the YSZ-AgCuTi3-Crofer sample in wt.-%

No	Ag	Cu	Ti	Fe	Ni	Cr	Zr	O
1	-	-	-	-	-	-	64.2	35.8
2	86.5	5.4	7.8	0.3	-	-	-	-
3	87.8	5.1	7.1	-	-	-	-	-
4	6.1	92.0	-	1.9	-	-	-	-
5	6.4	92.7	-	0.9	-	-	-	-
6		1.2	9.4	68.6		20.8		
7	-	2.0	11.7	67.9	-	18.4	-	-
8	-	1.4	1.8	74.7	-	22.1	-	-

Mechanical properties

The samples brazed with the Nb0.73CoCrFeNi2.1 EHEA reached an average joint strength of 97±7 MPa, whereas the shear strength of the AgCuTi3 joints was only 52±2 MPa. Hence, the joint strength is ~2 times higher compared to the conventionally brazed joint.

As shown in Figure 10, the EHEA brazements fractured mainly in the HfO_2 reaction layer. Whereas a complete remnant layer of HfO_2 still adheres to the YSZ ceramic, only brittle, partly broken fragments of the reaction layer remain at the steel side. Furthermore, the bright niobium-rich phases feature several cracks close to the fracture mirror, indicating a distinct brittleness of this phase. In contrast, the quaternary dark phase remained free of cracks. Hence, the metallization of the YSZ ceramic has to be modified in order to fully exploit the potential of the EHEA as a brazing filler alloy.

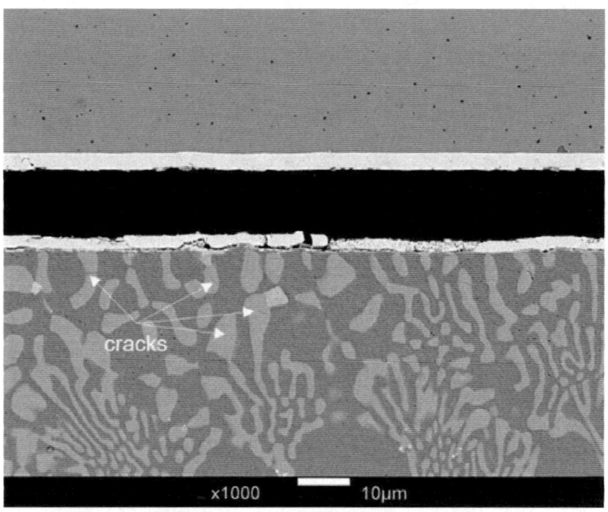

Figure 10: Fractured YSZ-Crofer sample using $Nb_{0.73}CoCrFeNi_{2.1}$ filler alloy

In contrast to the joints with the EHEA filler alloy, the joints brazed with AgCuTi3 fractured mainly in the brazed area close to the Crofer substrate, as shown in Figure 11. Additionally, cracks have formed at the ceramic / braze interface and lead to a severe formation of small cracks, which proceed perpendicularly to the shearing direction, initiating from the fracture mirror.

Figure 11: Fractured YSZ-Crofer sample using an AgCuTi3 filler alloy

Conclusions

The current research work demonstrates that eutectic high-entropy alloys can be used for high temperature applications like sealing of SOFCs. Sound joint between Crofer22APU and Hf-metalized YSZ ceramic were obtained using the Nb0.73CoCrFeNi2.1 EHEA. The eutectic microstructure of the brazing foil in the as-cast condition, consisting of a FCC structured solid solution and a Laves phase, was maintained in the joint area after brazing. A strong diffusion between the substrate and the braze alloy could be detected by EDS measurements. This diffusion leads to a significant change of the composition of the two phases in the joint area. Due to slow cooling rates compared to the smelting of the foils, a coarser structure was found in the joint.

The average shear strength of the EHEA-joints is 103 MPa and is about 2 times higher compared to YSZ-Crofer brazements using the AgCuTi3 filler alloy. In this regard, the joint strength of the EHEA was limited by the hafnium metallization and expect of few short cracks in the Laves phase, no significant detoriation of the EHEA in the brazing seam was detectable. Therefore, other metallization materials should be investigated in order to fully exploit the potential of the EHEA for sealing applications in SOFCs.

Acknowledgments

The authors thank Dr. Iris-Aya Laemmerhirt for her linguistic support with this paper.

References

[1] Hornbogen E., *et. al,* Metalle, Springer Berlin Heidelberg (Berlin, Heidelberg, 2016)

[2] Lu Y., Jiang H., Guo S., et al. "A new strategy to design eutectic high-entropy alloys using mixing enthalpy", *Intermetallics*, Vol. 91 (2017), pp. 124–128.

[3] Raphel A., Kumaran S., Kumar K.V., *et al*, "Oxidation and Corrosion resistance of AlCoCrFeTi High Entropy Alloy," *Materials Today: Proceedings* Vol. 4 (2017), pp. 195–202.

[4] Kai W., Cheng F.P., Liao C.Y., *et al*, "The oxidation behavior of the quinary FeCoNiCrSi x high-entropy alloys," *Materials Chemistry and Physics* (2017)

[5] Cantor B., Chang I., Knight P., et al. "Microstructural development in equiatomic multicomponent alloys," *Materials Science and Engineering*: A. 2004;375-377:213–218.

[6] Yeh J.-W., "Recent progress in high-entropy alloys," *Annales de Chimie Science des Matériaux*, Vol. 31, (2006), pp. 633–648.

[7] Lu Y., Dong Y., Guo S., *et al*, "A promising new class of high-temperature alloys: eutectic high-entropy alloys," *Scientific Reports*, Vol. 4 (2014), Article Number 6200.

[8] Fergus J.W., "Sealants for solid oxide fuel cells," *Journal of Power Sources*, Vol. 147 (2005), pp. 46–57.

[9] Miracle D., Miller J., Senkov O., *et al*, "Exploration and Development of High Entropy Alloys for Structural Applications," *Entropy*, Vol.16 (2014), pp. 494–525.

[10] Yeh, J.-W., *et al*, "Nanostructured High-Entropy Alloys with Multiple Principal Elements: Novel Alloy Design Concepts and Outcomes," *Advanced Engineering Materials*, Vol. 6 (2004), pp. 299–303.

[11] Rodríguez-López S., Comesaña R., del Val J, *et al*, "Laser cladding of glass-ceramic sealants for SOFC," *Journal of the European Ceramic Society*, Vol. 35 (2015), pp. 4475–4484.

[12] Huo W., Zhou H., Fang F., *et al*, "Microstructure and properties of novel CoCrFeNiTa x eutectic high-entropy alloys," *Journal of Alloys and Compounds*, Vol. 735 (2018), pp. 897–904.

[13] Wani I.S., Bhattacharjee T., Sheikh S., *et al*, "Ultrafine-Grained AlCoCrFeNi 2.1 Eutectic High-Entropy Alloy," *Materials Research Letters*, Vol. 4 (2016), pp. 174–179.

[14] Kendall K., Kendall M., High-Temperature Solid Oxide Fuel Cells for the 21st Century, (2016), pp. 1–24.

[15] Porter D.A., Easterling K.E., Phase transformations in metals and alloys, Chapman & Hall, (London, 1993)

[16] Gao MC, Yeh J, Liaw PK, *et al*, High-entropy alloys: Fundamentals and applications, Springer International Publishing, (Cham, 2016)

[17] VDM-Metals GmbH "VDM® Crofer 22 APU ", data sheet, received: 06.12.2017, 09:41 GMT+1

High Performance robot brazing with real time laser vision tracking and inspection

R. Simoneau, J.-P. Boillot, C. Demers-Roy, S. Latour
SERVO-ROBOT INC., St-Bruno, QC, Canada, J3V 6E1
jp.boillot@servorobot.com

Abstract

Advanced non-contact 3D laser-vision systems offer a definite advantage over contact based solutions, notably, because they combine real-time quality control of the brazing process using joint profile and bead geometry data with adaptive tracking capabilities. Online tracking and inspection of the brazed joint is carried out by two 3-D powerful high-resolution laser-vision systems directly interfaced through a high speed data communication link to the robot control unit.

Introduction

The laser brazing process is widely used for assembly of car components such as roofs and tailgates. In these applications, very high quality brazed beads are required to cope with high mechanical performances and cosmetic requirements. To achieve such high quality brazing of galvanized steel and aluminum car body components, the laser brazing beam must perfectly be positioned in the joint by the robot which must track its position precisely and adapt to its fit-up conditions at very high speed.

Furthermore, the brazing process which offers more freedom to car designers involves new joint configurations that cannot rely on contact wire based seam tracker which does not perform well due to the joint cross sections that are not appropriate to guide the brazing wire reliably at high laser power and high speed. Advanced non-contact 3D laser-vision systems offer a definite advantage over contact based solutions, notably, because they combine real-time quality control of the brazing process using simultaneously joint profile and brazed bead geometry data with adaptive tracking capabilities.

Online tracking and inspection of the brazed joint is carried out by two 3-D powerful high-resolution laser-vision systems directly interfaced through a high speed data communication link to the articulated robot control unit.

The brazing Process of Car Components

Base materials and filler wire metals

Laser brazing has been found in recent years to be one of the most cost efficient processes for producing aesthetic apparent joints of car components such as roofs and tailgates [1] [2]. It is used for joining zinc coated steels, Al-Si coated high strength steels and any one of these steels with an aluminum alloy sheet of around 1 mm thickness [3].

CuSi3 (3%Si) filler wire is generally used for joining galvanized or aluminized steel; Cu-Zn is also used. Al-Si, Al-Zn and Zn-Al wires are used for joining steel to aluminum or aluminum to aluminum.

Figure 2 shows the typical configuration for laser brazing a car roof flare joint. The leading filler wire, (diameter of 1.2 - 1.6 mm) is fed in front of the molten pool with an angle of 30 to 45 degrees above the joint surface. It is fused by a defocused laser beam (2-3 mm laser spot diameter) perpendicular to the joint surface in conduction mode.

The usual keyhole mode cannot be used since the underneath joint material must not be melted by the brazing process. A small and shallow keyhole could be used to melt faster the filler wire since much of the laser power is reflected on the solid wire surface. This is specially the case for Cu and Al wires, for which the laser diodes (800 – 900 nm) offer better absorption than the fiber laser (1060 nm) due to their lower wavelength. Coated or roughened wire surface can also increase laser absorption [4].

The laser beam also preheats and cleans the joint surface in order to improve wetting of the fused metal with the joint surface [5]. Poor preheating of the joint area in front of the wire, hindered by the shadow of the wire itself, leads to intermittent wetting and defective appearance of the brazed bead.

Many solutions and patents have been proposed to solve this problem: longitudinal scanning [6], two-beam laser brazing [7] [8] [4] two or more laser beams [9] [10] [11] [12] with oscillations [13], triple and twin-spot laser [14] [15] [16] coaxial beam shaping and wire feeding [17] and leading laser beam in front of the filler wire [18] [19].

Joint Configurations

Figure 2 and Figure 2 show a typical flange car roof joint laser brazing configuration. The defocused laser beam is used to melt the filler wire, heat the molten pool and the joint borders. For a joint between aluminum and steel, the aluminum plate may be fused when its melting point is closed to the one of the filler metal wire, as shown on Figure 2. Various configurations of lap and flange joints are used for different components and car designs.

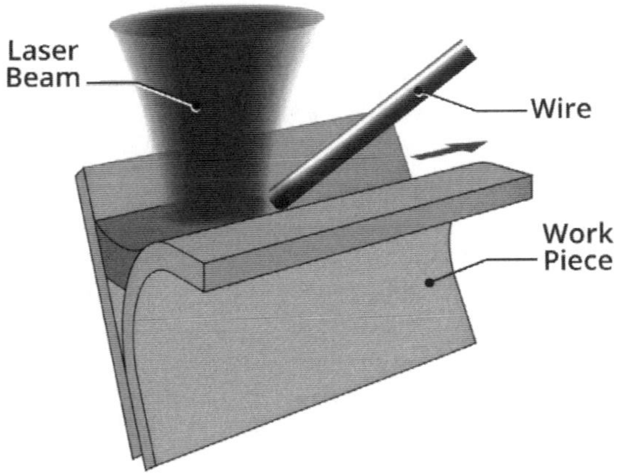

Figure 1. Typical car flange joint laser brazing.

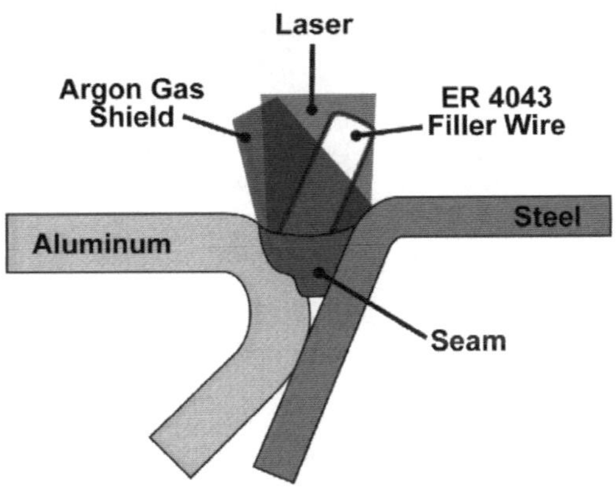

Figure 2. Laser welding-brazing of a steel-Al flange joint.

High Speed Triangulation Laser Cameras

Figure 3 shows a dedicated triangulation laser camera used for high speed tracking of laser brazed joint. The red laser line can acquire the joint profile at speed up to 2000 profiles/s with a resolution of 20 μm with a field of view of up to 16X16 mm.

Figure 3. High-speed high-resolution tracking laser brazing camera.

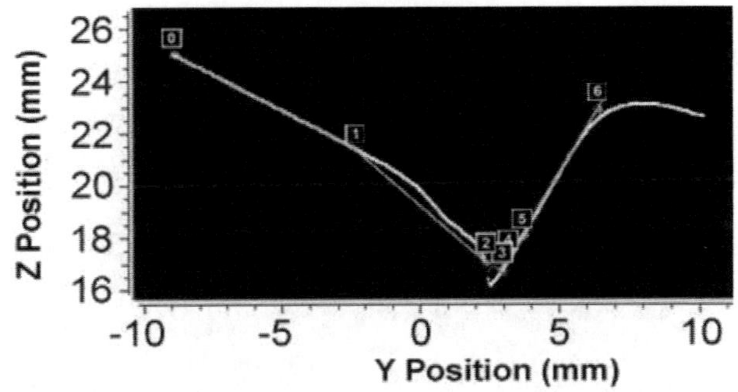

Figure 4. Tracking break points detected on a typical car joint.

On the joint profile acquired by the laser camera, many break points can be defined by the tracking algorithm. The wire tip and the laser spot can be accurately positioned relative to these break points in real time during the brazing process.

High Speed high resolution robot tracking

Laser brazing is carried out with high precision robots capable of moving the laser head at speed up to 10 m/min with an accuracy of ± 0.1 mm. We present below experimental tracking data obtained with a high precision robot on a curved lap joint using the Robot Error Compensation technique developed by Servo-Robot.

- **Part:** Lap joint (thickness = 1 mm) with a curved trajectory in the XY plane of the robot (curve radius = 180 mm)
- **Laser tracking system (tracking and TCP):** frame rate = 200 Hz
- **Application modules:** trajectory generator and Robot Error Compensation
- **Robot:** KUKA KR30 HA with KRC4 controller (communication cycle time = 12 ms)
- **Travel speed:** 9 m/min
- **Final process accuracy obtained:** ± 0.075 mm with robot error compensation; ± 0.3 mm without robot error compensation.

Figure 5 presents the graph of the trajectory acquired by the vision system using the laser-camera responsible of the seam tracking process. On this graph, there are five data plots (only one curve is visible since all data are superposed):

- Vision data which is the acquired position of the joint in the XY plane
- Filtered vision data is the above data with artifacts filtered out
- Robot data is the TCP of the robot along the trajectory (this is the curve in evidence on the figure)
- Correction data is the corrected trajectory (new TCP positions to apply)

Figure 6 presents the process accuracy without robot error compensation. It shows that even if seam tracking is applied, the robot is unable to follow the corrections completely, especially when it enters / exits the curved section on the part (around 150 and 250 mm on the graph).

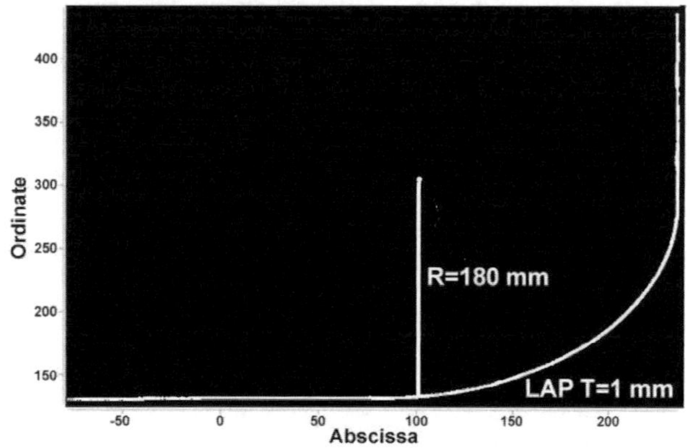

Figure 5. Trajectory used for the test (acquired by the laser-camera) – Abscissa = X coordinates; Ordinate = Y coordinates (mm).

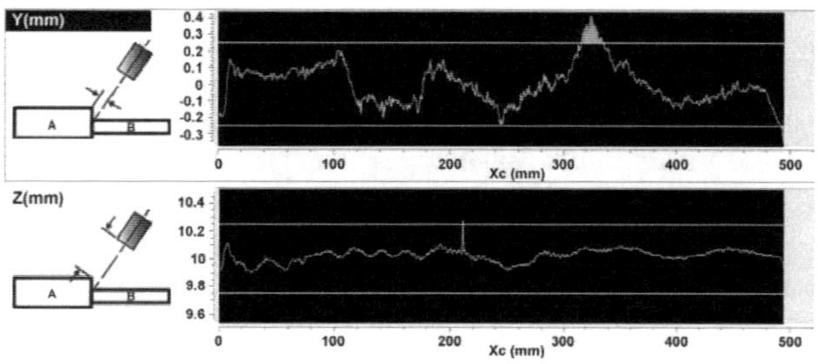

Figure 6. Process accuracy without robot error compensation – Y represents the lateral TCP error and Z the vertical error.

This path following error comes from many sources, but the fact that the robot cannot plan all the dynamics with the seam tracking corrections coming into play is the most important source.

In order to obtain the required ± 0.1 mm process accuracy required in laser brazing, a special module has been developed which is able to analyze and correct the robot error, an error which is found to be repeatable for a given trajectory even in presence of the typical offsets corrected by the seam tracking process.

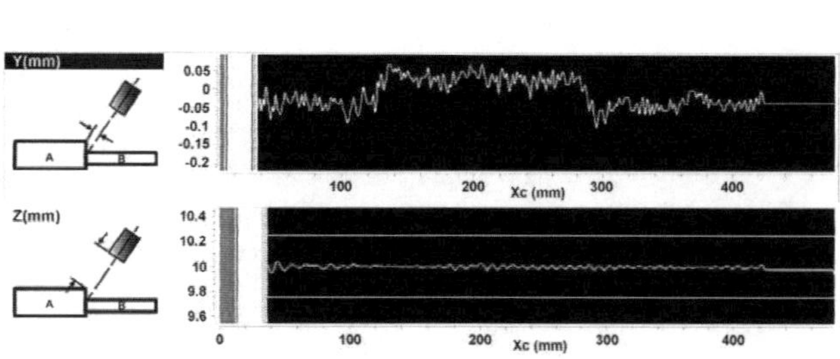

Figure 7 Final process accuracy (laser-camera at TCP) – Y represents the lateral TCP error and Z the vertical error.

Figure 7 shows the process accuracy results obtained with the same test conditions, but this time with the robot error compensation applied. This time, the final process accuracy falls within the required range of ±0.1 mm.

Car Brazed Joint Robot Tracking

Figure 8 shows a typical robot brazing set-up with the laser head, wire feeder, tracking and inspection cameras mounted on a high precision robot arm.

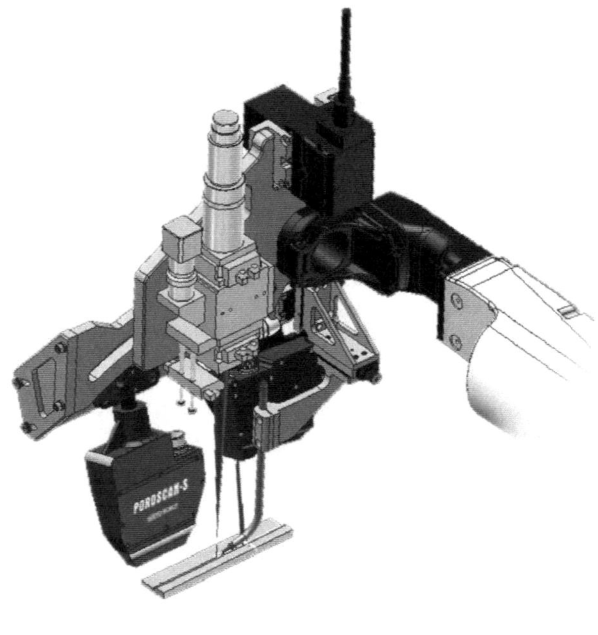

Figure 8. Set-up of a brazing laser head with wire feeder, tracking and inspection cameras mounted on a robot arm.

Initial wire tip TCP calibration and positioning

In order to obtain the required precision in positioning the wire tip and laser spot on the joint to be brazed, an accurate spatial calibration of the robot wire tip, the laser spot and the tracking camera within the spatial reference frame of the workpiece is used. And as pre-requirement the initial robot TCP calibration has to be very accurate to better than 0.1 mm. For that purpose Servo-Robot has developed an automatic TCP calibration method with an active target based system which is carried out by the robot itself without any manipulation by a human operator. This insures the most accurate and reproducible calibration, which is required for car joint robot laser brazing. After this accurate robot calibration, the measurement of the initial positioning of the wire tip, laser spot and tracking laser camera is realized in the following 3 steps with the help of an accurate target plate and 2D imaging techniques. During calibration the wire tip is replaced by a conical steel point and the laser beam spot is visualized making use of the pilot laser from the high power laser.

Step1: Find the wire tip position in a referenced target frame using the 2D camera

Step2 : Adjust the laser position on the same referenced target frame making use of automated pilot laser spot detection on the 2D image

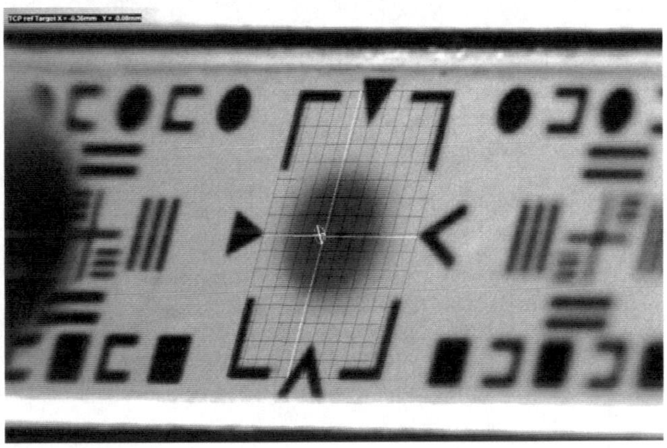

Step3 : Find the tracking laser line center position in the same referenced frame using geometric features on the target

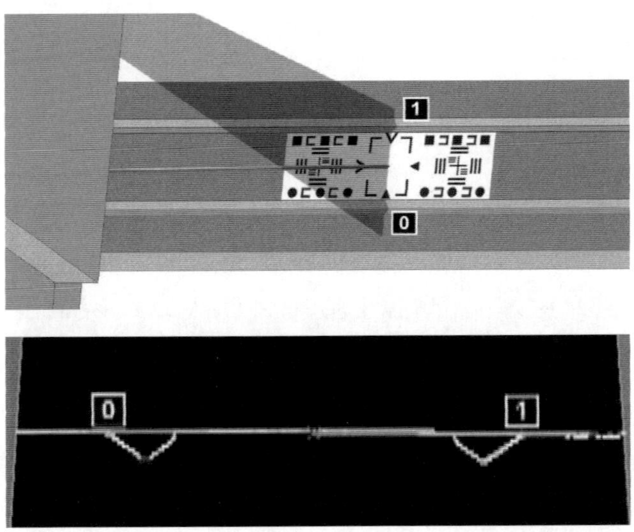

Figure 9. Wire tip and laser spot position calibration 3 steps.

Tracking a car roof joint

In order to respect the takt time of the car production line, the brazing process of the long car roof joints has to be carried out at quite high speed, in the range of 4 to 8 m/min. The high speed laser camera measures the joint profile at high frequency, up to 2000 profiles/s, every 0.5 mm. The tracking point on the joint is defined relative to the profile break points found by the tracking algorithm, as shown on Figure 4.

Process Control and Monitoring

The complete control system, shown in Figure 10, controls in real time the position of the laser spot and of the wire tip on the brazed joint. It also monitors in-line, the brazed bead profile and the molten pool temperature with the LPCS2 unit which includes an infrared pyrometer. This LPCS2 unit includes also a back reflection sensor to monitor the back reflection of the high power laser radiation to detect process instabilities.

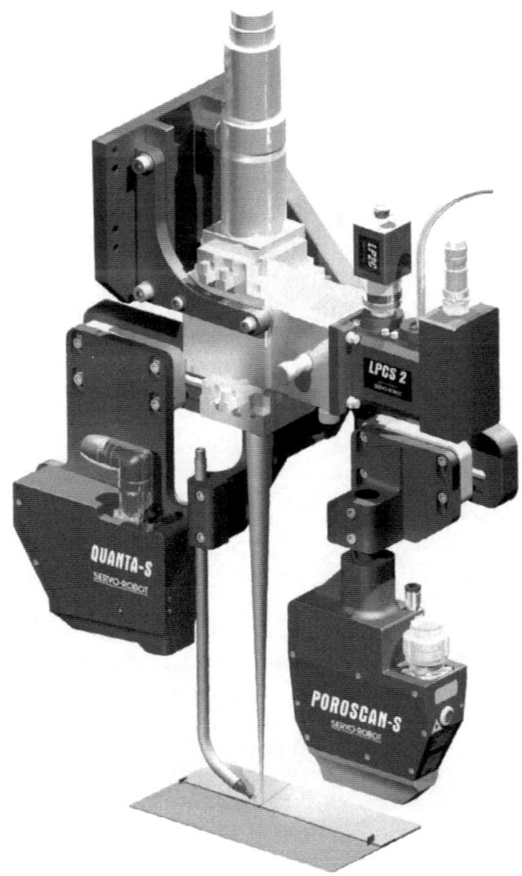

Figure 10. Laser brazing head with tracking, inspection and LPCS2 monitoring systems.

Inline Robotic Inspection

The inspection laser camera, (POROSCAN-S) shown in *Figure 11* includes a dual field of view that allows to measure very small defects, (> 0.1 mm), porosities or spatters with a reduced field of view which is automatically centered on the brazed joint using the dual sensor technology. The use of shorter wavelength blue laser (405 nm) improves the precision of the camera. From the successive joint profiles measured before and after brazing, 3D maps of the joint before and after brazing (*Figure 12*) are built and overlaid with the help of the position data received from the robot. From these two 3D maps, the brazed bead is extracted and its position and external geometrical features are computed on selected cross-sections. Features such as width, convexity/concavity, porosity diameter and depth, mismatch and height difference relative to the adjacent joint plates, as illustrated in *Figure 13*and *Figure 14* can be measured. **Table 1** shows typical bead inspection criteria used as defined in *Figure 14*.

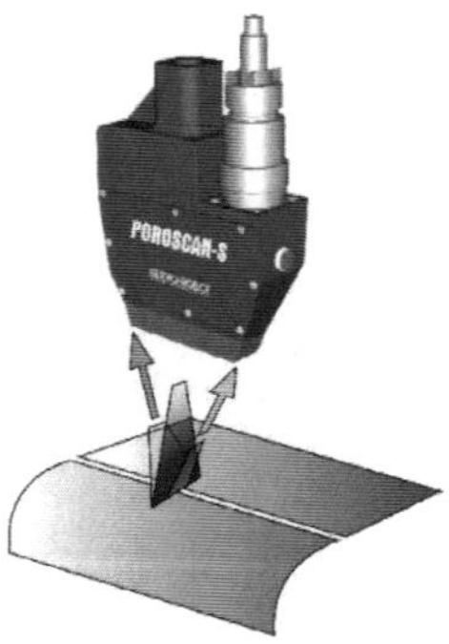

Figure 11. Dual sensor high resolution inspection laser camera.

Figure 12. 3D map of inspected laser brazed bead.

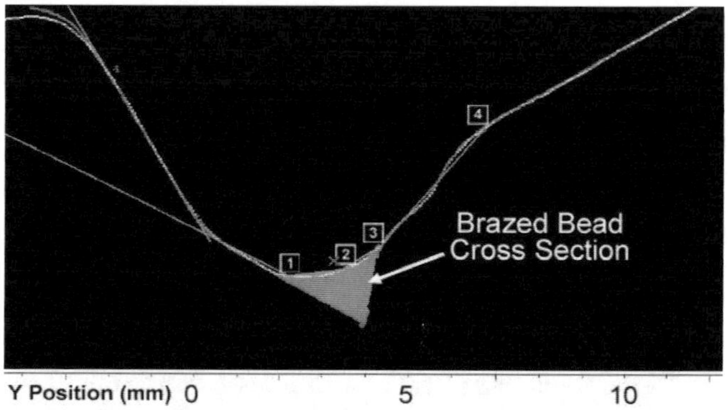

Figure 13. Cross-section of a brazed bead obtained from joint profiles measured before and after brazing.

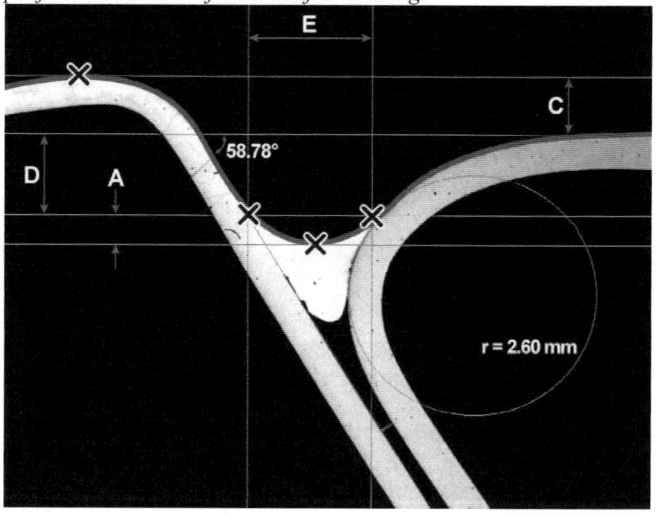

Figure 14. Cross-section of a brazed car flange joint showing typical inspection criteria.

Table 1. Typical bead inspection criteria

	Feature	Min. size (mm)	Max. size (mm)
A	Concavity	0	0.6
B	Porosity	0.2	NA
C	Mismatch	1.5	5
D	Height difference	0	1.2
E	Bead width	2.0	4.0

Conclusions

Laser brazing has been found most convenient to produce strong, sealed and aesthetic joints of visible car components such as roof and tailgates. High precision robots and triangulation laser cameras have been coupled to achieve high speed brazing of these complex joints with non-contact tracking and inspection. Special techniques have been developed to achieve accurate initial positioning of the laser spot and brazing wire tip at the right position on the joint. The comparison of joint profiles measured by the laser cameras before and after brazing also allows inline measurement of the main features of the brazed joint.

This robotic brazing technology was made possible by the integration of many recent technical developments such as high power lasers and laser-diodes, high speed and powerful microprocessors, high precision heavy duty robots, high speed high resolution laser cameras and sophisticated software. The synergistic cooperation of these technologies allows the production of lighter, less costly, more durable and more appealing cars.

Acknowledgments

The authors would like to acknowledge the contribution to this development of many co-workers at Servo-Robot Inc. and also within many car manufacturers.

References

[1] A. Luft, «Diode lasers in car body construction,» *Laser Technic J.,* vol. 10, no. 5, 2013.

[2] A. Luft, «Diode lasers and remote welding (and brazing),» *LaserTechnic J.,* vol. 11, no. 5, 2014.

[3] G. Filliard et al, «Industrial fluxless laser weld-brazing process of steel to Al at high brazing speed,» *J. Manuf. Proc.,* vol. 25, 2017.

[4] H. Kuwahara and al, "Vehicule roof structure and method for manufacturing". Patent US20170233014, Honda patent, 18 August 2017.

[5] S. Ernst, «Studies of laser brazing with regard to the quality influencing parameters,» U. Loughborough, D. Thesis, 2015.

[6] M. Heitmanek et al, «Laser brazing with beam scanning, experimental and simulative analysis,» *Phys. Procedia,* vol. 56, 2014.

[7] C. Mittelstadt et al, «Two-beam laser brazing of thin sheet steel for automotive using CuSiMnSn,» *Phys. Proc.,* vol. 56, 2014.

[8] H. Hornig, "Joining device for non-positive joining by means of a filler material using sensors, 2012". Patent WO2011042292A1, Scansonic, 2012.

[9] P. C. Wang and al, " Multi-heat source laser brazing system and method". Patent US20060261045, GM patent,, November 2006.

[10] N. Palmquist et al, «Laser brazing improvement with twin spot». Brevet US7875827, Volvo patent,, January 2011.

[11] J. Berndl et al, « Joining device for non-positive joining by means of a filler material using sensors». Brevet US20120267349, Scansonic, BMW patent, October 2012.

[12] M. Masahito and al, "Brazing device and brazing method which is capable of excellently brazing". Patent JP2014147954A, Honda patent, August 2014.

[13] F. Esaka et al, «Process and apparatus for welding with two or more laser oscillated beams». Brevet US5841097, Toyota patent, November 1998.

[14] Luft, A.; Laserline;, «Triple spot laser brazing,» *Industrial Laser Solution,* March, 2017.

[15] Laserline, «OTS Optics,» 2017. [En ligne]. Available: https://www.laserline.de/en/processing-optics/ots-optics/brazingoptic.html.

[16] IPG, "Trifocal YLS-5000-BR+Brochure," 2017. [Online]. Available: http://www.ipgphotonics.com/en/products/lasers/high-power-cw-fiber-lasers/1-micron/yls-6000-br.

[17] O. Pütsch et al, «Coaxial beam shaping for laser materials processing,» *Laser-Photonics,* January, 2018.

[18] C. Mittelstadt et al, « Novel approach to increase the energy-related process efficiency and performance of laser brazing,» *Phys. Procedia,* vol. 83 , 2016.

[19] F. Vollertsen and al, "Development for laser joining with high-quality seam surfaces," *Lightweight. Design,* vol. 5, 2017.

Estimation of corrosion resistance of ferritic stainless steel brazed joint by electrochemical method

M. Iwata, Y. Shinoda, Y. Hoshima, R. Shimada, and Y. Miyazawa
Tokai University, Hiratsuka, Kanagawa, Japan
Tel: +81-463-58-1211, E-mail: ymiyazawa@tokai-u.jp

Y. Bizen
Hitachi Metals, Ltd., Minato-ku, Tokyo, Japan

Abstract

Brazing of stainless steel is necessary to develop EGR cooler. Moreover, the use of ferritic stainless steel and Ni based brazing foil are expected in the fabricating field. However, formation of Cr-B intermetallic compounds and Cr-depleted layer beside brazed layer were confirmed when the stainless steel and the brazing foil were brazed in our previous works. Furthermore, deterioration in corrosion resistance around brazed layer was a concern due to the phases. Therefore, we aimed to estimate the corrosion resistances of each phase of brazed joints using electrochemical method. Anodic polarization curve measurement, in-situ observation of corroding microstructures, and SVET analysis were done for the brazed joints in this study. As the result of anodic polarization curve measurement, we consider the measurement is useful to estimate corrosion resistance of brazed joints due to reasonable values were obtained. Moreover, the results of the in-situ observation and SVET analysis showed possibility to estimate corrosion resistance of each phase of brazed joints.

Introduction

These days, manufacturing processes of heat exchanger and EGR cooler for automobiles are investigated energetically to take countermeasures of exhaust gas regulations. Stainless steel has been employed as the material because of its toughness and corrosion resistance; brazing of stainless steel is necessary for such the products. Moreover, the use of ferrite stainless steel that is relatively cheap as the base material and Ni based brazing foil that can reduce the quantity of the filler metal is expected in the fabricating field. However, there are only a few studies about brazing of ferrite stainless steel with Ni based brazing foil.

In our previous works, formation of intermetallic compounds of B, which was contained in brazing foil, and Cr, which was contained in stainless steel, was confirmed [1]. That caused Cr-depleted layer beside brazed layer. Moreover, the deterioration in the corrosion resistance around brazed layer was a concern due to the Cr-B compounds. Nevertheless, effect of the phases of brazed joint on corrosion resistance has not investigated in detail due to conventional test methods that evaluate full of the phases in a brazed joint using weight loss or tensile strength after exposing to corrosive environment.

In this study, ferrite stainless steel was brazed with Ni based brazing foil. Then we aimed to estimate the corrosion resistances of each phase especially the differences among base metal, filler metal, Cr-B compound, and Cr-depleted layer produced due to the creation of the Cr-B compound using electrochemical method.

Experimental Procedure

Base metal and brazing filler metal

SUS444 and MBF20 were employed as a base material and a filler metal, respectively. Moreover, SUS316L, MBF67, and MBF601 were used to discuss corrosion resistance of different phases. Table 1 shows chemical compositions of the stainless steels. Table 2 shows chemical compositions of the filler metals.

Table 1 Chemical compositions of the stainless steels

JIS	AISI	Chemical compositions (mass%)						
		Fe	Si	Mn	Cr	Ni	Mo	C
SUS444	444	Bal.	1.00	1.00	17.00-20.00	-	1.75-2.50	≦0.025
SUS316L	316L	Bal.	1.00	2.00	16.00-18.00	10.00-14.00	2.00-3.00	≦0.08

Table 2 Chemical compositions of the filler metals

Filler metals	Chemical compositions (mass%)							Temperature (℃)	
	Ni	Cr	Fe	Si	B	Mo	P	Solidus	Liquidus
MBF20	Bal.	7.06	3.43	4.92	3.01	-	0.01	969	1024
MBF67	Bal.	25.38	0.08	1.43	0.49	1.09	5.91	890	970
MBF601	43.18	15.97	32.02	0.92	0.35	1.27	6.05	959	1024

Specimens and brazing conditions

Single lap joints were used to estimate corrosion resistance and to observe cross-sectional microstructures at the brazed joints. Figure 1 shows the schematic diagram of the single lap joint.

Brazing was done under vacuum atmosphere. The brazing temperature was 1050℃, and the brazing time was 5 minutes. Making differences in the thickness of the brazed layer and formation of the compound layer, brazed joints holding three times the amount of brazing foils to original were also prepared.

Furthermore, base metals, heat-treated base metals, and filler metals that were heat-treated on base metals were prepared to compare their corrosion resistance with the brazed joints.

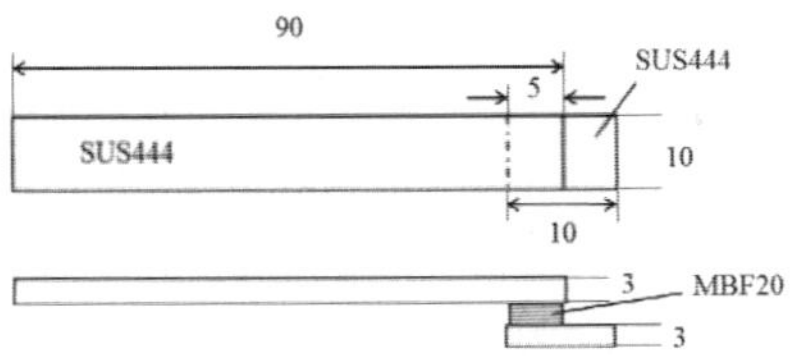

Fig. 1 Schematic diagram of the single lap joint

Anodic polarization curves measurement

Anodic polarization curves were measured using three-electrode electrochemical cell shown in Fig. 2 with a potentiostat. Test solution was 0.05 mol% H_2SO_4. Ag/AgCl reference electrode was used, and Cu electrode as counter electrode. The distance between working electrode and counter electrode were 40 mm.

Each specimen was masked with Kapton tape to control the test area as shown in Fig. 3. Cathodic reduction applying -0.7 V for 10 minutes was done before the test. 5 minutes later, corrosion potential (E_{corr}) was measured and anodic polarization was started from the potential. The voltage sweep rate was 50 mV/min. Maximum applied voltage was 2.0 V higher from E_{corr}.

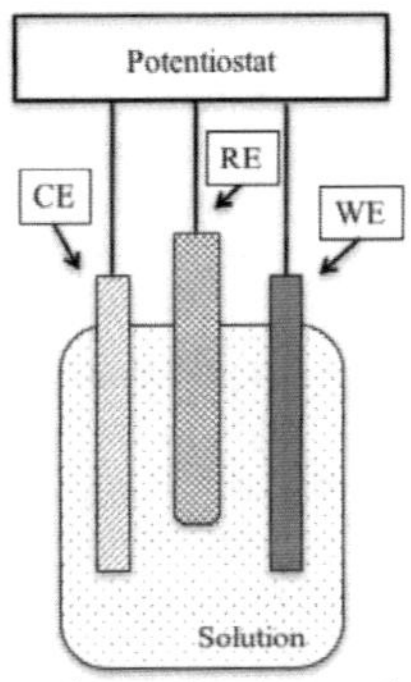

Fig. 2 Three-electrode electrochemical cell

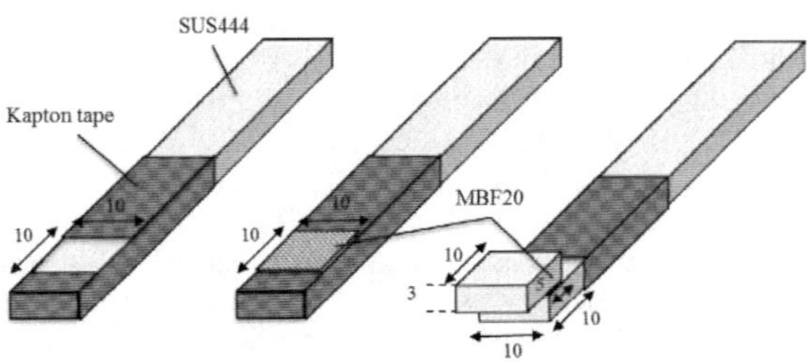

Fig. 3 Masking of the specimens
(a: Base metal, b: Filler metal, c: Brazed joint)

In-situ observation of corroding microstructures of brazed joints

The corrosion behavior of brazed joint was investigated by the in-situ observation. Polarization curve was measured coincident with observation of the cross-sectional microstructures. The method for measuring polarization curve was same as stated above. The voltage application was suspended in stages to observe the microstructures.

SVET analysis

Potential gradient on cross-section of brazed joint was investigated by scanning vibrating electrode technique (SVET). The test solution was tap water.

EPMA analysis

Elemental distribution on cross-sectional microstructure of brazed joints was analyzed using EPMA.

Results and discussion

Estimation of corrosion resistance

Figure 4 shows typical results of the polarization curve measurement. The base metal showed more stable passivation tendency than the filler metal regardless of whether it had been heat treated or not. Polarization curves of the brazed joints were shown between the curves of base metal and filler metal. When the polarization curves of heat treated base metal and filler metal were summed up with consideration to the ratio of its test area occupied with each one, the curve of brazed joint holding three brazing foils showed an almost same curve with the calculated curve as shown in Fig. 5. That indicated the whole of the brazed joint could work as anode. Therefore, anodic polarization of whole of brazed joint was confirmed with voltage sweeping in this method.

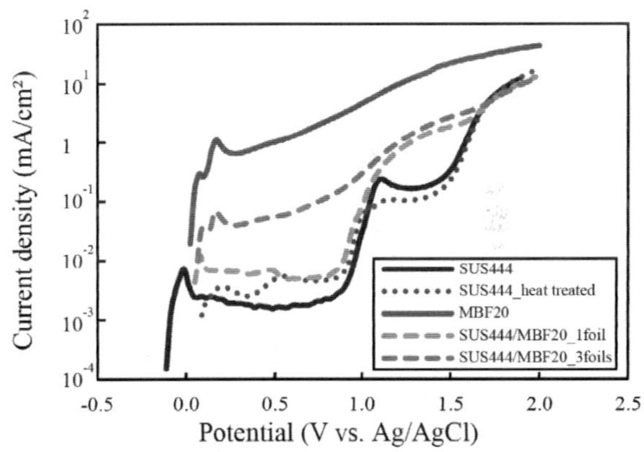

Fig. 4 Results of the polarization curve measurement for SUS444/MBF20 brazed joint

Table 3 shows values of E_{corr} measured before the polarization curve measurement for each specimen. That of the heat treated base metal was the most noble potential. Grain growth of the base metal was expected with the heat treatment, and it decreased grain boundary which was concerned as weak corrosion resistance area on the test surface. Consequently, the heat treated base metal showed more noble E_{corr} than the original base metal. Moreover, the difference in E_{corr} between the base metal and the filler metal was assumed to depend on differences in their chemical compositions.

Slops of the initial stages of the polarization curves which were considered as active stages were calculated to estimate polarization resistance of the heat treated base metal and the filler metal. That of heat treated base metal was quite larger as shown in Fig. 6.

When the brazed joint is put in actual corrosive situation with no other metals, anodic and cathodic polarization is concerned in the joint. The filler metal should be worked as anode and corroded in that case due to the difference in E_{corr} from the heat treated base metal. Furthermore, progress of Galvanic corrosion based on the polarization of the filler metal as anode and the base metal as cathode in the brazed joint with time was considered due to the differences in the polarization resistance and the stability of passivation film.

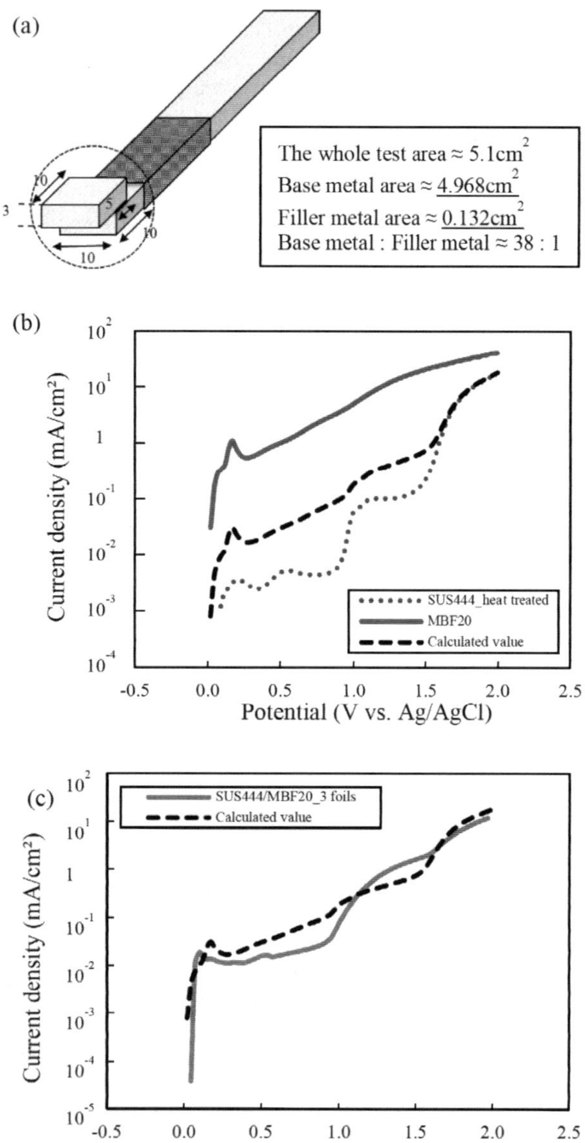

Fig. 5 Estimation of anodic polarization curve of the brazed joint (a: Test area of SUS444/MBF20 brazed joint, b: Calculated curve for the brazed joint, c: Calculated and experimental curves for the brazed joint)

Table 3 E_{corr} measured before the polarization curve measurement

	E_{corr} (V)
SUS444_heat treated	0.065
SUS444/MBF20_1 foil	0.050
SUS444/MBF20_3 foils	0.038
MBF20	-0.002
SUS444	-0.112

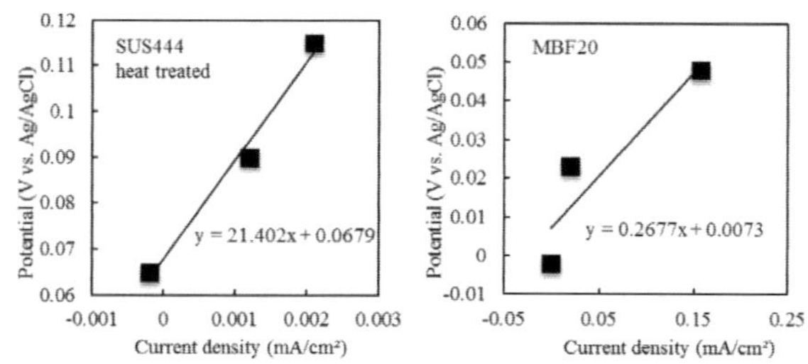

Fig. 6 Estimation of the polarization resistance

In-situ observation of corroding microstructures of brazed joints

Figure 7 shows cross-sectional microstructure of the SUS444/MBF20 brazed joint holding three foils obtained in the in-situ observation. In initial stage of the polarization, corrosion around the Cr-B compounds was observed. Cr-depleted layer caused by formation of the Cr-B compounds was concerned as the cause of the corrosion. In passivation stage, corrosion was not progressed significantly. When the polarization was progressed further, the phases eroded in the order of Cr-depleted layer, residual eutectic, Ni primary crystals, and then grain boundary of the base metal.

SVET analysis

Figure 8 shows cross-sectional microstructure of a brazed joint and potential gradients on the surface in each stage. The left and right images in Fig. 8 show potential distribution and the x-axis scanning data on the surface, respectively. Polarization of each phase was not observed significantly by 50 cycles (27 hours). After 53 cycles (28.5 hours), the filler metal layer showed a little noble potential temporarily, and then it shifted to less noble potential than the base metal soon. Since the polarization, they kept the tendency toward each polarity. That meant the filler metal should work as anode, and the base metal as cathode in the case of Galvanic corrosion at the brazed joint. Due to the results of E_{corr} measurement shown in Table 3, the filler metal was expected to work as anode consistently. However, the results of potential gradient showed the behavior that the base metal worked as anode in intermediate stage of the SVET analysis. The test surface was the cross section; therefore, we thought potential of the base metal on the test surface was different from that of heat treated base metal obtained in Table 3. The values of E_{corr} were slight different between the heat treated base metal and the filler metal; thus, we considered the filler metal showed more noble potential in this measurement in the initial stage. However, the potential had shifted to less noble and started working as anode around 57 cycles (30.5 hours) due to the difference in polarization resistance.

The less noble area around the brazed layer was not as liner as the actual filler metal layer observed using optical microscope. Furthermore, less noble areas were partially observed at the base metal. Therefore, we assumed that the potential gradient depended on each phase and the formation of Cr-B compounds made neighboring phases weak in corrosion resistance in this case. Results of EPMA analysis that focused on these phases were described in next section.

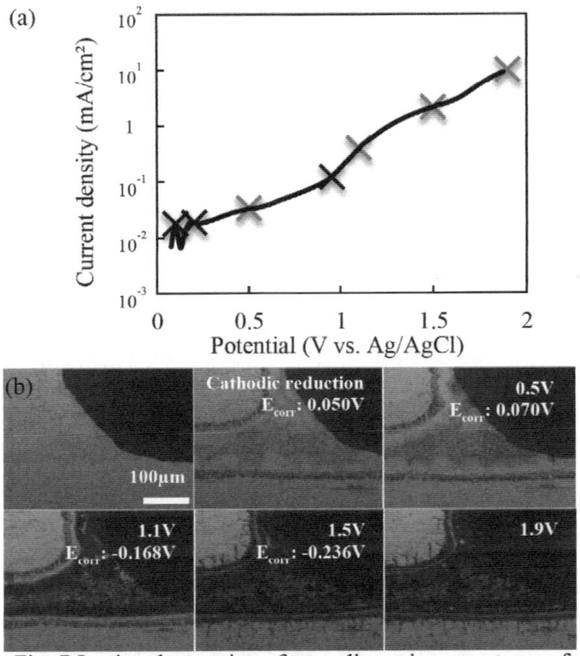

Fig. 7 In-situ observation of corroding microstructures of SUS444/MBF20 brazed joint (a: Polarization curve of the brazed joint holding three foils, b: Cross-sectional microstructures of the brazed joint)

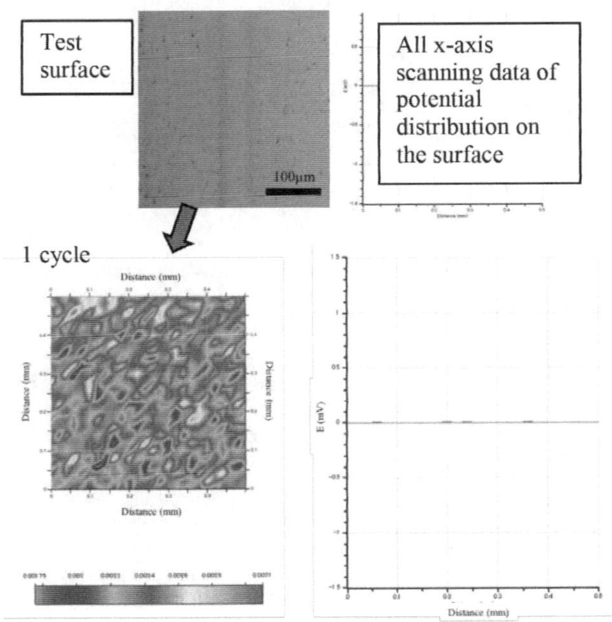

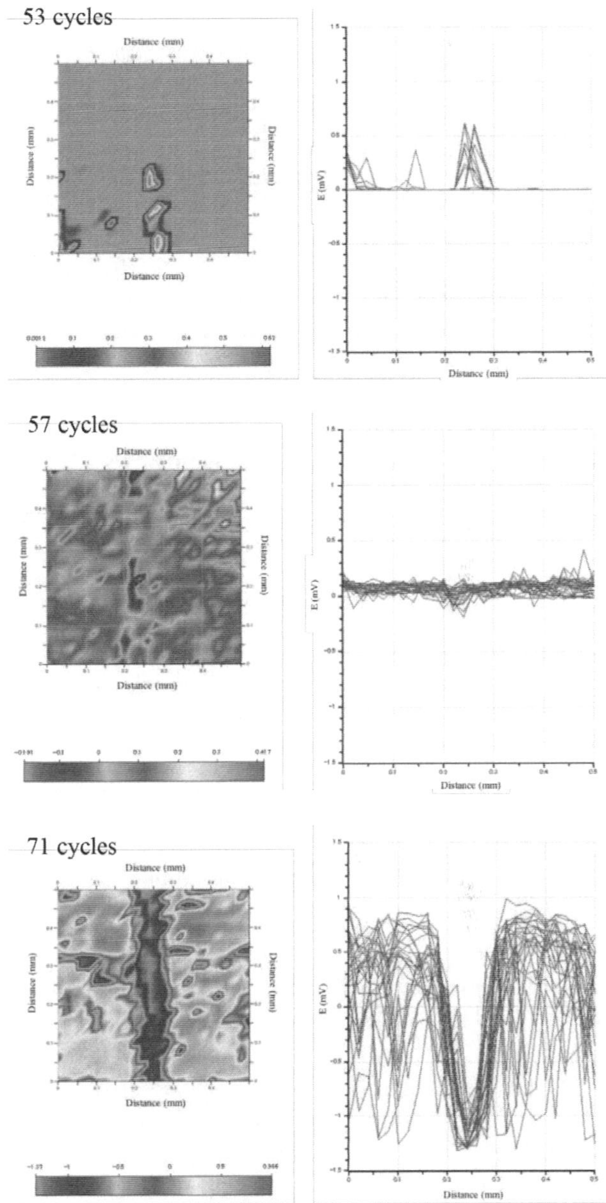

Fig. 8 Cross-sectional microstructure of a brazed joint and potential gradients on the surface

EPMA analysis

Figure 9 shows elemental distribution on microstructure of a brazed joint holding three foils. Formation of Cr-depleted layer around brazed layer was confirmed from the results. The area was less noble than base metal, but more noble than filler metal according to after 71 cycles potential distribution in Fig. 8. The area had been concerned as the least noble; therefore, we thought the Cr-depleted layer was not distinguished from Cr-B compound clearly due to SVET resolution.

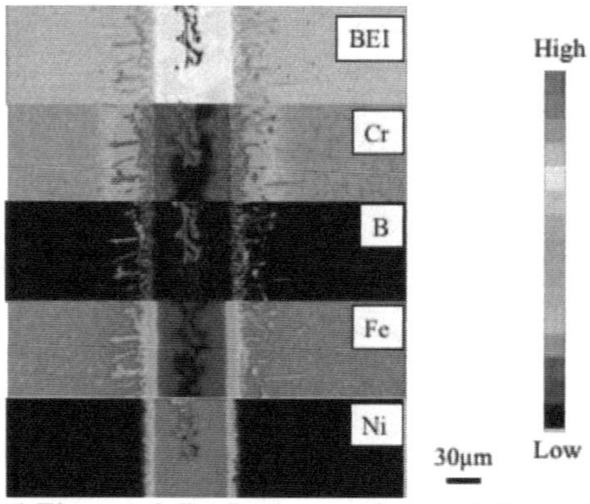

Fig. 9 Elemental distribution on phases of a brazed joint holding three foils

Comparison between SUS444/MBF20 and SUS316L/MBF20 brazed joints

Figure 10 shows results of polarization curve measurement for SUS316L/MBF20 brazed joint. Differences in the polarization curves among the base metal, the filler metal, and the brazed joint showed similar tendency to that of SUS444/MBF20 brazed joint shown in Fig. 4. Moreover, Fig. 11 shows comparison in the polarization curves with SUS444/MBF20 brazed joint. In general, SUS444 has better corrosion resistance than SUS316L due to its pure ferrite phase [2]. SUS316L showed higher current density that was associated with lower corrosion resistance than SUS444; however, SUS316L/MBF20 brazed joint showed lower current density than SUS444/MBF20 brazed joint.

Figure 12 shows result of the in-situ observation for SUS316L/MBF20 brazed joint holding three foils. Almost no differences in the corrosion behavior from that of SUS444/MBF20 brazed joint were observed. Therefore, we thought the differences in polarization curves between the brazed joints were affected by amount of filler metal area exposed on the test surface especially fillet.

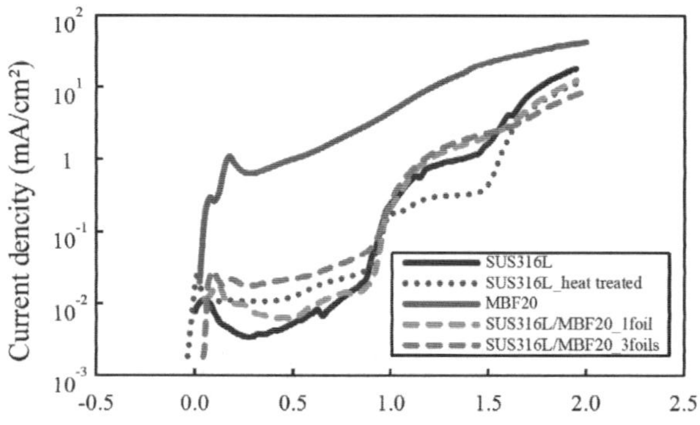

Fig. 10 Results of polarization curve measurement for SUS316L/MBF20 brazed joint

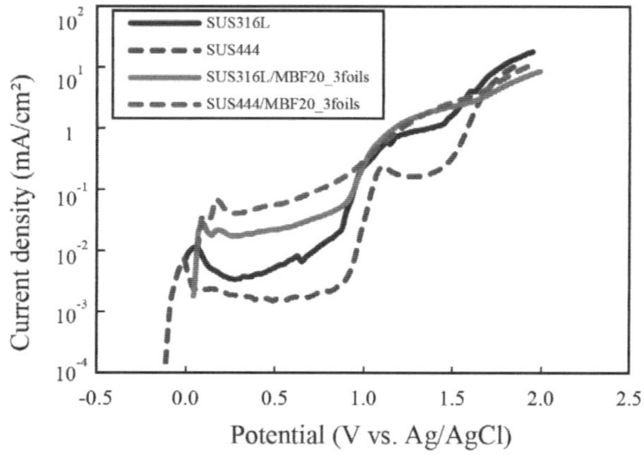

Fig. 11 Comparison in the polarization curves between SUS444/MBF20 and SUS316L/MBF20 brazed joints

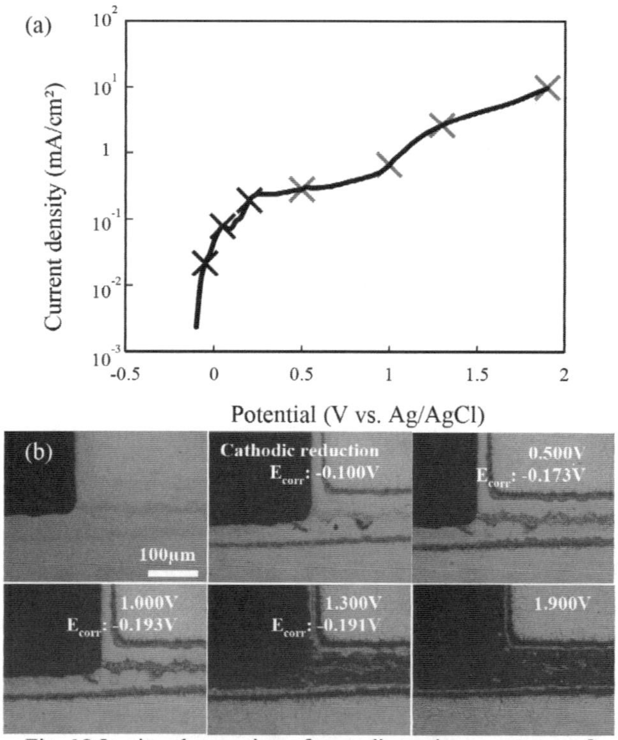

Fig. 12 In-situ observation of corroding microstructures of SUS316L/MBF20 brazed joint (a: Polarization curve of the brazed joint holding three foils, b: Cross-sectional microstructures of the brazed joint)

Comparison among SUS444 brazed joints with MBF20, MBF67, and MBF601

Figure 13 shows results of polarization curve measurement for MBF20, MBF67, and MBF601. These curves had different shapes, but that of MBF601 showed lower current density overall. We had thought that differences in corrosion resistance among the filler metals would be caused by their Cr content; however, we could not understand the differences in the polarization curves with only the Cr content. Increase of B content in filler metal decrease its corrosion resistance in general [3]. Therefore, we considered the difference between MBF20 and the others was caused by its B content. On the other hand, we thought the difference between MBF67 and MBF601 was caused by their P or Fe content.

Figure 14 shows polarization curves of SUS444 brazed joints with MBF20, MBF67, and MBF601. The differences among these curves had similar tendency with that of the filler metals shown in Fig. 13. Moreover, Fig. 15 and Fig. 16 show results of the in-situ observation for SUS444 brazed joint holding three foils of MBF67 and MBF601, respectively. Their corrosion behaviors were quite different from that of SUS444/MBF20 brazed joint. In comparison with SUS444/MBF20 brazed joint, corrosion of base metal of these two brazed joints was not observed. Therefore, we considered less noble phase caused by existence of filler metal or interfacial reaction at brazed layer affected on the corrosion behavior and corrosion resistance.

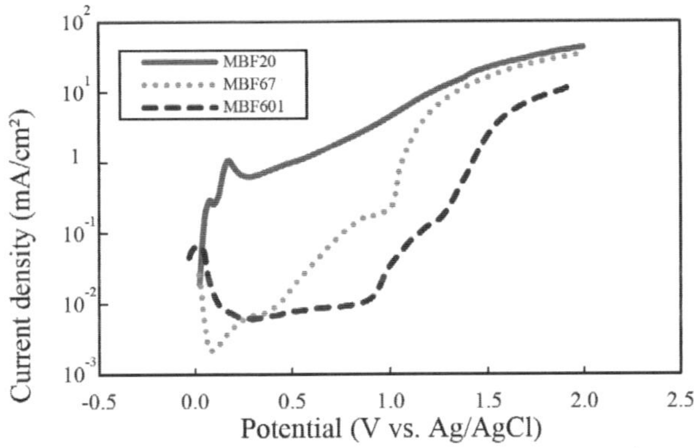

Fig. 13 Results of polarization curve measurement for MBF20, MBF67, and MBF601

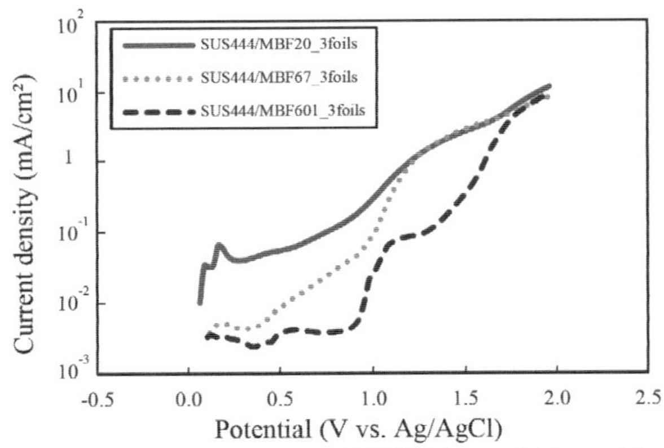

Fig. 14 Polarization curves of SUS444 brazed joints with MBF20, MBF67, and MBF601

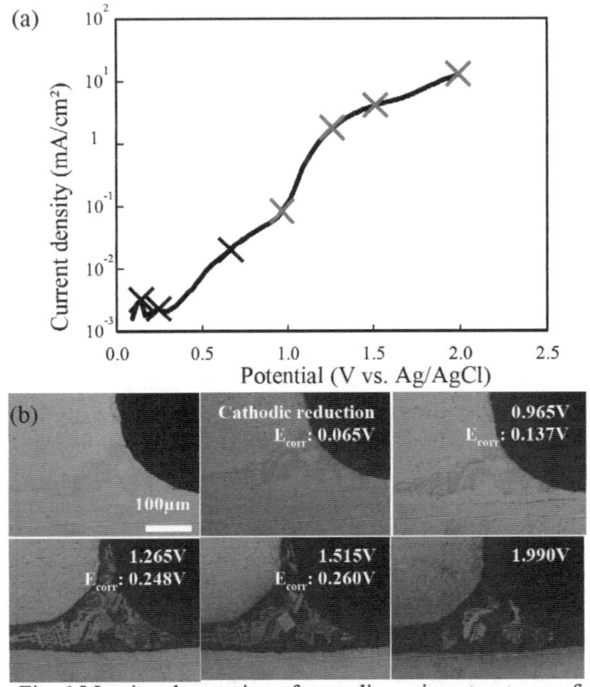

Fig. 15 In-situ observation of corroding microstructures of SUS444/MBF67 brazed joint (a: Polarization curve of the brazed joint holding three foils, b: Cross-sectional microstructures of the brazed joint)

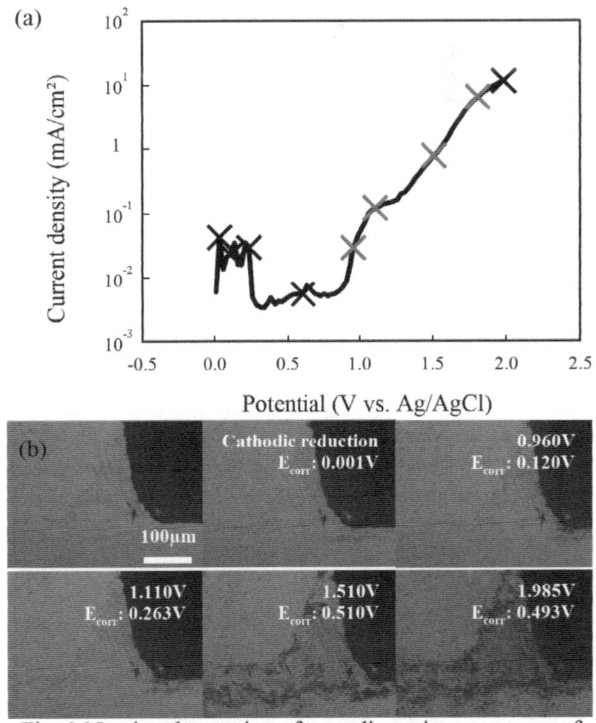

Fig. 16 In-situ observation of corroding microstructures of SUS444/MBF601 brazed joint (a: Polarization curve of the brazed joint holding three foils, b: Cross-sectional microstructures of the brazed joint)

Conclusions

Corrosion resistance of stainless steel brazed joints were investigated using electrochemical methods. Following results were obtained:

- Anodic polarization curves of brazed joints were obtained using a three-electrode electrochemical cell.
- The anodic polarization curves of brazed joints showed reasonable values; therefore, we consider the measurement is useful to estimate corrosion resistance of brazed joints.
- From the results of the in-situ observation and SVET analysis, we suggest it is possible to estimate corrosion resistance of each phase.

References

[1] Ohta, K. et al, Preprint of the National Meeting of Japan Welding Society (in Japanese), Vol. 95 (2014), pp. 264-265.

[2] Azuma, S. et al, "Localized Corrosion of SUS 316L/SUS 444 Powder Mixed Duplex Stainless Steel," Boshoku Gijutsu (in Japanese), 38 (1989), pp. 218-223.

[3] Matsu, K., Welding Technology (in Japanese), Vol. 59 (2011), pp. 54-57.

Influence of Gap Clearance on Brazing Joint Strength for Ni and Fe Based Filler Metals

Eddie Hedin
Höganäs AB, Höganäs, Sweden
Eddie.Hedin@hoganas.com, +46721460868

Abstract

The effect on brazing joint strength as gap clearance varies has been evaluated in this report. It is a known phenomenon that some brazing filler materials form stronger bonds when the gap clearance decreases, and this investigation adopts a practical approach at evaluating some of the modern Ni-based filler metals to determine if - and how much - strength increases as gap clearance decreases for those materials. Preliminary results show that joints brazed with the Ni-based fillers: BNi-2, BNi-5, BNi-7 seem to increase in strength as gap clearance goes under 50-60μm. This increase in strength may be caused by changes to the microstructure which are known to occur in those gap clearance regions. It has been confirmed in this report that at least for BNi-2, there is a significant shift in failure mode of the brazing joint as the microstructure changes. For large gaps (<100μm) the fracture occurs in the brittle centreline of the joint, and for smaller gaps (~50μm) a sort of delamination occurs in the interface between the base material and the filler metal.

No strength increasing effect has been observed for the filler metals Ni613 and BrazeLet® F300, however that is likely because there have been no tests at low enough gap clearances yet. More data is being gathered and analysed during the spring to compliment the preliminary results shown in this report.

Introduction

There are many different types of possible joint configurations when brazing components or test specimens. In practice, the joints produced in a component are often very complex and thus it is difficult to know the strength of the bond without performing some sort of destructive testing.

Certain rules of thumb have been established by the brazing community, such as the example for the single lap shear joint: "if the thickness of two parts to be joined is: "t", then a brazing overlap of "6t" will create a bond stronger than the materials being joined". Rules of thumb can be very helpful at times, but sometimes a deeper knowledge of how different parameters affect joint strength may be advantageous. This report will attempt to explain the influence of gap clearance on joint strength specifically in a butt joint configuration.

A lot of work has already been put into characterizing the influence of gap clearance on joint strength, where a notable example would be the results from Handy and Harman Laboratories in the 1930's as is discussed by Dan Kay in his article: "The famous joint strength vs joint clearance chart" [1]. This chart is shown in Figure 1 below.

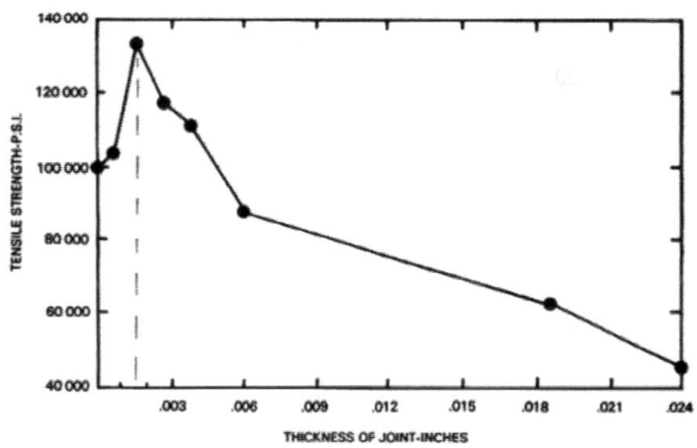

Figure 1. Joint clearance vs joint strength [1].

There's a peculiarity with the graph above which is worth mentioning, specifically the left part, which suggests a drop-off in joint strength as gap clearance becomes very small (somewhere below 0,002in / 50μm). As is discussed by Dan Kay in his article, this is likely not true in the sense that it is not gap clearance directly that causes a decline in strength. There are other parameters at play, such as the increasing influence on strength from voids and impurities as gap clearance decreases.

Either way it is an accepted notion in the brazing community that strength does indeed increase as gap clearance decreases, regardless of any potential disagreements concerning the cause. The work in this paper has been aimed towards re-evaluating this phenomenon with a practical approach for the (relatively speaking) newer Ni-filler metals: BNi-2, BNi-5, BNi-7 and Ni613 as well as Höganäs patented iron based alloy BrazeLet® F300. These materials add another level of complexity to the problem, as Ni-filler metals tend to form brittle eutectic centerlines in certain gap clearance intervals.

Material and Method

In this report, joint strength was investigated according to the standards in Table 1.

Table 1. Standards used in sample preparation and tensile testing.

EN10002-1 **Metallic materials – Tensile testing** – Part 1: Method of test at ambient temperature
EN12797 **Brazing – Destructive tests of brazed joints**

5 materials were evaluated in this investigation, see Table 2.

Table 2. Filler metal materials evaluated in this report (ISO 17672:2016 - Brazing - Filler metals).

Höganäs	AWS	ISO	Composition
BrazeLet® Ni2	BNi-2	Ni620	Ni82CrSiBFe
BrazeLet® Ni5	BNi-5	Ni650	Ni71CrSi
BrazeLet® Ni7	BNi-7	Ni710	Ni76CrP
BrazeLet® Ni613*	-	-	Ni60CrPSi
BrazeLet® F300	-	-	Fe39CrNiCuPSi

*Ni613 is a licensed product, patented by Fukuda Metal Foil & Powder co.,ltd.

Test specimen

The chosen test specimen geometry is the butt joint [3]. Two halves made from 316L stainless steel were brazed together with a specific gap clearance. Two different ways were used to reach specific gap clearance. The first is as recommended by EN10002-1, the second is Höganäs' own adaptation to the method in order to improve control over gap clearance.

1. Machined steps.
   30μm and 100μm steps were used.

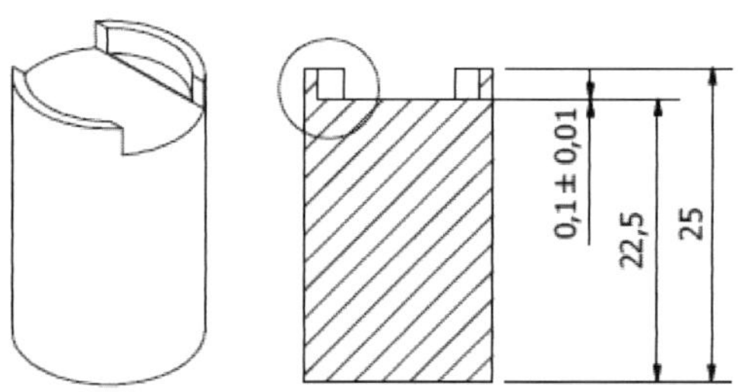

Figure 2. Bottom half of test specimen.

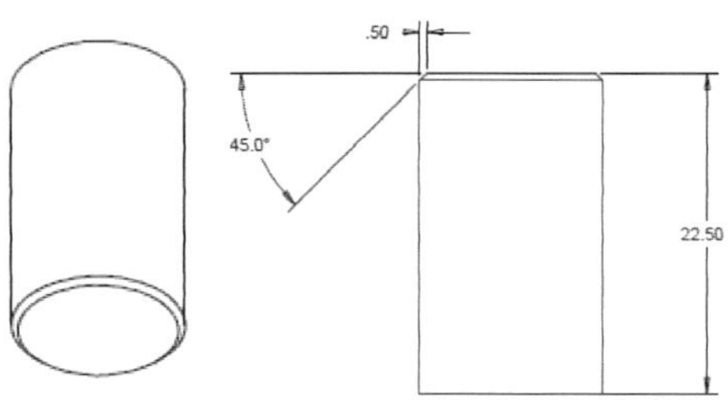

Figure 3. Top half of test specimen.

To illustrate the difficulties with machining steps for determining the gap clearance, Figure 4 and Figure 5 show a 30μm and 100μm step respectively, as cut and observed in an optical microscope. As the step becomes smaller, it becomes increasingly difficult to manufacture due to tolerances in the machining operation. In order to produce test specimens with even smaller gap clearances, another method, which is described below, was used.

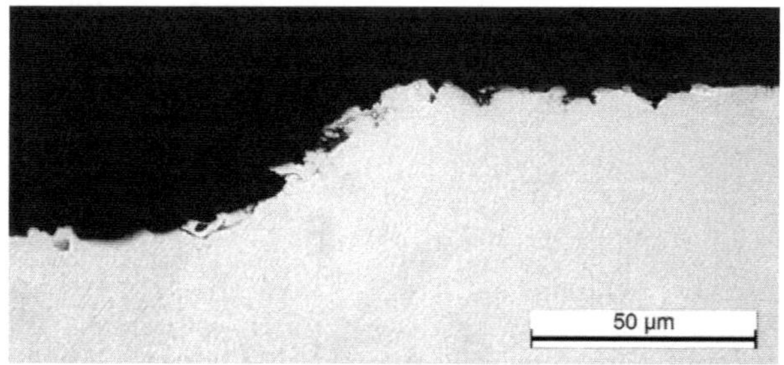

Figure 4. 30μm machined step.

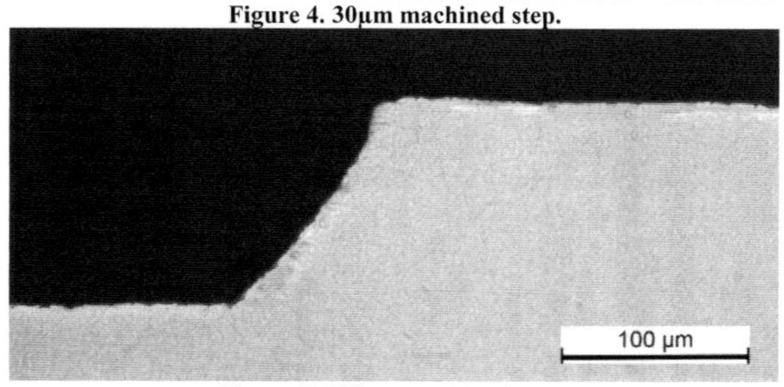

Figure 5. 100μm machined step.

2. Wire displacers made from 316 stainless steel.
   Stainless wires/thread were used as spacers to reach specific gap clearance. The bottom piece shown in Figure 6 did not have a step machined.

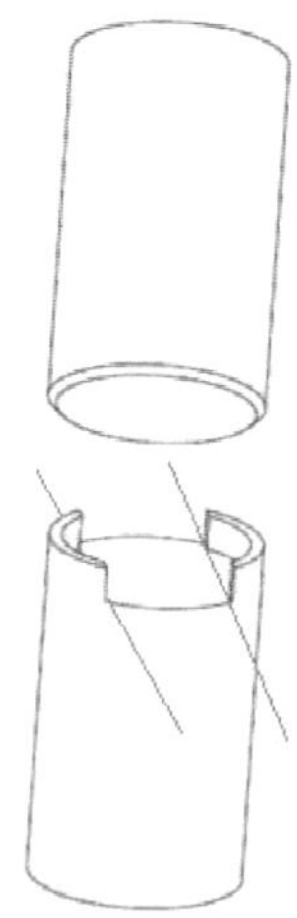

Figure 6. Gap clearance achieved through the use of thin (316 stainless steel) wire inserted between the two test pieces.

Table 3 shows the different thickness of wire used.

Table 3. Wire thickness used to reach specific gap clearances.

#	Wire thickness [μm]
1	"0" (no wire)
2	25μm
3	41μm
4	78μm
5	300μm

Specimens were placed in a fixture and brazing filler metal was applied as a paste as shown in Figure 7. The paste contained powder particles sieved to <63μm.

Figure 7. Sample prepared for brazing. Paste applied outside the gap between the pieces.

Figure 8. Sample after brazing.

4 test specimens were made per material and gap clearance.

Brazing process

Brazing was performed in a vacuum furnace under high vacuum (~10^{-5} mbar) at the recommended brazing temperature for each filler metal:

Table 4. Summary of tested filler metals and brazing temperatures used.

Filler metal	Brazing temperature
BNi-2	1050°C / 1922°F
BNi-5	1150°C / 2102°F
BNi-7	970°C / 1778°F
Ni613	1090°C / 1994°F
F300	1100°C / 2012°F

The temperature ramp was 9°C/minute and a 30 minute hold at 450°C was implemented to guarantee that the binder in the paste had burnt off completely. Indirect cooling in vacuum (switched off heating elements) was used in all tests.

Tensile test

After brazing, the assembly was machined to a tensile rod according to the drawing in Figure 9 and tested axially in a Zwick/Roell 10kN tensile testing machine. The tests were performed in room temperature according to EN12797 [4]. The diameter of the thin section has been changed to 4.0mm from 4.5mm (4.5mm is recommended by EN12797). This change was made to guarantee that the tensile testing machine would be capable of fracturing the specimens.

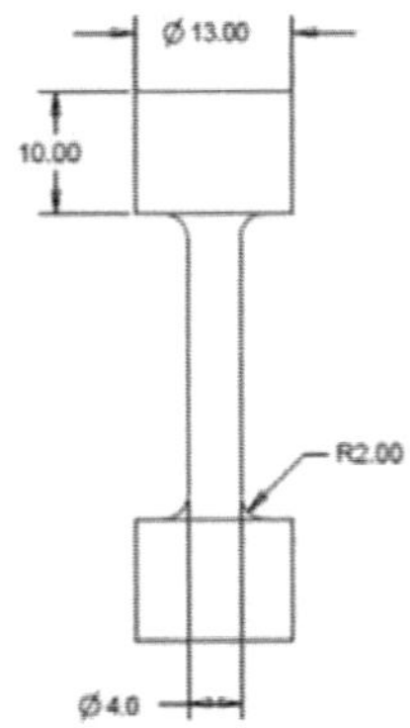

Figure 9. Tensile test specimen drawing.

Before tensile testing, gap clearance was evaluated non-destructively in a Lecia DM6000 light optical microscope. The gap clearance of each individual specimen was recorded and corresponded against the ultimate tensile strength of that specific tensile test.

Lastly, the fracture surfaces were investigated in an attempt to determine the mode of failure in each joint. The tested tensile specimens were fitted back together using Loctite 407 super glue - then mounted, polished and observed in a Leica DM6000 light optical microscope.

Homogeneous stainless steel 316L reference

Homogeneous 316L tensile test specimens were also produced. The geometry was the same as for the brazed samples (see Figure 9). These specimens were used to compare the strength of brazed samples to the actual strength of the base material.

Micro hardness measurements

As a complimentary investigation, Vickers hardness was measured using a Matsuzawa MMT-7 micro hardness tester to show the large difference in hardness and thus brittleness between different phases in the microstructure of the materials. Large gaps (>100μm) were tested across their thickness. Indentations were made in 40μm intervals.

Results

The results presented at this time in the report are from the preliminary trials and as such, are incomplete. A large-scale investigation is ongoing as of the time of writing. Several tests at different gap clearances will be performed per material in order to fill in the gaps in the diagrams below.

Strength vs gap clearance

The strength values from the tensile tests have been normalized against the highest value observed for each material. This means that the y-axis shows strength between 0-1 where 1 is the maximum observed strength for each material. This makes it possible to compare the "behavior" of each material.

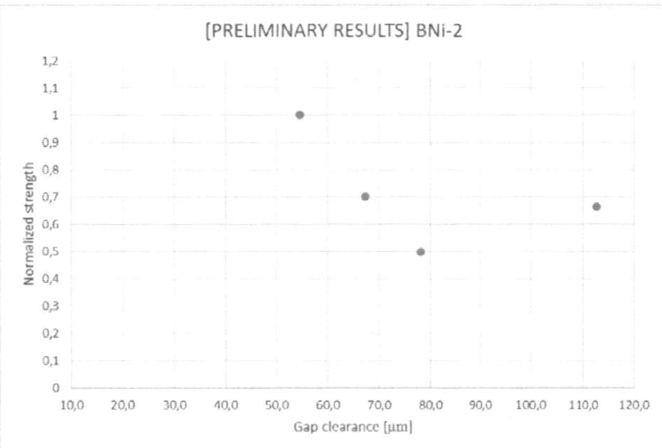

Figure 10. Strength vs gap clearance for BrazeLet® Ni2.

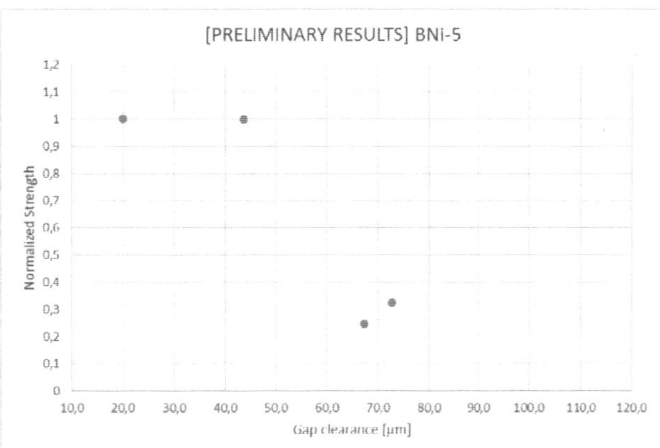

Figure 11. Strength vs gap clearance for BrazeLet® Ni5

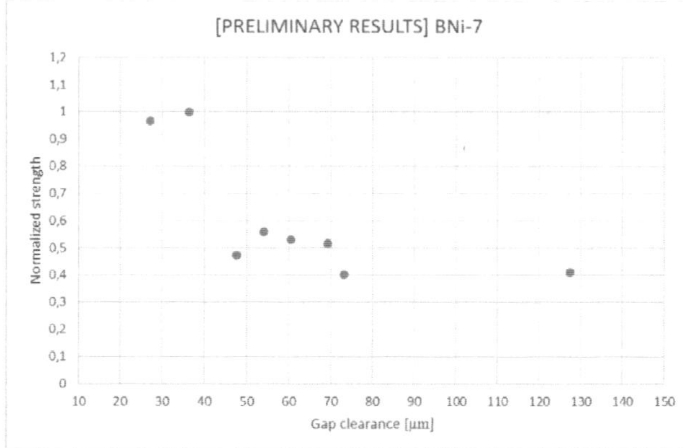

Figure 12. Strength vs gap clearance for BrazeLet® Ni7.

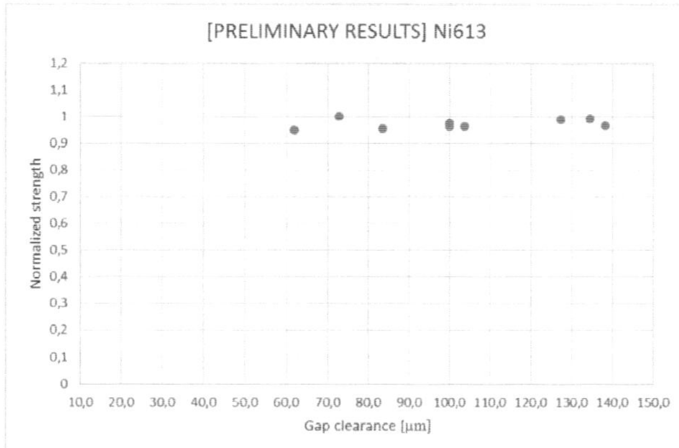

Figure 13. Strength vs gap clearance for BrazeLet® Ni613

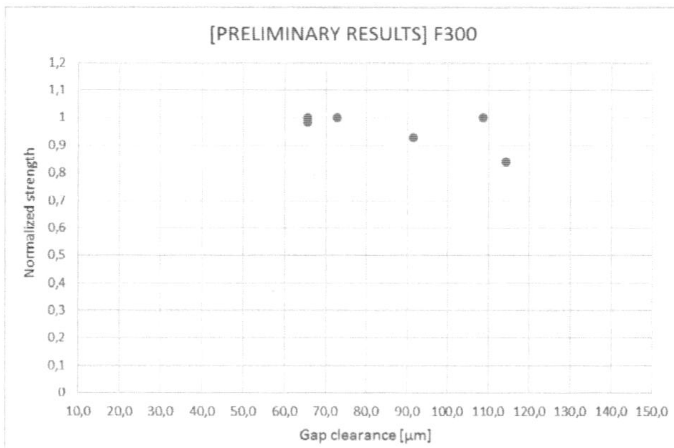

Figure 14. Strength vs gap clearance for BrazeLet® F300.

Fracture surface investigation

Different modes of failures were observed depending on the microstructure of the brazing joint. As an example, with BNi2 when the gap is large (>100μm) the fracture occurs in the continuous brittle centerline (dark phase). However, when the brazing gap clearance was smaller (~50μm) the failure mode that was observed is best described as a form of delamination, compare pictures in Figure 15.

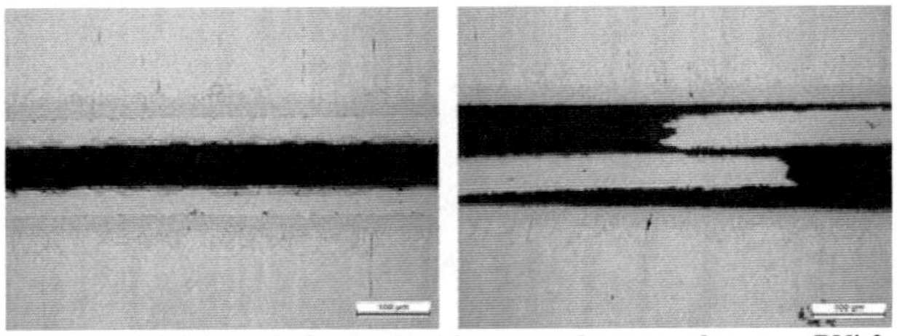

Figure 15. Difference in fracture mode as gap clearance changes – BNi-2.

Stainless steel 316L strength

RESULTS PENDING - will be part of the finished paper.

Large gap fracture modes

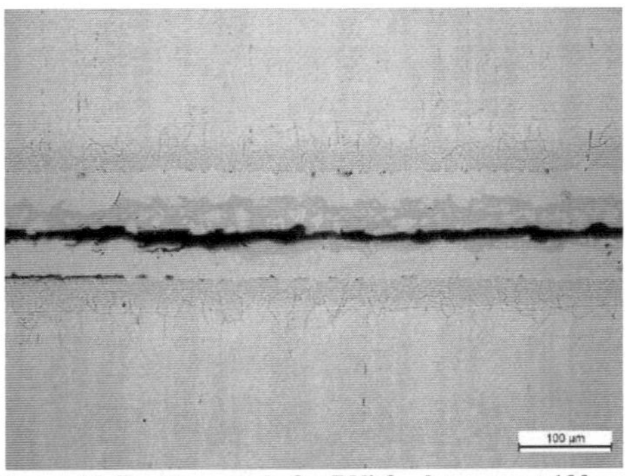

Figure 16. Fracture mode for BNi-2 – large gap >100μm.

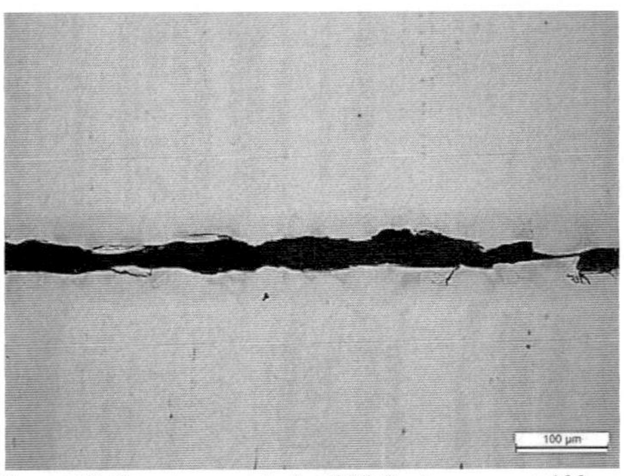

Figure 17. Fracture mode for BNi-5 – large gap >100μm.

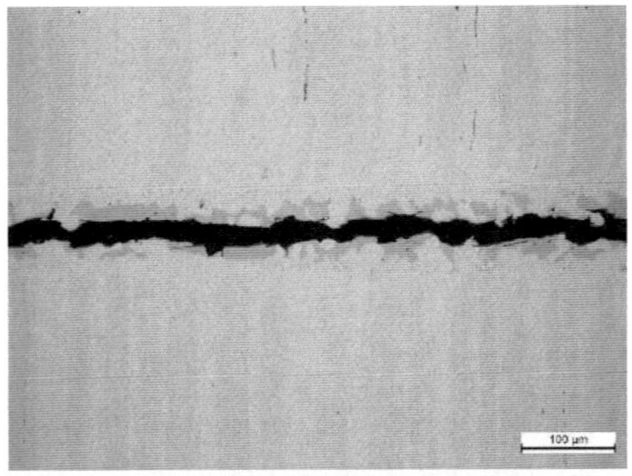

Figure 18. Fracture mode for BNi-7 – large gap >100μm.

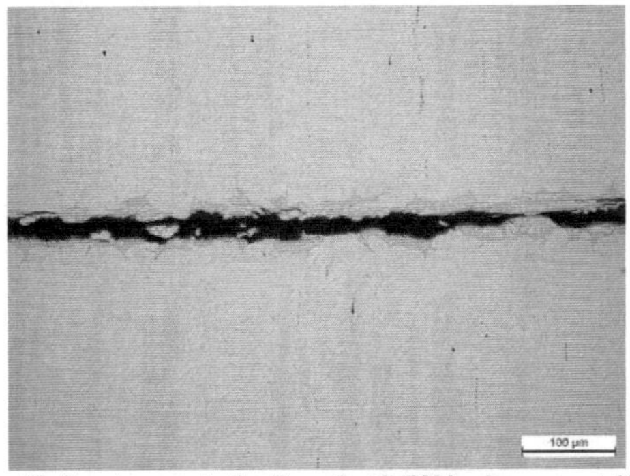

Figure 19. Fracture mode for BrazeLet® F300 – large gap >100μm.

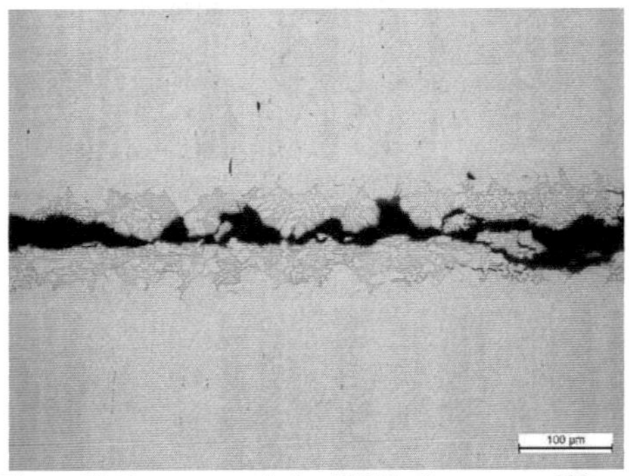

Figure 20. Fracture mode for Ni613 – large gap >100μm.

Small gap fracture modes

RESULTS PENDING - will be part of the finished paper.

Micro hardness

The results from the micro hardness tests show that there is a significant difference in hardness for the different phases. Figure 21 shows the microstructure of Ni613 in a large gap (>100μm) and the occurrence of 3 visually distinctive phases. Even by only looking at the images from the hardness test (Figure 22 and 23), it is obvious that there is a difference in brittleness of the phases marked 1 and 3. Clearly visible cracks have propagated from the edges of the indentation tool for the brittle phase.

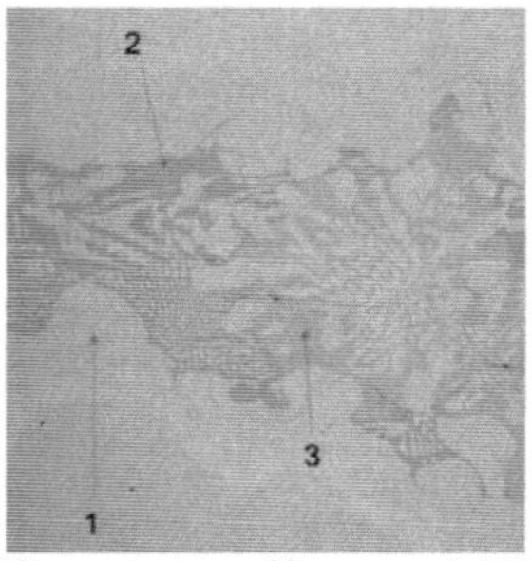

Figure 21. Microstructure of large gap (>100μm) Ni613.

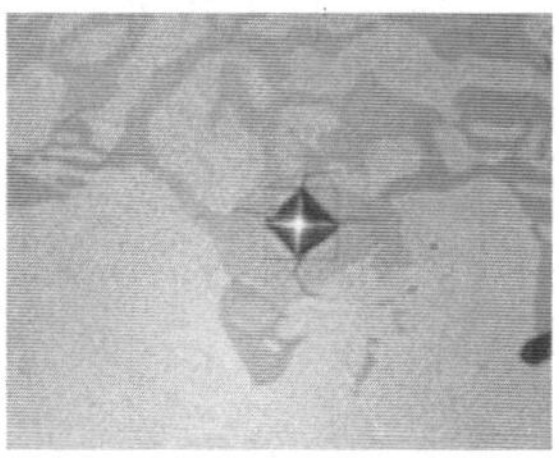

Figure 22. Indentation mark of phase 3 as shown in Figure 21.

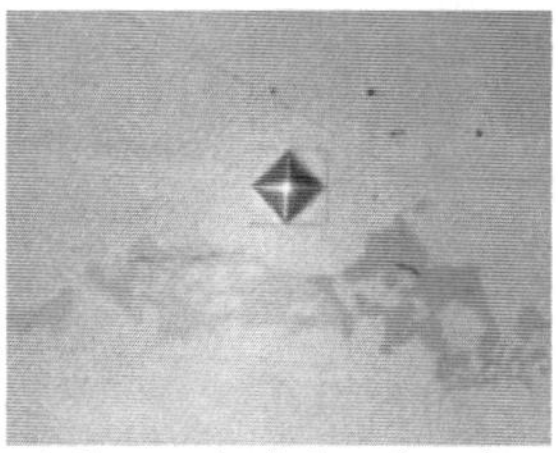

Figure 23. Indentation mark of phase 1 as shown in Figure 21.

A hardness profile of all the evaluated materials in this investigation is shown in Figure 24 below. It is clear that all the materials form hard phases in the centreline, to some extent, when the gaps are large (>100μm).

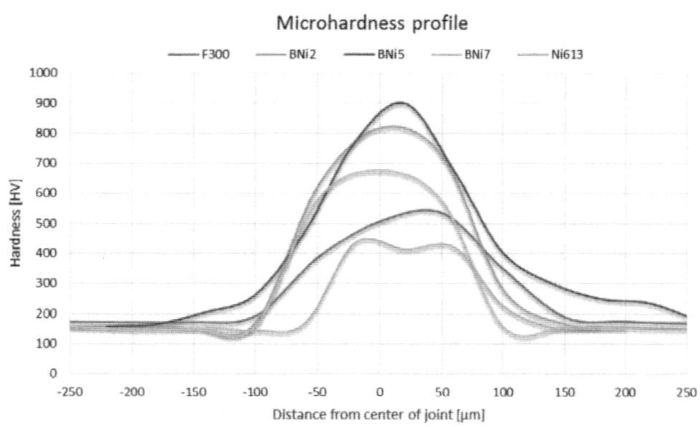

Figure 24. Microhardness profile for the tested materials BNi-2, BNi-5, BNi-7, Ni613 and F300.

Discussion

It is difficult to say anything with certainty at this time but judging by the preliminary results, there seems to be an influence of gap clearance on joint strength for at least some of the tested materials. The observed increase in strength may be entirely related to microstructural changes as it has been shown that there is a large difference in hardness of the phases that are produced when using these materials in large gaps (<100μm). As the failure mode also seems to change, at least for BNi-2, a preliminary conclusion is that the change in strength is purely due to changing microstructure. However, more tests are needed, especially in the lower gap clearance intervals in order to say anything for certain.

References

[1] The famous joint strength vs joint clearance chart – Dan Kay - https://vacaero.com/information-resources/vacuum-brazing-with-dan-kay/1300-the-famous-joint-strength-vs-joint-clearance-chart.html. Time of source: 2018-01-11.
[2] ISO 17672:2016 - Brazing - Filler metals
[3] EN10002-1 Metallic materials – Tensile testing – Part 1: Method of test at ambient temperature
[4] EN12797 Brazing – Destructive tests of brazed joints

Ni-Cr-Si-P and Ni-Cr-Si-B high corrosion resistance brazing of stainless steel: a new interpretation of single lap-joint (lap-shear) data.

David J. Kemmenoe
Cornell University Department of Mechanical and Aerospace Engineering, Ithaca, NY USA
djk293@cornell.edu

Eric A. Theisen and William Coughlan
Metglas Inc., Conway, SC USA

Shefford P. Baker
Cornell University Department of Material Science and Engineering, Ithaca, NY USA

Introduction

Nickel brazing of stainless steel has become the go-to solution to the problem of manufacturing thin sheet steel plate-fin heat exchangers that frequently operate at high temperatures in corrosive environments. An industry trend over the past 15 years has been to shift from Ni-Cr-Si-B filler metals to Ni-Cr-Si-P due to their increased corrosion resistance and lower brazing temperatures. This shift in composition has come with the unfortunate side effect of multiphase microstructures that frequently degrade joint mechanical properties.

In this study, we braze 316 stainless steel (SS) with Metglas Brazing Foil (MBF) 51 as a representative of the Ni-Cr-Si-B family and MBF 67 as a representative of the Ni-Cr-Si-P family to study the mechanical properties of these braze joints. The composition and melting characteristics of MBF 51 and MBF 67 are given in Table 1.

We begin by characterizing the macroscale mechanical properties of these two filler metal families. To do this, we use the single lap-joint (SLJ) shown in Fig. 1, also known as lap-shear joint, and follow AWS C3.2. We then go one step deeper by considering how the real braze stress state differs from the commonly used average shear interpretation of SLJ data. We do this on a qualitative basis with basic mechanics principles. We then examine the microstructure in detail through imaging and indentation to explain trends in the SLJ data.

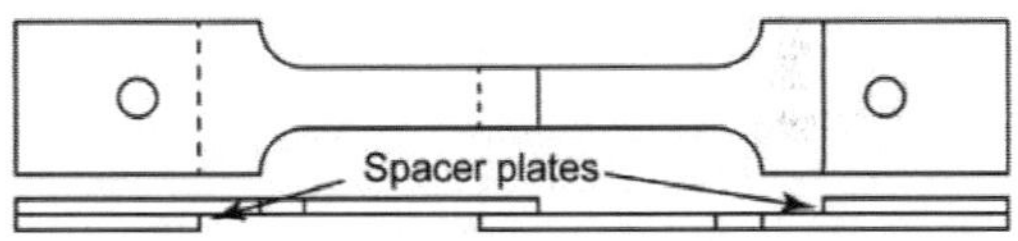

Figure 1: Brazed single lap-joint sample with spacer plates.

Experiments & Results

Three sets of SLJ samples were brazed under high vacuum (10^{-4} Torr). One set each of MBF 51 and MBF 67 SLJ's were brazed for 60 minutes at 1195°C. A second set of MBF 67 SLJs were brazed at 1090°C for 15 minutes. These samples were loaded into a standard tension test machine using spacer plates (Fig. 1) to account for the offset nature of SLJs and pulled to failure at a constant displacement rate of 1 mm/min.

Brazed cross-sections were polished and then imaged using scanning electron microscopy (SEM) to interrogate the structure of each joint. In Fig. 2A we see that thin MBF 51 joints form a single-phase braze microstructure and second phase precipitates at the SS grain boundaries. The MBF 67 microstructure in Fig. 2B is multiphase and lacks the grain boundary precipitates seen in Fig. 2A.

Table 1: Composition and melting characteristics of brazing foils.

MBF	AWS	Nominal Composition, wt %								Melting Temp. °C (°F)	
		Cr	Fe	C	B	Si	P	Mo	Ni	Solidus	Liquidus
51	BNi-5b	15.0	-	0.06	1.4	7.25	-	-	~76.3	1030 (1886)	1126 (2058)
67	-	25	<1	-	-	1.5	6	1.5	~64.5	890 (1634)	970 (1778)

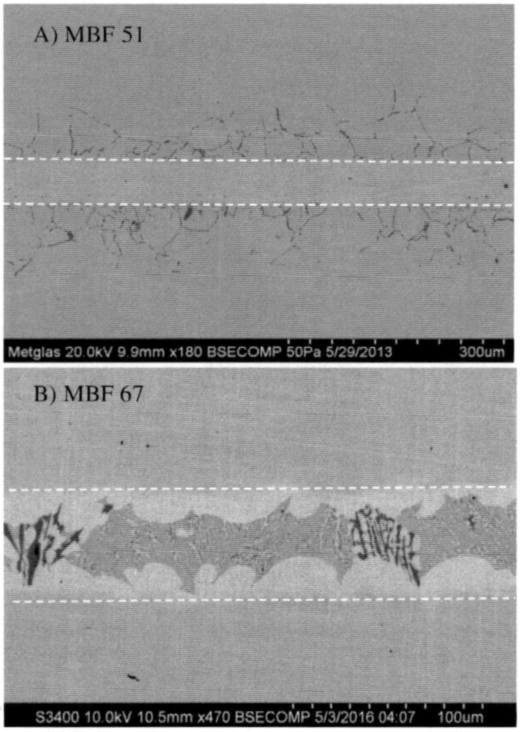

Figure 2: SEM images of MBF 51 (A) and MBF 67 (B). Dashed white lines indicate the boundary between the braze joint and the base metal.

Discussion

The primary differences in microstructure between MBF 51 and MBF 67 are based on the difference in diffusivity of B compared to P. It is well known that B diffuses along SS grain boundaries with a net flux out of the joint. The removal of B raises the remaining material's melting point enabling isothermal solidification of a single-phase Ni solid solution within the braze. The B in the SS grain boundaries reacts with Cr from the SS to form the second phase precipitates decorating the SS grain boundaries in Fig. 2A. This concentration of Cr leaves the surrounding SS depleted in Cr and lowers the corrosion resistance. P on the other hand has minimal flux across the base metal boundary. This results in a multiphase microstructure because P has minimal solubility in the Ni solid solution.

We find that the brazed SLJ stress state is more complex than the common interpretation of average shear. The stress state can be further complicated by effects based on base metal material properties.

Conclusions

Brazed SLJ data can only be explained with braze microstructure when appropriate considerations are taken for how the SLJ geometry and base metal properties affect the braze stress state. When these stress state considerations are properly accounted for, a new interpretation of brazed SLJ data can be developed which reveals new information that was not apparent with the traditional average shear interpretation.

Effect of active element on crack occurrence in Al_2O_3/SS 430 brazed joints

Hoejun Heo [1], Jeongyong Lee[1], Keeyoung Jung [2], Youngson Choe [3], Chungyun Kang [1]

1 Department of Material Science and Engineering, Pusan National University, South Korea; hjheo@pusan.ac.kr, kangcy@pusan.ac.kr

2 Energy Storage Materials Research Group, Research Institute of Industrial Science and Technology (RIST), South Korea; keeyoung.jung@rist.re.kr

3 Department of Polymer Science and Chemical Engineering, Pusan National University, South Korea choe@pusan.ac.kr

Abstract

Effect of active element (Ti and Zr) of Cu based filler alloy on crack occurrence of Al_2O_3/SS 430 brazed joint, which was utilized in sodium sulfur battery, was investigated to propose an advanced high temperature brazing alloy and joining process. Newly designed Cu-7Al-3.5Ti and Cu-7Al-3.5Zr model filler alloys, which were not dissolved into molten sodium, were used to bond ceramic-metal heterogeneous joints at 1000°C (solid-liquid state bonding), respectively. Reaction layer of Al_2O_3/filler alloy interface according to active element was characterized to show the relationships with crack occurrence. The results showed that the cracks of the brazed joints at 1000°C using Cu-7Al-3.5Zr filler alloys were not observed whereas the specimens using Cu-7Al-3.5Ti filler alloys was found with edge crack of the Al_2O_3 ceramic.

Introduction

The sodium-sulfur (NaS) battery has been widely focused as the representative electrochemical devices for high capacity energy storage systems (ESS)[1]. Molten sodium and sulfur has been used as anode and cathode, respectively, and β/β"-Al_2O_3 as solid electrolyte was utilized to divide them delivering a sodium ions (Na^+) at high temperatures about 270~350°C. The battery has a number of attractive features such as high charge/ discharge efficiency, a high energy density, and long cycle life[2]. Many researchers in in Japan, China and U.S.A. has accomplished variable technologies about sodium sulfur battery[3-6].

A high capacity NaS battery (500~1,200Wh) have main three heterogeneous joints such as ceramic-ceramic, metal-metal, and metal-ceramic joints. Among them, the ceramic-metal joint have been mainly applied by using thermal compression bonding (TCB) state of the art technology, which joined ceramic and metal components with active filler alloy(ABA) by applying a high temperature and pressures [7].

TCB joint have two barriers to guarantee high reliability of the tubular NaS battery. One is a high bonding strength to resist thermal stress and the other is 15 year corrosion resistance against molten sodium which contacts with filler alloy during operation. Thermal stresses applied on the TCB joints caused by different coefficients of thermal expansion (CTE) of sealing pieces have been studied to select an appropriate material for TCB joint[8,9]. Since another consideration is a corrosion behaviors against molten sodium attack, commercial aluminum alloys and modeled alloys were compared to enhance the corrosion resistance and a popular active brazing alloys, Ag-Cu-Ti, was immersed in to measure its corrosion kinetics under liquid sodium environment [10]. In other words, it's important to prevent the heterogeneous joints from unexpected failure [11].

Al-1Mn alloy (Al 3003) has been used as Al sealing bracket, and Al-Si-Mg alloys (Al 4004, Al 4104 and Al 4032), which has about 530°C eutectic temperature, have been generally used as active brazing alloy for TCB joint in sodium sulfur cells. However, in case that β/β"-Al_2O_3 was failed during operation, the battery would rapidly go up to 500°C due to reactions with Na and S [11-13]. Then, Al-based bracket and filler alloy could not ensure the safety of the TCB joints against an extra explosion. Therefore, a newly developed material for high temperature resistance was needed to provide its high temperature safety. Cu-based filler alloy and SS 430 could be applied due to its high melting point (above 780°C) and low CTE ($11.0*10^{-6}$ mm/K) properties, respectively.

In this study, effect of active element of a new filler metal (Cu-7Al-3.5Ti and Cu-7Al-3.5Zr) on crack occurrence in Al_2O_3/SS 430 joints has been studied to have a high bonding strength. Design of newly developed filler alloy was introduced, and microstructure of the brazed joint was investigated to reveal a relationship with the crack occurrence.

Experiment

The Cu-Al-Ti and Cu-Al-Zr model alloys were fabricated by casting in an induction furnace in $1.0*10^{-2}$ torr. The melted specimens were poured into a rod shape tool with a diameter of 25.4mm and a length of 100 mm. Compositions of the

Table 1 Glow discharge spectrometer (GDS) compositional results of modeled filler metals used in this study.

Filler alloy	Cu	Al	Ti	Zr	Analyzed
Cu-7Al-3.5Ti	Bal	6.9	3.5	-	GDS
Cu-7Al-3.5Zr	Bal	6.9	-	3.4	GDS

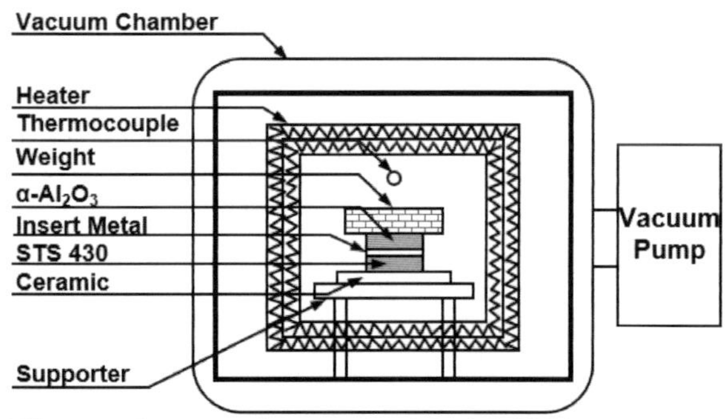

* Vacuum furnace of approximately 1.4x10^{-4} torr

Figure 1 Schematic illustration of bonding experiment

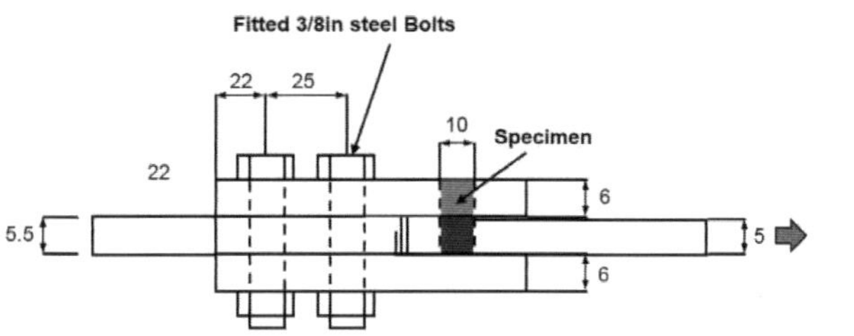

Figure 2 Schematic illustration of shear strength device, ASTM B565-04

alloys were analyzed by glow discharge spectrometer (GDS) as shown in Table 1.

The filler alloys were composed of Cu-Al-Ti and Cu-Al-Zr alloying system and its morphology is a sheet in 10mm diameter * 200μm thickness. Cylindrical rod of α-Al_2O_3 (99.7% purity) and SS 304 (10mm diameter, 5mm thickness) were prepared as work pieces. Brazing was conducted in high vacuum (under better than 1.0*10^{-6} torr)furnace as shown in Figure 1. The Specimens were brazed at 1000°C followed by an holding for 10minutes. Cooling was conducted for about 8 hours. Shear strength were measured by a tensile machine with jig ASTM NO.B565-4 as shown in Figure 2. Crosshead speed of the test was about 0.5mm/min.

The brazed specimens were cross-sectioned to the bonding interface and polished. Microstrucutre was characterized by filed emission scanning electron microscope (FE-SEM), X-ray diffraction (XRD), energy dispersive spectrometer (EDS) and field emission electron probe micro analyzer (FE-EPMA) to characterise the reaction layer between alumina and filler alloys.

Results and Discussion

Alloy design of the filler metal

In order to choose an element of the filler metal used in NaS battery, corrosion resistance against molten sodium(Na) should be observed. Among a number of elements, Al, Cu, Ti and Zr was chosen that they are not liquefied forming a liquid solution in rich-Na area (molten sodium) at 350°C (red stitched line) and 500°C (green stitched line) as shown in Figure 3. The phase diagrams of Cu-Na(a), Al-Na(b), Ti-Na(c) and Zr-Na(d) were displayed to reveal that molten sodium(Na) and each element are respectively existed without any solutions.

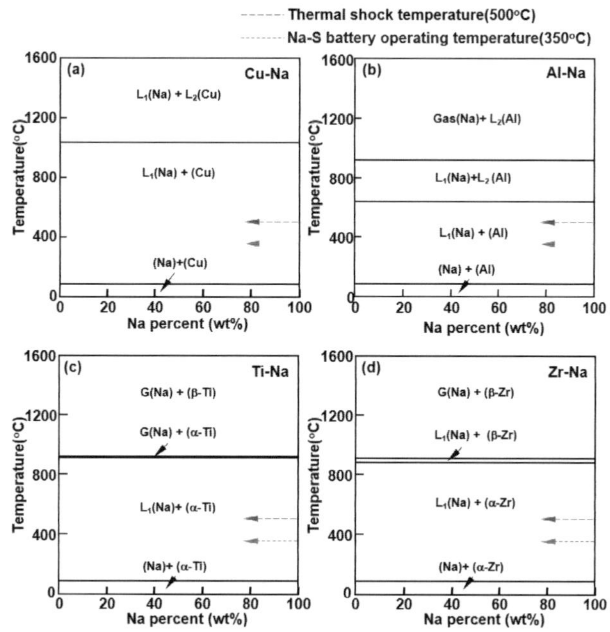

Figure 3 Equilibrium binary phase diagrams of (a) Cu-Na (b) Al-Na (c) Ti-Na and (d) Zr-Na

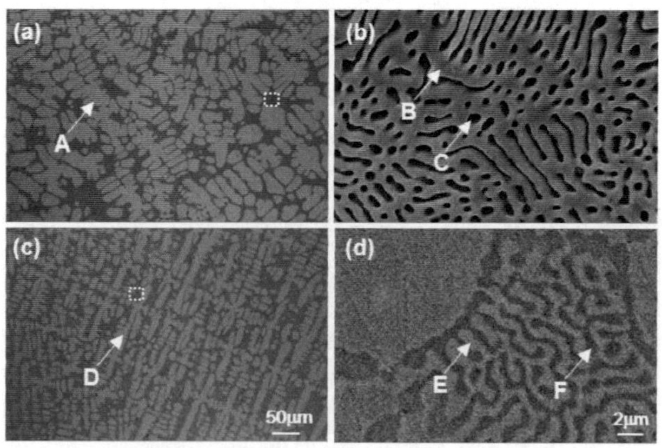

Figure 4 Optical micrographs of microstructure of the filler alloy, (a) : Cu-7Al-3.5Ti, (c) : Cu-7Al-3.5Zr and (b, d) : magnified SEM images for □ area.

Table 2 EDS spot analysis (at.%) of the point in Fig. 4.

	Cu	Al	Ti	Zr	Possible Phase
A	87.9	12.1	-	-	α-Cu
B	87.7	12.3	-	-	α -Cu
C	51.3	25.9	24.2	-	$AlCu_2Ti$*
D	86.6	13.4	-	-	α -Cu
E	87.7	12.2	-	-	α -Cu
F	56.7	20.1	-	23.2	$AlCu_2Zr$**

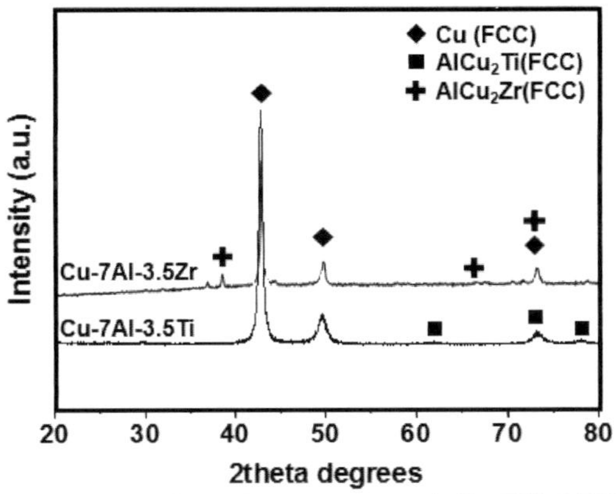

Figure 5 X-ray diffraction patterns of Cu-7Al-3.5Ti and Cu-7Al-3.5Zr modeled filler alloy.

Cu element was selected as a base matrix due to its higher melting point above 600oC. Ti and Zr active elements, which enhanced the chemical reaction with α-Al_2O_3 ceramic forming an oxide layer such as TiO_2, Ti_3Cu_3O, ZrO_2, ZrO, MgO and $MgAl_2O_4$ on the Al_2O_3 surface. Al was added to decrease the solidus and liquidus point of the Cu based filler metals.

Chemical compositional range was modified depending on solidus and liquidus melting point using Thermo-calc TCW 5 computational simulation tool. Al was added under 7wt.% due to formation of Al_4Cu_9 intermetallic compound, thereby 7wt.% was fixed in this study. Ti and Zr were limited under 6wt.% to prohibit high hardness.

Cu-7Al-3.5Ti and Cu-7Al-3.5Zr filler alloys were immresed into molten sodium at 350°C and 500°C for 50hr, and it was shown that the corrosion prodcut was not present

Figure 5 shows the optical microstructure (OM) of as-cast Cu-7Al-3.5Ti and Cu-7Al-3.5Zr filler alloys used in this study and its magnified seconday electron (SE) images were also suggested. EDS spot analysis (at.%) of the point in Fig. 4 and XRD anlaysis of filler metal were suggested in table 2 and figure 5. It's determined that lots of primary Cu were existed as dendritic shapes with eutectic phases comprised of copper and intermetallic compound $AlCu_2Ti$ phases.

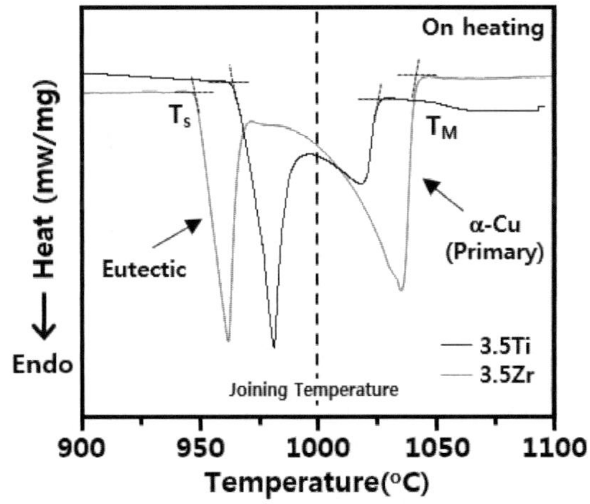

Figure 6 Differential scanning calorimetry (DSC) curve of Cu-7Al-3.5Ti and Cu-7Al-3.5Zr at the temperature range of 900~1100°C

Figure 6 shows the Differential scanning calorimetry (DSC) curves of Cu-7Al-3.5Ti and Cu-7Al-3.5Zr filler alloys at the temperature range of 900~1100°C. Melting temperature range of the of Cu-7Al-3.5Ti and Cu-7Al-3.5Zr filler alloys were 965~1020°C and 945~1045°C, respectively. Left knee point on left side and right knee point on right side indicated that soldius and melting temperature, respectively. Cu-7Al-3.5Zr has lower solidus and higher liquidus temperature than Cu-7Al-3.5Ti. This result infers that joining temperature, 1000°C, let the filler alloys exsist as mushy state (solid phase and liquid phase).

Crack occurrence of the brazed joints

Figure 7 shows cross-section macrostructures of the specimens bonded at 1000°C using Cu-7Al-3.5Ti (M.P. 965~1020°C) and Cu-7Al-3.5Zr (M.P. 945~1045°C) filler metals. The crack was indicated by the arrow, and the crack occurred at the edge of the joint. Crack was occurred at Cu-7Al-3.5Ti specimen, not Cu-7Al-3.5Zr specimen. In other words, Cu-7Al-3.5Ti joint is more susceptible to crack than Cu-7Al-3.5Zr joint.

It is known that the cracks were generated at the edge of the ceramics in ceramic-metal joint. This kind of crack is called edge crack, which is generally caused by the residual stress formed by the mismatch of thermal expansion coefficient between ceramics and metal. From these results, it was confirmed that crack-free joints can be obtained when Cu-7Al-3.5Zr filler alloy is used to braze ceramic and metal at 1000 °C (solid-liquid state bonding).

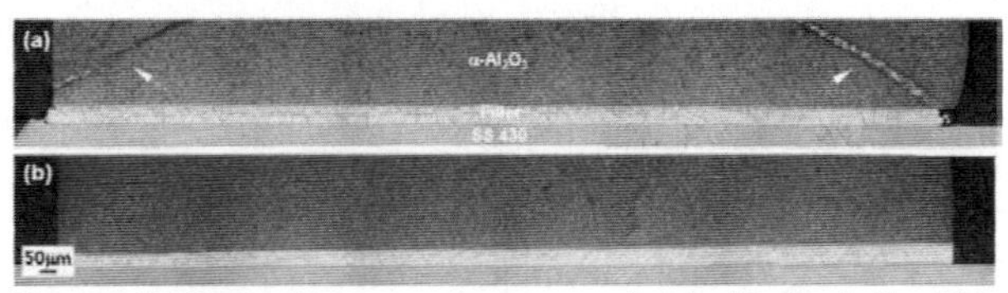

Figure 7 Optical macrostructure of (a) Al_2O_3 / Cu-7Al-3.5Ti / SS 430 brazed at 1000°C Al_2O_3 / Cu-7Al-3.5Zr / SS 430 brazed at 1000°C

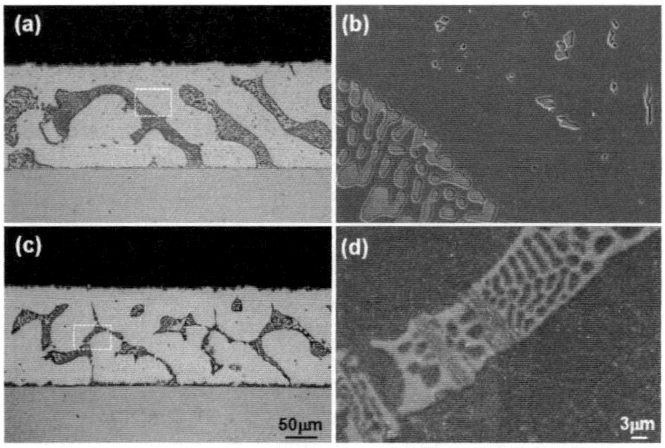

Figure 8 Optical micrographs (OM) of a cross-section of the specimen brazed at 1000°C using (a) Cu-7Al- 3.5Ti and (c) Cu-7Al- 3.5Zr filler alloy, and (b, d) its magnified SEM images.

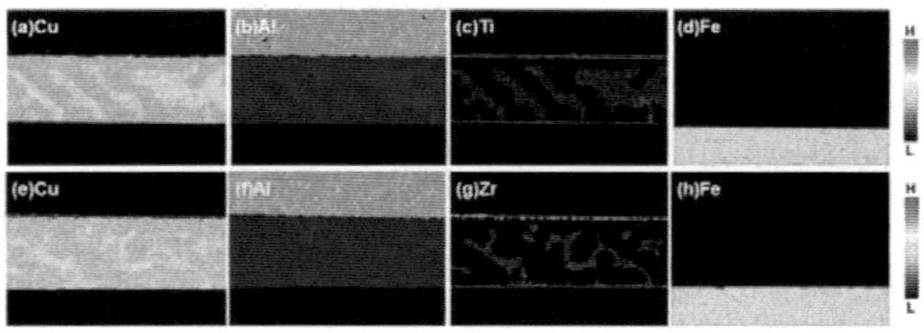

Figure 9 Elemental mapping of Cu, Al, Ti(Zr), and Fe for the joints brazed at 1000°C using (a~d) Cu-7Al-3.5Ti and (e~h) Cu-7Al-3.5Zr filler alloy.

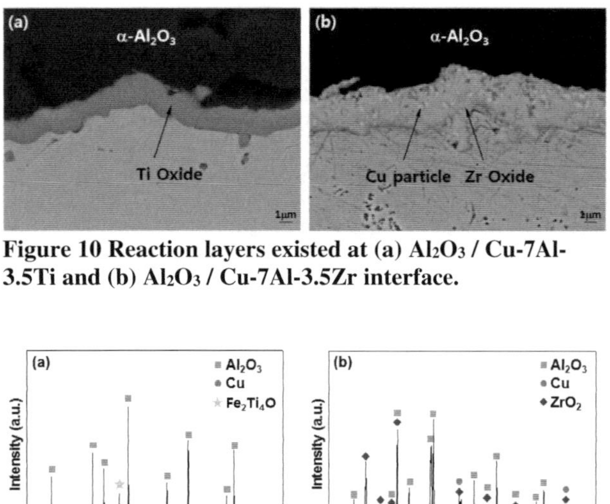

Figure 10 Reaction layers existed at (a) Al_2O_3 / Cu-7Al-3.5Ti and (b) Al_2O_3 / Cu-7Al-3.5Zr interface.

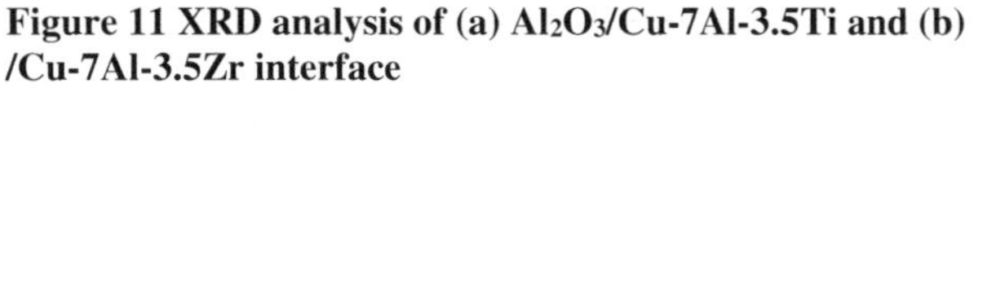

Figure 11 XRD analysis of (a) Al_2O_3/Cu-7Al-3.5Ti and (b) /Cu-7Al-3.5Zr interface

Microstructure of the brazed joints

Brazing were conducted at 1000°C, which is a glass bonding (between α-Al_2O_3 and β''-Al_2O_3) temperature, then ceramic-metal and ceramic-ceramic bonding process could be conducted simultaneously by increasing productivity. At 1000oC, ceramic-metal joint was brazed between solidus and liquids temperature, which is called "solid-liquid bonding". This process was adopted to reduce residual stresses by decreasing volume change of liquid related with shrinkage stress. Also, more pressure was loaded to induce the wetting behaviour with little liquid.

To understand why crack occurrence phenomenon was different depending on the filler alloy, microstructure was observed. Figure 8 shows the optical microscopies of the brazed joints between SS 430 and α-Al_2O_3 using Cu-7Al-3.5Ti and Cu-7Al-3.5Zr filler alloy. FE-EPMA elemental mapping of Cu, Al, Ti(Zr), and Fe for the joints was shown in Figure 9. It's confirmed that eutectic phases were distributed inside filler alloy. In addition, Ti-rich and Zr-rich layers on both α-Al_2O_3 and SS 430 interfaces were continuously existed. Fe elements was not detected in filler alloy, which was seemed that Fe was not diffused without forming intermetallic compound.

Figure 10 represents SEM images of reaction layer along Al_2O_3/filler alloy interface which has diffrerential morphology. A flat layer was existed at the interface made from Cu-7Al-3.5Ti filler alloy, whereas rough layer with a bit of Cu on the inside was found at the interface racted with Cu-7Al-3.5Zr alloy. To identify the layers, XRD analysis was used from the surfce between Al_2O_3 and Cu-Al-Ti filler alloy as shown in Figure 11. The peaks suggested the presence of Cu, Al_2O_3 and $FeTi_2O_4$ was observed in Figure 10 (a) and Cu, Al_2O_3 and ZrO_2 was observed in Figure 10 (b). Cu peak was distinctly found in the specimens brazed with Cu-7Al-3.5Zr, which means that the volume of Cu phases were high at the interface.

It is inferred that presence of Cu particles help it release the residual stresses occurred at the interface, thereby which prevents the specimen from cracking.

Conclusions

Effect of active element (Ti and Zr) of Cu based filler alloy on crack occurrence of Al_2O_3/SS430 brazed joint was investigated to propose an advanced high temperature brazing alloy and its new joining process. According to the results in this work, the following conclusions can be obtained.

1. Cu-7Al-3.5Ti and Cu-7Al-3.5Zr alloy was suggested as a reliable brazing ally which is not dissolved into molten sodium and melted at about 500°C. Melting temperature range of the of Cu-7Al-3.5Ti and Cu-7Al-3.5Zr filler alloys were obtained as 965~1020°C and 945~1045°C, respectively.

2. Cracks of the brazed joints at 1000°C using Cu-7Al-3.5Zr filler alloys were not observed whereas the specimens using Cu-7Al-3.5Ti filler alloys was found with edge crack of the Al_2O_3 ceramic.

3. A flat $FeTi_2O_4$ reaction layer was found at the interface of the specimen brazed with Cu-7Al-3.5Ti filler alloy, whereas rough ZrO_2 layer with Cu particles was existed at the interface reacted with Cu-7Al-3.5Zr alloy. It is seemed that Cu particles help it release the residual stresses occurred at the interface.

References

[1] J.T. Kummer, N. Weber, U.S. Patent 3,413,150, 1968.

[2] Chloride Silent Power Limited, Sodium Sulfur Electric Vehicle Battery Engineering Program, 1993. DOE report, DE-AC04e88AL54303.

[3] T. Oshima, M. Kjita, A. Okuno, "Development of Sodium - Sulfur Batteries", Int. J. Appl. Ceram. Technol. 1 (2004) pp. 269-276.

[4] R. Bauer, W. Haar, H. Kleinschmager, G. Weddigen, W. Fischer, "Stress analyses for the glass joints of contemporary sodium sulfur batteries", J. Power Sources 1 (1976/77) pp. 109-126.

[5] Z. Wen, J. Cao, Z. Gu, X. Xu, F. Zhang, Z. Lin, "Research on sodium sulfur battery for energy storage", Solid State Ion. 179 (2008) pp. 1697~1701.

[6] J.L. Sudworth, A.R. Tilley, "The Sodium Sulfur Battery", Chapman and Hall, London, 1985.

[7] Z. Wen, J. Cao, Z. Gu, X. Xu, F. Zhang, Z. Lin, "Research on sodium sulfur battery for energy storage", Solid State Ionics 179 (2008) pp. 1697–1701

[8] K. Jung, S. Lee, Y.-C. Park, C.-S. Kim, "Finite element analysis study on the thermomechanical stability of thermal compression bonding (TCB) joints in tubular sodium sulfur cells", Journal of Power Sources 250 (2014) pp. 1-14

[9] K. Jung, H.J. Heo, J.H. Lee, Y.C. Park, C.-Y. Kang, "Enhanced corrosion resistance of hypo-eutectic Al–1Mg–xSi alloys against molten sodium attack in high temperature sodium sulfur batteries", Corrosion Science, 98 (2015) pp. 748-757

[10] K. Jung, S. Lee, G. Kim, C.-S. Kim, "Stress analyses for the glass joints of contemporary sodium sulfur batteries", Journal of Power Sources, 269 (2014) pp. 773-782

[11] H.J. Heo, K. Jung, C.-Y. Kang, "Corrosion behavior of Ag-Cu-Ti brazed joints against molten sodium" IBSC conference proceedings (2015)

[12] J.K. Min, M. Stackpool, C.H. Shin, C.-H. Lee, "Cell safety analysis of a molten sodium–sulfur battery under failure mode from a fracture in the solid electrolyte", Journal of Power Sources, 293 (2015) 835-845.

[13] J.K. Min, C.-H. Lee, "Numerical study on the thermal management system of a molten sodium-sulfur battery module", Journal of Power Sources, 210 (2012) 101-109.

Characterizing Crack Propagation Life for Joints by Abnormal Magnetic Signals

Changliang Shi, Weixue Tang, Hao Zhan
Guangdong Industrial Analysis and Testing Center, Guangzhou, Guangdong, China
barrystone@163.com

Abstract

In the geomagnetic field, stress can induce spontaneous magnetic signals in ferromagnetic materials, which can be potentially applied to estimate the joint reliability. In this research, the abnormal magnetic signals were measured during dynamic tension test on the ferromagnetic joint specimens. The relationship between fatigue crack propagation and abnormal magnetic signals was studied. The results indicated that the gradient of abnormal magnetic signals was linear with abnormal peak, which were the key parameters for characterizing crack propagation life of joints. Then the fitting of gradient and fatigue cycles was done referring to the Paris equation and a simple model was derived.

Introduction

At present, most key components of machinery and equipment are made of ferromagnetic materials. Meanwhile, many steel structures components are joined by welding, brazing and soldering. The residual stress and defects in the joints are the key factors which affect the service performance and service life of the mechanical parts. With the development of the application of welding, brazing and soldering technology, the evaluation of joint reliability is becoming more and more important. Over the years, the joint reliability is generally guaranteed by the welding procedure, which is not adequate. The evaluation of joint reliability, especially the fatigue life, is still short of effective and rapid testing method [1].

Because the welding components are subjected to repeated loads, fatigue fracture is prone to occur. Fatigue failure is one of the main failure modes for the joints. Because the fatigue failure has a certain suddenness and contingency, it often leads to major accidents. Therefore, the fatigue life prediction is becoming more and more important. The various defects in the joints will be produced in the process of welding, brazing and soldering. Under the action of fatigue load, these defects are easily becoming the source of crack, which leads to the fatigue fracture of the component. However, according to the fracture mechanics, it is found that even if there are cracks in some components, as long as we can monitor the crack growth and ensure that the crack will not expand to the critical size, the component can also be safely used. Therefore, it is of great significance to study the fatigue crack propagation life for the joint. In the field of engineering, conventional mechanical testing methods are very difficult to complete the expansion of fatigue crack monitoring tasks. The nondestructive testing method is very suitable for monitoring the fatigue crack propagation [2-5].

The metal magnetic memory technology was promoted by the Russian scholar Dubov in 1997 for the first time. The abnormal magnetic signals, which were affected by the geomagnetic field and stress, could be detected. The novel nondestructive testing method can be used to detect the stress concentration and surface defects of ferromagnetic material [6]. Compared with traditional nondestructive testing methods, the technique does not need to pretreat workpiece and can be used when the workpiece is in service. Meanwhile, the stress concentration area or micro defects can be detected in order to early warning of fatigue failure [7-9].

The abnormal magnetic signals were detected during dynamic tension test on the ferromagnetic joint specimen in the paper. The relationship between abnormal magnetic signals and fatigue crack propagation was studied.

Experiment

Specimens

The base metal of specimens are 45 steel, which is a kind of medium carbon steel. The base metal were joined by welding in order to make the joint meet enough size. Then the specimens were polished. Many factors, such as the machining process and heat treatment condition, intensively affect the initial magnetic signal, so the specimens were under inductive demagnetization before experiments in order to study the relationship between stress and abnormal magnetic signals in the initial state.

Experimental instruments

The fatigue test was carried out by Instron 8810 hydraulic servo test machine, whose dynamic load error 71.0%. Tension–tension fatigue tests of constant amplitude (sinusoidal waveform) were performed. The metal magnetic memory testing device is the EMS-2003 intelligent magnetic memory detector manufactured by Eddysun, and the detection probe is a two channel pen probe. The normal component *Hp(y)* of the scattering magnetic field intensity was detected by the instrument. In order to avoid the manual error, the magnetic memory detection probe is clamped on the three-dimensional electronic control displacement platform of non-ferromagnetic material, and the step distance is 2.5 u m. The detection schematic diagram is shown in Fig.1.

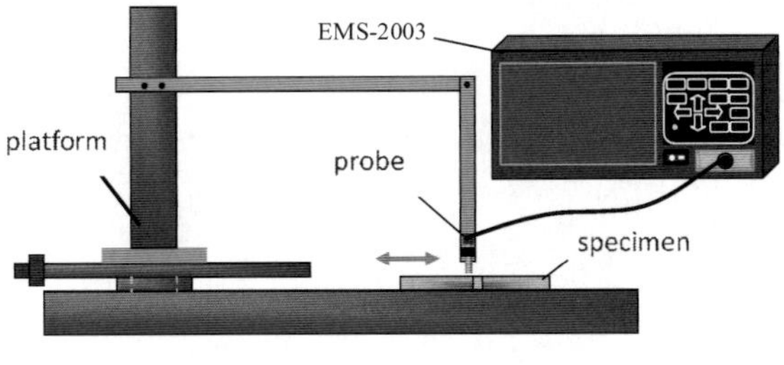

Fig.1 detection schematic diagram

The fatigue crack propagation was monitored by Olympus i-speed high speed camera, and the length of crack can be measured, the specimen is shown in Fig.2.

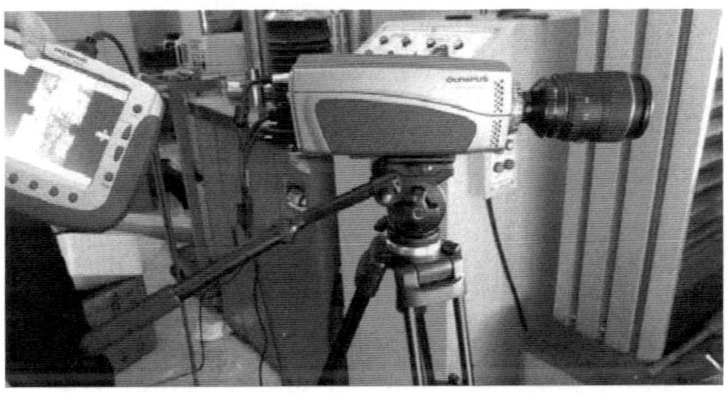

Fig.2 Olympus i-speed high speed camera

Testing procedure

The flat U notched specimens (*Kt*=5) were selected to carry out the tension fatigue test. Under the cyclic loading (220MPa, sine wave, stress radio *R* is 0.1, loading frequency *f* is 10), the number of cyclic loading increased sequentially. The specimen was removed from the fatigue testing machine and placed in the north and south direction on the platform after loading to the predetermined number of cycles. In the middle part of the test piece, 5 parallel lines, whose interval is 4.5mm and length is 60mm, was drawn as the detection paths, shown in Fig.3.

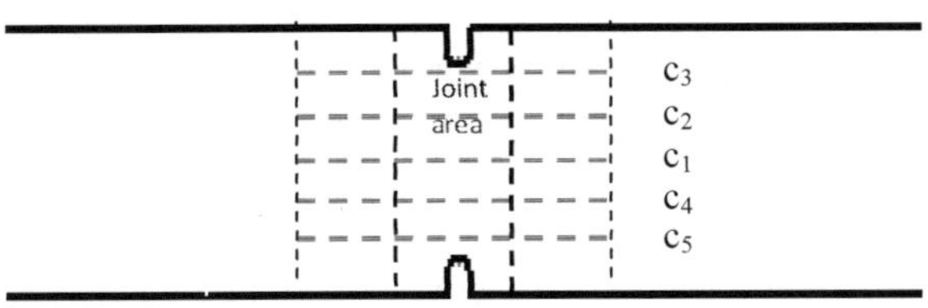

Fig.3 Scanning lines of the fatigue tensile testing specimen with notch

The probe was gripped on a non-ferromagnetic 3D electric scanning platform with a lift-off value of 0.5mm, the testing speed was 10mm/s, and the sampling interval is 30ms. After detection, the specimen was loaded again to a higher preset circle value, the above procedure was repeated until the specimen failed.

Discussion

The experiments were carried out in laboratory condition and the environmental magnetic field is dominated by the relatively constant geomagnetic field. The abnormal magnetic signals *Hp(y)* in different detection paths were detected after 500 fatigue cycle times, shown in Fig.4.

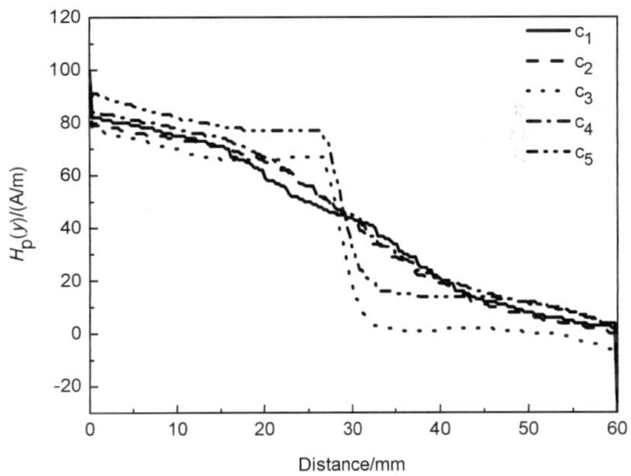

Fig.4 Abnormal magnetic signals $H_p(y)$ in different paths after 500 cycle times

The results indicate that the gradient k of the abnormal magnetic signal *Hp(y)* curves increases dramatically near the *U* notch in the joint area, and the gradient k of abnormal magnetic signals will increase more when it is closer to the *U* notch. According to the material mechanics, the stress concentration degree will increase near the *U* notch. The gradient k of the abnormal magnetic signals increases when the stress increases [8], so the gradient k increases dramatically near the *U* notch in the joint area and the stress concentration of the specimen can be determined qualitatively according to the gradient k of the *Hp (y)* detection curve. It can be seen that the change of abnormal magnetic signals *Hp(y)* in c_5 detection path is the most obvious in Fig.4. The following analysis will be only around the results in c_5 detection path in order to facilitate the analysis and make it representative.

The abnormal magnetic signals *Hp(y)* in c_5 detection path of the specimen is shown in Fig.5. The results indicate that the abnormal magnetic signals *Hp(y)* increase when the fatigue cycle times increase. The gradient k of the abnormal magnetic signal *Hp(y)* curves increases dramatically near the *U* notch in every fatigue cycle time and initiate the peak of *Hp(y)*. Therefore, the maximum value k_{max} of the gradient and the peak-peak value $\triangle H_{pa}(y)$ of abnormal magnetic signals $Hp(y)$ could be the key parameters to reflect the relationship between the fatigue cycle time and the magnetic signals.

Firstly, the maximum value k_{max} of the gradient was studied. The relationship between k_{max} and fatigue cycle time N was shown in Fig.6.

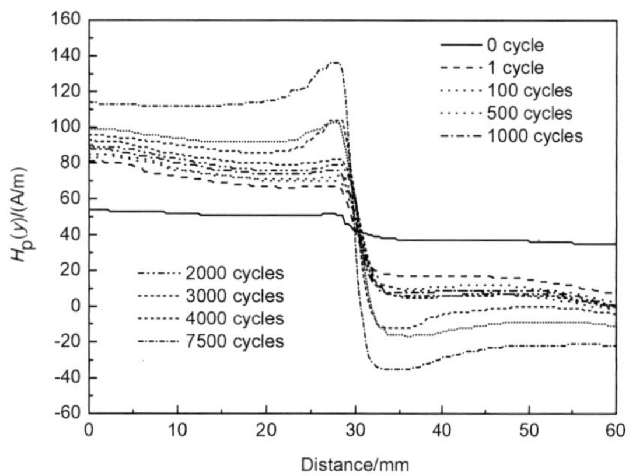

Fig.5 Abnormal magnetic signals $H_p(y)$ of specimen in c_5 path after different fatigue cycle times

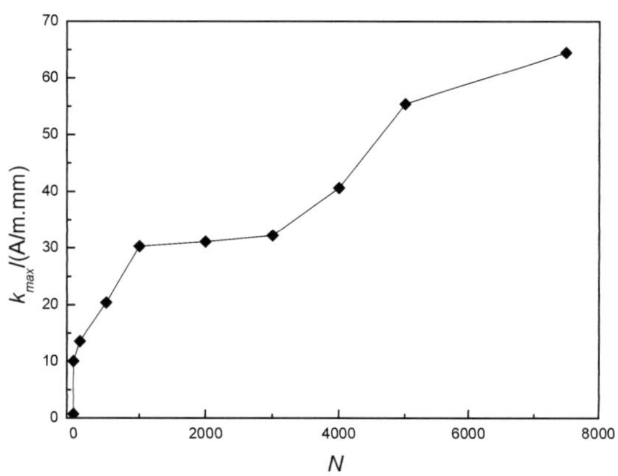

Fig.6 Relationship between k_{max} and fatigue cycle time N

The results indicate that the maximum value k_{max} of the gradient apparently increase at first when the fatigue cycle time increase. But when the cycle number increase to a certain extent, the maximum value k_{max} trends in the stable. With the increase of fatigue cycle time, the micro fatigue cracking was detected by the high speed camera at 4000 cycle times, then the maximum value k_{max} increase dramatically to the maximum value until the specimen cracked at 8000 cycle times. Meanwhile, there is also the same relationship between peak-peak value $\triangle H_{pa}(y)$ and the fatigue cycle time N.

According to the fracture mechanics [4], fatigue crack length will increase with the increase of fatigue cycle times. At the same time, the preliminary results indicate that the maximum value k_{max} and peak-peak value $\triangle H_{pa}(y)$ also increased with the increase of fatigue cycle times. Therefore, there must be some relationship between peak-peak value $\triangle H_{pa}(y)$ and the length of fatigue crack a. In order to analyze the correlation of them, a fitted curve of peak-peak value $\Delta H_{pa}(y)$ and the length of fatigue crack a was shown in Fig.7.

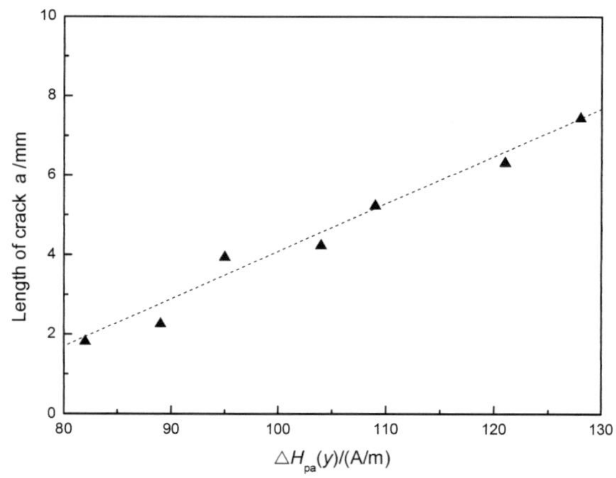

Fig.7 Fitted curve of $\Delta H_{pa}(y)$ and a

It can be seen that the peak-peak value $\triangle H_{pa}(y)$ of abnormal magnetic signals increase gradually with the increase of length of fatigue crack a. Then, the experimental data are fitted, and the fitting equation is shown by Eq.(1) as following.

$$a=0.11\triangle H_{pa}(y)-5.16 \tag{1}$$

According to the Eq.(1), there is a linear relationship between the peak-peak value $\triangle H_{pa}(y)$ and the length of fatigue crack a. In this research, the Eq.(1) is simplified to be Eq.(2) as following in order to be easy to analyze.

$$a=A\triangle H_{pa}(y)+B \tag{2}$$

where A and B are constant, affected by material and stress.

At present, the fatigue crack growth life is mainly predicted by the fracture mechanics method [7]. In engineering practice, the method, which is most commonly used to predict the fatigue crack growth life, is the classical Paris formula, shown in Eq.(3) as following.

$$N=\frac{1}{C(f\Delta\sigma\sqrt{\pi})^m(0.5m-1)}\left(\frac{1}{a_0^{0.5m-1}}-\frac{1}{a_c^{0.5m-1}}\right) \quad m\neq2$$

$$N=\frac{1}{C(f\Delta\sigma\sqrt{\pi})^m}\ln\left(\frac{a_c}{a_0}\right) \quad m=2 \tag{3}$$

where C and m are constant, calculated by fatigue crack propagation test. f is the geometric correction coefficient.

Because this fatigue test was carried out in constant amplitude condition, the Eq.(2) could be used in Eq.(3), shown in Eq.(4) as following.

$$N=\frac{1}{C(f\Delta\sigma\sqrt{\pi})^m(0.5m-1)}\left(\frac{1}{(A\Delta H_{pa}(y)_0+B)^{0.5m-1}}-\frac{1}{(A\Delta H_{pa}(y)_c+B)^{0.5m-1}}\right)\quad m\neq 2$$

$$N=\frac{1}{C(f\Delta\sigma\sqrt{\pi})^m}\ln\left(\frac{(A\Delta H_{pa}(y)_c+B)}{(A\Delta H_{pa}(y)_0+B)}\right)\quad m=2 \qquad (4)$$

where $\Delta H_{pa}(y)_0$ is the origin $\Delta H_{pa}(y)$, $\Delta H_{pa}(y)_c$ is the critical $\Delta H_{pa}(y)$.

In order to verify the accuracy of the Eq.(4), the test parameters are used as geometric correction coefficient f is 1.12, maximum stress is 220MPa, minimum stress is 22MPa, m is 2.64 and C is 2.86×10^{-12}. The test parameters were brought into the Eq.(4), and the calculated values of the fatigue crack propagation life could be obtained, which compared with the measured values, shown in Fig.8.

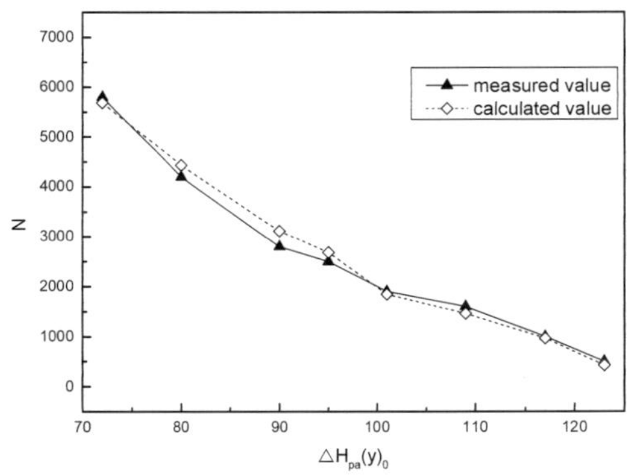

Fig. 8 Measured values and calculated values for propagation life of the joint

According to the results of Fig.8, it is known that the calculated value of the model is in good agreement with the actual measured value, and the maximum error is less than 10%. The Eq.(4) is obtained in the conditions of this paper, and there are more restrictive conditions. Although the method is obtained in laboratory conditions, it can direct the prediction for fatigue crack propagation life of ferromagnetic joints by the abnormal magnetic signals, and it can greatly promote the development and perfection of the joint reliability for welding, brazing and soldering.

Conclusions

The gradient k and the peak-peak value $\triangle H_{pa}(y)$ of the abnormal magnetic signals increase with the increase of fatigue cycle times. The maximum value k_{max} of the gradient and the peak-peak value $\triangle H_{pa}(y)$ of abnormal magnetic signals are the key parameters to reflect the relationship between the fatigue cycle time and the magnetic signals. There is a linear relationship between the peak-peak value $\triangle H_{pa}(y)$ and the length of fatigue crack a. Combined with the classical mechanics Paris formula, a predictive model for fatigue crack propagation life of ferromagnetic joints was established.

Acknowledgments

This work was financially supported by Public Research and Capacity Building Project of Guangdong in China (No. 2015A030401073), Environment and Capacity Construction Project of Guangdong Scientific Research Platform (No.2016GDASPT-0201, No.2016GDASPT-0316) and Innovation Ability Construction of Key Scientific Research Institutions of Guangdong Academy of Sciences (No. 2017GDASCX-0114).

References

[1] B.S. Xu, W. Zhang, S.C. Liu, et al. Remanufacturing Technology for the 21st Century [C]. Proceedings of the 15th European Maintenance Conference, Gothenburg, Sweden, 2000: 335~339.

[2] X.L. Zheng. On some basic problems of fatigue research in engineering [J]. Int J Fatigue 2001; 23:751–766.

[3] D. B. Garcia, A. F. Grandt. Application of a Total Life Prediction Model for Fretting Fatigue in Ti–6Al–4V [J]. International Journal of Fatigue. 2007, 29(7): 1311~1318.

[4] K. C. Liu, J. A. Wang. An Energy Method for Predicting Life Crack Orientation and Crack Growth under Multiaxial Loading Conditions [J]. International Journal of Fatigue. 2001, 23: s129~s134.

[5] H. Agerskov. An Analytical Model for Fatigue Life Prediction Based on Fracture Mechanics and Crack Closure [J]. Construct. Steel Res.1996, 37(3): 229~261.

[6] A. A. Dubov. A Study of Metal Properties using the Method of Magnetic Memory [J]. Metal Science and Heat Treatment. 1997, 39(9-10): 401-402.

[7] A. Tesch, R. Pippan, H. Döker. New Testing Procedure to Determine da/dN–ΔK Curves at Different, Constant R-values using One Single Specimen [J]. International Journal of Fatigue. 2007, 29(7): 1220-1228.

[8] L.H. Dong, B.S. Xu, S.Y. Dong, et al. Variation of stress-induce magnetic signals during tensile testing of ferromagnetic steels [J]. NDT&E Int 2008;41:184–189.

[9] S. T. Jiang, W. Li. Condensed Matter Magnetic Physics [M]. Beijing Science Publishing Company, 2003: 102~120.

Influence of the grain size of alloy and superalloy additives on green tapes and preforms for gas turbine repair brazing

S.M. Uddin, H. Krappitz
Innobraze GmbH, Esslingen, Germany
minhaz.uddin@innobraze.de

K. Bobzin, M. Öte, S. Wiesner
Surface Engineering Institute, RWTH Aachen University, Aachen, Germany
wiesner@iot.rwth-aachen.de

Abstract

Blades and vanes of gas turbines are thermally, chemically and mechanically stressed during operation. The subsequent material removal from the surfaces results initially in a reduction of the efficiency and finally in the failure of the turbine. These turbine damages can be repaired by brazing processes with polymer bonded flexible green tapes or binder-free pre-sintered preforms consisting of brazing filler metal and Ni-based superalloy powders.

As the shrinkage and the porosity of the repair products are relatively high after brazing, packing density and homogeneity of the green tapes as well as preforms can be improved by the variation of grain size and proportion of brazing filler metals and superalloy powders. The influence of the grain size distribution on the porosity of green tapes made of BNi-9 and the superalloy Mar M 247 is analyzed and the microstructure of brazed coating is investigated. Moreover, the melting behavior of the tapes in dependence of brazing temperature and time is analyzed by thermal analysis and in melting experiments at brazing process condition.

Introduction

Key aspects of the repair processes for worn or corroded turbine components are recovery of the original microstructure, sealing of cracks and compensating material loss. Conventional welding processes reach their limit easily as complex geometries and materials with numerous alloying elements are hardly weldable. In these cases repair brazing processes based on the transient liquid phase (TLP) bonding or the wide gap brazing are advantageous [1, 2]. Both processes lead to strong joints without significant occurrence of brittle intermetallic compounds and to a remelting temperature above the brazing temperature. Ni-based filler metals for TLP bonding contain elements lowering the melting point such as Si or B. During the brazing process, these elements diffuse into the substrate changing the composition of the filler metal. Consequently, the solidus temperature raises and the molten filler solidifies isothermally. After a subsequent homogenization step, the microstructure and mechanical properties of the brazed joint are equal to the base material [3]. If the joint gap is too large or if there is no actual gap as in repair surface brazing to build up material, the melting point lowering elements can only partly diffuse out of the filler metal. Consequently, the solidification does not occur isothermally but by lowering the temperature below solidus temperature. Due to the high amount of metalloids in the filler metal, brittle intermetallic phases form. In order to avoid the formation of these undesired phases in wide gap brazing, additives in form of substrate particles are added to the filler metal. These superalloy additives have several functions. They bridge the wide gap between the faying surfaces and thus induce the capillary filling pressure necessary to fill the gaps and voids. Additionally, they absorb the melting point lowering elements leading to an isothermal solidification of the filler and a joint without undesired intermetallic phases [2].

An important factor for the quality of the repair brazed area is the careful selection of not only the filler metal and the melting point lowering elements but also that of the superalloy additive. The focus should be on the composition, particle size and shape of the additive powder matching the chosen filler metal. Pores and intermetallic phases can be the result of limited flowing and diffusion processes due to an unadapted choice of grain size or mixing ratio of filler metal and superalloy additive [2, 4, 5, 6, 7, 8].

The material mixture for repair brazing can be applied as pastes, as stiff, sintered or hot isostatic pressed preforms or as flexible, polymer bonded and unsintered green tapes. While the application of pastes is also flexible in irregularly formed joining gaps, it has the disadvantage that the evaporating organic binder leads to the formation of pores and to the shrinkage of the tape up to 50 % during brazing [9]. In contrast, the relatively high density of presintered preforms and green tapes results in improved dimensional stability.

In order to produce dense preforms and tapes, a small amount of organic binding agent is mixed with blended superalloy as well as brazing filler metal. The organic binding agent is known to decompose in gaseous products at temperatures up to T = 450 °C leaving pores in the tapes. When molten, the brazing alloy infiltrates these cavities because of its high capillary action and forms cohesive bonds to the superalloy particles and the substrate [10].

However, an insufficient compaction of the powder or a local agglomeration of big particles in the production process of the preforms and tapes can cause a low packing density with a high porosity of $P > 5\ \%$ prior to brazing [11]. Also a high amount of organic binding agent can lead to extensive pore formation during the heating process. Although, even a high amount of pores in the preforms and tapes can partly be filled with molten filler metal during brazing but it will result in strong volume shrinkage of up to $\Delta V = 10\ \%$ of the preform [9].

If the mixing ratio of filler metal und superalloy additive or the ratio of the particle sizes is not appropriate, the melting process of the preform or tape can be inhomogeneous, leading to a microstructure with numerous voids and a limited material bond between filler and substrate material or superalloy particles. In industrial practice, a porosity of repair brazed areas up to $P = 10\ \%$ has to be accepted in consequence of the inhomogeneous microstructure of the preforms and tapes preventing a higher repair quality [9]. This porosity can reduce the mechanical properties of the repaired component and thus lead to premature failure of turbines. However, by optimizing the grain size of superalloy as well as brazing filler metal powders with the addition of a small amount of organic binding agent, the percentage of porosity in the preforms and tapes as well as in the coating after brazing can be limited to a very low amount of around $P = 1\ \%$ [7].

A specially developed adhesive can be used to fix the tapes to the base metal surface before brazing. The function of this specially developed adhesive is to ensure adhesion between tapes and substrates before brazing, to attain similar strength at high temperature, to sustain metallurgical compatibility and to ensure a good wetting during brazing [9]. On the other hand, to overcome the differences in the thermal expansion behaviour of filler as well as additive metal and to ensure good interfacial bonding between base metal and filler metal, some sustainable weight should be applied on the tapes before brazing.

This study focuses on green tapes containing the superalloy Mar M 247 and the filler metal BNi-9. The dependence of the brazing properties on grain size distribution of the tapes is analyzed. Thus, particles with different grain sizes are used for tape production. In this paper, the results of powder as well as tape analyses are shown. The melting behavior as well as the microstructure development of the green tapes in dependence of grain size distribution as well as brazing temperature and time is determined.

Experimental

The basis of the green tapes for repair brazing were powders made of the superalloy Mar M 247 (MM247) and the brazing filler metals BNi-9. 60 wt.-% of superalloy powders were blended together with 40 wt.-% of brazing filler metal powders and mixed with small amount of polymer bonded organic binding agent. The amount of binder agent is approximately $m_{binder} = 1.5$ wt.-% depending on the grain size of the powders to form tapes with a nominal thickness of $t_{nom} = 1.5$ mm. Different combinations of grain size of superalloy as well as brazing alloy powders were chosen in order to achieve optimum packing density of the tapes and to minimize the percentage of porosity in the coating after brazing. The chemical formula of the used powders is listed in Tab. 1.

	Superalloy MM247 [12]	Brazing filler metal BNi-9 [13]
Ni	Balance	Balance
Cr	8.2	13.5 - 16.5
Mo	0.6	
Fe		max. 1.5
Ta	3	
Co	10	0.1
Al	5.5	
Ti	1	
Hf	1.5	
W	10	
B		3.25 - 4.0
< 0,5 wt.-%	C	C, P

Table 1: Chemical composition of superalloy MM247 and filler metal BNi-9

In order to understand the influence of the grain size distribution on the properties of the green tapes and repair brazing, superalloy and filler metal powders with different grain sizes were used for tape production. For example, tape T1 should consist of superalloy particles with a targeted grain size of $d_{T1; MM247} = 45$ µm and of filler metal particles with a grain size of $d_{T1; BNi\text{-}9} = 106$ µm. The targeted grain sizes are D[v, 0.9] values. This diameter derived from the volumetric particle distribution is defined as the value where $V = 90\ \%$ of the particle diameters are smaller. The variation of the particle size of the filler metal and the superalloy resulting in the tape variants T1 to T6 is listed in Table 2.

	Tape variants					
	T1	T2	T3	T4	T5	T6
Targeted grain size MM247 [µm]	45	63	106	45	63	106
Targeted grain size BNi-9 [µm]	106	106	106	45	45	45

Table 2: Tape variants with targeted maximum particle size of superalloy MM247 and filler metals BNi-9

In order to determine, if the targeted grain size is correct, the actual grain size distribution of the brazing filler metal and superalloy powders was determined by the particle image analyzer system Morphologi G2 (Malvern Instruments GmbH, Herrenberg, Germany). During analysis, a two-dimensional image of all particles is recorded and transformed into a circle with the same area. The diameter of this circle is assumed to be equivalent to the diameter of the particle. For examination of the grain size distribution, the volume distribution of the equivalent circle diameter is plotted. The equivalent volume mean diameter D[4, 3] and the diameter D[v, 0.9] are determined. Additionally, the convexity, circularity and elongation of the particle have been examined as an indicator for their shape. The values of these factors range between 0 and 1. A perfectly spherical particle with a smooth surface would exhibit convexity and circularity values of 1 and elongation of 0.

For determining the density of a tape, a circular form with a diameter of Ø = 40 mm was used to punch out the tape of a particular thickness. The thickness of the tape is determined by averaging four thickness values measured at different positions of the tape. The volume of the circular tape has been calculated from its thickness and diameter. Subsequently, the density of the tape has been calculated by dividing weight and volume of the circular shape tape.

The filler metal has an eutectic composition with a melting point of $T_m = 1{,}055$ °C [13]. For conventional brazing processes, a brazing temperature of around $T_{brazing} = 1{,}100$ °C is recommended. The melting behavior of the brazing filler powder and of the green tapes was analyzed by combined differential scanning calorimetry (DSC) and thermal gravimetry (TG). The experiments were performed in the thermal analysis system SETSYS Evolution (SETARAM Instrumentation, Caluire, France) in fine vacuum at a pressure of $p = 5 \cdot 10^{-3}$ mbar. The heating rate was $\dot{T} = 10$ K/min. At first, the maximum temperature was chosen to be $T = 1{,}100$ °C corresponding to the recommended brazing temperature of the filler metal. Additional measurements on the tapes T3 and T4 were conducted with longer holding times or at higher brazing temperatures according to Table 3. Thus, the influence of temperature and time on the melting and solidification behavior could be assessed. The sample weight of powders and tapes was in the range of $24 \text{ mg} < w < 45 \text{ mg}$ due to the different densities. The evaluation of the results is carried out on the basis of mass specific data.

	$T_{brazing}$	$t_{brazing}$
Brazing process BP 1	1,100 °C	30 min
Brazing process BP 2	1,100 °C	60 min
Brazing process BP 3	1,150 °C	30 min

Table 3: Parameter set of the thermal analyses and brazing processes

In order to analyze the melting behavior, the microstructure development and the shrinkage of the tapes without the overlaying influence of the substrate, pieces of green tapes were placed on alumina substrates and heat-treated in the high-vacuum furnace MOV 553T (PVA TePla AG, Wettenberg, Germany). The pieces with an area of around $A = 5 \times 5$ mm^2 were cut out of a larger tape by means of a cutter knife. After this, a small deviation of the tape thickness compared to the nominal thickness of $t_{nom} = 1.5$ mm was detected. Measurements with a digital measuring slide after cutting in smaller pieces showed an actual thickness of around $t = 1.4$ mm. This is possibly due to a minor deformation of the tape during cutting. The parameter sets of the brazing processes investigated in this study are listed in Table 3.

Subsequently to the brazing process, the thickness of the now stiff tape parts was measured again. Then, the parts were metallographically prepared. The microstructure of the longitudinal section was documented by means of the optical light microscope Zeiss Axiophot. Additionally, the porosity of the brazed tapes was determined using the corresponding software Axiovision.

Results and Discussion

Figure 1 shows the values of the convexity, circularity and elongation of the superalloy MM247 and brazing filler metal BNi-9 particles. As the values didn't show a dependence on the actual material or the grain size, the results are merged in one plot. The median value of the convexity is $\overline{X}_{med, Con} = 0{,}995$ indicating a smooth particle surface. The high value of the circularity of $\overline{X}_{med, Cir} = 0{,}924$ and the low value of elongation of $\overline{X}_{med, Elo} = 0{,}094$ suggest that the particles are mostly spherical. Based on the results it can be expected that the shape and topology of all powders is similar. Thus, an overlying influence by the particle shapes on the properties of the tapes and brazed joints can be neglected in the following analyses.

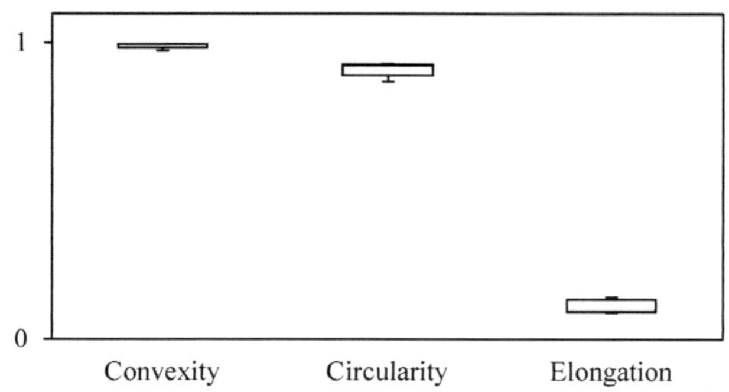

Figure 1: Convexity, circularity and elongation of the superalloy and brazing filler metal particles

The volumetric distribution of the circle equivalent diameter of the powders used for the tape production is shown in Figure 2. The distribution of the MM247 powder with the targeted grain size of $d = 63$ µm is noteworthy, as it is very narrow. Although, the grain size distributions for the others powders are relatively broad, the different targeted distributions can be distinguished. However, the overlapping of the grain size distribution of the superalloy powders with $d = 63$ µm and $d = 106$ µm has to be considered for the evaluation of the tape microstructure.

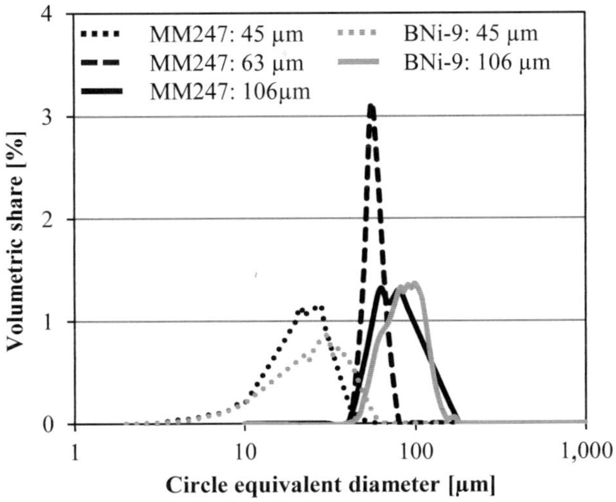

Figure 2: Volumetric distribution of the circle equivalent diameter of superalloy and filler metal particles

In order to compare the targeted and the actual grain size of each powder, the equivalent volume mean diameter D[4, 3] and the diameter D[v, 0.9] is listed in Table 4. It can be seen that the diameter D[v, 0.9] enclosing V = 90 % of the particles in the respective powder is very close to the targeted grain size. However, the equivalent volume mean diameter D[4, 3] differs notably from the targeted particle size. Only for the MM247 powder with the targeted size of d = 63 µm, the mean value of $\bar{d}$ = 57 µm is quite close to the targeted diameter value.

	Targeted grain size	D[4, 3]	D[v, 0.9]
MM247	45 µm	20.9 µm	31.0 µm
	63 µm	56.7 µm	65.5 µm
	106 µm	76.2 µm	102.5 µm
BNi-9	45 µm	27.5 µm	43.3 µm
	106 µm	85.4 µm	113.5 µm

Table 4: Equivalent volume mean diameter D[4,3] and diameter D[v, 0.9] where V = 90 % of the volume has smaller diameters of the superalloy and brazing filler metal powders

The results of the thermogravimetry show that the weight of the tapes decreases when heated up in vacuum up to T = 1,100 °C, Figure 3. Two distinctive steps of weight loss can be observed in the temperature range of 275 °C < T_I < 325 °C and 380 °C < T_{II} < 460 °C. These steps indicate the decomposition and/or evaporation of the mostly organic compounds of the binder. Nevertheless, the weight loss is relatively low with around Δm = 4 % up to the melting point of the filler metal. No notable difference between the green tape variants can be observed.

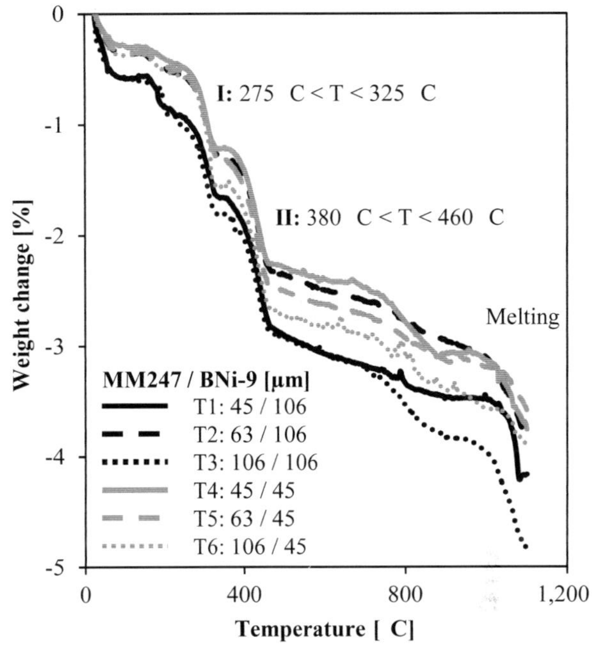

Figure 3: Weight change of the green tapes T1 to T6 in vacuum in the temperature range of 20 °C < T < 1,200 °C

The results of the DSC measurements of the BNi-9 powders with different grain sizes as well as of the tapes T1 to T6 are shown in Figure 4 for the heating and the cooling process. If not otherwise stated, the maximum temperature of the analysis was T = 1,100 °C and the holding time at that temperature was t = 30 min according to brazing process BP 1.

The onset temperature for the melting process of all powders and tapes is T_{onset} = 1,060 °C. The DSC curves of the filler metal BNi-9 show a broad endothermic melting peak indicating that the used alloy does not have an exact eutectic composition. The melting range of the fine-grained powder is 1,060 °C < $T_{m,\ BNi\text{-}9;\ 45\ \mu m}$ < 1,086 °C while that of the coarse-grained powder is broader ending at $T_{liq,\ BNi\text{-}9;\ 106\ \mu m}$ = 1,110 °C.

The tapes show a more complex melting behavior as the endothermic melting peak is overlain by an exothermic reaction. This indicates that the molten filler metal reacts with the superalloy additive. While the actual peak positions differ, this combined melting and reaction process finalizes at around 1,080 °C < T_{tapes} < 1,085 °C, probably corresponding to the completion of the melting of the filler metal. The peak positions of the combined melting and reaction process vary slightly for the tape variants but no consistent trend can be identified.

It can be noted that the curve progression of tape T4 with the smallest size of both superalloy and filler metal particles differs from the results of the other tapes. The reaction intensity of tape T4 is notably lower and the reaction interval of 1,060 °C < T_{T4} < 1,069 °C is very narrow. It can be assumed that the fine-grained filler metal is distributed more homogeneously in the tape and thus, reacts faster with the multiplicity of fine-grained superalloy particles around them.

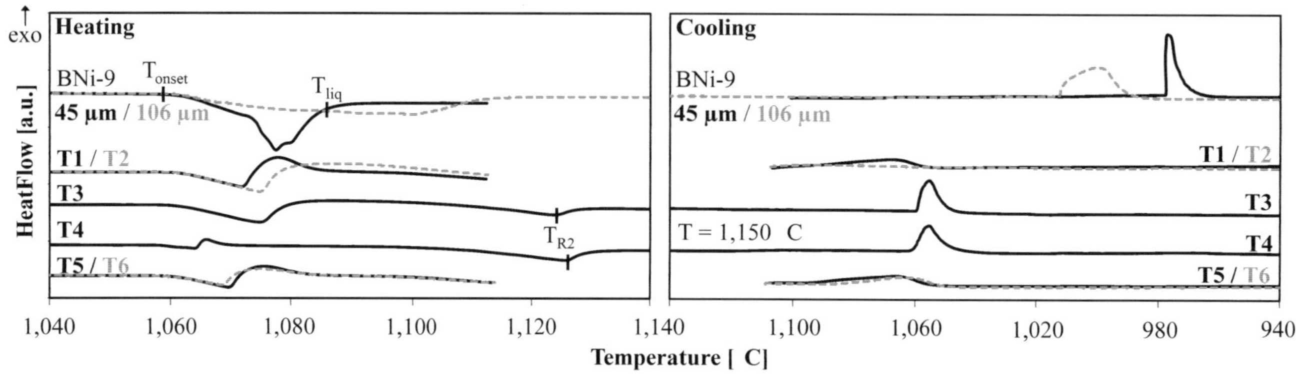

Figure 4: Results of DSC measurements of filler metal BNi-9 and tapes T1 to T6 during heating (left) and during cooling (right)

While the melting process is complete at temperatures above T = 1,100 °C even for the coarse-grained filler metal, in the DSC curves of the tapes T3 and T4, which were additionally heated up to T = 1,150 °C, a small second endothermic peak with a peak temperature of around T = 1,125 °C can be observed. As the superalloy MM247 does not melt in this temperature range, this peak could possibly indicate the melting of the phases formed by the reaction between filler metal and superalloy. Above this temperature, no further peaks could be observed.

The solidification of the filler metal takes place in the temperature range of 1,012 °C > $T_{S,\ BNi\text{-}9;\ 106\ \mu m}$ > 986 °C and of 978 °C > $T_{S,\ BNi\text{-}9;\ 45\ \mu m}$ > 963 °C in dependence of the grain size, respectively. In contrast, the solidification of the tapes occurs at higher temperatures. The tapes analyzed according to brazing process BP 1 up to T = 1,100 °C show broad and low-intensity solidification peaks in the temperature range of 1,095 °C > $T_{tapes;\ BP\ 1}$ > 1,045 °C. This leads to the assumption that the elements lowering the melting point have diffused out of the filler metal and into the solid superalloy particles. Thus, the melting and solidification temperature increases.

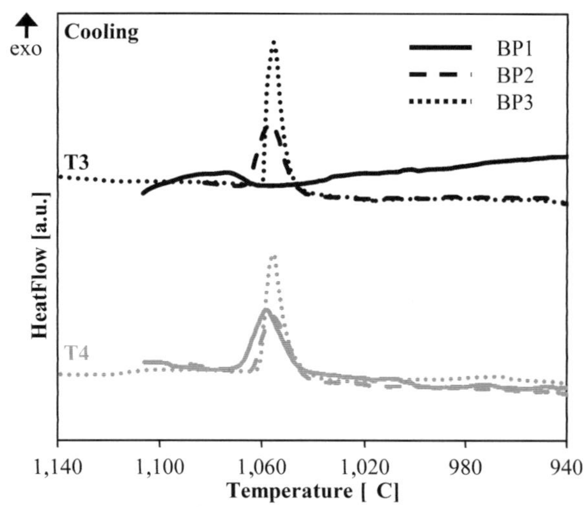

Figure 5: Results of the DSC measurements of tapes T3 and T4 during cooling after the brazing processes BP 1 to BP 3

The tapes T3 and T4 were also analyzed with the parameters of the brazing processes BP 2 and BP 3, Table 3. While the heating curves of the respective tape variant coincide, differences in the DSC curves during cooling in dependence on the brazing processes can be observed, Figure 5.

In contrast, the tapes analyzed with BP 2 and 3 show more defined solidification peaks between 1,065 °C > $T_{tapes,\ BP\ 2/3}$ > 1,045 °C. This indicates that the phase composition of the tapes has grown more homogeneous due to the increased energy input of the processes BP 2 and 3. It can be expected that a higher percentage of the superalloy particles in the tape was at least partially molten. Thus, the bigger amount of molten material probably solidifies at a defined temperature leading to a high and narrow solidification peak.

After the heat treatment, the tapes are not flexible anymore. In dependence of the original grain size distribution, the surface of tapes made of coarse-grained particles, e.g. T3, is rougher than that of e.g. T4 made of fine-grained particles. It has to be noted that only the samples after heat treatment at T = 1,150 °C show evidence of a global melting of the material due to their homogeneous metallic appearance. This is in accordance with the above mentioned assumptions that especially the process BP 3 leads to a more homogeneous melting of the tapes including also the superalloy particles.

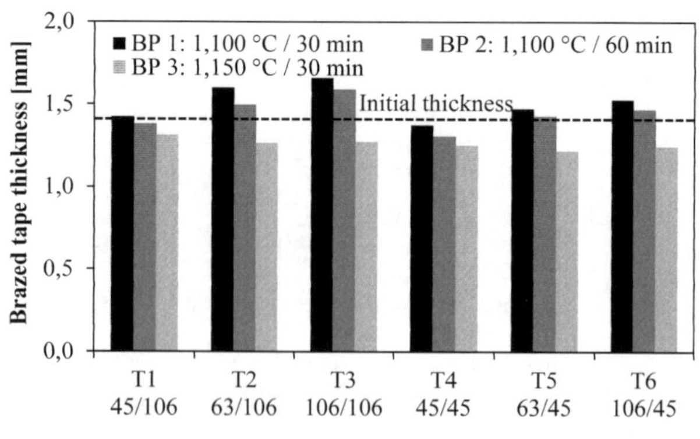

Figure 6: Thickness of the tapes T1 to T6 after brazing process BP 1 to BP 3; e.g. "45/106": grain size of superalloy and filler metal, respectively

The thickness of the tapes after the different brazing processes is shown in Figure 6. After the processes with a maximum temperature of T = 1,100 °C, the thickness of the tapes T1 to T3 is higher than that of the tapes T4 to T6. It can also be observed that the thickness increases from T1 to T3 and from T4 to T6, respectively. This indicates an influence of the grain size of the filler metal and additive superalloy. In comparison with the initial tape thickness of around t = 1.4 mm, a slight thickness reduction after the heat treatment at T = 1,100 °C can only be noted for T1 and T4 which are the tapes with the smallest grains. Based on these results, as a first assumption it can be stated that tapes made from small superalloy particles could tend more to shrinkage. The influence of the grain size of the filler metal is less pronounced but shows the same trend.

In contrast, the tapes heat treated according to process BP 3 with a maximum temperature of T = 1,150 °C exhibit a tape thickness notably lower than the initial thickness of the tape with $t_{tape, BP 3} = 1{,}26 \pm 0{,}03$ mm. It is also relatively constant for all tapes showing no dependence on grain size. This is possibly resulting from the assumed homogeneous und more complete melting of the tape material.

Figure 7 shows the longitudinal sections of tapes T1 to T6 heat treated according to BP 1 at T = 1,100 °C for t = 30 min. While it can be concluded that a fraction of the tape material is molten, the microstructure is not homogeneous but characterised by a network of connected, roundish particles. It can be assumed that the network is formed by the superalloy particles connected by the molten filler metal.

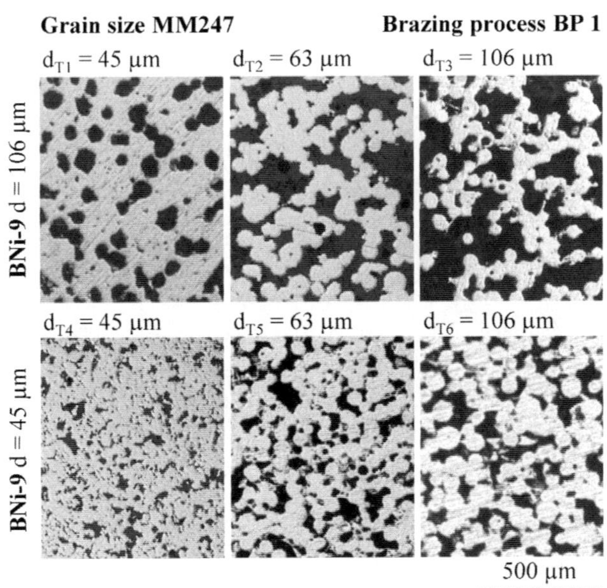

Figure 7: Longitudinal sections of tapes T1 to T6 heat treated according to BP 1 at T = 1,100 °C for t = 30 min

In dependence of the grain size, differences in the microstrucutre of the heat treated tapes can be observed. The tapes made with the coarse-grained filler metal, T1 to T3, are characterised by bigger voids while the tapes containing fine-grained filler metal particles, T4 to T6, seem denser. Additionally the shape of the voids changes from spherical to a more polygonal form when reducing the size of the filler metal. Independent of the used filler metal powder, an increasing diameter of the superalloy MM247 particles lead to an increase in porosity. This correlates with the observations about the tape thickness shown in Figure 6. The increase of the porosity is probably accompanied by a higher tape volume after the heat treatment.

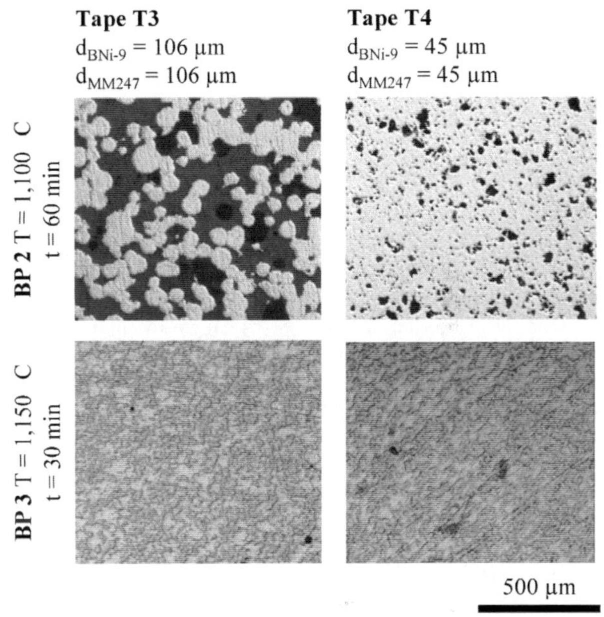

Figure 8: Longitudinal sections of tapes T3 to T4 heat treated according to BP 2 at T = 1,100 °C for t = 60 min and to BP 3 at T = 1,150 °C for t = 30 min

The tapes T3 and T4 were selected for further investigations due the very different grain sizes of the used filler metal and superalloy particles and the resulting differences in melting behavior and microstructure. The microstructure of the tapes after heat treatments according to the processes BP 2 and 3 are shown in Figure 8. In comparison to the results observed in Figure 6, it can be seen that a longer holding time results in a slightly denser microstructure. Nevertheless, the network made of particles can still be distinguished.

Only after the process BP 3 with a maximum temperature of T = 1,150 °C, the particle network is complete replaced by a homogeneous microstructure indicating a complete melting process of the tape material. Both tapes show a very similar multi-phase microstructure with a minor amount of small spherical voids. Further analyses will be performed to determine if the grain size of the particles in the tape influences the quantity and size of the voids as well as the phases.

Conclusions

The analysis of powders of the filler metal BNi-9 and the superalloy MM247 confirmed that mostly spherical particles were used for the tape production. The maximum grain sizes were identified to be close to the targeted diameters although the size distribution was mostly relatively broad. Nevertheless, the different powders can be distinguished by their targeted grain size. Nevertheless, at least a slight overlaying effect has to be considered when analyzing the influence of different grain sizes on the properties of the tapes. Thermogravimetric analysis showed that the tapes made of filler metal and superalloy particles loose only weight of around $\Delta m = 4\ \%$ due to evaporation of organic binder components. Thus, no increased porosity due to binder evaporation has to be expected in the tape.

By means of DSC measurements differences in the melting behavior of filler metal powders and tapes could be detected. The tapes showed a more complex behavior below the typical brazing temperature of BNi-9 of $T = 1{,}100\ °C$. The overlapping of the melting process by an exothermal reaction is possibly due to interactions between the molten filler metal and the superalloy leading to the formation of new phases. When increasing the temperatures above $T = 1{,}125\ °C$, an additional endothermic peak can be observed due to the melting process of the newly formed phases. The result is a dense and homogeneous microstructure while the tapes heat treated at lower temperatures showed a highly porous particle network connected by the filler metal. A notable influence of the grain sizes on microstructure and tape shrinkage could only be detected for the heat treatments at a temperature of $T = 1{,}100\ °C$ but further analyses will show if there are grain size effects also for the process at higher temperatures.

These observations are a first step to define optimal brazing parameters adapted for using the newly developed tapes with different grain sizes in repair brazing for turbine components. For actual repair brazing, the interactions of the tapes with the base metal have to be considered. Therefore, the tapes have to be attached to the base metal surface by means of a special adhesive [7]. Additional, some sustainable weight has to be applied on the top of tapes with adhesive addition in order to overcome differential thermal expansion behaviour of materials during brazing. Suitable parameters for tape application and their impact on the joining quality are currently developed.

Acknowledgments

The research is promoted by the Federal Ministry of Economic Affairs and Energy according to a resolution of the German Bundestag. The authors are thankful for this support.

References

[1] R. Acharya, R. Bansal, J. J. Gambone, S. Das, "A Couple Thermal, Fluid Flow, and Solidification Model for the Processing of Single-Crystal Alloy CMSX-4 Through Scanning Laser Epitaxy for Turbine Engine Hot-Section Component Repair (Part I)", *Metallurgical and Materials Transactions B*, Vol. 45B (2014) pp. 2.247-2.261

[2] X. Huang, W. Miglietti, "Wide Gap Braze Repair of Gas Turbine Blades and Vanes – A review", *Journal of Engineering for Gas Turbines and Power*, 134.1 (2012) 010801

[3] W. F. Gale, D. A. Butts, "Transient liquid phase bonding", *Science and Technology of Welding and Joining*, 9, 4 (2004) pp. 283-300

[4] L. C. Lim, W. Y. Lee, O. Lai, "Nickel base wide gap brazing with preplacement technique Part 2 – Formation mechanisms of macrovoids", *Materials Science and Technology*, Vol. 11, Iss. 10 (1995) pp. 955-960

[5] E. Lugscheider, V. Dietrich, J. Mittendorf, "Wide Joint Clearance Brazing with Nickel-Base Filler Metals", *Welding Research Supplement* 67.2 (1988) pp. 47s-52s

[6] E. Lugscheider, T. Schnittny, E. Halmoy, "Metallurgical aspects of additive-aided wide-clearance brazing with nickel-based filler metals", *Welding Research Supplement* (1989) pp. 9s-13s

[7] S.M. Uddin, H. Krappitz, J. Barnickel, I. Reinkensmeier, "Repair brazing and protection of titanium turbine components using polymer bonded tapes", *Proc LÖT 2016*; Aachen, Germany, June 2016, DVS-Band 325, pp. 78-83

[8] J. Zhang, Z. Zhang, J. Sun, X. Liu, "Effect of Gap Filler on Microstructure of Wide-gap Brazing Seam", *Materials Transactions, JIM*, Vol. 41, No. 8 (2000) pp. 1073-1076

[9] D. Fortuna, Sulzer Metco, "Gas Turbine Repairs with Binder-free Braze Alloys", *Sulzer Technical Review*

[10] M. Koschling, H. Krappitz, "Wear-resistant coatings close to the final contour built up by brazing by means of hard material and brazing mats", *Schweißen & Schneiden,* Heft 9/93 (1993)

[11] S.K. Tung, L.C. Lim, "Wide gap brazing prepacks of nickel base braze mixes", *Materials Science and Technology*, Vol. 11 (1985) pp. 949-954

[12] R. Przeliorz, J. Piątkowski, "Thermophysical properties of nickel-based cast superalloys", *METABK*, 54, 3 (2015), pp. 543-546

[13] DIN EN ISO 17672 "Hartlöten – Lote", Version 2010-11

Qualification of brazed ferritic stainless steel joints for corrosive applications

Tillmann, W.; Manka, M., Wojarski, L.; Becker, T.
Institute of Materials Engineering, Technical University Dortmund, Germany, Dortmund
Corresponding author:Matthias.Manka@udo.edu

Abstract

Brazing of stainless steel such as AISI 304 is a frequently used technique to produce high strength joints with a moderate corrosion resistance in exhaust gas systems. The usage of the ferritic stainless steel AISI 430 could reduce the material costs by half while maintaining comparable mechanical strength properties.
In this regard, the joint quality of vacuum brazed AISI 430 with the filler alloys BNi-2, and B-Ni67CrPSiB-945/1040 were analyzed. For this purpose, miscellaneous brazing parameters were utilized and assessed concerning the resulting microstructure and the tensile shear strength. According to the test sheet VDA 230-214, the corrosion resistance of the joints was investigated using the condensate K2.2 and an aging temperature of 400 °C. The best mechanical strength properties of 313.1 MPa were achieved for AISI 430/BNi-2 joints with a heating rate of 50 K/min, a brazing temperature of 1050 °C, a dwell time of 2 minutes, and with free vacuum cooling. The results of the mechanical properties and corrosion resistance were compared with the commonly used AISI 304/BNi-2 joint.

Introduction

In recent years, the use of ferritic stainless steels for structural applications has attracted strong interest due to their good strength and durability properties in conjunction with a favorable purchase price [1]. A comparison of the price development between the most common austenitic- AISI 304 and ferritic steels AISI 430 reveals a cost advantage in the range of half by using AISI 430 [2]. Nevertheless, reliable design parameters are still missing for a consistent development of new applications [1]. Especially in the field of brazing, only few journal contributions thematizes scientific investigations of brazed ferritic steel joints. Winiowski and Majewski used AISI 430 as base material for their investigations of new fluxes for a torch brazing process with the silver brazing alloy paste AG 203. For an overlap configuration with an undefined overlap, they achieved tensile shear strengths of approximately 200 MPa [3]. Further investigations of Winiowski and Rozanski disclose the potential of the filler alloys AG 401, B-Ag68CuSn-730/755, and B-Ag65Cu28Sn5Ni2 applied as a 100 μm thick foil and brazed in a vacuum atmosphere [4]. The superior spreading properties as well as the best average shear strength of 196 MPa were achieved by using B-Ag65Cu28Sn5Ni2 as a filler alloy, brazed at 1000 °C for 15 minutes [5]. In comparison to miscellaneous brazing processes with austenitic stainless steels, only few publications are available in which corrosion resistant nickel is used as filler alloy for the brazement of ferritic steels. At the IBSC 2012, Miyazawa et al. presented a detailed examination of the brazed ferritic steel X2CrCuTi18 with three different nickel filler alloys that contained boron, silicon or phosphor as melting point depressors [6]. They observed within their brazing process at 1080 °C for 30 minutes in an argon gas atmosphere that only a limited diffusion of phosphor into the ferritic steel took place with the phosphor-containing filler alloy. In further studies Miyazawa et al. investigated the microstructural results of brazed ferritic steel X2CrMoTi18-2 joints, using different amorphous nickel filler alloys by systematically reducing their boron content [7]. The results of this investigation evince that a boron content below 1 mass-% prevents a penetration of boron into the ferritic steel. Using the brazing filler alloy B-Ni67CrPSiB-945/1040 to braze different austenitic stainless steels as well as the ferritic steel AISI 444, Marsilius and Hartmann examined the element distribution of chromium, boron, and phosphor inside the brazing seam for different brazing parameters within a temperature range of 1100-1150 °C and dwell times of 10 to 60 minutes. They determined that at short dwell times, almost no phosphor diffuses into the base material. However, due to the large diffusion zone of approximately 50 μm, a phosphor embrittlement may occur [8]. Compared with similar investigations of AISI 304/BNi-2 joints [9, 10], boron penetrations along the grain boundaries can reduce the mechanical and especially the corrosion resistance of the brazing seam. Besides the boron content within the filler alloy itself, the formation of so-called chromium borides can be influenced by the brazing process. Utilizing a scientifically administrated experimental design, Tillmann et al. investigated the influence of process relevant parameters on microstructural formations and the resulting tensile shear strength of AISI 304/BNi-2 joints. The highest tensile shear strength of 260 MPa was obtained with a brazing gap of 50 μm, a heating rate of 50 K/min, and a brazing temperature of 1050 °C for 2 minutes in a vacuum atmosphere of approximately 10^{-5} to 10^{-6} mbar [11]. Further examination of Schmiedt et al. dealt with a detailed investigation of the influence of the corrosion outsourcing according to VDA 230-214 on the mechanical properties of AISI 304/BNi-2 joints using the synthetic exhaust gas condensate K2.2 [12]. In contrast to the existing base for a purposeful usage of brazed austenitic steel joints in corrosive environments, the knowledge concerning the use of significant cost-effective ferritic steels remains insufficient. Therefore, this study deals with the investigation of the brazing parameters to braze the most common ferritic steel AISI 430 with BNi-2 and B-Ni67CrPSiB-945/1040, focusing on their metallurgical as well as remaining strength properties after being exposed to extreme corrosive environments.

Experimental

The experimental procedure was divided into a two-step approach. Starting with fundamental investigations of an appropriate brazing process by varying the most relevant brazing parameters according to Tillmann et al. [11], the corrosive damage of the joints was assessed by the shear strength after conducting the dip and dry test according to VDA 230.214 [13].

All investigations were conducted with the ferritic stainless steel X6Cr17 (AISI 430) as the base material. Commonly utilized BNi-2 and B-Ni67CrPSiB-945/1040 (VZ2177) were used as brazing filler metals. Both filler alloys were provided as amorphous solidified foils, in case of BNi-2 with a thickness of 50 µm and for the second alloy with a thickness of 25 µm. The composition of the materials used, including the melting temperatures of the filler alloys, are summarized in Table 1.

Table 1: Composition of the used materials [14 - 16]

Materials	X6Cr17	BNi-2	VZ2177
	[wt.-%]	[wt.-%]	[wt.-%]
Ni	-	Bal.	Bal.
Fe	Bal.	3	-
Cr	17	7	25
Si	1	4.5	1.5
B	-	3.2	0.5
P	0.04	-	6
Mn	1	-	-
C	0.08	≤ 0.08	-
S	0.02	-	-
$T_{sol.}$ [°C]		970	945
$T_{liq.}$ [°C]		1000	1040
Foil thickness		50 µm	25 µm

Prior to the brazing experiments, the samples were laser cut into the geometry shown in Figure 1. The rejuvenation at the brazing area was specifically applied due to the appropriate high strength properties of the joints. The force was initiated by a pin, which was positioned into the drilled hole of the sample and fixed in the specimen holder. With regard to the brazing process, all samples were prepared by using a silicon carbide grinding paper with a grit of 1200 and underwent a subsequent cleaning process in an ultrasonic bath for 5 minutes. All brazings were conducted in the vacuum furnace provided by the company Torvac in vacuum atmospheres within the range of $5 \cdot 10^{-4}$-$5 \cdot 10^{-5}$ mbar. For the microstructural observation, cross-sections of the joints were prepared using metallographic techniques. The analysis itself was conducted by using a light- as well as a scanning electron microscope (SEM), which is equipped with an energy dispersive X-ray spectroscopy (EDS). In order to identify the elementary contribution of different phases within the brazing seam, both, the microstructure as well as the fractured surface were analyzed in backscattering mode.
To measure the tensile shear- and the shear strength properties, the uniaxial testing device of the company Mohr & Feldhoff was utilized. The following calculation of the strength properties was conducted with the real joining surface, which was measured with the light microscope after mechanical testing.
As mentioned in the introduction, no available data exists for the brazing of the ferritic stainless steel AISI 430 with the nickel filler alloys BNi-2 or VZ2177. Therefore, for both possible solutions, three completely different brazing procedures were tested and qualified according to their microstructure as well as to the resulting tensile shear strength using the single lap geometry shown in Figure 1a. With the aim to reduce scatter effects and to gain further strength data, a double lap configuration (Figure 1b) was chosen for further corrosion- as well as for the shear strength tests.

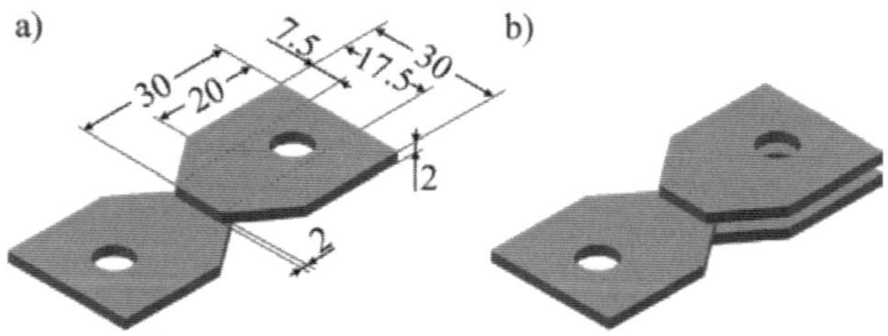

Figure 1: Sample geometry: a) single lap; b) double lap

Dip and dry corrosion test

The corrosion test according to VDA 230-214 was conducted in the self-made device visualized in Figure 2. With the aim to investigate the worst operating conditions for the joints, the maximal temperature permitted for the steel was chosen as the aging temperature and set to 400 °C. For the corrosive medium, the synthetic condensate K2.2 was selected. In the VDA test sheet, the condensate K2.2 is defined as a mixture of 0.035 ml/l acetic acid, 0.039 ml/l formic acid, and 1650 mg/l NaCl, dissolved in water.

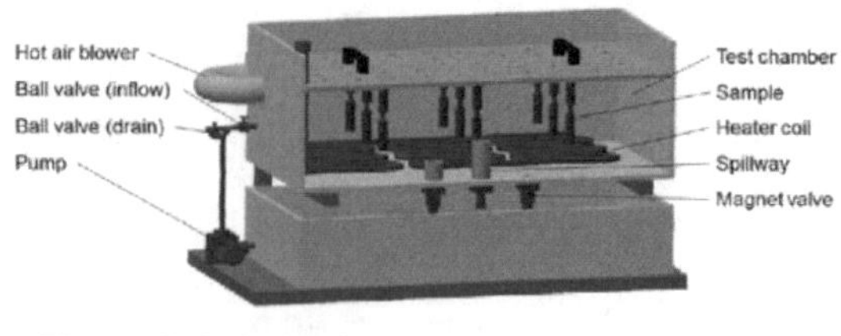

Figure 2: Schematic image of the corrosion test rig

Results and discussion

In order to guarantee a constant vacuum atmosphere, even for fast heating rates, a holding time was integrated at 400 °C for 30 minutes. Furthermore, in order to adjust a constant sample temperature, an additional holding time for 10 minutes at 900 or 950 °C was realized. For the AISI 430/BNi-2 joints, the brazing gap was set to 50 µm and the brazing temperature to 1050 °C. The heating rates were varied between 10-50 K/min and an additional annealing process for 2 hours was applied at 1000 °C. According to the microstructure as well as the resulting average tensile shear strength of 3 samples of 293.2 MPa, the best properties were achieved for a fast heating rate of 50 K/min, a short dwell time of 2 minutes and free vacuum cooling.

In case of using VZ2177, the 25 µm amorphous brazing foil was applied twice so that the brazing gap was set to 50 µm. The brazing temperature was varied between 1060-1080 °C. Meanwhile the heating rates, as well as the annealing process, were varied in the same manner as for the BNi-2 joints. As a result of these investigations, the best average tensile shear strength of 186.8 MPa was achieved at a brazing temperature of 1060 °C, a short dwell time of 2 minutes, and a downstream annealing process. According to the resulting microstructure shown in Figure 3, a more or less continuous intermetallic phase band appeared in the center of the brazing seam. This region seems to be dominated by an unavoidable Fe_3P phase (spectrum 2), which also could be observed in Müller et al.'s investigations, using the austenitic steels AISI 304 and AISI 316L with a comparable but boron-free brazing filler alloy [17].

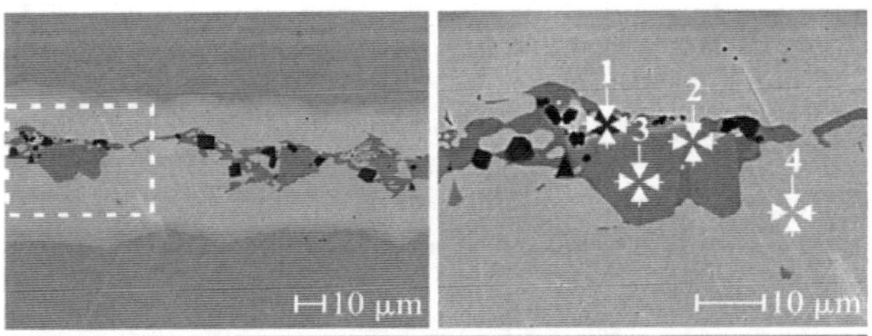

Spectrum	Fe	C	Si	Mn	P	Cr	Ni
1	1.63	3.18	-	-	9.85	83.04	2.50
2	9.68	5.49	0.07	0.21	21.17	32.94	30.44
3	2.70	3.65	5.01	-	5.37	47.90	35.36
4	23.01	2.90	1.41	0.27	-	17.98	54.43

Figure 3: Microstructure of the AISI 430/VZ2177 joint with a brazing gap of 50 µm

In order to decrease the segregation of the continuous phosphor band in the center of the brazing seam, the foil thickness of the brazing material VZ2177 was reduced by half for the pursuing experiments. Furthermore, the single lap geometry was changed to a double lap configuration as shown in Figure 1b, which permits the investigation of the shear strengths and is less sensitive for fluctuations caused by the small sizes of the samples. Additionally, the temperature for the downstream aging process was adjusted to 850 °C.

Based on the drawn conclusions from the determination of the brazing parameters, Figure 4 represents the microstructure of the brazed AISI 430/BNi-2 joint in the initial condition. In the center of the brazing seam (Pos. 5) a nickel-dominated phase can be observed, which contains approximately the composition of the brazing filler metal used. Due to diffusion processes, this region is infiltrated by iron. In general, the brazing seam appears ductile without the formation of brittle phases. The microstructural appearance at the diffusion zone resembles the well-known behavior of brazed chromium containing stainless steels like AISI 304 with BNi-2. Due to uneven diffusion rates, the phenomenon of Kirkendall-voids can be observed as dark points at the border area. Furthermore, most likely chromium borides have formed in a dark needle-like structure within the diffusion zone. In this context, it must be said that the detection of boron and carbon with EDS is limited due to their low atomic number. Anyway, the tendency of a carbon penetration from the base material into the brazing seam appears doubtless.

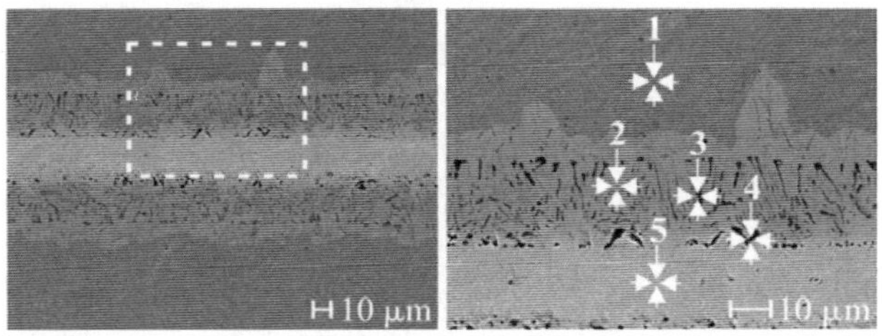

Spectrum	Fe	C	Si	Mn	Cr	Ni
1	76.62	7.46	0.27	0.40	15.26	-
2	81.24	3.62	0.37	0.28	14.13	0.36
3	70.06	4.41	0.33	-	24.69	0.51
4	39.11	7.05	2.96	-	10.82	40.07
5	17.91	6.77	5.38	-	9.82	60.13

Figure 4: Initial microstructure of the AISI 430/BNi-2-joint

Concerning the AISI 430/VZ2177 joint, the resulting microstructure in Figure 5 appears less brittle compared to the brazing joint utilized with a larger brazing gap of 50 µm visualized in Figure 3. The reduction of the brazing gap severs the continuous phosphor band and could lead to higher strength properties. Although, the formation of the isolated phosphor containing phases can be observed at position 4 or 5 and could be the reason for potential metallurgical notches. Most likely, the formation of Fe_3P-phases can be observed there. These phases are surrounded by a FeNiCr-rich phase, which seems to have a relatively ductile behavior. Because of the contrast ratio of the used materials, a clear brazing zone could not be detected by the SEM. However, the EDS measurements disclose a clear separation from the base material due to the rapid decrease of nickel and the strong growth of the iron content at position 1.

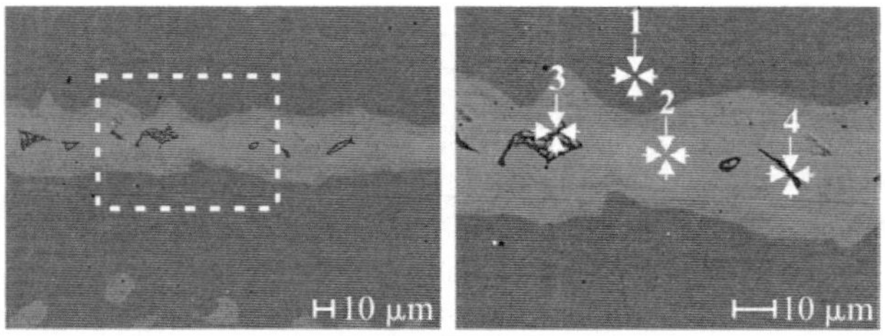

Spectrum	Fe	C	Si	P	Cr	Ni
1	76.74	4.97	0.44	0.38	16.65	0.82
2	49.91	5.65	0.90	0.30	15.66	27.58
3	43.14	9.64	0.23	8.53	26.27	12.20
4	44.35	7.86	-	9.13	27.83	10.83

Figure 5: Initial microstructure of the AISI 430/VZ2177-joint

In accordance with the test sheet VDA 230-214, the samples were weighed, in order to measure the corrosive damage, every week. Therefore, Figure 6 visualizes the percentage weight loss of 3 removed samples in comparison to the preceding week. It can be observed that no huge differences between both used filler alloys exist. Generally, the weight lost increases slowly up within the first week and was relatively stable for the following weeks. The corrosive damage is slightly worse for the filler metal BNi-2 and has a maximum value of 0.36%.

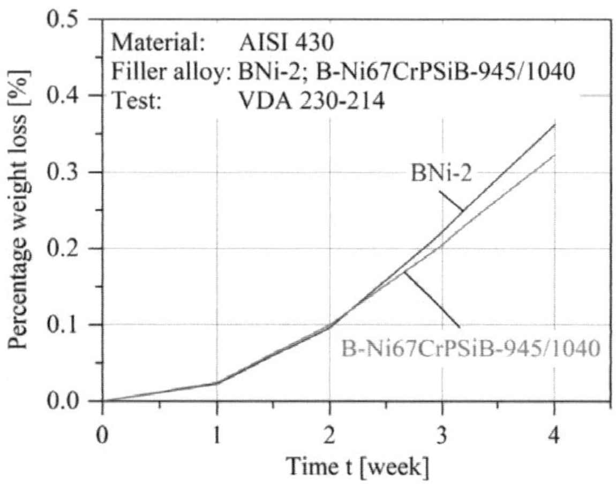

Figure 6: Percentage weight loss based on the weight of the previous week

If we compare the corrosive rate based on the weight of the initial condition, a very interesting phenomenon can be observed in Figure 7. While the corrosion rate behaves similar to the weight loss for the first two weeks, the corrosion rate seams to achieve a maximum at a value of 0.4 mg/a·mm² for the AISI 430/VZ2177 joint and a slightly higher value of approximately 0.45 mg/a·mm² for the compound brazed with BNi-2. In general, the performance of the joint AISI 430/VZ2177 is less vulnerable under the tested conditions.

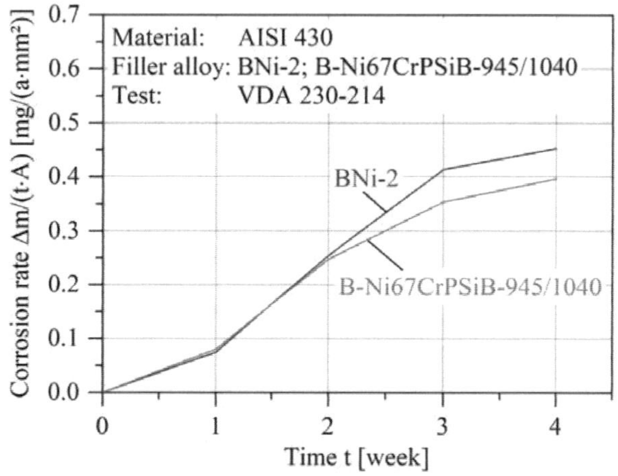

Figure 7: Corrosion rate based on the weight at initial condition

Besides the corrosive investigations, the mechanical properties of the joints were measured in the initial state as well as after each test week, (Figure 8). Basically, it can be said that joints brazed with VZ2177 show slightly better mechanical properties at all investigated conditions compared to the AISI 430/BNi-2 joints. An interesting observation can be made by comparing the initial state with the first week. Although the corrosive load leads to a minor decrease of the shear strength of the AISI 430/BNi-2 joint, the test conditions provoke an increase of the shear strength for the AISI 430/VZ2177 joint. A possible explanation of this behavior might be the annealing process for 5 hours at a temperature of 400 °C and the thus reduced stress level of the joint. The shear strength reached a maximum average value of approximately 250 MPa. In comparison to the reached average tensile shear strength of 186.8 MPa, presented for the larger brazing gap of 50 µm, this is an improvement of about 34% for a supposedly worse load. The subsequent outsourcing caused a decrease down to a value of 180 MPa in the third week, which is therefore lower than expected.

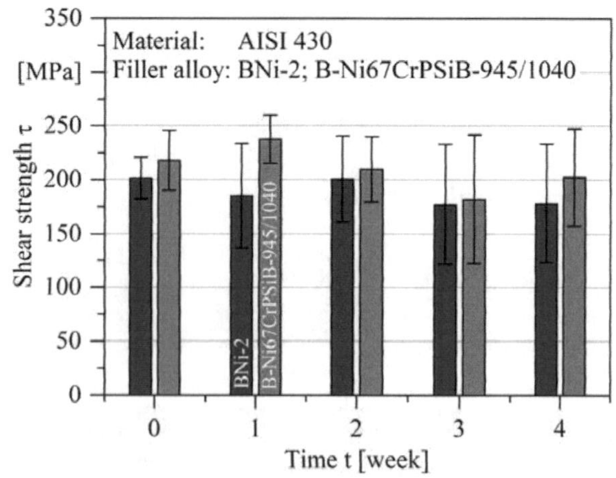

Figure 8: Shear strength properties at initial state and after each week of the dip and dry test

In case of the AISI 430/BNi-2 joint, the maximum shear strength was reached at the initial state with a value of 200 MPa. The maximum reduction of the shear strength could also be recognized in the third week and has, with a value of approx. 175 MPa, the same level of the AISI 430/VZ2177 joint. Therefore, this compound seems to be less sensitive to the corrosive dip and dry test.

Comparing the fractured surfaces of all investigated samples, no differences could be observed. The fracture of AISI 430/BNi-2 joints always occurred at the border area between the diffusion zone and brazing seam (Figure 9a) in the region where the Kirkendall-voids (Figure 4) were present. This goes in line with the shear strength properties of the joints, which exhibit only a decrease of about 10 % after 4 weeks of corrosive outsourcing according to the VDA 230-214 test sheet.
In contrast to that, the fracture path of the AISI 430/VZ2177 joint always occurred in the center of the brazing seam, Figure 9b. Although the formation of the seemingly brittle continuous phase formation could be significantly reduced by adjusting a brazing gap of only 25 µm, the isolated Fe_3P-phases are the weakest point in this compound. Therefore, the fracture propagation is almost unbiased by the corrosion as well and represents a rather inherent problem of the used filler metal.

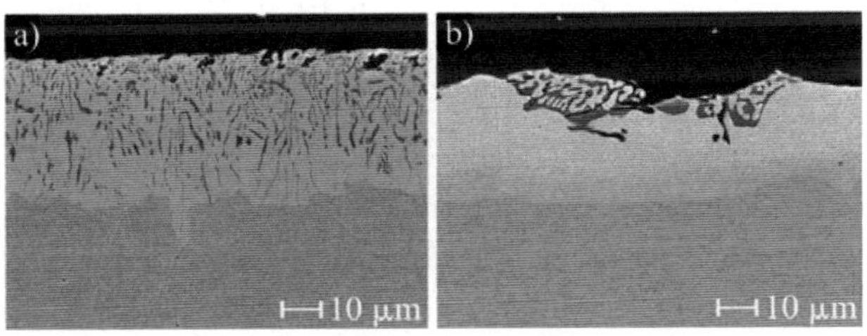

Figure 9: Fracture path for the samples of test week 2 a) AISI 430/BNi-2; b) AISI 430/VZ2177

Conclusions

The aim of the present study was to investigate the behavior of the ferritic stainless steel AISI 430, brazed with two different nickel filler alloys under corrosive conditions according to the test sheet VDA 230-214. From the procedures, following conclusions can be drawn:

1. The reduction of the brazing gap of the filler alloy VZ2177 from 50 to 25 µm severs the formation of a brittle continuous band in the center of the brazing seam.

2. With a heating rate of 50 K/min, a brazing temperature of 1050 °C, a dwell time of 2 minutes, and a free vacuum cooling, an average tensile shear strength of 293.2 MPa and a shear strength of 200 MPa were achieved by using BNi-2 as brazing filler alloy in form of a 50 µm amorphous foil.

3. With a heating rate of 50 K/min, a brazing temperature of 1060 °C, a dwell time of 2 minutes, an annealing process for 2 hours at 850 °C, and a free vacuum cooling, an average shear strength of 220 MPa was achieved by using VZ2177 as brazing filler alloy in form of a 25 µm amorphous foil. An average tensile shear strength of 186.8 MPa was achieved by using a foil with a thickness of 50 µm and an annealing temperature of 1000 °C.

4. For the synthetic condensate K2.2 used and the outsourcing temperature of 400 °C, the AISI 430/BNi-2 joint seems to be slightly more sensitive concerning to the weight loss and the corrosion rate. An almost constant corrosion can be observed from test week 4 on for both filler alloys used.

5. The dip and dry test has only a minor effect on the shear strength properties of both joints. A reduction of 10% for AISI 430/BNi-2 and 7% for AISI 430/VZ2177 can be observed.

6 The fracture for AISI 430/BNi-2 occurred along the border area between the diffusion zone and brazing, seam most likely because of the formation of Kirkendall-voids. For the joint AISI 430/VZ2177, the fracture appeared through the center of the brazing seams due to isolated phosphor phases. Corrosion has no influence on the crack propagation of both joints.

Acknowledgment

The authors thank Thomas Hartmann and the company Vacuumschmelze GmbH & Co. KG for their great support by providing the brazing filler alloys for our investigations.

References

[1] Cashell, K. A., Baddoo, N. R.: Ferritic stainless steels in structural applications. Thin-Walled Structures, 83, (2014), pp. 169–181

[2] NN: Legierungszuschlag - Übersicht, 2018. http://legierungszuschlag.info/, retrieved: 14.01.2018

[3] Winiowski, A.; Majewski, D.: New fluxes for brazing stainless steel and titanium and its alloys. Welding International, 28, (2011), pp. 1–4

[4] Winiowski, A., Rózanski, M.: Impact of Tin and Nickel on the Brazing Properties of Silver Filler Metals and on the Strength of Brazed Joints Made of Stainless Steels. Archives of Metallurgy and Materials, 58, (2013) 4, pp. 1007–1011

[5] Winiowski, A., Różański, M.: Vacuum brazing of stainless steel with silver binder containing nickel and tin. Welding International, 28, (2013) 12, pp. 947–952

[6] Miyazawa, Y., Saito, K., Fukikoshi, T., Ariga, T., Takizawa, T.: Brazing of ferritic stainless steel and microstructure at the brazed joint. In: Gourley, R., Walker, C. (ed.): Proceedings of the 5th international brazing and soldering conference, April 22th to 25th, 2012, Las Vegas, Nevada USA. Materials Paer: ASM

[7] Miyazawa, Y., Ishihara, M., Ota, K., Bizen, Y.: Brazing of Ferrite Stainless Steel with Ni-based Amorphous Brazing Foils, (2015)

[8] Marsilius, M., Hartmann, T.: Influence of boron and phosphor containing nickel based brazing alloy on different base materials. In: Brazing, high temperature brazing and diffusion bonding. LÖT 2016: Lectures and posters of the 11th international conference, June 7th to 9th, 2016, Aachen, Germany. DVS-Berichte, Band 325. Düsseldorf: DVS Media GmbH, (2016), pp. 227–232

[9] Castano, C., Blanco, A., Carnota, J., Sotelo, J., Gonzalez, M., Verdera, D., Porto, E.: Study of brazing filler metals with low Nickel content applied to the austentic steel joint by brazing in vacuum furnace. IBSC, International Brazing and Soldering Conference, 4. Materials Park: ASM International, (2009), pp. 119–124

[10] Wielage, B., Hoyer, I.: Nickel-base standard brazing solder with improved corrosion resistance for corrosion-loaded components. Schweißen und Schneiden, 60, (2008), 7/8, pp. 401–403

[11] Tillmann, W., Wojarski, L., Manka, M., Walther, F., Wittke, P., Schmiedt, A.: Examination of the influencing factors on the microstructure of AISI 304/ BNi-2 joints. In: Brazing, high temperature brazing and diffusion bonding. LÖT 2016: Lectures and posters of the 11th international conference, June 7th to 9th, 2016, Aachen, Germany. DVS-Berichte, Band 325. Düsseldorf: DVS Media GmbH (2016), pp. 233–238

[12] Schmiedt, A., Manka, M., Tillmann, W., Walther, F.: Influence of pre-corrosion caused by synthetic exhaust gas condensates of the automotive industry on tensile and fatigue properties of brazed AISI 304/BNi-2 joints. In: Vormwald, M., Beier, H. T., Breidenbach, K. (ed.): Proceedings of the 5th Symposium on Structural Durability, May 17th to 18th, 2017, Darmstadt, Germany. Technische Universität Darmstadt - Institut für Stahlbau und Werkstoffmechanik (2017), pp. 1–13

[13] Verband der Automobilindustrie e. V. (VDA): VDA Test sheet 230-214 - Resistance of metallic materials to condensate corrosion in exhaust gas-carrying components, (2016)

[14] DIN EN ISO 17672; 2017. Hartlöten- Lote

[15] NN - ACCIAI SPECIALI TERNI: 1.4016. data sheet

[16] Vacuumschmelze GmbH: VZ2177 - Data sheet, Vacuumschmelze, 2016

[17] Müller, W. H., Wilden, J., Schmorl, B., Stahn, O.: Calculation of Thermodynamically Optimized Temperature-/Time-Cycles in Nickel-Based Brazing Joints. PRZEGLĄD SPAWALNICTWA, 88, (2016), 8, pp. 65–70

Fracture Mechanics Test Methods for Ni-base Superalloy Braze Joints

Bryan Riggs, Jacob Wildofsky, Boian Alexandrov, Avraham Benatar
The Ohio State University, Columbus, Ohio, U.S.A.
riggs.106@osu.edu, wildofsky.2@osu.edu, alexandrov.1@osu.edu, benatar.1@osu.edu

Ray Xu
Rolls-Royce Corporation, Indianapolis, Indiana, U.S.A.
Ray.Xu@Rolls-Royce.com

Abstract

In this study, the fracture toughness of Inconel 718 Ni-base superalloy braze joints with a BNi-2 filler metal was assessed using four different fracture mechanics test methods. Experimental results are provided and differences between test methods are discussed. Inconel 718 base metal fracture toughness properties are also measured to provide a base line for comparison with braze joint test results.

Introduction

Voids and defects are commonly found in braze joints due to the inherent nature of the brazing process [1]. As brazing continues to be an efficient and cost effective joining technology, it is becoming increasingly important to account for voids or defects when present in joints. When defects are detected in bulk materials or welds (often using non-destructive inspection such as X-ray or ultrasonic), fracture mechanics can be used to estimate the stress intensity factor (K_C) at the crack tip and compare it to the critical stress intensity factor (K_{IC}) fracture toughness of the material where K_{IC} is a function of the flaw size, stress and material properties. In this approach, if the calculated K_C exceeds the K_{IC} of the material, failure will occur. Fracture mechanics has been incorporated into industry standards such as API 579 or BS 7910 to generate failure assessment diagrams as part of a fitness-for-service assessment [2]. To calculate the fracture toughness of the material(s) for FFS assessment standard procedures have been developed, such as those listed in ASTM E399 or ASTM 1820 are applied [3]. Typical test methods include the compact tension (CT) test or three- and four-point bend tests to experimentally determine the critical stress intensity factor or fracture toughness. With these properties known, fitness-for-service assessments can be used to determine maximum flaw size and failure criteria in components where flaws have been detected. While this approach has been successfully implemented for weld or bulk metals, the same approach has yet to be applied to braze joints.

There have been recent research efforts to adopt a similar approach for braze joints. Leinenbach et al. used a DCB test specimen to measure the critical stress intensity factor (K_{IC}) of martensitic stainless steels brazed with an Au-18Ni braze alloy [4]. Gan et al. used a V-notch type of SENT test to measure the fracture toughness of vanadium/stainless steel braze joints [5]. Ghovanlou et al. used a similar SENT and SENB type specimen to measure fracture toughness of A36 steel braze joints with two copper filler metals [6]. Weiss et al. tested Inconel718/BNi-5 braze joints (similar to the material combination used in this study) using a traditional four-point SENB test geometry [7]. K_{IC} values reported in this study ranged from 21.1-28.2 ksi√in. A summary of some of the publications in this area is provided in Table 1. In general, fracture toughness values range from 10 ksi – 70 ksi depending on the base metal/braze alloy combination.

While efforts in this area have shown promise, there is still a lack of published data in the literature and a standard test method for measuring braze joint fracture toughness is not currently available. This is partially due to the wide range of materials that have been used in these studies. Therefore, the

Table 1: Summary of several publications on fracture mechanics of braze joints.

Author	Base Metal	Braze Alloy	Test Method	K_{IC} (ksi*in$^{1/2}$)
Leinenbach et al.[4]	Martensitic Stainless Steel	Au-18Ni	DCB	43.0 - 46.1
Weiss et al.[7]	Inconel 718 (full HT)	BNi-5	SENB	21.1 - 28.2
Moorhead et al.[8]	Alumina, PSZ, Ti	Cu-44Ag-4Sn-4Ti	DCB (modified)	2.5 - 10.7
Ghovanlou et al.[6]	A36 steel	Copper (BCu-1)	SENB	19.74
		Copper (BCu-1)	SENT	42.32
Gan et al.[5]	SS 304, V-5Ti-5Cr	AuNi	SENT (V-Notch)	18.20
		MBF80 (BNi-9)		1.55
		AuNiPd		1.91
		Cu		10.01
Philips et al.[9]	SS 304	Ni-P	DCB (modified)	1.0 [kJ m^{-2}]

Table 2: Formulas used for K_{IC} calculations for four test methods used in this study.

Test Method	K_{IC} formula	Geometry Factor - f(a/W)
SENT[5]	$K_{IC} = \sigma_c\sqrt{\pi a}\, f\left(\frac{a}{W}\right)$	$f\left(\frac{a}{W}\right) = 1.12 - 0.23\frac{a}{W} + 10.6\left(\frac{a}{W}\right)^2 - 21.7\left(\frac{a}{W}\right)^3 + 30.4\left(\frac{a}{W}\right)^4$
DCB[4]	$K_I = \frac{F}{Bh^{0.5}}(8 + 13.25\frac{a}{h} + 12\left(\frac{a}{h}\right)^2)^{0.5}$	N/A
CT[3]	$K_{IC} = \frac{F}{B\sqrt{W}} * f\left(\frac{a}{W}\right)$	$f\left(\frac{a}{W}\right) = \left(2 + \frac{a}{W}\right)\left[0.886 + 4.64\frac{a}{W} - 13.32\left(\frac{a}{W}\right)^2 + 14.72\left(\frac{a}{W}\right)^3 - 5.6\left(\frac{a}{W}\right)^4\right] / \left(1 - \frac{a}{W}\right)^{3/2}$
SENB[6]	$K_{IC} = \frac{F}{B\sqrt{W}} * f\left(\frac{a}{W}\right)$	$f\left(\frac{a}{W}\right) = \frac{3(S_{outer} - S_{inner})}{W}\left(\frac{a}{W}\right)^{\frac{1}{2}}\left[1.99 - \left(\frac{a}{W}\right)\left(1 - \frac{a}{W}\right)\left(2.15 - 3.93\left(\frac{a}{W}\right) + 2.7\left(\frac{a}{W}\right)^2\right)\right]$

purpose of this work was to compare test results from four different fracture mechanics test methods for the Inconel 718/BNi-2 braze joints. After a review of the technical literature, four test geometries were selected for study: Single Edge Notch Tensile (SENT), Double Cantilever Beam (DCB), Compact Tension (CT), and Single Edge Notch Bend (SENB) or four-point bend specimen.

Experimental Procedure

Materials selected for study were a nickel-based superalloy, Inconel 718 and a conventional Ni-Cr-B based braze filer metal, AWS BNi-2. Two rounds of testing were carried out for the specimen types of SENT, DCB, CT and SENB, and each type of the specimens included 5-6 duplicates tested in each round. Braze joint test specimens were created by EDM machining Inconel 718 plates. The DCB and V-Notch geometries and manufacturing procedures have been previously reported [10]. For CT and SENB test specimen basic dimensions are shown in Figure 1. For brazing, BNi-2 amorphous foil was placed between two machined halves and graphite fixtures were used to hold the assembly in place. Pre-cracks were created by applying brazing stop-off on both halves in the area of the crack. This method was found to be easier than using a fatigue pre-cracking method such as those described in ASTM E399 due to the brittle nature of the braze joint. Analysis of the crack tip showed a sharp crack with a notch radius less than half of the joint gap at the start of each braze joint. Furnace brazing was carried out in high vacuum (about 10^{-5} Torr) at 1065°C for 10 min followed by a full Inconel 718 solution and age heat treatment. Joint gaps were found to be typically 0.001-0.002 inch. Metallographic cross-sections of the braze joints carried out in a separate study showed typical joints contained approximately 20-40% volume fraction of centerline intermetallic phases.

Inconel 718 base metal SENT and SENB test specimens were also fabricated from the same 0.5" plate for comparison with the braze joints test results. These samples were EDM machined to the same dimensions as the final brazed specimens and put through the same furnace brazing and heat treatment. Fatigue pre-cracking was applied for these samples to ensure a sharp crack tip prior to testing.

All specimens were tested to failure using an MTS 810 tensile tester at a constant displacement rate of 0.04 in/min. Force vs. crosshead displacement curves were generated for each test. The SENT specimens were pulled in tension using flat serrated grips, CT and DCB specimens were tested using clevis grips with a pin and the SENB tests were carried out using a four point bend fixture similar to the ASTM E1820 standard.

Equations used to calculate K_{IC} based on the peak load measured for the four test geometries have been provided in the technical literature and are listed in Table 2. In all cases K_{IC} is a function of the crack size (a), the sample geometry (w=width,

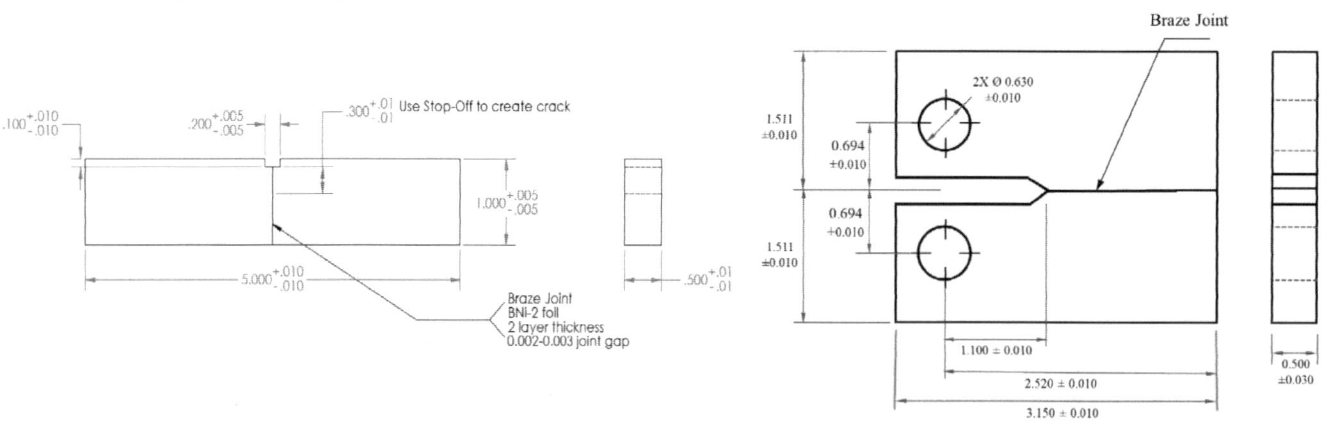

Figure 1: SENB (left) and CT (right) geometries used for fracture mechanics tests.

h=height, B=thickness) and applied load. For CT and SENB specimens Inconel plate thickness of 0.5" was used for both with an average crack size (a) to width (W) ratio of 0.45 for the CT tests and 0.10 (Round 1) and 0.40 (Round 2) for the SENB tests, respectively.

Results and Discussion

Braze Joint Test Results

The test results for the SENT braze test specimens are shown in Figure 2. All failures occurred in the BNi-2 braze joints with an average calculated K_{IC} of 43.38 ksi√in. Force vs. displacement curves indicated a brittle fracture occurred in these samples before any crack propagation was observed and failure occurring immediately at the peak load. There was a significant reduction in the calculated K_{IC} between Round 1 and Round 2. Cross-sections of both braze joints revealed that Round 2 joint gaps were larger than expected approximately 0.003-0.004 and contained a higher volume fraction of brittle, centerline intermetallics compared to Round 1 specimens. Centerline intermetallics such as CrB and Ni_3B borides are often found in near-eutectic braze alloys such as BNi-2 and have been shown to be detrimental to joint mechanical properties which may explain the lower K_{IC} measured in Round 2. These results indicated microstructure and joint gap can influence K_{IC} in braze joints.

DCB and CT test results are shown in Figure 3 and Figure 4, respectively. The average calculated K_{IC} was found to be 14.43 ksi√in using the DCB test method and 16.43 ksi√in for the CT tests. Force vs. displacement curves showed most samples failed immediately at the peak loads similar to the SENT ones indicating brittle failure mode. Three CT braze joints tested in Round 2 showed crack propagation prior to failure. Cracking was also heard in test samples that did not

V-Notch Single Edge Notch Tension:

Test Round	Average K_Q (ksi√in)
1	50.53
2	32.64
Total	43.38

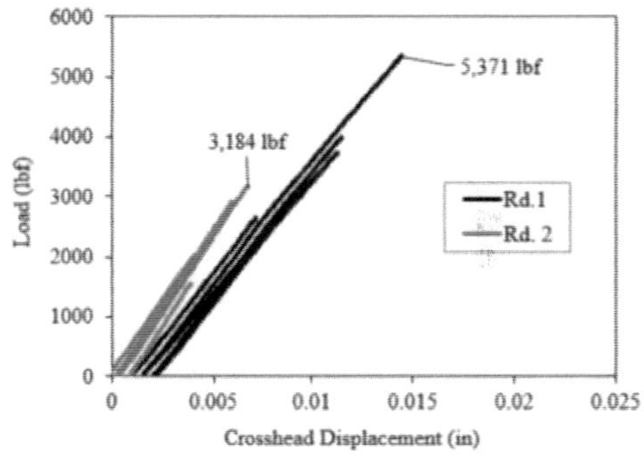

Figure 2: Experimental test results for V-Notch (SENT) Inconel 718/BNi-2 braze joints.

Double Cantilever Beam:

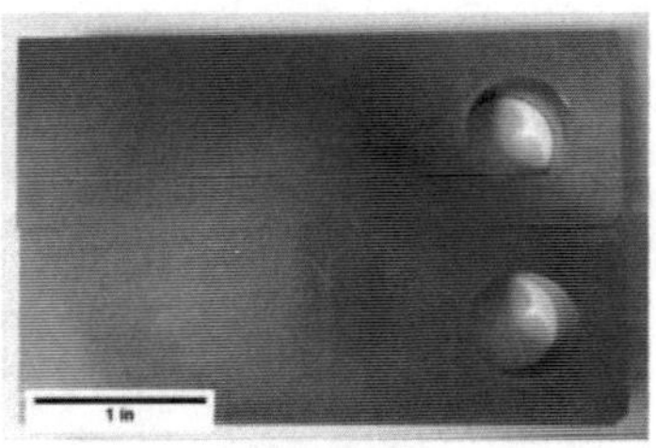

Test Round	Average K_{IC} (ksi√in)
1	14.88
2	13.86
Total	14.43

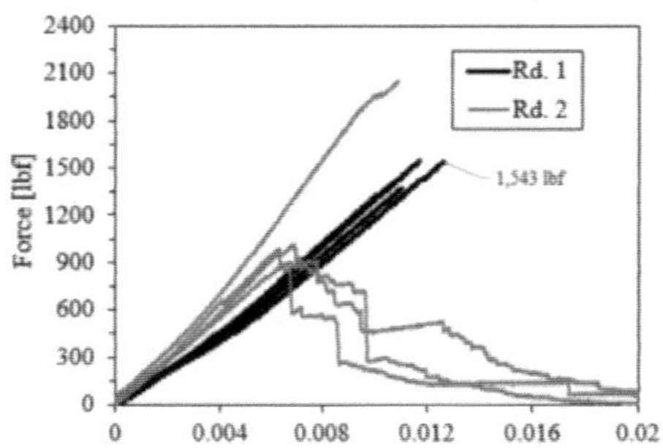

Fig

Compact Tension:

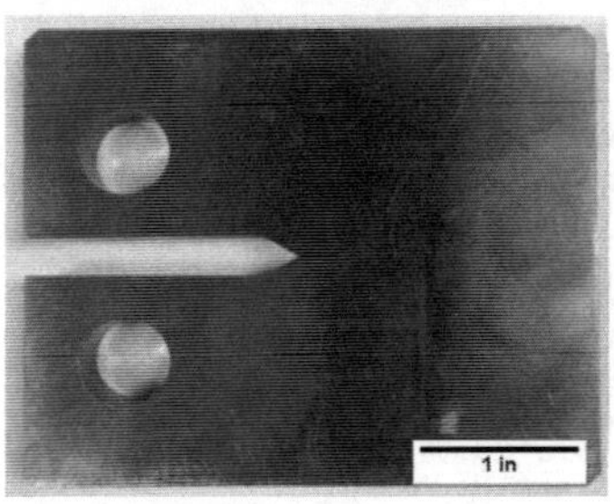

Test Round	Average K_Q (ksi√in)
1	12.78
2	20.99
Total	16.43

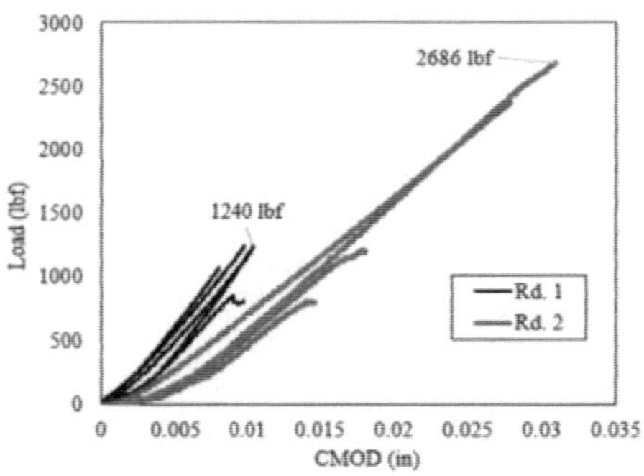

Figure 4: Experimental test results for Inconel 718/BNi-2 braze joints using CT test method.

Single Edge Notched Bend:

Test	K_{IC} (ksi√in)
1	23.50
2	22.70
Total	23.10

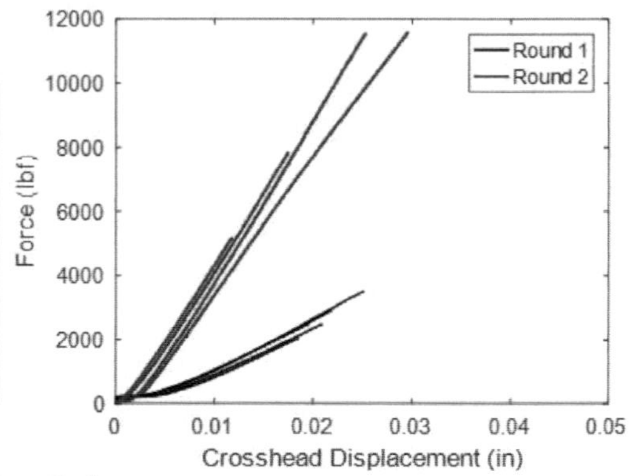

Figure 5: Experimental test results for Inconel 718/BNi-2 braze joints using SENB test method.

Average K_{IC} (ksi√in)
165.03

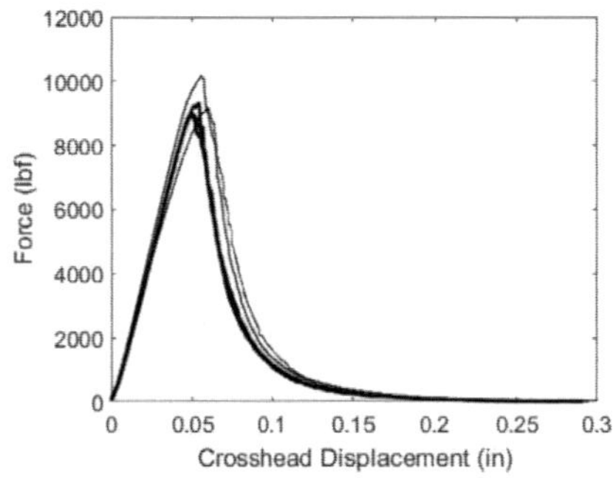

Average K_{IC} (ksi√in)
167.11

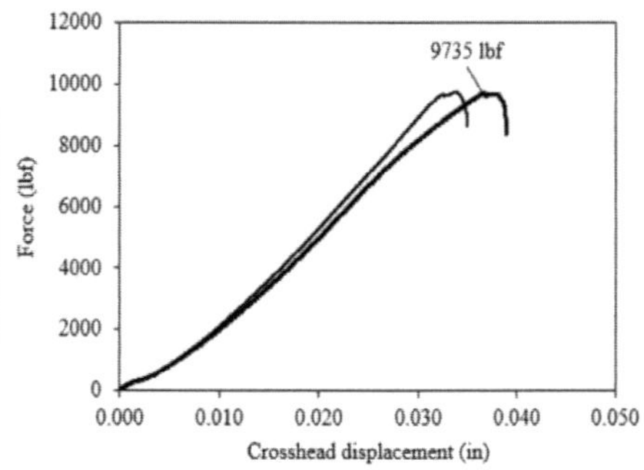

Figure 6: Experimental test results for SENB (left) and SENT (right) Inconel 718 base metal.

show propagation prior to reaching the failure load starting at approximately 90% of the failure load. Weiss et al. reported similar results measuring acoustic emission during fracture tests of Inconel 718/BNi-5 braze joints and attributed these to cracking along the centerline intermetallic phases [7]. The average K_{IC} of the braze joints from DCB testing was found to be significantly lower than the 43-46 ksi√in reported by Leinenbach et al. using the same test method. However, this might be expected as the Au-18Ni has been shown to form a more ductile, single phase braze joint without continuous brittle intermetallic phases compared to Ni-Cr-B braze alloys [4]. In general, the average DCB and CT test results were more consistent between Round 1 and Round 2 compared to the SENT test results. There was still some scatter for the DCB and CT test results that were attributed to changes in joint gaps and quality of braze joint.

The SENB test results for the BNi-2 braze joints are shown in Figure 5 along with force vs. displacement curves from Round 1 and Round 2 tests. The average K_{IC} was 22.70 ksi√in for a/w=0.1 and 23.50 for a/w=0.4, respectively. These results indicate the calculated fracture toughness of the BNi-2 braze joints is independent of crack size. All samples failed in a brittle fracture at the peak load through the braze joint. Similar to the DCB and CT test specimens cracking was audible prior to failure. These results are within the range of K_{IC} measured by Weiss et al. using a similar geometry to test the fracture toughness of Inconel 718 with a similar Ni-based braze alloy, BNi-5. While BNi-5 contains silicon rather than boron as the primary melting point depressant element this suggests good repeatability between studies and manufacturing/brazing techniques is possible.

Base Metal Test Results

Figure 6 shows test results for Inconel 718 base metal using SENB (left) and SENT (right) test specimens. The average experimental K_{IC} was found to be 165.03 ksi√in for the SENB specimens and 167.11 ksi√in for the SENT specimens. For SENB testing the crack size to width ratio (a/w) was 0.4 the same used for Round 2 braze joint tests. Base metal SENT dimensions were the same as the braze joint test specimens [7]. As expected, base metal tests were much more consistent in terms of peak load and calculated K_{IC}. Force vs. displacement curves also indicate a much more ductile failure especially for the SENB test specimens where the loads slowly decreased once the peak load was reached. This was an indication of crack propagation through the Inconel 718 rather than an immediate brittle fracture.

Another important observation was that there was very little difference in the calculated K_{IC} value between the SENB and SENT test methods for the base metal specimens. Whereas, braze joint tests showed the average K_{IC} from SENT testing was significantly higher than the K_{IC} calculated by the other three methods tested. This is likely due to the tri-axial stress that develops in braze joints loaded in tension due to the restraint imposed by the surrounding base material. The SENB, CT and DCB specimens on the other hand apply more of a bending crack opening force on the braze joint that could reduce the amount of tri-axial stress and allow for more local deformation of the joint material. Similar results were reported by Ghovanlou et al. which both SENT and SENB tests were performed for the same copper braze joints. Additional analysis using finite element models or equivalent is recommended to better understand this phenomenon.

Conclusions

This paper provides a comparison of four different fracture mechanics test methods for measuring the fracture toughness (K_{IC}) of braze joints. Experimental results for SENT, DCB, CT and SENB test methods were provided for Inconel 718/BNi-2 braze joints. Inco 718 base metal tests were also carried out using the SENT and SENB specimens for comparison. The results showed some scatter in test data based on the quality of the initial braze joints (joint gap, presence of centerline intermetallics, defects/voids, etc.). Average fracture toughness values obtained through two rounds of testing indicated the CT and DCB test specimens gave the most conservative K_{IC} values of 16.43 and 14.43 ksi√in, respectively. Pure tensile loading conditions in the SENT tests resulted in significantly higher K_{IC} values compared to bending loading conditions applied in the SENB, CT and DCB tests. The average K_{IC} for the Inconel 718 braze joint material was approximately 1/10th that of the Inconel 718 base metal fracture toughness of 166 ksi√in. The base metal K_{IC} was unaffected by the loading condition applied in the test with SENT and SENB average results coming within 2 ksi√in of each other.

Acknowledgments

This research has been supported by the Rolls-Royce Corporation and performed within the Manufacturing and Materials Joining Innovation Center (MA2JIC), a U.S. National Science Foundation, Industry/University Cooperative Research Center.

References

[1] American Welding Society (AWS), Brazing Handbook, Miami, FL: American Welding Society, 2007.
[2] British Standard, "BS 7910: Guide to methods for assessing the acceptability of flaws in metallic structures, ISBN 0 580 45965 9," Technical Committee WEE/37, London, 2005.
[3] ASTM E399-12e3, Standard Test Method for Linear-Elastic Plane-Strain Fracture Toughness K_{IC} of Metallic Materials, ASTM International, West Conshohocken, PA, USA, (2012).
[4] Leinenbach, C., Schindler, H.J., Başer, T.A., Rüttimann, Wegner, K., Quasistatic Fracture Behavior and Defect Assessment of Brazed Soft Martensitic Stainless Steel Joints, Engineering Failure Analysis 17 (2010), pp. 672/682.
[5] Gan, Y.X., Aglan, H.A., Steward, R.V., Chin, B.A., Grossbeck, M.L. Microstructure-fracture toughness relationship of vanadium alloy/stainless steel brazed joints, Journal of Nuclear Materials 299 (2001), pp. 157/164.
[6] M. K. Ghovanlou, H. Jahed and A. Khajepour, "Mechanical reliability characterization of low carbon steel brazed joints with copper filler metal," vol. 528, no. 6146-6156, 2011.
[7] B. Z. Weiss, H. D. Steffens, A. H. Engelhart and B. Wielage, "Static and Dynamic Crack Toughness of Brazed Joints of Inconel 718 Nickel-Base Alloy," vol. October 1979, no. 288-295, 1979.
[8] Moorhead, A.J. and Becher, P.F. Adaptation of the DCB test for determining fracture toughness in ceramic materials. Oak Ridge, Tennessee, USA, Journal of Materials Science 22 (1987), pp. 3297/3303.
[9] Philips, N.R., He, M.Y., Evans, A.G., A wedge fracture toughness test for intermediate toughness materials: Application to brazed joints, Acta Materialia 56 (2006), pp. 4593/4600
[10] Gan, Y.X., Aglan, H.A., Steward, R.V., Chin, B.A., Grossbeck, M.L. Microstructure-fracture toughness relationship of vanadium alloy/stainless steel brazed joints, Journal of Nuclear Materials 299 (2001), pp. 157/164.
[11] B. Riggs, B. Alexandrov, A. Benatar and R. Xu, "Evaluation of DCB and V-Notch Fracture Mechanics Test Methods for High Temperature Brazed Joints", LÖT 2016 11th International Conference on Brazing, Aachen, Germany, June, 2016.

Corrosion Resistance of High-Temperature Intermetallic Nickel Alloy with Palladium Brazing Filler Metal

S. Maksymova, V. Miasoid, L. Nyrkova
E. O. Paton Electric Welding Institute of the NAS of Ukraine
E-mail: maksymova@paton.kiev.ua

Abstract

Corrosion resistance is an important factor for operation of brazed joints of high-temperature nickel alloys in the air with high humidity. Corrosion potentials of base metal, brazing filler metal and brazed joints were investigated for express-evaluation of electrochemical inhomogeneity of the joints.
This paper presents the results of comparative investigations of corrosion resistance of deposited (on Ni_3Al – based alloy substrate) brazing filler materials of Pd-Ni-(Me) system containing different amount of germanium as well as the relative value of brazing filler metals heat resistance after isothermal holding at increased temperature.
The sample after experiments on brazing filler metal spread in vacuum was used for the investigations. Data on corrosion resistance of intermetallic nickel alloy and deposited brazing filler metal, received by means of measurement of corrosion potentials in 3% solution of NaCl, showed that decrease of germanium concentration from 5 to 1.5% in spread of brazing filler metals of Pd-Ni-Cr-(Me)-Ge system over nickel alloy substrate (based on Ni_3Al) results in displacement of corrosion potential to more positive area as well as improvement of high-temperature resistance that rises corrosion resistance.

Introduction

Brazing and brazing filler metals of different alloying systems [1, 2] are used to produce permanent joints (repair) of high-temperature nickel alloys, including intermetallic ones on Ni_3Al base). Nickel-based brazing filler metals alloyed with other chemical elements are referred to the most widespread. They provide set temperature interval of melting, wetting pf base metal and necessary mechanical properties to brazed joints [2, 3]. Boron and silicon are usually used as depressants for nickel brazing filler metals [1-4]. On the one hand, they reduce melting temperature and provide good wetting of the base metal. On the other hand, they promote the formation of fusible brittle eutectic compounds in the brazed welds and on boundaries of base metal grains that have a negative influence on mechanical properties at long-term operation under conditions of increased temperature. Reduction of a number of brittle phases allows carrying additional continuous heat treatment [2], but increases energy consumption and is not always acceptable in development of different structures.
Formation of brittle fusible phases can be eliminated using brazing filler metals with solid solution structure, developed based on binary metallic systems characterizing with complete solubility. Palladium-nickel system is referred to one of such systems. It is characterized with the presence of solid solutions minimum temperature of melting at 40% nickel concentration [5]. Further decrease of melting temperature is possible by means of application of additional alloying using other elements, including silicon and germanium [6, 7]. Both elements have limited solubility in nickel, but germanium has a larger area of solubility in nickel and palladium in comparison with silicon [6]. Application of brazing filler metals of this system provides high mechanical properties to brazed joints of high-temperature nickel alloys. In this connection, there is a necessity in the evaluation of corrosion resistance of brazed joints operated under atmospheric conditions for a long time.
This work presents the results of comparative investigations of corrosion resistance of base metal and brazing filler metals of Pd-Ni-(Me) system, deposited on the substrate of high-temperature nickel alloy based on Ni_3Al.

Experiment

Nickel intermetallic alloy Ni_3Al (Ni-8Al-18Mo) was used for experiments. The samples from experiments on brazing filler metals spread (Table 1) on the base metal substrates (nickel intermetallic alloy) were used as deposited brazing filler metal. Evaluation of corrosion resistance of intermetallic nickel alloy based on Ni_3Al and deposited brazing filler metal was carried out by measurement of their corrosion potentials in 3% solution of NaCl.

Table 1. System of brazing filler metals and heating temperature

No.	Alloying system	Temperature, °C
1	Pd-Ni-Cr-(Me)-1,5Ge	1235
2	Pd-Ni-Cr-(Me)-3Ge	1235
3	Pd-Ni-Cr-(Me)-5Ge	1225

This method is used for express-evaluation of electrochemical inhomogeneity of different zones of welded joints for preliminary evaluation of their primary failure at contact with the corrosion-aggressive medium.
The sample before measurement was degreased using magnesium oxide, rinsed in running and then in distilled water, dried by filter paper. Due to the small size of the samples, a method of drop potential measurement was used. The potentials measured in 3% NaCl relatively to silver-chloride reference electrode at room temperature using high-resistance voltmeter during 1 hour (until stationary value).

Potential of a layer of brazing filler metal being spread over the substrate from high temperature alloy based on nickel aluminide was compared with the substrate potential.
The high-temperature resistance of brazing filler metals was studied by gravimetric method after isothermal oxidation on air at increased temperature. The samples were weighted accurately to firth character after coma on VLR-200 scales then they were heated in a muffle furnace in open air at 900 °C temperature during 100 hours.
Heating was carried out to set temperature with further holding for 6-8 hours and till complete cooling, after what the cycle was repeated.
Mass measurement of investigated samples was performed using the formula:

$$\Delta m = m_1 - m_0,$$

where Δm - measurement of mass of samples after testing,
m_1 –mass of sample after heating;
m_0 –mass of sample before heating;
Specific increment (g/m^2) was calculated on the formula: $\frac{\Delta m}{S}$,
where S is the area sample surface (m^2),
Δm is the change of sample mass after testing.
Microstructure and elemental composition of separate phases were investigated using scanning electron microscope Tescan Mira 3 LMU, equipped with energy-dispersive spectrometer Oxford Instruments X-max 80 mm^2 with INCA control packet. In-situ measurements made up to 1 μm. Distribution of elements and exposure of microsection was carried out in back-scattered electrons (BSE) allowing investigation of micro sections without etching as well as etching.

Results and discussion

Visual examination of received samples shows good spread and wetting of the base metal. The results of performed investigations showed that the potential of base metal makes -0.357 V, the values of potentials of deposited brazing filler metals with different content of germanium have more positive values in comparison with the base metal. It should be noted that reduction of germanium content from 5 to 1.5 % promotes a shift of corrosion potential in the field of more positive values (Fig. 1).

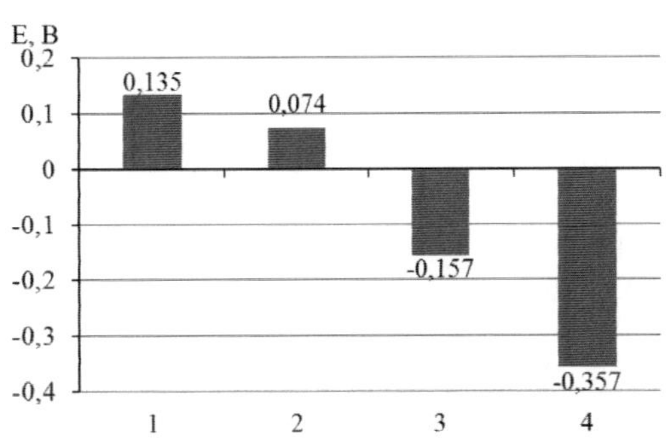

Fig. 1. Histograms of corrosion potentials of deposited brazing filler metals: Pd-Ni- Cr-Me-1,5Ge (1); Pd-Ni–Cr-Me-3Ge (2); Pd-Ni-Cr-Me-5Ge (3) on substrate of Ni_3Al and base metal (4).

This indicates the formation of a protective film on the brazing filler metal surface, which serves as a barrier between the base material and environment. One of the constituent elements of these brazing filler metals is chromium, which is characterized by high stability and have protective coatings [8].
It is noted that the largest difference of corrosion potentials is reached at 1.5% germanium content that indicates its higher corrosion resistance in comparison with brazing filler metals containing 3 and 5% of germanium.
Al_2O_3 oxide is formed in the oxidation of intermetallic nickel alloy based on Ni_3Al that is typical for high-temperature nickel alloys [8]. It is also characterized by stability, but have lower stability in comparison with chromium oxide. NiO oxide formation takes place at lager rate in comparison with chromium and aluminum oxides, but it has lower protective properties that explain lower corrosion resistance of the base metal in comparison with brazing filler metals.
Following the analysis of the received results of microstructure examinations, there is a correlation between corrosion resistance of the brazing filler metals and their structural state. In the initial state brazing filler metal No. 1 consists of the nickel-based solid solution and separate single inclusions of phase based on palladium Pd_xGe_y (Fig. 2, a, b). The letter forms eutectic with a solid solution.
The increase of germanium concentration in the initial brazing filler metal to 5% results in an increase of phase based on palladium, enriched with germanium, the rise of the amount of eutectic constituent and decrease of the amount of solid solution (Fig. 2, c).
Spreading of palladium brazing filler metal over the nickel alloy substrate based on Ni_3Al preserves the same tendency of structure formation, namely solid solution based on nickel and palladium phase, but in this case, it is enriched with aluminum, not germanium. Interaction of liquid brazing filler metal and solid base metal takes place in process of spread and wetting of the base metal.
Diffusion processes, promoted by the presence of concentration gradient (on aluminum) at the interphase boundary and high diffusion activity of aluminum, results in the formation of palladium-based phase enriched with aluminum. In accordance with binary state diagrams [6], Pd_xAl_y phase containing an insignificant amount of other elements is formed in the brazing metal.
A similar formation of brazed seams is observed in brazing of titanium aluminides with titanium brazing filler metals containing eutectic constituent.
Aluminum diffuses from base metal in the brazed welds, which results in formation in the brazed weld of phase typical for base metal [9].
Investigation of high-temperature brazing filler metals was also directed on the study of their capability to withstand chemical failure at an increased temperature under air environment effect. Increase (decrease) of the mass of tested sample due to the formation of oxide films on the surface was used as a quantitative characteristic of high-temperature strength.

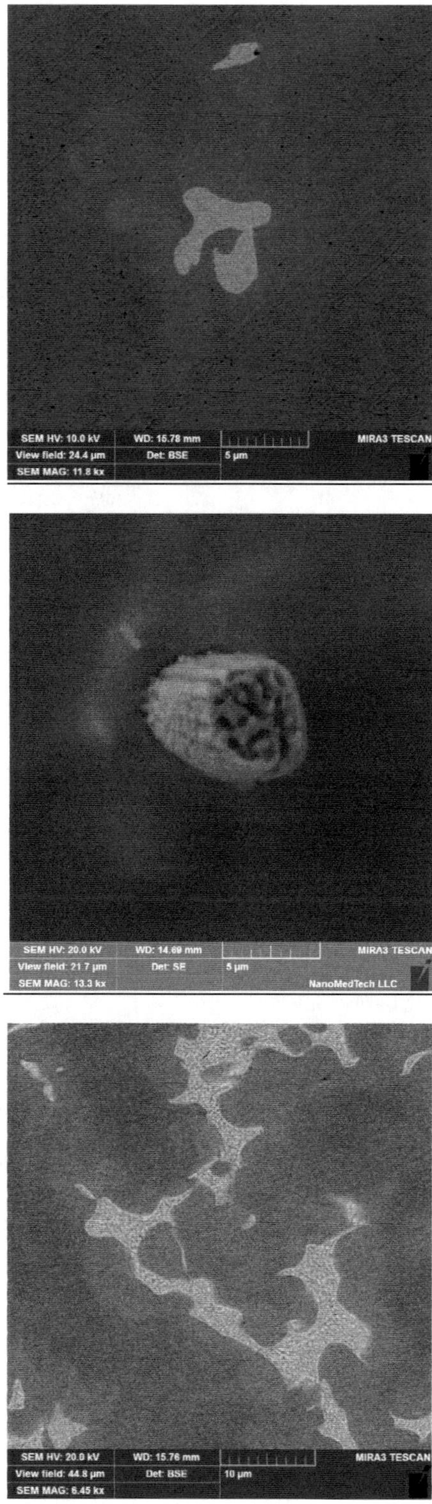

a

b

c

Fig. 2. Microstructure of brazing filler metals containing: 1.5% (a, b) and 5% (c) of germanium

It was assumed that the lower variation of sample mass in isothermal holding at increased temperature the higher its corrosion resistance.
Figure 3 represents the dependence of mass increase of brazing filler metals with different germanium content.

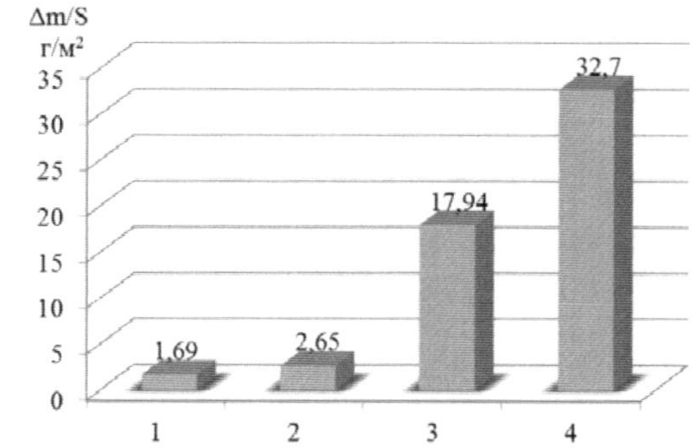

Fig. 3. Mass increase of brazing filler metals with different content of germanium (wt. %): 1 – 1.5% ; 2 – 3%; 3- 5% and 4 – base metal

It can be seen from received data analysis that reduction of germanium concentration in the brazing filler metal to 1.5% results in decrease of sample mass increase in isothermal holding (at 900 °C temperature), that indicates improvement of high-temperature strength in comparison with brazing filler metal containing 5% of germanium (Fig. 3).
Metallographic and local micro X-ray spectrum examinations of the brazing filler metals (after isothermal holding) determined on the brazing filler metal surface (No.3) presence of two layers (Fig. 4, b) different in chemical composition and color. A layer based on chromium (98%) is formed on the outer surface and under it that based on germanium with the following composition 53,65 Ge-42,05 Pd-28, 4,30Ni (Fig. 4).
Obtained results of investigations show that reduction of germanium concentration in the brazing filler metals results in an increase of the amount of solid solution and decrease of the amount of Pd_xGe_y phase, which during spreading of the brazing filler metal over nickel aluminide is transformed in Pd_xAl_y. Reduction of the amount of this phase in the spread brazing filler metal promotes displacement of corrosion potential of brazing filler metal in the more positive area.

Conclusions

Received investigation results show that spread of brazing filler metals of Pd-Ni-Cr-Me-Ge system over nickel alloy substrate (based on Ni_3Al) with a decrease of germanium concentration from 5 to 1.5% promotes displacement of corrosion potential of the deposited metal in ore positive area that rises corrosion resistance.

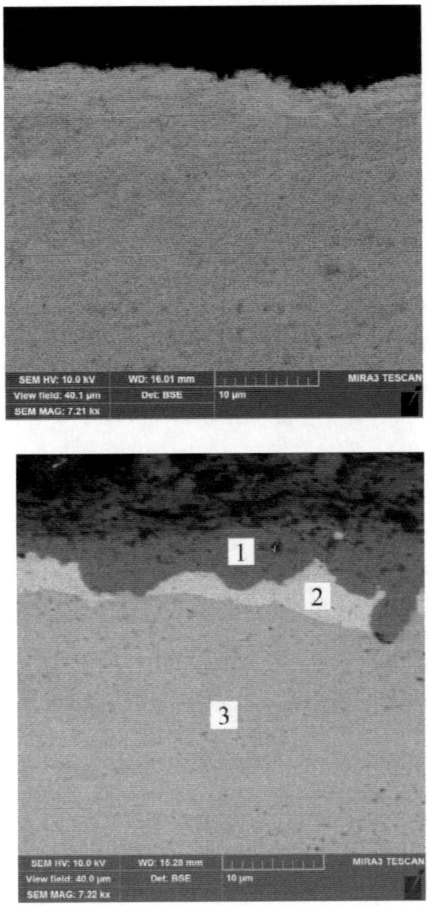

Fig. 4. Surface of brazing filler metal No.1 (a) and No. 3 (b) after isothermal holding at 900 °C temperature

Microstructure and micro X-ray spectrum examinations showed that reduction of germanium concentration in the initial brazing filler metal results in an increase of the amount of solid solution and decrease of the amount of palladium phase enriched with germanium. Pd_xAl_y phase is formed instead of Pd_xGe_y in the spread of brazing filler metals over the substrate of intermetallic nickel alloy (based on Ni_3Al).

Brazing filler metals containing 1.5% of germanium are characterized with the least mass loss, and respectively, highest high-temperature resistance. This correlates with the results of potentiometry.

References

[1] Khorunov, V.F., Maksimova S.V., "Production and application of rapidly quenched brazing alloys," Welding International, 20 (5) (2006), pp. 405.

[2] Rabinkin A. Advances in brazing. Science, technology and applications. (Cambridge, 2013), pp. 121-159.

[3] Petrunin I.E., Handbook on brazing. (Moscow, 2003). P. 480.

[4] Schwartz, M.M. Brazing. (Ohio, 1987), P. 455.

[5] Massalski T. B., Binary Alloy Phase Diagrams (Ohio, 1990), CD.

[6] Lyakishev N. P. State diagrams of double metallic systems, Vol. 3, (Moscow, 1999). P. 448.

[7] Sidorov V. V., "On mechanism of effect of microadditives on oxidation resistance of high-temperature nickel alloys." *Metallovedenie i termicheskaya obrabotka metallov.* No1, (1995), pp. 33-36.

[8] Khorunov V.F., Maksymova S.V. Advances in brazing. Science, technology and applications. (Cambridge, 2013), pp. 85-120.

[9] Maksymova, S.V., Khorunov, V.F., Myasoed, V.V. "Influence of germanium as a depressant on the structure and intervals of melting of palladium alloys.". *Metallofizika i Noveishie Tekhnologii, V.* 37(6) (2015), pp. 803-816.

DIRECT ALUMINUM SOLDERING

By William F. Avery and Dr. Yehuda Baskin
Superior Flux & Mfg. Co., Cleveland, OH, USA

Abstract

The techniques for soft soldering to aluminum are described, and include liquid flux and solder, paste flux and solder, solderpaste, and cored-wire solder. Many heating methods can be used and include hot plate, convection oven, soldering iron, torch, induction, and resistance. Cleaning aluminum has advanced from using dangerous and corrosive chemistries to using nonhazardous chemicals. Aluminum soldered connections must withstand various environmental conditions including exposure to salt. The strength of the soldered connection depends on the aluminum alloy being soldered and on the solder alloy used.

Introduction

Aluminum oxidizes with extreme rapidity; it is almost impossible to remove its ever-present oxide coating more rapidly than it reforms. This oxide, aluminum oxide Al_2O_3, is extremely tenacious; it melts at 2050°C / 3722°F. Aluminum oxide is neither melted nor reduced by temperatures that melt the metal. However, oxides normal to mill-finished aluminum alloys are removed during soldering by specialty aluminum cleaners and soldering fluxes[1].

There are many elements used to combine with aluminum to make different alloys. These elements are used to enhance given properties of the aluminum. Silicon is added to lower the melting point and, more importantly, improve the extrusion properties of aluminum alloys. Magnesium is added to improve the strength of the aluminum alloy; the same is true for zinc and titanium. The problem is that the addition of most of these elements decreases the interaction with pure aluminum and in the case of silicon, magnesium, and titanium, can significantly reduce the solderability of those particular alloys, because of the inherent poor solderability of these elements[2].

The major consideration when selecting a solder is "will it make a metallurgical bond with the aluminum"? If this does not occur, the bond to the aluminum will fail. The common solders that have been found to properly bond to aluminum contain zinc, silver, and copper. Cadmium will bond to aluminum, but its highly toxic nature precludes its use. Neither tin nor lead will alloy to aluminum as a solder, so a simple tin-lead solder should be avoided when soldering directly to aluminum. Typical solders that work well for aluminum soldering include tin-silver, tin-copper-silver, tin-copper, and tin-bismuth-silver. A low-cost alloy such as a simple plumbing alloy like tin-antimony will not work, because neither tin nor antimony will bond metallurgically to aluminum. Any solder will work if one first pre-tins the surface with a proper solder, then solder over the pre-tinned surface. Properly soldering to aluminum means making an intermetallic bond between the solder alloy used and the aluminum surface. In the example below, 91/9 tin-zinc solder is used to solder to aluminum. The solder joint was cross-sectioned, polished, then placed in a scanning electron microscope to show the intermetallic bond between the solder alloy and the aluminum surface. The intermetallic connection (**Figure 1**) is actually a new compound that is neither solder nor aluminum[3].

Figure 1[4]

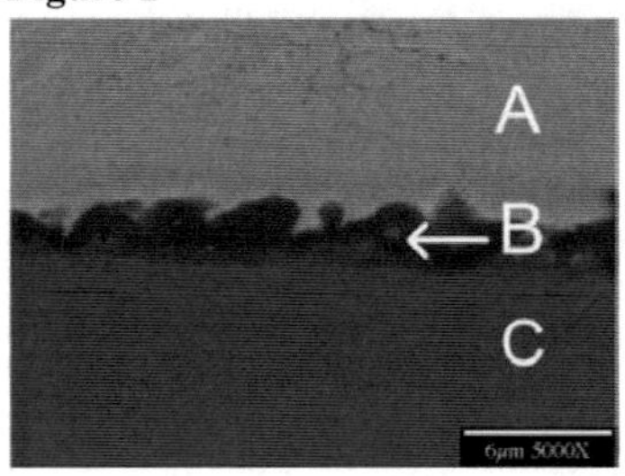

Un-plated Aluminum Surface Soldered with 91/9 Tin-Zinc

Description:
A – 91/9 Tin-Zinc Solder
B – Zinc/Aluminum Intermetallic Zone
C – Aluminum Base Metal

Results and Discussion

METHODS FOR SOLDERING TO ALUMINUM

Liquid flux and solder - The liquid flux for aluminum soldering is a mixture of organic amines and inorganic fluoroborate salts and typically has the consistency of honey, whose color may range from amber to deep brown. Other chemicals such as alcohols are sometimes added to modify the viscosity of the flux for those operations that require a less viscous flux (**Figures 2 and 3**). Organic flux formulations are designed for temperatures of 177°C to 316°C / 350°F to 600°F. Above this temperature, the flux begins to carbonize at an ever increasing rate. At 316°C / 600°F, deterioration and charring are so rapid that organic flux can only be used with ultra-rapid soldering techniques, such as induction soldering[5].

Figure 2

Liquid flux applied to surface between two aluminum parts with a solder foil present.

Figure 3

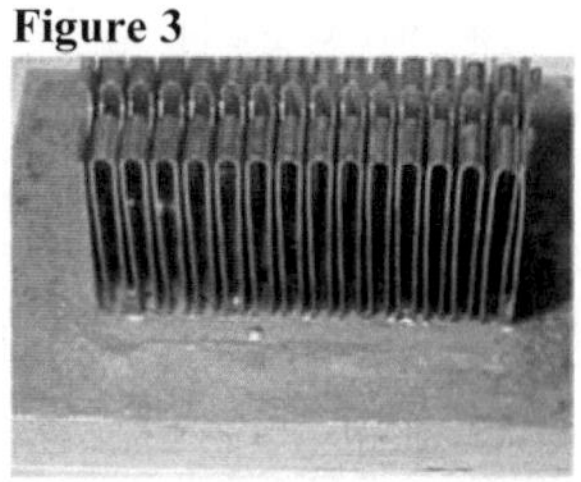

After hot plate heating, solder has flowed and has connected the aluminum parts.

Paste flux and solder - When chemical binders are added to the liquid flux, it transforms the state of the flux, making it a paste (**Figure 4**). The paste is readily dispensable by a needle, which allows for more accurate placement of the flux[6].

Figure 4

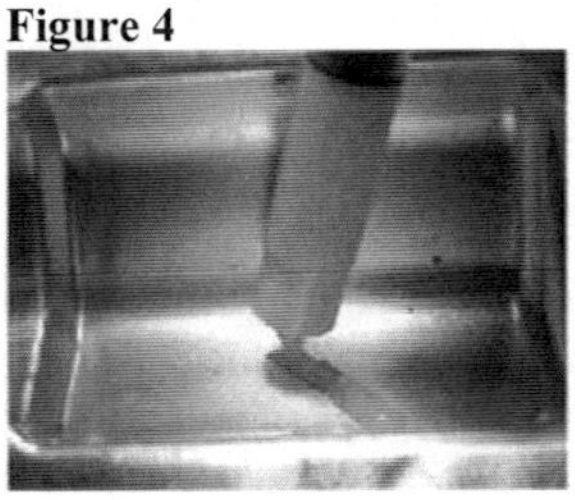

Paste flux dispensed to metal surface

Flux-cored solder wire – When the organic flux formulation is further modified to have higher solids content it will be more viscous, instead of being liquid at room temperature. When this material is heated it will liquefy, which allows it to be injected into a solder core. Once the wire is cooled, the flux core becomes hard, which allows the wire to be used as a flux-cored solder wire. The chemistry of this modified flux solid is activated for soldering aluminum at the 280°C to 380°C / 536°F to 716°F temperature range. The flux-cored aluminum wire solder is different than typical flux-cored wire solder for copper, in that the aluminum itself must be heated to the activation temperature before melting the core wire solder on the hot surface (**Figures 5 and 6**). The chemistry of the flux will not allow it to be melted onto a soldering iron, as is often done with flux-core soldering of copper surfaces[7].

Figure 5

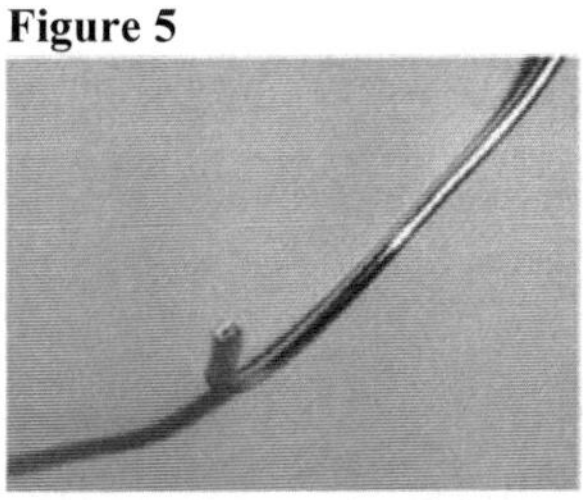

Wire solder with aluminum core flux in the center of wire.

Figure 6

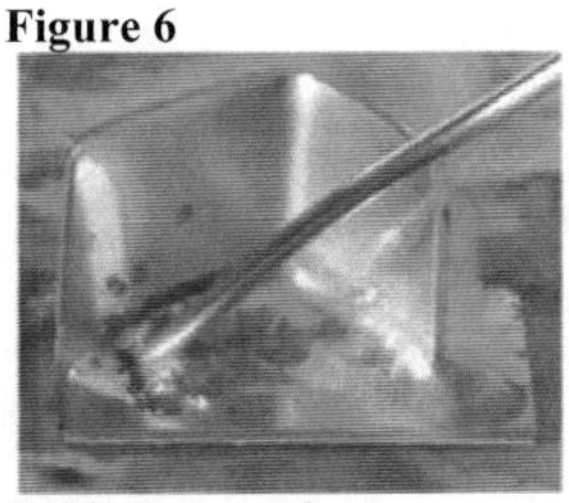

Soldering aluminum to aluminum with wire solder containing core flux.

Solderpaste - One advantage of the organic type of flux is that it can be converted to a solderpaste, by incorporating solder powder and binders (Figures 7 and 8). The same temperature limitations, 180°C to 316°C / 350°F to 600°F, apply to most versions of this solderpaste. A high-temperature solderpaste has been developed, which has been formulated to have activating temperatures in the 280°C to 380°C / 536°F to 716°F temperature range, and can withstand a temperature of 420°C / 788°F[8].

Figure 7

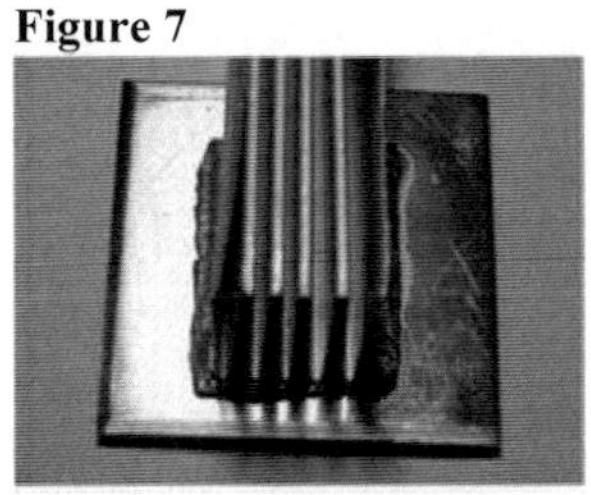

Aluminum fin placed on solderpaste dispensed on copper

Figure 8

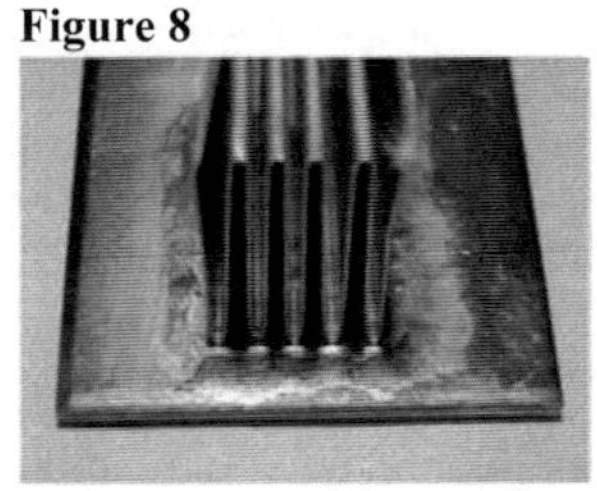

After solder reflow fins are soldered to copper

HEATING METHODS FOR SOLDERING ALUMINUM

Hotplate – A hot plate may be used to bring aluminum parts up to soldering temperature. Such a device can vary from being: thermostatically controlled electric devices, non-controlled electric units, and even metal plates set over gas jets. Parts to be soldered are cleaned, fluxed and positioned. The solder may be preplaced or face-fed. In the latter event, the parts themselves must have sufficient weight to remain immobile while the solder contacts them, or they must be kept in position by suitable fixtures. Heat is applied, and when the solder melts and flows through the joints, the heat is turned off. If the heat is not automatically controlled, but is manually shut off, it is generally done a little before the joint is complete to prevent overheating[9].

Convection oven – Furnace soldering lends itself to both small-lot and high volume production. As many assemblies and joints as can fit into the furnace may be soldered at one time. If the batch furnace's pass through rate is too low, a semi-automatic continuous furnace is substituted. Parts too large or massive to be evenly heated by other means may be soldered with minimum distortion in a furnace. A furnace is also useful for soldering complex and intricate parts with joints that cannot be easily heated after assembly by other techniques. Solder reflow via conveyor belt oven is a proven and widely used procedure for consistent reliable soldering. Furnace soldering is excellent for long solder joints, producing highly controlled solder fillets, neater joints and more efficient use of solder. Any furnace capable of encompassing the assembly to be soldered, bringing it up to the melting temperature of the solder used and holding it there for the necessary time may be used. It may be heated by any means including infra-red lamps, convection heated air, and static heating coils. Soldering furnaces should be power-ventilated. Flux vehicle vapors, such as are given off by alcohol and other organics, are highly combustible and explosive when mixed with air. These vapors can be sometimes eliminated by drying the flux assembly in air prior to soldering. Flux fumes, produced during soldering, cannot be as easily avoided. They must be vented outside the plant with attention to local antipollution laws, and may require the installation of scrubbers[10].

Induction – Joints soldered by this method are brought to soldering temperature by high-frequency electrical currents induced in the faying surfaces by an inductance (coil) positioned nearby. Heating is localized, fast and generally accomplished in a few seconds. There are no open flames, and power is consumed only during the soldering cycle. Once soldering parameters have been ascertained, soldering by induction is an extremely accurate and repeatable operation, which can be automated. Post induction soldering cleaning is easier than post torch soldering cleaning, because there are fewer combustion products to react with the flux and form insoluble residues. However, the equipment required is expensive compared to that necessary for other soldering methods[11].

Resistance – With resistance soldering an intense heat can be rapidly developed directly within the joint area and in a tightly controlled manner, since the heat is created by passing a current through the metal being heated. This allows a faster ramp-up to the required solder melt temperature and minimizes thermal travel away from the solder joint. This helps minimize the potential for thermal damage to materials or components in the surrounding area. Heat is only produced while each joint is being made, making resistance soldering more energy-efficient. Resistance soldering equipment, unlike conduction irons, can be used for difficult soldering applications where significantly higher temperatures may be required. This makes resistance soldering comparable to flame soldering in some situations. When the required temperature can be achieved by flame or resistance methods, the resistance heat is more localized because of direct contact, whereas the flame will spread the heat, thus warming a larger area.

Torch – This is the simplest way to provide soldering heat. It is low in cost, portable, and suited to production work as well as single assemblies and repairs. Its flame is hot enough to be used readily with all solders, and its output can be varied to accommodate small and large assemblies. Although any source of heat can be used for torch soldering, commercial torch soldering is generally accomplished with the same type of torch, controls, and gases that are used for fusion welding. Conversion to aluminum merely requires a change in torch nozzles. Since the chemicals used to make the relatively low-temperature fluxes, solder pastes, and cored-wire solder are partially organic, they will ignite if exposed to an open flame. For this reason, torch heat should only be used to heat the surface before applying the soldering media. After this any torch heat applied must be directed away from the joining area, either behind the metal or far away from the metal to prevent igniting the flux component[12].

Soldering Iron – A soldering iron may be used; it may be heated electrically or by a gas flame. However, the weight of the "copper" and its temperature is most important. It must be large enough and hot enough to bring the joint and much of the adjoining metal up to soldering temperature in a fairly short time. The use of an iron is therefore limited by the size of practical existing irons and the mass of the work pieces to be soldered. Generally, soldering irons are used only with soft and intermediate solders and only on relatively low mass materials. To secure maximum heat transfer and to prevent undesirable alloying, the tip of the iron should be tinned with the solder that is to be used[13].

CLEANING ALUMINUM

Aggressive cleaners – These can be either caustic (basic) or strong acid solutions. A hazardous and corrosive caustic "deoxidizer" may be made by mixing a 5% solution of sodium hydroxide and water, generally kept at 67°C / 150°F. The strong acid cleaning systems (of nitric acid or a combination of nitric acid and hydrofluoric acid) is always followed by multiple water rinses to stop the activity of the strong acid cleaner on the aluminum. The problem with such cleaning solutions is that they are dangerous to handle and the raw chemicals to make them are extremely hazardous. After cleaning, water is used to wash the chemicals involved away from the cleaned aluminum.

Hot water – This method is used with the chemistry of the organic "honey" flux. If the parts are immediately washed in hot water, the flux can generally be removed. However, the use of hot water will not effectively remove this type of flux residue if the residues are not immediately removed after soldering.

Non-hazardous cleaners – A new chemical cleaner has been developed, using citric acid chemistry, that is not like the strongly corrosive and hazardous cleaners previously used, and which works well on the organic "honey" type of soldering chemistry. This cleaner is nonhazardous and can be handled with much greater ease than the dangerous caustic or acidic cleaners[14]. This cleaner is also safer to the environment.

Figure 9

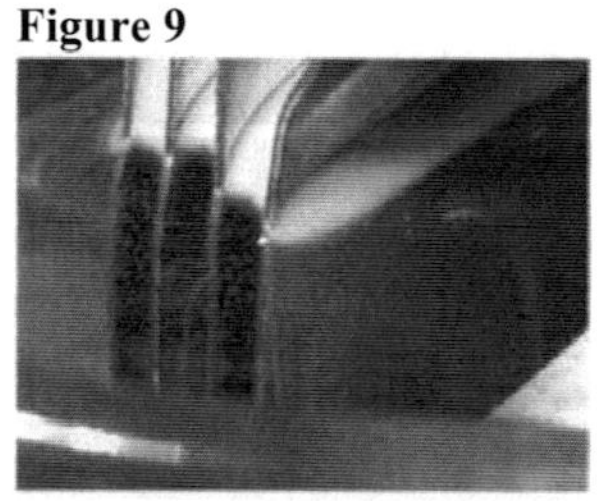

Nonhazardous cleaner removing oxides from aluminum.

The formation of small bubbles shows the reaction occurring.

Figure 10

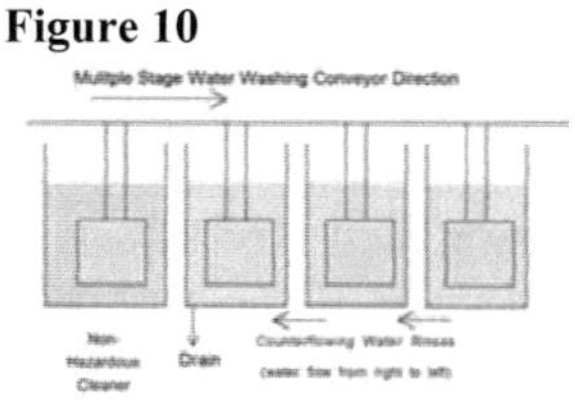

Cleaning schematic for multiple stage nonhazardous cleaner with counter-flowing water rinses to make the last tank of water as clean as possible.

ENVIRONMENTAL EXPOSURE

Exposure to moisture – A major consideration in making any aluminum soldered connection is: "how viable is the solder connection"? Experience has shown that just making a connection where the solder has appeared to flow on the aluminum is not sufficient to prove that the connection is truly sound. Sometimes after creating the connection, the solder can de-bond over a short period of time. When this occurs, the proper amount of attention has often not been given to the conditions needed to make a viable solder joint. If the wrong solder is used (no element to create the intermetallic bond to the aluminum) there will be no true soldered connection between the bulk solder and the aluminum. Such a connection cannot last, because true soldering has not occurred. If insufficient heat is used to create the joint, a cold joint will occur and aluminum is far less forgiving to having the proper conditions for soldering than copper. To prove if a soldered connection has worked, a week long exposure to a high humidity environment (80-95% humidity at 40°C / 104°F) will separate sound solder connections from false ones. Only a connection with a true intermetallic bond will survive the week-long humidity test.

Exposure to salt – Another major consideration for applications, particularly automotive and marine applications, is the ability for soldered aluminum connection to withstand exposure to a salt environment. This is not an issue for brazed connections (made with a low melting aluminum alloy containing, for example, 12% silicon) or a so-called "hard solder" containing a unique high melting solder alloy made of zinc and aluminum. However, when it comes to conventional soft solder alloys used to solder aluminum that may be tin-zinc, tin-silver, tin-copper, tin-silver copper, or tin-silver-bismuth, testing has shown that all of these solders, as well as everything else of standard solder types tested, failed to withstand long exposure to 2-10% salt solutions, after a one to three weeks' immersion at room temperature. Some of the proposed exotic protection schemes to prevent salt exposure, such as coating the part after soldering or pre-tinning the aluminum before soldering, were not practical and did not protect the aluminum soldered connection.

A new solder alloy, ALUSAC-35[15], based on tin-silver-copper, but having other proprietary elements added has proven that it can withstand the difficult room-temperature immersion test. Indeed, this new solder alloy also withstands the ASTM B117 salt fog test[16] (the automotive industry elevated temperature salt environment test). This test is in an enclosed chamber with a solution of 2-10% salt water that is heated to 40°C / 104°F and is sprayed constantly on the soldered parts in the chamber for a given time period (usually 24, 48, or 96 hours). The requirement is that the solder joint does not break apart under these conditions. The new solder alloy has a relatively high melting temperature, 341°C / 646°F).

STRENGTH OF SOLDERED JOINT

Aluminum to copper – The copper to copper solder bond is stronger than any of the combinations below. That is not surprising considering how easy copper is to solder and copper's easy ability to make an intermetallic bond to the solder. The testing used two flat sheets soldered to make a lap joint (**Figures 11 and 12**). However, when considering copper to aluminum, these also can be a relatively consistent bond, but it tends not to be as strong as copper to copper, because on strips of the same size, copper is stronger than aluminum. When the tensile test is done, many times the break in the connection is actually in the body of the aluminum, not in the solder bond.

Figure 11

Copper to Copper Lap Joint

Figure 12

Aluminum to Aluminum Lap Joint

Figure 13

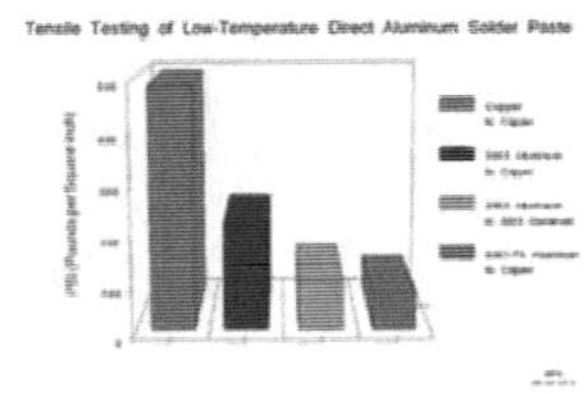

Figure 14

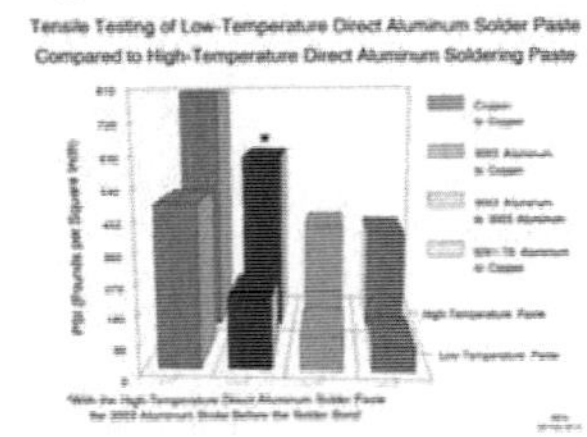

Aluminum to aluminum – This is always the weakest connection if all factors are equal. Aluminum is harder to solder than copper and two aluminum parts being soldered represents the hardest test. In the chart above (**Figure 13**) alloy 3003 aluminum to copper is much stronger than 3003 aluminum to 3003 aluminum.

Effect of aluminum alloy on final joint strength – The aluminum alloy chosen is an extremely important factor in how strong a given soldered connection will be. As can be seen in the charts above (**Figures 13 and 14**), 3003 aluminum to copper bond is much stronger than the 6061 aluminum to copper bond. As a matter of fact, the 6061 aluminum to copper bond is slightly weaker than the 3003 aluminum to 3003 aluminum bond, which is surprising.

Effect of soldering method on final joint strength – By soldering technique, there is the consideration of what chemistry is used, what solder alloy is used, and the soldering temperature employed. The charts above (**Figures 13 and 14**) show a comparison of low-temperature direct aluminum solder paste (using a low-temperature solder alloy), to a relatively high-temperature direct aluminum solder paste, using a different flux chemistry and solder alloy. These differences have a great effect on the strength of the solder bond achieved[17].

Conclusions

Direct aluminum soldering eliminates the need to plate the aluminum before soldering. When soft soldering aluminum, it is important to consider the challenges of the tenacious aluminum oxide layer, the differences in aluminum alloys, and choosing the right solder alloy to properly bond to aluminum. There are different techniques now available to soft solder aluminum, including liquid flux and solder, paste flux and solder, solder paste, and cored-wire solder. Similarly, there are many heating methods that will work on soldering aluminum. Once the soldering is completed, there are relatively simple and nonhazardous cleaning methods for removing flux residues.

Soldered aluminum connections need to be evaluated for being truly complete after exposure to moisture. If an intermetallic bond has not been made between the solder and the aluminum, exposure to moisture will break the solder bond. Another major consideration is the relative strength of the soldered connection, especially compared to the copper to copper bond. The copper to aluminum bond is stronger than aluminum to aluminum bond. The aluminum alloy used is a controlling factor in the ultimate strength of these connections. Finally, a new specialty solder, ALUSAC-35, has been developed that is capable of withstanding the ASTM B117-16 salt fog test.

Acknowledgments

The authors would like to thank the Aluminum Association for giving us the opportunity to work on the Aluminum Soldering Handbook, and helping us to further develop the important area of aluminum soldering.

References

[1-14] Aluminum Soldering Handbook, 6th Edition, February 2017, The Aluminum Association, Edited by William Avery and Dr. Yehuda Baskin, Chapter 1, page 15

[15] Haruka Nishino and Takuro Fukami, "An Innovative Approach to Soldering Aluminum with ALUSAC-35", U.S. Tech, September 2017

[16] ASTM B117-16 "Standard Practice for Operating Salt Spray (Fog) Apparatus", ASTM International

[17] William F. Avery, "Low-Temperature Direct Aluminum Soldering Paste", IMAPS - 8th International Conference and Exhibition on Device Packaging, March 2012

Basics of Induction Heating Related to Brazing and Soldering Applications

W. Adam Morrison
Ajax Tocco Magnethermic Corporation, Albertville, AL USA
amorrison@ajaxtocco.com; 256-279-1220

Abstract

Induction heating is a very established heating technology that many still consider cutting edge or new technology with a great deal of mystery and art associated with its application. The theory of induced electrical currents via electromagnetic induction was proven by Michael Faraday in 1831. The technology was first commercialized for industrial heating applications by Dr. Edwin Finch Northrup in 1918. Although induction heating technology is approaching 100 years of commercialization a great deal of questions and misunderstanding of its application still exist. One such example is a common assumption that induction heating will only heat magnetic components, when in fact induction heating can be used to heat any electrical conductive material to any temperature below its vapor pressure. It is the intent of this presentation to help engineers and tradesmen not directly involved with the induction heating market to gain a greater understanding of how induction heating works as related to brazing and soldering processes. Topics include conversion efficiency, calculating heat times, calculating power requirements, proper frequency selection, heating inductor design and a review of specific brazing and soldering case studies. Readers will be better equipped to specify induction heating system requirements as related to their own unique application challenges.

Background

Both brazing and induction heating are well established technologies. Brazing dates back to ancient Egypt. Though not quite as old as brazing, induction heating as a process is over 100 years old. One can imagine the volumes of data that has been presented for both processes, yet there are still many who do not fully understand how to apply either technology. When the two processes are combined the mystery only deepens. It is important to consider both brazing and induction heating, as separate and different processes in order to understand and apply them. Brazing is a joining process, while induction heating is a heating process that is often used as a single step of the brazing process.

Brazing as a basic definition is the joining of two metallic components through the use of a molten filler metal that melts at a temperature greater than 840F (465C). Soldering by definition is the same as brazing, only that the filler metal melts at a temperature less than 840F. Brazing creates bonds between the base metals and the filler metal at the atomic level (alloying), while soldering is an intermetallic bond (mechanical).

Adhesive bonding is also a common method of metals joining that often employs the use of heat via an induction heating. Adhesive bonding is purely a mechanical bond. Adhesive bonding by itself warrants a separate discussion/topic regarding the physics associated with this process. However, all 3 processes can benefit from the understanding of how to apply induction heating technology as related to either of the joining processes.

Induction heating is a thermal process in which an electrically conductive material is heated via induced electrical current flow within the conductive material. The induced electrical current is created within the conductive material by exposure to a changing (AC) magnetic field. The changing magnetic field is generated by a varying electrical current (AC) being passed through an electrical winding (coil/inductor). Heating can be very rapid and very focused, thus providing significant benefit to the brazing process as compared to other heating technologies.

Electromagnetic induction was first observed by Michael Faraday in the year 1831. He learned that electrical energy can be transmitted from one conductor to another, if the conductors are in close proximity and if current flowing in the primary conductor is alternating (EPRI, 1998, p.1). This concept is the exact way that electrical transformers operate. American physicist Joseph Henry also discovered electromagnetic induction independently in 1831. The unit for inductance, the Henry, was named after Joseph Henry.

It wasn't until 1918 that the first practical application of the phenomenon was used as an industrial thermal process. Dr. Edwin Northrup developed the wireless fireless furnace in 1918 to melt lead for making lead shot. Dr. Northrup helped found Ajax Induction Heating Corporation. Later in the 1930's the Ohio Crankshaft Company (OCCO) applied the technology to selectively harden large crank and cam shafts to combat the rising costs of high alloy steels. Stronger cranks and cams were necessary as engine horsepower increased. Induction heating technology affects all aspects of our lives. Everything from underwires for women's undergarments to mission critical aerospace parts to food products and even the latest wireless charging devices are all a result of induction heating technology.

Brazing in Greater Detail

Brazing as previously stated is a process of joining two closely fitting metallic components by the introduction of a molten filler metal at a temperature above 840F. There are many factors that can influence the success and strength of a braze joint such as service environment, joint geometry, base metals, filler metals, heating method, production requirements fluxes and post braze inspections. As one can appreciate the matrix for brazing variables is quite large. For this reason a significant investment in understanding the process should be made by those not well versed in the technology, including hiring a qualified brazing consultant. Due to the many combinations of variables associated with the brazing process, this paper will predominantly focus on induction heating technology as related to brazing and metals joining.

Intended Service Environment.

The first step for considering any joining process, specifically brazing is to consider the end use or service environment of the component being manufactured. The service environment will dictate base metal selection, filler metal selection, part geometry, heating method as well as a few other key process parameters. For example, it would not be a good idea use a copper casting as tool holder/shank for a carbide cutting tool due to the ductility and limited strength of copper. Obviously one would choose a stronger more heat resistant base materials such as iron or steel to be used for the shank. This is just one of many such considerations when specifying a brazing the process. Similarly, iron would not be an ideal material for a water plumbing component or a medical grade cutting instrument, etc... The success of any brazing process ultimately starts on the drawing board. A very good and repeatable induction heating process can not compensate for an inadequate part design or a poorly planned brazing process.

Filler Metal Selection.

As previously noted, the service environment will influence the choice of filler metal used to join the base metals. The selection of filler metal will influence the heating method, oxide prevention (fluxes and or protective atmospheres), joint geometry, method of filler introduction, and the amount of filler used. Staying with the carbide cutting tool example, proper filler metal selection will be key to producing a cutting tool that has adequate strength to minimize the occurrence of the carbide being damaged or separated from the shank during use. It is important to select a filler metal that will readily "wet" both the carbide and the iron shank. Wet or wetting refers to the ability of a filler metal to adhere and spread across the surface of a base metal in a continuous even layer (ASM Metals Handbook, Volume 6, Welding Brazing and Soldering, page 20). The ability to wet a base metal is critical for proper alloying to take place between the filler metal and base metal. Further, the filler metal should provide some level of ductility to provide shock absorption for the carbide thus minimizing fracturing of the carbide during use. Most silver bearing alloys in the AWS BAg series of filler metals are ideal for brazing of carbides to steel (BAg-3 is commonly used). Due to the high vibration environment that most cutting tools operate in, combined with the relative brittleness of carbides, BAg-3 alloy by itself is not an ideal solution for carbide brazing. BAg-3 does not have the necessary ductility to optimize tool life of the finished braze joint. The ductility of the braze joint can be improved by the use of a silver clad copper filler metal, sometimes referred to as tri-metal alloy. Tri-metal alloys will have a thin layer of copper with each side of the copper plated with the desired filler metal choice (BAg-3). This alloy is typically introduced as a preplace shim/foil between the carbide and shank. During the brazing process, only the silver cladding is melted which joins the carbide to the copper foil and the copper foil to the iron/steel shank. Since the copper is not melted it retains its parent metal properties and provides a ductile layer between the carbide and shank. This is just one example of how intended use can influence filler metal choice. Some high temperature environments may dictate the use of a copper filler metal in which a traditional silver alloy would fail due to its lower melting point (1270F for BAg-3 vs 1950F for a typical copper alloy). The type of filler metal selected will ultimately influence the type of protective atmosphere as well as the joint geometry.

Protective Atmospheres and Fluxes

Most metals react freely with oxygen to form surface compounds, oxides, such as iron oxide (FeO^2) in the case when iron is exposed to oxygen. We commonly refer to iron oxide as rust. Oxides are not easily wetted by filler metals and are a very serious concern for any brazing process. Further the bond between the oxide and the base metal is relatively weak. Any braze joint with excessive oxide inclusion will ultimately result in premature failure. For this reason, oxides must be removed prior to the brazing process. The rate at which oxides form is rapidly accelerated when the base metal is heated. Oxide formation during the brazing process must be prevented. Any minor oxides present prior to brazing must be dissolved during the brazing process. The use of a chemical flux and/or protective atmosphere is required to protect the brazing surface from oxide formation and to remove any minor surface oxides while the brazing process is being performed.

A protective atmosphere is void of oxygen. Without oxygen, oxides can not form. Some typically used atmosphere gasses are argon, nitrogen, and hydrogen as well as others. Information regarding proper selection of an atmosphere gas needs to be reviewed for base metal compatibility and safety (proper ratios of hydrogen and oxygen can be explosive). Some base metals can become brittle or develop other adverse conditions such as embrittlement when exposed to certain gasses at elevated temperatures. Hydrogen gas in the form a disassociated ammonia is often used due to it's oxide reducing properties, which will remove oxides at elevated temperatures. Another common practice is the use of a vacuum environment for brazing. Not only is a vacuum void of oxygen, the low partial pressure of oxygen and moisture will cause oxides to disassociate at elevated temperatures. A vacuum environment is essentially a form of controlled atmosphere.

Another method of creating a controlled atmosphere is the use of a chemical flux. A flux is a chemical compound containing mixture of a wetting agent and one or more of the following materials: boric acid, borates, fluorides, and fluoborates. (ASM Metals Handbook 9th Addition, Volume 6, Welding, Brazing and Soldering, page 971). Chemical fluxes can be in powder, liquid or paste form. Often powdered fluxes are mixed with water or alcohol into a paste consistency, allowing them to be brushed onto the braze area of the parts being joined. When heated, the flux releases gasses that displace the oxygen from the area where the flux is applied. The flux also dissolves any minor oxides that may have been present prior to starting the brazing process. Further the flux reduces the surface tension of the molten filler metal to promote free flow and capillary action of the filler metal. The most common chemical flux used with induction heating processes is AWS spec FB3A, which is active between 1050F and 1600F.

The underlying key concept is to protect the joint area from oxidation during the brazing process and remove any residual oxides present prior to brazing. This can be accomplished via vacuum, oxygen free gasses or chemical fluxes. Another key concept to remember is that the parts should be thoroughly cleaned immediately prior to brazing. Under no circumstances should a flux or reducing atmosphere be relied upon to compensate for poor part preparation prior to brazing. There is a limit to the amount of oxides that can be reduced by a chemical flux and the amount of oxides that are disassociated in a reducing environment is a function of time. Excessive heat times can create additional metallurgical issues that could result in premature failure of the braze joint. The necessity for precleaning parts is especially true for aluminum base metals, which oxidize almost immediately in ambient air. Aluminum oxides are very stable and very difficult to remove in a reducing atmosphere or with a chemical flux.

Joint Geometry

The choice of base metals, filler metal and flux will have a significant influence on joint design as will the intended use/service environment of the brazement and the method of heating for brazing. There are essentially only 2 styles of braze joint configurations. The butt joint and the lap joint. All other joints are just variations of these two joint configurations.

The butt joint is simply a joint created when two end planes of two base metals are butted end to end to each other. The braze area is only comprised of the surface area between the two butted ends. The surface area of a butt joint can be increased through a simple modification called a scarf joint in which the mating surfaces are beveled.

The other basic joint configuration is the lap joint. The lap joint is created when two parallel surfaces are stacked on top of each other. An example would be the ends of two steel sheets/strips placed on top of each other. Another example would be a smaller tube inserted into the ID of a larger tube. The lap joint is a stronger joint configuration than a butt joint due to the greater percentage of braze joint surface area created. It should be noted that the optimal amount of overlap is 3-6 times the thickness of the thinnest member. For example if a lap joint were created from 0.010 inch thick steel, the length of the overlap would be between 0.030 and 0.060 inches, which will create a joint that is at least as strong as the parent base metal of the thinnest member. Longer joints provide minimal improvements in joint strength and can be difficult to create consistent braze quality due to the distance the filler metal has to flow, creating the potential for voids and flux inclusions.

It would be easy to oversimplify braze joint geometry. From a brazing process viewpoint, there are numerous considerations such as joint gap, fixturing, placement of filler metal, type of filler, type of flux, and heating method. A separate paper could be written exclusively about induction braze joint design.

The gap between the two parts will vary depending on the type of filler metal, flux and thermal expansion of the parts (if dissimilar metals are being joined). Gaps between components being joined can range from a slight interference fit to 0.010 inch. For most induction brazing applications, the gap will range from 0.002 to 0.005 inches, which is the range for optimal joint strength. Smaller gaps can prevent the flux from being forced out of the joint by the molten filler metal, creating flux inclusions that will reduce joint strength and/or lead to premature corrosion within the joint. Smaller gaps and interference fits are acceptable for fluxless brazing applications such as when brazing in an inert atmosphere using a copper filler metal. Larger gaps will lack proper capillary action to completely fill the joint area. Joint strength starts to approach the yield strength of the filler metal as the gap increases. A properly designed joint will have a yield strength that is close to the weakest base metal in the joint. The joint gap at the brazing temperature should be considered when engineering a braze joint. Thermal expansion will obviously increase the gap between the parts at the brazing temperature. Considering the gap at the brazing temperature is especially important when brazing dissimilar metals.

Ideally the joint design would be self-fixturing, thus minimizing the amount of tooling required to hold the components in position during the brazing process. In most cases some form of part fixturing is required and can not be avoided. Additional discussion will be provided regarding tooling design in relation to induction heating.

Most induction heating processes use a fixed geometry inductor. The braze joint design should take in consideration the geometry of the heating inductor. The components should be able to be assembled within the inductor and the finished brazement should be able to be removed from the inductor after brazing.

The method of applying the filler metal to the joint during the brazing process will influence joint design. The filler metal can be hand fed by a wire, it can be pre-applied as a paste, or as preformed ring, flat shim, pellet/slug or even clad directly onto the parts. All the before mentioned methods of filler metal introduction have been successfully used with induction heating as a brazing heat source. Ideally a preplaced filler is used for any brazing process. Preplaced filler metal provides greater consistency in the amount of filler used. It also speeds up the process because the filler is being heated simultaneously as the base metal. Preplaced filler metals also allow the opportunity for the filler metal to be placed within the interior of the joint, insuring complete filler metal coverage. Preplaced filler metals also allow for automation of the braze process. If preplaced filler metals are used the joint should be engineered to provide a location for the filler to rest during the heating process. As an example a "shelf" should be provided for preplaced paste type filler metals. Without the shelf, the filler metal paste would fall off the joint before reaching the brazing temperature.

Part Fixturing

Significant investment has been made in specifying the base metals, filler metal, flux, joint geometry and heating method. Proper fixturing to hold the parts during the brazing process is also critical to the overall success of the process. Some key design features for fixturing include:

- Provide proper component alignment
- Allow for thermal expansion
- Easily loaded/unloaded
- Compatible with the chosen heat source (flame, resistance, induction, etc…)
- Heat resistant
- Wear resistant
- Chemical resistant
- Minimal contact with components
- Low thermal conductivity
- Minimal thermal mass

Ceramic and stainless steel are most commonly used. Ceramics provide great temperature resistance, chemical resistance and wear resistance. Ceramics also have low thermal conductivity. However ceramics are brittle and are prone to fracture. Stainless steel is heat resistant, chemical resistant and wear resistant. Stainless steel has relatively low thermal conductivity. It provides good toughness and wear resistance. Sometimes a combination of these materials are used to fabricate fixtures capitalizing on the benefits of both materials. Fixtures can be as simple as a single alignment pin, or a complex arrangement of clamps, springs, pins and/or weighted centers. Fixtures should allow normal physical growth of the parts being heated. Parts that are restricted during heating can be distorted or damaged.

Fixturing in relation to the use of induction heating for brazing processes should also be none magnetic to minimize stray heating of the fixture. Acceptable none magnetic materials would be 300 series stainless steel, and aluminum. The high conductivity of aluminum can be addressed by the addition of smaller more thermally stable ceramic inserts that make direct contact with the parts being brazed. Not only is aluminum none magnetic, but it is also highly electrically conductive. The high electrical conductivity of aluminum greatly reduces any potential eddy current heating of stray induced currents within the fixture.

To further minimize stray heating of a holding fixture, the distance between the fixture and any portion of the inductor assembly should be a minimum of 2 inches. Sometimes it is not possible to have 2" of space between the inductor and the holding fixture, in which case it may be necessary to add a water cooled path to the fixture. There are also other design techniques that can be implemented to stop or minimize induced currents within a holding fixture, such as physical breaks, and EMF shielding. The scope of this paper does not allow detailed discussions of these techniques. Qualified induction heating suppliers know and understand how to implement best practices related to fixture design.

Method of Heating

There are several heating methods to consider for a brazing process. Each heating technology has its own advantages and disadvantages. This paper is written to better understand how to apply induction heating technology, but before one can fully understand and appreciate induction heating technology, we need to briefly examine a few of the more common heating technologies used for brazing. Other heating technologies include:

- Flame Brazing
- Resistance Brazing
- Furnace Brazing
- Dip Brazing
- Induction Brazing

The decision to pick a particular heating technology can be influenced by several factors such as:

- Utility costs
- Production rates
- Utilization
- End Use
- In situ processes
- Capital investment costs
- Heat sensitive areas of brazement
- Availability of skilled labor

Open Flame/Torch Brazing

Perhaps the most common heating method for brazing is open flame (torch). Torches require relatively low skill level to operate, have a low acquisition cost, have a low cost of operation, easy to integrate, and are compatible with almost any metal to be joined. Torches allow for single piece part flow.

Although a torch requires little skill level to operate from a basic on/off and heating principle, considerable skill and time is required for an operator to be able to manipulate the torch properly for brazing. The skill level required for hand held torch brazing is similar to that of a skilled artisan. Torches are often integrated into automated work cells to eliminate the operator skill factor.

Tooling and fixturing to hold parts during brazing is often more simple for torch brazing than other heating methods. Torches for the most part are point sources of heat allowing specific areas of an assembly to be heated. One key advantage of gas torches is that a gaseous flux can be injected into the oxy fuel mixture to protect the parts from oxidation during heating and thus eliminate the need for a chemical flux in some cases. Post cleaning of torch brazed parts is often required. Time and costs associated with post cleaning should be considered.

Due to torches being a gas combustion process, they are susceptible to environmental changes such as atmospheric pressure, temperature and humidity. This is perhaps one of the biggest disadvantages of open flame brazing in that they require constant adjustment to maintain a consistent brazing process. Another disadvantage is energy consumption. Although bottled gasses are relatively inexpensive, a torch is burning during production whether a part is present or not and thus wasting energy. Another issue to be considered is operator comfort and the additional load that a torch or other combustion process places on facility environmental systems such as HVAC and the need to ventilate the work area. Open flames also present their own obvious fire and burn risks that should be considered.

Furnace/Oven Brazing

Furnace and oven brazing are popular heating processes for joining components at high production rates. The primary difference between an oven and furnace is the temperature each operates at. Ovens operate below 1000F while furnaces operates above 1000F. Oven's are typically limited to low temperature aluminum brazing and lower temperature soldering applications. Continuing to focus on brazing processes, we will limit our discussion to only furnaces. Just note that mechanically speaking ovens and furnaces operate the same way. Most furnace brazing systems operate near 2000F and are primarily used to braze steel components using a 95% copper filler metal. The heating environment within the furnace is a hydrogen rich atmosphere that displaces oxygen and provides an ideal reducing atmosphere that will remove surface oxides during heating. Flame curtains are typically at the entrance and exits of a furnace to prevent oxygen from entering the furnace and to burn off any excess hydrogen that may escape the furnace. Brazed parts exit the furnace brighter than when they entered. Parts are fed into, through and out of the furnace on a metal chain conveyor.

As stated previously belt furnaces are popular due to their ability to continuously process parts, allowing high production rates. A single furnace can also process a wide variety of part geometries small enough to fit through the entry and exits of the furnace. Irregular shaped parts that may be difficult to fixture are sometimes good candidates for furnace brazing. Furnace brazing is very controllable in that they operate at a constant temperature, minimizing the chance of overheating a part. Stainless steel fuel rail assemblies are commonly processed in belt furnace.

Due to the constant moving conveyor and the volume of parts that are typically processed by a furnace it is not practical to invest in complex part holding fixtures which can be expensive and eventually will degrade over time in a furnace environment. Parts processed in a furnace should ideally be self fixturing to eliminate the need for a holding fixture. Other options for holding components together for furnace brazing include spot welding, staking, and wire banding.

Brazing in a furnace heats the entire assembly vs only the joint area. This can be an advantage or a disadvantage. The high brazing temperature will anneal the assembly removing stresses in the part which may be advantages for end use characteristics or post processing after brazing. Heating of the entire part may not be desired or permitted. In cases where selective heating is required using a copper filler metal, controlled atmosphere brazing with induction heating, dip brazing or resistance should be considered.

Furnaces have a relatively high acquisition cost along with relatively high maintenance costs and operating costs. Furnaces have to run continuously, even between shifts, to prevent premature wear associated with large temperature swings. Furnaces also consume significant energy in the form of natural gas. It is critical that a furnace have a high utilization rate to realize a return on the investment. Furnaces consume a significant amount floor space due to the length of conveyer required which provides the proper time for heating, followed by adequate time for cooling. Steel parts should not be introduce to an oxygen bearing environment until after the parts have cooled below 400-500F, or discoloration will occur that may not be desirable. Furnaces radiate a significant amount of heat that must be exhausted from the operator area to maintain a comfortable safe working environment. Some state and local environmental codes may prevent the installation of a furnace or impose additional environmental protection systems to be installed.

Dip Brazing

Dip brazing is a brazing process in which parts are conveyed through a hot molten salt bath. The molten salt heats the braze joint(s), melts the filler and brazes the parts. Dip brazing shares many of the same pros and cons as furnace brazing, such as the ability to operate at high production rates, and great temperature control. One key advantage of dip brazing over furnace brazing is that dip brazing has the ability to only heat a portion of the part by only submerging the joint area within the molten salt. Fluxes can be combined with the salt bath to clean and protect the part from oxidation during heating. Heating rates for dip brazing is typically 4 times faster than a belt furnace (ASM Materials Handbook, 9th Addition, Volume 6; Welding Brazing and Soldering, page 989) Most of the disadvantages discussed for belt furnaces apply to dip brazing such as acquisition cost, operating cost, maintenance, floor space, part fixturing, etc...

Resistance Brazing

Like flame brazing and induction brazing, resistance brazing can be used as a point source of heating, which is one of the primary reasons it is selected as a brazing process. Very small delicate parts such as eyeglass frames are primary candidates for resistance brazing.

Resistance brazing is the heating of two electrically conductive components via an electrical current that is passed through the joint area via contact with two electrodes strategically placed on either end of the braze joint. The joint area becomes a part of an electrical circuit. Depending on the resistivity of the components being joined the heat for brazing may primarily come from the base metal (highly resistive parts like steel) or from the contact electrodes themselves when brazing low resistance base metals like copper. Resistance heating requires pressure to be applied to the braze area, which can be used to "seat" the parts after the filler metal is melted. Resistance brazing only uses energy when a part is being heated. Resistance brazing allows for single piece part flow. Resistance brazing can be used with chemical fluxes or within a controlled atmosphere environment. Due to contact contamination issues resistance brazing is best suited for applications that do not require a chemical flux such as in a controlled atmosphere, but more commonly when self fluxing filler metals such as SilPhos alloys (AWS BCuP) are used.

Acquisition costs for resistance brazing systems are relatively low. The equipment used for resistance brazing is very much the same as that used in resistance welding. The technology is well established and readily available from many sources. Operating costs are very low. Resistance heating is almost 100% efficient and energy is only used when a part is being heated. There are no combustion gasses that need to be exhausted. Floor space requirements are minimal, with some systems being table top mounted. The technology is very scalable and some very large parts have been successfully brazed with resistance heating. Resistance heating can be used

in sensitive operating environments. Due to the selective heating nature of resistance heating, heat can be contained within specific areas of a part and not affect adjacent sensitive areas.
Due to physical contact being necessary with the brazement and the electrodes, part manipulation during the heating process, such as rotation, is not practical. Further the physical contact and pressure applied by the electrode may leave undesirable cosmetic markings on the parts being joined. Proper routine maintenance of the electrical contacts and electrical interconnect are critical for consistent braze quality. Although production rates are much slower with resistance technology as compare to furnace or dip brazing there are significant positive attributes that make it a valid consideration for a brazing process.

Induction Brazing

A very brief review of induction heating will be provided in this section of this paper. A more detailed presentation of induction heating technology as related to brazing will be provided once we complete our general overview of brazing in general.
Induction brazing is the use of induction heating technology to heat metallic components for the purpose of melting a filler metal for brazing. Induction heating is a process in which an alternating magnetic field is generated within an inductor/winding. Metallic parts (magnetic or non-magnetic) placed within this alternating magnetic field have electrical currents induced into them via transformer action. The induced currents within the part cause the part to generate internal heating due to the resistance of the material to the flow of the induced electrical current. The heat generated in the part via induced current is the same heat generated when a part is subjected to electrical current flow via the resistance heating method previously described. The key difference between induction heating and resistance heating is that physical contact with the heated parts is not made when using induction heating technology. Since physical contact is not required, parts can be rotated or moved in other ways during the brazing process.
Induction heating rates can be extremely rapid (fractions of a second) or very long (hours) depending on the process requirements. Further, induction heating can be very selective, heating only small portion (a few millimeters) of a part. Induction heating can also be used to heat entire assemblies in single piece part flow or continuous conveyor type production. The choice to heat a small portion of a part vs the entire part depends entirely upon the process requirements. Induction heating is very adaptable to a wide variety of part geometries.
Induction heating is very efficient, almost as efficient as resistance heating with typical efficiencies ranging between 40-90%. Induction heating like resistance heating only use energy when a part is being heating. The heat is instant on and instant off, therefore not wasting any power during standby or delays in production. There are no preheat/warm up procedures for starting an induction heating system.
Induction heating is very controllable, repeatable and scalable. Temperature rises of a few 10's of degrees to several thousands of degrees can be provided with induction heating technology. Induction heating has the ability to achieve temperatures not easily possible with torches, resistance, and/or furnaces. Unlike combustion heating methods, induction heating is not affected by environmental changes such as dewpoint, pressure and ambient temperature.
As with any heating system, induction heating systems do require regular maintenance to maintain optimum performance, however the maintenance required is much less than other heating technologies discussed in this paper. Overall system reliability is very high with typical mean time between failures being 20,000 hours or more.
Acquisition costs for induction heating equipment is generally much higher than alternate comparable (BTU rating) heating technologies. The investment costs are often quickly offset by energy savings, increased production rates, reduced scrap, and adaptability. Induction heating systems are easily adapted to new part geometries by simply changing the geometry of the heating inductor connected to the output of the induction heating power source. Costs for induction heating equipment can range from a few thousand dollars ($2,000 for a 1kW system) to several million dollars ($10,000,000 for a 20MW system). Induction brazing systems typically range from 3kW to 50kW with the majority being in the 10kW to 20kW range. A typical 10kW system can produce up to 150 pounds of steel per hour at 1450F. If you consider a typical braze joint will comprise less than 0.5 inch^2 of material, 150 pounds would represent over 1000 braze joints per hour at 100% utilization. The ROI for an induction heating system can be very short.
Induction heating is compatible with flux and fluxless brazing applications. It is very common for induction heating to be used within a vacuum or a positive pressure inert gas environment. Selective heating of automotive hydraulic components joined with a copper filler metal at 2000F is a common use of induction brazing technology. In addition to being compatible with a with a wide range of brazing processes. Multiple heating steps can be performed with a single induction heating system via the use of an electrical transfer switch that will allow power to transferred to a 2nd or 3rd inductor which can be used to heat a different portion of a part or another part all together. Within a single work cell and with a single induction heating system it would be possible to braze at one coil, quench the part and then transfer power to second coil for a 2nd braze joint or possibly to perform a localized heat treat. Generally speaking, induction heating systems do not consume a lot of space. Most induction brazing systems are table top mounted.
Induction heating combines many of the desirable attributes provided by resistance heating, furnace brazing, torch brazing and dip brazing. It is not the silver bullet for all brazing applications, but it certainly should be a primary consideration when deciding which heating technology to use for a new brazing process. In order to have a successful brazing process, the proper selection of the heating source, along with joint design, fixturing, base metal selection filler metal selection, flux requirements, and quenching, are critical and must be compatible in order to produce quality brazed assemblies.

Quenching and Cleaning

After the parts have been assembled, fluxed, heated and brazed, the parts must be cooled for safe handling and/or successive processing steps. The quenching process is equally as important, possibly more important than any of the other discussed steps in the brazing process. Improper quenching can lead to fractured parts, brittle parts, trapped flux, none uniform

metallurgy, poor aesthetics and other potential negative characteristics.
Quenching is simply the method used to remove heat from a heated part for further processing. The method of quenching can be as simple as a long cool down area that will allow sufficient time for the parts to cool via natural convection. Quenching can also use forced cooling methods, such as compressed air, fans, water, oil, and even sand. The key to any quenching process is consistency in order to have a repeatable quality process.
Some quenching methods are not compatible with certain brazing processes. For example it would not be possible to use a liquid quench inside a vacuum chamber or a belt furnace. The quenching method for a furnace of vacuum process is simply TIME.
The timing of the quench is important. Generally speaking, you want the molten braze to have set/solidified before using a forced quench media. Quenching too early can cause the molten filler to be ejected from the joint. Quenching too early and also lead to foreign debris such as scale or flux being inserted into the joint area. Quenching too early can also lead to fractured parts due to the rapid change in part temperature. This is especially true for carbide brazing and more so for larger carbides.
It is often advantageous for flux brazing processes to apply a forced quench (water) to the parts immediately after the filler has solidified and significant heat still resides within the part. The rapid cooling of the heated part will help break up and remove scale and crystalline flux residues. Again, be mindful that certain parts can not be quenched too soon due to potential fracture concerns. Sometimes liquid quenches are heated to slow down the rate of cooling, still providing the before mentioned scale and flux residue removal, while minimizing the potential for part fractures. Heated oil is yet another forced method of quench that is less aggressive than heated water due to oil having a much lower specific heat and conductivity.
Fans and compressed air can be used to speed up the cooling process when liquids are too aggressive or incompatible with the brazing environment. Fans and compressed air can be used in positive pressure systems and furnaces.
Sometimes even natural convection can be too aggressive of a cooling medial for brazed parts. This is the case for larger carbide tools or dissimilar metals with significant differences in thermal expansion. To further slow down the cooling process, such parts are often dipped into hot sand to cool before being further processed.
As one can appreciate a separate paper can be written about quenching. It is not the intent of this paper to provide any greater detail related to the subject but just use this opportunity to help reinforce the fact that brazing is very complicated process that is often over simplified during the initial design phase.
To further complicate the brazing process, one also has to consider post cleaning and inspection of the parts. Quenching alone may not be enough to adequately remove scale and flux residue. The quenching process itself may leave undesirable residue. After brazing, parts may require painting, plating or some other surface processing. Perhaps the end use can not tolerate any foreign matter within the assembly such as an automotive fuel rail or medical device. It is important that the method of quenching and the method of post cleaning be compatible with post processes and end use requirements. There are numerous post cleaning methods to consider such as grit blasting, acid dip, vibratory bowl, solvent wash, sanding, and hot tank immersion.

Post Braze Inspection

Now that the entire brazing process from assembly to cleaning is completed we are left with the final most important step of the brazing process, post braze inspections. Due to braze joints residing between the surfaces of 2 or more close fitting components it is not easy to simply visually inspect the surface a part for quality. However this is often the only option unless potentially expensive and time consuming equipment and/or methods are employed. Each participant in a brazing process must evaluate critical final product characteristics related to its fitness for use, potential liability, cost of repairs/warranty, serviceability, etc. and then design a written inspection procedure that can be used to guarantee quality. Part of the quality verification process may involve other checks and inspections throughout the brazing process along with establishing upper and lower operating limits for flux volume, filler volume, energy input, machining, etc.
Early in the design and development stages of a brazing process it is likely necessary that destructive testing methods and accelerated life cycle testing be used to verify a brazing process. Tensile testing, cutting, polishing, metallographic inspections, compression, torque testing, etc… may be necessary. Even after the design is initial confirmed it may be necessary that periodic samples are pulled from production for destructive testing to make sure the process is not experiencing unseen issues.
It is possible to design into the assembly certain mechanical characteristics that will help visually verify a part is properly brazed. One such design feature is called a witness hole. A witness hole is small hole strategically placed in the outer brazed component. The witness hole is typically placed in the braze joint area likely farthest away from where the filler metal was introduced to the joint. Capillary action will pull and distribute the molten filler uniformly throughout the braze joint. The witness hole will allow a quick visual inspection to see if the filler metal had indeed been pulled all the way down and around the area where the witness hole is located. Sometimes the part geometry itself can provide a "built in " witness area such as in the case of brazing a tube to a threaded fitting. In this case if the filler was applied at the top of the joint during brazing and filler could be seen all the way around the tube at the bottom/inside the fitting, then one could assume complete fill had occurred. Especially if there is no evidence of unmelted filler or a cold joint at the top of the joint.
A cold joint is evident by filler metal that has partially melted or does not have a nice smooth fillet. Although fillets provide very little if any strength enhancement of a braze joint, they are good evidence of even heating and proper wetting of the base metals being heated. A cold joint or a joint with poor wetting will have filler that is not melted smoothly. It will be lumpy, non uniform and likely be balled up on the surface where the filler was introduced to the joint.
In cases where it is not possible to witness the complete flow of the filler and it is necessary to verify each part has been properly brazed such as in critical aerospace assemblies, it may be

necessary to employ X-ray inspections to see through the surface of the parts and visualize the flow of the filler metal within the joint. X-ray inspection equipment requires a significant investment in capital, personal training and inspection labor.
Other critical braze joint inspections take place in the refrigeration market where refrigerant leaks can lead to excessive pollution, reduced energy efficiency, and costly repairs. Such joints are often pressurized with helium and then a "sniffer" is used to detect any leaking helium. Helium molecules are very small compared to ambient air and typical refrigerants. A seal that will not leak helium is considered air tight and satisfactory for refrigeration.
Certain mission critical electrical assemblies that are brazed get 100% inspection by measuring the electrical resistance of a braze joint. This is accomplished by passing a precise known amount of current through a braze joint and then precisely measuring the resulting voltage drop across the joint area. If the braze is of poor quality, the measured voltage drop will be higher than normal.
It is ultimately up to each practitioner of a brazing process to determine what is necessary to insure individual quality and be dedicated to operating within the bounds of established parameters. As I have said before, brazing is a very complex process due to the many different variables and inputs. It is absolutely critical that every step of the process from beginning to end be defined, managed and periodically verified. Otherwise ultimate final quality will be a guess at best and could put unnecessary risk to environment, property and life.
Basics of Induction Heating

Basics of Induction Heating

The concept of induction heating has already been described in this paper. The key concept is that the principle is the same as how an electrical transformer works. The heating inductor is the primary winding and the part being heated is the secondary. Another key concept is that induction heating can be used to heat any electrically conductive material, even glass. Molten glass will carry an electrical current and can be directly heated with induction once introduced into an induction heating coil in the molten state. With sufficient power and the correct operating frequency, the temperature of the molten glass can be increased for further processing such as the manufacture of fiberglass insulation.
Frequency of operation refers to the rate at which the electrical current and corresponding magnetic field reverses direction, measured in cycles per second, also know as the Hertz. Most induction heaters for brazing processes operate within the kHz range (1000's of cycles per second). Induction heaters can operate from a few 10's of Hz to several Million Hz (MHz). A deeper discussion of frequency selection will be provided later in this paper.

Hysteresis Heating

As previously stated induction heating can heat any electrically conductive material. There is a common misconception that induction heating is only effective on magnetic parts. There is one key difference associated when heating a magnetic part vs a none magnetic part. Magnetic parts like iron and carbon steel have magnetic particles within their atomic structure called magnetic domains. These magnetic domains are sort of like small bar magnets. These bar magnetics physically align themselves with the polarity of the magnetic field applied to the part. As the field reverses, the domains reverse direction. This constant reversing of direction results in internal friction heating, similar to the heat you feel when you rapidly rub your hands together. The faster the rubbing, the more heat is produced. Faster rubbing would equate to a higher operating frequency (or 1kHz vs 25kHZ). Higher frequencies do yield slightly higher energy conversion efficiencies. This heating through magnetic domain direction reversal is know as hysteresis heating. Hysteresis only occurs in magnetic materials and is the most efficient form of induction heating. Hysteresis heating can exceed 90% conversion efficiency.
When sufficiently heated, magnetic parts will lose their magnetism. This is called the Currie point (typically around 1400F for iron). Once a magnetic part reaches its Currie point, hysteresis heating stops. Without proper frequency and/or power selection the heating all together will stall at 1400F. This is another reason some novices believe induction heating will not heat above 1400F. With proper frequency and power selection it is possible to heat to any temperature up to the vapor point of the material being heated. Tungsten is routinely heated to over 6000F with induction heating for the purpose of manufacturing welding electrodes.

Eddy Current Heating

The heating that takes place beyond the Currie point is eddy current heating. This is the transformer type energy transfer previously described, also known as "I squared R losses" derived from Ohm's law (V=I*R). Eddy current heating is what allows induction heating to heat any electrically conductive material, including gold, copper, silver, aluminum and many other materials. The efficiency of Eddy current heating will range depending on the alloy being heated from 50% to as low as 5% conversion efficiency. The induced electrical current is called an eddy current, just like the eddy currents you may see swirling on the surface of river or stream. Eddy currents in a stream of water circulate on the surface of the stream of water and not the full depth of the water. Induced electrical eddy currents act the same way in metal objects. They only circulate on the surface of the part. The higher the frequency of operation (Hz) the more shallow the eddy currents are. The depth at which the current penetrates the surface is called the reference depth and represents the depth at which 63% of the induced current is flowing.
Previously we mentioned proper frequency selection. There is a relationship to the operating frequency and the depth of electrical current penetration beyond the surface of the part. This is called the reference depth of heating.

Current Reference Depth

It's important to consider depth of heating when choosing an induction heating system. If the depth is excessive very little heating will occur due to cancellation of eddy currents. Cancellation occurs with eddy currents of opposite directions from opposing sides of the part come in contact with each other.

As a general rule the depth of current should be no greater than ½ the thickness of the part being heated. In the case of tubular objects, the frequency should be no greater than ½ the OD of the tube regardless of the wall thickness. Ideally the material thickness (or diameter if a tube) would be 4 times greater than the reference depth for maximum efficiency. At the very minimum the reference depth should not exceed the radius of the part being heated.

Resistivity

Although frequency of operation is the primary influencer of depth of heating, so too are the electrical resistivity of the part being heated and its magnetic permeability. The formula for calculating reference depth is $D=3160*\sqrt{(\rho/\mu f)}$. ρ is the electrical resistivity of the material being heated in micro ohm inches. μ is the relative magnetic permeability in henries/meter. f is the frequency of operation in Hertz (Hz). From the formula you can see that the depth of heating increases as the electrical resistivity increases. Conversely the depth decreases as the magnetic permeability or operating frequency increases.

Magnetic Permeability

Magnetic permeability is the ability of a material to absorb and retain magnetic flux. This would be similar to the ability of a sponge being able to absorb and retain water. The greater the ability to retain flux the higher the permeability. None magnetic materials all have a permeability of 1, while a good magnetic material like iron has a permeability of 5000. Carbon steel has a relative magnetic permeability of 100. It is important to consider the permeability at the brazing temperature due to the currie point of some materials. Remember that steel loses is magnetism at approximately 1450F at which point its permeability changes from 100 to only 1. This would result in a 10 fold increase in the depth of current penetration at temperatures above currie. If the wrong frequency is selected, then heating will stop at the currie point and will not increase regardless of how much additional power is applied to the part.

Depth of Heating

We previously noted that the depth of penetration will increase with increases in electrical resistivity. As a material is heated the resistivity increases, thus it is also important to consider the depth of heating at the brazing temperature. For example, carbon steel has a resistivity of 6μΩ-inches at 68F and 40μΩ-inches at 1400F, which would represent a 2.58X increase in the depth of current penetration. The same concerns with selection apply with resistivity as those discussed for magnetic relative permeability.

Notice we have repeatedly said "depth of current penetration" and **NOT** "depth of heat penetration". The laws of thermal dynamics are not avoided with induction heating. The heat generated within the part is not limited to the depth of current penetration. The **depth of heating** will increase via conduction and is purely a time and thermal conductivity function. The additional depth of heating resulting from soak/heat time can be approximated by the formula $D=\sqrt{0.0015t}$, where D is the depth in inches and t is the time in seconds.

Heat flow in to adjacent areas will still have to be considered if there are sensitive areas that do not need to be heated. The best way to minimize heat flow is to use the shortest possible heat time or active heat sink/removal techniques.

Due to the tendency of alternating currents to collect on the surface of a part, it is also possible to select a frequency that is too high. An operating frequency that is too high will cause the current to collect in corners of parts such as the corners of a hex nut assembly. This will create localized hot spots that could be sufficiently high to melt the base metal before the filler metal is melted. In order to prevent overheating, less power has to be used to reduce the power density, which will extend the heat time for the braze process. A frequency that is too high will require longer soak times for joints with thicker cross sections. An excessively high operating frequency may produce excessive temperature gradients within the joint area resulting in poor alloying with the inner most component.

Although much has been discussed in relation to choosing an operating frequency, most brazing applications are easily and efficiently performed within a frequency range of 10-50kHz. It's not often that the frequencies are below this range, but higher frequencies in the 200-400kHz range are regularly used. Frequency optimizes efficiency, but at the end of the day, it is power that gets the work done.

Induction Heating Power (kW's)

Power is the amount of work performed over time, or the energy transferred over time. Most induction heaters are rated by their ability to deliver power, typically rated in kilowatts (kW). Power offerings for induction heating power supplies are very wide ranging. Industrial grade induction power supplies can be found from as low as 1kW to over 20MW (1,000W to 20,000,000W). For brazing applications, most brazing processes fall within the 5kW to 50kW range, but several hundred kW for larger applications is not uncommon.

Selecting the proper kW rating is as important as proper frequency selection. A kW rating that is too low will lack the power to achieve the brazing temperature, take too long to braze, and/or have excessive heat affected zone (HAZ). A kW rating that is too large will lack the necessary resolution for stable process control, overheat parts, and is a waste of capital resources to acquire and maintain.

Ideally the kW rating of the power supply would be 130% of the estimated power required to achieve a defined brazing process. This provides adequate room for process adjustment, and optimizes capital costs. Power supplies with ratings exceeding 5X the required power for a defined process should never be used due to the likelihood of not being able control the process.

There are several factors in determining the required power for a brazing process. Allowed heat time, target temperature, starting temperature, heated mass, and base metal composition are the primary factors.

Base metals vary in density, specific heat, electrical resistivity, and relative magnetic permeability as well as other defining characteristics, but these 4 traits are the most important to calculating power requirements for a process.

BTU's Pounds Degrees Watts

If we know the density of the base metal being heating along with the volume of the material being heated, we can estimate the mass of the heated area. At this point in time the target temperature and the starting temperature have been defined which will provide us with a delta-T (change in temperature). We can multiply the specific heat of the metal by the delta-T and the mass of the heated area to arrive at the total number of BTU's required to achieve the necessary brazing temperature. It is an easy conversion to convert BTU's to kW-Hrs. (kWH) using the conversion factor of 3412BTU/1kWH. This value would not include conductive losses to adjacent areas, radiation or convection losses, nor the overall energy conversion efficiency of the induction set up. A further discussion about these individual base metal traits and how they related to estimating induction heating power requirements is provided in the following.

Density

Estimating the mass of the heated area, not necessarily the entire mass of the object, is the first step of the process. An engineering drawing or good physical measurements are required. If drawings are not available and or the cross section of the heated area is very complex, it is perfectly OK to weigh the part(s) and ratio out the mass of the heated area. An additional safety factor beyond the 1.3X power rating factor previously noted will be necessary to have room to adjust the process. If we know the cross sectional area and length of the area to be heated for brazing we can estimate the volume of material. We can multiply the volume of the heated mass by the material density to determine the mass of the heated area. For example, 0.5 cubic inches of copper will weigh 0.16 pounds, based on the density of copper being 0.321 pounds per cubic inch. Below are a few common base metals and their densities.

- Ferrous metals 0.284 $lbs/inch^3$
- Aluminum 0.1 $lbs/inch^3$
- Copper 0.321 $lbs/inch^3$

Specific Heat

Specific heat refers to the amount of energy (BTU) required to raise the temperature of 1 pound of material by 1 degree Fahrenheit (F). This is typically expressed at BTU/lbs/F. Water for example, requires 1BTU/lbs/F. Some heat content values for common base metals is provided below.

- Ferritic materials 0.12 BTU/lbs/F
- Aluminum 0.22BTU/lbs/F
- Copper 0.09BTU/lbs/F

Energy

Multiplying the specific heat of the base metal being joined by the mass of the heated area and the required temperature rise, the total amount of energy in BTU's is determined. Let's continue with the previous copper example in which we determined the heated mass was 0.16 pounds. We will assume we require a 1400F temperature rise for a brazing process. We know that copper has a specific heat of 0.09BTU/lb/F. Multiplying (0.090BTU/lb/F)*0.16lb*1400F, we arrive with a value of 20.16 BTU's. This is BTU-hours. Most brazing processes are in seconds, so we need to convert this value to BTU-seconds by multiplying by 3600, which will provide us with 72,576 BTU-seconds. We need to convert the BTU-seconds to kW-seconds in order to work towards determining out power requirements. There are 3412 BTU's/kW. So divide 72,576/3412. The energy required to the part is 21.27kW seconds. This would mean that 21.27kW delivered for 1 second would do the job, or 10.63kW for 2 seconds or 0.7kW for 30 seconds would also work. For rest of this process analysis, we will use an assumed heat time of 10 seconds. This would require 2.12kW of power, **to the part!**

Efficiency

Remember our discussion about hysteresis heating and eddy current heating. Copper is none magnetic, so only eddy current heating occurs. Further copper is very conductive and will require a lot more current to generate heat internally. Remember the heating via eddy currents is I^2R losses. The lower the value for R the more current is required for the same temperature rise. Copper is one of the single most difficult materials to heat with induction in terms of efficiency. Induction heating efficiency when heating copper is typically only 15% and could be much worse, depending on the application. A summary of conversion efficiencies for common base metals heated with induction heating is provided below.

- Pure iron 95-98%
- Medium carbon steel 85-90%
- 316 stainless steel 45-50%
- Aluminum 35-40%
- OFC Copper 5-15%

Selecting a power supply size

We will assume 15% efficiency for this process analysis in which we are heating a ½" copper cube for brazing. We have determined the energy required **to the part** is 21.27kW seconds and that our heat time is 10 seconds. Simply divide 21.27kW-seconds by the assumed efficiency (15%) and we see that 141.8kW-seconds is required from the power source to achieve our desired results. Divide by the heat time of 10 seconds and the actual power to be delivered from the power supply is 14.18kW. Remember we recommended a 30% safety factor for process variability. Multiply by 1.3 and the minimum kW rating of the power supply would be 18.4kW. 20kW, 25kW and

30kW are common power supply offerings from most reputable induction heating OEM's.

Induction Heating Power Supply Explained

An induction heating power supply has been referenced numerous times within this paper. What is an induction heating power supply? An induction heating power supply is a power conversion and control circuit responsible for converting and/or controlling the 60Hz AC power available from the local utility company. Most industrial scale induction heaters operate from 3 phase 60Hz power, either 230V or 480V (in North America). 380V and 400V 50Hz power is typical for Asia, Europe, Africa, Australia, China, Russia and other regions of the World. Since an induction heating coil is single phase device, it is necessary to convert the 3-phase power to single phase. It is also often necessary to convert the frequency of operation from the mains frequency of 60Hz to something significantly higher. Most brazing operations take place between 10kHz and 50kHz. In addition to converting the mains 3 phase power to a higher single phase power it is also necessary to accurately control the power delivered to the inductor. So power conversion and control are the main purposes of an induction heating power supply.

The power conversion usually takes place in the basic format of AC to DC to AC. First the 3-phase voltage is passed through a 3-phase rectifier where it is rectified and filtered into clean DC voltage. This DC voltage is then fed into an AC inverter circuit comprised of transistors, diodes and capacitors. This transistors alternately turn on and off to alternately charge/discharge capacitors within the inverter network to convert the DC back into a single phase AC output. The rate at which the transistors turn on/off determine the operating frequency of the power supply. The output of the inverter is then fed to a tuning network, which can be internal or external to the induction heating power supply. The tuning network consists of additional capacitors, inductors and/or transformers. The purpose of the tuning network is to match the output impedance (resistance) of the power supply to the input impedance of the heating inductor and part being heated. Without the tuning network it would be necessary to engineer and build a new power supply for every application, which is not practical and would be very expensive. So, OEM's design power supplies to be compatible with a range of applications based on frequency, current, voltage and power output capabilities. The transformers within the tuning network either step up or step down the voltage from the inverter to match the requirements of the inductor. The inductors and/or capacitors within the tuning network are required to adjust the power factor/phase relationship between the inductor voltage and inductor current for optimal power transfer.

Heating Inductors

Like the power supply we have mentioned the heating inductor numerous times within this paper. We briefly described an inductor previously as an electrical winding in which an AC current is passed for the purpose of generating an electromagnetic field with a specific patter for the purpose of heating an electrically conductive object.

When most people think of an induction heating coil they envision a simple round profile winding made from small diameter copper tubing, like a coil on a whiskey still. Although this would be one form of an inductor, there are many more to consider. Heating inductors can be made from hollow tubing, solid rod, flexible cable, machined billet and even 3D printed. In all cases they are typically made from pure copper. The choice of which form of copper the inductor is made from is mostly determined by the process requirements and sometimes the budget and ability of the user to build their own inductors. Most inductors are water cooled due to the part temperature and the amount of current flowing within the inductor.

Heating inductor design

Inductors can be in many different configurations such as solenoid, hair pin, pancake, transverse, ID, channel, clamshell, and machined profile. Only the solenoid inductor completely surrounds the part. It is a common misconception that the inductor has to completely surround the part. The fact is that an inductor can heat from the outside, the inside, one side, two sides or 3 sides. Inductors are generally profiled to the shape of the part and/or the area of the part to be heated.

Besides the part being heated, the most important part of an induction heating process is the inductor. A poorly designed inductor will create a lot of scrap, use excessive power, have poor repeatability, require post working the parts, or may not work at all. Weather you build your own inductors or purchase them from an induction OEM it is vital that the inductor be properly designed and documented.

Key inductor design attributes

Proper documentation is critical to producing exact spares. Inductors are electromechanical devices. They do experience expansion and contracting from the applied magnetic forces. They experience rapid temperature changes many times per hour. They are exposed to corrosive chemicals such as fluxes, solvents, and other organic compounds. Inductors may experience a lack of water flow and burn up. Inductors may come in contact with a part during heating in which an arc is created that burns the inductor. Automated material handing may become out of adjustment and impact the inductor, deforming it. Inductors can and do fail. Inductors are not a one-time purchase, but should last 10's of thousands of cycles.

The majority of inductors are made by forming round or rectangular tubing around a form/jig using heat. Often the tubing is packed with a removeable filler such as steel shot to prevent the tube from collapsing during the forming process. The process of forming a quality inductor is very much an art, very similar to that of a skilled blacksmith. To watch a skilled coil builder perform his craft is a real treat! Although tempting, I will not divulge into a full discussion of heating inductor fabrication. Just know the induction heating process is only as good as the inductor being used for the process. It is not uncommon for inductors to cost between $5,000 and $10,000 for one unit.

Key design attributes of an inductor are as follows:

- Allows for easy insertion and removal of heated parts

- Is compatible with the tooling holding the heated parts
- Has adequate copper size to accommodate the kW rating of the induction heater
- Provides uniform heating of the braze joint area
- Retains its shape and position during use
- Properly insulated to protect from contamination, and accidental contact with the parts being heating
- Low inductance leads for efficient power transfer
- Properly documented for duplication

Water cooling systems

Most heating inductor and induction heating power systems are water cooled due to the power density, high electrical current values and extreme temperature changes experienced by the tooling. The water is typically provided in a controlled manner by a closed loop water cooling and pumping system.

These water systems come in many formats from simple fan cooled radiators, to water cooled heat exchangers and refrigerated chillers. In all cases there is typically a closed loop reservoir that contains a clean coolant that meets the induction OEM's coolant specs based on their unique system design. It is important that an induction heating OEM's water specifications be followed very closely to maximize the life of the equipment. Poor water quality is the number one cause of maintenance for all induction heating systems. Water recirculating systems can be outfitted with a host of sensors to automatically monitor the quality and condition of the cooling system. Inlet and outlet temperature, flow rate, water conductivity, water levels, etc... can all be controlled with modern day controls.

Controls

Modern day controls can be integrated throughout the induction brazing process to accurately and consistently control the brazing process. Modern induction heating systems employ numerous microprocessors and are able to provide reams of data regarding about the performance characteristics of a process. Filler metal volume measurements, quench flows, min and max temperatures, cooling rates, energy delivery, etc... are just a small sample of the data that can be collected. Modern day induction heaters can even detect if a part is out of position or of the wrong composition based on the phase angle relationship of the current and voltage in the induction coil. Further this data is easily monitored from remote locations over corporate networks and the internet. This level of communication and data acquisition capability on a single part by part basis is very attractive to those with critical brazing processes. Individual process data for individual parts can be stored for later use. It is entirely up to the end user to determine which parameters of their process need monitoring and then take advantage of all the data streams available to collect this data. Individual part processing and verification along with extreme repeatability are just a few of the hallmarks for induction heating for brazing applications.

Finite Element Analysis

Not only is induction heating very controllable and repeatable, it is also very predictable. Through the use of Finite Elemental Analysis (FEA) software and proprietary simulation software packages, induction OEM's can accurate predict heat patterns, temperature rises, temperature distribution, stress, strain and even metallurgical changes in parts before a single part is physically heated. Super computing power is now a reality for larger induction OEM's. This processing power greatly reduces engineering and process development time and the ultimate lead time for new installations by weeks and months. Pending the complexity of your brazing process it may be worth investing in and FEA project to optimize a new process or an existing process to reduce energy use, improve quality and/or performance characteristics.

Summary

Continuous improvements in quality and performance are goals for any process engineer. Both brazing and induction heating can be used accomplish these goals weather be using brazing over welding to reduce the weight of an assembly or the use of induction heating to speed up a brazing process. This paper has presented a wide range of information regarding these two industrial processes that are often used together. A person can obviously select any one of the many concepts regarding these two processes, as discussed in this paper, and discover a significant amount of additional information to be study. It is my hope that from this paper, that the reader is less intimidated about the use of induction heating and will give this wonderful heating technology serious consideration for a brazing or other joining process. Once the basics are understood, users can rely upon various industry experts within the induction heating community to help solve any unique challenges they may be having. Induction heating experts love to be heard and are more than willing to assist those who want to know.

The Mechanical Performance of Sn-Pb Solder Joints on LTCC Substrates

P. Vianco, S. Williams, A. Kilgo, and B. McKenzie
Sandia National Laboratories, Albuquerque, NM USA
ptvianc@sandia.gov; 505-844-3429

M. Walsh, W. Price, and E. Guerrero (ret.)
Nuclear Security Complex, Kansas City, MO USA

Abstract

The assembly of ceramic components often use soldering technologies to attach metal structures to the ceramic base material. Because many suitable solder alloys do not readily wet-and-spread on ceramics, a metallization layer is deposited on the latter to support wetting and spreading by the molten solder for completion of the joint. The metallization layer must be sufficiently robust to retain its integrity though the soldering process as well as not negatively impact the long-term reliability of the joint. A study was performed to evaluate the mechanical properties of solder joints made to a 0.200Ti/W-4.0Cu-2.0Pt-0.375Au (μm) thin film metallization deposited on low-temperature co-fired ceramic (LTCC) base materials. The solder joints were made with the 63Sn-37Pb solder (wt.%, abbreviated Sn-Pb). A pin pull test was developed to measure the tensile strength of the solder joint as a function of soldering parameters. Failure mode analysis was a critical metric for assessing the roles of interfaces, bulk solder, and the ceramic on mechanical performance. The Sn-Pb solder joints experienced a nominal strength loss with increased severity of the soldering process parameters. The strength decline was attributed to changes in the solder joint microstructure and not a direct degradation to the thin film structures.

Key words: Thin film conductor, low-temperature co-fired ceramic, pin pull strength, soldering process conditions.

Introduction

Ceramic substrates have traditionally been specified for printed circuit boards (PCBs) that must meet the performance requirements of radio-frequency (RF) electronics or to minimize the thermal mechanical fatigue (TMF) degradation of solder joints belonging to large ceramic components (e.g., capacitors, filters, etc.). Today, these PCBs are also being targeted for use in high-temperature electronics that provide in-situ monitoring of oil, gas, and geothermal wells (Fig. 1). Down-hole temperatures are typically 150 – 200°C, but can peak at 300°C. Similar conditions exist for the very high reliability electronics that are aboard deep-space probes exploring the planet, Venus, or the Sun [1 – 3].

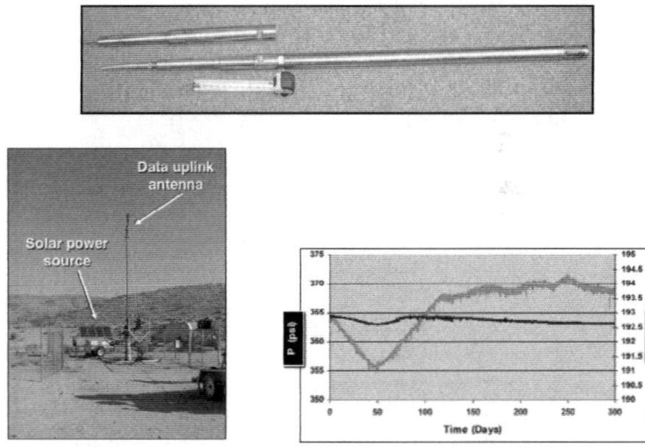

Figure 1 Photographs show the Navy/Sandia Coso down-hole tool and test site. The graph shows the extreme pressure and temperature conditions under which the tool operated without failure for nearly one year.

Electronics based on ceramic PCBs are referred to as *hybrid microcircuits (HMCs)*. Traditionally, HMCs used an alumina ceramic substrate, an example of which, is shown in Fig.2a. Traces and bond pads were created from pastes that were a combination of metal particles, typically Ag, Au-Pd, or Au-Pt-Pd alloy; a glass frit; and organic binders. The circuit pattern was formed by printing the paste through a screen or stencil sheet that duplicated that pattern. Once the paste was printed on the ceramic surface, it was dried to drive off the organic binder and then fired at 850 – 900°C for 15 – 30 min in air. The firing step sintered the metal particles together to form an electrically conductive layer; it also caused the glass frit to melt and diffuse to the metal layer/ceramic surface where it bonded the two materials together. Lastly, the ceramic substrate was assembled with components using the same solder processes as are used with traditional laminate PCBs, creating the HMC assembly shown in Fig. 2b.

Alumina substrates are generally limited to having circuitry on one or both surfaces because of the difficulty of creating internal conductor layers. Signal transmission between the two sides is achieved by edge clips or holes that are mechanically drill through the substrate. If the substrate is sufficiently thin (< 0.5 mm), the holes can be formed by laser drilling. The hole or *via* walls were covered with the same thick film conductor and fired to create a continuous electrical path. These "through-ceramic" holes provide adequate reliability for the circuit; but, they can add significant manufacturing costs to the product.

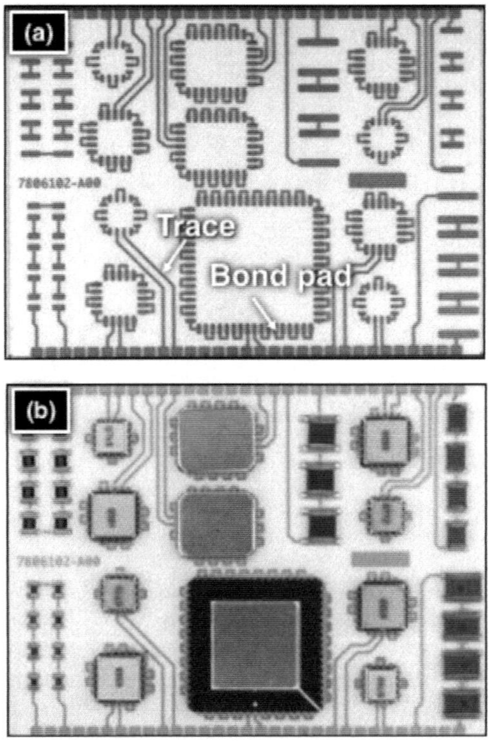

Figure 2 (a) Photograph shows an alumina substrate having traces and pads formed from a Au-Pt-Pd thick film. (b) HMC assembly is shown with the components soldered to the bond pads.

Unfortunately, even the above two-layer HMC technology based on screen printed, thick film conductors, has reached its engineering limits with respect to supporting the further miniaturization and increased functionality of electronic products. The answer to the two-layer limitation was the development of low-temperature, co-fired ceramic (LTCC) substrate technology. The LTCC ceramic attains a multilayer construction similar to laminate PCBs using the stack-up of individual "tape" layers, each having a thick film, circuit pattern printed and dried on it. The layers, which include vias that connect them together, are pressed and fired at elevated temperatures (800 – 850°C) to complete a bond that is both electrically conductive and mechanically strong.

Although LTCC provided an answer to the need for multiple substrate layers, thick film conductor technology has become the bottleneck towards further miniaturization of HMC electronics. The screen printing-drying-firing process has reached its limits with respect to providing minimum line widths and spacings that can support solder processes and provide interconnections that meet long-term reliability requirements. The reliability limitation became particularly apparent in the development of thick film/LTCC PCBs for multi-chip module (MCM) sub-assemblies [4].

An alternative approach is to replace thick film conductors with thin film layers. The thin films are deposited by physical vapor deposition (PVD) techniques – evaporation or sputtering. The glass adhesion layer is replaced with a thin film metallization such as titanium (Ti) or chromium (Cr). On top of the adhesion layer is deposited the *solderable finish*. The solderable finish is the layer to which the solder joint is made. Common solderable finishes include copper (Cu), nickel (Ni), and platinum (Pt). The solderability of the layer is maintained by adding a *protective finish* on top of it. Often-used protective finishes include gold (Au), silver (Ag), and tin (Sn). While the adhesion layer is typically only fractions of a micron thick, the other layers can have thicknesses that vary over several microns, depending upon the soldering process, signal transmission properties, and long-term reliability requirements.

The particular advantage of thin film conductors is that the corresponding circuit pattern can be made with increasingly finer features to support further miniaturization. Circuit definition can be achieved by one of three methodologies: (a) shadow masks, (b) photoimaging as is similarly used in the microelectronics industry, and (c) laser ablation. The use of shadow masks provides a rapid means to define the metallization pattern. However, its drawback is a limitation on feature definition due to the combination of minimum mask feature dimensions and proximity of the mask to the substrate surface.

The photoimaging process can realize considerably finer lines and spaces. Ultraviolet light passes through a glass mask to define the pattern on a photosensitive film or *resist* covering the thin film metallization. The unexposed film is removed by chemicals; the exposed metallization is removed by plasma etching, leaving behind the desired circuit pattern under the exposed photosensitive layer. The latter layer is removed by a different chemical removal step. Drawbacks include capital equipment costs; time duration to plasma etch away the thin film, which increases with the latter's thickness. Also, prolonged etching times can expose the substrate surface to temperatures of 150 – 200°C.

The third technique is laser ablation [5]. The particularly advantage comes with thicker film layers because it reduces both process time and temperature rise at the substrate while still providing feature resolution equivalent to that of photoimaging techniques. Besides capital equipment costs, its other drawbacks include metal and substrate spatter as well as surface micro-cracks and general roughness that may impact high-frequency, RF signal performance. See Fig. 3 for an example of a laser ablated surface.

Hybrid microcircuit products have been developed at Sandia National Laboratories using a thin film-on-LTCC technology. The metallization stack was as follows: 0.200 μm Ti adhesion layer, 4.0 μm copper (Cu) signal layer, 2.0 μm Pt solderable finish, and 0.250 μm Au protective finish. The relatively thick metallization required plasma etching as the photo definition process. Titanium was selected as the adhesion layer based upon its success on alumina ceramics. An extensive study was conducted that is similar to that reported below, which examined the adhesion strength of the Ti-Cu-Pt-Au system when soldered with the 63Sn-37Pb solder (wt.%, abbreviated Sn-Pb) and 95.5Sn-3.9Ag-0.6Cu (SAC396) Pb-free solder [6]. The thin films withstood exposure to molten solder as well as long-term, solid-state aging at temperatures as high as 170°C and time periods as long as 200 days.

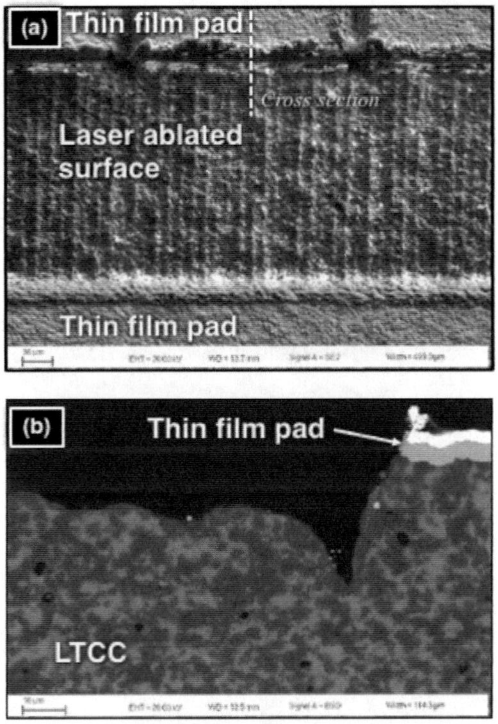

Figure 3 (a) SEM photograph shows laser ablated, LTCC surface used to define the thin film pads. (b) SEM image shows the cross section between the ablated (left) and non-ablated (right) region; the latter is under the thin film.

Nevertheless, the microanalysis of metallographic cross sections uncovered localized separations between the Ti layer and the LTCC, specifically, at the Ti/SiO_2 interface. This anomaly was sensitive to the conditions of the LTCC surface, which were difficult to control as part of the overall process. As a result of these findings, a task was initiated to explore alternative adhesion layers having improved bonding to the LTCC surface. Candidate layers included chromium (Cr), zirconium (Zr), silicon (Si), and the titanium/tungsten (Ti/W) alloy. Chromium, Zr, and Si can be deposited by evaporation. The deposition of Ti/W layer required establishing a new radio-frequency (RF) sputtering capability. This report describes the results of a study, the goal of which was to assess the processibility and long-term reliability of the 0.200 Ti/W – 4.0 Cu – 2.0Pt – 0.375Au metallization for Sn-Pb solder joints made to LTCC.

A second degradation scenario is the effect of *solid-state aging* on the joint strength. The results from those tests and microanalyses will not be presented in this report in the interest of brevity. Those findings will be provided in a follow-on text.

Experimental Procedures

Materials
The thin film stack used in this study was as follows:

- 0.200Ti/W – 4.000 Cu – 2.000 Pt – 0.375 Au (μm)

The Ti/W film served as the adhesion layer. The Ti/W ratio was determined to be $Ti_{1.5}/W_{3.5}$ using spectral analysis techniques, the details of which, are provided in the results section. The Cu layer supported electrical signal conduction. The Pt layer served as the solderable finish. The 0.375 μm thick Au layer was the protective finish. The Ti/W film was deposited by RF sputtering. The Cu and Pt layers were also deposited by RF sputtering in the same chamber without a break in the vacuum condition. The Au layer was evaporated on the Pt layer in a second chamber. The limited oxidation of the Pt layer allowed the test pieces to be exposed to the air environment prior to the Au deposition. Initial attempts were made to deposit the Ti/W film by RF sputtering; break the vacuum; and move the test pieces into the evaporation system to deposit the Cu, Pt, and Au layers. However, the Cu layer exhibited very poor adhesion to the Ti/W film using this process sequence.

The LTCC substrates were fabricated using Dupont™ 951 tape (Dupont is a registered trademark of the E.I. Dupont de Nemours Corp., Wilmington, DE). The test vehicle and pin used to perform the pull test are shown in Fig. 4. The post-fired substrate thickness was 0.914 mm. The test specimen had an x-y footprint of 25.4 mm x 25.4 mm. The substrates were not subjected to additional surface preparation steps (e.g., polishing) other than those routinely performed at the manufacturer. This surface condition is typically the starting point from which thin film circuits were created with the LTCC substrate. A pattern of nine test sites was used to assess thin film adhesion. The pattern was created by sputtering the Ti/W layer through an aperture mask followed by the Cu and Pt then lastly, evaporation of the Au layer through the same mask.

Copper pins were soldered to each pad using the 63Sn-37Pb solder (wt.%, abbreviated Sn-Pb). A specially designed fixture held the LTCC substrate and pins for the soldering process (Fig. 5). Solder paste was dispensed onto each pad and the substrate was placed onto the fixture base plate (Fig. 5a). The Cu pins were fed through a frame that held them in an upright position. The frame was attached to the base plate after which, the pins were lowered onto the solder paste deposit (Fig. 5b). The entire assembly was placed into the furnace to complete the soldering process. Once the fixture had cooled to room temperature, the test specimen – substrate-plus-soldered pins – was cleaned of flux residues and inspected for solder joint quality (Fig. 5c).

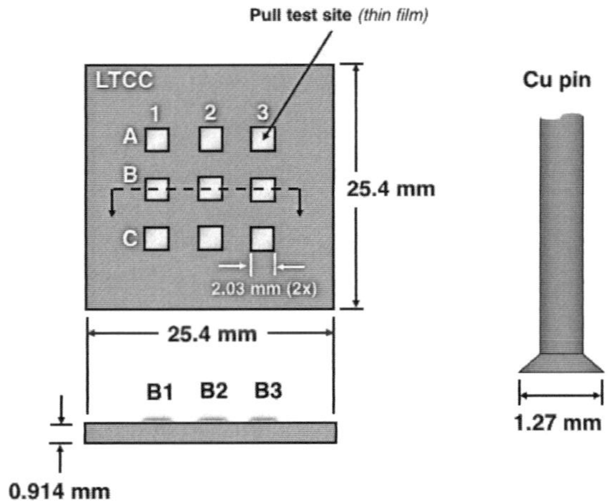

Fig. 4 LTCC pull test specimen and the Cu pin

This study evaluated the solder joint strength as a function of the *soldering process*. Potential degradation mechanisms include (a) incomplete dissolution of the Au layer; (b) excessive leaching of the Pt layer that exposed the Cu layer to dissolution by molten solder; and (c) excessive intermetallic compound (IMC) layer development between the solder and Pt thin film layer. In addition, exposure of the Ti/W-Cu-Pt-Au to the high temperatures of the soldering process could potentially lead to delamination between individual layers or an adhesion loss between the Ti/W layer and LTCC surface.

The soldering process variables included exposure time to the molten solder and the molten solder temperature. The time durations were:

- 15s, 60s, and 120s.

These values took into account the range of automated soldering processes. The molten solder temperatures, which are listed below, were likewise based upon typical soldering assembly processes:

- 215°C, 240°C, 260°C, and 290°C

Duplicate samples were fabricated and tested for each combination of solder temperature and time duration.

All pull tests were performed in a mechanical test frame. The displacement rate was 10 mm/min. Eight of the nine solder joints were pull tested per sample, leaving one joint for cross section and microanalysis. The pull strength was represented by the mean and the standard deviation of the sixteen (16) data points.

Failure mode analysis was also performed on the test samples. There were four predominant failure modes were:

- *Solder failure:* crack path that remains in the solder and/or associated IMC layers;
- *Thin film delamination*: a separation of the Ti/W layer from the LTCC surface.
- *Thin film peeling*: separation along the interface(s) between layers within the thin film stack.
- *Divot failure*: loss of bulk LTCC from under the joint area

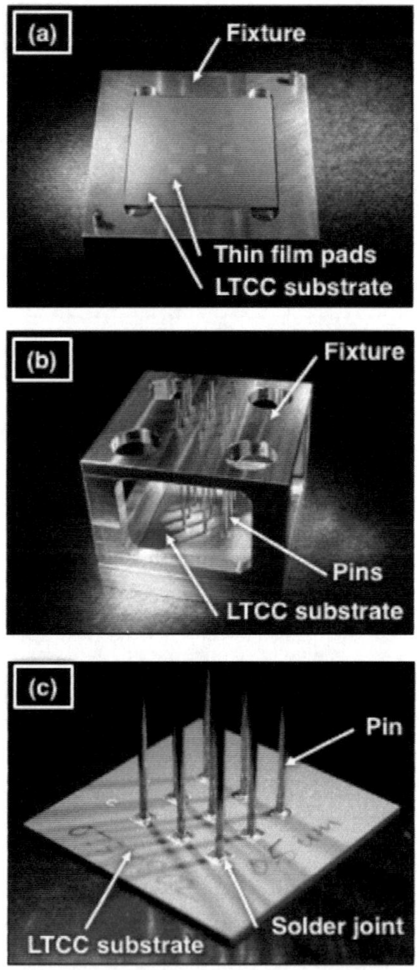

Fig. 5 (a) LTCC substrate is located into the base plate of the fixture. (b) Fixture is assembled with the LTCC substrate; solder paste; and pins in place. The entire assembly would be placed into the furnace to make the joints. (c) Photograph of the final test specimen.

Scanning electron microscope images will be presented that illustrate these failure modes in the discussion, below.

The failure mode analysis methodology began with each pull-tested site being assigned one of the four failure modes that predominated the fracture surface. A further discrimination was not made between minor failure modes because, in general, there was one mode that controlled the fracture for each pin pull test. The failure modes were added up and the percentage determined by dividing that count by sixteen (16), which is the total number of test sites.

Results and Discussion

Spectral analysis of the Ti/W-Cu-Pt-Au thin film

The Ti/W-Cu-Pt-Au thin film was evaluated using scanning transmission electron microscopy (STEM). The specimen was prepared by the focused ion-beam (FIB) cutting process using Ga ions. The STEM image is shown in Fig. 6. The Cu and Pt layers developed with largely columnar grain structures. A small demarcation line was observed near the top of the Pt layer. The columnar grain structure appeared to be continuous across the line, which implies that the source of the artifact did not significantly interrupt the deposition process and, as such, did not introduce a material defect in the Pt layer.

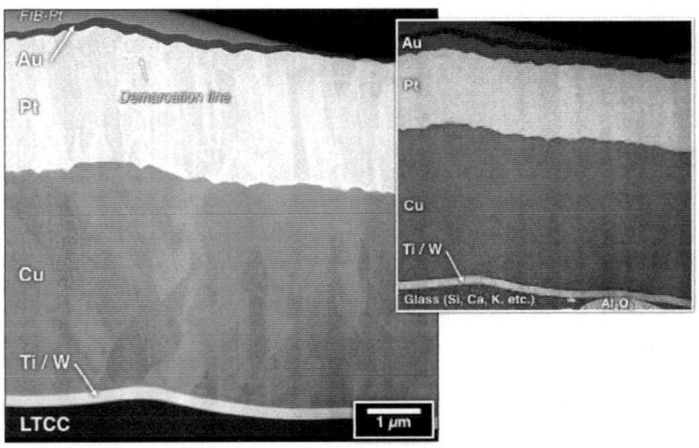

Fig. 6 STEM image shows the as-deposited, Ti/W-Cu-Pt-Au layer. A faint demarcation line is identified in the layer. The inset picture is the annular spectral image that identifies the elements by color, including the different phases in the LTCC.

The inset picture is an annular spectral image that identified the elemental construction of the layer: blue, Au; green, Pt; red, Cu; cyan, Ti/W; magenta, the silica-based glass component of the LTCC; and yellow, the alumina particles also within the LTCC base material. An important observation is that the Ti/W layer exhibits excellent adhesion to the silica-based, glass component of the LTCC base material, which was a shortcoming of the Ti adhesion layer. Besides a potentially improved intrinsic interaction between the Ti/W layer and glass, the higher energy, RF deposition process may also have contributed to an improved adhesion performance by partially sputtering the LTCC surface. Interdiffusion was also absent between the individual layers. Lastly, quantitative data, which were obtained from this analysis, determine the Ti/W layer composition to be $Ti_{1.5}/W_{3.5}$.

A high magnification image is shown in Fig. 7 of the Ti/W layer. A faint demarcation line was also observed, which as was the case with the Pt layer, above, was likely caused by a brief interruption of the deposition process. Although the gray tone implies a slightly lower material density, discontinuities such as voids and cracks were not associated with the demarcation line. General porosity was limited to less than 5 nm size. The layer had excellent adhesion to the LTCC substrate. Excellent adhesion was also confirmed between the Cu and Ti/W layers.

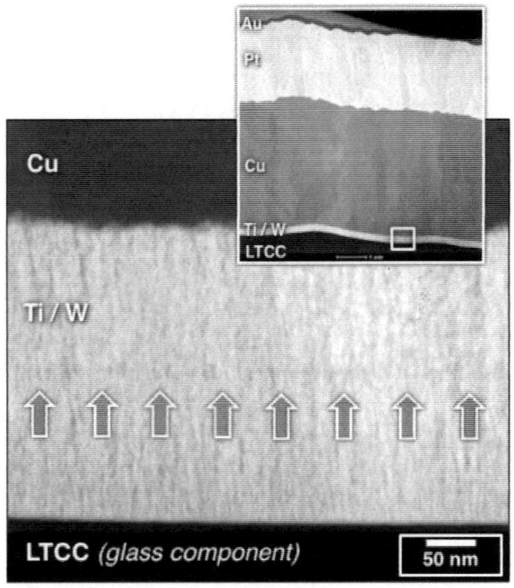

Fig. 7 High magnification, STEM image shows Ti/W-Cu-Pt-Au layer at the location of the film. A faint demarcation line is identified in the Ti/W layer by the magenta arrow.

Effect of soldering process parameters on *pull strength*

The pull strength data are shown in Fig. 8 as a function of the process temperature. The blue circles, magenta squares, and red diamonds represent the 15s, 60s, and 120s data, respectively. The pull strengths are presented in load (lb.). Tensile stresses were calculated at minimum and maximum load values bounding the data. The very high pull strengths reflect the plain strain effect resulting from the confined geometry created by the thin gap and the wide footprint of the joint. The strength enhancement is demonstrated, by comparing the data in Fig. 8 with the tensile strength of bulk Sn-Pb solder, which is 5.8 – 7.3 ksi (40 – 50 MPa) at a similar displacement rate [7]. A minor, yet statistically significant difference, was observed between the 15 s interval versus the 60s and 120s soldering times, but for only the 240°C and 260°C soldering temperatures. The short error bars indicate the excellent reproducibility of the strength values.

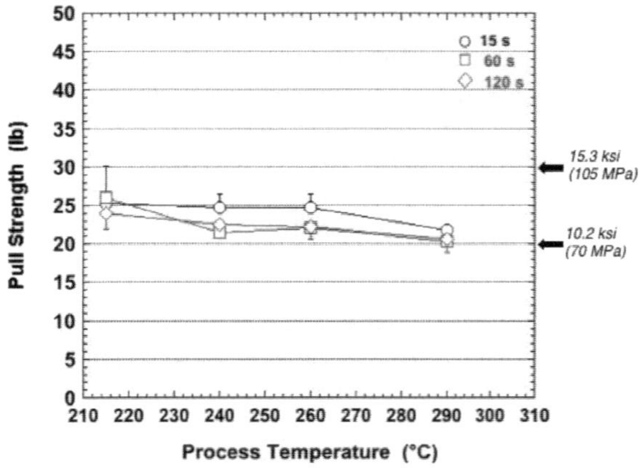

Fig. 8 Pull strength is shown as a function of process temperature for each of the three process time intervals. The tensile stresses were calculated for the 20 lb. and 30 lb. loads to illustrate these limiting cases – those values are shown at the right of the graph.

Effect of soldering process parameters on *failure mode*

The failure mode results are presented in Fig. 9 for the test vehicles assembled by soldering for 15 s. At the soldering temperature of 215°C, 50% of the test sites exhibited failure within the solder joint (blue bar); the remainder of the failures were evenly split (25%) between thin film delamination at the Ti/W-LTCC interface (yellow bar) and fracture by the LTCC divot mode (red bar).

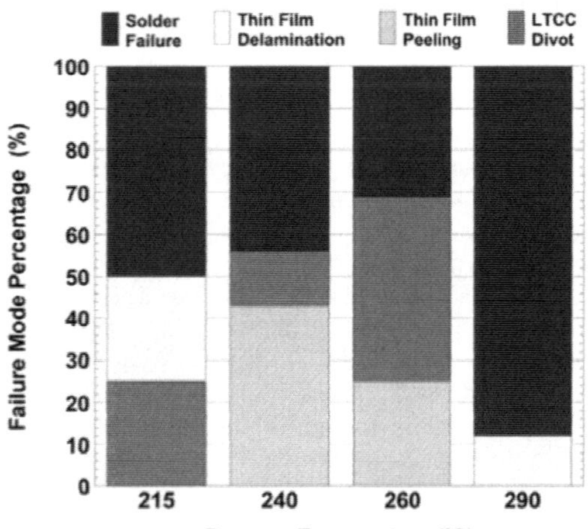

Fig. 9 Bar chart shows the percentage of the observed failure modes as a function of solder temperature for a soldering time of 15 s. Each of the four failure modes is distinguished by the color of the bar per the legend above the plot.

An increase of the soldering temperature to 240°C introduced the thin film peeling mode, which replaced the thin film delamination mode and a portion of the solder failure and LTCC divot modes. The thin film peeling mode remained at the higher soldering temperature of 260°C, but to a lesser degree as the LTCC divot mode increased in frequency. The LTCC divot failures have been attributed to residual stresses caused by the mismatch of coefficients of thermal expansion (CTE) between the Cu pin, Sn-Pb solder, and LTCC [6]. This point will be further discussed later in the report. Lastly, when the soldering temperature was raised to 290°C, ninety-percent (90%) of the test sites showed solder failures; the remaining failures were thin film delamination at the Ti/W-LTCC interface.

The failure mode data in Fig. 9 were compared to the pull strength data in Fig. 8. The pull strengths remained essentially unchanged across the 215°C, 240°C, and 260°C soldering temperatures. Yet, a significant variation occurred to the failure modes. This trend implies that, by-and-large, the intrinsic strengths were similar between the microstructural features associated with each failure mode.

When the soldering temperature was raised to 290°C, ninety-percent (90%) of the test sites showed the solder failure mode; the remaining failures were thin film delamination of the Ti/W-LTCC interface. The increased presence of the solder failure mode was accompanied by a small decrease of pull strength as shown in Fig. 8. This trend implies that, in the event that there is a strength loss by the solder joints, the likely culprit can be found in the solder joint microstructure and not the adhesion properties accompanying the thin film layers.

At this point, an analysis is made of the physical metallurgies that characterized each of the failure modes in Fig. 9. The solder failure (blue bar) is illustrated by the scanning electron microscope (SEM) images in Fig. 10. Figures 10a and 10b show the fracture surfaces on the LTCC pad and pin sides of the solder joint, respectively. This joint was made using the process conditions of 215°C and 15 s. Three distinctive regions were identified on the fracture surfaces by the letters: "A,", "B", and "C." Their physical metallurgies were determined with the assistance of the energy dispersive x-ray (EDX) technique.

Region A had a relatively smooth topography in Fig. 10. That topography resulted from intergranular fracture as shown by the higher magnification SEM image in Fig. 11. The accompanying EDX elemental maps indicate that the fracture surface was comprised primarily of Au, Cu, and Sn. Although intermittent areas showed only the Sn signal (white circles), by-and-large, the fracture surface showed all three elements to overlap with one-another (cyan circles). The same surface topography and elemental distributions were repeated on the pin side of the fracture (region A, Fig. 10b).

The failure mode morphology of region A resulted from fracture *through* the (Cu, Au)$_x$Sn$_y$ intermetallic compound (IMC) layer that formed at the solder/Cu pin interface. The Au layer, which topped the thin film stack on the LTCC substrate, dissolved into the molten Sn-Pb solder. While the solder was molten, as well as following solidification, a driving force developed for Au to diffuse to the solder/Cu interface where it was incorporated into the IMC layer. Note that the contribution by solid-state diffusion had taken place *at room temperature*.

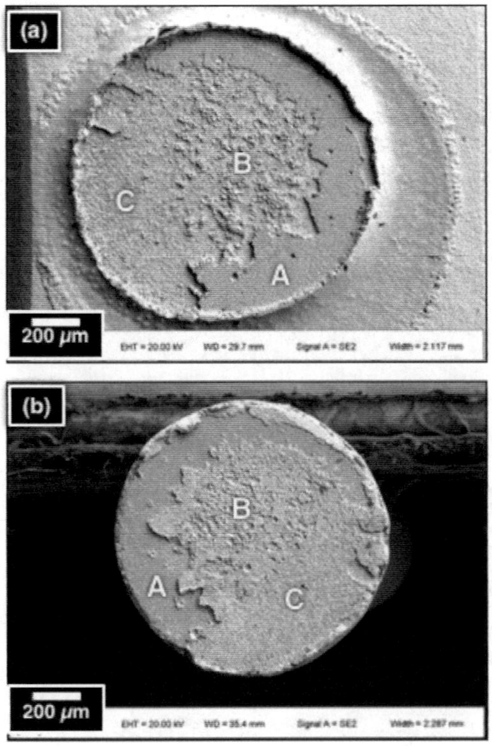

Fig. 10 SEM photographs show the solder failure mode, fracture surfaces belonging to (a) the thin film pad on the LTCC substrate and (b) the corresponding pin. The letters "A," "B," and "C" denote the three surface topographies. The soldering process parameters were 215°C and 15 s

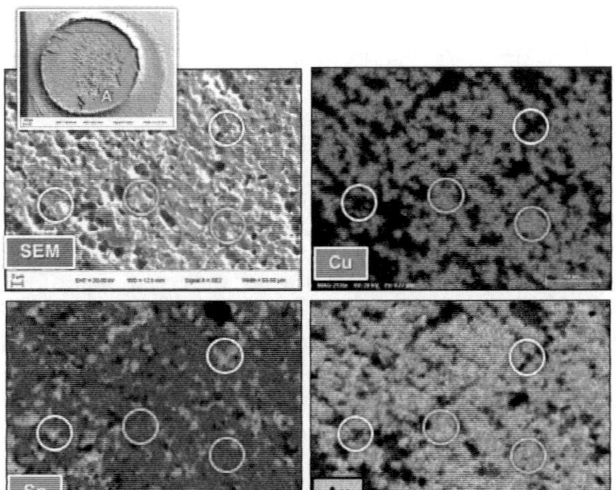

Fig. 11 High magnification, SEM image of the region A (see the inset picture) is shown, together with the EDX maps of Cu, Sn, and Au that predominated the fracture surface (process conditions: 215°C and 15 s). These data were obtained from the pad (LTCC) side of the failure.

The regions B and C were very similar. Their relative heights versus that of region A in Fig. 10 indicated that the fracture surface was closer to the solder/thin film interface. The primary difference between regions B and C was that region B included a greater percentage of failure within the ductile Sn-Pb solder that generated the cup-and-cone fracture surface relief. Therefore, further discussion of this fracture behavior will focus on region C.

A high magnification, SEM image is shown in Fig. 12 of the region C. The EDX maps show the fracture surface to be comprised predominantly of Pt and to a lesser extent, Sn. Lead (Pb) was also observed, which reflected an intermittent fracture in the adjacent, Sn-Pb solder. At first glance, the data suggest that fracture occurred along a Pt_xSn_y IMC layer that formed at the solder/Pt interface after the Au had been dissolved away. However, the yellow circles show that there is *minimal coincidence* between the Pt and Sn signals, which implies that the fracture path did not have a significant association with a Pt_xSn_y IMC layer.

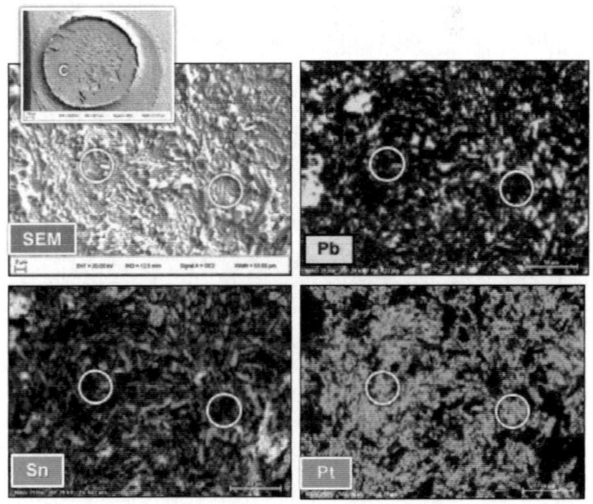

Fig. 12 High magnification, SEM image of the region C (see the inset picture) is shown, together with the EDX maps of Sn, Pt, and Pb on the fracture surface (process conditions: 215°C and 15 s). These data were obtained from the *pad (LTCC) side* of the failure.

In order to confirm the above hypothesis, SEM and EDX data were analyzed of the region C on the *pin side* of the fracture. Those results are shown in Fig. 13. *Platinum was not detected on the fracture surface.* However, there was a significant Au presence. The Pb signal (white circles) is due to localized fracture in the adjoining Sn-Pb solder field as noted earlier. The remaining fracture surface was comprised primarily of overlapping Sn and Au signals. These findings, when coupled with those of the fracture surface on the pad side (Fig. 12) indicate that the Au was not fully dissolved from atop the Pt layer by the molten solder. Rather, some Au remained and reacted with Sn to form a Au_uSn_v IMC layer. The C-region failure mode originated from fracture along the interface between the Au_uSn_v IMC layer and the Pt layer. Given that region C comprises a significant fraction of the solder failure surface, Au_uSn_v IMC/Pt interface fracture does not necessarily indicate that the latter poses an intrinsic weakness to the joint.

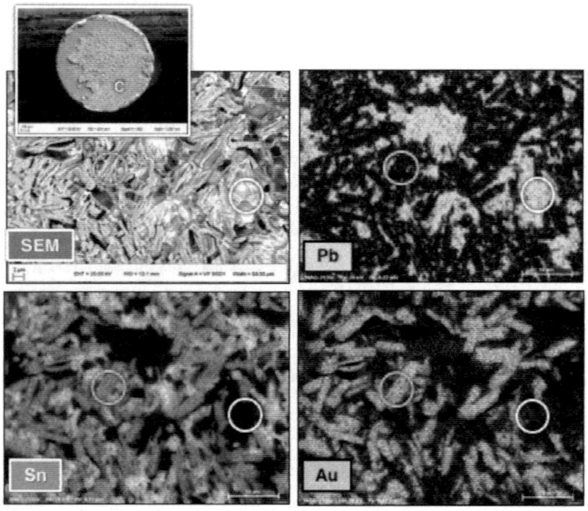

Fig. 13 High magnification, SEM image shows the region C (see the inset picture) together with the EDX maps of Sn, Au, and Pb (process conditions: 215°C and 15 s). These data were obtained from the *pin side* of the failure.

The divot failure mode is illustrated by the complementary SEM images in Fig. 14. Failure initiation occurred in the bulk LTCC material, which is indicated by the letter A in both images. As the LTCC fracture expanded from the pad center (Fig. 14a), it transitioned into a the "fractal" pattern (B) that is the combination of thin film delamination and thin film peeling. As the crack approached the outer edge of the joint, a second transition took place whereby separation was entirely thin film delamination (C). This behavior was documented as an "LTCC divot" failure mode because it predominates the LTCC fracture. Also, given the rapid nature of the LTCC fracture, the remaining fracture surface may not reflect the intrinsic failure mode of the joint in the absence of the LTCC fracture.

The thin film peeling failure mode is illustrated by the SEM images in Fig. 15 and, specifically, location A. This sample was fabricated at 240°C and 15 s. The LTCC pad and pin sides of the fracture are shown in (a) and (b), respectively. A small divot (B) and associated "fractal pattern" (C) were observed in both images. However, the latter fracture paths were secondary to the thin film peeling mode.

The fracture surface appeared to be relatively featureless at the magnifications used in Fig. 15. A high magnification, SEM image is shown of the fracture surface in Fig. 16 (LTCC side). The corresponding EDX maps, together with those from the pin side of the fracture, confirmed that failure took place at the Ti/W – Cu interface. Residual Cu was not observed on the exposed Ti/W film to within the detection limits of the EDX technique.

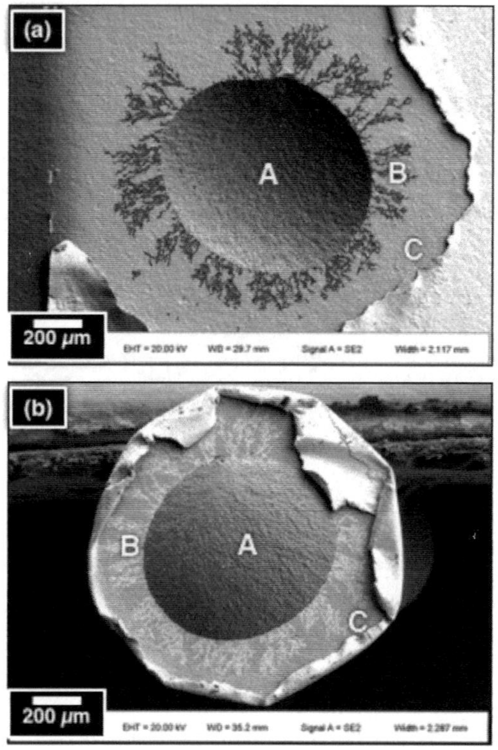

Fig. 14 SEM photographs show the complementary fracture surfaces of (a) the thin film pad on the LTCC substrate and (b) the pin that together illustrate the "LTCC divot" failure mode. The A region is the divot, itself; the B region is a fractal pattern comprised of a mixture of both the "thin film peeling" and "thin film delamination" failure modes The C region is solely the thin film peeling mode. The soldering process parameters were 215°C and 15 s.

At the higher magnification level of Fig. 16, the fracture surface showed significant relief. The Ti/W layer is only 0.200 µm thick so that it replicates the surface topography of the LTCC substrate. The limited thickness of the Ti/W layer also permitted the x-ray signal to sample the chemistry of the LTCC underneath it. The protrusions of the film correspond to the alumina phase of the LTCC (Al map). The glass phase is represented by the Ca signal because the Si signal, which would be generated by the primary constituent of the glass material, silica (SiO_2), is overlapped by the W line of the Ti/W film.

The failure mode percentage data appear in Fig. 17 that were obtained from test sites soldered for 60s. The thin film peeling appeared at the soldering temperature of 215°C; it was no longer observed at 260°C as recorded at 15 s (Fig. 9). The failure mode percentages were nearly identical at 240°C. At the soldering temperatures of 260°C and 290°C, there was a significant increase of the solder failure mode at the expense of the other three modes. This trend implies that, with increasing soldering temperature, the interfaces associated with the thin film structures and the bulk LTCC (divots) had lesser roles in the fracture behavior while the solder joint microstructural features – bulk solder and IMC reaction layers – had greater effects on mechanical performance.

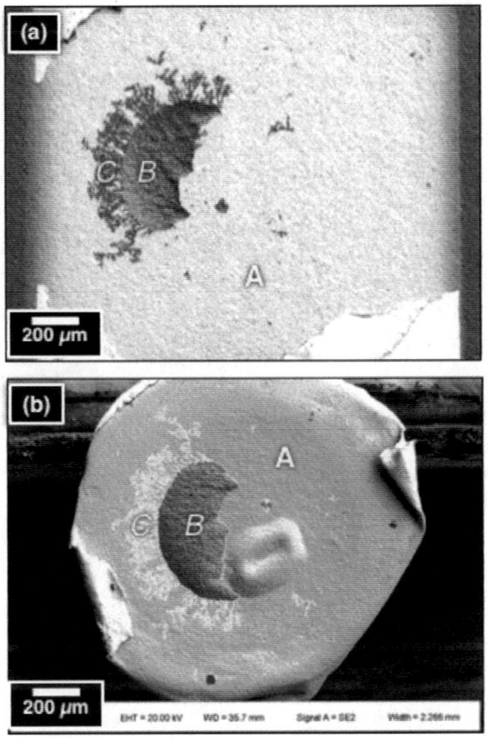

Fig. 15 SEM photographs show the complementary fracture surfaces of the (a) thin film pad on the LTCC substrate and (b) the pin showing the "thin film peeling" failure mode. The A region identifies the film peeling mode. The secondary fracture paths were the LTCC divot (B) and the fractal patterns of thin film delamination and thin film peeling (C). The soldering process parameters were 240°C and 15 s.

The details of the fracture surface morphologies associated with the failure modes in Fig. 17, were generally similar to those discussed in association with Figs. 10 – 16.

The failure mode data in Fig. 17 (60 s) were compared to the pull strength results in Fig. 8. The pull strength decreased slightly between 215°C and 240°C, which happened to coincide with the increase of the thin film peeling mode from 12% to 50%. However, nearly the same failure mode percentages were observed at 240°C, 15 s (Fig. 9), but without the lower pull strength associated with 240°C and 60 s data (Fig. 17). Thus, a conclusive correlation cannot be drawn between low solder joint strength and the thin film peeling mode. A similar conclusion can be made with respect to the thin film delamination failure mode. At the soldering temperature of 215°C, increasing the soldering time from 15 s to 60 s caused this failure mode to increase from 25% to 38%. Yet, the pull strength did not change, significantly, between those same conditions. In summary, the analysis of failure mode data, which were obtained for 15 s (Fig. 9) and 60 s (Fig. 17), when compared to the pull strength results (Fig. 8), indicated that a direct correlation cannot be made between the failure mode and strength metrics for the soldering temperatures, 215°C, 240°C, and 260°C soldering temperatures. This trend implies, once again, that the particular fracture paths associated with each failure mode have similar intrinsic strengths.

Fig. 16 High magnification, SEM image of the region A (see the inset picture) is shown, together with the EDX maps of W, Al, and Ca (process conditions: 240°C and 15 s). These data were obtained from the LTCC pad side of the failure.

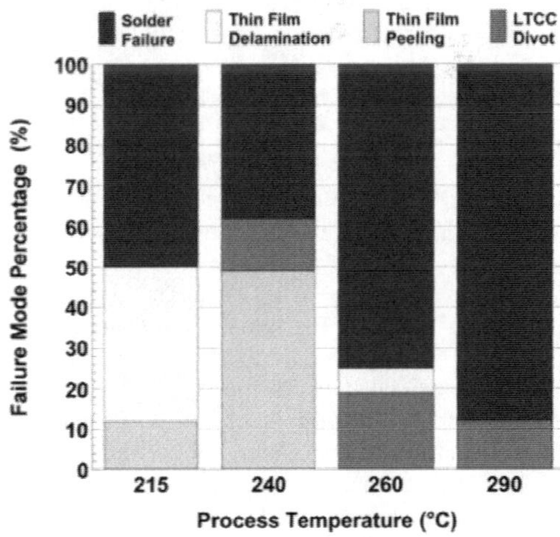

Fig. 17 Bar chart shows the percentage of the observed failure modes as a function of solder temperature for a soldering time of 60 s.

However, the failure mode data in Fig. 17 (60 s) for the 290°C soldering temperature, like that in Fig. 9 (15 s), showed a reduce pull strength and 88% occurrence of the solder failure mode. This comparison indicates that changes to the Sn-Pb solder and associated IMC reaction layers, which control the solder joint failure mode, have the ability to reduce the pull strength of the joint. Moreover, those features are expressly a function of the soldering conditions.

The details of the fracture surface morphologies associated with the failure modes in Fig. 17, were generally similar to those described with respect to Figs. 10 – 16.

The failure mode percentages are shown in Fig. 18 for the longest soldering time of 120 s. The thin film peeling mode was observed at the soldering temperatures of 215°C and 240°C; it increased at the former temperature and decreased at the latter temperature when compared to the 60 s solder time results in Fig. 17. Moreover, the solder failure mode increased, significantly, at the two lowest soldering temperatures when compared to either of the two shorter soldering times. At the higher soldering temperatures of 260°C and 290°C, the solder failure mode retained its dominance at 120 s as was the case at 60 s. But, there was also an increased frequency of LTCC divot failures with the longer soldering time.

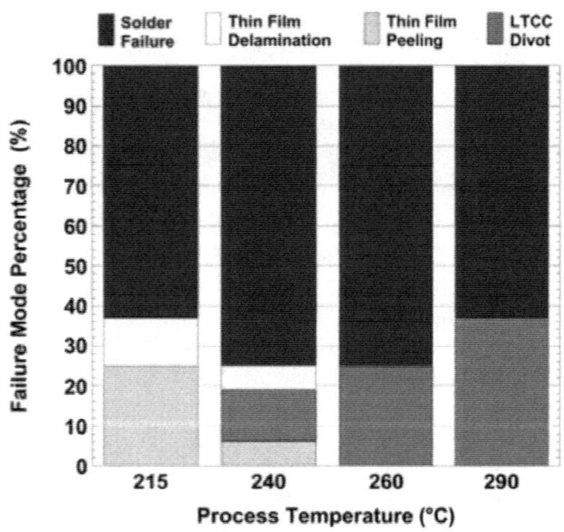

Fig. 18 Bar chart shows the percentage of the observed failure modes as a function of solder temperature for a soldering time of 120 s.

The failure mode trends in Fig. 18 were compared to the pull strength data in Fig. 8. A reduced pull strength was observed at 215°C that coincided with the increased prevalence of the solder failure mode. The pull strength was unchanged at the soldering temperatures of 240°C and 260°C as was also the case for the dominance of the solder failure mode. In addition, the increased frequency of LTCC divots between these temperatures was not reflected by a change to the pull strength. A slight drop of pull strength was observed at 290°C. However, the slight increase of the LTCC divot failure mode (Fig. 18) cannot be confidently asserted as the cause. Rather, the solder failure mode with its dependence on the Sn-Pb microstructure and IMC layers likely caused the drop in the pull strength.

A closer examination was made of the fracture surfaces associated with the four failure modes between soldering process conditions. Even at higher magnifications, the details of the fracture surface morphologies changed very little for the thin film delamination, thin film peeling, and LTCC divot modes as a function of either soldering temperature or time. Therefore, the "appearance" and "disappearance" of these three failure modes versus the soldering parameters are not correlated to intrinsic changes to the thin film layers, their interfaces, or the properties of the bulk LTCC.

On the other hand, the fracture surfaces associated with the solder failure mode did change as a function of the soldering parameters in two manners. First, the proportion of the fracture area represented by region A (Fig. 10) increased with severity of the soldering conditions. Recall that region A results from separation in the IMC layer that formed at the solder/Cu pin interface. This trend is illustrated in Fig. 19, which shows the surfaces on the pad (LTCC) side of the fracture path for the range of soldering conditions: (a) 215°C, 15 s; (b) 215°C, 60 s; and (c) 290°C, 120 s. Note that region A has increased at the expense of the combined regions, B plus C.

Second, the details of the fracture surface topography belonging to region A changed as a function of soldering conditions. This behavior is illustrated in Fig. 20, which shows the region A fracture surface generated by the following soldering parameters: (a) 215°C, 60 s and (b) 290°C, 120 s. The smooth topography in Fig. 20a has changed to a rougher topography in Fig. 20b. The latter surface was attributed to an increased growth of the (Cu, Au)$_x$Sn$_y$ IMC layer caused by the prolonged exposure to molten solder.

A review was made of the EDX data, the goal of which was to determine whether the topographical changes to region A as illustrated in Fig. 20, were likewise accompanied by chemistry changes to the associated (Cu, Au)$_x$Sn$_y$ IMC layer. *A decrease was observed in the Au signal relative to that of Cu with the increased severity of the soldering conditions.* In the absence of cross section analyses, it could only be hypothesized that the higher soldering temperature allowed the Au to reach a higher concentration in the Sn-Pb solder, which reduced its concentration in the (Cu, Au)$_x$Sn$_y$ IMC layer formation. Because the reduction in the Au content of the IMC layer would cause the latter to become *less* brittle, the increased propensity for fracture at this interface (region A) was attribute to the increased thickness of the IMC layer.

In summary, the solder failure mode, the propensity of which increased with the severity of the soldering conditions, was correlated to the reduction in solder joint pull strength. Furthermore, the loss of pull strength was associated with an increased proportion of the fracture surface occurring at the solder/pin interface (region A). The reduce pull strength of the latter interface coincided with a thickening of the (Cu, Au)$_x$Sn$_y$ IMC layer taking place at the region A fracture surface. It is important to reiterate that the strength loss observed in Fig. 8 was relatively small and certainly would not portend a degradation to the performance or long-term reliability of the joint.

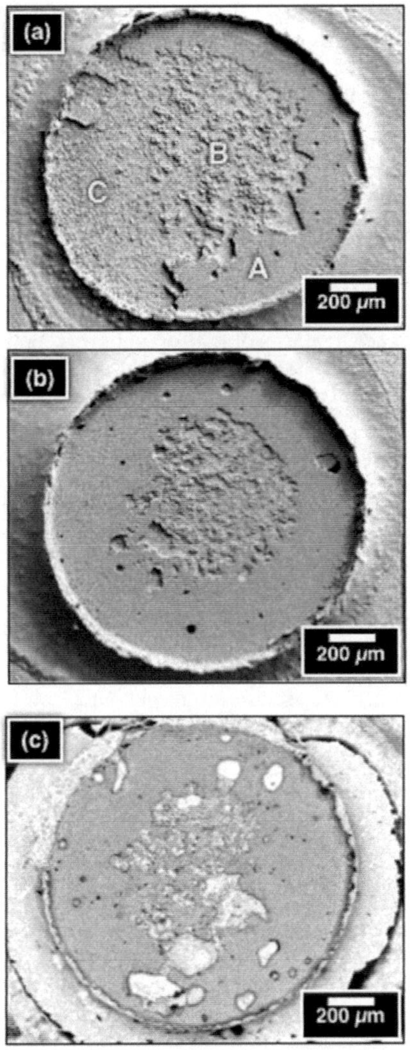

Fig. 19 SEM photographs illustrate the changes to the fracture surfaces and, specifically the proportions of regions A, B, and C as a function of soldering process conditions: (a) 215°C, 15 s; (b) 215°C, 60 s; and (c) 290°C, 120 s.

Program goal: a robust thin film technology

The long-term objective of this program was to develop a robust thin film technology for LTCC-based electronic products. Although the solder failure mode appeared to control the pull strength of the Sn-Pb joints, it required the presence of the Cu pin. A Cu base material may not always be present in an HMC solder joint. Therefore, additional consideration was given to the other three failure modes in order to understand their individual behaviors.

Once goal was to ascertain the ability of the Ti/W adhesion layer to withstand the range of soldering process conditions required to assemble HMC products. In the case of the *thin film delamination mode*, the SEM and EDX analyses showed that the Ti/W film, like its Ti counterpart, exhibited a slightly better bond to the alumina phase (particles) than to the silica-based glass phase of the LTCC material. Nevertheless, these differences of adhesion were subtler than was the case of the Ti layer and, as such, did not significantly impact the solder joint pull strength.

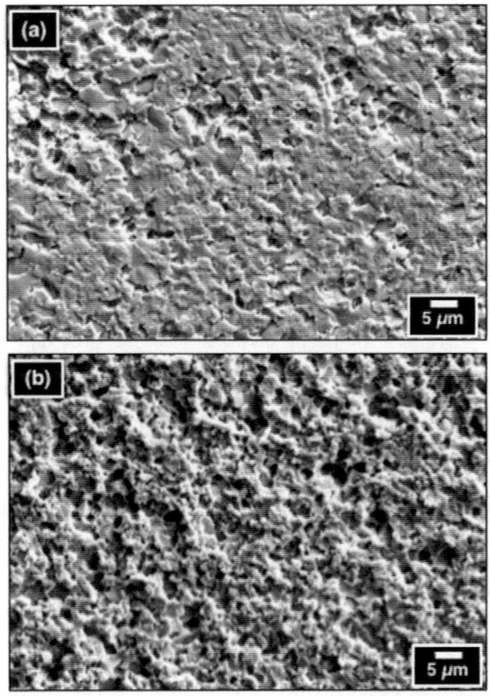

Fig. 20 SEM photographs shows an increased roughness to the fracture surface topography of region A as a function of soldering conditions: (a) 215°C, 60 s and (b) 290°C, 120 s.

The *thin film peeling failure mode* represented a separation between the Ti/W adhesion layer and the Cu layer. Although depositing the Cu layer without breaking vacuum improved its adhesion to the Ti/W surface, this measure reduced, but did not completely eliminate, its occurrence. Interestingly, thin film peeling appeared to diminish with increased severity of the soldering process conditions. This trend suggests that the additional thermal exposure improved adhesion at the interface between the Cu and Ti/W layers. Perhaps an ancillary thermal treatment would increase the bond strength of that interface. Nevertheless, the thin film peeling fracture path was not sufficiently pervasive nor weak enough to pose a risk to the mechanical strength of the solder joint.

Lastly, *the LTCC divot failure mode* occurred persistently in the pull tests, but without an obvious trend as a function of soldering process parameters. Moreover, the occurrence of LTCC divots did *not* correlate with the pull strength trends. Clearly, the strength of the LTCC ceramic was, at least, comparable to the intrinsic strengths of the structures that contributed to the other three failure modes.

Further consideration was given to the LTCC divot failure mode. The tensile strength of LTCC is in the range of 39 – 51 ksi, which exceeds the nominal stresses generated by the pull loads in Fig. 8 by nearly three-fold [8]. The failure mode analysis did not indicate that the test substrates contained a repeatable flaw in the LTCC material. Therefore, a stress concentration must have been generated in the substrate such that, when combined with the stresses of the pull test, caused the total stress in the LTCC to exceed its intrinsic strength.

The hypothesis was mentioned earlier that residual stresses can result from the mismatch of CTE values between LTCC (5 – 8 ppm/°C) and those of either the Sn-Pb solder (21 ppm/°C) or the Cu pin (17 ppm/°C) [7, 9]. Radial *tensile* stresses are generated under the solder joint footprint *after its solidification* as the Sn-Pb alloy and Cu pin contracted to a greater degree than did the LTCC material. A detailed finite element analysis was outside the scope of this study. Nevertheless, it is hypothesized that the pull load placed on the joint superposed additional tensile stresses on the pre-existing residual stresses so that the total tensile stress exceeded the local tensile strength of the LTCC, resulting in the divot failure mode.

The CTE mismatch residual stresses can only develop after solidification of the solder (183°C). Per se, the residual stresses were not sensitive to the specific soldering temperatures because they exceed 183°C because the Sn-Pb alloy would still be molten and could not support a mechanical load. The same trend was observed with the LTCC divot failure mode – that is, its occurrence was largely insensitive to the soldering temperature. This correlation supports the premise that residual stresses, which arose only after the Sn-Pb solder had solidified at 183°C, contributed to the divot formation. Also, the LTCC divot failure mode was not sensitive to soldering time. The latter observation, when coupled to the soldering temperature insensitivity, implies that the strength of the LTCC material was not explicitly degraded by the harshness of the overall soldering conditions so as to be a cause of the divots. Lastly, the residual stresses and LTCC divot failure mode were not of a magnitude or prevalence, respectively, to control the solder joint performance and/or reliability.

Conclusions

1. A study was undertaken to develop a 0.200Ti/W-4.0Cu-2.0Pt-0.375Au (μm, abbreviated Ti/W-Cu-Pt-Au) thin film metallization system to serve as the conductor on low-temperature, co-fired ceramic (LTCC) substrates for high-reliability hybrid microcircuit (HMC) electronic products.

2. A pin pull test was used to assess the tensile strength of solder joints made to the Ti/W-Cu-Pt-Au thin film with the 63Sn-37Pb (wt.%) filler metal. The assembly process, which attached Cu pins to the thin film pads, was based on the matrix of the soldering temperatures of 215°C, 240°C, 260°C, and 290°C and soldering times of 15 s, 60 s, and 120 s.

3. The pull strengths were well in excess of calculated values based on the tensile strength of Sn-Pb solder due to the plain strain effect of the solder joint geometry. The strength values experienced a modest loss from 25 lb. to 21 lb. between 215°C and 290°C. Strength differences were within the experimental error as a function of soldering time at either of these temperatures. At 240°C and 260°C, strengths were slightly higher at 15 s versus 60 s and 120 s, but only just outside the error bars of the respective data sets.

4. Failure modes were dominated by fracture within the solder joint and specifically, at the solder/thin film or solder/Cu pin interfaces. The slight loss of pull strength was correlated to an increased prevalence of failure in the $(Cu, Au)_xSn_y$ intermetallic compound (IMC) layer at the solder/Cu pin interface.

5. The thin film delamination and thin film peeling failure modes had a negligible role because they did not correlate to the trends in the pull strength. The thin film peeling failure mode, which was caused by separation along the interface between the Ti/W and Cu layers, appeared to actually diminish with increased severity of the soldering process.

6. The LTCC divot failure mode was also a minor contributor to the fracture behaviors. Its occurrence shifted slightly to the higher soldering temperatures, but without a consistency that could be correlated to the pull strength data. The divots were attributed to the presence of tensile residual stresses generated by the CTE mismatch of the materials after solidification of the Sn-Pb solder joint.

Acknowledgments

The authors wish to thank Brian Wroblewski for his thorough review of the manuscript and Paul Kotula for the spectral analysis. Sandia National Laboratories is a multi-mission laboratory managed and operated by National Technology and Engineering Solutions of Sandia LLC, a wholly owned subsidiary of Honeywell International Inc. for the U.S. Department of Energy's National Nuclear Security Administration under contract DE-NA0003525.

References

1. Amalu, E., Ekere, N., and Bhatti, R. "High Temperature Electronics: R&D Challenges and Trends in Materials, Packaging, and Interconnection Technology," *Proc. Sec. Inter. Conf. on Adaptive Science and Tech.*, IEEE (New York, NY, 2009) 147 – 154.
2. Nowak, D. and Dziedzic, A. "LTCC Package for High Temperature Applications" *Microelect. Rel.*, Vol. 51 (2011) pp. 1241 – 1244.
3. Del Castillo, L., Schatzel, D., Tudryn, C., Hatake, T., Chen, Y., Jojharradi, M., and Kolawa, E. "Extreme Environment Electronic Packaging for Venus and Mars Landed Missions, *Proc. Inter. Planetary Probe Workshop 4,* NASA, Pasadena, CA, June 24 – 29, 2006.

4. Vianco, P., et al., "An Analysis of the Pull Strength Behaviors of Fine-Pitch, Flip Chip Solder Interconnections Using a Au-Pt-Pd Thick Film Conductor on Low-Temperature, Co-Fired Ceramic (LTCC) Substrates," *Sandia Report* SAND2008-5736 (Sandia National Laboratories, Albuquerque, NM; 2008).
5. Girardi, M., Peterson, K., Vianco, P., and Wieliczka, D. "Laser Ablation of Thin Films on Low-Temperature Co-fired Ceramic," *J. Microelect. and Elect. Pack.* Vol. 12 (2015) pp. 72 – 79.
6. Vianco, P., et al., "Establishing a Ti-Cu-Pt-Au Thin Film-on-Low Temperature Co-Fired Ceramic (LTCC) Technology for High Temperature Electronics," *Proc. Surf. Mount Tech. Assoc. Inter. Conference*, Surface Mount Technology Association (Edina, MN; 2015), www.smta.org.
7. Vianco, P., Soldering Handbook, 3rd Ed. AWS (Miami, FL; 2000) p. 197.
8. Bermejo, R., "Mechanical Properties of Low Temperature Co-fired Ceramics: Testing Methodologies for Strength Characterization," *J. of Microelec, Elec. Comp., and Mater.*, Vol. 42 (2012) pp. 254 – 259.
9. Sakamoto, S., Fujita, S., Sugimoto, Y., and Tanaka, N., "Mechanical Strength of Low-Temperature Co-Fired Ceramic Multi-Layered Substrate," J. Ceramic Soc., of Japan. Vol. 125 (2017) pp. 569 – 573.

Achievement in soldering of thermo-electric cell with zinc-based alloy

Igor N. Pashkov, Evgeny O. Shargayev,
PBSU, Moscow, Russia
paikaspp@gmail.com

Tatyana A. Bazlova, Viacheslav E. Bazhenov
National University of Science and Technology "MISiS", Moscow, Russia
v.e.bagenov@gmail.com

Abstract

Increasing of working temperature of thermo-electric cells is one of the most important goals. Solution of this problem is closely related with the temperature of soldering of thermoelement (Bi-Sb-Te) with copper bus. The choice of zinc-based Zn-4Al alloy and replacing of copper bus on aluminum one has been done. The soldering was carried out at temperatures 440-480°C in argon atmosphere and samples were loaded by pressure 1.7 MPa. Experiments elicited a fact that quality joining could be done if thermoelement had a protective layer of Al with thickness more than 30-40 micrometers. In this case base material dissolution or damage was not observed. Cold spray method with Al powder was used for coating. The most important thing was the condition of Al bus and Zn-4Al foil surface as so as roughness of it. It determined fracture of oxide layers during loading and allowed joint formation. There was not any porosity in joints that usually are typical for Zn based alloys. Influences of temperature, load pressure, thickness of Al coating on joint forming and microstructure have been studied as well.

Introduction

The problem of thermo-electric cell operating temperatures increasing, as well as the used solder and the product itself cost depreciation, is rather actual at present. Rising of operating temperatures is associated with the soldering temperature of the cell parts. The thermo-electric cell based on Peltier element [1] represents Peltier refrigerating system, consisting of connected in series semiconductors of p- and n-type, forming p – n and n – p junctions (Fig. 1.). Each of such junctions has a thermal contact to one of two heat exchangers. Because of the electric current flow of certain polarity, the temperature gradient is formed between heat exchangers of the thermo-electric cell.

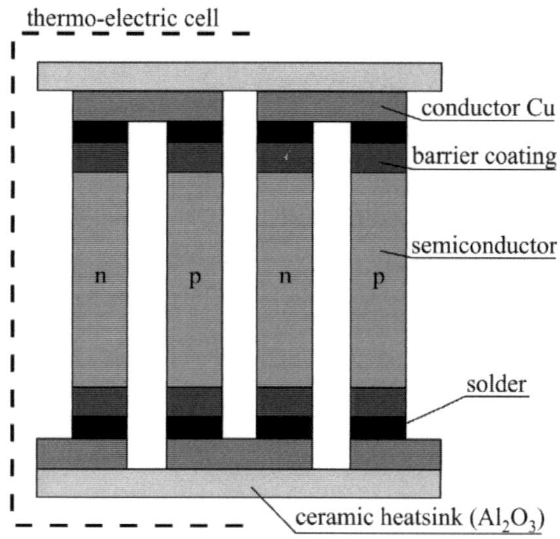

Fig. 1. Thermo-electric cell scheme.

At present time, various low-temperature solders are used while the thermo-electric cell soldering. The most widespread solders are Sn-based alloys: Sn - Sb ($t_{melt.}$ = 230°C) and Sn - Au ($t_{melt.}$ = 280°C). Soldering is implemented in the protective gas furnaces. Usually solder is applied as paste, consisting of a solder powder and a flux with binder. Bi-Sb-Te-based semiconductor material has the nickel barrier coating with the gold immersion layer for the interdiffusion of atoms prevention between the semiconductor part and the solder alloy, and for the thermo-electric cell durability increasing, as well. During the soldering process, bonding occurs between the copper conductors and the barrier coating of the thermo-electric cell. The degradation temperature of semiconductor material is about 560°C. Therefore, there is still a margin of solder temperature for the thermo-electric cell operating temperature increasing. However, there is no suitable solder alloys for the specified materials with melting temperatures above 300°C, except Zn-based solders. However, these solders have the restricted application and basically are applied for soldering of aluminum and aluminum with copper using flame soldering [2].

Possibility of thermo-electric cell soldering with Zn-based solder alloys

Zn-based solder alloys working temperature is within interval of 420-500°C. The joint structure formation, while soldering with Zn-based solders, is considered explicitly in articles [3]. It is known that the soldering temperature rising and the soldered joint width increasing result in the shrinkage porosity appearance in the soldered joint. Therefore, while soldering with the filling elements of Zn-based solders, it is necessary to strive towards the lowering of the soldering temperature and of the gap between soldered parts. The last is feasible easily while thermo-electric cells soldering, as far as from the constructive point of view, generally, the plain elements are bonded.

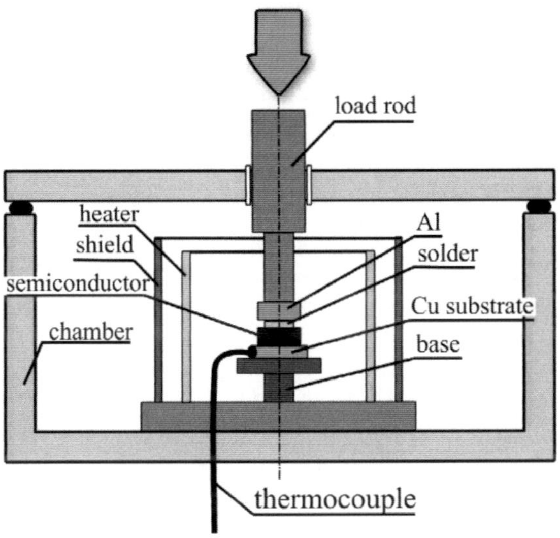

Fig. 2. - Scheme of chamber and sample assemblage for soldering.

Changes are necessary in a thermo-electric cell design for the aluminum surfaces soldering implementation. Firstly, the copper buses should be replaced by the aluminum ones. Secondly, the aluminum coating should be applied on the semiconductor part surface. Then, while such soldering, the bonding will occur between the aluminum coating and the aluminum bus. However, the basic problem of soldering of aluminum is the surface oxide layer fracture provision to ensure the wetting of aluminum by solder alloy. Due to high activity and problems with fluxes removal while soldering of aluminum, the flux-free soldering is considered to be the only alternative.

There are several methods of such mechanical treatment application; these are the ultrasonic treatment, which application has been explored and described already [4], and the pressure application – the method, generally applied at pressure welding and at diffusion welding of aluminum. The pressure application essence consists in creating a local plastic deformation of aluminum surface in the place, where soldering occurs. Oxide layer fractures, and in a presence of a solder melt, the wetting of aluminum surface with Zn-based solder occurs. On the other hand, during the pressure application the solder melt starts moving while melting, that also creates favorable conditions for the oxide layer fracture due to friction. Access for oxygen from outside is absent, that also influences favorably on the possibility of aluminum surface wetting with solder melt. The pressure value at soldering process should not exceed the breaking point of semiconductor material to prevent its deformation and fracture. Soldering in the protective gas furnace has been offered by analogy with the existing soldering technology of thermo-electric cells.

Thus, for the possibility verification of the thermo-electric cell parts bonding by means of Zn4Al (Zn – 4 wt. % Al) solder, it is necessary to solve the following problems:

- To research the solderability of thermo-electric cell with aluminum bus through the aluminum coating on a semiconductor surface;
- Researching of temperature and pressure influence on the soldered joint formation.

Materials and methods

Base material and preparation of solder

Ferrotec Nord Corporation thermo-electric cells of Bi-Sb-Te system, size 4.55x4.55x2.1 mm, with the applied anti-diffusion barrier coating of Ni and Sn and the gold immersion coating were produced for exploration.

The aluminum coating was applied on initial samples by cold spray coating method [5], 40 μm thickness. The aluminum powder with particle sizes from 20 up to 150 μm was used for spraying. It is necessary to pay attention to a series of features of the given method: firstly, the obtained coating surface is rough; secondly, the surface is not sintered, as far as the mechanical interaction is implemented; and thirdly, there are oxides and some quantity of ceramics (up to 5-10 wt. %) in the applied aluminum coating.

The foil of Zn4Al alloy, 200 μm thickness was taken as a solder. Embedded elements were cut out of the foil in the form of squares 4x4 mm size. Before the embedded elements cutting and the solder placement, the foil tape was exposed to grinding with an abrasive paper with 60-80 μm grain sizes.

The aluminum bus 4.55x4.55x2 mm size was exposed to grinding before the bonding assemblage also.

Soldering of samples

Samples were assembled and placed in a vacuum chamber according to the scheme at Fig. 2. Soldering was carried out in argon atmosphere and at heating from 420 °C to 480 °C. The chamber was evacuated to a residual pressure of 1.3 Pa prior to argon back filling. Samples without aluminum coating and coated ones were exposed to soldering, as well.

The pressure application was implemented by certain weight-loading placement on a rod, which through vacuum feed-through leaned on the sample. The pressure value on the sample made 1.7 MPa. At such pressure, no deformation of the sample at temperatures up to 500°C occurs.

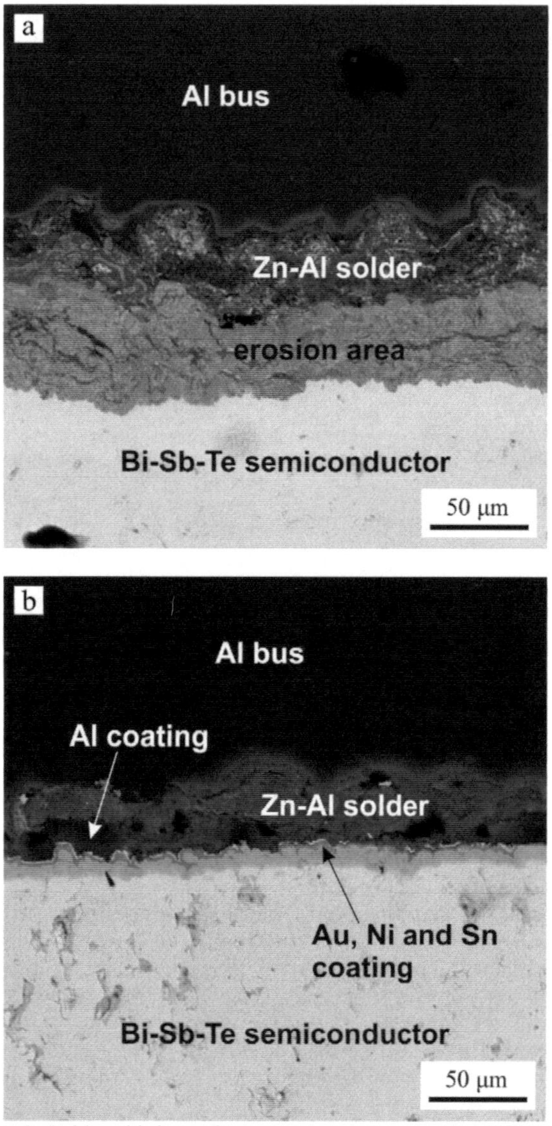

Fig. 3. Soldered joints microstructure at temperature 440°C of the sample
a) without coating and b) with aluminum coating 40 μm thickness.

After soldering, the samples were sliced, and metallographic specimens were prepared for observation. Metallographic observation of the obtained soldered joint structure and the element distribution were carried out using TESCAN VEGA 3 SBH scanning electron microscope with the energy dispersive X-ray analysis system of Oxford Instruments Advanced AZtecEnergy.

Results and discussion

The influence of aluminum coating

The melting temperature of solder makes 392°C. However, up to 440°C the soldered joint formation does not occur. Starting with the temperature of 440°C, we succeeded to solder together the samples both with an aluminum coating and without it, as well. It is worth noticing, that due to the excess solder pressing-out because of pressure application, it was a success to avoid the shrinkage porosity formation inside the soldered joint. However, the metallographic observations have shown, that the barrier coating dissolution of nickel with tin by Zn-based solder alloy does occur, and the noticeable erosion of semiconductor material is observed (Fig. 3). In this connection, the further experiments were carried out with aluminum coating only, which is not dissolved entirely within the temperature range of 440-480°C.

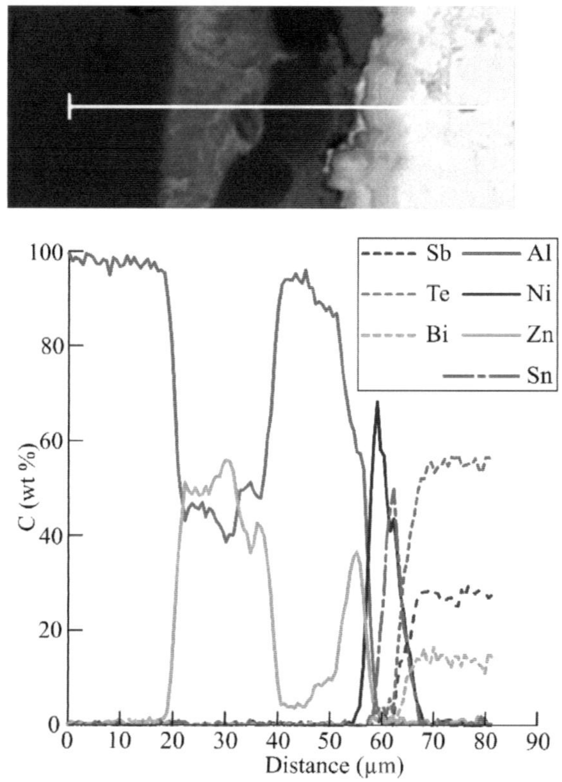

Fig. 4. Microstructure and element distribution of samples with 40 μm aluminum coating at soldering temperature 440°C.

In the presence of aluminum coating, alloy melt interacts with it. Along with dissolution of aluminum, the solder temperature rises, its ability to migrate further deep into the sample decreases.

Thus, at samples where the soldering was implemented without aluminum coating, the widening of mutual dissolution zone is observed, and solder alloy dissolves the thermo-electric cell, that resulted in semi-conductor material fracture within all the investigated temperatures range.

Soldering temperature influence

Apparently from metallographic analysis results (Fig. 4 and 5), the decrease of solder alloy layer width in the soldered joint is observed while the soldering temperature increasing.

According to the phase diagram, the solubility of aluminum in Zn4Al solder melt increases several times in 440-480°C temperature interval. The intense dissolution of aluminum by Zn-based solder alloy occurs. Prior to soldering, the aluminum content in a soldered joint is 4 %, after soldering is 15-20 %.

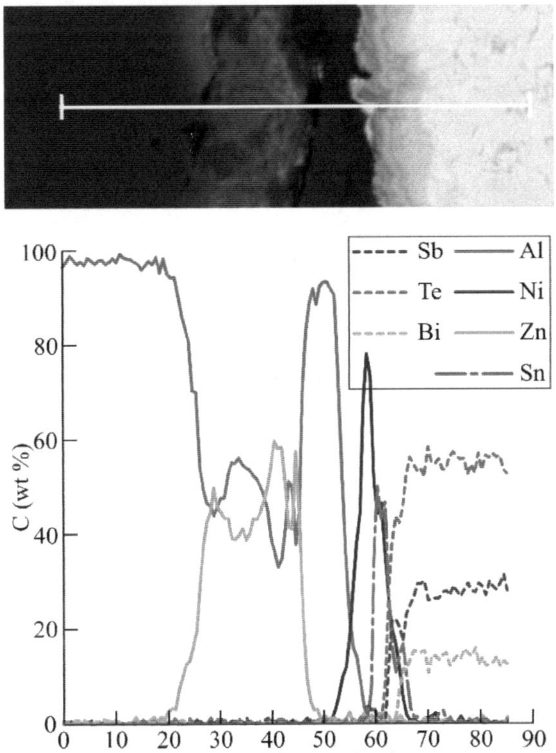

Fig. 5. Microstructure and element distribution of samples with 40 μm aluminum coating at soldering temperature 480°C.

The wide zone of base material and coating dissolution is observed. However, the pressure application results in the excess melt alloy removal out of soldered joint zone that restricts the ability of solder melt to dissolve aluminum. So, the thickness of aluminum coating on a semiconductor part makes ~20 μm at temperature 440°C (Fig. 4), and while soldering at 480°C it makes ~10 μm only (Fig. 5).

Thus, the soldering temperature increasing leads to solder layer decreasing in a soldered joint and to its pressing-out towards the sample periphery. Semiconductor erosion is not observed. To provide the prevention of solder with base material interaction, the aluminum coating thickness should be increased conceivably.

From the analysis of EDS maps (Fig. 6), implemented with the help of energy dispersive X-ray analysis, it follows, that aluminum oxides, which were on a surface of samples, and also in cold spray coating volume on semiconductor material, are distributed evenly throughout the solder and coating volume. Defects in soldered joint are not observed. It confirms the fact, that at pressure application process and at solder melt presence, the oxide film fraction and the new melt-aluminum boundary formation occur. Solder is spreading on pure aluminum surface, and oxide particles are distributed evenly throughout the joint volume.

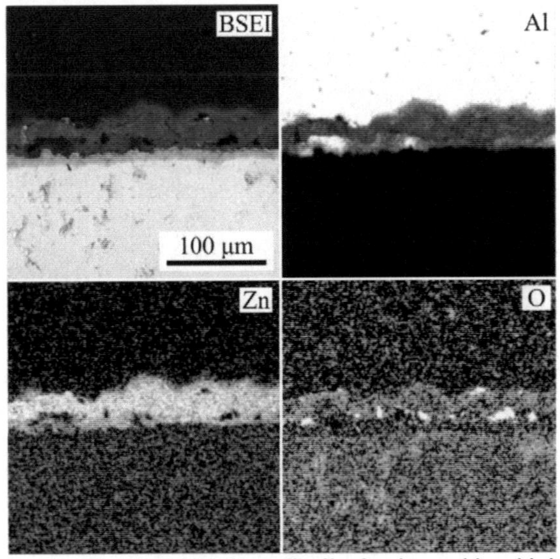

Fig. 6. EDS maps of elements distribution in a soldered joint. Soldering temperature is 460°C.

Conclusions

1. The semiconductor material with the aluminum bus bonding with the help of flux-free soldering with Zn - 4 wt. % Al solder is possible in the 440 – 480 °C temperature interval with the pressure application, which fractures oxide films and allow attaining the dense homogenous soldered joints.
2. Absence of Al coating leads to dissolution of Ni/Sn coating by solder and to the semiconductor material erosion within all temperature range.
3. The Al coating width should be not less than 40 μm for the joint stability attaining at soldering temperature of 440-480°C.
4. The soldering temperature rising leads to the layer thinning of solder in the soldered joint and to its pressing-out towards the periphery of sample.

Acknowledgments

This research has been supported by Ferrotec Nord Corporation.

References

[1] Riffat, S.B., Ma, X., "Thermoelectrics: a review of present and potential applications," *Applied Thermal Engineering*, Vol. 23, No. 8 (2003), pp. 913-935.

[2] Dai, W., Xue, S., Lou, J., Lou, Y., Wang, S., "Torch brazing 3003 aluminum alloy with Zn-Al filler metal," *Trans. Nonferrous Met. Soc. China*, Vol. 22, No. 1 (2012), pp. 30-35.

[3] Bazhenov, V.E., Pashkov, I.N., Pikunov, M.V., Cheverikin, V.V., Anohin, A.A., "Interaction of Zn and Zn–4Al, Zn–15Al (wt-%) solder alloys with aluminum,"

Materials Science and Technology, Vol. 32, No. 8 (2016), pp. 752-759.
[4] Nagaoka, T., Morisada, Y., Fukusumi, M., Takemoto, T., "Selection of soldering temperature for ultrasonic-assisted soldering of 5056 aluminum alloy using Zn–Al system solders," *Journal of Materials Processing Technology*, Vol. 211, No. 9 (2011), pp. 1534-1539.
[5] Moridi, A., Hassani-Gangaraj, S.M., Guagliano, M., Dao, M., "Cold spray coating: review of material systems and future perspectives," *Surface Engineering*, Vol. 30, No. 6 (2014), pp. 369-395.

Analysis of coatings deposited by brazing and cold spraying for casting aluminum/steel hybrids

*K. Bobzin, M. Öte, S. Wiesner, L. Gerdt**
Surface Engineering Institute, RWTH Aachen, Germany
*gerdt@iot.rwth-aachen.de

Abstract

The application of hybrid lightweight constructions in the automotive industry is triggered by the striving for an efficient and sustainable use of energy and resources. In this regard, techniques for joining dissimilar materials are developed and refined. Pressure die casting provides an innovative method for manufacturing aluminum cast/steel sheet hybrid components, which are of a particular interest for the automotive sector. In the current study, Cu and Al-based coatings are applied on the steel sheet by means of thermal spraying and brazing to enhance the bonding between aluminum melt and steel inserts during the die casting process. Additionally, the influencing factors on the bonding quality are determined. The microstructure at the interface of aluminum/steel compounds is investigated by light microscopy as well as scanning electron microscopy. The mechanical properties of the joints are examined by shear tests. The high pressure die casting (HPDC) experiments show that Al-based coatings and Cu coating enable to achieve a solid bond between melt alloy and steel insert. However, the bond strength between the cold gas sprayed (CGS) Al-based coatings and steel insert should be enhanced to withstand the high dynamic loadings during the casting process.

Introduction

For decades, steel and aluminum remain the most widely used materials in the automotive industry. Especially the hybrid structures of both metals are coming to the fore as an important research topic. The main aim is to combine the good mechanical properties of the steel with the low density of aluminum to reduce weight and thus reduce fuel consumption and CO_2 emissions in the transport sector. However, the joining of aluminum and steel is a challenging task due to strongly different thermophysical properties of both materials and possible formation of brittle intermetallic phases. Besides welding and brazing, high pressure die casting is an alternative technique for joining dissimilar metals. Casting of grey cast iron liners with aluminum [1] or Al inserts with Mg alloy [2] in the engine crankcase is already state of the art. Pressure casting of aluminum onto a steel insert has been also at the focus of research at RWTH Aachen University in recent years [3,4]. Further improvement of the compound properties is still required, as the problem of the gap formation at the aluminum/steel interface has not been completely solved. Due to enormous cooling rates after pressure casting and insufficient wetting of steel by aluminum, it is difficult to achieve a metallurgical bonding at the interface.

In order to enable a gap-free material bonding between the steel and the cast aluminum alloy, coatings of AA7075 and Cu applied onto the steel insert are investigated in the current study. The main challenge is the application of coatings with a reliable adhesion on the steel substrate which shall facilitate metallurgical bonding with the aluminum melt during the pressure casting. In the pressure casting process the steel insert is heated up to a temperature of T = 400 – 520 °C, while the temperature of the molten aluminum is about T = 700 °C [4]. Therefore, the Al-based alloy AA7075 with a low solidus temperature was selected to enable a partial melting of the coating during HPDC to achieve a metallurgical bonding with cast alloy. Copper is widely used as a brazing filler metal for joining different components made of steel. In the current approach the copper coating applied by means of deposition brazing should provide a better wetting of the surface to facilitate the formation of a solid joint between Al cast alloy and steel.

Experimental

In Figure 1 the principle scheme of the surface pretreatment of the steel inserts for HPDC is shown. The main aim was to achieve a solid metallurgical bond between the cast alloy and coatings.

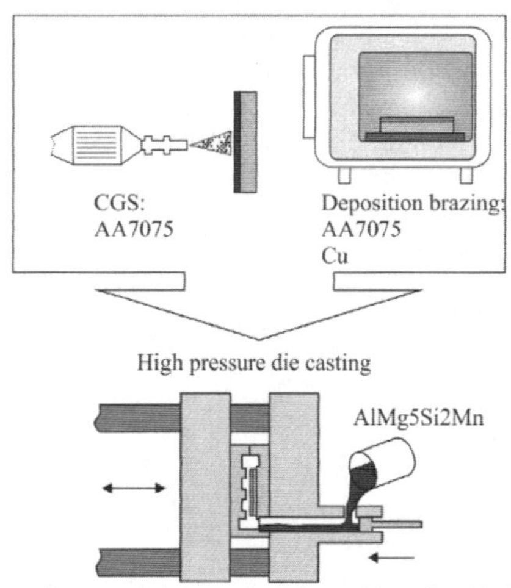

Figure 1: Scheme of the coating deposition for high pressure die casting

In the current study, cold-rolled deep-drawing steel DC04 was selected as the substrate material, which is used in body construction. Steel sheets with the dimensions of $l \times b \times t = 106 \times 24 \times 2\ mm^3$ were used as inserts to produce aluminum-steel hybrid components. The composition of the steel, coating materials and Al cast alloy is given in

Table 1. The coating deposition of AA7075 was performed by means of Kinetics 8000 Cold spray system from Oerlikon Metco (Pfäffikon, Switzerland).

Table 1: Chemical composition of the steel, cast alloy and coatings (wt.-%)

	DC04	AA7075	AlSiMn9	Cu
Si		0.4	9.0	
Mn	0.4	0.3	0.4	
C	0.08			
S	0.03			
P	0.03			
Cu		1.6	0.03	99.9
Zn		5.5		
Mg		2.5	0.08	
Fe	bal.		0.15	
Al		bal.	bal.	

The spray parameters for AA7075 coating are shown in Table 2. Foil of AA7075 with a thickness d = 200 µm was deposited by means of vacuum brazing at T = 690 °C with a holding time of t = 5 min as a reference for the cold sprayed coating. Copper coating was applied as foil with a thickness of d = 100 µm by means of vacuum brazing at T = 1,100 °C with a holding time of t = 10 min. Die casting trials were conducted by means of the high pressure die casting cell Frech DAK 450 Vacural by Oskar Frech GmbH + Co. KG (Schorndorf, Germany). In order to characterize the deposited coatings they were metallographically prepared and examined in terms of their microstructure. A microhardness tester Fisherscope HM2000 by Helmut Fischer GmbH (Sindelfingen-Maichingen, Germany) was used for the examination of the hardness of the reaction zone at the interface coating/steel with a load of m = 5 g. In order to determine the chemical composition of the possible reaction layers at the interfaces between steel, coating and aluminum cast alloy after HPDC, an energy dispersive spectroscopy analysis was performed by means of the scanning electron microscope Zeiss Leo 1530 with the EDS device Bruker Quantax 200.

Table 2: Spray parameters for cold gas spraying of AA7075

Parameter	AA7075
Gas temperature [°C]	490
Process gas flow [m^3/h]	60
Carrier gas flow [m^3/h]	3.0
Spray distance [mm]	25
Transverse movement velocity [mm/s]	200

Results and Discussion

The AA7075 coating could be deposited by cold spraying on the steel substrate without previous surface treatment. The feedstock material exhibited a good plasticity resulting in a tight bonding between particles and good adhesion on steel surface, Figure 2a. The coatings of AA7075 have a porosity of about p = 0.5 %. The thickness of the coating was of about d = 120 µm.

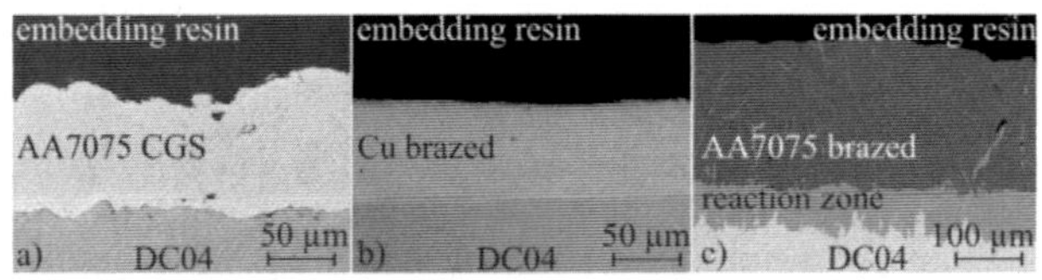

Figure 2: Micrographs of the coatings after deposition a) AA7075 CGS, b) Cu deposition brazed and c) AA7075 deposition brazed

In contrast to the thermally sprayed AA7075 coating, the brazed copper layer is metallurgically bonded to the steel substrate. Furthermore, it has a homogeneous microstructure and a smooth surface, Figure 2b. The thickness of the Cu coating was about d = 100 µm. Deposition brazing of AA7075 foil resulted in a strong reaction of Al and Fe with the formation of a thick reaction zone of AlFe intermetallic phases, Figure 2c. As already stated, these kind of intermetallic compounds have a negative effect on mechanical properties of the compound. The samples with deposition brazed AA7075 coating were nevertheless used for the HPDC trials as the reference for the cold sprayed coatings.

High pressure die casting

After the HPDC, the steel/Al cast compounds were first analyzed metalographically in order to characterize the quality of the bonding at the interfaces between Al cast and coating as well as between coating and steel. A metallurgical bond between Al melt and AA7075 CGS coating could be partially achieved, Figure 3a. The application of copper coating resulted in a solid bonding almost along the whole interface. The brazed AA7075 coating did not withstand the HPDC process as the compound failed directly after the samples were removed from the mold. The cracks propagated at the interface between reaction zone and AA7075 coating. However, metallurgical bonding could be achieved at the interface between Al cast and coating, Figure 3c. The micrographs show the sections of the compound with a solid metallurgical bonding.

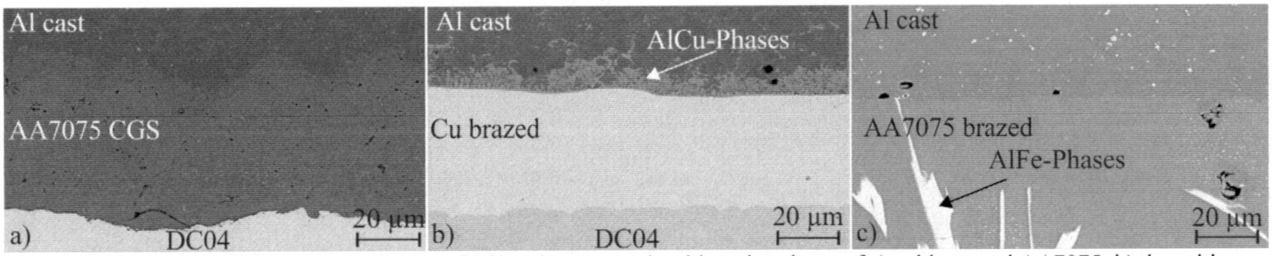

Figure 3: SEM images of hie pressure die cast of Al/steel compounds with an interlayer of a) cold sprayed AA7075, b) deposition brazed Cu, c) deposition brazed AA7075

In order to describe the interaction at the cast/coating/steel interfaces, SEM/EDS analysis was performed. Figure 4 shows the results of this analysis for the brazed AA7075 coating. The interface between the Al cast and the coating can be reconstructed in particular by following the silicon and magnesium content, due to the differences in the chemical composition of both alloys. A low gradient of the Mg and Si content at the interface between coating and cast indicates the diffusion of these alloying elements and, thus, a good metallurgical bonding. However, the quality of the bond was inhomogeneous along the whole interface. A solid, gap free bond was achieved at approximately 50 % of the whole boundary surface. The AlFe phases, which were observed in the coating, were formed during the deposition brazing process.

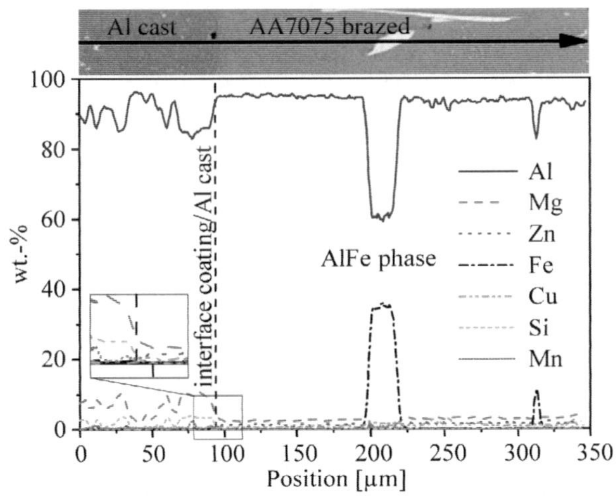

Figure 4: SEM/EDS analysis of the compound with brazed AA7075 coating

For the AA7075 coating deposited by cold gas spraying, neither diffusion at the interface between steel and coating nor the formation of AlFe intermetallic phases after HPDC was observed. However, considering the low gradient of the Si and Zn content at the interface between coating and cast, a diffusion of both elements could be assumed, Figure 5. This fact points out a solid metallurgical bonding between Al cast and coating. A gap free joint between coating and Al cast could be achieved at approximately 60 % of the whole interface. Furthermore, a densification of the coating occurred during the HPDC process, possibly due to the sintering effect at the interfaces between the splats in the coating as described in [5]. Only some small single pores could be detected in the microstructure of the coating after casting.

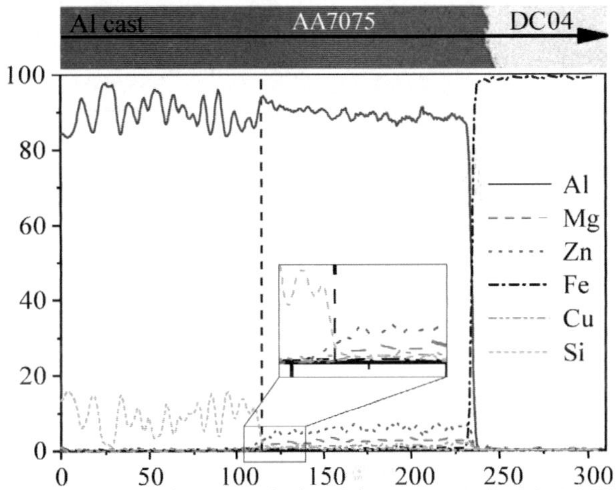

Figure 5: SEM/EDX analysis of the compound with AA7075 CGS coating

Figure 6 shows the results of the SEM/EDS analysis for the compound with Cu coating. The diffusion of the copper into the cast alloy and the formation of AlCu phases was observed. The whole diffusion zone with a high Cu-content was of about d = 20 µm. The influence of the formed reaction zone on the mechanical properties of the compound should be further investigated.

Figure 6a shows the microstructure of the reaction zone. At the interface of the Cu coating, a thin layer with a thickness of approximately d = 3 µm with a chemical composition indicating the existence of the Al_3Cu intermetallic phase has formed. Above this thin layer, the microstructure of the reaction zone consists of Al-AlCu eutectic. The morphology of this area in the reaction zone is similar to that observed by Zhou et al. for Al-Cu joints produced by means of a low heat input pulsed double-electrode gas metal arc welding [6]. However, the thickness of the intermetallic phases and the whole reaction zone formed after the HPDC was significantly lower.

Hardness at the bond zone

In order to characterize the mechanical properties of the bond zone, microhardness measurements were conducted. Figure 7 shows the hardness of the compound at the interface steel/coating/Al cast for brazed Cu coating and AA7075 CGS coating. The gradient of the hardness between steel, coating and al cast is significantly smaller in the case of the Al-based coating, due to a higher hardness of AA7075 CGS compared to brazed copper. Furthermore, the differences related to the hardness of the steel near the interface with the coating have to be mentioned.

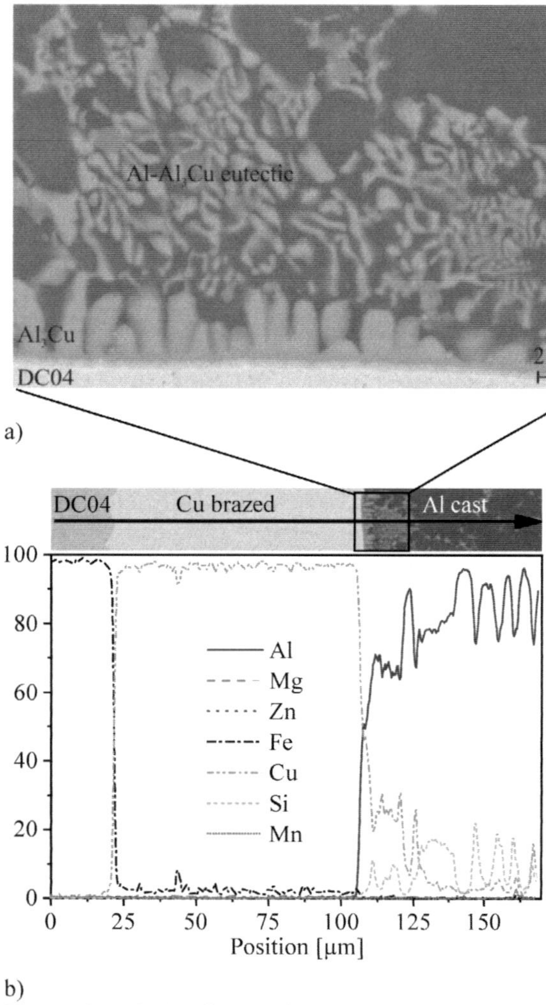

Figure 6: a) SEM micrographs of the reaction zone of the compound with Cu coating and b) SEM/EDS analysis of the compound

Higher microhardness of the steel in the case of AA7075 CGS coating can be firstly explained by work-hardening effect occurred during the cold gas spraying by high velocity impact of the powder particles. On the other hand the heat treatment during the deposition brazing of the Cu coating results in a softening of the cold rolled steel. As the initial hardness of the steel sheet HU = 1445 ± 22 N/mm² between the values of the state after deposition brazing and after cold spraying both explanation are valid. In the case of Cu coating, the microhardness at the interface between coating and cast alloy reaches its maximum of HU = 1,860 ± 90 N/mm² due to the copper diffusion into the Al cast and formation of the reaction zone, which did not occur in the case of the AA7074 CGS coating.

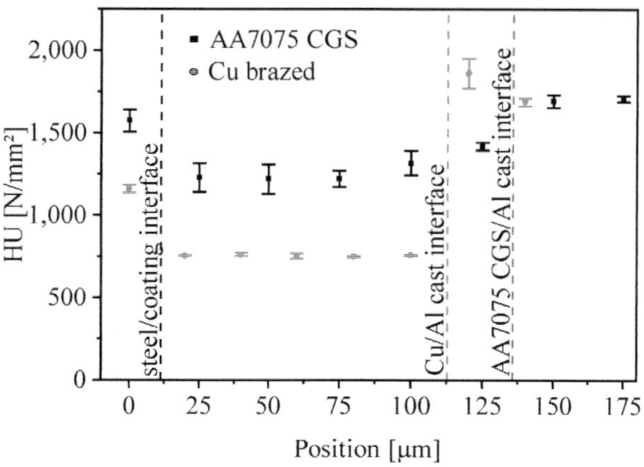

Figure 7: Microhardness at the interface steel/coating/Al cast

Conclusions

Three different coatings on a steel insert for hybrid die casting of steel/aluminum compounds were investigated. In the case of the brazed Cu coating, a metallurgical bond could be produced almost along the whole interface between coating and cast alloy. A thin reaction zone with a chemical composition of Al_3Cu intermetallic phases has formed at the interface between Cu coating and Al cast, due to the diffusion of the Cu into the aluminum melt. For the Al-based coating of AA7075, a partial bonding to Al cast could be achieved, both for deposition brazed and cold gas sprayed coatings. However, brazed interlayer failed during HPDC process due to the thermally induced cracks in the region of the AlFe intermetallic phases, which were formed during the deposition brazing. After application of the cold gas sprayed coating, no formation of the intermetallic compounds was observed at the interface between coating and steel. The microhardness distribution in the bond zone was homogenous for AA7075 CGS coating, while the hardness of the copper coating is significantly lower compared to the Al cast. The possible effect of such a gradient on mechanical properties of the compound will be further investigated by tensile shear tests. Furthermore, the corrosion behavior of die cast Al/steel hybrid compounds will be analyzed.

Acknowledgments

This research is supported by the German Research Foundation (DFG) under contract No. BO 1979/39-1. Special thanks are addressed to our project partners from the institute of metal forming (IBF) and foundry institute (GI) from RWTH Aachen University.

References

[1] Nisslé, S., Dassler, S., Noster, U., Mundl, A. (2008). Verbundguss, Neue Auslegungsmöglichkeiten für Gussteile durch Funktionsintegration oder lokale Verstärkung. Tagungsband der 5. Ranshofener Leichtmetalltage. LKR-Verlag, Ranshofen, pp. 35–44.

[2] Fent, A. (2006). Magnesium-Aluminium-Verbundkurbelgehäuse- Einsatz von Neutronen in der Bauteilentwicklung. Garching.

[3] Oberschelp, C. (2012). Hybride Leichtbaustrukturen für den Karosseriebau - gusswerkstofforientierte Anwendungsuntersuchungen für das Druckgießen. Diss., Gießerei-Inst. der RWTH, Aachen.

[4] Lao, B. (2013). Druckgegossene Metallhybridstrukturen für den Leichtbau-Prozess, Werkstoffe und Gefüge der Metallhybriden. Diss., Gießerei-Inst. der RWTH, Aachen.

[5] Bobzin, K., Öte, M., Wiesner, S., Gerdt, L., Bührig-Polaczek, A., Brachmann, J. (2018). Effect of Alloying Elements on Growth Behavior of Intermetallic Compounds at the Cold-Sprayed Coating/Steel Interface During Immersion in Aluminum Melt. Inter Metalcast **249**/1-2, 167.

[6] Zhou, X., Zhang, G., Shi, Y., Zhu, M., Yang, F. (2017). Microstructures and mechanical behavior of aluminum-copper lap joints. Materials Science and Engineering: A **705**, pp. 105–113.

Joining of Aluminum and Stainless Steel for Heat Exchanger Applications

Hui ZHAO[1,*], *Stefan ELBEL*[1,2], *Pega HRNJAK*[1,2]

[1]*Creative Thermal Solutions, Inc.*
2209 North Willow Road, Urbana, IL 61802, USA

[2]*University of Illinois at Urbana-Champaign*
Department of Mechanical Science and Engineering
1206 West Green Street, Urbana, IL 61801, USA

* *Corresponding Author Email: hui.zhao@creativethermalsolutions.com*

Abstract

Aluminum alloys are used extensively in heat exchanger fabrication owing to the apparent advantanges, such as high thermal conductivity and low density. However, when an aluminum heat exchanger is installed in a system, such as in the HVAC&R applications, there is often the need to join the heat exchanger to other components that are made of dissimilar metals, such as stainless steel (SS) and copper pipes. In this study, the feasibility of using controlled atmosphere brazing technology for joining dissimilar metals (Al/SS) in HVAC&R systems is investigated. It is well known that the formation of intermetallic compound between Al-based filler metal and SS base metal is inevitable during brazing. The key factor to ensure a high quality brazed joint is to control the IMC formation and minimize the IMC layer thickness. 304 stainless steel and AA3003 aluminum alloys are brazed with selected Al based filler metal with the assistance of non-corrosive fluxes. An indepth analysis of filler metal wetting behavior and flowability are presented. The brazed joint quality between dissimilar materials is examined metallurgically. The purpose of this study is to explore proper brazing conditions and filler metal selections for joining aluminum and stainless steel in HVAC&R applications.

Introduction

Heat exchangers, such as evaporators and condensors, are essential components for HVAC&R systems. The current trend is to replace the traditional round copper tube heat exchangers with all aluminum microchannel heat exchangers, owing to many advantages of the microchannel heat exchangers: (1) high thermal performence; (2) light weight; and (3) low material cost. One challenge for a wider application of the aluminum heat exchangers is to join aluminum with dissimilar materials, such as stainless steel and copper alloys in a HVAC&R system. Brazing is the technology often used for joining dissimilar metals. The main difficulties in joining dissimilar metals such as aluminum and stainless steel is the intermetallic compound (IMC) formation at the interface which may reduces the joint strength. Some studies of joining aluminum with stainless steel by brazing have been reported [1-5]. Based on the considerations of brazing temperature and material compatibilities, Al based filler metal, such as Al-Si binary alloys are often selected for such brazing applications. The Al-Si alloys are the most commonly used filler metals for aluminum brazing because of the excellent compabibility with many aluminum alloys for brazing. In the case of using Al-bearing filler metal for brazing the stainless steel, the formation of a Fe-Al compound layer at the solid/liquid interface will take place. The existence of the IMC layer reduces the strength of a brazed joint. An example of a crack developed between the Al-Si filler metal and the stainless steel base metal after brazing is illustrated in Fig.1. An apparent intermetallic layer is attached to the filler metal.

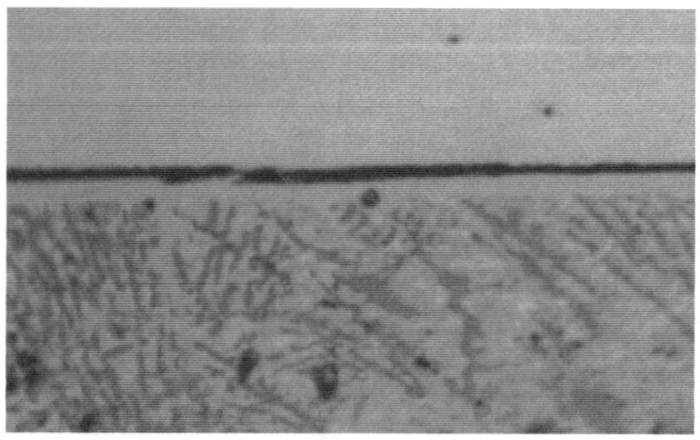

Figure 1. A crack found in a brazed joint between aluminum and stainless steel.

The key for a successful brazing of aluminum and stainless steel is to minimize the growth of the IMC layer. Controlled atmosphere furnace brazing of aluminum and stainless steel using eutectic Al-Si filler metal has been studied. It was found that excessive dwell time at brazing temperature will lead to thicker IMC layer and more complication of reaction product structures [2]. It is reported that induction brazing with Al-Si filler metal can effectively reduce the IMC layer thickness [3]. Fast heating is preferrable but not always applicable, therefore, precoating of the stainless surface with a thin layer of nickel beofre brazing has been reported [6]. In recent years, studies of using low melting tempature Zn-Al alloys to join aluminum and stainless steel by flame brazing have been reported[1,4,5]. Due to the Al composition in the filler metal, the IMC layer still forms at the stainless steel and the filler metal interface. Recent studies show that Zr addition in Zn-15Al as filler metal can reduce the thickness of intermetallic layer at joint interface to improve brazed joint strength [5].

In this study, wetting behavior and flowability of Al-Si filler metal during brazing of alumium and stainless steel are studied. Aluminum based filler metal AA4047(Al-12Si) is used. Furnace brazing with a controlled atmosphere is adopted. Wetting behaviors of the filler metal on different substrates are observed in-situ using a transparent furnace as well as a heating stage facility. The brazed joint quality between dissimilar materials is examined metallurgically. Cross section samples are prepared and examined using an optical microscope to study the interactions between filler and base metals.

Materials and Experiments

Base alloys and filler metal

For the sessile drop wetting test, small rods (~1mm length) of AA4047 (Al-12Si), cut from a roll of 0.8mm diameter wire, is used as filler metals. The Al-12Si alloy has a melting temperature range of 577-582°C. Coupon plates of 304 stainless steel and AA3003 (~10x10 mm) aluminium are used as base metals. The wetting behavior of filler metal may be affected by the surface morphology of the base metal [7,8]. In addition to the smooth surface of the commercially available base metal coupons, a very rough surface feature was generated by sandblasting on the base metal surfaces for the wetting tests. Figure 2 illustrates a comparison between the cross section images of a smooth SS304 surface and a rough SS304 surface.

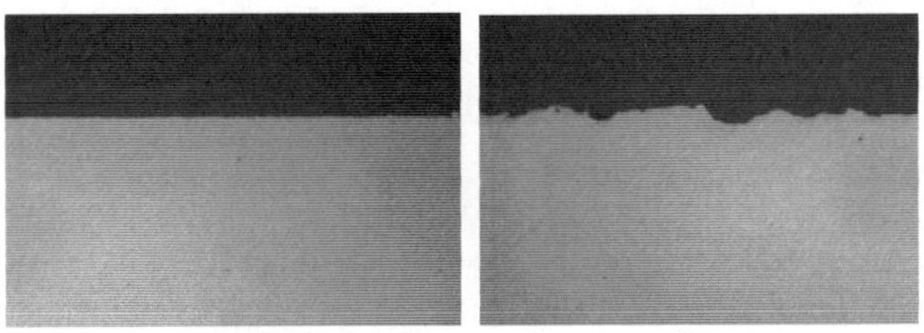

(a) (b)

Figure 2 Cross section veiw of the stainless steel surface: (a) smooth surface; (b) rough surface.

Joining dissimilar metals such as Al/SS in a refrigeration system is often needed in tube-to-tube connections. A flowability test is used to visualize the capillary flow of liquid filler metal during brazing. In this study, a partial circular ring (~20mm in diameter) is made of the AA3003 alloy foil (0.25mm thick), which is sitting on a SS304 coupon plate (~25x25x0.8mm). A small piece of filler metal (Al-12Si alloy, 2mm diameter, ~ 3mm in length) is placed at one end of the metal ring. Upon melting, the capillary force will drive the liquid filler metal flow through the gap between the dissimilar metals. A picture of a flowability test sample (Al/Al brazing) is illustrated in Figure 3. The noncorrosive NOCOLOK®[1] flux is applied for brazing.

Figure 3 Experimental set-up for flowability test

Experiments

The sessile drop wetting tests are performed using a heating stage test facility. The stage is made of stainless steel. Inside the stage chamber, a mini-heating coil is embedded in a ceramic holder. Inert gas atmosphere is maintained by continous flow of high purity nitrgon. A quartz glass window provides visual access through a microscope to the wetting and speading behavior of the molten filler metal.

The flowability tests are performed using an in-house built transparent furnace. The transparent furnace wall provides real-time observation of the brazing process. A video camera was used to capture the filler metal melting and joint formation process. Continuous flow of nitrogen protects the metal surfaces from oxidation during heating.

To minimize the thickness of the IMC layer, a brazing temperature that is slightly above the melting temperature (577-582°C) of the filler metal with no dwell time is applied for most of the tests. The heating rate for the sessile drop wetting test is around 30℃/min, and the heating rate in the flowability test using a transparent furnace is 25℃/min to ensure a uniform temp distribution during heating.

Results and discussion

Wettability of Al-Si filler on SS304 and AA3003

It is known that AA4047 filler metal shows excellent wettability on the AA3003 alloy in a controlled atmosphere brazing condition. Figure 4 illustrates pictures of the filler metal on the AA3003 surfaces during a sessible drop wetting test. In Figs 4(a), the AA3003 coupon has an "as-is" surface that is relatively smooth. The liquid filler metal shows good wetting on the smooth surface and the final wetting area after solidfication measures more than 20mm^2. A rough surface feature on the AA3003 is also generated by sandblasting. A

[1] NOCOLOK® is a registered trademark of Solvay Fluor GmbH, Hannover, Germany

very fast and extensive spreading of the liquid filler metal on such a surface is observed during the sessile drop wetting test, see Figs 4(b). The filler metal covers the substrate as a thin film. The increase of the surface roughness on the AA3003 surface has significantly enhanced the wetting under the current test condiditons.

(a) (b)

Figure 4 Wetting test on the AA3003 surface

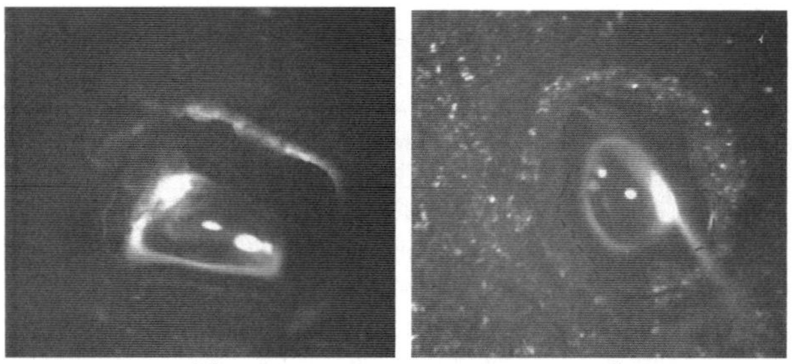

(a) (b)

Figure 5 Wetting test on the SS304 surface

Figure 5(a) illustrates the wetting test of the filler metal on the smooth SS304 surfaces. During real time observation, it was found that upon melting, the liquid filler metal forms a sphere initially, and gradually starts wetting and spreading on the base surface. The measured wetting area, after solidification, is around 5.5mm^2. For the wetting test performed on the rough SS304 surface (see Fig. 5(b)), no apparent improvement of wetting is observed. The bulk liquid covers even less area than on the smooth surface. However, it is interesting to note that a smaller halo extented beyond the bulk liquid, indicating the further spreading of the liquid filler metal in the valleys formed on the rough surface. The red coutour on Fig. 5(b) illustrates the boundary between the bulk filler metal and the halo ring. The sessile drop test indicates that the increase of surface roughness on the SS surface does not provide apparent enhancement on the wettability but may have hindered the spreading.

Flowability tests

The sessile drop wetting test has shown that Al-Si filler metal wets both the SS304 and AA3003 under the controlled atmosphere brazing conditions. However, the wettability on the SS304 surface is inferior to the wettability on the AA3003 surface. To study the join formation process in furnace brazing of AA3003 and SS304, the flowability test is performed using the transparent furnace. Figure 6 illustrates pictures extracted from the video that captures the flow of liquid filler metal during brazing of an AA3003 ring and a SS304 coupon plate. Both substrate surfaces are smooth with no flux applied. The filler metal is covered with a thin layer of flux.

When the sample is heated to a temperature above the flux melting point, an extensive spreading of the molten flux on both metal surfaces is observed, see Fig.6(b). The spreading of the flux on the AA3003 surface is essential to remove the tenacious oxide layer on the aluminum surfaces. Upon melting of the filler metal, an excellent flowability of the filler metal around the ring is illustrated in Figs. 6(c)&(d). The cooling starts immediately after the filler metal fills the joint gap to limit the IMC layer formation.

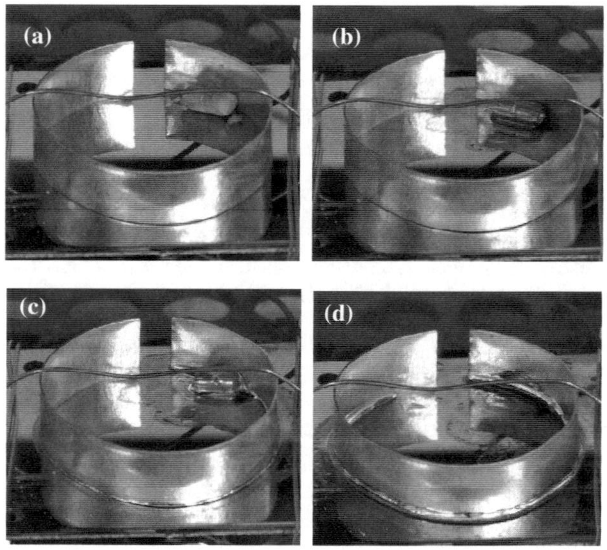

Figure 6 Flow of liquid filler metal between AA3003 ring and SS304 plate (smooth surface)

One brazed sample is metallurgically cut at the middle position of the ring, and polished to reveal the microstructure at the joint interface. The optical microscope images of the interface between the filler metal and the SS304 surface are presented in Figure Fig. 7. As expected, a reaction layer formation at the interface can been observed, as illustrated in Fig.7(b) between the dashed lines. However, the thickness of this layer is very small (less than 10 μm), indicating that a relatively low brazing temperature and fast joint formation process may effectively reduce the growth of the IMC layer. An in-depth study of the IMC composition, thickness and the correaltions with the brazing parameters will be performed in the future study.

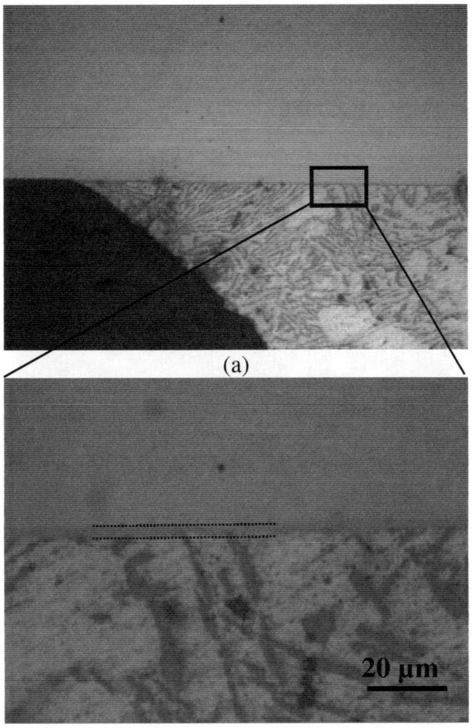

(b)

Figure 7 Microscope images of filler metal/SS304 interface

Another flowability test is performed on a sample with a rough SS304 surface, as illustrated in Fig.8. In this test, flux is only applied on the filler metal as in the previous test with a smooth SS304 surface. During the brazing test, it is interesting to observe that the molten flux spreads extensively on the rough SS304 surface, as illustrated by the red dashed lines on Fig.8(b). However, upon melting of the filler metal, the molten filler metal did not show a good flowability. The liquid filler metal accumulates at the initial position and shows limited wetting on the AA3003 surface. A small amount of liquid filler metal was able to flow into the gap between the ring and plate, but stops at a small distance at around the 3 O'clock position. In this case, the increased surface roughness seems to hamper the flow of the filler metal. The gap between the AA3003 ring and the SS304 plate is largely unbonded. The limited wetting of the liquid filler metal on the AA3003 and the extensive spreading of the flux on the SS304 may indicate that there is not sufficient flux wetting of the aluminum surface due to the depletion of flux by the rough stainless steel surface. Without the destruction of the oxide layer on the aluminum surface, good wetting will not take place. Therefore, no joint was formed between majority part of the AA3003 ring and SS304 plate. To further explore the hypothesis, a flowability test was performed with the same combinations of the A3003 ring and a SS304 plate with a rough surface. An excessive flux amount is applied on all sufaces of the ring, the plate and the filler metal. The results are illustrated in Fig.9.

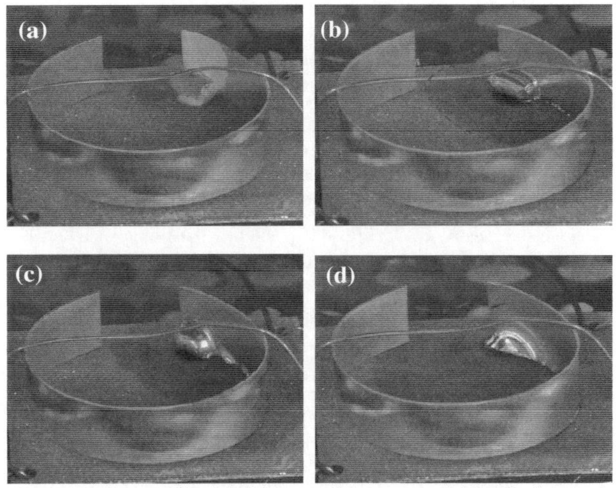

Figure 8 Flow of liquid filler metal between AA3003 ring and SS304 plate (rough surface)

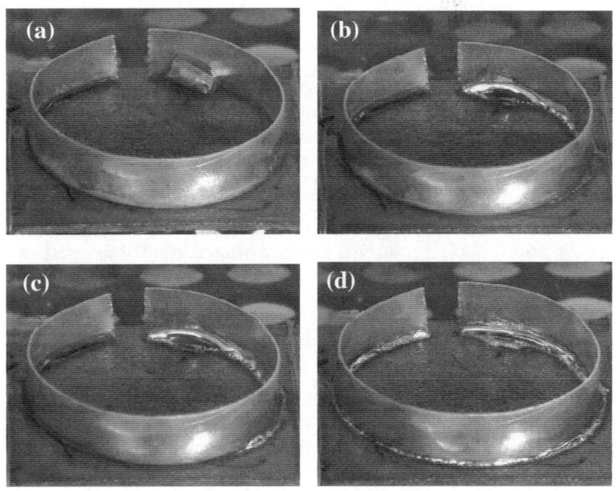

Figure 9 Flow of liquid filler metal between AA3003 ring and SS304 plate (rough surface and excessive flux)

Apparently, in this test, the excessive amount of flux ensures good coverage on both the aluminum and the stainless steel surface. Upon melting, the liquid filler metal immediately starts flowing through the gap between the ring and plate. However, the flow seems to make a stop at a 5 O'clock position due to some obstacle (see Figs. 9(b)&(c)). A continue heating to a higher peak temperature (600 °C) seems to promote the flow. It is interesting to observe that the liquid filler metal overcomes the unknown obstacle at around 595 °C and quickly fills the entire gap (see Fig.9(d)). The test shows that more flux amount is needed for an increased surface roughness of base metal. And the increase of surface roughness on the SS304 does not improve the wettability and flowability of the liquid Al-Si filler metal.

Conclusions

In this paper, wetting and flow behaviors of Al-Si filler metal during furnace brazing of AA3003 to SS304 alloys are experimentally studied. The influence of base metal surface features and flux application are explored. It was found that the increase of surface roughness on the SS304 surface does not enhance the wettability of AA4047 filler metal on the base surface. Such an alteration of the surface may even hamper the capillary flow during brazing, and requires a sufficiently large amount of flux application. The AA4047 filler metal wets the SS304 metal, but the wettability is inferior to the wettability on the AA3003 surface. Excellent flowability is estabilished between the AA3003 and the smooth SS304 substrate. Low brazing temperature and a fast joint formation process may effectively reduce the growth of the IMC layer between the filler metal and the stainless steel surface.

References

[1] Yang, J.L., Xue, S.B., Xue, P., Lv, Z.P.,Dai, W. and Zhang, J.X. , " Development of novel CsF-RbF-AlF3 flux for brazing aluminium to stainless steel with Zn-Al filler metal", *Materials & Design*, Vol. 64 (2014), pp. 110-115.

[2] Roulin, M., Luster, J.W.,Karadeniz, G. and Mortensen, A., "Strength and structure of furnace brazed joints between aluminium and stainless steel", *Welding Journal*, 05 (1999), pp 151s-155s.

[3] Fedorov, V., Weis, S. and Wagner, G., "Mechanical and microstructural behavior of brazed aluminum / stainless steel mixed joints" , *IOP Conf. Series: Materials Science and Engineering*, Vol. 118 (2016), 012003.

[4] Yang, J.L., Xue, S.B., Xue, P., Lv, Z.P.,Long, W.M., Zhang, G.X., Zhang, Q.K and He, P., " Development of Zn-15Al-xZr filler metals for brazing 6061 aluminium alloy to stainless steel", *Materials Science & Engineering A*, Vol. 651 (2016), pp. 425-434.

[5] Yang, J.L., Xue, S.B., Xue and Sekulic, D.P., "An impact of Zirconium doping of Zn-Al braze on the aluminum-stainless steel joints integrity during aging", *Journal of materials engineering and performance*, Vol. 26 (2017), pp 358-365.

[6] Brazing and Soldering Manual, Zhang, Q.Y. and Zhuang H.S. Eds,China Machine Press, (Beijing, 2008), pp505 (in Chinese)

[7] Zhao, H., Elbel, S. and Hrnjak, P., "Influence of Surface Morphology on Wetting Behaviors of Liquid Metal during Aluminum Heat Exchanger Fabrication", *Proceedings of the 15th Purdue Refrigeration Conference*, University of Purdue, West Lafayatt, IN, July 2014.

[8] Zhao, H., Elbel, S. and Hrnjak, P., "Joining dissimilar materils in heat exchanger manufacturing process", 5th International Congress on Aluminum Heat Exchanger Technologies for HVAC&R, Düsseldorf, Germany, May 2017.

Brazing of Titanium alloys by amorphous foils BNi-2 and BNi-5: shear strength of brazed joints

Alexander E. Shapiro, Leonid A. Shapiro

Titanium Brazing, Inc. Columbus, OH

ashapiro@titanium-brazing.com

Abstract

Application of nickel-based filler metals in order to substitute for expensive titanium-based alloys, for brazing Titanium Grade 2 and Grade 5 is discussed. The standard BNi-2 (Ni-7Cr-3Fe-4.5Si-3.1B wt.%) and BNi-5 (Ni-19Cr-10Si wt.%) filler metals in the form of amorphous foils 60 microns thick were used. Shear strengths of titanium brazed joints made by amorphous foil of BTi-5 and powder of BTi-1 were chosen as criteria to compare with joints made by nickel-based filler metals. Mechanical testing showed that BNi-5 forms brazed joints comparable by strength to titanium-based filler metals, while joints made by BNi-2 have lower strength, which can be improved by diffusion heat treatment performed within the same thermal cycle. Besides, there is a negative circumstance in applying BNi-2 or BNi-5, -- the filler metals require brazing temperatures that are higher than alpha-beta phase transformations of base titanium alloys. Therefore, strength data of base metals after heating above 1000°C are also presented.

1. Introduction

Conventional high-temperature brazing filler metals for joining titanium materials are represented by three alloy families: (1) Ti-Cu-Ni, (2) Ti-Zr-Cu-Ni, and (3) Ag-Cu. Titanium-based fillers are sometimes modified by refractory metals to improve heat resistance of brazed joints. Silver-copper eutectic can be alloyed by tin and indium to decrease brazing temperature, as well as by a small amount of titanium in order to activate wetting of some base materials. Despite significant differentiation in compositions, all above mentioned filler metals are characterized by high cost. Expensive filler metals essentially increase production costs of such brazed structures as titanium honeycomb panels, fin-plate heat exchangers, cryogenic systems, aerospace devices, etc. In order to decrease production costs of titanium structures, we evaluated standard nickel-chromium brazing filler metals such as BNi-2 and BNi-5. Prices of these filler metals are lower by a factor of 3-4 than prices of conventional titanium brazing filler metals.

In the past, application of nickel-based filler metals was limited due to high brazing temperatures: 1010-1100°C for BNi-2 and 1150-1200°C for BNi-5. These temperature ranges are higher than α-β transus temperatures of Titanium Grade 2 (955°C) and Titanium Grade 5 (995°C).

It would be reasonable to test strength of brazed joints and evaluate the risk of losing strength of base metals after brazing titanium alloys by nickel-based filler metals. At least, we can provide the industry with strength data information that will help to evaluate a possible applications of nickel-based standard filler metals from technical and economical points of view.

The following brazing experiments and mechanical testing of shear strength of brazed joint demonstrated that application of nickel-based filler metals is reasonable in many applications.

2. Experimental

Titanium Grade 2 (CP Titanium) and Grade 5 (Ti-6Al-4V) alloys were used as based metals, and BNi-2 and BNi-5 were used in the form of amorphous foils 60 microns thick. Shear strengths of titanium brazed joints made by amorphous foils of BTi-5 and powder of BTi-1 were chosen as criteria to compare with Ni-Cr-based filler metals. Compositions and melting ranges according to AWS A5.8M/A5.8:2011 of all tested filler metals are presented in Table 1.

Table 1

Compositions and melting ranges of tested filler metals

Brazing filler metal	Composition, wt.%	Melting range	Brazing temperature
BNi-2	Ni-7Cr-3Fe-4.5Si-3.1B	971-999 °C	1010 °C
BNi-5	Ni-19Cr-10Si	1079-1135 °C	1200 °C
BTi-1	Ti-15Cu-15Ni	902-950 °C	1010 °C
BTi-5	Ti-20Zr-20Cu-20Ni	848-863 °C	920 °C

A specimen and tooling for shear test by applying compression are shown in Fig. 1-2. Three to five specimens of every combination of the base and braze metals were fabricated and subjected to mechanical testing in order to determine shear strength. Polished cross-sections of brazed joints were etched using the 50%-diluted Kroll's reactive and then were examined using an optical microscope to reveal solidification microstructures.

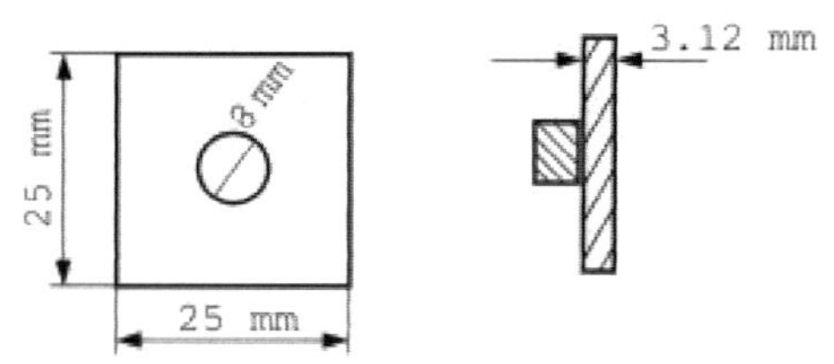

Fig. 1 Specimens used for measuring shear strength of brazed joints: a design [5] for the compression test. Position of a brazed joint to be tested is colored in red

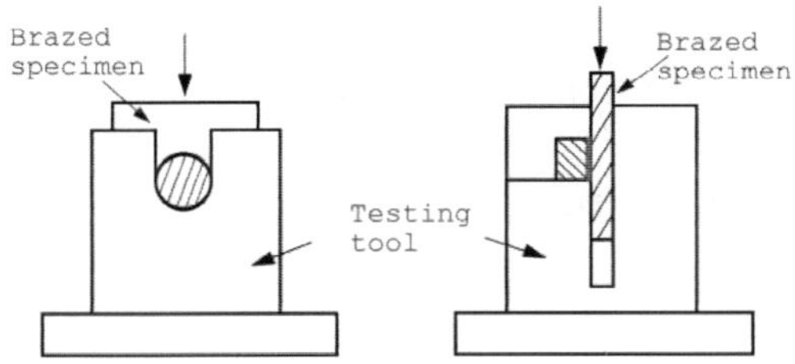

Fig. 2 Scheme of testing and a tool for measuring shear strength of brazed joints by applying compression [5]

Brazing was performed in vacuum $5.5x10^{-5}$ Torr. Brazing foils were placed between the support and the plug (Fig. 1), that was compressed by a "dead weight" ~250 g/cm^2 during heating and cooling in a vacuum furnace. The heating rate was 430°C/hour up to brazing temperature with a step for 20 min at 750°C to equalize the temperature of all brazed parts. Holding time at the brazing temperature was 10 min. Cooling was carried out with the furnace down to 260°C, followed by cooling to room temperature by purging dry argon.

The type or pattern form of the load-displacement diagram was used for the evaluation of the mechanical behavior of tested brazed joints. The type represented by Fig. 3 was considered as ductile behavior of brazed joints, while the type represented by Fig. 4 - as brittle behavior.

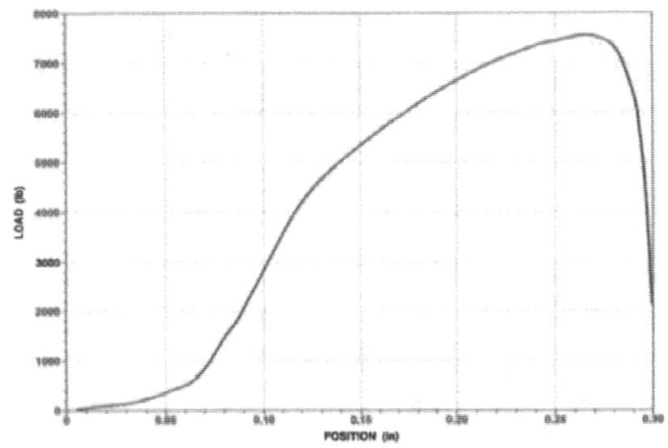

Fig. 3 The load-displacement diagram of Titanium Grade 2 joint brazed by BTi-5 amorphous foil at 920°C, 10 min: Ductile pattern.

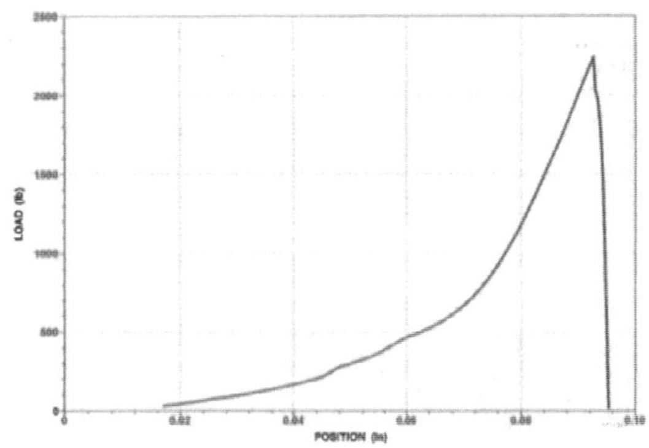

Fig. 4 The load-displacement diagram of Titanium Grade 2 joint brazed by BNi-2 amorphous foil at 1010°C, 10 min: Brittle pattern.

3. Results and discussion

Shear strengths of brazed joints of titanium alloys are presented in Tables 2 and 3. Comparison of strength data for brazed joints of Titanium Grade 2 (Table 2) showed :

(a) Shear strength of brazed joints made by nickel-based filler metal BNi-2 is lower than that of joints brazed by titanium-based filler metals BTi-1 or BTi-5,

(b) The loss of strength values is critical for application of BNi-2, where brazed joints lose over 50% of shear strength if compared with Bti-5, and over 60% - if compared with BTi-1;

(c) Shear strength of titanium joints made by BNi-2 can be significantly improved by increasing brazing temperature and time, e.g., rising of brazing temperature from 1010°C to 1080°C resulted in growth of the strength from 213 MPa to 349 MPa (Table 2). This result indicate that the strength of brazed joints highly depends on diffusion processes, and as a consequence, a post-braze diffusion heat treatment can be effective to improve mechanical characteristics of brazed joints made by nickel-based filler metals;

(d) An effect of high brazing temperature was not found for titanium joints brazed with BNi-5: increasing temperature from 1080°C to 1200°C resulted in the gain of strength only by 9.5%. We suggest that in this temperature range the grain growth and strength loss of the base metal is the main factor to affect the strength of titanium brazed structure. Change from brittle mechanical behavior to ductile (Table 2) supports this conclusion. Comparison of strength data (Table 3) for brazed joints of Ti-6Al-4V alloy (Titanium Grade 5) showed :

(a) Shear strength of brazed joints made by nickel-based filler metal BNi-2 is lower than that of joints brazed by titanium-based filler metals BTi-5. However, the strength of Ti-6Al-4V alloy joints is higher by a factor of 2.25 than that of brazed joints of Titanium Grade 2;

(b) The loss of strength values is not critical for application BNi-2, where brazed joints lose only 19% of shear strength if compared with BTi-5 and 26% - if compared with BTi-1;

(c) Again, maintaining the brazing temperature from 1010°C to 1080°C resulted in the increase of the strength to 541 MPa, which is comparable with strength of joints brazed by BTi-5. This means that low-cost nickel-based filler metal BNi-2 can substitute for expensive BTi-5 for joining Ti-6Al-4V base metal considering the fact that both filler metals reveal a brittle type of failure;

(d) The shear strength of Ti-6Al-4V alloy joints made by another nickel-based filler metal BNi-5 is higher by 3% than that of joints made by BNi-2 and comparable with the joints brazed bi BTi-1. So, BNi-5 can also be recommended to substitute for expensive titanium-based filler metals.

Table 2

Shear strength of Titanium Grade 2 brazed joints and type of their mechanical behavior

	BNi-2	BNi-2	BNi-5	BNi-5	BTi-1	BTi-5
Brazing temperature	1010°C, 10 min	1080°C, 20 min	1080°C, 20 min	1200°C, 10 min	1010°C, 10 min	920°C, 10 min
Form of brazing filler metal	Amorphous foil 60 μm thick	Amorphous foil 60 μm thick	Amorphous foil 60 μm thick	Amorphous foil 60 μm thick	Powder -170 mesh (-90 μm)	Amorphous foil 75 μm thick
Shear strength of brazed joints, MPa	211.0-216.4 213.6	336.4-362.1 349.3	294.0-391.4 342.7	338.1-418.1 379.0	506.9-566.75 536.8	469.7-471.9 470.8
Type of the load-displacement diagram	Brittle	Brittle	Brittle	Ductile	Ductile	Ductile

Table 3

Shear strength of Titanium Grade 5 (Ti-6Al-4V) brazed joints and type of their mechanical behavior

	BNi-2	BNi-2	BNi-5	BTi-1	BTi-5
Brazing temperature	1010°C, 10 min	1080°C, 20 min	1200°C, 10 min	1010°C, 10 min	920°C, 10 min
Form of brazing filler metal	Amorphous foil 60 μm thick	Amorphous foil 60 μm thick	Amorphous foil 60 μm thick	Powder -170 mesh (-90 μm)	Amorphous foil 75 μm thick
Shear strength of brazed joints, MPa	422.4-538.0 480.2	493.4-588.6 541.0	599.4-613.4 606.4	619.8-684.6 652.2	562.7-615.0 588.8
Type of the load-displacement diagram	Brittle	Brittle	Brittle	Brittle	Brittle

Microstructures of brazed joints of Titanium Grade 5 alloy made by BNi-2 and BTi-5 are shown in Figures 5-6. Both nickel-based and titanium-based filler metals form similar microstructures of brazed joints that are characterized by: (a) Mixture of solid solution grains and eutectic in the middle of joint metal, (b) Fully recrystallized diffusion zones that have a Widmanstätten-like crystals, and (c) No intermetallic compounds at the interfaces of the base metal. Slightly different appearance of joint metal patterns in Fig. 5 and 6 can be explained by presence of copper and nickel in the composition of the BTi-5 filler metal, while BNi-2 is copper-free and contains only nickel. As proved in [6], formation of wide diffusion zones with the Widmanstätten-like microstructure improved the strength of brazed joints.

4. Strength of base metals

As brazing temperatures by both tested nickel-based filler metals are higher than 1000°C that is above $\alpha \leftrightarrow \beta$ transus of titanium Grade 2 and Grade 5 alloys, it is important to check if base metals lose strength after brazing. Base metals were heat-treated according to the brazing thermal cycles and tested by tension of standard "dog-bone" flat specimens [7]. Table 4 shows values of tensile strengths of titanium base metals before brazing and after heating to 1000°C.

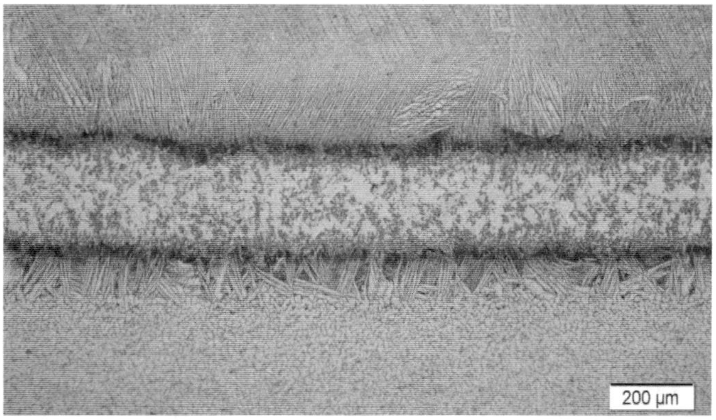

Fig. 5 Microstructure of Ti-6Al-4V alloy joints brazed by nickel-based filler metal BNi-2, x100

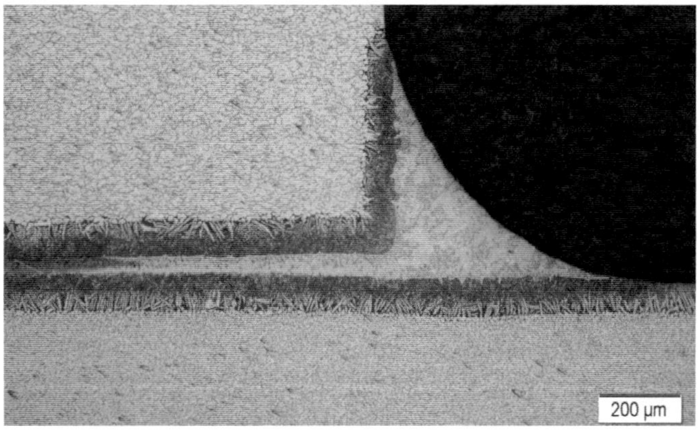

Fig. 6 Microstructure of Ti-6Al-4V alloy joints brazed by nickel-based filler metal BTi-5, x100

Table 4

Tensile strength of titanium base metals after heat treatment according to brazing thermal cycles

Thermal cycle	Titanium Grade 2		Titanium Grade 5 (Ti-6Al-4V)	
	Yield strength, MPa	Tensile strength, MPa	Yield strength, MPa	Tensile strength, MPa
Before brazing	460	524	1045	1070
Brazing at 900°C	346	525	954	997
Brazing at 1000°C	331	419	871	966

Thus, yield strength of Titanium Grade 2 dropped from 460 to 331 MPa, which means that the base metal lost ~28% of its strength due to overheating to 1000°C. We can expect further loss due to brazing at 1080°C. The same time, Ti-6Al-4V alloy lost 17% in yield stress and only ~10% in ultimate tensile strength that can be admissible in some applications in view that the strength of brazed joints made by nickel filler metal BNi-5 is comparable not only with BTi-5 but also with the mostly used titanium filler metal BTi-1 (Ticuni®70) .

Both nickel-based filler metals BNi-2 and BNi-5 can be recommended for vacuum brazing titanium alloys, and also as low-melting components of composite filler metals used for healing cast defects of titanium or superalloy parts. Placed on the bottom of such defects as pin-holes, outside shrinkage pores, outside the local porosity area, these low-melting components spread on the defect surface and infiltrate the high-temperature, nonmelted component of the composite brazing filler metal which is located above.

5. Conclusions

5.1 Both standard nickel-based filler metals BNi-2 and BNi-5 in the form of amorphous foils are suitable for brazing Titanium Grade 5 (Ti-6Al-4V) alloy in vacuum in the temperature range 1080-1200°C.

5.2 Despite the loss of shear strength of brazed joints by ~19%, the low-cost foil BNi-2 can substitute for expensive titanium- and zirconium-containing amorphous foil BTi-5, because the shear strength of Ti-6Al-4V brazed joints is still fairly high, and besides, it can be improved by diffusion heat treatment.

5.3 The low-cost filler metal BNi-5 can substitute for both titanium-based filler metals BTi-1 and BTi-5:-- the strength values of brazed joints are comparable for all these filler metals. Possible loss of strength of the base metal is a point of discussion when brazing by BNi-5 at the high temperature range.

5.4 Both nickel-based filler metals BNi-2 and BNi-5 are also suitable to substitute for expensive BTi-5 amorphous foil for brazing Titanium Grade 2 alloy at the temperature 1080°C but they are not recommended to substitute for the high-strength BTi-1 filler metal that is also used for joining Titanium Grade 2.

6. References

1. Rabinkin, A., Liebermann, H., Pounds, S., Taylor, J., Reidinger, F., and Siu-Ching, L. 1991. Amorphous Ti-Zr-base Metglas®, brazing filler metals. *Scripta Met. et Mater.* 25, 399–404

2. Onzawa T., Suzumura, and Ko M. W. 1990. Brazing of titanium using low-melting-point Ti-based filler metals, Welding Journal, vol. 69, No. 12, 462s-467s.

3. Shapiro A. E. and Rabinkin A. 2003. State of the art of titanium-based brazing filler metals, Welding Journal, vol. 82, No. 10, 36-43

4. Shapiro A.E. Brazing of conventional titanium alloys. 2016. ASM Metals Handbook, 1-25.

5. Matsu K., Miyazawa Y., Nishi Y., Ariga T. 2007. Brazing and interfacial reaction of CP titanium with Ti-Zr-based filler metals, Materials Transactions, vol. 48, No.5, 1055-1059

6. Botstein O. and Rabinkin A. 1994. Brazing of Ti-based alloys with amorphous Ti-25Zr-50Cu

filler metal. Material Sci. and Eng. A188: 305–315.

7. Shapiro A., Niezgoda S., Hanes B., Brooks D., Zhang X.-G. 2017. Microstructure and strength of titanium after heat treatment at different temperatures in the range of 680-1000°C, The Ohio State University, Materials Science Dept., Capstone Project

Diffusion Bonding of Aluminium a Review

Gemse, Felix; Dahms, Steffen; Jahn, Simon
ifw Jena - Guenter-Koehler-Institut fuer Fuegetechnik und Werkstoffpruefung GmbH, Jena, Germany

Abstract

Aluminium alloys are excellent materials for heat transfer (high heat conductivity) in lightweight components (low density). With these properties aluminium is the ideal material for heat sinks or heat exchangers. Nowadays, further requirements are added, for example structural functions. But not every aluminium alloy can fulfil this boundary condition. Often parts consist of two halves, which have to be joined together. Not all aluminium alloys are suitable for brazing because of alloying elements, such as Magnesium. With the solid state joining process diffusion bonding, the limitations can be extended. For the diffusion bonding process, it is important to get metallic surfaces in close contact. The natural oxide layer decreases the diffusion between the components to join. Besides concepts to treat the oxide layer, requirements for diffusion bonding are presented, too. In this paper, the state of the art of aluminium alloy diffusion bonding is given, as well as current applications. Lightweight heat exchangers are used for cooling energy conversion devices like inverter or rectifier for electrical current (e.g. in wind power plants). Other applications of diffusion bonding are the joining of aluminium based materials with superior properties like aluminium metal matrix composites and multi material joints, e.g. bonding of Al to Cu or steel.

Introduction

Aluminium alloys are an excellent material wherever lightweight and high thermal conductivity matter. Due to its low density combined with a high thermal conductivity aluminium could reach a specific thermal conductivity above copper or even silver. Additionally the material costs are considerably lower. Another feature of aluminium and its alloys is the spontaneous formation of a stable oxide layer on its surface which gives this ignoble metal an outstanding chemical resistance against corrosive media. With all this properties aluminium is an ideal material for cooling and heat exchange applications.

In the recent years the requirements for such parts raise by means of higher packing density and compactness. To reach this, parts with integrated cooling channels were developed. The common solution to create inside structured parts is to mill channels in several sheets and join them to a compact unit. But it is a huge challenge to join aluminium without a loss of its properties in the joining area.

One of the most promising techniques to achieve this is the diffusion bonding process.

Applications

Like mentioned before, aluminium is an almost perfect material for heat exchanging solutions at temperatures between -270 °C and +150 °C [1, 2]. Especially in electronic components the power density and with it, the thermal losses increase. Therefore the commonly used air cooling is not sufficient to keep those components in tolerable operation temperatures. This is the reason why more and more electronic applications use water cooling due to its higher cooling capacity and a reduction of needed space and noise from ventilators. The diffusion bonding technique allows production of cooling plates laminated of cut sheets or plates with milled cooling channels like shown in **Fig.1**. Compared to brazed or sealed coolers, Diffusion bonded heatsinks combine the properties of the aluminium material itself, without concessions to chemical or corrosion resistance. Furthermore machining the final shape after bonding is not a problem so sophisticated structures with inner cooling can be realized.

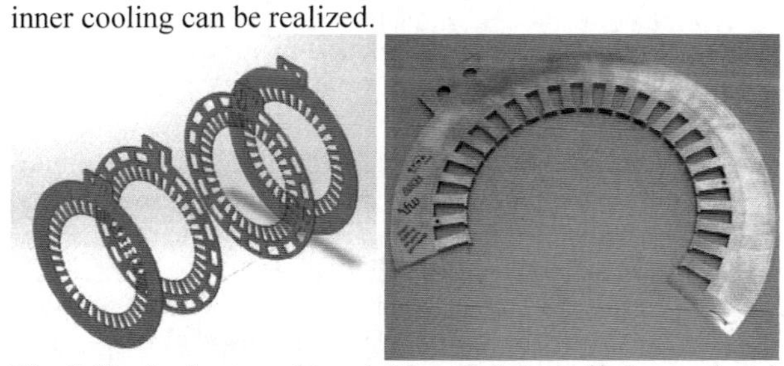

Fig. 1 Single sheets and bonded heatsink (cutoff) for an electric motor with integrated heatsinks [27]

A second field of application is compact heat exchangers, made of a stack of many sheets with channels inside. Compact heat exchangers provide a large surface for heat transfer internally and small outer dimensions. Especially for low temperature application those heat exchangers reach huge conversion rates due to the high heat conductivity.

Alloys for diffusion bonding

In many applications diffusion bonding competes with the well-established brazing or non-thermal joining processes like adhesive bonding. That is why diffusion bonded parts are used were the demands exceed the capabilities of other joining technologies. The main limitation in this field is the brazeability of some alloys especially when the brazing has to be done fluxless. That is the reason why the range of diffusion bonded parts is almost reduced to alloys of the 6000 series [4]. Those alloys provide a very good balance between heat conductivity and mechanical strength **Fig.2**. They are hardenable so the loss of strength caused by the bonding process could be reverted. Their chemical composition provides good machinability and a high resistance against corrosion. But due to their magnesium content the brazeability is limited. An alloy outside the 6000 series which is used for diffusion bonded products is the AlMg3 alloy (AW 5754). Although the material properties could be considered as average this alloy is interesting wherever material costs are in focus.

In the past investigations with high strength alloys of the 7000 series were made. These alloys are of special interest for aerospace applications [5, 6]. But nowadays the application focus of these alloys is in structural parts. Therefore joining technologies like fusion welding or non-thermal joining processes like clinching or adhesive bonding are more efficient [4; 7].

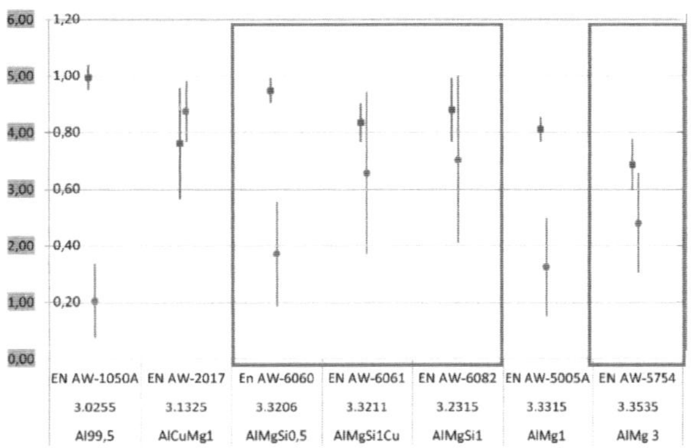

Fig. 2 Comparison of yield strength (red) and heat conductivity (blue) of different aluminium alloys with Al99,5 as base (= 1,00) [3]

Comparatively new classes of aluminium alloys are aluminium matrix composites (AMC). These materials are reinforced with up to 70% of ceramic fibers, whiskers or powder particles to achieve better material properties, especially higher wear resistance, increased strength and further increase of thermal conductivity. Due to the variety of reinforcement materials and their content in the matrix it is possible to tailor material properties in a wide range [8; 9 10]. With the development of such AMCs investigations on diffusion bonding of these materials were carried out [11, 12, 13]. Fundamental research on the behaviour of the reinforcement during the bonding process is conducted. The target applications lie in space and military applications.

Diffusion bonding

Diffusion bonding is a solid-state welding process wherein a joint is formed by interdiffusion between the faying surfaces to form a bond on an atomic level.

To allow interdiffusion bonding is carried out at elevated temperatures between 50% and 90% of the melting temperature of the material. Usually diffusion bonding is done in inert atmospheres like argon or in vacuum [14]. The last one has the advantage of an additional cleaning effect by removal of fluent contaminations and gases from the surfaces. One boundary condition for diffusion is a close contact between the faying surfaces. This is realized by applying a defined load onto the parts, high enough to allow a smoothing of the asperities on the surface by creeping processes but low enough to prevent extensive macroscopic deformation of the whole part. The bond formation takes place by volume diffusion between the both faying surfaces, driven by the surface energy. The main steps of bond formation are shown in **Fig. 3.**

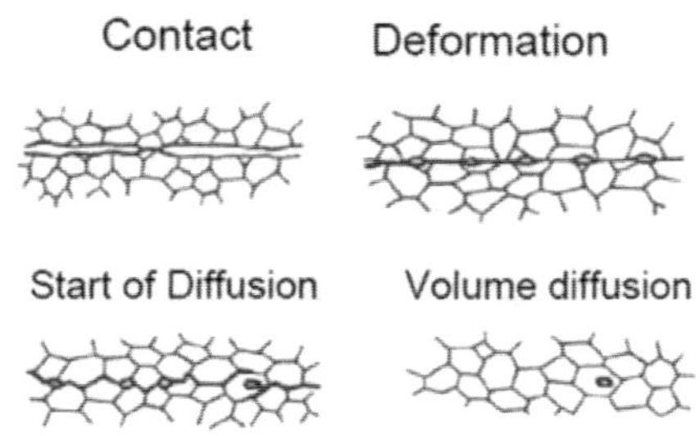

Fig. 3 Process steps of bond formation [26]

In the result diffusion bonding leads to an aerial joint without interlayers or filler materials which has almost the same properties as the bulk material.

One major requirement to achieve this are faying surfaces which are totally clean from residues like dirt, grease and oxides, because they act like diffusion barriers which lead to unbonded areas within the joint and hence lower the quality of the bond.

Especially the demand of oxide free surfaces is almost unreachable for aluminium alloys. The high reactivity of pure aluminium leads to reformation of the oxide layer immediately after the blank metal has contact to air or water (moisture).

Surface treatment

The main challenge in diffusion bonding of aluminium is to overcome the diffusion barrier formed by the stable oxide layer on the surface.

This Al_2O_3 layer has a high chemical resistance and a melting point approx. three times higher than the aluminium below. In the following approaches to get faying surfaces with defined conditions are introduced

Mechanical

The first and possibly easiest way for surface treatment of aluminium is the mechanically removal of the oxide layer. Many authors describe different methods of grinding, milling or brushing the surface. Mechanical cleaning leads only to a temporary removal due to fast reformation. Ortloff [15] stated that the kind of treatment has an important impact on the reformation speed and morphology of the rebuilt layer. So many authors suggest brushing the surface as the best method because of the low thermal influence to the material and the formation of small sharp roughness tips [15, 16]. The deformation of those tips during the bonding process causes a breaking of the oxide layer and leads to pure metal contacts between the faying surfaces.

Other approach is to prevent the reformation of the oxide layer. Wernick [17] describes grinding in organic fluids to shield the metal from oxygen and keep the time between grinding and bonding as short as possible to keep re-oxidation low.

Furthermore procedures are described to cover the prepared surface with sealant which closes the surface up from the environment. This sealant is thermally crackable and could be removed during the heating phase inside the bonding furnace. But especially for larger parts the removability without residues is questionable [18, 19].
A Solution to overcome the problems with reformation of oxides and sealants it to machine the surface in inert atmosphere and keep the samples in it until the bonding process is finished. The way to achieve this is the connection of a glove box to the diffusion bonding furnace. Inside the box and in argon atmosphere the surface preparation of the samples is done. Through a lock the samples were transferred into the furnace chamber were diffusion bonding is done in a vacuum atmosphere. This setup guarantees a total shielding from oxygen and water. Bonding experiments showed that due to the lack of the oxide layer, the bonding temperature could be reduced by 200°C for some alloys. Besides the advantages the huge effort for the whole bonding system is to mention. The furnace needs a second door which acts as lock, between glove box and furnace chamber a transfer unit is necessary which allows the transport and the accurate positioning of the sample in the furnace and of course all the lock and the movement device must withstand the high bonding temperatures. All this might be the reason why the actual known systems are rather small sized and used for research applications [29].

Chemical Approaches

Besides the mechanical methods chemical cleaning is an option to remove/reduce the oxide layer. Most important are etchants based on caustic soda or caustic potash which is able to solute the oxide layer but both chemicals relief a film of aluminium hydroxide on the etched surface which again could be removed by acids [17, 18, 25]. Acids like sodium dichromate are also able to dissolve the oxide layer they relief reactive layers e.g. chromium on the surface that also impede the diffusion bonding process.
At least all chemical procedures demand a final rinsing step with water to neutralize and remove the etchant. This supports the reformation of oxide.

Physical Approaches

Some workgroups utilized a focused ion beam of Argon to remove the oxide combined with the transfer of the prepared samples under vacuum to the bonding chamber. A feature of this technique, besides the cleaning effect, is an activation of the surface by distortion of the near surface microstructure. This allowed a significant reduction of the bonding temperature and a high quality bond [20].
With respect to process time of several hours for ion cleaning and the necessary vacuum equipment the production scale feasibility has to be considered.

Interlayers

An approach to partially circumvent the problems of removal and reformation of the oxide layer is the usage of interlayers like Copper, Nickel, Silver or Zinc [16; 21]. Most of the investigated interlayers cannot improve the bond strength significant due to the formation of brittle intermetallics in die bond zone. A promising approach is the use of a thin (1 µm – 10 µm) Copper interlayer which forms an eutectic at 548 °C [22; 23, 24] followed by isothermal solidification of the reaction layer due to a shift of the melting point by solving aluminium from base material into the melt. The presence of a liquid phase aids the fracture of the oxide layer. This process is known as transient liquid phase bonding (TLP). Although TLP bonding provides high bond strengths the usage of a different material may influence the corrosion resistance by means of bimetallic corrosion.
A special technique which differs from the above mentioned described in [30]. In this case Gallium is utilized as interlayer. Gallium forms a eutectic with the aluminium, as soon as both metals get in contact. This could be done by simply whipping the liquid Gallium over the aluminium surface. The Al-Ga Eutectic forms below the oxide layer so the oxide could be whipped off, while leaving a Gallium-rich coating on the surface. The Coating prevents from re-oxidation. Due to its low melting temperature, its chemical properties and low concentration the Gallium diffuses into the aluminium during the bonding process without a detectable effect to the material. As most of the actual solutions, this process is very effective on smaller surfaces due to the handling of the liquid gallium [30].

Process Parameters

The common parameters to define a diffusion bonding process are the parameters load, soak time and bonding temperature. As for all diffusion bonding processes these parameters depend strongly on the chemical composition of the alloy which should be bonded, the surface condition and the acceptable deformation in pressing direction. Typical values for aluminium alloys are around 500 °C to 550 °C , 30 - 60 min soak time and a pressure (load per area) of 2.5 to 5 MPa [15, 16, 19]. In a certain range the process parameters are interchangeable for example higher temperature and low pressure. Other criteria are part geometry and the capability of the bonding equipment. Since the development of the diffusion bonding process one aim is to reduce the bonding temperature. Besides the above mentioned surface treatment procedures the extreme increase of the bonding pressure was investigated [25]. With this approach bonding temperatures around 300 °C could be realized but these technologies can be considered as a special case of friction welding due to high deformation rates. And of course the demand on stability of the bonding equipment is clearly higher.

Conclusions

Amir Shirzady titled one of his publications with "Aluminium - perhaps most difficult to bond material" [28].
Diffusion bonded aluminium components have an unrivalled potential for heat transfer applications in lightweight products. The chemical resistance and long term stability is considerably high because of the absence of filler materials. With this range of properties diffusion bonding is can compete or even beat other joining technologies. That is the reason why the range of diffusion bonded parts especially in the field of heat exchange applications is a still growing market. Nevertheless diffusion bonding of aluminium is challenging due to its oxide layer and the ongoing research on how to deal with, show that a general and reliable solution is still not available. Furthermore the development of new aluminium based materials is progressing and demanding adapted joining technologies.

References

[1] http://www.linde-engineering.com/internet. global.lindeengineering.global/en/images/P_3_2_e_12_150dpi 19_5772.pdf, (2016)

[2] Kvackaj T. Aluminum alloy, Theory and applications, Intech Open (2011)

[3] Jahn, S. et al. Cooling system for electric power trains – diffusion bonded heatsinks, Proceeding of the 8th International Congress Aluminum Brazing (2014)

[4] http://aluminium.matter.org.uk/content/html/ ger/default.asp?catid=214&pageid=2144417055#plane; 2016

[5] Heinz, A., Recent development in aluminium alloys for aerospace applications, Materials Science and Engineering: A, Vol. 280 (2000) Iss. 1, pp. 102 – 107

[6] Merati, A., Eastaugh, G., Determination of fatigue related discontinuity state of 7000 series of aerospace aluminium alloys, Engineering Failure Analysis, Vol. 14, (2007) pp. 673 – 685

[7] Pujari K.S.; Patil, D.V. A Review on GTAW Technique for High Strength Aluminum Alloys (AA 7xxx series); Intl. Journal of Engineering & Technology, Vol. 2 (2013)

[8] Surapp, M.K.; Aluminum matrix composites: challenges and opportunities; Sadhana; Vol. 28 (2003); pp. 319-334

[9] Torralba J.M. et. al., P/M aluminium matrix composites: an overview; Journal of Materials Processing Technology; Vol. 133 (2003) Iss. 1-2; pp 203-206

[10] Ibrahim, I.A. et al. Particulate reinforced metal matrix composites – a review, Journal of Materials Science, Vol 26 (1991) pp1137-1156

[11] Zhang X.-P. et.al, Investigation on diffusion bonding characteristic of SiC particulate reinforced aluminum metal matrix components (Al/SiCp-MMC), Composites Part A: Applied Science and Manufacturing, Vol 30 (1999) Iss.12, pp 1415 - 1421

[12] Zang G. et al. Joining of Al_2O_3p/Al composites by transient liquid phase (TLP) bonding an a novel process of active-transient liquid phase (A-TLP) bonding; Materials Science and Engineering Vol. 488,(2008), Iss. 1-2, pp 146 - 156

[13] Muratoglu M. et al. Investigation on diffusion bonding characteristics of aluminum metal matrix composites (Al/SiCp) with pure aluminum for different heat treatments. Journal of Materials Processing Technology, Vol. 178 (2006) pp. 211 – 217)

[14] Kazakov, N.F.; Diffusion Bonding of Materials. Pergamon Press, 1985

[15] Ortloff S., Diffusionsschweißen hochfester Aluminiumlegierungen, Diss. Herbert Utz Verlag (1995)

[16] Shirzadi A.A. Diffusion Bonding Aluminum Alloys and Composites – New Approaches and Modelling, PhD thesis,

[17] Wernik, S. et al. Die Oberflächenbehandlung von Aluminium, Eupen Lenze Verlag (1974),

[18] Method of fluxless brazing and diffusion bonding of aluminium containing components; Patent US 3937387 A; (1976)

[19] Huang, Y. et al. Diffusion bonding of superplastic 7075 aluminium alloy, Materials Science and Engineering: A, Vol. 266 (1999) pp. 295 – 302

[20] Haiyan Ch. et al; Low-temperature diffusion bonding of pure aluminium; Applies Physics A; Vol. 113 (2013); Issue 1; pp 101/104;

[21] Jian, Z. et al; An investigation on diffusion bonding of aluminium and magnesium using a Ni interlayer; Materials letters 83 (2012); 189-191

[22] Shirzadi A.A., Wallach E.R.; New approached for transient liquid phase bonding of aluminium based metal matrix composites; Material Science and Technology; Vol. 13 (1997); Iss. 2.

[23] ASM Handbook Vol. 3, Alloy Phase Diagrams, (1992), page 2.44

[24] Urena, A. et al. Diffusion Bonding of an Aluminum-Copper Alloy reinforced with silicon Carbide Particles (AA2014/SiC/13p) using Metallic Interlayers; Scripta Materialica, Vol. 35 (1996) Iss. 11, pp. 1285 -1293

[25] http://www.sgersing.de/pdf/SFT041.pdf

[26] Owszarski, W.A., et al. Application of Diffusion Welding in the USA, Welding Journal, Vol.22, (1981)

[27] Köhring P. et al. Ein neuartiges Kühlsystem für elektrische Fahrzeugantriebe, Proceedings Workshop hochtemperaturfügen (2012)

[28] Shirzadi et al., Aluminum – perhaps most difficult material to bond; Surface and Interface Analysis 31 (2001); 609-618

[29] Habisch, S., Mayr, P. In-line surface treatment and diffusion bonding, Symposium of World Experts of Diffusion Bonding, oral presentation 20.06.2017

[30] Shirzadi A.A. and Saindrenan G. (2003): New method for flux-free diffusion brazing aluminium alloys using liquid gallium (UK Patent 2386578), Science and Technology of Welding and Joining, Vol. 8, No. 2. pp. 149-153.

Mg tolerance in controlled atmosphere brazing with different TRILLIUM® claddings

Torkel Stenqvist, Slawomir Koscielski, Richard Westergård
Gränges Sweden AB, Slottsvägen 1-3, 61281, Finspång, Sweden

Abstract

A study relating to controlled atmosphere brazing (CAB) of Al with Mg containing core is disclosed. Lab scale clad sheet cores are made with varying Mg contents, an alloying element which cause brazing difficulties using normal CAB flux of the Al-K-F type. The cores are clad with the flux containing TRILLIUM® Solid or TRILLIUM® Lean Al-Si composite filler materials. These new fillers are compared with normal CAB filler with varying flux amounts applied using normal slurry fluxing. The brazing tolerance to Mg in the core alloy is evaluated in brazing under a variety of heating conditions. The Mg tolerance for all materials is increased with heating rate to brazing temperature, and the higher amount of flux either from inside or externally applied to the cladding. It is concluded that especially the TRILLIUM® Solid has better tolerance to Mg compared to normal slurry fluxed sheet. The tolerance for Mg for TRILLIUM® Lean is surprisingly large, considering the small amounts of flux maintained in the cladding. Also, the small amounts of flux in TRILLIUM® Lean clad brazing sheet reduces or eliminates industry issues with post-brazed flux residue in heat exchangers in low Mg core sheet.

Introduction

The controlled atmosphere brazing (CAB) of Aluminium use a Al-K-F flux and clad brazing sheet. The flux reacts with Mg ("poisons" the flux) [1], thus rendering it less efficient. Hence, the braze cladding is normally free from Mg while core alloys with up to 0.4 wt-% exist. To braze these, for CAB, highly Mg alloyed materials require very high loads of flux, or special Cs-bearing flux types. The ability to use Mg in clad brazing sheet is very appealing as it is a very efficient strengthening element, both when in solid solution in the alloy as well as in the form of strengthening precipitates e.g. Mg_2Si.

Gränges introduced the TRILLIUM® brazing material several years ago; this is a filler alloy with flux incorporated in the cladding material itself. The TRILLIUM® cladding is supplied on brazing sheets, thereby removing all need for fluxing and the problems associated (waste water, housekeeping, floor space, environmental and health issues, etc.) with that process. Brazing sheet where the complete cladding is made of TRILLIUM® plates, TRILLIUM® Solid, shows an extremely robust and repeatable brazeability also in very adverse atmospheric conditions [2, 3]. Presently a version where only a part of the filler metal is the TRILLIUM® composite, TRILLIUM® Lean, has been introduced in response to industrial demands for lower flux residue.

Early developments of TRILLIUM® Solid indicated that the brazing tolerance for Mg was surprisingly good, but the extent was not fully established until recently. As the development TRILLIUM® Lean was made it too had to be tested for Mg tolerance, and a comparison with ordinary surface fluxing is needed to put the results into proper context.

Experimental details

Materials

Chemistries and denotations

In Tables 1-3 the materials tested here are seen. It comprises materials clad with TRILLIUM® Solid, TRILLIUM® Lean and corresponding normal clad materials.

For the cores, lab cast materials with AA3003 base chemistry were used, and Mg was added to reach 0.1, 0.2, 0.3, and 0.4 wt-%. Materials clad with TRILLIUM® Solid were also tested with cores containing 0.25 and 0.35 % of Mg. As non-TRILLIUM® filler cladding, AA4045 was used. The TRILLIUM® cladding materials were based on AA4045 with a brazing flux incorporated into it. The angle was made of an unclad 0.42 mm AA3003 in temper H16. The TRILLIUM® material was sampled from an extruded bar with 25 mm thickness made in full scale production. The basic chemistries are seen in Table 1. The flux used in the TRILLIUM® was of the $Al_{1-3}KF_{4-6}$ type.

Table 1. Chemical composition of the AA4045, TRILLIUM® and the angle materials.

Material	Si	Fe	Cu	Mn	Mg	K
AA4045 (clad and TRILLIUM® Lean)	10.1	0.19	<0.01	0.02	<0.01	---
TRILLIUM® Solid	9.5-10.5	0.15	0.05	0.01	0.01	0.6-1.6
AA3003 – angle	0.18	0.50	0.11	1.1	0.035	---

Table 2 Chemical composition of cores 1-side clad with 10% TRILLIUM® Solid.

Alloy AA3003	Si	Fe	Cu	Mn	Mg
+0Mg	0.5	0.5	0.11	1.1	<0.01
+0.1Mg	0.51	0.62	0.11	1.22	0.10
+0.2Mg	0.51	0.51	0.12	1.17	0.21
+0.25Mg	0.51	0.35	0.13	0.65	0.26
+0.3Mg	0.46	0.43	0.11	1.11	0.30
+0.35Mg	0.50	0.47	0.11	1.12	0.35
+0.4Mg	0.50	0.49	0.11	1.17	0.41

Table 3. Chemical composition of the core samples for TRILLIUM® Lean and clad materials for slurry fluxing.

Alloy AA3003	Si	Fe	Mn	Cu	Mg
0Mg	0.13	0.52	1.1	0.10	0.03
+0.1Mg	0.12	0.5	1.11	0.1	0.10
+0.2Mg	0.12	0.51	1.13	0.1	0.20
+0.3Mg	0.12	0.48	1.10	0.09	0.30
+0.4Mg	0.12	0.49	1.09	0.09	0.40

Sheet build-up and lab production

All sheet materials were lab cast after mixing alloying ingredients in a proper order at temperatures of ~710-740 °C and cast in a book mould. The resulting ingots and the extruded TRILLIUM® bar were scalped to a thickness of 14 mm. Then the materials were subjected to a simulated preheating by 10 hours linear heating to reach 500 °C, soaked for 2 hours and then cooled to RT. The cores and claddings were then rolled to appropriate gauges for cold clad rolling, and then softened at 250 °C for 2 h.

The TRILLIUM® Lean claddings were made by combining an AA4045 lid, an AA4045 base and a TRILLIUM® plate. The lid was always made to be 10 % and the TRILLIUM® and base thicknesses were made to vary the flux load (g/m^2) in the combined TRILLIUM® Lean plate.

Then, when all claddings and cores were ready they were combined in cold clad rolling to provide clad brazing sheet, rolled to 0.1, 0.2, 0.3, 0.5, 1.0, and 1.5 mm, giving several different flux loads. Then, materials thinner than 0.5 mm were given a final partial annealing at 250 °C for 2 h to reach an H24 temper. Material of 0.5 mm or thicker were given a full soft annealing at 380 °C for 2 h. After production, all samples were ultrasonically degreased in ethanol. The build-up and microstructure of a TRILLIUM® Lean clad plate is exemplified in Fig. 1.

All TRILLIUM® Solid materials were made with 10 % cladding, rolled to 0.3 mm thickness and partially annealed at 250 °C during 2 h.

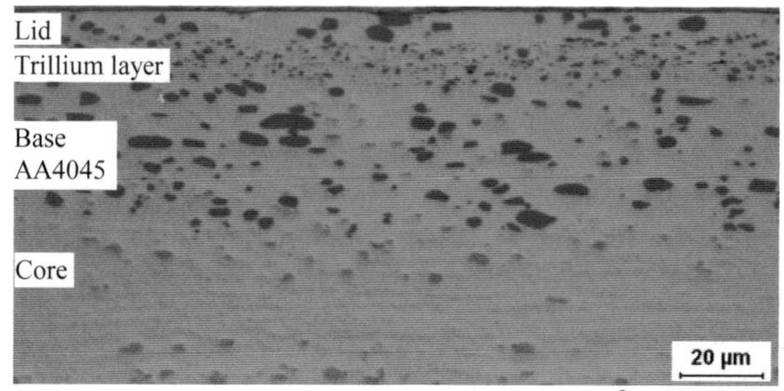

Figure 1. Microstructure example of TRILLIUM® Lean clad sheet.

Evaluation techniques

Brazing

Brazing tests were made in angle on coupon geometry on clad coupons in a laboratory glass furnace. The angles had a milled contact side. All samples were degreased in ethanol before brazing.

The atmosphere consisted of pure N_2 with < 3 ppm O_2 and a dew point of < -60 °C, but the furnace is not completely air tight and some oxidation of the samples and the flux during the heating may occur. The linear heating rate from room to brazing temperature 600 °C was either 5, 15, 30, or 60 °C/min.

Fluxed samples

Flux slurries were mixed using Nocolok100 with either isopropanol or deionized water with 0.15 % Antarox BL-225 wetting agent. The flux concentrations in the slurries were calculated to give the correct flux coverage using an automatic pipette sampling 20 µl from the agitated flux slurry. Samples were fluxed to 1, 2, 4, 6, or 8 g/m^2. A hot plate was used to dry the samples with water as carrier. Isopropanol wets the surface well and the flux is spread over all the surface with a little wiggling of the sample. After some seconds the alcohol has evaporated, leaving the flux. For the water carrier the slurries form a droplet on the surface. In one set of experiments the flux mixture was spread out over the surface manually, Fig. 2, in another set the droplet was put on one side of the coupon and allowed to dry leaving a flux agglomeration; here one leg of the angle was set on the agglomeration and the other was on the bare surface, Fig. 3. The leg on the flux will have an abundance of flux, and always join; the test was rather how well the flux spread and facilitated the joining of the other leg.

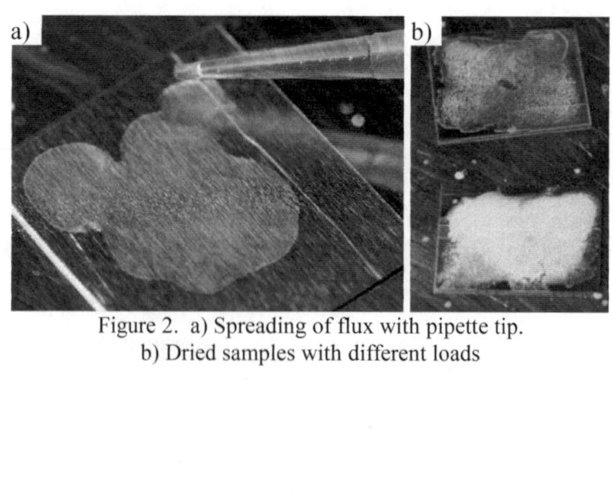

Figure 2. a) Spreading of flux with pipette tip.
b) Dried samples with different loads

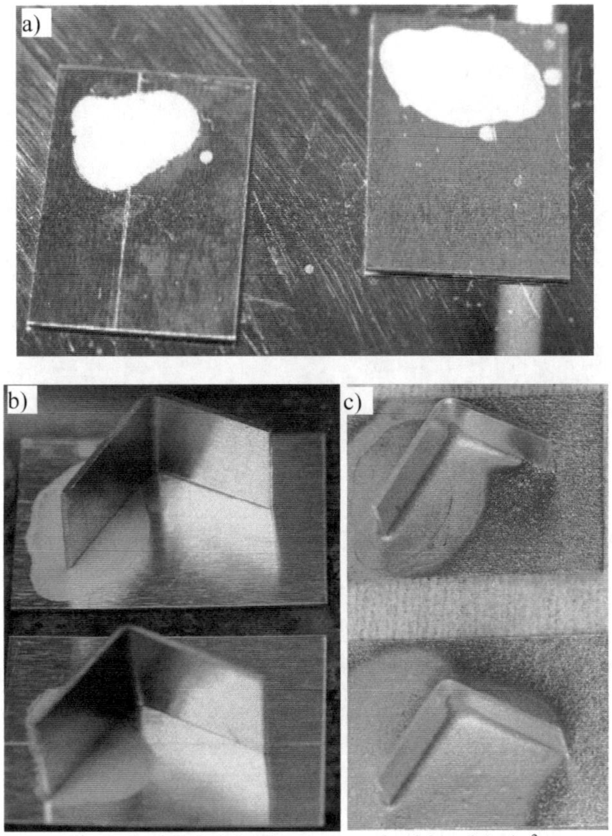

Figure 3. a) Dried droplets, average load:2 and 4 g/m^2;
b) The samples before brazing
c) The same samples with angle after brazing.

Analyses

After brazing the length of the fillets on each side of the legs were graded from 0 to 100 % joint filling. TRILLIUM® Lean, with several different gauges had different amounts of filler metal available for joint filling, giving a variety of joint sizes; however, no regard was taken to the size of the fillet, more than that the fillet was readily visible.

Samples of all materials made here were cut in the longitudinal direction before and after simulated brazing, hot mounted, ground, polished and etched in 2 % HF(aq.) to observe the particle microstructures using light optical microscopy. TRILLIUM® Solid materials were cut transverse the angle near the middle of the angle legs, cold mounted, then ground and polished. The joint areas were then measured using image analysis.

Spark excited optical emission spectrometry (OES) was used to analyze the composition of cores and non-TRILLIUM® claddings. TRILLIUM® containing claddings were analyzed using glow discharge optical emission spectrometry (GDOES) to analyze the chemical content as a function of the depth below the original sheet surface. It was specifically used to measure the amount of K in the material and, assuming the flux is present as $KAlF_4$, the amount of flux/m^2 was calculated.

Scanning electron microscopy with energy dispersive spectroscopy (SEM-EDS) was used to study the surfaces of the brazed materials.

Results

Joint formation using standard fluxing of conventional brazing sheet

Typical brazed parts are exemplified in Fig. 4 and the cross-sections are illustrated in Fig. 5. The summarized braze results of all the fluxed samples as a function of flux load and Mg content are shown in Fig. 6. The expected general trends of decreased brazeability with increasing Mg content in the core and decreased heating rate are clear. Further, with the spread-out flux the two flux bearing agents are very similar.

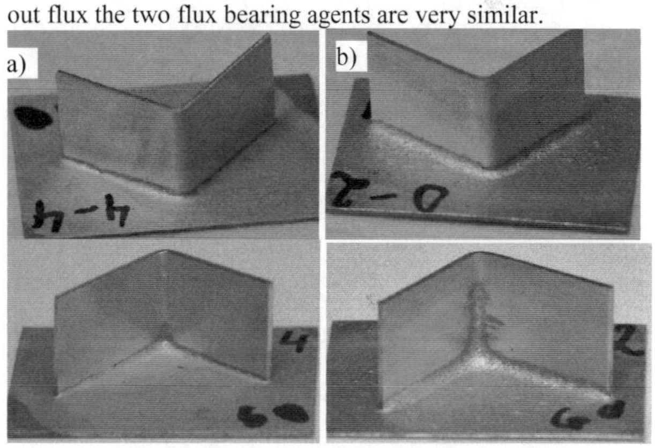

Figure 4. Typical brazed joints, isopropanol fluxing, 60 °C/min heat up rate.
a) Smaller joint: 0.4 wt-% Mg, 4 g/m^2 flux,
b) Large joint: 0 wt-% Mg, 2 g/m^2 flux.

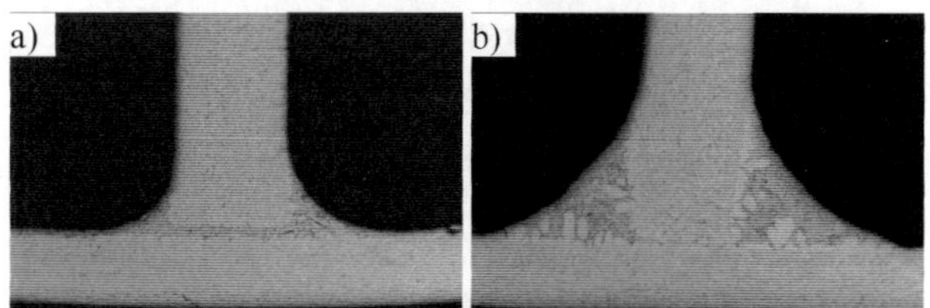

Figure 5. Cross-sections
a) Smaller joint: 0.4 wt-% Mg, 4 g/m^2 flux
b) Large joint: 0 wt-% Mg, 2 g/m^2 flux

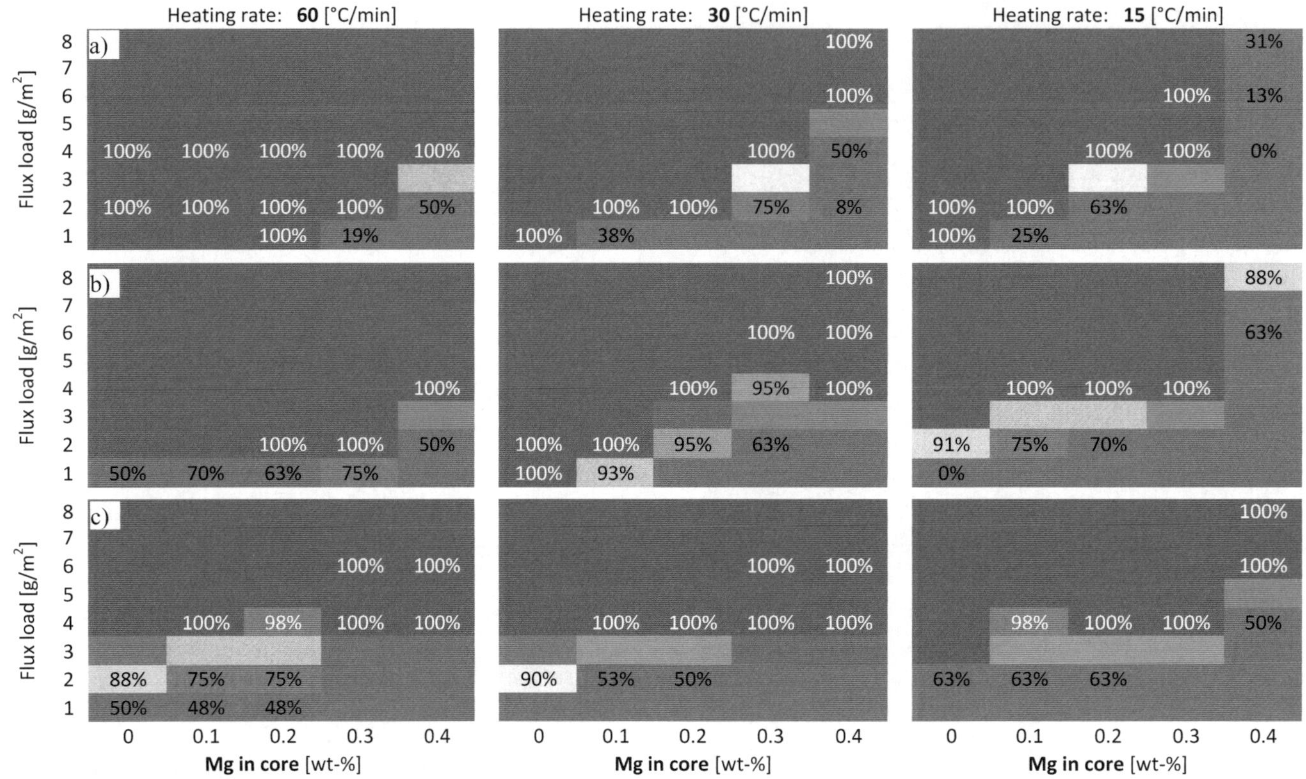

Figure 6. Fluxed samples, brazing results (ratio of filled length) at different heating rates.
a) Flux isopropanol slurry. b) Manually spread flux water slurry. c) Dried droplet of flux water slurry.

Joint formation in TRILLIUM® Lean

Typical brazed part is exemplified in Figure 7 and the cross-section is seen in Figure 8.

Figure 7. Braze joint of 0.3 mm TRILLIUM® Lean with flux load 0.27 g/m^2. Heating rate 60 °C/min.

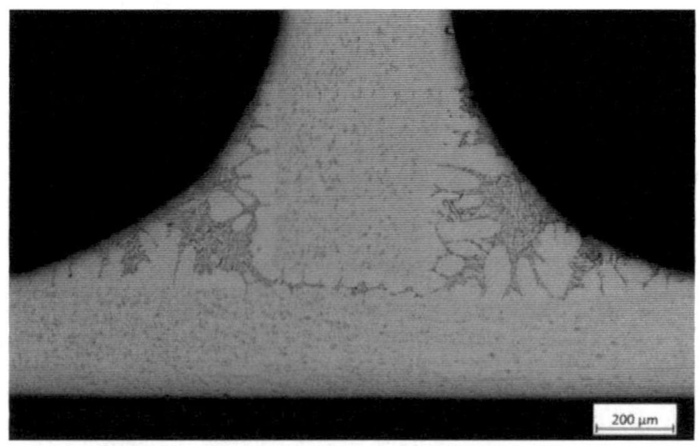

Figure 8. Cross-section, sample from Fig. 7.

The summarized braze results of all the TRILLIUM® Lean samples are shown in Figure 9, sorted after flux content, Mg in core, and heating rate. The brazeability decreases with increasing Mg content in the core and decreased heating rate as expected.

Flux load [g/m²]	Heating rate: 60 [°C/min], Mg in core 0 [wt-%]	0.1	0.2	0.3	0.4	Heating rate: 30 [°C/min], Mg in core 0 [wt-%]	0.1	0.2	0.3	0.4	Heating rate: 15 [°C/min], Mg in core 0 [wt-%]	0.1	0.2	0.3	0.4
4.1					100					100			100	100	100
2.8				100	100			100	100	100			100	50	88
2.4			100	100	100			100	100	75			100	50	13
1.8			100	100	0		100	100	75	0			88	0	
1.4			95	50				80	0	0		100	100		
1.3			100	38			100	100	0			100	0		
1.1	100	100	45	50		100	100	25				100	0		
0.88	100	100	63			100	98	0			100	98			
0.82		100	56			100	100	50	0			100			
0.65	100	100	0			100	98				100	100			
0.59	100	98	75			100	75				100	50			
0.47	100	50	0			100	61					0			
0.41	88	75				100	88				100	3			
0.27	75	50				98	85				100				
0.23	88	50				8					0				
0.19	50					0					25				
0.18	0					0					0				

Figure 9. TRILLIUM® Lean, brazing results (ratio of filled length) at different heating rates. Sorted after flux content.

Joint formation in TRILLIUM® Solid

The flux inherent in the filler corresponded to 2.1 g/m². The example appearance of cross-sectioned joint is shown in Fig. 10. The joint size variation with Mg is seen in Fig. 11. Joints with size < 0.8 mm² gave stitching and discontinuities which worsened as joints became smaller. The brazed joint size is reduced as Mg content in the core increases and as the heating rate to brazing temperature decreases.

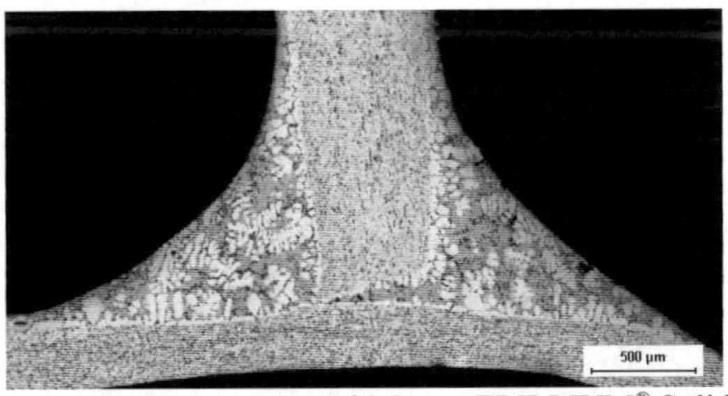

Figure 10. Cross-section of 0.3 mm TRILLIUM® Solid, 0.3 wt-% Mg brazed with a heating rate of 60 °C/min.

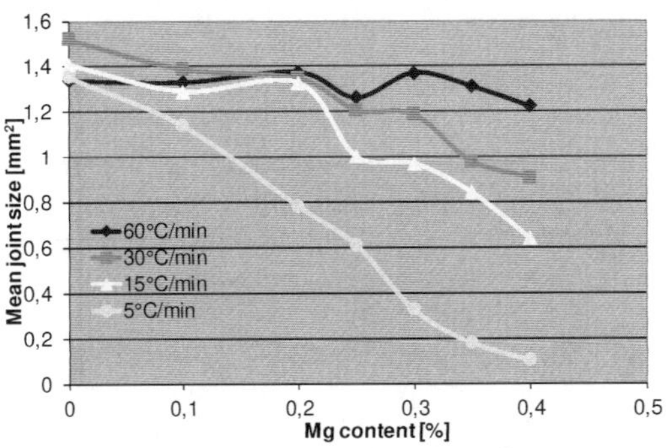

Figure 11. Brazed joint size as a function of Mg content and heating rate when brazing TRILLIUM® Solid.

All Mg containing core materials, regardless if it was TRILLIUM® or slurry fluxed claddings, gave surface residue with needle-like, "poisoned" flux residue whereas the Mg-free materials gave flux residue without such particles, see Fig. 12. Also, small flux amounts gave little surface flux residue whereas much flux in or on the cladding gave more residue.

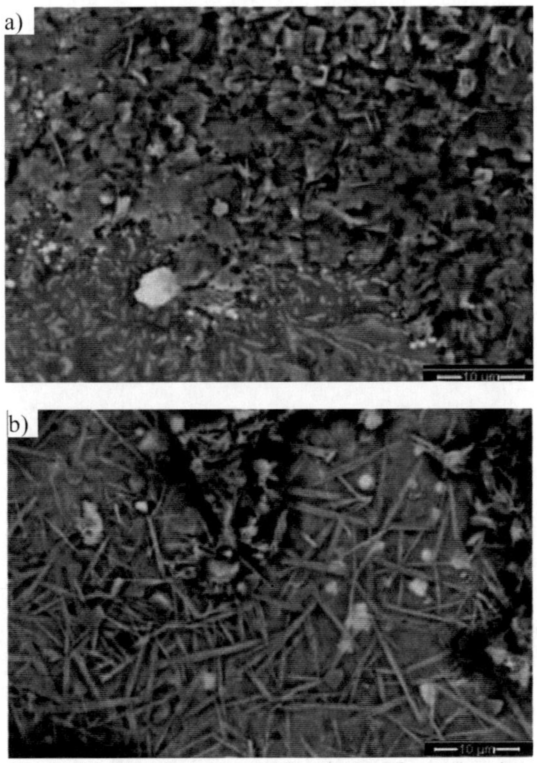

Figure 12. The appearance of brazed surfaces showing a) surfaces with undisturbed flux residue on Mg-free cores b) poisoned flux residue on Mg-containing cores.

Discussion

Limitations with angle-on-coupon brazing and glass furnace set-up

For successful angle on coupon trials to be made it is important that the coupon is as flat as possible. Here, the coupons were made using lab rolling and the flatness is neither especially good nor repeatable and the gap size between angle and coupon varies. This causes the wettability to differ along the nominal contact between angle and coupon. In several cases poor joints are made because of a non-optimal test geometry between these two components. Also, the angle is important and here the contact surface and geometry between angle and coupon varies from a microscopical perspective. This means that variations in the contact surface between angle and coupon may influence how wetting and joint formation may occur, and the symmetry of the formed joint. Further, there is no external pressure applied to the angle, only its weight.

The glass furnace itself also influences how wetting and joint formation takes place. The specimen table has a small temperature gradient, which is carried over to the surface of the coupon. If a part of the braze in the angle/coupon contact is molten and wetted earlier than others it may affect how joint growth occurs, with the already formed joint attracting the molten filler as it becomes available. This may cause an asymmetric joint size along the legs of the angle.

The atmosphere quality is very important in CAB. The glass furnace used here is displacement purged using a very clean gas (< 3 ppm O_2, dew point < -60 °C). However, the mixing of the clean gas and the original air is imperfect and even a 10 min purge before heating starts may leave significant residual O_2 and water vapor in the brazing atmosphere. The furnace has also open vents. This will make wetting and joint formation more difficult.

The above considerations must be taken into account when interpreting the outcome of the brazing trials. One can safely assume that an industrially produced material using state-of-the-art components provides as good or better result when brazing heat exchangers.

Slurry fluxed samples

At the high heat rate for low Mg contents 1 g/m^2 flux is sufficient using isopropanol while the water carrier needs 2 g/m^2. The poor results for the latter is not consistent with the brazing at the intermediate rate, where these samples showed better results. It seems as other random factors (the actual flux content in the 20 μl sampled could vary due to the agitation, there is a small uncertainty in the exact volume sampled, the degreasing of the sample, the contact with the angle, etc.) have influenced the result, and it can be expected that such differences have more importance at lower flux levels.
Both test series with spread-out flux failed to braze the material with 0.4 wt-% Mg with even 8 g/m^2 flux in the low heating rate. However, this was possible if the flux was applied as droplets, an apparent difference. This behavior can be explained, as there is a **local** fluxing, where the actual concentration is higher than the nominal load. Figure 13 shows the difference for the two droplet samples with 0.2 wt-% Mg brazed at 30 °C/min with 2 and 4 g/m^2 flux, respectively. With a low load of flux, the flux does not spread much; the actual original droplet edge is still visible in the flux residue. With a higher load, the flux wet the surface and flow along the length of the opposite angle leg and then spread out from there. Thus, the local flux load at the joint site was above the 8 g/m^2, but 4 g/m^2 average load was necessary for the flux to migrate.

For lower average flux loads the flux does not spread, and hence the brazeability of the second leg is poor. All samples with about 50 % filled joint length or less have not shown any flux spreading.

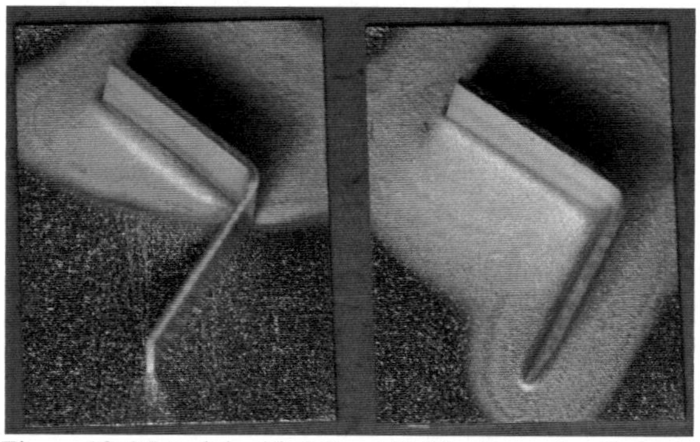

Figure 13. Material with 0.2 wt-% Mg. Braze heating rate 30 °C/min. Flux load 2 and 4 g/m^2.

TRILLIUM® Lean

With no Mg in the core the brazeability is very good, well formed repeatable joints can be achieved with 0.23 g/m^2 flux in a fast braze cycle, however some results with low flux are erratic as seen comparing the different heating rates. Since a number of different gauges have been used there is more variation than with the slurry fluxed samples, for example the clad thickness – which is the distance the Mg atoms have to diffuse to reach the surface and react with the flux. The varying lid thickness is also a factor to be considered, a thicker lid gives a larger migration distance to the surface once the flux has molten. With these lab samples similar flux levels have been obtained in different ways. A more detailed inspection indicates a better brazing with a thicker TRILLIUM® layer, at the same flux coverage; however, more experiments are needed to verify this.

It is clear that very low flux levels braze well with low Mg contents. A core content of 0.1 wt-% Mg requires about 0.6 g/m^2 flux, and with 0.2 wt-% Mg the flux needed increase to about 1.3 g/m^2. At higher Mg levels more than 2 g/m^2 are required, but comparing with ordinary fluxing this is rather low. Since the flux in the TRILLIUM® Lean is spread out the best comparison is with the spread flux, Figures 7 and 8, where 1 g/m^2 flux gave varying results for low Mg and 4 g/m^2 was required for high levels using a good heating rate. For the slow heating rate TRILLIUM® Lean brazed 0.4 wt-% Mg cores with 4 g/m^2 flux, while 8 g/m^2 was not sufficient with spread flux.

TRILLIUM® Lean was designed to replace current production, and presently low Mg contents are favored by the heat exchanger producers. The angle on coupon test is conservative, but it is clear that a TRILLIUM® Lean strip with 0.6 g/m^2 flux could be used for all applications where the Mg level is less than 0.1 wt-%. These low flux levels likely reduce or eliminate industry issues with residual flux after brazing. For a normal "Long-life" material (0.25 wt-% Mg) about 2 to 2.5 g/m^2 flux is likely needed when brazing in our glass furnace.

TRILLIUM® Solid

For the rapid industrial-like heating rates the tolerance to Mg is obviously very good. This is because at high heating rates less Mg has time to diffuse out into the filler and react with the flux.

As seen above, standard fluxed materials have a varying performance depending on the flux distribution. However, no sample with evenly distributed flux manage to braze 0.4 wt-% Mg core material at 60 °C/min with loads less than 4 g/m^2. The 2.1 g/m^2 flux in TRILLIUM® Solid thus works much more efficiently. This suggests that the detrimental phase and morphology flux-Mg reaction product forms slower, or at least differently, when the flux is inside the filler than when it is applied on the surface.

Present results versus industrial applicability

In a real application the circumstances for assembly, surface intimacy and joint formation differ markedly from those encountered in this study. Firstly, the surfaces from industrially rolled and formed materials are comparatively flat. Secondly, the joint geometries are better designed to make wetting easier and to provide a significant capillary pressure once wetting has occurred. Thirdly, industrial brazing use fixtures to apply a small force to the joint that makes surfaces adapt slightly relative each other, good for joint quality. Fourth, modern industrial brazing atmospheres are probably better than the one in the glass furnace used here. These are all examples of features that make industrial brazing easier and more reproducible. For example, industrial mass production of an as yet undisclosed commercial heat exchanger has been successful using TRILLIUM® Lean with a flux load $<\sim 0.15$ g/m^2. When brazing takes place under ideal conditions (atmosphere, cleanliness, fixtures etc.) the TRILLIUM® Lean has proven to braze with flux loads as low as ~0.050 g/m^2 [4].

With the above in mind one should consider the presently developed results to be very conservative.

From discussions with industrial representatives, it appears that the lower benchmark in industrial slurry fluxing is about 2 g/m^2 for successful brazing of heat exchangers of Mg-free materials. When applying flux using slurry fluxing it ends up everywhere, not only on the braze clad surface, which means that all unclad surfaces are also covered, thus there is more flux available for braze joints formation.

Joint formation vs. heating rate

It is obvious that the heating rate has a marked influence on the joint formation ability of TRILLIUM® on Mg containing cores and the same behavior is seen in normally fluxed brazing sheet. Faster heating provides larger tolerance than slower heating does from a joint size and joining quality perspective. This general behavior is expected since faster heating provides a shorter time for oxide growth and for Mg to diffuse into the braze and eventually reach the braze surface.

Mg exhibits a well documented affinity for molten flux and the proposed reaction between flux and Mg results in MgF and $KMgF_3$ that have higher melting temperatures than aluminium and also malign morphologies that interfere with brazing. Since flux is consumed by these reactions there is less left to undertake oxide removal and wetting, thus making fillet formation more difficult. The compounds may not only inhibit wetting but also make fillet growth more awkward. A brazing sheet surface that contains contaminants will make it more difficult for a growing fillet to dislodge it and move it aside. That a heating rate-joint formation relationship is seen here is caused by the fact that the Mg rapidly diffuses from the core through the solid cladding, and even more rapidly when it is molten. A rapid heating rate provides a smaller amount of Mg in the cladding and near the superficial oxide than a slower heating rate. It also provides excessive growth of MgO and spinel type oxides on or near the superficial surface. It is known from vacuum brazing that such oxides may be as detrimental as excessively thick Al_2O_3 when it comes to preventing wetting and joint growth.

Conclusions

The faster the heating rate and the higher the flux load, the better the Mg tolerance.

For the slurry fluxed samples:

- A threshold amount of flux (>2 g/m^2) exists for migration of molten flux.
- On high Mg samples an uneven flux distribution may increase local concentration and facilitate joint formation.
- Evenly spread flux performs better on low Mg samples than unevenly fluxed samples.
- Evenly spread 2 g/m^2 flux is sufficient on low Mg samples in normal conditions.

TRILLIUM® Lean braze:

- Mg-free material with flux levels below 0.6 g/m^2. This will give low amounts of flux residues after brazing.
- 0.3 wt-% Mg material with <2.5 g/m^2 flux.

TRILLIUM® Solid braze 0.4 wt-% Mg material with 2.1 g/m^2 flux.

References

[1] EAA Aluminium Automotive Manual – Joining – Brazing, *European Aluminium Association (updated 2015).*

[2] Yu, C. N., Hawksworth, D., Liu, W. & Sekulic, D. P., "Al brazing under severe alterations of the background atmosphere: A new vs. traditional brazing sheet", *5th International Brazing and Soldering Conference, IBSC 2012, April 22, 2012 - April 25, 2012, 2012 Las Vegas, NV, United states*. ASM International, 347-354.

[3] Yu, C. N., Sekulic, D., Hawksworth, D., "Impact of the background atmosphere on the aluminum brazing joint formation", *Proc. ASME 2013 Int. Mech. Eng. Cong.&Expo.*, IMECE2013-64776, Nov. 15-21, 2013, Cal., USA.

[4] Westergård, R., Koscielski, S. and Oskarsson, A., "TRILLIUM® - a new aluminium brazing technology", *2nd Conference & Exhibition on Light Materials, 08-10 Nov. 2017, Bremen, Germany*

Microstructural evolution of SiC joints soldered with the assistance of ultrasonication using Zn–Al filler metals

Bingzhi Wu, Jiuchun Yan
State Key Laboratory of Advanced Welding and Joining, Harbin Institute of Technology
Harbin 150001, P.R. China
bingzhw@163.com

Abstract

SiC ceramics were successfully soldered assisted by ultrasounds. Two kinds of filer metals of non-eutectic Zn–5Al–3Cu and eutectic Zn–5Al alloys were used. The effects of ultrasonic action on the microstructure and mechanical properties of the soldered joints were investigated. The results showed that ultrasounds could promote the wetting and bonding between the SiC ceramic and filler metals within tens of seconds. In the case of Zn–5Al–3Cu solder, a fully grain-refined structure in the bond layer was obtained as the ultrasonic action time increased. This may lead to a substantially enhancement of strength of soldered joints. In the case of Zn–5Al solder, the shear strength of the soldered joints was only ~ 102 MPa when the ultrasonic action time was shorter, and fractures occurred in the brittle lamellar eutectic phases in the center of the bond layer. Along with the increase of ultrasonic action time, the lamellar eutectic phase in the bond layer of SiC joints could be completely transformed to a fine non-lamellar eutectic structure. Meanwhile, the grains in the bond layer were obviously refined. Those results led to the remarkable enhancement of the shear strength of the joints (~ 138 MPa) using the Zn–5Al solder, which had approached that enhancement using the Zn–5Al–3Cu solder. The enhanced mechanical properties of joints were attributed to the significant refinement of the grains and the change in the eutectic structure in the bond layer. Prolonged enhanced heterogeneous nucleation triggered by ultrasonic cavitation is the predominant refinement mechanism of the bond metals of the SiC joints.

Introduction

Currently, as one of the most promising ceramic materials, SiC ceramics are becoming increasingly attractive for applications in several industries such as the aerospace, automotive, energy, electronics and optical device [1-3], especially for high-performance microelectronic packaging due to their remarkable mechanical properties, excellent thermal conductivity and low thermal expansion coefficient. In practical applications, SiC ceramics must be firmly joined to themselves or other metals. Therefore, reliable joining of SiC ceramics is absolutely essential to widen its applications.

Brazing is the most commonly used method to join ceramics. The majority of work on brazing of SiC ceramics to themselves or metals has been performed by using mainly the filler metals of Ag–Cu–Ti [4], Co-based [5], Fe–Si [6] and Ni–Cr [7] alloys. Nevertheless, the brazing process cannot be applied in electrical and electronic engineering field due to their high brazing temperature ranging from 800 to 1200 °C, which can cause thermal damage to electronic components. Therefore, the joining at relatively low temperature is in demand for the application. The Sn-based lead-free solders possess very low melting points, and they are commonly used as joining materials for the application in current electronic packaging industry [8, 9]. However, due to the low melting lower than 250 °C of Sn-based solders, they are unable to meet the requirements of the application in the power electronics devices performed at elevated temperatures, which makes further investigation of soldering processes highly necessary.

In comparison with the widely used Sn-based solders, the Zn-Al alloys was one kind of high temperature solders and possess a moderate melting temperature above 350 °C, higher corrosion resistance, excellent thermal and electrical conductivities and superior mechanical properties, making them reasonable candidate high-temperature solder alloys used extensively in die-attach, power semiconductor, and optical device packaging, flip-chip packaging [10, 11].

In our present study, Zn–5Al and Zn–5Al–3Cu alloys were selected as the filler metals to join SiC ceramics by ultrasonic-assisted air soldering. The objective of this work has been to identify the microstructural evolution of the joints and its relationship with corresponding mechanical properties and fracture modes. Such a study is expected to be helpful in developing high-reliability SiC ceramic joints for high-density electronic packaging application.

Experimental procedures

Pressureless sintering SiC ceramic used in this study was provided by Shanghai Unite Technology Co. Ltd. Shanghai, China. The SiC ceramic was sectioned with a diamond blade to obtain samples with dimensions of 40 mm× 10 mm× 3 mm. Prior to soldering, all of the joining surfaces were polished by 1 μm diamond paste, and then was ultrasonically cleaned in acetone for 10 min. The Zn–5Al (wt%) eutectic alloy and Zn–5Al–3Cu (wt%) alloy were used as filler metals. High purity metals of Al (99.99%), Zn (99.99%), and Cu (99.999%) were used for alloy preparation. The principle of ultrasonic-assisted soldering is schematically represented in Fig. 1a. The SiC samples were placed in a single overlap configuration with a ~20 mm overlap length. Then the samples were heated to 420 °C in air at a rate of 50 °C/min then ultrasonic vibrations was applied (Fig. 1b). The amplitude of the ultrasonic vibration was ~3.5 μm, as measured by a laser doppler vibrometer (Polytec OFV-505/5000, Germany). The ultrasonic action time was referred to as one of the main process parameters, and ranged from 5 s to 20 s. Subsequently, the soldered couples were held for 3 min at soldering temperature, and then cooled in air to room temperature.

After joining, the soldered joints were mounted in epoxy resin and then were cut perpendicular to bonding interface for the cross-sectional microstructure analysis. In addition, the soldered joints were cut paralleled to bonding interface until the bond layer was exposed for the electron backscattering diffraction (EBSD) microstructure analysis of solder matrix. EBSD measurements were obtained using commercially available software (supplied by the company HKL Technology APS). The joints were then observed by scanning electron microscope (SEM) (FEI; Quanta 200FEG) equipped with an energy-dispersive X-ray spectroscopy (EDS) BRUKER AXS system. The grain size of the bond layer was assessed in EBSD mode. The shear strength tests were measured by shear testing using a specially designed fixture as illustrated in Fig. 1c. The soldered specimens were compressed at a constant speed of 0.5 mm/min at room temperature. The shear strength was calculated by the load at the fracture divided by the nominal area of the joint. To ensure reliable strength results, five samples at a certain parameter were examined to calculate the average value of the joint as well as the standard deviation of the shear strength. After shear testing, the fracture path and surfaces were all investigated via SEM.

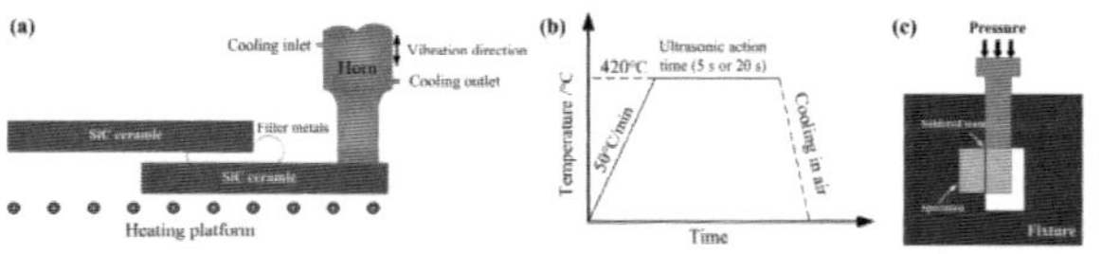

Fig. 1. Schematic diagram of the (a) ultrasonic-assisted soldering principle; (b) thermal profile in the soldering process; (c) shear test configuration.

Microstructural examination of joints

Fig. 2 shows the representative microstructures of the joints using the Zn–5Al filler metal and Zn–5Al–3Cu with an ultrasonic action for 5 s. As shown in Fig. 2a and b, the joints are free from obvious defects such as voids and cracks. The further magnification observations of the cross-sectional interface show that no obvious reaction products could be observed at the interface (Fig. 2c and d). It can be seen that the bond layer of joints using Zn–5Al solder (Fig. 2e) mainly constituted by η-Zn phase (light) and a large area of Zn–Al eutectic interdendritic structures (textured pattern). According to the Zn–Al binary phase diagram [12], primary η-Zn phase precipitates from the molten solder at first during solidification. When the soldering temperature is below 382 °C, eutectic transformation occurs and α+η lamellar eutectics are established from the remaining liquid enriched with a solute segregation. For the Zn–5Al solder, the η-Zn phase was in a grain-like morphology, while the eutectic phases are mainly lamellar structure in the center of the bond layer and only a small amount of non-lamellar structure (Fig. 2e).

For the Zn–5Al–3Cu solder, as shown in Fig 2f, η-Zn phases and eutectic phases were identified as the main phases in the bond layer. In addition to the lamellar Zn–Al eutectic phase, non-lamellar eutectics were also observed in the bond layer. According to the Zn–Al–Cu phase diagram, the non-lamellar eutectic phases were in the form of gray η-Zn matrix embedded with white rod-like or plate-like ε-$CuZn_4$ phase and dark whisker-like α-Al phase. The formation of $CuZn_4$ phase was due to the strong reaction activity between Zn and Cu elements. The lamellar structure was composed of dark α-Al phase and gray η-Zn phase. The result was similar to the results in ref. [10]. Compared with the joint soldered with Zn–5Al filler metal that is shown in Fig. 2a, it can be seen that the volume fractions of primary η-Zn phases increased, but the amount of eutectic structure decreased. That is to say, the addition of Cu to Zn–5Al alloy could promote the heterogeneous nucleation of η-Zn phases, but suppress the formation of eutectic structures; thereby it will be helpful to improve the mechanical properties of Zn–Al based alloy.

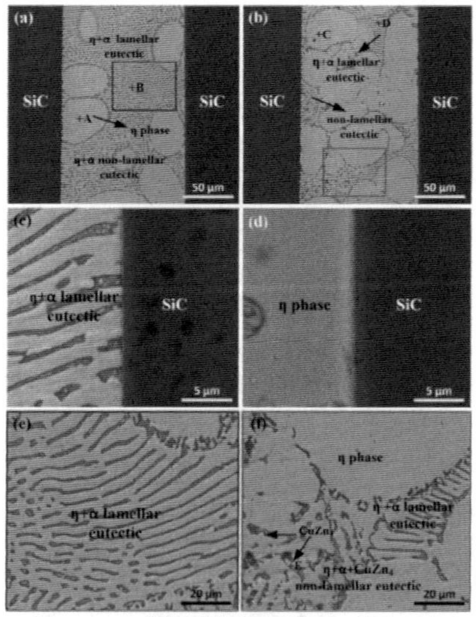

Fig. 2. Microstructures of the joints with an ultrasonic action for 5 s using (a, c) Zn–5Al solder and (b, d) Zn–5Al–3Cu solder; (e, f) the corresponding magnification views of bond layer.

Fig. 3 shows the representative microstructures of the joints using the Zn–5Al filler metal with an ultrasonic action times of 20 s. With an increase of ultrasonic action time, the morphology of eutectic structure varies considerably and the lamellar Zn–Al eutectic structure in the bond layer completely disappeared when using Zn–5Al solders (Fig. 3a). Instead, more fine and non-lamellar eutectic phases occurred in the bond layer. As shown in Fig. 3b, no unsoldered gaps were observed at the soldered interface. Fig. 3c shows the high magnification SEM micrographs of non-lamellar eutectic. It can be seen the non-lamellar eutectic structure is composed of secondary η–Zn phases (white) and α-Al phase (black). The size of the α-Al phase was smaller and more uniform in size compared to that with shorter ultrasonic action time.

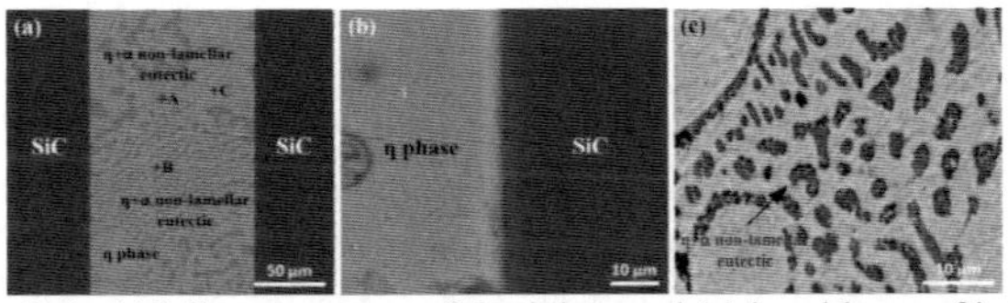

Fig. 3. (a, b) Microstructures of the joints and (c) bond layer of joints using Zn–5Al solder with an ultrasonic action for 20 s.

Fig. 4 shows the representative microstructures of the joints using the Zn–5Al–3Cu filler metal with an ultrasonic action times of 20 s. The large size of lamellar eutectic phases existed in the soldered seam could be completely eliminated. Only a small amount of fine and discontinuous eutectic phase can be observed (Fig. 4a and b). The ε-$CuZn_4$ embedded in the non-lamellar eutectic phase became particles-like after the ultrasonic treatment (Fig. 4c).

Fig. 4. (a, b) Microstructures of the joints and (c) bond layer of joints using Zn–5Al–3Cu solder with an ultrasonic action for 20 s.

Mechanical properties of the joints

The shear strength of the joints is shown in Fig. 5 as a function of the ultrasonic action time. For the joints using Zn–5Al solder, results indicated that joint strength increases with the increase in ultrasonic action time. When the ultrasonic action time was 5 s, the shear strength was about 102 MPa. The brittle lamellar Zn-Al eutectic phases were predominant in the joint. With an ultrasonic action time of 20 s, the lamellar eutectic phases in the center of bond layer were completely eliminated. The shear strength of joints could reach up to ~ 138 MPa when the ultrasonic action time was prolonged to 20 s. The noticeable improvement of shear strength of joints is mainly attributed to the change of eutectic structure.

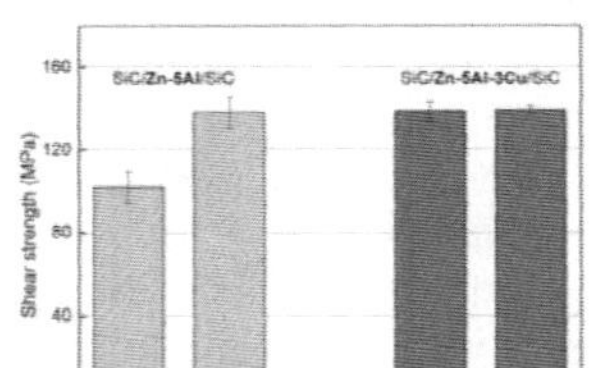

Fig. 5. Shear strength of SiC joints as a function of ultrasonic action time.

For the joints using Zn–5Al–3Cu solder with ultrasound for 5 s, the shear strength was ~ 139 MPa, which is close to maximum shear strength of the joint using the Zn–5Al solder. The improvement of the shear strength with the additional Cu element was achieved. The volume fraction of eutectic phases decreased and the non-lamellar eutectic phases occurred in the joint. The shear strength kept relative constant as the ultrasonic action time was prolonged to 20 s, which was equivalent to the highest shear strength soldered using Zn–5Al solder.

Fig. 6 shows the SEM images of fracture surfaces of failed joints using Zn–5Al solder with ultrasonic action time of 5 s and 20 s. When the ultrasonic action time was 5 s, uneven eutectic structures and flat η-Zn phases can be distinguished at the fracture surface (Fig. 6a). It indicated that the fractures mainly propagated in the eutectic structures in the bond layer, while bypassing the η-Zn grains. The brittle rupture of the joints is mainly due to the eutectic structure. Large volume fractions of lamellar eutectic phases existed at the center of the soldered seam cause excessive brittleness of joints, resulting in the low strength joints (~ 102 MPa). As shown in Fig. 6b, the fracture surface shows apparently cleavage fracture when the ultrasonic action time was 20 s. That is to say the fracture started and propagated completely within the SiC ceramic and away from the interface. As the ultrasonic action time increased, the brittle lamellar eutectic structures in the bond layer were effectively eliminated. Therefore, the shear strength of joints remarkably increases due to the change of microstructure with a further increase in ultrasonic action time (~ 138 MPa).

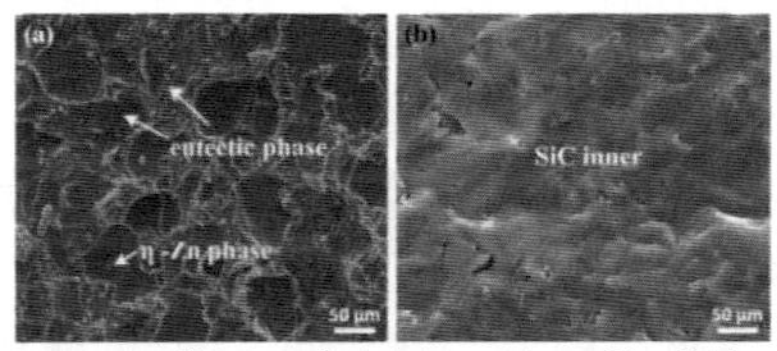

Fig. 6. Fracture surfaces of joints using Zn–5Al solders with ultrasonic action for (a) 5 s and (b) 20 s.

Fig. 7 shows the images of fracture surfaces of failed Zn–5Al–3Cu joints with ultrasonic action time of 5 s and 20 s. The addition of Cu element in Zn–Al alloy improved the microstructure of soldering seam by suppressing the formation of lamellar eutectic phases and promoting the formation of η-Zn phases, resulting in high strength of ~ 139 MPa. As shown in Fig. 7a, the fractures propagated within the SiC ceramics, indicating that the bonding strength of interface was actually higher than the results of shear strength of joints. As the bond layer is no longer the weak region of the joint, the shear strength of the joint could not be improved further by increasing the ultrasonic action time.

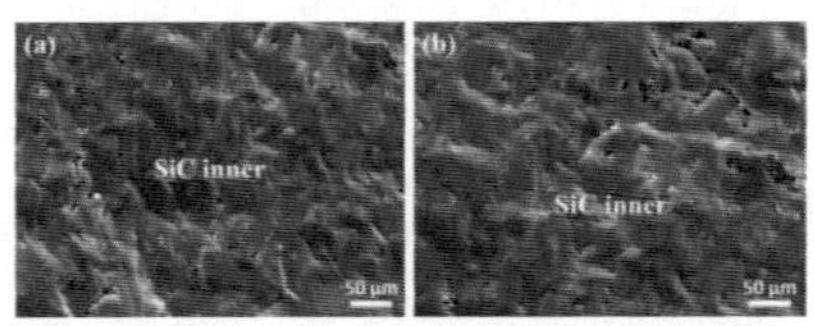

Fig. 7. Fracture surfaces of joints using Zn–5Al–3Cu solders with ultrasonic action for (a) 5 s and (b) 20 s.

Discussion

From the above results obtained we can see that the mechanical properties of joints can be obviously improved by ultrasound action. This depends mainly on the changes in microstructure of the joints caused by ultrasound. To further identify the effect of ultrasonic action time on the microstructure and grain refinement of joints quantitatively, the typical EBSD microstructures of the solder matrix without ultrasonic action and bond metals with ultrasonic action are comparatively displayed in Figs. 8 and 9. The effect of ultrasonic action on the grain structure was examined by EBSD analysis. Fig. 8 shows the typical EBSD microstructure of the Zn–5Al solders. The results show that non-dendritic spheroidal grains can be clearly observed in all samples. The sample without ultrasonic treatment also shows a non-dendritic structure, which may be due to the addition of conventional grain refiners of Al-5Ti-1B master alloys during preparation of the solders. By comparing EBSD microstructure of the solder without ultrasonic vibration (Fig. 8a), the grain refinement efficiencies vary gradually with the prolongation of ultrasonic action time. When the ultrasonic action time was 5 s, there was little change in the grain size of Zn–5Al alloy (Fig. 8b). Further increase of the ultrasonic action time to 20 s led to very marginal decrease of grain size of Zn–5Al alloy (Fig. 8c). As shown in Fig. 9, the longer ultrasonic action time resulted in the grain refinement of Zn–5Al–3Cu alloys, which is consistent with the experimental observation using Zn–5Al solders. It can be seen that the average grain size of Zn–5Al alloy was refined from 32 μm with ultrasonic action time of 5 s to 19 μm with ultrasonic action time of 20 s.

Fig. 8. EBSD microstructures of (a) Zn–5Al solders without ultrasonic action; (b, c) bond metals in joints with ultrasonic action times of 5 s and 20 s, respectively.

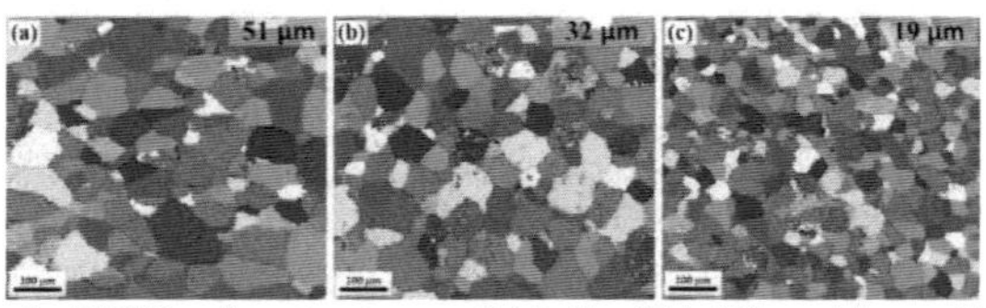

Fig. 9. EBSD microstructures of (a) Zn–5Al–3Cu solders without ultrasonic action; (b, c) bond metals in joints with ultrasonic action times of 5 s and 20 s, respectively.

Fig. 10 schematically presents the morphological evolution of microstructure of bond metals soldering using Zn–5Al and Zn–5Al–3Cu solders with different ultrasonic action times. As shown in Fig. 10a, with shorter ultrasonic action time (5 s), the volume fraction (VF) of the lamellar eutectic structure (E1) is up to 65%. Only a small amount of non-lamellar eutectic (E2) and Zn-rich phases exist in the bond metal. This result is basically consistent with the original microstructure of the Zn–5Al solder matrix. This suggests that the microstructure transformation may occur to a very limited extent with shorter ultrasonic action time. Increasing ultrasonic action time to 20 s favors the refinement of η-Zn grain size (Fig. 10b). Meanwhile, the lamellar eutectic structure completely disappears and becomes a non-lamellar eutectic structure composed of fine α-Al and η-Zn phases. This morphology differs greatly that with shorter ultrasonic action time. It has been reported that finer eutectic provides better mechanical properties than a coarse lamellar morphology in Zn–Al alloy [23]. For the Zn–5Al–3Cu solders, as illustrated in Fig. 10c, a large number of Zn-rich phases exited in the bond metal. The eutectic microstructure is characterized by lamellar (E3) and non-lamellar (E4) structures. When the ultrasonic action time is further increased to 20 s, as shown in Fig. 10d, the Zn-rich phases are obviously refined and lamellar eutectic structure completely disappears, which is consistent with the results of bond metal soldered using Zn–5Al solder with longer ultrasonic action time. The microstructure evolution strikingly improves the mechanical of bond metal.

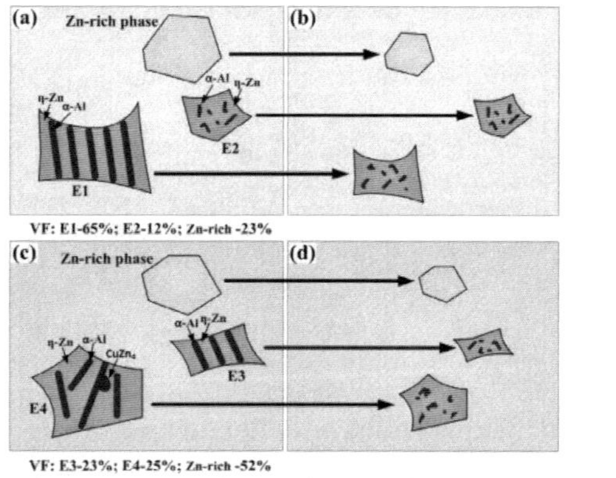

Fig. 10. Microstructure evolution diagram of bond metals soldered using (a, b) Zn–5Al solders and (c, d) Zn–5Al–3Cu solders with an ultrasonic action time of 5 s or 20 s, respectively.

From the above microstructure evolution, it can be summarized that a prominent structural refinement of both the Zn–5Al and Zn–5Al–3Cu solder matrix was all achieved after the ultrasonic treatment. The effect mechanisms of ultrasound on the microstructural refinement are discussed as following. According to previous studies [14, 15], several kinds of mechanism were proposed, including the dendrite fragmentation, the detachment of nucleus from mould walls and the promotion of heterogeneous nucleation. Since the ultrasonic vibration was terminated above the liquidus temperature in this study, both the dendrite fragmentation and the detachment of nucleus from the mould walls are not responsible for the microstructural refinement of alloy. Thus, the grain refinement can be explained by the heterogeneous nucleation theory of solid crystals induced by ultrasound in the molten solder during the soldering. The grain refinement of alloy is determined by the number of nuclei in unit melt. The influences of ultrasonic action time on the microstructure were contributed by the combined effects of acoustic cavitation and streaming induced by the ultrasound in the molten solder during soldering.

The cavitation-induced heterogeneous nucleation mechanism can be further divided into three different subcategories [15]. Firstly, the presence of ultrasound in the molten solder could induce pressure fluctuation and consequently, generate microbubbles in the liquid medium. The collapse of bubbles increases the pressure on the molten solder, resulting in the increasing of melting point as suggested by the Clausius-Clapeyron equation,

$$\Delta P = \frac{\Delta H}{T_m(V_L - V_S)} \Delta T$$

Where ΔP is the increased pressure in 105 Pa, ΔH is the latent heat of freezing in J, V_L and V_S are the specific volume of the liquid and the solid phase in cm^3/g, respectively, Tm is the equilibrium freezing point in K, and ΔT is the change in the freezing point. The great increase of localized ΔP and (V_L-V_S) provided by ultrasonic cavitation increases the localized melting point in the vicinity of cavitation bubbles, which is equivalent to raising the localized undercooling. Secondly, the formation, growth and adiabatic collapse of microbubbles may have caused the bubble-liquid interface to undergo changes from undercooling to superheating at high frequencies. Rapid undercooling and superheating would enhance the nucleation of crystals in the molten solder. Thirdly, non-metallic inclusions exist in the molten solder and enhanced wettability can be achieved by the formation of cavitation bubbles through the degassing of hydrogen gas trapped at the surface of inclusions. The wetted inclusions can act as effective nucleation sited during solidification after soldering process. Meanwhile, the homogenization of alloying elements and the distribution of the wetted inclusions over the whole melt can be improved with the aid of the acoustic streaming. So the acoustic streaming also plays a significant role in the refinement of the grains.

Base on the above-mentioned, it can be said that the increased heterogeneous nucleation due to the addition of ultrasonic effect is the reason responsible for this microstructural refinement when the ultrasonic treatment is performed above the liquidus temperature. One or more of these three mechanisms may be responsible for increased density of active nuclei with ultrasonic treatment, which might have resulted in finer microstructures observed in the present work. The change of microstructure is considered to be caused by the inadequate space for the primary arms growth along the non-dendritic tip because the growth of primary arms will be restricted when two grains encounter. The increase of the number of nuclei results in the increasing amount of η-Zn phase and α-Al phase. The increasing amount of η-Zn phases are response for the effective refinement of the η-Zn phase. As the α-Al phase mostly exists in the eutectic structure in Zn–Al alloy. Thus, the increase of α-Al phase leads to the increasing amount of eutectic, which triggers the refinement of eutectic simultaneously. In addition, the variations of the morphologies of η-Zn phase are also associated with the increasing of the eutectic structures. Because the fine eutectic phase distribute in the interdendritic regions between the growing dendritic structures of primary η-Zn phase, resulting in the inhibition of continued dendritic growth. As a result, the eutectic phase was effectively refined.

If the joints were soldered without an assistance of ultrasonic wave, the nuclei within the undercooled molten solder will develop into coarse grains. Under optimal ultrasonic treatment conditions, the entire microstructure of the solder metals could be grain-refined, yielding a homogeneous non-dendritic structure. When the ultrasonic action time is short (5 s), the grain refinement of alloys is not obvious. This may be due to the absence of sufficient ultrasonic cavitation effects on nucleation during soldering, thus the grain got an adequate growth space and finally achieved coarse non-dendritic structure. Longer ultrasonic action time results in more nuclei formed, and in this case the nuclei would have almost the same growth velocity in all directions. It helps to prohibit the growth of dendrites, and the primary particles finally become near-spherical. Thus, a large number of non-dendritic grains form and higher grain refinement efficiency is achieved when the ultrasonic action time is 20 s.

Conclusions

In this work, SiC ceramics were soldered using the non-eutectic Zn–5Al–3Cu and eutectic Zn–5Al filler metals by ultrasonic-assisted air soldering. The conclusions can be drawn as follows:

(1) The joints using Zn–5Al–3Cu solder had the highest shear strength of ~ 139 MPa due to the appearance of a large number of η-Zn phases in the bond layer. As the ultrasonic action time increases, the η-Zn grains were obviously refined and the lamellar eutectic structure was completely avoided in the bond layer. The average grain size of bond metals was decreased from 51 to 19μm with the ultrasonic action for 20 s, in comparison with the original solder metals without ultrasounds.

(2) The microstructure of bond layers using the Zn–5Al solder consisted of primary η-Zn phases and large amounts of lamellar eutectic phases when the ultrasonic action time was shorter. The shear strength of joints was ~ 102 MPa and fracture occurred in brittle lamellar eutectic. As the ultrasonic action time increased, the shear strength of joints could reach up to ~ 138 MPa, which is equal to that of joints using Zn–5Al–3Cu solder. Such an enhancement resulted from the elimination of brittle lamellar eutectic phases and the grain refinement in the bond layer.

(3) The improvement of mechanical properties of joints was closely associated with significant grain refinement and changing of eutectic structure of bond metals. The predominant refinement of bond layers with increasing the ultrasonic action time was ascribed to the increased nucleation sites for crystallization triggered by ultrasonic cavitation induced heterogeneous nucleation. This work demonstrated that ultrasounds can be applied as an effective way to refine the microstructure of bond metals of joints and to enhance the performance of joints.Text about conclusions reached resulting from the research.

Acknowledgments

The authors gratefully acknowledge the financial support from the National Natural Science Foundation of China (Grant No. 51435004).

References

[1] R. Kisiel, Z. Szczepański, "Die-attachment solutions for SiC power devices," *Microelectron. Reliab.*, Vol. 49, No. 6 (2009), pp. 627-629.

[2] R. Singh, "Reliability and performance limitations in SiC power devices," *Microelectron. Reliab.*, Vol. 46, No. 6 (2006), pp. 713-730.

[3] M.G. Nicholas, D.A. Mortimer, "Ceramic/metal joining for structural applications," *Mater. Sci. Technol.*, Vol. 1, No. 9 (1985), pp. 657-665.

[4] M.C. Halbig, B.P. Coddington, R. Asthana, M. Singh, "Characterization of silicon carbide joints fabricated using SiC particulate-reinforced Ag–Cu–Ti alloys," *Ceram. Int.*, Vol. 39, No. 4 (2013), pp. 4151-4162.

[5] H.P. Xiong, W. Mao, Y.H. Xie, B. Chen, W.L. Guo, X.H. Li, Y.Y. Cheng, "Control of interfacial reactions and strength of the SiC/SiC joints brazed with newly-developed Co-based brazing alloy," *J. Mater. Res.*, Vol. 22, No. 10 (2007), pp. 2727-2736.

[6] J.R. McDermid, M.D. Pugh, R.A.L. Drew, "The interaction of reaction-bonded silicon carbide and Inconel 600 with a nickel-based brazing alloy," *Metall. Trans. A.*, Vol. 20, No. 9 (1989), pp. 1803-1810.

[7] Y.W. Mao, S.J. Li, L.S. Yan, "Joining of SiC ceramic to graphite using Ni–Cr–SiC powders as filler," *Mater. Sci. Eng. A.*, Vol. 491, No. 1 (2008), pp. 304-308.

[8] S.K. Kang, A.K. Sarkhel, "Lead (Pb)-free solders for electronic packaging," *J. Electron. Mater.*, Vol. 23, No. 8 (1994), pp. 701-707.

[9] H. Ma, J.C. Suhling, "A review of mechanical properties of lead-free solders for electronic packaging," *J. Mater. Sci.*, Vol. 44, No. 5 (2009), pp. 1141-1158.

[10] N. Kang, H.S. Na, S.J. Kim, C.Y. Kang, "Alloy design of Zn–Al–Cu solder for ultra high temperatures," *J. Alloys Compd.*, No. 467, No. 1 (2009), pp. 246-250.

[11] T. Shimizu, H. Ishikawa, I. Ohnuma, K. Ishida, "Zn–Al–Mg–Ga alloys as Pb-free solder for die-attaching use," *J. Electron. Mater.*, Vol. 28, No. 11 (1999), pp. 1172-1175.

[12] H.Y. Li, Z.G. Li, Y. Liu, H.F. Jiang, "Effect of zirconium on the microstructure and mechanical properties of Zn–4% Al hypoeutectic alloy," *J. Alloys Compd.*, Vol. 592, (2014), pp. 127-134.

[13] W.R. Osorio, C.M.A. Freire, A. Garcia, "Dendritic solidification microstructure affecting mechanical and corrosion properties of a Zn4Al alloy," *J. Mater. Sci.*, Vol. 40, No. 17 (2005), pp. 4493-4499.

[14] J.D. Hunt, K.A. Jackson, "Nucleation of solid in an undercooled liquid by cavitation," *J. Appl. Phys.*, Vol. 37, No. 1 (1996), pp. 254-257.

[15] M. Qian, A. Ramirez, A. Das, "Ultrasonic refinement of magnesium by cavitation: clarifying the role of wall crystals," *J. Cryst. Growth*, Vol. 311, No. 14 (2009), pp. 3708-3715.

Study of material thickness variation in soldering dissimilar metals (Brass-Steel)

***Rehan Waheed*[a], *Hasan Aftab Saeed*[b], *Wasim Tarar*[c], *Khalid Mahmood*[d] *and Sajid Ullah Butt*[e]**
Department of Mechanical Engineering, College of Electrical and Mechanical Engineering, National University of Sciences and Technology, Islamabad, 44000, Pakistan
[a]*rehan.waheed@ceme.nust.edu.pk*, [b]*hasan.saeed@ceme.nust.edu.pk*,
[c]*wasimakram@ceme.nust.edu.pk*, [d]*khalid.mahmood@ceme.nust.edu.pk*, [e]*sajidullahbutt@yahoo.com*

Keywords: dissimilar metals, soldering, joint strength, EMI Shielding

Abstract

EMI shielded doors made from brass extruded channels need to be welded with shielded enclosures to attain optimum shielding performance. Control of welding induced distortion is a problem in welding dissimilar metals like steel and brass. In this research soldering of the steel-brass joint has been proposed to avoid weld distortion. The material used for brass channel is UNS C36000. The thickness of brass is defined by the manufacturing process i.e. extrusion. The thickness of shielded enclosure material (ASTM A36) can be varied to produce joint between the dissimilar metals. Steel sections of different gauges are soldered using (91% Tin, 9% Zinc) solder to the brass and strength of joint is measured by standard test procedures. It is observed that thin steel sheets produce a stronger bond with brass. The steel sections further require to be welded with shielded enclosure steel sheets through TIG welding process. Stresses and deformation in the vicinity of soldered portion is calculated through FE simulation. Crack formation in soldered area is also studied through experimental work. It has been found that in thin sheets deformation produced due to applied force is localized and has no effect on soldered joint area whereas in thick sheets profound cracks have been observed in soldered joint. The shielding effectiveness of EMI shielded door is compromised due to these cracks. The shielding effectiveness of the specimens is tested and results are compared.

Introduction

The fabrication of EMI shielded doors require a lot of welding to be done on door and frame sections. A key parameter in fabrication of these doors is dimensional accuracy. Joining of dissimilar metal i.e. steel with brass, and long seam welding passes the tolerance limit of whole assembly are disturbed due to distortion produced during welding process. In this research prospects have been studied to use soldering as joining process for the steel-brass joint. The soldering of metal sheets by soft leaded and lead free solders has been studied by many researchers. S. Weis et al [1] studied the soldering of high strength aluminum alloy. S. Weis suggested that in situation where thermal influence on base metal is critical soldering is more suitable than welding process. Brett Daly [2] also mentioned the advantage of brazing over welding process in situation where dissimilar metals have to be welded and welding induced distortion is critical. The strength of the joint is compromised in case of soldering. In situations where soldered joint is under fatigue loading, strength of joint becomes more important. Nourani et al [3] studied the fracture toughness of soldered joint by using a DCB and ARCAN specimens. Milan Brozek [4] study the strength of soldered joints on galvanized steel sheets and copper sheets of 1mm thickness. Milan tested the soldered specimens in universal testing machine and compared the strength of leaded solder and lead- less solder. Shangari.S [5] studied the failure of a soldered joint in nuclear power plant copper tubing. Shingari proposed that non-destructive testing techniques must be used to check the quality of joint as surface inspection only cannot predict the failure of joint under loading. Milosavljević et al [6] proposed a special solder for joining steel with brass. Darveaux et al [7] defines a constitutive relation for tin based solder alloys under thermal cyclic load. The strength of solder joint is weak as compared to the base metal strength. In this case joint strength is 3 to 4 times less than the base metal therefore reliability of the whole assembly needs to be investigated. Test criteria laid down in US unified facilities guide specifications [8] is used. Door static load test and door sag test are used as test points for checking the integrity of soldered joint. A CAD model of door is prepared and analyzed through FE simulations to select appropriate sheet thickness of door frame at design stage. Some samples of door frame section are prepared for experimental study. The tightening torque of door hinge bolts is measured and their effect on soldered joint is studied. The physical and mechanical properties of solder alloy Sn-9Zn is taken from the research work of A.A El-Daly et al [9].

FE Simulation

A CAD model of door section is prepared with door brass section profiles, the strength members of steel and the soldered fillet joint with its proper width. The assembly of door and frame section is shown in fig 1. The main emphasis of this research is the door frame section. Door frame is made up of UNSC 36000 brass extruded channel. The profile of the channel makes it a female section with its counterpart male brass section attached to the door. The female brass section is soldered with a box section made up of ASTM A36 steel. The thickness of steel section can be varied keeping in consideration commercially available sizes and material. The steel box section takes all the structural load of door assembly during operation i.e. opening and closing of door.

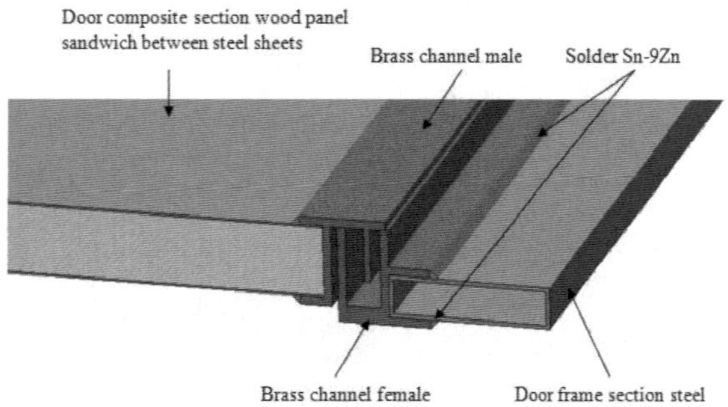

Fig 1 Sectional view of EMI shielded door assembly

The opening and closing of door requires a lot of force due to EMI shielding gaskets placed in the female brass channel of door frame. A three point locking mechanism is installed on door for smooth operation. The three point locking mechanism is not modeled here for simplicity of FE analysis. As per standard criteria defined in static load tese mentioned in US unified facilities guide specifications [8] the door assembly should sustain a pressure of 2KPa applied on door surface area with door placed horizontally such that it rests on door frame section and is open able in downward direction. The weight of whole door is also sustained by the door frame sections. The FE model of door is shown in Fig 2.

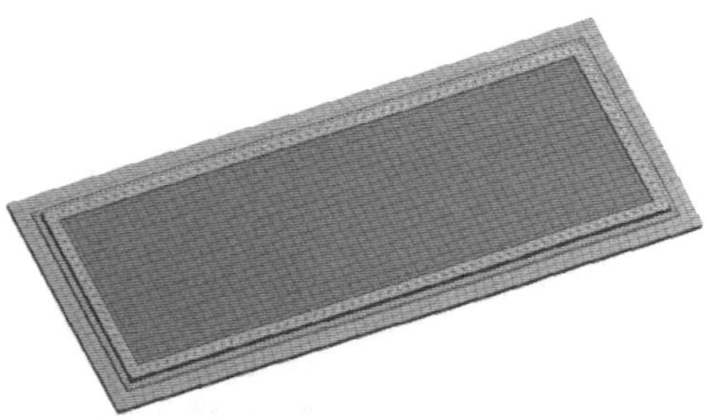

Figure 2. FE model of EMI shielded door assembly

The requirement of sag test is that the door is opened 90° and two weights of 23 Kg are hanged with the extreme end of door within a distance of 5 inches, in addition this weight should be kept hanging for 10 minutes. To simulate the sag test a transient structural analysis is performed with total step timing of 600 seconds. When the door is opened and test weights hanging with it all the load is transferred to door frame section through hinges therefore to simplify the FE model of sag test only door frame section is considered. The appropriate loads are calculated and applied to door frame section at the point of hinge bolts. The passing criteria as mentioned in unified facilities guide specifications for both the static load test and door sag test are that deformation if produced in any section of door assembly should be less than 2mm. In this research work the main emphasis is on the integrity of soldered joint therefore stress has been calculated in the vicinity of solder joint to check if this stress is within the rage of yield stress for Sn-9Zn solder alloy. In FE simulation three test cases are considered

Case1: Door frame section with 2mm thick steel box section
Case2: Door frame section with 1mm thick steel box section
Case1: Door frame section with 0.8 mm thick steel box section

All of the above mentioned cases are simulated and the results compared to find the optimum sheet thickness for dissimilar brass-steel soldered joint.

Results of FE Simulation

The equivalent von-misses stress for static load test on 2mm steel section is shown in fig 3. On the door panel maximum stress is in the center and at corners of the door. The male brass sections are cut at 45^0 and joined at the corner making a sharp edge. These sharp edges act as stress riser hence more stress is observed on these points. Care must be taken to solder the corners carefully. On door frame side maximum stress is in the center along the length of door. However these stresses are below the yield point of Sn-9Zn solder.

Fig 3 Static load test simulation of 2mm thick steel section

The maximum and minimum stress levels for the three cases are summarized in table 1. The maximum total deformation in all cases is given in table 2 for static load test.

Table 1 Static load test stress result

	Max Stress(MPa)	Min Stress (MPa)
Case 1	2.203	325 e-6
Case 2	3.153	637 e-6
Case 3	5.960	289 e-6

Table 2 Static load test total deformation

	Max total deformation (m)	Min total deformation (m)
Case 1	3.156e-5	3.507e-6
Case 2	2.682e-5	2.981e-6
Case 3	9.496e-6	1.055e-6

The deformation found out in transient structural analysis for sag test in 1mm steel section is shown in fig 4. As the load is applied on the hinge area it is transferred to the adjoining solder joint and the brass female channel.

Fig 4 Deformation in 1mm steel section during sag test

The maximum and minimum stress level in transient structural analysis is given in table 3. The maximum total deformation of sag test is given in table 4.

Table 3 Door sag test stress result

	Max Stress(MPa)	Min Stress (MPa)
Case 1	4.522	0.502
Case 2	16.654	1.850
Case 3	18.524	2.058

Table 4 Door sag test total deformation

	Max total deformation (m)	Min total deformation (m)
Case 1	7.602e-7	8.447e-8
Case 2	1.265e-6	1.406e-7
Case 3	1.585e-6	1.762e-7

The total deformation found in both tests i.e. static load test and sag test for all cases is less than 2mm. The stress level found for all cases mentioned in table 1 and 3 is also less than the yield criteria of Sn-9Zn solder alloy. The stress is in elastic range and hence cannot produce any permanent deformation or cause initiation of crack in solder joint.

Experimental work

In fabrication of EMI shielded door the hinges, door lock assembly and door handle need to be bolted. The holes drilled for bolting purpose became suspect sites for electromagnetic radiation penetration. The attenuation level of whole shielded enclosure drops due to a tiny hole or penetration point left in shielding process. To cover the hinge bolts EMI shielding gaskets are placed under the nut and washer of hinge bolt. The bolt is then tightened to produce perfect electrical bonding between bolt washer and outer skin steel sheet. The hinge and bolt with EMI gasket is shown in Fig 5. A total number of three hinges are used for 7ft standard height of door. EMI gasket is placed on inner side of door frame to stop electromagnetic radiation from entering the shielded enclosure. Bolts of size M10 length 35mm are used for this purpose. Bolt holes are drilled 80 mm away from the solder joint as per requirement of hinge. A total number of three bolts per hinge at 50 mm distance from center to center are required. For test purpose only single bolt is installed in test sample.

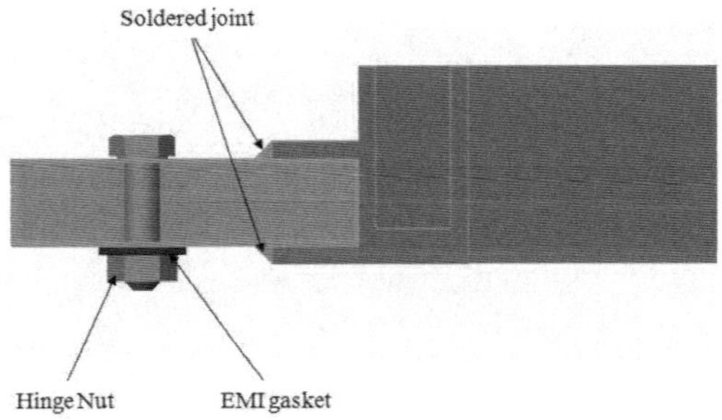

Fig 5 Door frame section with hinge bolt installed

In this research the bolt tightening torque is measured by tightening the bolt with a calibrated torque wrench and measuring the EMI shielding effectiveness through shielding effectiveness leak detector (SELD) equipment. The bolt tightening torque is found out to be in the range of 40 to 55 N.m for perfect electrical bonding. As the hinge bolts are very close to the soldered joint an experiment is performed to check the three different steel frame sections i.e. 2mm, 1mm and 0.8mm under the load of tightened torque. For this experiment small scale door frame sections are prepared. These sections have all the original dimensions accept length. The length of these sections is kept 100 mm. The samples are soldered properly with Sn-9Zn solder alloy. After soldering, holes for hinge bolts are drilled at appropriate location. The samples are then fastened in a test jig. Bolts are tightened on these samples with torque wrench. As starting point torque of 30 N.m is set on the torque wrench. The bolts are further tightened with increasing torque of 5 N.m and cracks in the soldered joint are observed. In 2mm thick steel sample the solder joint fails at a torque of 35 N.m. However in 1mm and 0.8mm steel sections no crack is observed up-to a torque of 70 N.m. At 70 N.m both the thin steel sections deform plastically under the bolt washer. This permanent deformation was measured and found to be more than 2mm in both samples. To counter the deformation in thin plates honey comb steel sections are placed inside the door frame box section. The honey comb came under compressive load during the bolt tightening process and did not allow any deformation to take place at the maximum torque of 70 N.m.

A complete EMI shielded door assembly is prepared with door frame section of 1mm thickness and soldered with Sn-9Zn alloy at steel –brass joint area. The door is then tested with shielding effectiveness test equipment in the frequency range of 10 KHz to 1 GHz as per US military standard 188-125 [10]. The shielding effectiveness of door was found satisfactory at all frequencies. A crack of size 30 mm length is deliberately produced on the hinge side of door. The door is again tested for EMI shielding effectiveness. This time the door does not meet the standard criteria. The graph of shielding effectiveness test is given in fig 6 along with the standard test criteria. At lower frequencies i.e. from 10 KHz to 1 MHz the attenuation values are above the standard curve, but at higher frequency attenuation level drops drastically to small opening in the crack area. Small cracks and pit holes in solder joint have a huge impact on shielding effectiveness of EMI shielded enclosure.

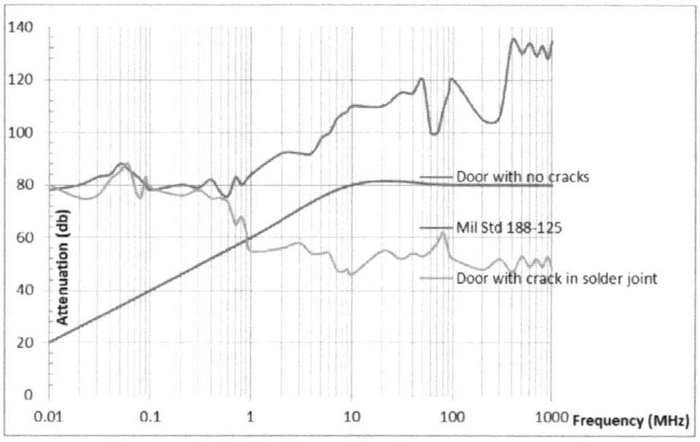

Figure 6 Shielding effectiveness test result

Conclusion

The integrity of soldered joint on dissimilar metals steel with brass is tested in context with its application on EMI shielded door. Sheet thickness variation is studied on frame side of door to check bonding of the joint. It is revealed in this study that thick sheet i.e. steel 2mm is acceptable for this application under tensile or shear loading but if bending load is applied on the frame section as shown in the experimental study of hinge bolt tightening torque, cracks will appear in the steel to solder alloy interface of the bond. The brass to solder alloy interface remains intact, but electromagnetic radiations find passage through these cracks and the shielding effectiveness of the whole enclosure is compromised. The thin steel plates remain intact in tensile, shear and bending loads. In the torque tightening test the thin sheets deform locally in the area of load application. The stress produced due to this local plastic deformation is not transferred to the solder joint. The joint remains intact even when the tightening torque is increased much higher than the required level.

Recommendations for future work

In this study EMI shielded door having solder joint along the periphery is tested under static load test and swing test. A further investigation of swinging door closure test may be performed to authenticate the integrity of solder joint and overall reliability of door. In addition these tests are for EMI shielded doors installed in static enclosures that are placed on ground during service life. For mobile shielded enclosures that are to be installed on vehicles for transportation fatigue test must be performed to check reliability of solder joint.

References

[1] Weis, S., Hoyer, I., & Wielage, B. (2008). Joining of high-strength aluminum-based materials with tin-based solders. WELDING JOURNAL-NEW YORK-, 87(3), 35.

[2] Daly, B. (2013). Basics of Brazing With Induction Heating. Weld. J, 92, 52-54.

[3] Nourani, A., & Spelt, J. K. (2015). Effect of processing parameters on fracture toughness of lead-free solder joints. Engineering Fracture Mechanics, 142, 64-78.

[4] Brožek, M. (2013). Soldering sheets using soft solders. Acta Universitatis Agriculturae et Silviculturae Mendelianae Brunensis, 61(6), 1597-1604

[5] Shangari, S. (1992). Soldering in nuclear power plants. Paper presented at the 73rd American Welding Society annual meeting.

[6] Milosavljević, A., Todorović, R., Kostov, A., & Todorović, L. SPECIAL BRASS SOLDER FOR STEEL PARTS HARD SOLDERING.

[7] Darveaux, R., & Banerji, K. (1992). Constitutive relations for tin-based solder joints. IEEE Transactions on Components, Hybrids, and Manufacturing Technology, 15(6), 1013-1024.

[8] Unified facilities guide specifications section UFGS-8 56 46.20 20 (2006) Radio frequency shielded enclosures, welded type

[9] El-Daly, A., & Hammad, A. (2010). Elastic properties and thermal behavior of Sn–Zn based lead-free solder alloys. *Journal of Alloys and Compounds, 505*(2), 793-800.

[10] MIL-STD-188-125-1 17 July 1998 "High-Altitude Electromagnetic Pulse (HEMP) Protection For Ground Based CI Facilities Performing Critical, Time-Urgent Missions, Part 1 Fixed Facilities"

Fabrication of Ni-alloy Metal Sheet by Laser Powder Bed Fusion

Toshi-Taka Ikeshoji, Makiko Yonehara

Fundamental Technology for Next Generation Research Institute, Kindai University, Higash Hiroshima, Japan

Ikeshji.tt@hiro.kindai.ac.jp

Kazuhiro Nakamura

Technology Research Association for Future Additive Manufacturing, Tokyo, Japan

Hideki Kyogoku

Faculty of Engineering, Kindai University, Higashi Hiroshima, Japan

Abstract

Using laser powder bed fusion (LB-PBF), it was fabricated metal sheets of Ni-base alloy which is a brittle material and is difficult to form a foil by the rolling process. The surface of Ni-base alloy powder bed was irradiated by the laser whose spot size was ø0.1mm and the laser power 100—200 W with scanning in the scanning speed 600—1200 mm/s. The Ni-alloy thin sheets were successively fabricated. The surface morphology and the microstructure of its cross section were analyzed.

Introduction

The Ni-based alloys are usually lack of ductility, especially, the Ni-based brazing filler metals contains the minor element like B and Si, they are difficult to form in thin sheet by rolling. For the brazing, the interface gap is usually as small as several 0.1 mm. The brazing sheet to insert in the interface gap is better to have such a small thickness. Currently, the thin sheets of Ni-base alloy, however, can only be obtained by the single roll quenching method to form the amorphous ribbon.

The laser beam powder bed fusion (LB-PBF) is currently used to build the metal 3D object from wide variety of metal powders. During LB-PBF process, the metal powder is molten by the scanning laser. And it can produce the thin wall structure whose thickness is almost 0.5 mm with Ni-base alloys e.g. Inconel 718, Inconel 625. Therefore, using the LB-PBF process, the thin metal sheet could be formed, and the metal sheet might be utilized as brazing filler metal. The feasibility to produce thinner metal sheet as to 0.1 mm is not, however, clear.

In this research, using LB-PBF, the feasibility of fabrication of meal sheets of Ni-base alloy, Inconel 718 is investigated. In the first step it is conducted to obtain the suitable parameters by the numerical analysis involving the melting and solidification phenomena of metal powder. Then using the parameters obtained by the analysis, the metal sheets attempted to fabricate by LB-PBF. The parameters investigated are the laser power and the scanning speed.

Feasibility Study by Numerical Simulation

It was conducted to check the feasibility to form the metal foil by the laser beam powder bed fusion (LB-PBF) using the transient heat transfer analysis with the melting and solidification by FEM[1]. The calculated region was a 5.50× 1.85×0.5 mm rectangular parallelepiped, which was assumed to filled with Inconel 718 powder. The powder layer thickness was assumed to be 0.050 mm but it did not have effect to analysis because the analysis area was totally filled by the powder.

The scanning length is 5 mm and 10 times scan was simulated. The building conditions were the laser power of p = 150 W, the laser spot size of d = Ø0.1 mm, the the laser scanning speed of v = 900 mm/s, the hatching pitch h = 0.15 mm. The initial temperature of powder and bulk plate was set to 50° C. The laser power distribution was assumed to Gaussian distribution. On the side and bottom surfaces, it was applied the heat conduction condition of 20 W/(K·m) and the temperature outside of those surfaces was set to 50° C.

The temperature depent physical values for Inconel 718 powder was used involving the volumetric spefic enthalpy, the thermal conductivity, and the density. The volumetric spefic enthalpy H_v^b calculated from the the bulk density ρ^b kg/m^3 and specific enthalpy h_m^B J/kg [2,3].

$$H_v^b(T) = \begin{cases} -1.1600 \times 10^9 + 2.3613 \times 10^6\, T + 1.3623 \times 10^3\, T^2 \\ \quad -3.5435 \times 10^{-2}\, T^3 - 1.1272 \times 10^{-5} \\ \qquad 400 < T < 1000 \text{ K} \\ -2.8141 \times 10^9 + 5.4716 \times 10^6 T - 1.4051 \times 10^2 T^2 \\ \qquad 1000 < T < 1528 \text{ K} \\ -4.4581 \times 10^{10} + 3.76667 \times 10^7 T - 3321.76 T^2 \\ \qquad 1528 < T < 1610 \text{ K} \\ -2.5345 \times 10^9 + 6.5088 \times 10^6 T - 1.8983 \times 10^2 T^2 \\ \qquad 1610 < T < 2100 \text{ K} \end{cases}$$

The heat conductivity λ W/(K·m) is

$$\lambda^b = \begin{cases} 1.1224 + 0.0165T & T < 1528 \text{ K} \\ 4.8985 + 0.0136T & 1600 < T < 2100 \text{ K} \end{cases}$$

For the powder, the density ρ^p kg/m^3 was estimated using the bulk density ρ^b and void ratio of powder ϵ.

$$\rho^p(T) = (1 - \epsilon)\rho^b(T)$$

The volumetric specific enthalpy of powder H_v^P J/m^3 was decided in similar manner.

$$H_v^P(T) = (1 - \epsilon)H_v^b(T)$$

The heat conductivity of powder is estimated using Kunii-Smith model[4].

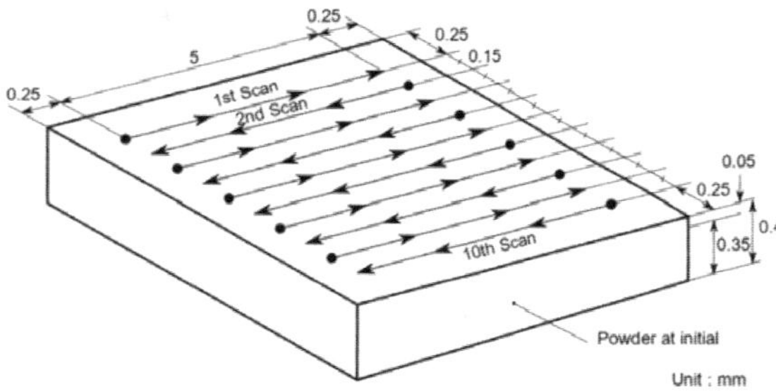

Fig. 1 Analyzed region for feasibility study.

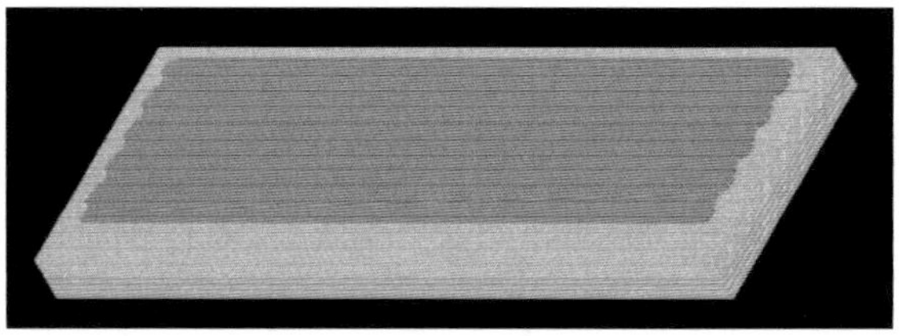

Fig. 2 Melted and solidified region expected by the transient heat transfer analysis (purple part).

$$\lambda^{\mathrm{P}}(T) = \left\{ \epsilon\left(1 + \beta\frac{h_{\mathrm{rv}}D_p}{k_{\mathrm{g}}}\right) + \frac{\beta(1-\epsilon)}{\left(\frac{1}{\phi} + \frac{h_{\mathrm{rs}}D_p}{\lambda^{\mathrm{g}}}\right)^{-1} + \gamma\left(\frac{\lambda^{\mathrm{g}}}{\lambda^{\mathrm{b}}}\right)} \right\} \lambda^{\mathrm{g}}$$

To simulate the melting and solidification of powder, finite elements' material properties were changed by their temperature values. For a finite element with powder material property, when its temperature value raised up to the liquidus of Inconel 718, *i.e.* 1610 K, its material property of enthalpy and thermal conductivity changed from the property for powder to for bulk.

A result of the analysis is shown in Fig.2. The purple part is the once-melting-and-solidified powder part. The light blue part is the powder part and added to see whether the hole was in the solidified part. On the top surface, no hole was observed. That means A sheet of foil might be able to form by the LB-PBF.

Forming Foil by Laser Beam Powder Bed Fusion

Experimental Details

Sheets of Inconel 718 foil were attempted to form using the laser beam powder bed fusion (LB-PBF). The foil's dimensions were 10 × 10 mm. One corner of each specimen was cut-off by 2 × 2 mm to indicate the direction in the building chamber of LB-PBF machine. No support structure was added under the foil. The powder box was filled by multiple squeezing process prior to the laser irradiation. Every layer thickness, t mm, was 0.05 mm. The accumulated powder thickness was more than 5 mm over the base plate.

Tab. 1 Process parameters and specimen number. Values in the second line in the cell is the ordinary energy density, J/mm³.

		Scanning speed, mm/s		
		600	900	1200
Laser power, W	100	#1, #10 33.3	#2, #11 22.2	#3, #12 16.7
	150	#4, #13 50.0	#5, #14 33.3	#6, #15 25.0
	200	#7, #16 66.7	#8, #17 44.4	#9, #18 33.3

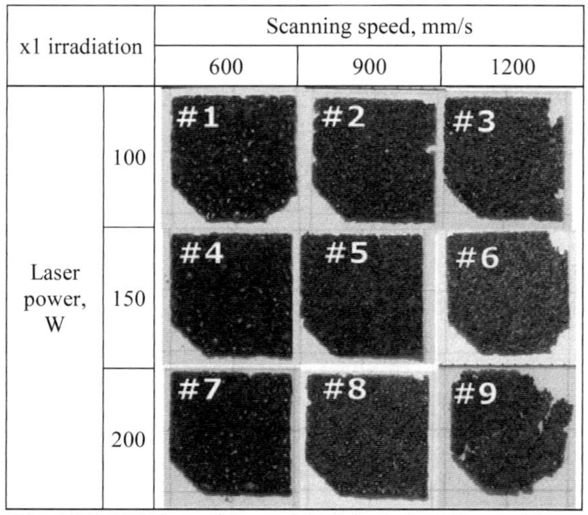

Fig. 3 LB-PBL formed Inconel 718 sheets by 1 times of laser irradiation.

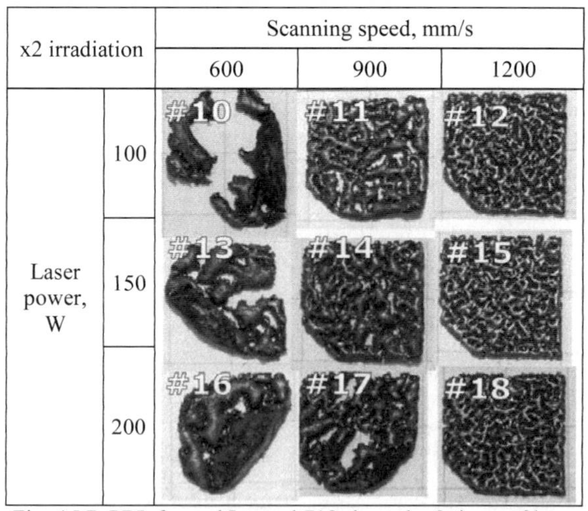

Fig. 4 LB-PBL formed Inconel 718 sheets by 2 times of laser irradiation.

The laser power, P W, was varied 100, 150, and 200 W, and the scanning speed, v mm/s, was varied 600, 900, and 1200 mm/s. The laser spot size was Ø0.1 mm. The hatching pitch, h mm, was 0.15 mm. The ordinary energy density, E J/mm^3, of laser irradiation was

$$E = P/(vht)$$

And in this research, the nominal energy density was defined as

$$E' = P/(vht) \times n$$

where n was the number of laser irradiations.

In this research, the laser was irradiated on the powder bed 1 or 2 times scanning the same geometry. First 9 specimens, #1 --- #9, experienced 1 time of laser abrasion process, and the next 9 specimens, #10 --- #18, experienced 2 times of laser abrasion process. The multiple laser irradiation was originally to reduce the surface roughness to expect the laser polishing effect.

Results

Fig.3 shows the sheets fabricated by LB-PBF. For most of the process conditions provided in this experiment, sheets were formed. But, for #3, #6 and #9, the sheet was brittle and easily broken apart during the handling, e.g. anchoring up by tweezers.

Specimen #1, #5 and #9 were fabricated with the same energy density value, E = 33.3 J/mm^3, but their outlook became different. #1 sheet had small holes inside. #5 was the firmest sheet. #9 was the weakest sheet among them. According to the brittleness and the surface morphology, no clear tendency was observed to the change in the energy density.

Fig.4 shows the sheets fabricated by 2 times successive irradiations. The melt alloy was agglomerated, especially for #10 with 100 W and 600 mm/s, #16 with 200 W and 600 mm/s, and #16 with 200 W and 600 mm/s.

The sheet's thickness values measured by a micrometer are shown in Fig.5. For the laser irradiation conditions with the scanning speed of 600 mm/s in 2 times laser irradiation, the fabricated specimens were not in the sheet form, and their thickness value is plotted as 0 mm (blank circles in Fig.5). With the laser power, the thickness values stay almost the similar when the scanning speed The sheet's thickness values varied from 0.5 mm to almost 1.5 mm. With the laser scanning speed, thickness values decreased.

Observation of the process showed the agglomeration occurred in the 2nd laser abrasion. During the 2nd abrasion, the sheet formed during 1st abrasion was almost totally melt, and the molten liquid was shed on the powder bed. This was like the balling phenomena during the ordinary LB-PBF process [5]. But the difference was the substrate was the powder in this experiment. Usual balling is mentioned on the bulk material e.g. the base plate or the formally solidified surface.

This total melting might be caused by the difference in the heat conductivity between the powder and the bulk metal. The heat conductivity of Inconel 718 powder is 2 digits smaller than the its bulk. The heat input by 2nd abrasion quickly spread in the sheet formed by the 1st abrasion to melt the sheet totally.

If the contact phenomena are involved in the agglomeration of the melt alloy on the powder layer, the oxygen concentration in

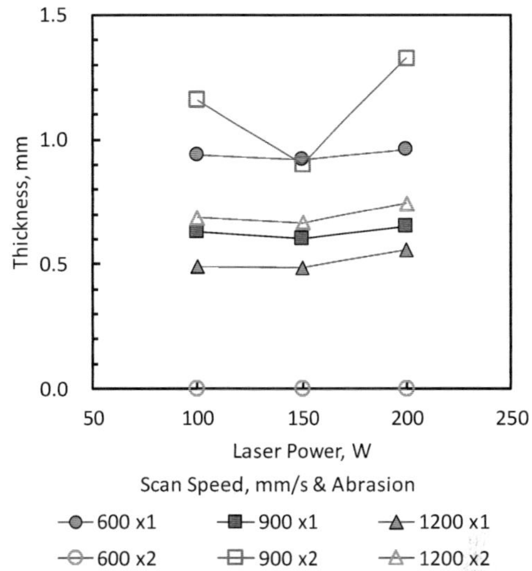

Fig. 5 Thickness of LB-PBL fabricated sheets vs the laser power.

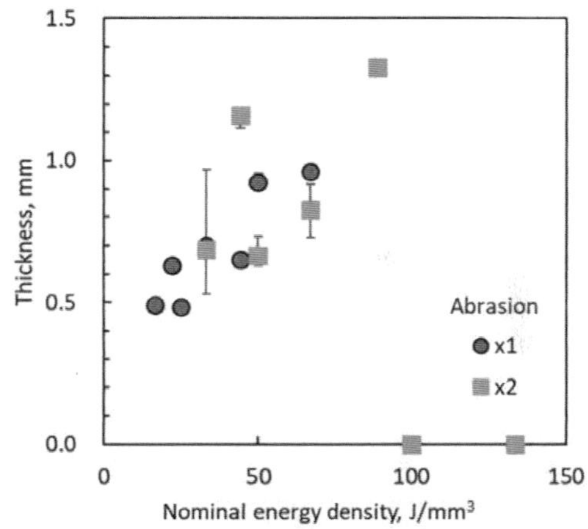

Fig. 6 Thickness of LB-PBL fabricated sheets vs the nominal energy density.

the atmospheric gas should be discussed. As shown in Fig.3 and 4, the surface of sheets was mostly oxidized which suggested the high oxygen concentration. Therefore, if the process were conducted in the atmosphere with low oxygen concentration, the sheet was formed with much wider window in the fabrication parameters.

The sheet's thickness values did not show the clear tendency with the laser power varied from 100 --- 200 W, but with the laser scanning speed varied from 600 --- to 1200. The nominal energy density, which is the index combining these parameters, was calculated (Fig.6). With increase in the nominal energy density, the sheet's thickness increased. But over 100 J/mm^3, the agglomeration of melt occurred to fail to form sheet. In the cases of the two times of laser irradiation, the thickness values tend to be larger, but this could be due to the rough surface of the sheets.

According to these results, the brazing filler sheet metal might be able to fabricate with LB-PBF for the Ni-based brazing filler, which has the similar compositions with Inconel 718. Usually, the brazing filler sheet has the thickness of around 0.5 mm. Under the laser parameters investigated in this research, it will be fabricated the thick sheets compared with the conventional brazing filler sheet. So, the parameter study under 20 J/mm^3 is required for thinner sheet fabrication.

Conclusions

Using laser powder bed fusion (LB-PBF), it was fabricated metal sheets of Ni-base alloy, Inconel 718, with in variation in the laser power and the scanning speed; the laser power 100—200 W, and the laser scanning speed 600—1200 mm/s. The following results were obtained.
(1) With one abrasion of laser, the Inconel 718 sheets could be obtained. But those sheets were brittle.
(2) Within the parameter range in this research, the tendency of the sheet quality to those parameters was ambiguous.
(3) With two times irradiation of laser, the sheet formed in the 1st abrasion melted totally and agglomerated.
According to the results, under the proper conditions, the sheet metal or the sheet brazing filler of Ni-based alloy is possible to fabricate. And the thickness will be around 1 mm.

Acknowledgments

This research was supported by JSPS KAKENHI Grant-in-Aid for Scientific Research (C) 17K06097.

References

[1] Ikeshoji, T.-T., H. Kyogoku, M. Yonehara, M. Araki and K. Nakamura, "Numerical Transient Heat Transfer Analysis of Multi Laser Track on Powder Bed of Ni-Based Alloy," Solid Freeform Fabrication 2016: Proceedings of the 27th Annual International Solid Freeform Fabrication Symposium – An Additive Manufacturing Conference. Austin, TX, USA (2016)398-405.
[2] Wilthan B, Schuetzenhoefer W, Pottlacher G. "Thermal Diffusivity and Thermal Conductivity of Five Different Steel Alloys in the Solid and Liquid Phases". Int J Thermophys,36,8(2015)2259-72.
[3] Pottlacher G, Hosaeus H, Kaschnitz E, Seifter A. "Thermophysical Properties of Solid and Liquid Inconel 718 Alloy". Scandinavian Journal of Metallurgy,31,3(2002)161-8.
[4] Kunii D, "Heat Transmission of Solid Particles," Transaction of JSME, 60,456 (1957) 54-61.
[5] Li, R., J. Liu, Y. Shi, L. Wang and W. Jiang, "Balling behavior of stainless steel and nickel powder during selective laser melting process." The International Journal of Advanced Manufacturing Technology 59(9-12) (2012) 1025-1035.

Investigation of joining process regarding the brazing of copper with aluminium

Dr. M. Lindemann, Dr. P. Koronakis, Dr. M. Ebbinghaus, M. Zabel, F. Gley
from ***Voestalpine Böhler Welding Fontargen GmbH, Eisenberg****, Germany*
Info@voestalpine.com , Tel. +49 (0) 6351 401-0

Abstract

Copper-aluminium compounds and assemblies have been used for years in the HVAC&R sector. Meanwhile, a variety of aluminium alloys has been further developed. Joining by brazing is an important criterion here.
The aim of the paper is on the one hand to consider the material selection taking into account brazing suitability, alternatives and process safety and optimization. The influencing factors are evaluated depending on the constructive design.
In addition to the base materials, the potential filler metals and fluxes are considered according to their product characteristics and processing properties.
In the further course, the test of the properties of the brazing joints using ZnAl filler metals will be described. Both destructive and non-destructive tests as well as corrosion tests are carried out.
The results are aimed at giving the user more confidence in the use of hybrid joints made of aluminium and copper base materials in Heating, Ventilation, Air-Conditioning and Refrigeration systems (HVAC&R). At the same time, support is provided regarding the selection of brazing processes and filler metals on the one hand and the evaluation of the brazed joint on the other.

Introduction

Examinations of brazed joints on copper and aluminium pipes using soft soldering and brazing filler metals for HVAC systems should reflect the state of the art and provide a recommendation for the reliable manufacture of the dissimilar joints.
Within the scope of this examination the testing of joint properties and of brazed joints using ZnAl solders was described. Furthermore, the results of the observations were summarised.

Dissimilar joints of copper and aluminium materials are predominantly used in industry as pipe/sleeve joints. For reasons of strength, plug-in or overlap joints are preferred. The main base materials used for refrigeration systems in HVAC are aluminium materials of the types EN AW 1050, 5005, 6060 and 3103.

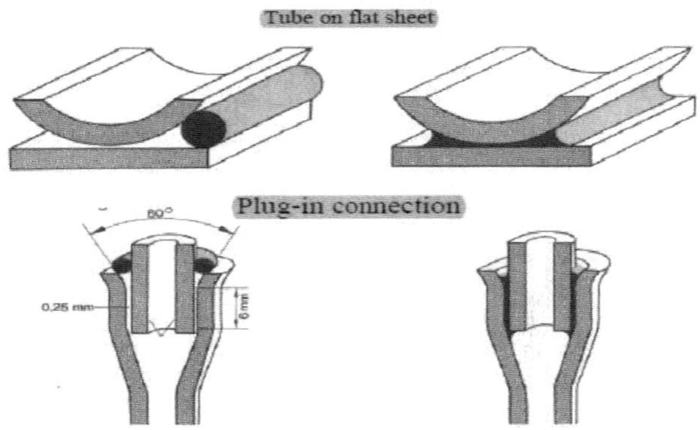

Fig. 1: Types of joint for aluminium pipes according to [1]

The goal is to provide manufacturers and fitters of aluminium and copper joints with a reliable method with which high-quality, reproducible and durable joints can be brazed. The focus is thereby on the solderability of dissimilar joints of aluminium and copper pipes in HVAC systems.

State of the art

Table 1 shows that joints between aluminium and copper can be manufactured according to the purpose of use with various welding methods. Friction and other pressure welding methods are described as being suitable for assemblies in the electrical industry and "brazed joints" are created through the generation of an intermediate layer referred to as "solder". During the welding process, brittle intermetallic phases may be created that can be avoided by coating the copper side with silver solder, zinc or tin. The brittle phases impair the strength characteristics of the joint.

Diffusion welding with very long holding times under pressure and temperature is technically possible, but requires very precise preparation of the pipes. The traditional soldering method is described at this time [2] as not yet being possible. **Table 1** (dated 1972) has been expanded by the corresponding state of development from 2017 (see "O" marking)

Table 1: Application possibilities of various methods of joining Cu with Al (dated 1972) [2].

Application / Process	Possible	Via intermediate layer	Not possible	Technically applied
TIG-/ MIG-welding		X		X
Electron beam welding	O		X (1972)	
Flash butt welding		X		X
Friction welding	O	X		
Friction stir welding	O			O
Cold pressure welding		X		X
Wedge welding	X			X
Ultrasonic welding	X			X
Diffusion welding	X	X		X
Shockwave welding	X			
Brazing	O		X (1972)	O
Gluing	O		X (1972)	O

In every case, welding methods necessitate the melting of the base material of at least one of the joint partners, in the course of whose connection intermetallic phases are created. The gluing process makes use of 1- and 2-component adhesives on an epoxy or polyurethane basis. Hot-, warm- and cold-curing adhesives can be used, depending on requirements. In the case of soldering, the base materials remain in the solid state; strengths are achieved among other things through the diffusion zones that form.

Diffusion layers between copper and aluminium are created in relation to the mass-weight ratio between aluminium and copper on reaching a temperature higher than 450°C/842°F. **Table 2** shows the hardening properties of the various phases and the base materials [3]. The hardening curve is shown in **Fig. 3** [3].

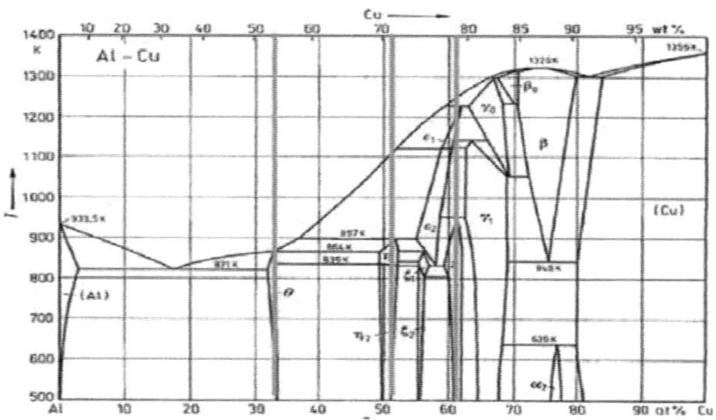

Fig. 2: Phase diagram - dissimilar joint: Al/Cu joint

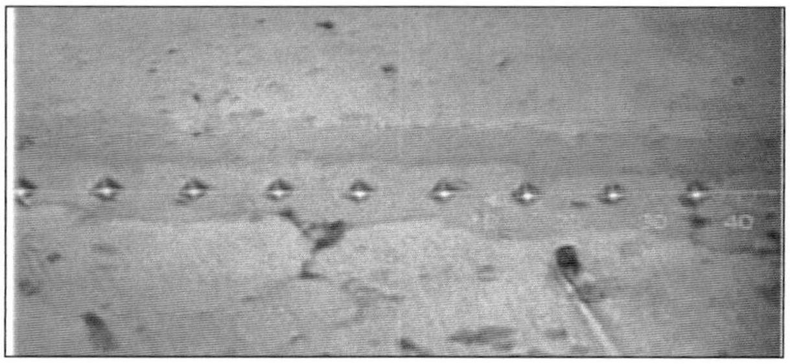

Fig. 3: Hardening curve (see also Table 2)

Table 2: Hardening values of the phase zones in Al/Cu dissimilar joints with diffusion welding [3]

Alloy	Weight % Cu	At. % Cu	Vickers hardness [HV]
Copper	100	100	158
θ-phase (Al_2Cu)	52	33	1050
η2-phase (AlCu)	71	51	168
γ2-phase (Al_4Cu_9)	82	66	172
Aluminium	100	100	56
Vickers hardness (test force 50 μN; penetration time 20 s)			

The method described by Mr Peter [4] for the joining of aluminium and copper pipes using induction heating without filler metals represents a further possibility to manufacture corresponding dissimilar joints. To this end, the copper pipe is pointed a little and inserted into the widened aluminium pipe. Both the copper pipe and the aluminium pipe are preheated in a heating process. On expiry of the heating time, the copper pipe is pressed rapidly into the aluminium pipe. This results in the solder being pressed out of the joint.

The reproducibility of this joint and its use under production or building site conditions are difficult. Therefore, flame soldering will continue to be important.

Soldered joints between evaporator (aluminium) and compressor (copper connection) have been used in refrigeration technology for many years. These are usually manufactured with a zinc-aluminium alloy which, on account of the liquidus temperatures of between 382 °C/720 °F and 450 °C/842 °F, belongs to the soft solders. An exception to this is the alloy A665 (S-ZNAl22-415°C/779°F – 475°C/887°F). With regard to the classification of the solders and ISO 857-2: 2007-03 [5], this ZnAl alloy is strictly speaking a hard solder, because the liquidus temperature is higher than 450 °C / 842 °F.

Soldering of copper / aluminium dissimilar joints.

Solderability is the capability of a component to be manufactured by soldering such that the joints meet the requirements [7]. A component is only solderable if the influencing factors (features) soldering suitability, soldering possibility and soldering reliability exist (**fig. 4**).

The following three sections will deal with these three influencing factors in order to answer the question of solderability.

Fig. 4: Dependence of solderability

Selection of suitable base materials, solders and fluxes (soldering suitability)

Table 3: Soldering suitability of aluminium alloys and the corresponding soft and hard solders

Alloy type	Mat. Nr.	Temperature C°/ °F	Brazing	Soldering
A407L	BAlSi-4	430-440 / 806-824	AlSi Alloys	ZnAl Alloys
A631	ZnAl2	585-595 / 1085-1103		
	Group 1			
Al 99,5	EN AW-1050A	647 / 1197		
Al 99,7	EN AW-1070A	659 / 1218		
Al 99,8	EN AW-1080A	669 / 1236		
Al 99,98	EN AW-1098	660 / 1220		
Al 99,0	EN AW-1200	658 / 1216		
	Group 2			
AlCu4Mg1	EN AW-2024	505-640 / 941-1184		
AlCu4PbMgMn	EN AW-2007	507-650 / 945-1202		
AlCu6BiPb	EN AW-2011	585-650 / 1085-1202		
	Group 3			
AlMn1Cu	EN AW-3003	643-654 / 1189-1209		
AlMn1Mg1	EN AW-3004	629-654 / 1164-1209		
AlMn1Mg0,5	EN AW-3005	630-654 / 1166-1209		
AlMn1	EN AW-3103	645-655 / 1193-1211		
AlMn0,5Mg0,5	EN AW-3105	640-655 / 1184-1211		
AlMn0,6	EN AW-3207			
	Group 5			
AlMg1	EN AW-5005			
AlMg5	EN AW-5019	630-650 / 1166-1202		
AlMg2Mn0,8	EN AW-5049	620-650 / 1148-1202		
AlMg2	EN AW-5051			
AlMg2,5	EN AW-5052			
AlMg4,5Mn0,7	EN AW-5083	607-649 / 1125-1200		
AlMg4	EN AW-5086	607-649 / 1125-1200		
AlMg3,5	EN AW-5154	641 / 1186		
AlMg4,5Mn0,4	EN AW-5182	638 / 1180		
AlMg2	EN AW-5251	610-650 / 1130-1202		
AlMg3Mn	EN AW-5454	602-646 / 1116-1195		
AlMg3	EN AW-5754	610-640 / 1130-1184		
	Group 6			
AlSiMg	EN AW-6005A	585-650 / 1085-1202		
AlMgSiPb	EN AW-6012	585-650 / 1085-1203		
AlMgSi	EN AW-6060	585-650 / 1085-1204		
AlMg1SiCu	EN AW-6061	580-640 / 1076-1184		
AlMg0,7Si	EN AW-6063			
AlMgSi(B)	EN AW-6101B	620-650 / 1148-1202		
	Group 7			
AlZn4,5Mg1	EN AW-7020	600-650 / 1112-1202		
AlZn5Mg3Cu	EN AW-7022	640 / 1184		
AlZn5,5MgCu	EN AW-7022	640 / 1184		
	Group 8			
AlFeSi(A)	EN AW-8011A	640-655 / 1184-1211		

Suitable
Challenging
Not suitable

Table 3 shows the soldering suitability of aluminium alloys and the temperature ranges of base materials and filler metals. In addition, the microstructure formations and the way in which they change in the case of thermal expansion must be considered.

Point 4.3.1.1 in DIN EN 14276-2 [8] "Pressure equipment for refrigerating systems and heat pumps" prescribes the base materials to be used for the soldering of pipes according to ISO 15608 [9] (Tables 4 and 5)

Table 4: permitted aluminium and aluminium alloys according to DIN EN 14276-2 [8]

Designation according to EN 15608 [9]	Groups according to **Table 1**
21	Group 1
22	Group 3

Table 5: permitted copper and copper alloys according to DIN EN 14276-2 [8]

Designation according to EN 15608 [9]	Copper / copper alloy
31	Pure copper
32	Copper-zinc alloys (brass / nickel silver)
33	Copper-tin alloys
34	Copper-nickel alloy
35	Copper-aluminium alloy

All copper and copper alloys according to **Table 5** are well suited to soldering. In practice, the base material (31) used most frequently is pure copper. The use of iron-alloyed copper materials (CuFe2P) is also possible according to DIN EN 14276-2 [8] point 4.3.1.2. This base material is well suited to soldering using hard and soft solders.

From a relatively small selection of hard and soft solders, in Europe four alloys have become established on the market.

Table 6: most important solder alloys for the manufacture of Al/Cu dissimilar joints

Fontargen brazing	ISO 3677 [10]	DIN designation	Hard solder (B) Soft solder (S)
A 407 L	B-Al88Si-575/585	AlSi12	B
A 633	S-Zn97Al3-430/450	ZnAl3	S
A 631	S-Zn98Al2-382/407	ZnAl2	S
A 665	S-Zn78Al22-415/475	ZnAl22	S/B

Tin-based solders, which can also be used for joints, have not been considered in this publication due to their low strengths.

The generally used types of flux are shown in **Table 7** according to their use for soft soldering or brazing. If soldering takes place under atmospheric conditions, the flux must be selected according to the aluminium alloy and the working temperature.

Table 7: the most important fluxes for the manufacture of Al/Cu dissimilar joints in Europe

Fontargen brazing	EN 1045 [11] / EN ISO 94541-1 [12]	Form of delivery	Hard solder (B) Soft solder (S)
F 400 MD	FL 10	Pasty (corrosive)	B
F 400 NH	FL 20	Powder (non-corrosive)	B
F 600 AL	2.1.2.A	Oil (corrosive)	S
F 600 ZA	not standardised	Pasty (non-corrosive)	S

As a guideline, the non-corrosive fluxes with a magnesium content of up to 0.7% and the corrosive fluxes with a magnesium content of up to 3% are usable in the aluminium alloys suitable for soldering.

Manufacturing by soldering (soldering possibility)

Apart from the possibilities for joining by welding with the problems described, the dissimilar joints of aluminium alloys and copper can be manufactured using the joining methods named in **Table 1**. The method to be used must be analysed in the individual case by the manufacturers and selected while taking into account all essential influencing factors. A decision-making matrix according to DIN 8514 [7] is useful here. An example is shown in **Table 8**.

Table 8: Example of a decision-making matrix with grading at one's own discretion (4 = very good, 2 = adequate)

		Flame		Induction		Resistance soldering	
Criteria	**Weight in %**	Grade		Grade		Grade	
Tolerances	10	2	0.2	2	0.2	4	0.4
Process reliability	40	3	1.2	4	1.6	3	1.2
Availability	10	4	0.4	2	0.2	2	0.2
Personnel costs	20	2	0.4	4	0.8	4	0.8
Procurement costs	20	4	0.8	2	0.4	2	0.4
Total	100		**3.0**		**3.2**		**3.0**

Constructive design of the joint (soldering reliability)

The soldering reliability essentially takes into account the constructive design and the stress condition of the soldered joint.

Table 9: factors influencing the soldering reliability according to DIN 8514 [7]

Constructive design	Stress condition
Position of soldered seam/soldering zone	Static or dynamic stress
Cross-sectional ratios	Stress condition in the soldering zone
Notch effect	Stress speed
Irregularities in the soldered seam	Stress temperature
	Stress medium

The most important factors influencing soldering reliability are given in **Table 9**.

Explicitly in the case of the dissimilar joint between copper and aluminium, the constructive design and the soldering possibility lead to the question: which of the joint partners is to be inserted into the other? In this regard, only subjective statements are made in practice. Both variants are possible; a connection between irregularities and their dependences has not been proven. Depending on the pipe diameter and wall thickness, insertion depths of between 10 mm and 20 mm have proven to be effective with dissimilar joints. The soldering rule of thumb "overlap length = 3 to 6 x component thickness" can be applied.

Checking the joint properties

Terms such as "pressure resistance" and "leak test" are omnipresent in refrigeration and air-conditioning technology. What these terms describe and how they influence the evaluation of a soldered joint is often misinterpreted. Point 6 "Testing and commissioning" in EN 378-2 [6] describes practicable methods for testing refrigeration systems and their components. However, the solderability should be secured prior to commissioning.

General testing during commissioning

The test of pressure resistance described in EN 378-2 [6] represents the stress condition in the soldering zone and thus has a direct influence on the soldering reliability.

The leak test, conversely, is merely a snapshot of the permeability in relation to a medium. The irregularities that form in the soldered seam are not considered. Deviations from the original test results may already have occurred after the transport of the components and must therefore be classified in the testing possibilities section. The pressure resistance and leak tests are non-destructive tests and are not suitable for testing a constant soldered seam quality.

An increase in irregularities in resoldered joints according to ISO 18279 [13] due to changing process parameters has not been proven.

Testing of soldered joints

Applicable testing methods are described in the DIN EN 12797 [14] and DIN EN 12799 [15] standards. These tests essentially serve the purpose of maintaining the soldering reliability of brazed joints.

Destructive tests are essentially unsuitable for manufacturing, but can be used beforehand to determine soldering reliability.

Table 10: Composition of the tests

Non-destructive tests DIN EN 12799 [15]	**Application suggestions**
Visual inspection	100% inspection possible with little effort; assessment of solder flow, flute shape, surface pores and flux residues; particularly efficient with inlaid solder;
Ultrasonic testing/thermography	Hardly used, high expenditure, difficult to correlate the irregularities. High degree of practical experience necessary;
Radiographic testing	Hardly used, high expenditure, difficult to correlate the irregularities. High degree of practical experience necessary;
Dye penetration testing	Little meaning; only surface cracks discernible;
Leak test / overpressure test	Normal practical test method; only continuous irregularities can be detected, observe test pressure
Destructive tests DIN EN 12797 [14]	**Application suggestions**
Tensile test	Rarely used because the soldered joint is usually exposed to shear stress;
Shear pull test	Often used; position of the breakage and the loadability of the joint are of great significance;
Metallographic examination	Often used in advance to determine the process parameters for soldering;

Hardness test	Detection of hardening in the microstructure; rarely used;
Peel test	Used very often, subjective statement about strength is possible; exact determination of the irregularities after breakage in the solder;
Bending test	Rarely used; indication of deformability of the assembly if used;

The individual test methods in **Table 10** will not be dealt with any further. Combinations of these have proven themselves in practice for testing the soldering reliability.

The corrosion test is not considered in the aforementioned standards and their test methods. Nevertheless, corrosion has without any doubt an influence on the soldering reliability, depending on the stress medium (see **Table 9**).

The effort and thus the costs for the tests to be carried out must be defined individually and specifically for the plant in question. In the past, many assemblies were not subjected to any tests. If there are no recall campaigns and complaint rates are very low, then test methods are hardly discussed in practice. Increasing numbers of items, larger dimensions and higher stresses in the sector of aluminium/aluminium and aluminium/copper joints demand a rethink.

The following empirical values and examinations carried out are intended to provide security in dealing with the material pairings using AlZn alloys.

Properties of soldered joints using ZnAl solders

Strengths

In addition to the forces applied and the stresses created, the shear pull test frequently used in soldering technology also provides a statement regarding where the failure of the joint will ultimately occur under load. Shear pull stresses on brazed (B-Al88Si-575/585) and soft-soldered (S-Zn98Al-382/407) aluminium/aluminium and aluminium/copper joints are listed as an example. The sample geometry used (**fig. 5**) deviates from that recommended in EN 12797 [14]. The solder gap was 0.1 mm.

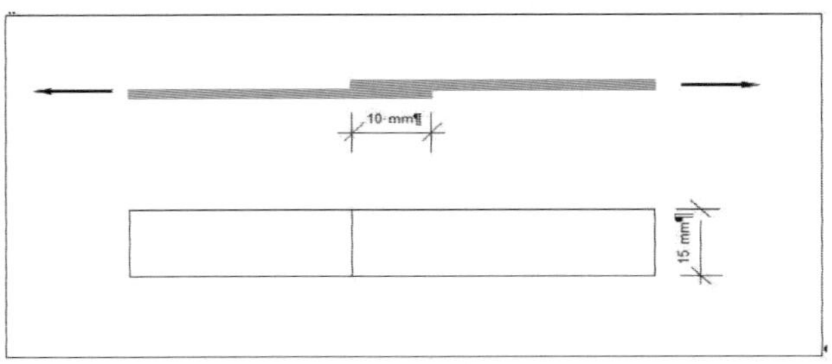

Fig. 5: Simplified illustration of the sample geometry according to [16]

Table 11: Shear pull stresses averaged according to Güldner Table 17 [16]

Solder filler metals	Base materials	Shear pull stresses
B-Al88Si-575/585	Al / Al	100.56 N/mm²
	Al / Cu	102.85 N/mm²
S-Zn98Al2-382/407	Al / Al	98.84 N/mm²
	Al / Cu	99.39 N/mm²

The breakage position in all samples was located in the aluminium base material, which indicates a high shear pull strength of the joint.

The bursting pressure test is not described in EN 12797 [12], but makes a good contribution to the appraisal of piping systems and their behaviour under pressure. The pressures to be applied are to be defined depending on the area of application of relevant standards.

Table 12: Mean bursting pressures according Güldner Table 11

Solder filler metals	Base materials	Bursting pressure
B-Al88Si-575/585	Aluminium / aluminium	165 bar
	Aluminium / copper	155 bar
S-Zn98Al2-382/407	Aluminium / aluminium	166 bar
	Aluminium / copper	173 bar

Plug-in joints were used (pipe 10x1). In each case, the breakage was located in the area of the aluminium base material with an insertion depth of 10 mm.

For quality monitoring during manufacturing, the peel test described in DIN 12797 is suitable for the fast determination of irregularities in soldered connections and for the subjective estimation of the strengths. Different evaluations are possible depending on the breakage position. The strength of the joint is confirmed if the breakage is located in the base material. If the breakage is located in the soldered joint, the peeling forces can be estimated subjectively on the one hand and an evaluation of the wetting and the irregularities is possible on the other.

Peel tests for the aluminium/copper joint are realised below as examples. **Figures 7 to 10** show the solder sample and the results of the peel test. The flux residues were removed with a brush after soldering. Following the shortening of the copper pipe, the aluminium pipe was clamped and a cut was made in the region of the copper pipe, before the latter could be peeled from the aluminium pipe with a pair of tongs (see **fig. 8**).

It is possible to make out some of the irregularities on the aluminium pipe in the soldered joint. **Figure 10** shows the circumference with the contained irregularities of the soldered seam on the copper pipe.

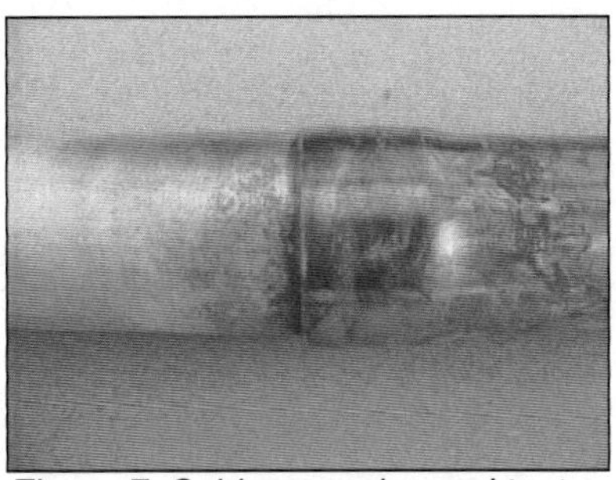

Figure 7: Solder sample, peel test

Fig. 8: Result of the peel test - overview

Fig. 9: Result of the peel test - aluminium pipe

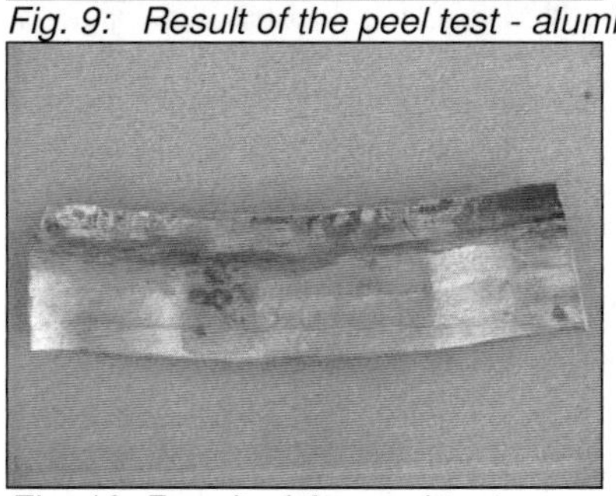

Fig. 10: Result of the peel test - copper pipe

Internal irregularities

The warm extraction divides opinion among experts in soldering technology and is no longer considered as a test in most standards, because it is indisputable that the test results have only a limited significance in relation to the detailed evaluation of several internal irregularities.

In many practical cases, however, the evaluation of the soldered joint is limited to selected error patterns that can be evaluated relatively well following the warm extraction. Examples of this are lacks of fusion and insufficient flow.

In the case of soldered joints that are not suitable for a peel tests on account of the breakage location (base material), the warm extraction is an inexpensive, practicable alternative for the estimation of the quality during the manufacturing process.

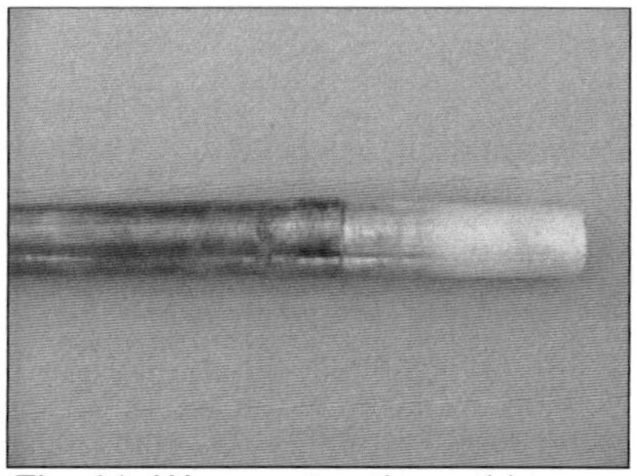

Fig. 11: Warm extraction solder sample

Figure 11 shows the solder sample for the warm extraction. The soldering parameters are appropriate for the soldering procedural instructions for the peel test.
A meaningful evaluation of the aluminium/copper soldered joint after warm extraction is only possible on the copper side (**figure 12**). A reliable determination of the irregularities is difficult on account of similar hues between the base material aluminium, as well as its alloys and the ZnAl filler metal.

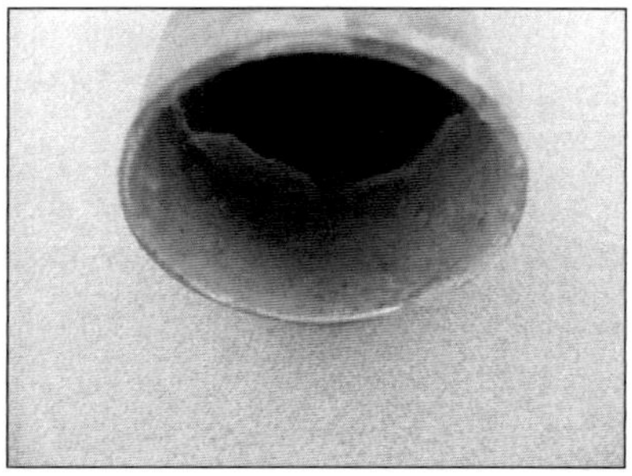

Figure 12: Result of the warm extraction - copper tube

Resistance to weather influences

The corrosion tests is indispensable for the aluminium/aluminium and aluminium/copper joints using ZnAL solders. If the joint is protected against moisture, for example with paint, this corrosion mechanism plays a subordinate role.

In the tests by Güldner, the mass loss as a consequence of corrosion was determined in a corrosion spray mist test (**Table 13**). The samples were tested over a period of 240 hours in accordance with DIN EN ISO 9227 (former 50021-SS) using the solder S-Zn98Al-380/420.

Table 13: Mean mass difference according to Güldner

Solder filler metals	Base materials	Mass difference
S-Zn98Al2-382/407	Al / Al	14 mg
	Al / Cu	139 mg

The subsequent leak test at 35 bar with a dwell time of 60 minutes did not produce any leaks. The extent to which the corrosion test is used depends on the usage case and the expected media.

Conclusion

The aim of this paper is to provide security in the manufacture of joints in refrigeration and air-conditioning systems using aluminium alloys as well as dissimilar joints of aluminium and copper alloys.

Ultimately it is the manufacturer of soldered assemblies who selects the joining process to be used and who is thus responsible for the joint. All positive substance joining methods (**see Table 1**) and positive locking pressing methods have advantages and disadvantages. A decision-making matrix is useful for the selection of a cost-effective and standard-compliant joining process (**see Table 8**).

Due to the flame soldering process commonly used for the manufacture and repair of pipe systems, the selection of a solder and the possibilities to test the soldered joint are illustrated below. Apart from the advantages, the secure handling of the process disadvantages is mentioned.

The solderability is determined by the soldering suitability, soldering possibility and soldering reliability. The soldering suitability requires the user to select the base materials. Before deciding on a solder filler metal, the soldering process must be selected - soft soldering or brazing. For historical reasons, strength characteristics of the joint partners are connected with both options. Indicative of this is the remark in DIN EN 378-2 [6], point 5.3.1.3, which states that soft soldering may only be used if strength plays a subordinate role. Presumably the remark refers to the group of Sn-based solder alloys whose strength and melting temperature ranges lie well below the Zn-based solders. In the brazing process (B-Al88Si-575/585) the material pairing Al/Al can still be manufactured well under manufacturing conditions. The joining process with the dissimilar Al/Cu joint becomes borderline as the wall thickness increases. The different thermal conductivities of Al and Cu that facilitate the melting of the aluminium alloy make the reproducible soldering possibility more difficult. The soft soldering process with Zn-Al solders, conversely, is advantageous with melting temperatures of between 380 °C and 450 °C. The joints soldered with ZnAl alloys can attain the strength values of brazed assemblies (**see Table 11**).

In parallel with the advantageous melting temperatures of ZnAl solders, the soldering possibility for repairs should be mentioned. A skilled solderer can manufacture the soldered joints with a decreased temperature difference of 100°C/212°F The disadvantages with regard to the corrosion problem can be countered structurally. The available testing methods help irregularities to be detected at an early stage and thus soldering reliability to be ensured in order to be able to manufacture reproducible, inseparable and utilisable soldered joints.

In practice soldered joints are protected against moisture by thick surface coats and laid in areas with low dynamic or cyclic stress. In addition to the testing methods mentioned, which not only serve quality assurance purposes, but also and mainly the legal security of responsible persons, the results of decades of application experience are incorporated into the question of soldering reliability. However, these empirical values require complete documentation of the manufacturing conditions and the technological implementations. Equally important is the training and advanced training of skilled workers for the manufacture and self-checking of joints. Amongst other things, procedural instructions for soldering should exist in some easily-accessible form.

References

[1] Aluminium Vereinigte Metallwerke Ranshofen Berndorf AG: Data sheet 16 Soft soldering and brazing.
Issued: Date of issue unknown

[2] Overview of the welding methods for joining copper with aluminium: Horst Erhardt: ZIS notifications.
Issued: Volume 1 / 1972

[3] Prof Bergmann TU Ilmenau / DKI documents:
Lecture. Issue of documents TU: 03.11.2011

[4] Induction soldering - principles - plant technology - inductors - effective soldering technologies, H. J. Peter:
Expert Verlag. Issued: 2013

[5] DIN ISO 857-2:
Welding and allied processes - Vocabulary - Part 2: Soldering and brazing processes and related terms Issued: March 2007

[6] EN 378-2:
Refrigerating systems and heat pumps - Safety and environmental requirements
Design, construction, testing, marking and documentation.
Issued: 2008

[7] DIN 8514:
Brazeability, solderability.
Issued: 2006-05

[8] EN 14276-2:
Pressure equipment for refrigerating systems and heat pumps.
Part 2: Piping
Issued: August 2007

[9] ISO 15608:
Guidelines for a metallic materials grouping system.
Issued: July 2000

[10] DIN EN ISO 3677:
Filler metal for soldering and brazing - Designation
Issued: December 2016

[11] DIN EN 1045:
Brazing - Fluxes for brazing -
Classification and technical delivery conditions:
Issued: August 1997

[12] DIN EN ISO 94541-1:

Soft soldering fluxes - Classification and requirements -
Part 1: Classification, labelling and packaging
Issued: July 2016

[13] DIN EN ISO 18279:
Brazing - Imperfections in brazed joints
Issued: 2004-04

[14] DIN EN 12797:
Destructive tests of brazed joints.
Issued: 2000-12

[15] DIN EN 12799: Non-destructive examination of brazed joints. Non-destructive examination of brazed joints
Issued: 2000-07

[16] Mr Güldner, degree thesis:
Qualification of a soldering process for the manufacture of aluminium and aluminium/copper pipe connections in the production of industrial heat exchangers:
KTV/02/2008

Preventing active braze filler metal run-out using conformal ALD coatings

Charles Walker, Dennis De Smet, Ron Goeke, Bonnie McKenzie, Paul Vianco
Sandia National Laboratories, Albuquerque, New Mexico, USA

Sandia National Laboratories is a multi-mission laboratory managed and operated by National Technology and Engineering Solutions of Sandia, LLC., a wholly owned subsidiary of Honeywell International, Inc., for the U.S. Department of Energy's National Nuclear Security Administration under contract DE-NA0003525.

Abstract

Active brazing is a permanent metallurgical joining method in which highly reactive brazing filler metals are utilized to directly braze metals to nonmetals. Due to limitations of the active brazing filler metal spreading adequately over a nonmetal surface, the filler metal must be preplaced between the two faying surfaces of the brazement. When heated above the liquidus temperature, the liquid filler metal is normally contained within the brazement by capillary attraction. With certain metal-nonmetal base material combinations, however, chemical reactions at the brazement faying surfaces may lead to excessive braze filler metal flow and render the brazed assembly useless. Nanometers thick conformal coatings, deposited by atomic layer deposition (ALD) onto the metal surfaces, can modify the surface chemistry to eliminate excessive filler metal flow. Unlike other means used to prevent excessive filler metal flow, such as braze stop-off, thin ALD coatings do not hinder next assembly processes, require post-braze cleaning or alter the base material mechanical properties. This effort focuses determining the thickness bounds of conformal ALD coatings that fully prevent excess braze filler metal flow while maintaining joint integrity. Test results and analysis of 94% alumina ceramic ASTM F-19 tensile buttons [1] having Kovar® [2] alloy interlayers that have been actively brazed in a high-vacuum furnace using silver-based active filler metal alloys will be presented along with conclusions, and recommendations.

Introduction

Background
Brazing metals to nonmetals such as oxide ceramics is typically accomplished by either modifying the nonmetal faying surface to promote wettability, i.e. metallizing, or using active braze filler metals: filler metal alloys containing small weight percentages of Group IV metals, such as titanium, zirconium or hafnium [3,4].

The metallization process most commonly used is a multistep process beginning with the application of a molybdenum-manganese (Mo-Mn) slurry that is fired onto the nonmetal faying surface to be brazed in a reducing atmosphere at temperatures ranging from 1475°C to 1600°C. Subsequent processing applies a nickel plating layer which is sinter-fired in a reducing environment at approximately 800°C. The resulting metallization bi-layer total thickness is on the order of 12.7-15.2μm (400-500μin). Following the metallization process the nonmetal and metal substrates are fixtured as required, the braze filler metal is applied, and brazing thermal cycle is performed to accomplish the braze.

The active filler metal brazing process is accomplished by preplacing the active filler metal between the properly fixtured or positioned metal and nonmetal faying surfaces and the brazing thermal cycle is performed. Figure 1, Conventional brazing and Active Brazing Processes Compared, displays the two brazing processes. While the active brazing process appears to be much simpler to accomplish, in certain cases, the active element material used can lead to problems not observed when using a Mo-Mn/Ni plate metallization.

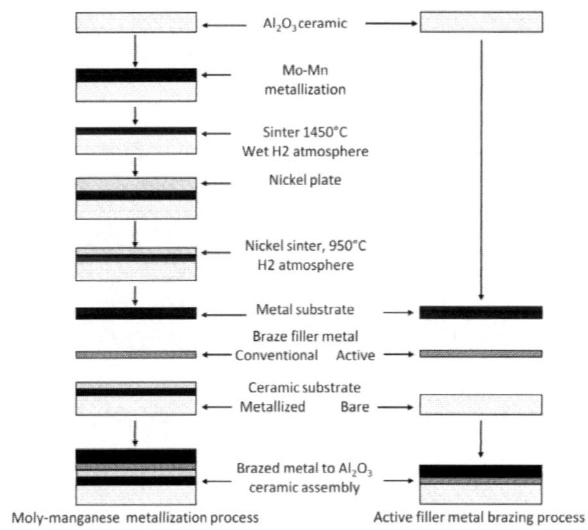

Figure 1. Conventional and Active Brazing Processes Compared

Two typical problems observed when developing and active brazing process are insufficient wetting of the nonmetal surface and excessive wetting of the metal surface. In the first case, the highly reactive Group IV metal in the active filler metal reacts with excess oxygen or in some cases hydrocarbons in the furnace atmosphere, essentially poisoning the braze filler metal, which prevents the formation of the necessary metal-oxide or metal-carbide bonds at the nonmetal interface. The second issue, excessive wetting or spreading of the liquid braze filler metal, has been demonstrated to occur when the active element reacts with the nonmetal substrate, freeing elements that may forms compounds which reduce the surface tension of the metal faying surfaces.

There are commercially produced materials available that can prevent excess brazing filler metal flow. These materials, commonly referred to by users as stop-off, braze stop-off or braze-stop, are typically fine oxide ceramic powders in a volatile solvent carrier that are applied to portions of the metal braze assemblies where filler metal flow is undesirable. Post-braze powder residues from these stop-off materials are in many cases unacceptable and must be removed, using mechanical or chemical methods, sometimes a combination of both. In instances where post-braze cleaning of stop-off materials cannot be used, other methods to reduce filler metal spread/wetting such as: decreased peak brazing temperature and/or time, reduced filler metal volume, joint geometry changes, surface roughness, etc., are attempted. In our specific case, these methods were equally ineffective as a solution.

Figure 2. ASTM F19 tensile button sample, 08082015F401-2, displaying excess filler metal flow

Vianco et al. [5] reported that while excessive filler metal flow occurred when brazing Kovar to alumina ceramic, no excessive filler metal flow or runout was observed when Ag-1Cu-2Zr braze filler metal was used to fabricate Kovar-Kovar assemblies, or alumina-alumina assemblies. In the presence, of aluminum, however, it was demonstrated that excessive filler metal flow does occur on the Kovar substrates. All evidence suggested that the required aluminum must come from the alumina ceramic. The atomic aluminum, therefore becomes available primarily due to the reduction-oxidation (redox) reaction $3Zr+2Al_2O_3 \Rightarrow 4Al+3ZrO_2$, which is spontaneous at the brazing temperature. This "free" aluminum is the product of a redox reaction between zirconium and alumina in which the liquid zirconium metal reduces the aluminum oxide, forming the more stable zirconium oxide. Elemental spectral analysis of the Ag-1Cu-2Zr/Kovar interfaces of Kovar-alumina ceramic assemblies reveal that multiple aluminide compounds are formed when Al and AlZr react with the base metal alloy constituents, iron, nickel and cobalt [5].

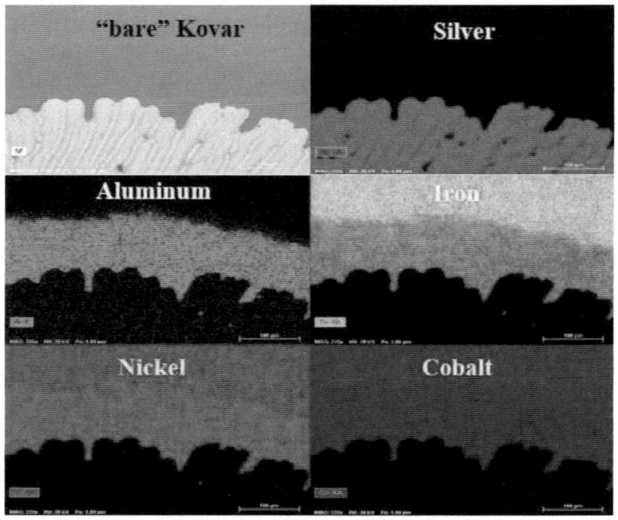

Figure 3. Sample 08082015F401-2 displaying aluminum wave-front on Kovar surface

Figure 2 shows a brazement having a region of excess braze filler metal flow on the Kovar surface of a fractured ASTM F19 Tensile Sample. As shown in the scanning electron microscope image and corresponding electro-dispersive spectroscopy image (EDS) maps, Fig. 3, there is a thin layer of aluminum (orange) layer on the Kovar surface in front of the leading edge of the primarily silver (blue) solidified braze filler metal that has reacted with the Kovar substrate. The braze filler metal, nominally 97Ag-1Cu-2Zr weight percent, contains no aluminum, indicative that the aluminum in the resultant braze joint came from the alumina ceramic. Atomistic modelling simulations [6] indicated that the $3Zr+2Al_2O_3 \Rightarrow 4Al+3ZrO_2$ redox reaction and subsequent iron aluminide reactions occur within tens of nanoseconds, leading to rapid spreading of the liquid filler metal. quickly, often within fractions of a second. The simulations indicated that the addition of aluminum to the liquid filler metal greatly enhanced the spreading ability when in contact with the Kovar substrates.

After these previous experimental and modelling results, potential paths to limit or eliminate iron aluminide formation were determined. One path would mask the metal surface with a thin, barrier layer coating. The second path would impede the aluminide reactions through adding alloying elements to the Ag-1Cu-2Zr braze filler metal or possibly reducing the amount of active element material in the braze filler metal. Because the second path would require an entire filler metal development effort at no little expense, the above-mentioned barrier layer path was selected to be performed first. The results of this effort are the subject of this manuscript.

The following requirements were determined necessary for a barrier layer: The selected layer must be conformal to surface features, limit surface wetting or spread, not hinder the braze joint performance (i.e. joints are hermetic and have high tensile strengths) remain intact and not compromise next assembly function. To address these requirements a very thin (1nm – 20nm) of alumina was deposited on the Kovar surface using an atomic layer deposition (ALD) process (Fig. 4).

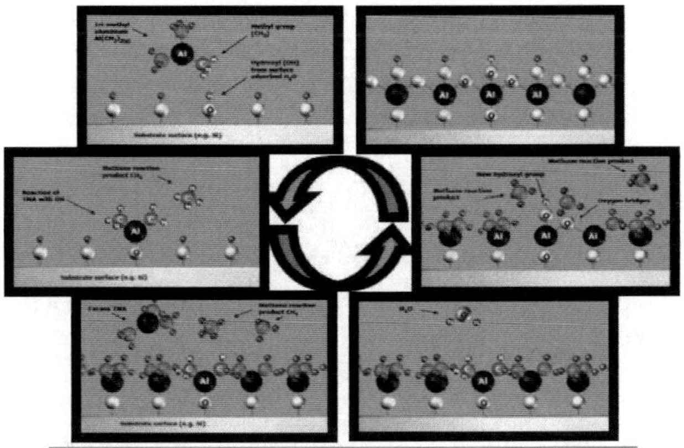

Al_2O_3 CVD - reaction:

$2Al(CH_3)_3(g) + 3H_2O(g) \rightarrow Al_2O_3(s) + 6CH_4(g)$

ALD Al_2O_3 half-reactions: (200°C ~ 1.1 Å/cycle)

$Al(CH_3)_3(g) + Al\text{-}OH^* \rightarrow Al\text{-}OAl(CH_3)_2^* + CH_4(g)$

$2H_2O(g) + \text{-}Al(CH_3)_2^* \rightarrow \text{-}Al(OH)_2^* + 2CH_4(g)$

Figure 4. Atomic Layer Deposition (ALD) alumina ceramic process

ALD is a chemical gas phase thin film deposition process. The film is deposited by introducing the chemical reactants sequentially and utilizing "self-limiting" adsorption reactions (chemisorption) to control the thickness of the deposited film to a single reaction layer. Repeating the reaction cycle results in a layer by layer highly conformal film growth. For the specific process used to grow the alumina oxide in the reactor used for these experiments, Al_2O_3, films, one complete cycle occurred every 32 seconds and a film thickness deposition rate of 1.1Å (0.011nm) per cycle was achieved.

Experimental Plan

Proposed Experiments

A series of three experiments is planned that will evaluate the effectiveness of using ALD Al_2O_3 coatings, in various thicknesses, on Kovar substrates when brazed to 94% alumina ceramic using the active braze filler metal Ag-1Cu-2Zr. A portion of the Kovar substrates will include "dimples", actually small mounds or bumps, approximately 25µm (0.001") tall on the faying surfaces. Though not completely effective in preventing braze filler metal runout for this system of components, dimples are effective in maintaining braze joint thickness uniformity, which can improve joint tensile strength.

The first experiment will be to braze Diamonite cylinders to "flush mount" Kovar discs, forming a chamber that can be leak tested. The flush mount disc substrates have nominally the same outer diameter as the Diamonite cylinders (25.4mm). The geometry allows the effectiveness of the coating to prevent the formation of large fillets (i.e. blobs) on the inner brazement interface and excessive flow on the outer Kovar substrate edges to be determined. The input variables are dimpled or undimpled Kovar; bare surface or 10nm thick ALD Al_2O_3 coated surface. Observed/Measured responses are: 1) runout/excess flow and 2) joint hermeticity. Runout or excess flow is defined as filler metal solidified on the inner or outer surfaces or edges of the Kovar alloy beyond the joint proper that can be observed without magnification. Hermiticity is defined as having total helium leak rates ≤5E-13 Pa-M^3s^{-1} (≤5E-12 atm-cm^3s^{-1}). These criteria are critical for next assembly joining processes as well as product performance. Three replicates of each unique combination will be fabricated, 12 samples total. Tensile button joint tensile strength data is collected after the samples are mounted into proper fixtures and are tested on a MTS test frame. A displacement rate of 5.1E-05m-s^{-1} (0.002in-s^{-1}) is used to fracture the tensile samples.

The second set of experiments will be to fabricate and test ASTM-F19 tensile buttons using the Diamonite alumina ceramic for the tensile button material and oversized Kovar interlayers. Oversized Kovar tensile button interlayers allow the runout, excessive wetting, and unwanted spread to be easily evaluated. The input variables are dimpled or undimpled Kovar; bare surface or ALD Al_2O_3 coated (5nm, 10nm, or 20nm). Measured responses are: 1) runout/excess flow; 2) hermeticity; and 3) brazed joint tensile strength.

The third set of experiments will be to fabricate and test ASTM F-19 tensile buttons using the ceramic 94ND10, a proprietary 94% alumina ceramic. For these samples, the input variables are ceramic resintering atmosphere (Wet-H_2 or Air); dimpled or undimpled Kovar; bare surface or ALD Al_2O_3 coated (1nm, 5nm, 10nm, or 20nm). Measured responses are: 1) runout/excess flow; 2) hermeticity; and 3) joint tensile strength. The input variables for the three sets of experiments are tabulated in Table 1.

	Ceramic Tensile Buttons		Kovar Interlayers			
Experiment Set	Material	Resintering Atmosphere	Bare	Bare with Dimples	ALD Al_2O_3 Coating Thicknesses	Dimples & ALD Al_2O_3 Coating Thicknesses
1	Diamonite	Wet H2	Y	Y	10nm	10nm
2	Diamonite	Wet H2	Y	Y	5nm, 10nm, 20nm	5nm, 10nm, 20nm
3	94ND10	Wet H2 or Air	Y	Y	1nm, 5nm, 10nm, 20nm	1nm, 5nm, 10nm, 20nm

Table 1. Materials and Input variables for Proposed Experiments

Sample preparation

The 94% alumina ceramic cylinders and tensile buttons were cleaned prior to brazing using a 3-step solvent cleaning process followed by resintering at high temperature. The 3-step solvent cleaning process is performed as follows: 1) vapor degrease in Lenium-ES solvent; 2) rinse in acetone; 3) rinse in isopropyl alcohol. The resintering process is accomplished in either wet hydrogen (28°C dew point) at 1500°C for 60 minutes, or air-fired at 1575° for 120 minutes. depending on the experiment input variables. The 1.25mm thick Kovar discs and Kovar tensile button interlayers and the 50µm thick Ag-1Cu-2Zr active brazing alloy preforms were cleaned using the above mentioned 3-step solvent cleaning method. Additionally, the Kovar substrates were subjected to a pickling acid etch process (deionized water/hydrochloric acid) followed by a bright-dip (acetic/nitric/hydrochloric acid solution) followed by rinsing in deionized water.

The ceramic cylinder samples for Experiment 1 were assembled into laser-machined alumina ceramic fixtures, while the alumina tensile buttons were assembled into Standard ASTM F-19 tensile button fixtures. Masses providing joint preload forces of approximately 30g/cm^2 (200g/in^2) were positioned onto each fixtured assembly. Each set of furnace brazed samples included two "dummy" assemblies, similar in mass to each type of component being brazed, into which work thermocouples were mounted to monitor the brazing process temperatures. The dummy ASTM F19 alumina tensile buttons utilized have machined holes to position the work thermocouples tips near the braze joint interfaces, thereby allowing more accurate monitoring of the peak brazing temperature and time at or near the braze joint proper.

Performing the braze

The samples were brazed in an oil-free, top-loading, cryogenic-pumped, high-vacuum furnace. Included among the standard compliment of controllers and monitoring instruments is a NIST-traceable ionization gauge to monitor/insure the proper pressures are achieved and maintained during the brazing cycles. Typical absolute pressures range from 5E-07 - 1E-06Torr during the peak brazing temperature soak segment.

The temperature cycle used was as follows:

15°C/min from ambient to 925°C, soak 15 minutes;
10°C/min from 925°C/ to 985°C, soak 5 minutes;
25°C/min from 985°C to 925°C, soak 0 minutes;
Furnace cool (uncontrolled) to ambient.

Results Experiment 1, Braze Flush-Mount Assemblies

Observed Response: Excess braze filler metal runout

Two of the three Diamonite cylinders brazed to bare Kovar failed the excessive runout criteria. One of the three samples brazed to the dimpled bare Kovar failed the visual criteria.

Figure 5. Active-brazed Diamonite Cylinders to flush-mount flanges

However, both sets of the 10nm thick ALD Al_2O_3, dimpled and non-dimpled, passed the visual inspection with no failures. Samples having excess flow or runout can be seen in Fig. 5. All failures occurred on the assemblies using bare Kovar, with or without dimples, the two left-most columns. Figure 6 is a close-up view of the excess braze filler metal flaws commonly observed when brazing alumina ceramic to bare Kovar with silver-based active braze filler metal alloys.

Figure 6. Excess braze filler metal flow occurring on flush-mount Kovar flanges brazed to Diamonite

Measured Response: Brazed assembly hermeticity

All 12 samples (100%) successfully passed the mass spectrometer helium leak test, having no detectable leaks (NDL).

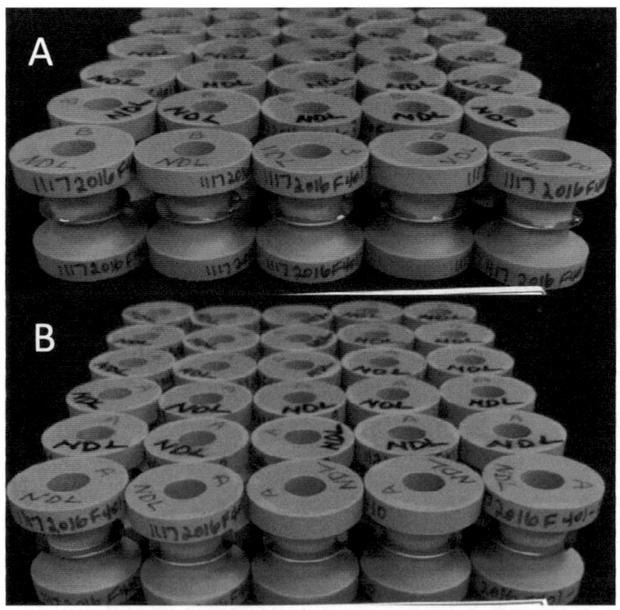

Figure 7. Diamonite tensile button assemblies A: Uncoated Kovar (30 of 30 with excess flow); B: 10nm ALD Al_2O_3 coated Kovar (0 of 30 with excess flow).

Results Experiment 2, Braze Diamonite Tensile Buttons to Oversized Kovar

Observed Response: Excess braze filler metal flow/runout
No runout or excess braze filler metal flow was observed on any of the braze joints having an ALD Al_2O_3 coating ≥5nm thick. 210 of 210 brazements for 105 tensile button assemblies passed the runout test. All tensile button assemblies having bare Kovar interlayers exhibited excess runout. 35 of 35 assemblies failed the runout test with 65 of 70 total brazements exhibiting excess filler metal spread/flow.

Tensile button assemblies using either bare (uncoated) Kovar or Kovar having a 10nm thick ALD Al_2O_3 coating are shown in Fig. 7. Note that in all 30 assemblies in the upper image (5A) have excessive flow the upper side (shown) brazements as well as the lower side (hidden) brazements. Conversely, no excess filler metal flow was observed on wither the upper sides or under sides of the 30 active brazed assemblies having the ALD coatings. Eight unique configurations of the Diamonite to Kovar to Diamonite ASTM F19 tensile buttons and excess filler metal flow results are displayed in Table 2. The numbers in parentheses () indicate the ratio of the quantity of samples with no runout (passing condition) compared to the quantity of samples brazed.

Atmosphere / Dimples	Bare Kovar	5nm ALD Al_2O_3	10nm ALD Al_2O_3	20nm ALD Al_2O_3
Wet H2 / no dimples	(0/30)	(30/30)	(30/30)	(30/30)
Wet H2 / dimples	(0/5)	(5/5)	(5/5)	(5/5)

Table 2. Diamonite/Kovar/Diamonite Brazed Tensile Button Excess Filler Metal Flow Results

As seen in Table 2, results like those displayed in Fig. 7 were observed for Diamonite tensile button assemblies having as little as 5nm thick ALD Al_2O_3 coatings or as much as 20nm thick coatings. In certain instances, small dimples (0.25μm-38μm) are formed into the metal substrates to maintain the desired brazement thickness and uniformity. Quantities of 5 brazed tensile button samples, each using dimpled Kovar with uncoated or ALD Al_2O_3 coatings in the same thicknesses used previously were fabricated and tested. These results are also tabulated in Table 2. The results clearly demonstrated the effectiveness of using thin film coatings of ALD alumina to eliminate excessive braze filler metal flow for this system of materials.

Measured Response: Diamonite tensile button hermeticity
One of the 140 Diamonite/Kovar/Diamonite active brazed tensile buttons was discovered to be non-hermetic. All brazed tensile button assemblies fabricated using bare, uncoated Kovar interlayers were hermetic (100%). Thirty uncoated samples, each having 2 brazements per assembly or 60 total brazements, were brazed using undimpled Kovar interlayers and 5 samples (10 brazements) having dimpled interlayers.

One of the 105 brazed tensile button assemblies was found to have a leak on one brazement resulting in a 99.5% success rate: 209 of 210 total brazements had no detectable helium leaks. The one leak detected was on an assembly where the Kovar interlayer was coated with 5nm of ALD Al_2O_3. Previous experience brazing alumina ceramic / uncoated (bare) Kovar / alumina ceramic tensile buttons brazed with Ag-1Cu-2Zr have generally shown very high yields with respect to making hermetic brazements. These results confirm that ALD Al_2O_3 coatings do not jeopardize hermeticity. The 8 configurations of the Diamonite to Kovar to Diamonite ASTM F19 tensile buttons and helium mass spectrometer leak detection results are shown in Table 3. The numbers in parentheses () indicate the ratio of the quantity of samples that are hermetic or NDL, compared to the quantity of samples brazed.

Atmosphere / Dimples	Bare Kovar	5nm ALD Al_2O_3	10nm ALD Al_2O_3	20nm ALD Al_2O_3
Wet H2 / no dimples	(30/30)	(29/30)	(30/30)	(30/30)
Wet H2 / dimples	(5/5)	(5/5)	(5/5)	(5/5)

Table 3. Diamonite/Kovar/Diamonite Brazed Tensile Button Helium Leak Check Results

Measured Response: Diamonite tensile button strength
The tensile test results of the first sets of brazed tensile button samples were also positive with respect to including ALD alumina coatings to reduce braze filler metal excess runout. The Diamonite tensile button assemblies incorporating undimpled and uncoated Kovar interlayers had tensile strengths slightly higher, on average, than undimpled 10nm thick ALD Al_2O_3, 94MPa (13.5ksi) versus 83MPa (12.0ksi) with standard deviations of 17MPa and 19MPa respectively. For the brazed Diamonite tensile buttons incorporating dimples, however, the results were nearly equal for the uncoated interlayers and 10nm thick ALD Al_2O_3 coated interlayers: 100MPa (14.5ksi) average tensile strength with a standard deviation of 14MPa for the uncoated Kovar versus 102MPa (14.8ksi) with a standard deviation of 25MPa for the 10nm ALD coated samples. Note the four groups for which the tensile strength data is not displayed in Fig. 8, 5nm and 20 nm ALD alumina coated Kovar, with and without dimples. This subsequent population of 70 samples were fabricated approximately eight months following the first groups and

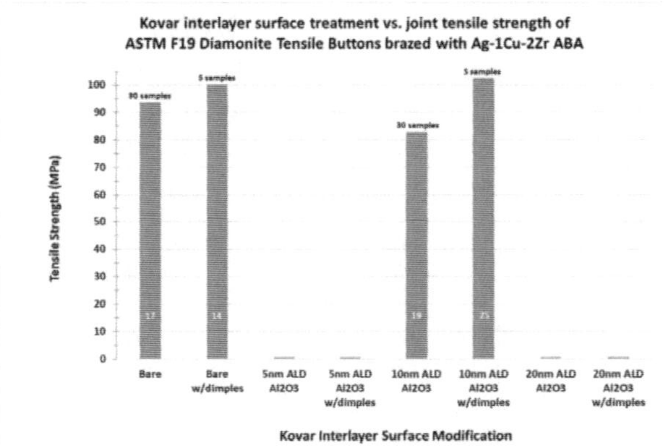

Figure 8. Diamonite/Kovar/Diamonite Brazed Tensile Button Average Tensile Strengths

resulted in average tensile strengths approximately 25-50% of the values for the bare and 10nm thick Al_2O_3, that are reported in Fig. 8, and are not displayed here. Comparison of furnace thermal-cycle data, post-braze metallurgical and SEM microscopy analysis performed on samples from these groups are inconclusive in determining cause(s) for the low strengths.

Results Experiment 3, Braze 94ND10 Tensile Buttons to Oversized Kovar

Observed Response: Excess braze filler metal flow/runout
No runout or excess braze filler metal flow was observed on 30 brazed tensile button assemblies (60 brazements) having an ALD Al_2O_3 coating ≥5nm thick for a yield of 100%. Conversely, 13 of 24, or 54%, of the tensile button assemblies having bare Kovar or ALD Al_2O_3 coatings <5nm thick exhibited excess runout. Of those 13 samples, runout occurred on 24 of the 26 total brazements.

Dimples did help prevent excess flow on the 94ND10 alumina tensile buttons when used on bare Kovar interlayers, while in the case of very thin (1nm thick) ALD Al_2O_3 layers did not help. The resintering atmosphere input variable, wet hydrogen at 1500°C for 60 minutes, or air at 1575°C for 120 minutes appeared to have minimal impact on the excess flow/runout response. Of the 13 samples exhibiting excessive braze filler metal flow, 6 were wet hydrogen fired, and the remaining 7 were air fired. The compiled excess filler metal flow results of the 18 unique configurations of the 94ND10/Kovar/94ND10 ASTM F19 tensile buttons are displayed in Table 4. Numbers in parentheses () indicate the ratio of the quantity of samples having runout compared to the total quantity of samples brazed. Only 3 replicates of each unique condition were used.

Atmosphere / Dimples	Bare	1nm	5nm	10nm	20nm
Wet H2 / no dimples	(3/3)	(1/3)	(0/3)	(0/3)	(0/3)
Wet H2 / dimples	(0/3)	(2/3)	(0/3)	(0/3)	--
Air-fired/ no dimples	(3/3)	(1/3)	(0/3)	(0/3)	(0/3)
Air-fired / dimples	(0/3)	(3/3)	(0/3)	(0/3)	--

Table 4. 94ND10/Kovar/94ND10 Brazed Tensile Button Excess Filler Metal Flow Results

Measured Response: 94ND10 tensile button hermeticity
The brazed 94ND10/Kovar/94ND10 tensile button assemblies had mixed hermeticity results: Only 48 out of 54 assemblies, 89%, were hermetic. An 11% failure rate of any 94% alumina (i.e. Sandi-94, Diamonite, 94ND10, etc.) active brazed to Kovar is unusually high. No non-dimpled tensile buttons failed the helium leak test, whether they were uncoated or had coatings of 1nm, 5nm, 10nm, or 20nm thick layers of ALD Al_2O_3. Because 100% of the non-hermetic tensile buttons were dimpled, it is believed that an error in the dimpling process (dimples too large) prohibited the 0.002" braze filler metal preform making a hermetic seal. The 18 configurations of the 94ND10/Kovar/94ND10 ASTM F19 tensile buttons and helium mass spectrometer leak detection results in Table 5. The numbers in parentheses () indicate the ratio of the quantity of samples that are hermetic or NDL, compared to the quantity of samples brazed. The firing atmosphere, wet hydrogen or air, appeared to have little impact upon the hermeticity outcome as each type of atmosphere had at least one failure. However, as previously noted, the all the failures were associated with dimpled Kovar interlayers.

Atmosphere / Dimples	Bare Kovar	1nm ALD Al_2O_3	5nm ALD Al_2O_3	10nm ALD Al_2O_3	20nm ALD Al_2O_3
Wet H2 / no dimples	(3/3)	(3/3)	(3/3)	(3/3)	(3/3)
Wet H2 / dimples	(1/3)	(3/3)	(2/3)	(2/3)	--
Air-fired/ no dimples	(3/3)	(3/3)	(3/3)	(3/3)	(3/3)
Air-fired/ dimples	(1/3)	(3/3)	(3/3)	(3/3)	--

Table 5. 94ND10/Kovar/94ND10 Brazed Tensile Button Helium Leak Check Results

Measured Response: 94ND10 tensile button strength
Of the eighteen unique material/processing conditions, the tensile strength results for eight of the unique sample configurations, those having dimpled Kovar interlayers, are displayed in Fig. 9. The average values for the brazed 94ND10 ceramic tensile button groups ranged from a low of 69.8MPa (10.1ksi), bare Kovar, air-fired ceramic, to 104.9MPa (15.2ksi), 5nm ALD Al_2O_3, hydrogen fired ceramic. Because each unique configuration had only three replicates, determining trends is more difficult than for the brazed Diamonite tensile buttons in Experiment 1 in which the unique material combinations contained either 5 or 30 replicates.

Consider, for example, the four unique interlayer surface treatment conditions, air-firing strengths were greater for the 1nm and 10nm thick Al_2O_3 coatings and the hydrogen-firing strengths were greater for the bare Kovar and the 5nm thick Al_2O_3 coating. The standard deviations are similar (15MPa) for all air-fired groups, with the hydrogen-fired samples being slightly more uniform (8MPa). Like that reported previously for the Diamonite tensile buttons, there were ten groups of 94ND10 tensile buttons fabricated with non-dimpled Kovar approximately twelve months following the first sets. The tensile strengths of these later fabricated samples were also approximately 25-50% of the values previously observed, and are not included in this document. Post-brazing sample inspection and analysis has yet to reveal the cause for these lower strengths

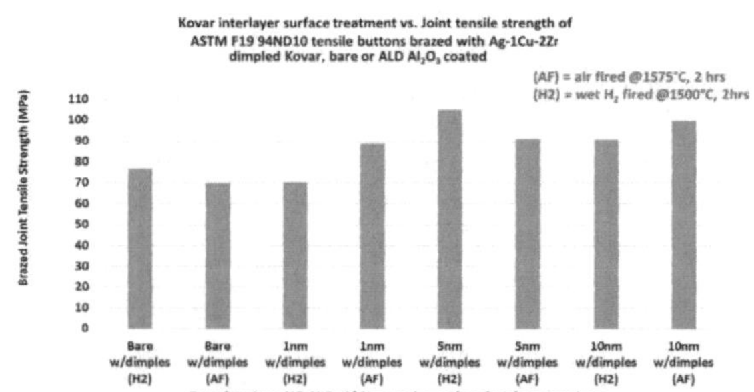

Figure 9. 94ND10/Kovar/94ND10 Brazed Tensile Button Average Tensile Strengths

Discussion

Observed Response: Excess braze filler metal runout

Excessive brazing filler metal flow, or runout, is typically a qualitative, rather than a quantitative, or measurable response: brazed samples either have runout or they do not. In many cases, a little, or even a large amount of filler metal spreading over a substrate surface has no detrimental effects. For the project related to these experiments, however, excessive braze filler metal flow cannot be tolerated, and is the primary cause for brazed subassembly failure. For that reason, any excess braze filler metal outside the braze joint proper observed without magnification is not allowable. All three experiments demonstrated the ability of thin coatings of alumina ceramic,

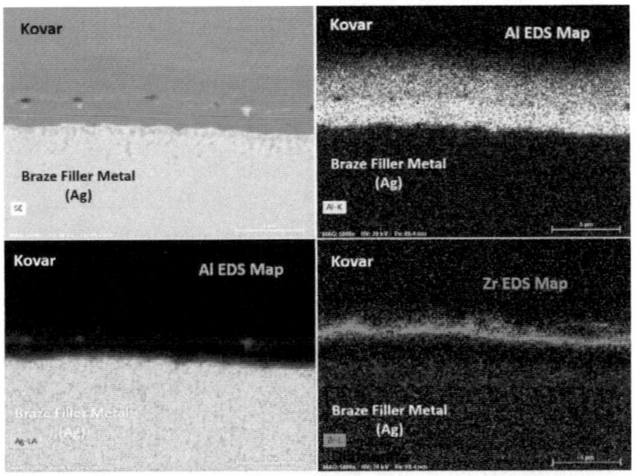

Figure 10. Diamonite/10nm ALD Al_2O_3 Kovar SEM Image and EDS Maps

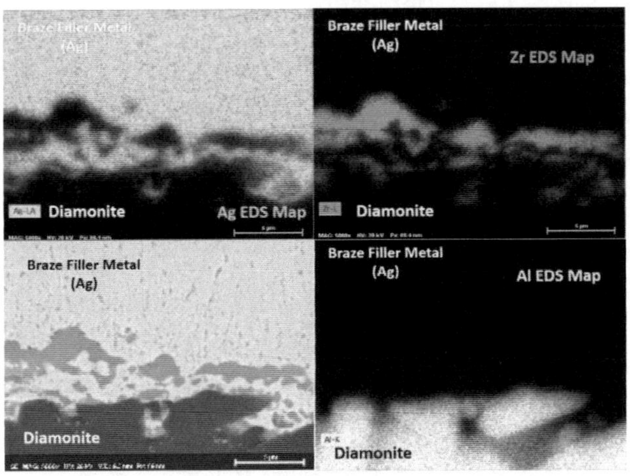

Figure 11. Diamonite/10nm ALD Al_2O_3 Kovar SEM Image and EDS Maps

Al_2O_3, to be effective in mitigating or eliminating altogether excessive filler metal flow and runout when using the active active brazing filler metal Ag-1Cu-2Zr. ALD alumina coatings ≥5nm thick were observed to be effective at eliminating runout 100% of the time, independent of the 94% alumina ceramic material or the resintering atmosphere utilized (276 out of 276 possible brazements). Samples brazed using dimpled bare Kovar substrates, were typically better performers than standard, non-dimpled, bare Kovar substrates. For example, Table 4 shows that when used with the 94ND10 alumina ceramic tensile button assemblies, e 100% of the dimpled bare Kovar substrates (6 of 6) passed the excess flow criteria, and 100% of the non- dimpled bare Kovar substrates (6 of 6) failed the evaluation. For the other two experiments, brazed Diamonite cylinders and tensile buttons, dimples were far less effective. None of the Diamonite tensile buttons using dimpled Kovar interlayers passed the excess flow inspection (100% failure), and 1 of the 3 Diamonite cylinders brazed to dimpled Kovar failed.

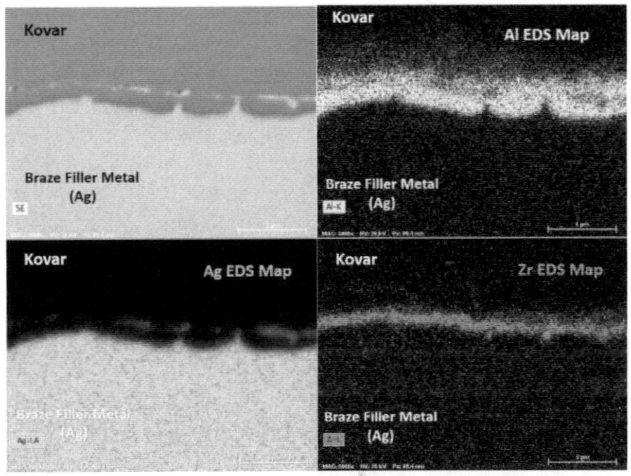

Figure 12. Diamonite/bare Kovar SEM Image and EDS Maps

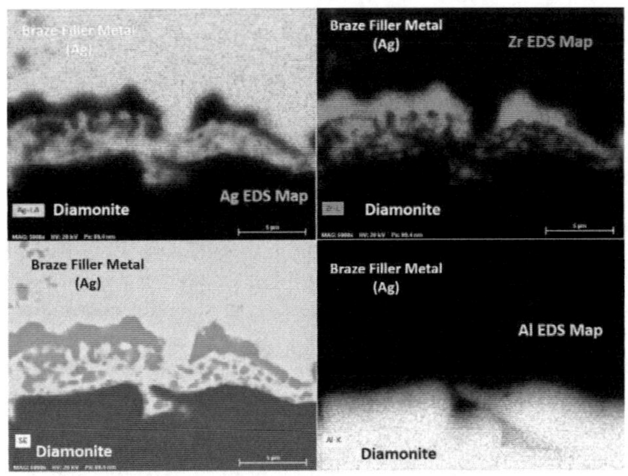

Figure 13. Diamonite/bare Kovar SEM Image and EDS Maps

Figures 10-13 illustrate the reactions occurring at the substrate interfaces when performing active brazes, and though displaying only Diamonite alumina ceramic brazements are representative of the 94ND10 alumina samples analyzed also. The figures have been positioned so that the SEM images and corresponding EDS maps for ALD Al_2O_3 coated Kovar and bare Kovar are displayed side by side. Figures 10 and 12 show the reactions at the Kovar interface while Figs. 11 and 13 show the reactions at the Diamonite alumina ceramic interface. Notice that in all four figures (Figs. 10-13) that there is a continuous band of zirconium (orange EDS maps) at both Kovar and Diamonite interfaces. These continuous layers confirm that there is adequate active element material, Zr in this case, to bond to two individual oxide faying surfaces. Recall the redox reaction mentioned previously, $3Zr+2Al_2O_3 \Rightarrow 4Al+3ZrO_2$. In this reaction, the 94% alumina ceramic, Diamonite in this case, is reduced by the atomic zirconium in the liquid braze filler metal, and the "free" atomic aluminum is brought into the liquid braze filler metal as a result. This highly reactive aluminum in the liquid filler metal is now free to react with exposed surfaces on the Kovar interlayer, forming aluminide compounds. As mentioned previously, these aluminide reactions drive the excessive wetting and spreading behavior on the bare Kovar interlayer surface. In samples having thin ALD Al_2O_3 layers deposited on the Kovar interlayers, the same interfacial reactions occur at the two faying surfaces with the zirconium in the liquid braze filler metal. In these samples, however, the alumina is reduced and the atomic aluminum also mixes with the liquid braze filler metal. At this point the aluminum is unable to react further as no "bare" Kovar surfaces are exposed, thus no runout occurs.

Measured Response: Brazed sample hermeticity

Of the 140 Diamonite/Kovar tensile buttons brazed, there was 1 failure, for a yield rate of 99.5%. The failure occurred on a sample using a 5nm thick ALD Al_2O_3 coated Kovar interlayer. The brazed 94ND10 tensile buttons did experience several failures during the helium leak testing. As mentioned preciously, these failures appear to be attributed to the dimpling process of the Kovar interlayers, as all 6 failures occurred on tensile buttons having dimpled Kovar interlayers.

Measured Response: Brazed sample strength

The measured tensile strengths were similar for both ceramic materials, Diamonite and 94ND10. For the Diamonite tensile buttons, dimples added to the Kovar interlayers tended to increase brazement tensile strength. In the case of 94ND10 tensile buttons, where only dimpled interlayer strengths were reported, tensile strength increased with increasing ALD Al_2O_3 thickness, independent of the resintering atmosphere used. As mentioned previously, the tensile button assemblies brazed and tested for Experiments 2 and 3 were fabricated in different calendar years with lengthy time intervals (12 & 8 months respectively) between builds. The Kovar interlayers used in later builds were from different material lots than those used previously. Comparison of the fracture surfaces of the different populations revealed that there were differences in appearance between the high-strength groups and the low-strength groups, especially with respect to the amount of well-bonded joint area: low-strength samples had much lower ceramic "pullout" than the high-strength samples. The later populations having lower tensile strengths included both bare Kovar and/or bare, dimpled Kovar, confirming that the poor tensile strength results are not related to the inclusion and use of the ALD Al_2O_3 coatings to prevent runout. Additional samples are currently being fabricated to aid in determining this strength discrepancy,

Conclusions

For all three experiments ALD Al_2O_3 was shown to be highly effective in eliminating excessive braze filler metal flow and/or runout on Kovar substrates when using the silver-based active filler metal, Ag-1Cu-2Zr. ALD Al_2O_3 coatings as thin as 1nm preventing excessive flow in 7 of 12 samples fabricated. When Al_2O_3 coatings were ≥5nm thick, however, the excess flow was eliminated 100% of the time. Additionally, the data also demonstrated that there was no decrease in yield due to loss of hermeticity or brazement tensile strength. Other analytical efforts are underway to gain more fundamental knowledge of the reaction chemistry at the atomic level. ALD Al_2O_3 coatings function well in this application and remain well adhered post-brazing. No loose particulates were observed. Additionally, though not the subject of this conference paper, next assembly processes (laser welding) were not detrimentally impacted. Future work to determine the effectiveness of these types of ALD coatings used with other active brazing filler metals is anticipated.

Acknowledgments

The authors wish to express their appreciation to the following people working in the Materials Characterization and Performance Department at Sandia for their excellent brazed sample preparation work enabling the materials analysis efforts: Alice Kilgo, Christina Profazi and Chad Taylor. Also, a special thanks to the personnel working in the Materials Mechanics and Tribology Department at Sandia, who were responsible for the tensile testing: Tom Crenshaw (Ret.), Todd Huber and John Laing.

References

[1] ASTM International, ASTM F19-11(2016), "Standard Test Method for Tension and Vacuum Testing Metallized Ceramic Seals," West Conshohocken, PA.

[2] Kovar is a registered trademark of Carpenter Technology Corporation, Reading, PA.

[3] Mizuhara, H and Oyama, T., "Ceramic/Metal Seals," Engineered Materials Handbook®, Vol. 4: Ceramics and Glasses, ASM International (Materials Park, OH, 1992)

[4] Walker, C. and Hodges, V.C., "Comparison of Metal-Ceramic Brazing Methods," *Welding Journal*, Vol. 87, No. 10, (2008), pp 43-50.

[5] Vianco, P., et al, "Understanding the Run-out Behavior of a Ag-Cu-Zr Braze Alloy When Used to Join Alumina to an Fe-Ni-Co Alloy," *Proc. 6th International Brazing and Soldering Conference. IBSC 2015*, eds. R. Gourley and C. Walker, CD-ROM. American Welding Society.

[6] Chandross, M., "Simulations of Active Brazing," *Thermec'*, Las Vegas, NV, Dec 2013.

Reduction of Residual Stress in the Brazing Joint of ZrB_2-SiC and TC4-TiB_w with Self-Toughened SiC Interlayer

Qinghua Feng, Peng He
State Key Laboratory of Advanced Welding and Joining, Harbin Institute of Technology, Harbin 150001, China

Abstract

In brazing ZrB_2-SiC composite ceramic to TC4-TiB_w titanium matrix composite, high residual stress generates in the ZS near the interface and deteriorates the joint strength. To overcome this problem, a novel design of joint architecture structure was introduced. The design involved a 3BC-SiC (SiC of 3D bridged network structure with continuous cavity) interlayer as substitution for bearing residual stress to increase the strength and reliability of joint. In brazing process, the 3BC-SiC, characterized with low coefficient of thermal expansion and high elastic modulus but large brittleness, was infiltrated by the liquid AgCu braze alloy. Thus the 3BC-SiC was toughened and the propensity to fracture of substitution interlayer to bear the residual stress was reduced effectively. As a result, the residual stress in the ZS matrix decreased, meanwhile the shear strength of the joint increased from 5.4MPa to 41.2MPa.

Introduction

ZS (ZrB_2-SiC composite ceramic) [1] and TTw (TC4 reinforced by TiB whiskers) [2] have been promising candidates for application in thermal mechanical structure. Reliable boning of ZS and TTw is the key issue for the exploitation of respective superiorities in an overall structure for thermal protection.

Active brazing [3] was the most efficient method to join the ceramic and metal, improving the wettability of the ceramic substrate. But the thermal residual stress from the coefficients of thermal expansion (CTE) mismatch and different mechanical responses between the couple of ceramic/metal deteriorates the strength of joint. Extensive approaches have been explored to relieve the residual stress.

Hard metal interlayers (W, Mo) with low CTE can block the residual stress transfer from metal to ceramic matrix. In contrast to conventional hard metals [4], ceramics possess relative lower CTE close to ceramic matrixes and higher elastic modulus for blocking the residual stress. However, the monolithic ceramic interlayers with high brittleness is easy to crack under residual stress as ceramic matrixes. The brazing of porous-Si_3N_4 to TiAl using AgCu filler [5] showed that the cracks occurs in the Si_3N_4 without infiltration close to the TiAl rather than infiltrated ceramic around the interface due to the accompanying toughening effect of infiltrating filler on Si_3N_4. It highlighted the toughening effect of ductile phases on brittle ceramic.

Enlightened by the toughening mechanism [6], a 3BC-SiC (SiC with 3D bridged network structure and continuous cavity) interlayer was introduced. In this study, ZS and TTw were brazed with AgCu filler metal activated by the dissolved Ti of TTw and the 3BC-SiC interlayer. In the brazing, the 3BC-ceramic was infiltrated and toughened by the liquid braze alloy to decrease the inherent brittleness. The further researches can focus on tuning the porosity of 3BC-ceramic to approach the CTE of ceramic substrate.

Experiment

The ZS ceramic containing 20 vol% SiC was cut into 4mm ×5mm × 2.5mm pieces. The TTw composite was cut into 4mm × 5mm × 3mm pieces. And the 3BC-SiC with the porosity of 30% was cut into 4mm × 5mm × 100μm plates as the interlayer. And their microstructures were all shown in Fig. 1a-c. All the surfaces to be brazed were polished with SiC paper up to grid 2000 and then ultrasonically cleaned with the acetone for 10 min. Subsequently, the brazed surfaces of ZS and TTw were coated with the Ag-28Cu (wt.%) powders using the terpineol. Then the 3BC-SiC was sandwiched into

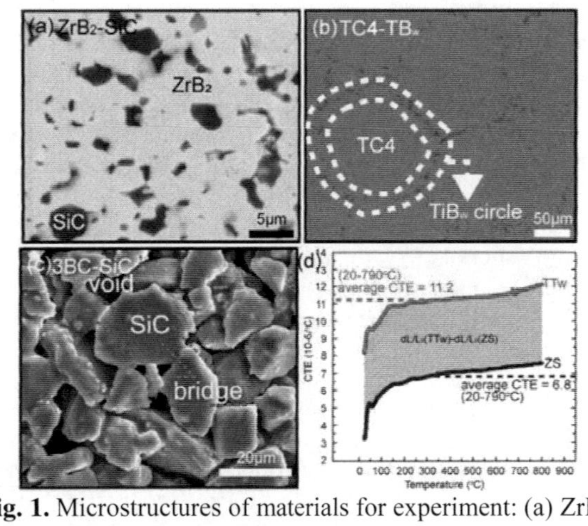

Fig. 1. Microstructures of materials for experiment: (a) ZrB_2-SiC ceramic, (b) TC4-TiB_w, (c) 3BC-SiC, and the (d) CTE mismatch of ZS and TTw

Table 1. **Average chemical compositions (points in Fig. 2) by EDS (at.%).**

Spots	Elements (at.%)								Possible phases
	B	C	Al	Si	Ti	Cu	Zr	Ag	
A	-	22.47	0.63	45.37	28.99	1.47	-	1.07	Ti-Si-C
B	-	-	1.75	-	0.34	6.03	-	91.88	Ag(s,s)
C	-	-	1.97	0.99	3.18	92.18	-	1.67	Cu(s,s)
D	-	-	0.41	0.35	19.76	77.64	-	1.85	$TiCu_4$
E	-	-	4.37	0.13	26.24	68.10	-	1.16	$TiCu_2Al$+ $TiCu_4$
F	-	-	1.46	0.16	38.34	59.13	-	0.90	Ti_2Cu_3
G	-	-	1.07	-	47.42	49.03	-	2.48	TiCu
H	-	-	5.74	-	62.11	30.70	-	1.45	Ti_2Cu
I	-	-	3.02	4.00	52.41	40.34	-	0.22	TiCu+Ti_2Cu

the coated ZS and TTw samples with the sequence of ZS/AgCu/3BC-SiC/AgCu/TTw.

The assembly was brazing at 810°C for 5min in a vacuum furnace (~10^{-6}Torr). After that, the specimen was cooled at 5°C min^{-1}. For comparison, a joint without the 3BC-SiC interlayer was brazed under the same condition.

The microstructure and chemical composition of joints were identified by scanning electron microscope (SEM, Quanta 200FEG) equipped with energy dispersive spectroscopy (EDS, AMETEK). The mechanical property of the joint was characterized by shear strength using an Instron-1186 universal testing machine. And he fracture surfaces were also observed by Quanta 200FEG SEM.

Discussion

The microstructure of the joint using 3BC-SiC interlayer brazed at 810°C for 5min (Fig. 2a-f) was free of cracks and voids (Fig. 2h), in which the AgCu alloy dissolved and was activated by Ti elements dissolving form TTw. The overall microstructure of the joint (Fig. 2a) can be divided into five characteristic regions: region I (interface between ZS and AgCu magnified in Fig. 2e), region II (zone between region I and III detailed in Fig. 2b), region III (3BC-SiC interlayer infiltrated with AgCu detailed in Fig. 2d), region IV (layer between region III and V magnified in Fig. 2c), region V (reaction zone of TTw and AgCu magnified in Fig. 2f). Additionally, possible phases in the region of I to V were identified by EDS, and the results were displayed in Table 1.

Around the ZS substrate, a Ti-Si-C (phase A) compound layer was formed adjacent to SiC from the reaction of SiC and Ti dissolved from the TTw matrix across the 3BC-SiC interlayer (Fig. 2e). Cu (s,s) (phase C) and Ag (s,s) (phase B) distributed close to ZrB_2 without obvious reaction products , but in interatomic bonding. Typical microstructure of the interface of TTw and brazing alloy was composed of five different reaction layers in order of $TiCu_4$ (phase D), $TiCu_2Al$+$TiCu_4$ (phase E), Ti_2Cu_3 (phase F), TiCu (phase G), Ti_2Cu (phase H)+Ti from brazing seam to TTw. And there were some TiB whiskers from the TTw distributing in these reaction layers, which indicated the dissolution of the TTw and the release of active Ti.

In the middle of joint, the 3BC-SiC interlayer was infiltrated by liquid filler metal containing active Ti with no voids (Fig. 2d). The infiltrated 3BC-SiC can be regard as the ceramic matrix composite of SiC skeleton and AgCu metal, with low CTE approaching ZS and high elastic modulus, and served as a substitution interlayer to bear the residual stress from the CTE mismatch of ZS/Tw, which was the key issue to obtain a joint without cracks caused by residual stress.

Remarkably, there was a double-ring structure distributing around the SiC particles: a Ti-Si-C layer (phase A) coated on the SiC particles and a TiCu+Ti_2Cu layer (phase I) around the Ti-Si-C layer. Ti-Si-C was the reaction product of SiC and active Ti. The active Ti was driven around the SiC particles to form a ring-like region which was rich in Ti, and the concentration of Ti leaded to that of the Cu, because that the partial enthalpy of Ti in the liquid Cu at infinite dilution is -10kJmol^{-1}. The concentration of Ti and Cu caused the generation of Ti-Cu compounds. The microstructure of the region IV, containing the Ag (s,s) (phase B) and Cu (s,s) (phase C), were primarily the same as region II.

To assess the effects of the 3BC-SiC interlayer on the microstructure and strength, another joint of ZS to TTw with no interlayer inserted (Figs. 2g) was brazed at the identical condition (810°C, 5min). It was worth noting that there was a crack (Figs. 2i) in the ZS ceramic, which was caused by high residual stress from the CTE mismatch (Fig. 1d) of ZS and TTw. And the crack in the ceramic substrate deteriorated the mechanical property. Thus, the shear strength of the joint increased from ~5.4 MPa to ~41.2 MPa after the insertion of the 3BC-SiC interlayer (Fig. 3a).

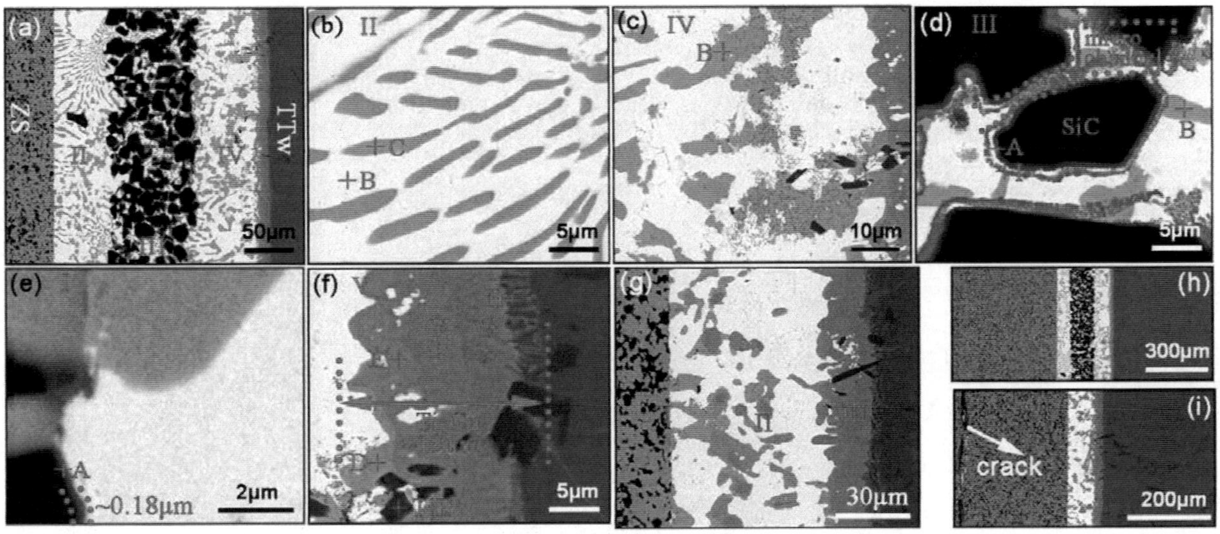

Fig. 2. Microstructure of the ZS/TTw joint with a 3BC-SiC interlayer: (a) integral joint, (b) connection layer between Zone I and Zone III, (c) connection layer between Zone III and Zone V, (d) 3BC-SiC interlayer infiltrated with braze alloy, (e) reaction layer adjacent to ZS, (f) reaction zone of AgCu/TTw interface, (h) joint and the adjacent ZS domain without cracks; microstructure without a interlayer for comparison: (g) integral joint, (i) joint and the adjacent ZS domain with a crack.

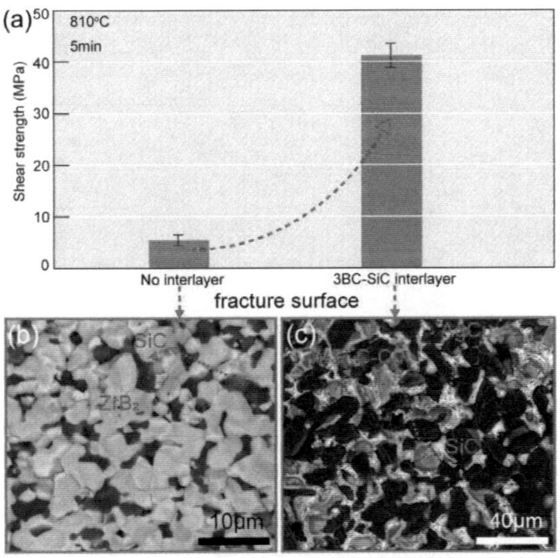

Fig.3. the shear strength of joints without and with a 3BC-SiC interlayer. SEM images of the fracture surfaces in the joints (b) without and (c) with an interlayer.

3BC-SiC interlayer raised the shear strength of the joint by transferring a majority of residual stress from the ZS around the interface to the self-toughened SiC interlayer, which was illustrated by the change of facture surfaces (Fig. 3b and c). After inserting the interlayer in the joint, and fracture surface changed from the ZS to the 3BC-SiC interlayer featuring with phases of SiC, AgCu and reaction layer (Ti-Si-C-Cu). The infiltrated 3BC-SiC was seemed as the composite of SiC and AgCu and its CTE and elastic modulus can be calculated through Schapery bound [7] and H-S bound expression [8] as 8.47×10^{-6}/K and 312GPa respectively. Thus the interlayer with lower CTE than TTw and higher elastic modulus than brazing seam, constituting the soft/hard/soft alternant structure with the AgCu filler between ZS and TTw, can hinder the transfer of residual stress to ZS and enhanced the shear strength of the joint in large measure.

Conclusions

In summary, ZS and TTw was successfully brazed with a 3BC-SiC interlayer without cracks. In the brazing process, the 3BC-SiC was infiltrated and toughened by the filler alloy. The toughened ceramic with hard skeleton characterized by high elastic modulus and low CTE can decrease the residual stress effectively. After the insertion of interlayer, the shear strength raising from 5.4 to 41.2MPa.

Acknowledgments

This research is supported by ''National Natural Science Foundation of China (NSFC, Grant numbers 51475103, 51474081)".

References

[1] Zhang, S. C. *et al*, "Mechanical properties of sintered ZrB2–SiC ceramics," *Journal of the European Ceramic Society*, Vol. 31, No. 5 (2011), pp. 893-901.

[2] Huang, L. J. *et al*, "Microstructurally inhomogeneous composites: is a homogeneous reinforcement distribution optimal ?," *Progress in Materials Science*, No. 71 (2015), pp. 93-168.

[3] Shi, J. M. *et al*, "Interfacial microstructure and mechanical property of ZrC-SiC ceramic and Ti6Al4V joint brazed with AgCuTi alloy," *Journal of the European Ceramic Society*, Vol. 37, No. 8 (2017), pp. 2769-2778.

[4] Zhou, Y. *et al*, "Interlayer selection and thermal stresses in brazed Si3N4− steel joints," *Materials science and technology*, Vol. 7, No. 9 (1991), pp. 863-868.
[5] Zhao, Y. X. *et al*, "Interfacial microstructure and mechanical properties of porous-Si_3N_4 ceramic and TiAl alloy joints vacuum brazed with AgCu filler," *Ceramics International*, Vol. 43, No. 13 (2017), pp. 9738-9745.
[6] Flinn, B. D. *et al*, "Fracture Resistance Characteristics of a Metal-Toughened Ceramic," *Journal of the American Ceramic Society*, Vol. 76, No. 2 (1993), pp. 369-375.
[7] Schapery, R. A., "Thermal expansion coefficients of composite materials based on energy principles," *Journal of Composite Materials*, Vol. 2, No. 3 (1968), pp. 380-404.
[8] Lasagni, F. *et al*, "Enhanced Young's Modulus of Al-Si Alloys and Reinforced Matrices by Co-continuous Structures," *Journal of composite materials*, Vol. 44, No. 6 (2010), pp. 739-755.

Microstructure and Strength of ZrB_2-SiC/TC4-TiB_w Joints Brazed with AgCu Alloy

Peng He, Qinghua Feng
State Key Laboratory of Advanced Welding and Joining, Harbin Institute of Technology, Harbin 150001, China

Abstract

Reliable joining of ZrB_2-SiC composite ceramic to TC4-TiB_w titanium matrix composite was accomplished by brazing with AgCu inactive alloy. The active Ti, dissolved from TTw to the liquid AgCu braze alloy, can react with ZS matrix and Cu elements in the braze alloy to form different compounds due to different brazing temperatures and holding times, the effects of which on the interface microstructure and mechanical strength of the joints were investigated. The results indicated that the dissolved Ti reacted with SiC to form Ti_3SiC_2 compound previous to the formation of TiB whiskers from the reaction with ZrB_2 with increasing the temperature; and the free Ti and Cu could react mutually to form a series of Ti_xCu_y phases adjacent to TTw and in the braze seam. The peak shear strength of ~158.4MPa was obtained at 790°C for 30min, when the fracture were originated from Ag(s,s) in the middle of joint, and then propagated along the path from ZS/braze interface to ZS matrix.

Introduction

ZS (ZrB_2-SiC composite ceramic) [1-3] and TTw (TC4 reinforced by TiB whiskers) [4-6] have been the promising candidates for application in thermal mechanical structure. Reliable boning of ZS and TTw is the key issue for the exploitation of respective superiorities in an overall structure for thermal protection.

Now several methods were used to join the couple of ceramic and metal components, in which the active brazing [7,8] was the most efficient method, improving the wettability of the ceramic substrates. But, the research on the brazing of both the ZS ceramic and TTw composite is rarely seen. Therefore, successful brazing of ZS and TTw is very important to the practical application of these two materials.

In this study, ZS and TTw were brazed with AgCu filler with good ductility for releasing the thermal residual stress from the coefficients of thermal expansion (CTE) mismatch between ceramic and metal. In the brazing process, the AgCu alloy was activated by the dissolved Ti from TTw and reacted with ZS and TTw to form the sound joints. The reaction products of the joint were identified. And the effects of brazing temperature and holding time on the microstructure and shear strength of brazed joints were investigated.

Experiment

The ZS ceramic and the TTw composite were cut into 4mm ×5mm × 2.5mm and 4mm × 5mm × 3mm pieces respectively, whose microstructures were all shown in Fig. 1a and b. All the surfaces to be brazed were polished with SiC paper up to grid 3000 and then ultrasonically cleaned with acetone for 10min. Subsequently, the brazing paste made up of the Ag-28Cu (wt. %) powders and the terpineol was sandwiched into the ZS and TTw with the sequence of ZS/Ag-Cu/ TTw. The assembly was heated to 790-920°C for 0-30min in a vacuum furnace (~10^{-6}Torr). After that, the specimen was cooled to room temperature at 5°C min^{-1}.

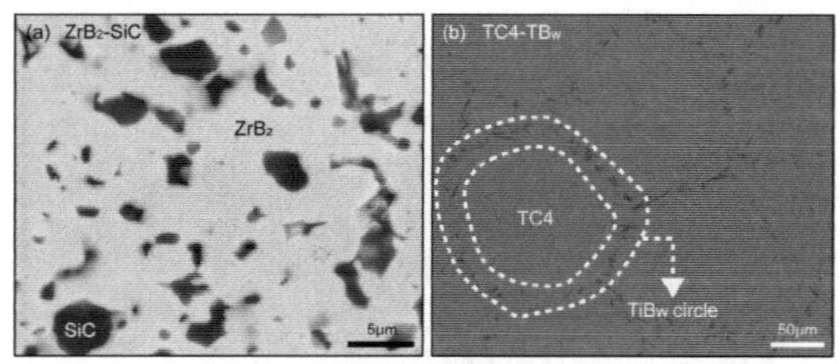

Fig. 1. Microstructures of substrates: (a) ZrB_2-SiC ceramic and (b) TC4-TiB_w.

The microstructure and chemical composition of joints were identified by scanning electron microscope (SEM, HELIOS NanoLab 600i) equipped with energy dispersive spectroscopy (EDS, AMETEK) and electron probe microanalyzer (EPMA, JXA-8230) equipped with wavelength dispersive spectroscopy (WDS). The mechanical property of the joint was characterized by shear strength using an Instron-1186 universal testing machine. And he fracture surfaces were also observed by HELIOS NanoLab 600i SEM.

Discussion 1 Typical microstructure of the brazed joints

Fig. 2 displays the typical microstructure of the ZS/TTw joint brazed using AgCu alloy at 810°C for 5 min. Fig. 2(a) shows the integral microstructure of the joint, indicating the successful joining of the ZS and TTw without voids and

Table 1. Average chemical compositions (points in Fig. 2 and 3) by EDS or WDS (at.%).

Spots	Elements (at.%)								EDS /WDS	Possible phases
	B	C	Al	Si	Ti	Cu	Zr	Ag		
A	2.69	31.68	0.51	14.75	49.08	0.21	-	1.08	WDS	Ti_3SiC_2
B	-	-	1.75	-	0.34	6.03	-	91.88	EDS	Ag(s,s)
C	-	-	3.50	-	2.11	91.81	-	2.57	EDS	Cu(s,s)
D	-	-	0.82	-	18.81	79.35	-	1.02	EDS	$TiCu_4$
E	-	-	7.77	0.29	24.48	66.24	-	1.22	EDS	$TiCu_2Al$
F	-	-	0.41	0.35	19.76	77.64	-	1.85	EDS	$TiCu_4$
G	-	-	8.14	0.09	24.45	65.63	-	1.68	EDS	$TiCu_2Al$
H	-	-	1.07	0.03	42.54	55.57	-	0.78	EDS	Ti_3Cu_4
I	-	-	1.07	-	47.42	49.03	-	2.48	EDS	TiCu
J	-	-	5.74	-	62.11	30.70	-	1.45	EDS	Ti_2Cu
K	49.10	-	0.05	0.06	47.44	3.29	-	0.06	EDS	TiB
L	32.41	7.63	0.35	0.03	57.48	1.48	0.62	-	WDS	TiB
M	-	19.26	2.27	1.23	64.81	10.34	1.35	0.74	WDS	$TiC+Ti_2Cu$
N	-	5.47	7.84	7.92	49.28	22.06	7.05	0.37	WDS	Ti_2Cu

cracks. Three characteristic zones, marked by I-III, were distinguished in the joint. To see the details of each region, the magnification of Zone I (reaction layer of the ZS/AgCu interface), Zone II (brazing seam) and Zone III (reaction layer of the AgCu/TTw interface) were illustrated in Fig. 2b-d, respectively. And the compositions of different reaction phases in each zone were measured by EDS or WDS, as listed in Table 1.

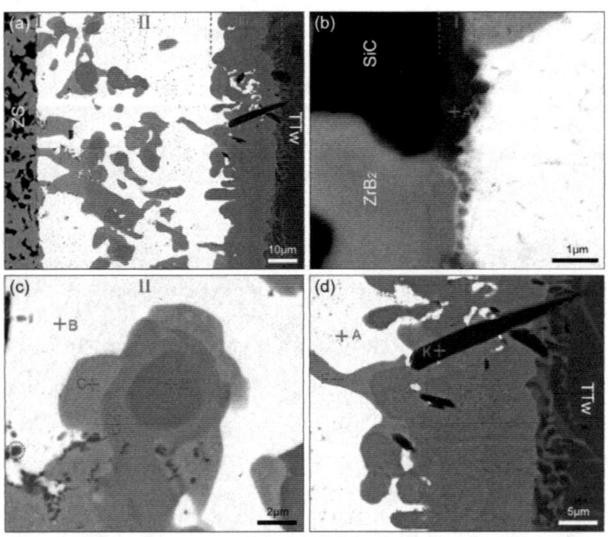

Fig. 2. Microstructure of the ZS/TTw joint: (a) integral joint, (b) reaction layer adjacent to ZS, (c) braze seam, (d) reaction zone of AgCu/TTw interface.

As shown in Fig. 2(d), typical microstructure of TTw interface (zone III) was a multilayer composed of five different reaction phases in sequence of $TiCu_4$ (phase F), $TiCu_2Al$ (phase G), Ti_3Cu_4 (phase H), TiCu (phase I), Ti_2Cu (phase J)+Ti from the brazing seam to the TTw substrate. Morever, there were some TiB whiskers from the TTw distributing in these reaction layers, which suggest the dissolution of the TTw and the release of Ti. Thus, the AgCu brazing alloy was activated by the dissolved Ti to react with the ZS matrix for sound joint. As seen in Fig. 2(b), in the ZS/AgCu interface area (zone I) a black reaction layer (phase A) formed adjacent to the SiC component of ZS substrate, which was identified as Ti_3SiC_2. And the formation of Ti_3SiC_2 has been reported. The Ti_3SiC_2 was the reaction product of Ti dissolved in the brazing alloy from the TTw and SiC substrate. Ag (s,s) (phase B) distributed close to ZrB_2 with no obvious resultant phases, but in sound interatomic bonding.

Fig. 2(c) shows the details of the brazing seam (Zone II). It can be identified that the brazing seam was a Ag(s,s) (phase B) layer dispersed by some Cu-containing blocks. And the Cu-containing blocks consisted of $TiCu_2Al$ (phase E), $TiCu_4$ (phase D) and Cu(s,s) (phase C) from the center to the outside successively which can be seemed as the Ag matrix composite with compounds containing Cu.

As analyzed above, the AgCu brazing alloy can be activated by the Ti dissolved from the TTw substrate, thus reacted with ZS and TTw matrix to form sound joint. The typical joint microstructure at 810°C for 5min was determined to be ZS/Ti_3SiC_2/Ag + Cu-contained blocks (Cu(s,s) + TiCu4 + $TiCu_2Al$)/Ti-Cu (TiCu4 + $TiCu_2Al$ + Ti_3Cu_4+ TiCu+ Ti_2Cu) layer.

Discussion 2 Effect of brazing temperature on the interfacial microstructure and shear strength of ZS/TTw joints

Fig. 3 shows the microstructure of the ZS/TTw joints brazed for 5 min at different temperature from 790°C to 920°C. It can be observed that the microstructure of brazed joints changed with the brazing temperature.

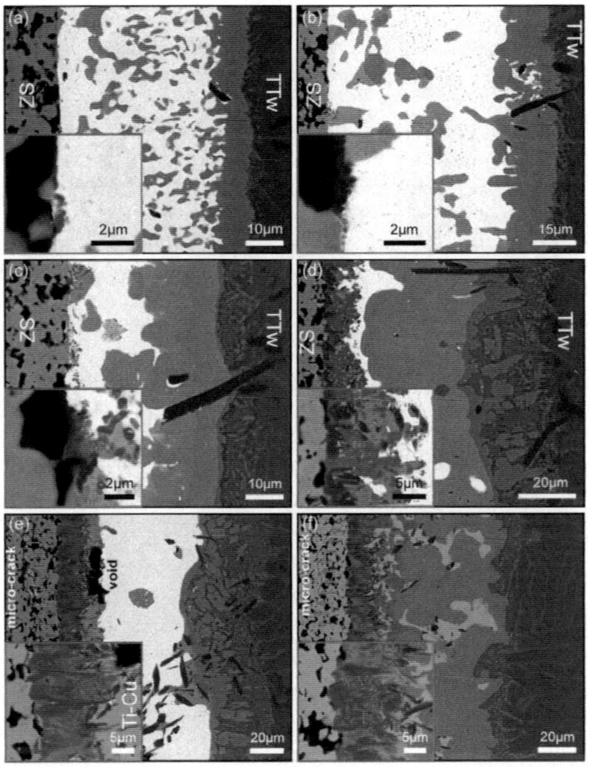

Fig. 3. Effect of brazing temperature on the microstructures of brazed joints: (a) 790°C, (b) 810°C, (c) 850°C, (d) 890°C, (e) 910°C, (f) 920°C.

When the joint was brazed at a low temperature range of 790°C to 850°C (Fig. 3a-c), the reaction layer adjacent to the SiC components of the ZS substrate was Ti_3SiC_2, and there was no obvious reaction phases within the interface of ZrB_2 and brazing alloy, including the joint brazed at 810°C described above. Further increasing the temperature in the range of 890°C to 920°C (Fig. 3d and f), the Ti_3SiC_2 disappeared and another two phases occurred beside the ZS. The new phases was identified by WDS as $TiC+Ti_2Cu$ (phase M) and Ti_2Cu (phase N) solved by some Si due to the fact that the Si atoms was easy to dissolve from the Ti_3SiC_2 to the brazing alloy. Moreover, there were TiB whiskers (phase L) generating in the ZS/AgCu interface result from the reaction of ZrB_2 and Ti dissolved from TTw.

On the other hand, with the increase of temperature the thickness of reaction layer of ZS/AgCu and AgCu/TTw elevated significantly due to the dissolution of Ti and the elevation of temperature. And meanwhile the thickness of the brazing seam decreased regardless of the joint brazed at 910°C (Fig. 3e). The abrupt increase of the brazing seam may be the result of improvement of wettability of AgCu on the ZS matrix. Due to the low reaction free energy of Ti and ZS, the Ti concentrated around the ZS, and with increasing the temperature the concentration of the dissolved Ti beside the ZS enhanced. Meanwhile the agglomeration of Ti resulted in that of the Cu, due to the fact that the partial enthalpy of Ti in the liquid Cu at infinite dilution is $-10kJmol^{-1}$. The concentration of both Ti and Cu gave rise to the formation of Ti-Cu compound adjacent to ZS. Moreover, the wettability of AgCu on Ti-Cu reaction layer was much higher than that of ZS resulting in the elevation of the layer of brazing seam. And with the increase of the temperature, the Ti-Cu layer occupied the brazing seam.

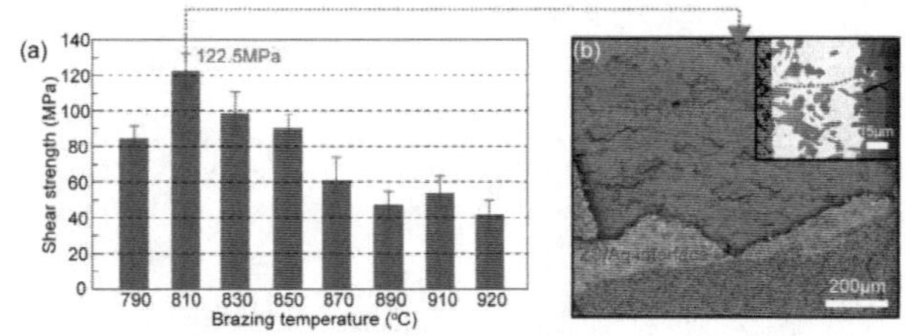

Fig. 4. (a) The shear strength of ZS/TTw joints at different brazing temperature and (b) the SEM image of the facture surface in the joint with the maximum shear strength.

The change of the shear strength was dependent on the microstructure of the joint. With increasing the temperature, the shear strength enhanced at first and then decreased, and the maximum value was 122.5MPa at the braze temperature of 810°C (Fig. 4a). In addition, the fracture surface of the joint of maximum shear strength occurred from the Ti-Cu reaction layer bordering the TTw and propagated along the ZS/AgCu interface then into the ZS substrate (Fig. 4b). There was a little increase of strength at 910°C for the elevation of Ag (s,s) layer with good ductility.

Discussion 3 Effect of holding time on the interfacial microstructure and shear strength

Based on the fracture surface analysis and to improve the joint strength more sufficiently, further researches will focus on brazing the ZS/AgCu/TTw couple at 790°C for long holding time, in which the thickness of the Ti-Cu brittle compounds layer within the AgCu/TTw interface decreased and the bonding of the interface of the ZS and TTw matrix was strengthened simultaneously compared to the joint brazed at 810°C.

Fig. 5 shows the microstructure of the ZS/TTw joints brazed at 790°C for different holding time of range from 0min to 30min. It can be seen that with extending the holding time there was no disappearance of Ti_3SiC_2 and the generation of TiB whiskers and the reaction layer increased meanwhile. Moreover, the effect of holding time on the reaction layer was slighter than that of brazing temperature.

Fig. 6a shows the room temperature shear strengths of the ZS/TTw joints brazed at 790°C for different holding time. The shear strength of joints increased with the elevation of the holding time and the maximum value of 158.4MPa was obtained at the holding time of 30min. And fracture surface of the joint of maximum strength changed into the path of Ag (s,s)-interface-ZS (Fig. 6b)

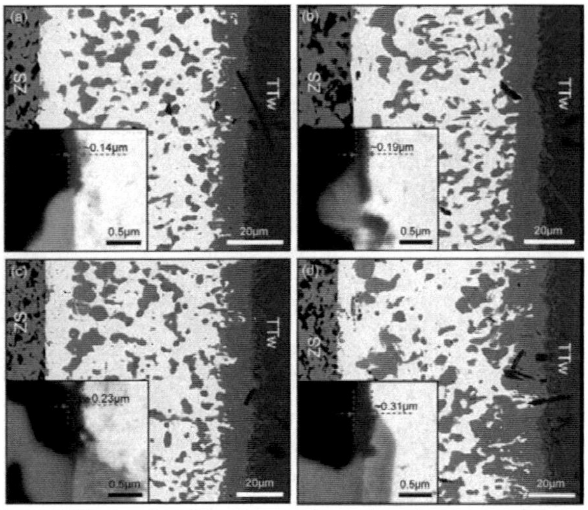

Fig. 5. Effect of holding time on the microstructures of brazed joints: (a) 0min, (b) 5min, (c) 15min, (d) 30min.

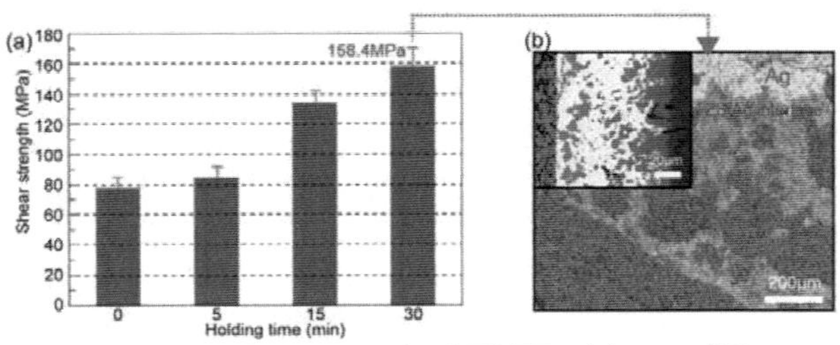

Fig. 6. (a) The shear strength of ZS/TTw joints at different holding time and (b) the SEM image of the facture surface in the joint with the maximum shear strength.

Conclusions

In summary, ZS and TTw was successfully brazed with AgCu alloy dissolved by Ti from TTw. With increasing the braze temperature, the reaction phases adjacent to the ZS substrate changed from Ti_3SiC_2 to $(TiC+Ti_2Cu)+Ti_2Cu+TiB_w$. The maximum shear strength of was obtained at 790°C for 30min.

Acknowledgments

This research is supported by ''National Natural Science Foundation of China (NSFC, Grant numbers 51475103, 51474081)".

References

[1] Zhang, S. C. *et al*, "Mechanical properties of sintered ZrB2–SiC ceramics," *Journal of the European Ceramic Society*, Vol. 31, No. 5 (2011), pp. 893-901.

[2] Zou, J. *et al*, "Thermoablative resistance of ZrB_2-SiC-WC ceramics at 2400°C," *Acta Materialia*, No. 133 (2017), pp. 293-302.

[3] Chamberlain, A. L. *et al*, "High-strength zirconium diboride-based ceramics," *Journal of the American Ceramic Society*, Vol. 87, No. 6 (2004), pp. 1170-1172.

[4] Huang, L. J. *et al*, "Microstructurally inhomogeneous composites: is a homogeneous reinforcement distribution optimal ?," *Progress in Materials Science*, No. 71 (2015), pp. 93-168.

[5] Huang, L. J. *et al*, "In situ TiB_w/Ti-6Al-4V composites with novel reinforcement architecture fabricated by reaction hot pressing," *Scripta materialia*, Vol. 60, No. 11 (2009), pp. 996-999.

[6] Huang, L. J. *et al*, "Room temperature tensile fracture characteristics of in situ TiB_w/Ti6Al4V composites with a quasi-continuous network architecture," *Scripta materialia*, Vol. 64, No. 9 (2011), pp. 844-847.

[7] Wang, H. *et al*, "Brazing mechanism and infiltration strengthening of CC composites to TiAl alloys joint," *Scripta materialia*, Vol. 63, No. 8 (2010), pp. 859-862.

[8] Shi, J. M. *et al*, "Interfacial microstructure and mechanical property of ZrC-SiC ceramic and Ti6Al4V joint brazed with AgCuTi alloy," *Journal of the European Ceramic Society*, Vol. 37, No. 8 (2017), pp. 2769-2778.

Brazing with Photochemical Etched Amorphous Filler Metal Preforms

W. Coughlan and Dr. E. Theisen
Metglas, Inc., Conway, SC USA
metglas@metglas.com, 1-843-349-6800

Abstract

Brazing with amorphous filler metal preforms allows for precise placement of braze filler metal, controlled volume and very little waste. The major processes used to manufacture amorphous filler metal preforms are mechanical stamping or thermal laser cutting. Recent developments in photochemical etching have created the opportunity to manufacture very thin braze performs with tightly controlled tolerances. This has driven increased demand, particularly for nickel-based alloys used by the aerospace market, for controlled application of brazing filler metal. Examination of the edge condition, its impact on wetting stainless steels, as well as benefits and considerations are discussed in this paper.

Introduction

Brazing filler metal preforms have been used in various industries for many years. Preforms using powders include tapes and pre-sintered powders, rolled and amorphous braze filler metal sheets can be stamped into preforms or laser cut to size. While each of these offers a different set of advantages the growth of more customization has led filler metal suppliers to look for alternative ways to add value to our customers. Photochemical etching or photochemical machining is a process developed in the 1960s as a rapid manufacturing process that give manufacturers a way to quickly design and manufacture complex shapes out of thin sheets (less than 0.06") [1]. Amorphous brazing foils thicknesses range from 0.00078" up to 0.003", which make them ideal candidates for utilizing the photochemical etching process to manufacture preforms.

Discussion

One of the major advantages of using amorphous brazing filler metals in a joining process is the ability to preplace the filler metals in only the areas that are required. This gives the designer the versatility of utilizing complex shapes which are specifically used in heat exchanger applications and allows optimal flow path designed to maximize heat transfer. Another benefit is the ability to automate your product assembly while minimizing waste as the total known volume can be tightly controlled with the geometry of the preform.
Nickel based amorphous brazing filler metal preforms are typically stamped or laser cut to size. While stamping offers a very robust process that can be very cost effective, tooling can be expensive (typically $5,000-20,000) and requires regular re-dressing due to the high hardness of the foils [2]. Laser cut is a good solution for custom designs and does not require tooling, however there are concerns about the edge condition as there will be some re-cast areas that can impact your stack-up tolerance and melting characteristics.

Photochemical etching utilizes a phototool that operates like a stencil by masking the area you want to remain after the etching process. Ultraviolet (UV) light is used to develop the image, which hardens the area exposed to the UV light. The softer photoresist left behind is washed away, leaving only the exposed metal to be chemically removed. The sheet is then sprayed with a heated acid etchant from both sides to dissolve the unwanted metal. Finally, the photoresist is stripped away and rinsed multiple times to ensure there is no remaining etchant [3]. Figure 1 shows a schematic depicting the process when etching from one side or both sides of the filler metal. Note that the edge condition is typically not perfectly straight due to the nature of the acid dissolving through the preform.

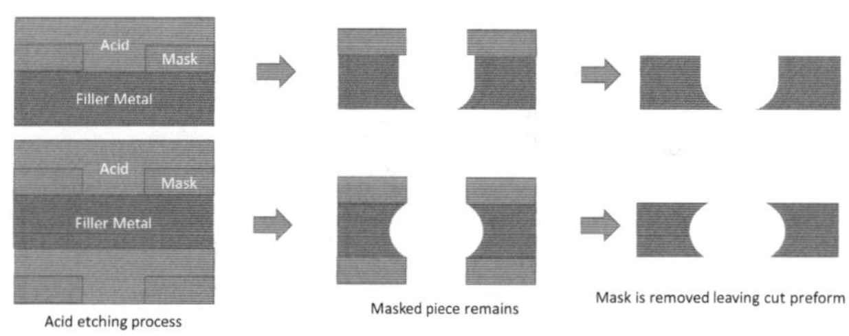

Figure 1: Schematic of etching process from one side or both sides of filler metal.

Figure 2 is an example of the intricate shapes that can be developed for amorphous brazing filler metal preforms using the etching process. The small holes at the edge of the samples are perforations and are left on the sheet. An automatic process can then remove each preform and place them, where required in the assembly.

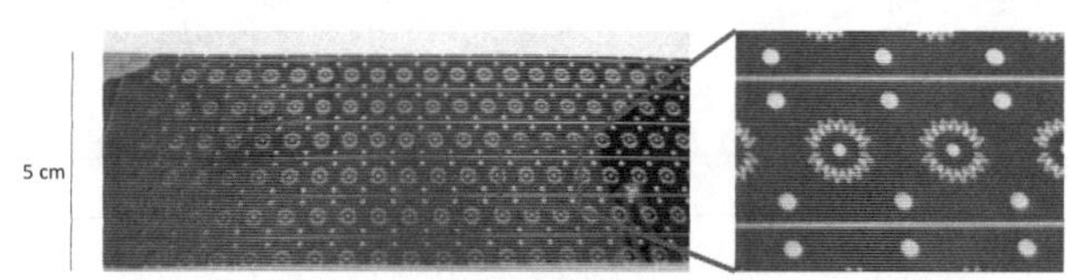

Figure 2: Example of detailed filler metal after etching process.

Scanning electron microscope images of the etched preform are shown in Figure 3. The higher levels of magnification show that the edge cut is slightly beveled. However, the edge condition of the filler metal preform on this small length scale is rarely significant in the brazing process. General tolerances are ± 15% of the base material thickness and a ± 0.001" location tolerance.

Most brazing processes do not require the level of complexity shown in Figures 2 and 3. Larger shapes are needed more often, such as the image of the preform with a triangular section removed in Figure 4. This type of braze preform is also ideal for chemical etching as it provides an alternative to stamping or laser cutting. In large volume production of simple preforms stamping is likely to be more cost effective. Shapes such as circular washers are commonly produced using stamping of the filler metal.

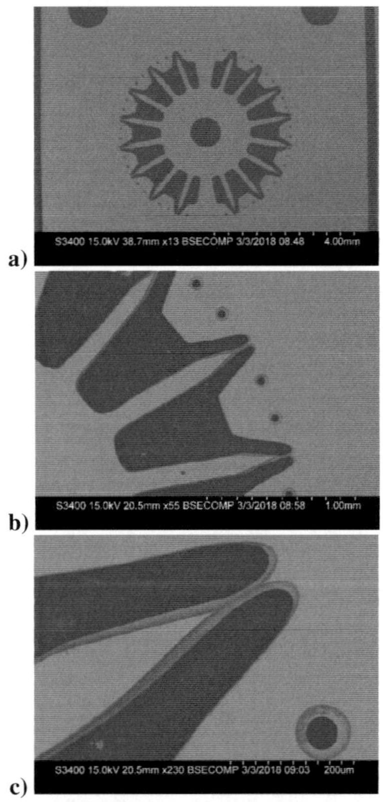

Figure 3: Example of complex shapes that can be manufactured using photochemical etching under increasing levels of magnification.

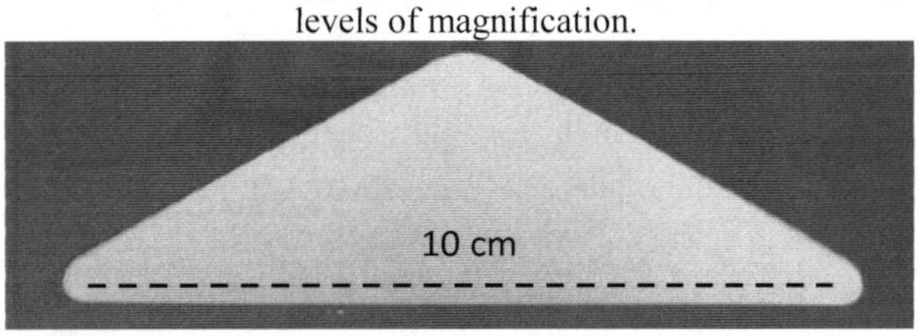

Figure 4: Example of simple shaped preform that can be manufactured using photochemical etching.

Conclusions

Photochemical etched amorphous brazing filler metal preforms give manufacturers the option of a process with inexpensive tooling, rapid production timing and ease of automation. The edge condition of these preforms is consistent, with no impact on stack-up tolerance, as there are no burs or re-cast layers. This also has no impact on wetting versus the virgin filler metal, as all the etchant is washed away, and there is no phase change in the preform.

Acknowledgments

A special thanks to Conard Corporation, Glastonbury, CT for their assistance in photochemical etching technology.

References

[1] D. M. Allen: CIRP Annals Volume 53, Issue 2, 2004, Pages 559-572

[2] A. Rabinkin: in Proc. IBSC 2003, "Overview: Brazing with (NiCoCr)-B-Si Amorphous Brazing Filler Metals: Alloys, Processing, Joint Structure, Properties, Applications"

[3] Conard Corporation: "The Comprehensive Guide to Photo Chemical Machining"

Aerospace Brazing – Yesterday, Today, and Tomorrow

Dan Kay
Kay & Associates, 4 Lawton Drive, Simsbury, CT 06070 USA
E-mail: dan.kay@kaybrazing.com, Phone: 1-860-651-5595

Abstract and Introduction

World War II seems to have been the catalyst for rapid development in "aero"space brazing around the world, and over the years since that time aerospace brazing has mushroomed significantly. This paper is designed to look at some of the aerospace brazing developments that began at that time in some of the early jet-fighters that actually saw service in WWII, and then to briefly explore how brazing has mushroomed in the years that followed, including its extensive use in both military and commercial aircraft, and its eventual use in missiles/rockets and ultimately in manned space-flight. These brief looks will also highlight developments in brazing technology over the years, leading to today's state-of-the art furnaces, fixture-materials, and techniques that will open up new doors to tomorrow's brazing technologies.

1940's.

World War II seems to have been the catalyst for rapid development of "aero-space" brazing around the world. Brazing had been quietly moving forward in the years prior to the big war, but the production needs to maintain a "war-footing" in Europe, America, and Asia quickly showed that brazing was the best method for many different, unique joining applications, such as those in radar, ordnance, marine use, and in aircraft engines.

Fig. 1 German Messerschmitt ME-262 used a high-temp silver-based brazing filler metal (BFM) in its jet engines

In Fig. 1 is a photo of the Messerschmitt ME-262 jet airplane used late in WWII by the Germans, in which it was reported that the Germans apparently used a high-temp silver-based brazing filler metal (BFM) they had developed (85Ag/15Mn) which could handle some (but not all) of the temperatures seen by those engine components. This unique BFM is still being used in the brazing industry here in The US today, being given brazing spec designations of AWS A5.8, Class BAg-23, and AMS 4766.

After the big war, many of those early German jet engines were examined in labs in England and the US, and improvements on that technology were sought. One such company here in the US, Curtiss-Wright Corporation, tried to duplicate, and then improve upon, the hollow sheet-metal turbine blades used in those Messerschmitt engines. Extensive testing by Curtiss Wright found, as did the Germans, that the silver-based BFM they were using did not have the thermal properties required, and a better high-temp BFM was needed.

A young welding-engineer at Curtiss-Wright, named Bob Peaslee, was interested in finding just such an alloy, and it fell into his lap one day when a thermal-spray salesman, who wanted to demonstrate a Ni-Cr hard-surfacing spray that had excellent hardness and high-temp capabilities, was visiting Bob in his office. Bob thought to himself that this hard-surfacing powder might just be the thing to replace the Ag-Mn alloy for hi-temp brazing. His subsequent braze-tests proved this to be correct. One of his early tests involved brazing not only stator components but also rotor components in a new Curtiss-Wright jet engine, using this Ni-Cr thermal-spray powder. It worked beautifully, the tests clearly showing the inherent strength and hi-temp capabilities of this new Ni-based brazing filler metal (BFM).

Fig. 2 shows a photo of one of the test rotors he brazed, and which was then high-speed tested in Curtis Wright's aero labs. Bets were taken as to how high a spin-test speed would cause the rotor to break apart. The rotor, normally operated at about 12,000 rpm was finally destroyed when it was spinning at almost 42,000 rpm! Fig. 3 shows what the destroyed parts looked like when re-assembled. Subsequent examination of all the component parts showed that none of the brazed joints had failed. Not one. Zero. Zilch. Nada. Every failure had occurred in the base metal, away from any of the brazed joints!

Fig. 2 Brazed rotor prior to high-speed spin-testing.

Fig. 3 Brazed rotor broke apart at almost 42,000 rpm, but not one of the brazed joints failed.

1950's and 1960's

Rapid development of brazing in jet engine and other aerospace applications continued through the 1950s. Then, in the 1950's brazing of honeycomb assemblies gained significant acceptance in both jet-engines and in wing and body panels in aircraft. As jet engines became more robust, operating at higher temps, and at higher rpms, sacrificial honeycomb seals had to be developed (such as that shown in Fig. 4), to allow spinning turbine blades to grow in length without hurting the engine walls or the blades themselves. As can be readily seen in the photo, there are hundreds of brazed

Fig. 4 Typical honeycomb brazement for air-seals in jet engines. Each honeycomb-cell is supposed to be brazed tightly to the curved metal substrate below the honeycomb.

joints in each length of honeycomb brazed to the curved substrate below it. Notice the small size of each honeycomb cell in that structure. BFM was often applied as powder poured carefully down into each cell after attaching the honeycomb to the substrate, or pressed (as tape) into the bottom of the honeycomb just prior to the honeycomb's attachment to the substrate. It was not until the late 1980's and 1990's that a new technique was developed for these honeycomb materials, in which the high-temp BFM was finally available in foil form, and thus able to manufactured as part of the honeycomb, as shown in Fig. 5. With that new innovation, assembly and brazing of honeycomb was made much simpler.

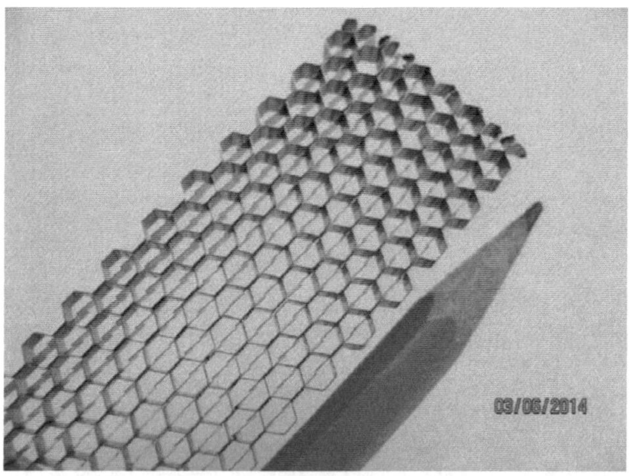

Fig. 5 Honeycomb material with BNi-2 BFM foil embedded right in the honeycomb as part of the structure. During brazing, the BFM foil melts, joining each cell to the substrate.

Honeycomb structural panels for wings and body parts were used extensively in aircraft such as the B-70, shown in Fig. 6, designed in the late 1950's and which flew many test missions through the early 1960's.

Fig. 6 The North American B-70 used a lot of brazing in its structure, including honeycomb panels for strong lightweight skin in wings and body panels.

These brazed honeycomb panels were assembled for brazing by placing strong, lightweight honeycomb panels between sheets of special high-strength stainless (PH 15-7 Mo).

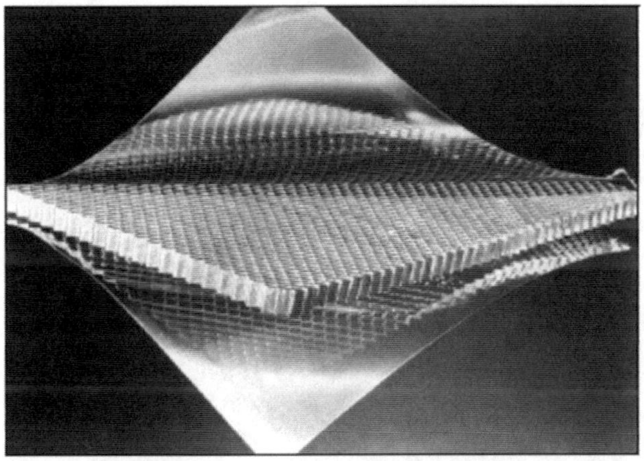

Fig. 6 Open view of the typical honeycomb structure of wing panels, the upper and lower thin sheets being brazed to each side of the honeycomb.

Some of this honeycomb-brazing technology found it way into the early development of manned spacecraft, such as Mercury, Gemini, and then in the Apollo missions to the moon in the late 1960's and 1970's, as shown in Fig. 7. The Command Module (CM) of the Apollo spacecraft used a lot of brazed honeycomb in its construction in order to create a very strong but lightweight structure.

Rocket-engines, too, benefited from brazing, as shown in Fig. 8. In the photo you can see that the bell of the engine nozzle is made from a series of tubular members, brazed side-by-side to form that bell-shaped structure. Fuel flows down through each of those tubes in these engine nozzles, and then back up through adjacent tubes not only to pre-heat the fuel for later combustion in the nozzle, but also to help keep the nozzle temperatures in control. Such nozzle designs make optimal use of brazing technology, using different kinds of BFMs, not only nickel-based BFMs, but particularly with the Au-Ni BFMs.

Fig. 7 The Apollo spacecraft used a lot of brazed honeycomb panels in the Command Module (CM), and the Saturn rocket used an extensive amount of brazing in the bells of the F-1 rocket engine nozzles.

Fig. 8 Individual fuel tubes brazed side by side to form the bell of this F-1 engine nozzle used to launch the Apollo/Saturn rocket to the moon.

1970's and 1980's

The 1970's brought us into the age of the Space Shuttle, and eventually to the International Space Station. By this time brazing was used in many more places than just launch vehicles, rocket engines, and aircraft structures. Another important use of brazing was to be found in some of the delicate instrumentation within the astronaut's space-suit backpacks, for example. The backpacks these astronauts wore, such as the one shown in Fig. 9, contained brazed heat-exchangers and related brazed tubular connections to properly control fluid flow (liquid and air) for their suits.

Fig. 9 Astronaut life-support systems make extensive use of brazing to keep astronauts safe and comfortable.

It's good to know, too, that so much of the technology used in making these fantastic aerospace developments possible has rolled over into wide-spread commercial use in our every-day lives today in our businesses and in our homes.

1990's and 2000's

Brazing applications in aerospace continue to expand each year, and the end is not in sight. Brazing developments such as those for surface-buildup applications on worn-out turbine blade components (like those shown in Fig. 10), in order to give those worn-out blades a new lease on life, so to speak; or some of the more exotic applications being developed, such as joining ceramics and metals via specialized brazing processes for unique applications in space-satellite "windows", are just examples of the innovations being created and developed in the wide world of aerospace brazing each day.

Ceramic bonding. Today, with the demand for assemblies that can operate at much higher temperatures for greater fuel efficiency, we are seeing the increased use of ceramic materials in aerospace applications. These assemblies require unique joining processes. In the last few decades, processes have been developed, using brazing technologies, that can enable ceramic materials to be bonded to metals and other ceramics (such as that shown in Fig. 11). These proprietary, patented processes show us that joining/bonding of a wide variety of diverse materials via brazing processes is possible, and will continue to be developed.

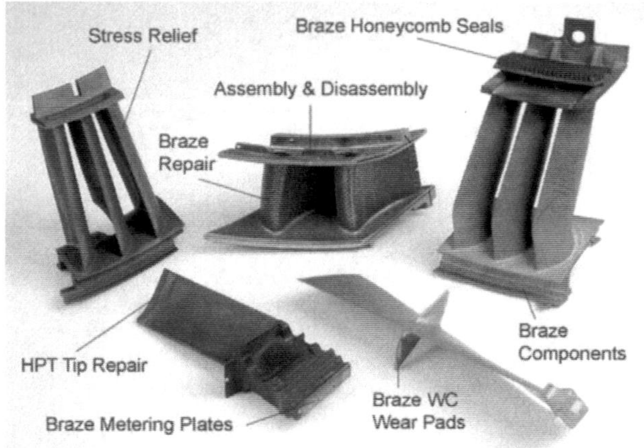

Fig. 10 Brazing is widely used for joining turbine blades/vanes to their substrates, for use in surface-buildup (via such things as pre-sintered braze preforms), among other uses.

Fig. 11 Proprietary ceramic-to-metal assembly, that was bonded/joined together using a very specialized, patented vacuum brazing process.

A caution here relative to joining/bonding ceramics using a brazing process. Please notice that I did NOT say that you can braze ceramics. I said that you can join/bond ceramics together using a brazing process. There is a significant difference between those two sentences. Brazing always involves the ability of the brazing filler metal (BFM) to "wet" the base material that is to be joined. But, BFMs do NOT like to bond to, or flow over, oils, dirt, grease or oxides! Thus, by definition, BFMs cannot braze ceramics. However, so-called "active brazing alloys" (ABA), which contain titanium, for example, can create complex ceramic structures (spinel-like structures) when the titanium in the ABA reacts with the binders in the ceramic (simplistically speaking, ceramics are essentially highly compressed ceramic powders with a special kind of binder holding the particles together). By reacting with, and bonding to the "binders" in the ceramic matrix, a very strong joint can be made. The process is carried out usually in a vacuum furnace, using these ABA alloys, and thus this procedure looks similar to a normal brazing process. But be careful, the actual BFM is not "wetting" the alumina particles, but is creating a complex bonding-mechanism to the intergranular binders within the matrix of the ceramic. I realize this is a very simplified explanation, but hopefully one that can help the reader to see why the words "joining" or "bonding" are used when referring to ceramics, rather than the word "brazing".

Tomorrow's Aerospace Brazing

What will tomorrow bring in the world of aerospace brazing? Brazing furnace technology continues to advance, as we see "continuous" (multi-chamber) vacuum brazing furnaces develop; vacuum furnaces that are getting bigger and bigger to handle larger and larger parts for "one-step" brazing; laser and microwave joining processes that are being explored and utilized in brazing more and more; and now even 3D-printed assemblies that are being produced and then brazed together.

Will brazing ever become outmoded and no longer be viable for tomorrow's joining needs? No, that will never happen. With constant growth and change occurring in the materials world, brazing processes will be here to stay. In fact, some materials being developed in today's metals-labs are so sensitive to cracking that they cannot be welded. But brazing processes, being more gentle and uniform in their heating of these materials, is proving to be a viable joining process for these complex new alloys.

Aerospace brazing is offering today many more opportunities than has ever been offered in the past. Where will YOU fit in? What will YOU do?

Brazing's future is bright: As a famous person once said: Let's take it "to Infinity... and beyond!"

Fig. 12 Brazing's future in aerospace is bright.

Brittle Brazing Filler Metals Made in Flexible Wire Forms

Ziani Abdelouahab
Morgan Advanced Materials, Hayward, California, USA
abdel.ziani@morganplc.com, Ph.: 510 491 1258

Abstract

A broad range of brazing filler metals (BFM) mainly available in paste or melt spun foil forms contain alloy elements which promote the formation of non-ductile, hard intermetallic phases. Typical BFM alloys falling within description are BFM's of the B-Ni series per AWS classification. Per those BFM's alloy constitution, the amounts of boron/silicon alloy element additions are such that the resulting alloy materials as obtained by conventional casting processes are hard and inherently brittle. Their ductility is severely impaired by boron/silicon intermetallic compounds precipitation during solidification of the cast product. That is mainly the reason why those BFM's are only available in powder/paste or amorphous foil forms. The powdered form is generally obtained by melt atomization techniques whereas the presence of glass phase formation promoters such as boron and silicon alloy elements enables their production in melt spun foils. In this work, we will be presenting a new process and raw materials approach to successfully producing those BFM's in flexible wire forms. Flexicore® is a Morgan Advanced Materials trade name designating a new class of wire products among the brittle BFM alloys. The work presented here is a case study of AMS 4777 (BNi-2) alloy made in flexible wire form.

1. Introduction

BFM's are mainly those containing one or more of the Ag, Au and Pd precious metals elements or those consisting of (Al, Mg)-, Co-, Cu-, Fe- and Ni-based alloys. BFM's alloys constitution has been optimized such that their solidus/liquidus temperatures can be varied within the practical process temperatures of brazing. In general, BFM's of the first group among Ag and Au alloys are ductile and adequate for cold forming operations and tolerant to large amounts of deformation at room temperature. Therefore, those alloys were easily formable into thin foils and wires for brazing products manufacturing. Those of the second group containing transition metal elements such as Co, Cu, Fe and Ni because of their higher melting temperature are usually alloyed with other elements mainly B, P and Si which act as melting temperature depressant through eutectic reactions. Among the second group of alloys, the B-Ni series listed under AWS classification existed since the 1950's and found broad applications through aerospace and automotive industries in honeycomb structures joining, Exhaust Gas Recirculation (EGR) coolers and heat exchanger components brazing.

Despite their early development in the late 1940's, products availability was limited to their paste or melt spun foils. The presence of B, P and Si of alloy elements in the BFM materials while reducing the solidus/liquidus temperature range to the required processing window for those alloys also severely impaired the ductility of those alloys. It is well known that under solidification and cooling regimes encountered in conventional foundry and metals casting operations, the cast ingot preforms used for strip rolling and wire drawing would inherently consist of brittle feedstock material due to heavy precipitation of borides, phosphides and silicides. Alternatively, P and the metalloids B and Si are known for being promoters of amorphous phases formation upon rapid solidification by melt spinning, a property that is exploited for the production of B-Ni alloys in high strength ribbons adequately usable for stamping brazing preforms. Similarly, given their relatively low melting temperature, the powder form usable for the production of brazing pastes is readily producible by melt atomization techniques.

There are few exceptions of wrought BFM's made into wires, one example is the BFM's of Cu-P alloy system. Due to their lower melting temperature range, Cu-P (AWS BCuP series) filler alloys were readily produced by hot extrusion and supplied in rods and wires up to 0.046" wire sizes [1]. The feasibility of hot extruding of Cu-P alloys near their solidus temperatures (< 600°C) at which material flow during forming is enhanced has made it possible for the braze alloy manufacturing industry to build adequate extrusion tooling that fairly withstood those temperatures at production scale. Besides the particular case of Cu-P alloys, it is worth mentioning those reported attempts made to produce melt spun wires [2]. Unlike planar melt spinning technique which is capable of producing foil at large volumes, the process disclosed in [2] is hardly scalable to industrial wire production for many obvious limitations.

Composite arc spray wires used in hard facing applications have been traditionally produced by roll/draw-type processes as described in [3]. This process uses a ductile cladding material strip that is rolled around a powder component material. The process was designed for volume production of wire at rather loose chemical composition tolerances and can merely achieve wire size reductions under 0.046". Furthermore, the core powder material is generally a blend of pure element powders comprising low and high melting point metals such as Al and Mo which can be simultaneously molten in a fast and high energy fusion process such arc spray/weld. In furnace brazing however, such a wide difference in melting temperatures among the constituents will inevitably cause early incipient melting of the metal with low melting point and result in a poor brazed joint.

In light of the foregoing background, and with a view of providing brazing wire products for those alloys prone to loss of ductility due to intermetallic phases precipitation, we developed a manufacturing solution to enable their production in flexible wire forms. The concept is generically designated under Morgan's Advanced Materials trade name Flexicore®. In a case study of BNi-2 (AMS4777) alloy, this paper outlines the development of a flexible wire form of the alloy with a reduced wire down to 0.025".

2. Flexible Wire Product Design

2.1. As-cast Phase Constitution of Alloy BNi-2

Figure 1 shows an SEM micrograph revealing the microstructure of an as-cast BNi-2. The microstructure consists of coarse angular-shaped chromium-boride intermetallic (black precipitates) scattered in a pseudo-eutectic structure comprising a (Ni, Cr, Fe)-silicide (dark grey areas) and a (Ni, Cr, Fe)-rich phase (light grey areas).

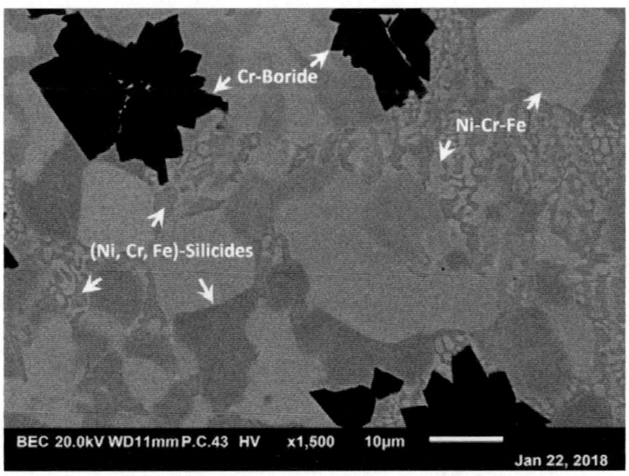

Figure 1. As-cast microstructure of alloy BNi-2

Clearly, an alloy like BNi-2 exhibiting a predominantly intermetallic-rich microstructure will have non-ductility to accommodate large forming deformations such as those to which other ductile metals and alloys are subjected during wire drawing operations.

2.2. Preform for Wire Drawing: Design Considerations

Specification for BNi-2 alloy composition in wt. % as listed in [4] is given as: Ni: Bal., Cr: 6.0-8.0, B: 2.75-3.50, Si: 4.0-5.0, Fe: 2.5-3.5. As a starting point for providing a feedstock preform which can be drawn into wire, the preform must be made of at least one ductile component. In a preferred arrangement, the ductile component could be sheath in which a powder of the non-ductile component can be encapsulated.

Chemical compositions of the sheath and the powder components can be subsequently tailored to allow for sufficient sheath wall thickness while maintaining the melting range of the powder component below the liquidus temperature of the sheath. Accordingly, weight fractions of sheath and powder components can be accurately adjusted to yield the nominal composition of BNi-2 alloy. For illustration, the relationship between the various parameters involved in the design of the preform are shown in Fig. 2. Wire drawing was carried out by conventional preform break down and bar size reduction processes, such as roll drawing, swaging, Bull block drawing, all performed at room temperature.

The generic approach disclosed in this paper is described in further details in Morgan Advanced Materials patent application EP2015/053486.

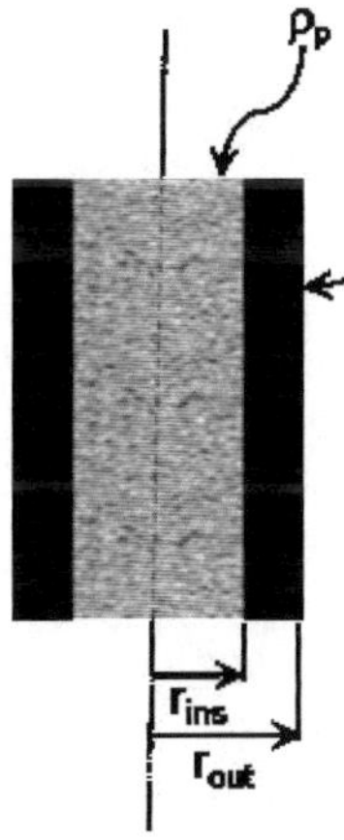

$$r_{ins} = \sqrt{\frac{f_w^p \rho_s}{f_w^s \rho_p + f_w^p \rho_s}}\, r_{out}$$

r_{ins} : Inside radius of sheath component

r_{out} : Outside radius of sheath component

ρ_s : Theoretical density of sheath component

ρ_p : Packing density of powder component

f_w^s : Weight fraction of sheath component

f_w^p : Weight fraction of powder component

Figure 2. Schematic representation of feedstock preform design for flexible wire drawing

3. Materials and Experimental

3.1. Sheath Alloy Options

At first glance, it seems trivial that Ni would be the material of choice for making the sheath component. Nickel strain hardening is slow and nickel maintains a fairly good ductility through many steps of heavy cold reductions. However, combining the remaining alloy elements into the powder component will result in an alloy powder with significant amounts of high melting point Cr-Boride compounds. Therefore, in order to prevent the risk of incomplete melting of the wire at the brazing temperature, the sheath was fabricated from a Ni-Cr alloy material.

3.2. Powder Alloys

As mentioned earlier, the powder component of the wire must not contain high melting point constituents. Likewise, elemental B powder admixing to other alloy powders in order to meet a target blend composition is undesirable. Instead, a pre-alloyed powder of a Ni-Si-B-Fe quaternary alloy made by atomization techniques would be a better approach. The powder component used in this study was produced by inert gas atomization. The powder was sieved through a 100 mesh screen, the resulting particle size distribution is shown in Fig. 3.

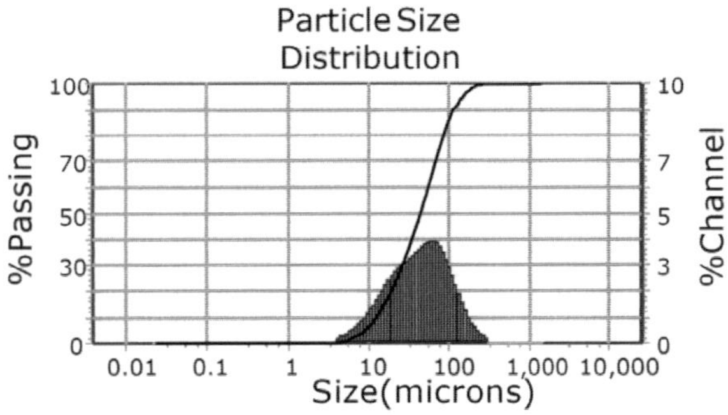

Figure 3. Particle size distribution of the as-received powder component used in this study

For BNi-2 alloy paste preparation, we used a -100 mesh inert gas atomized pre-alloyed powder of the five elements Ni-Cr-B-Si-Fe BNi-2 alloy commercially available, supplied by Praxair. The particle size distribution of the as-received powder is shown in Fig. 4. A binder comprising 80% deionized water, 15% glycerin and 5% of a hydrophilic derivative cellulose was mixed to the powder in a ratio of 15 wt. % binder : 85 wt. % powder.

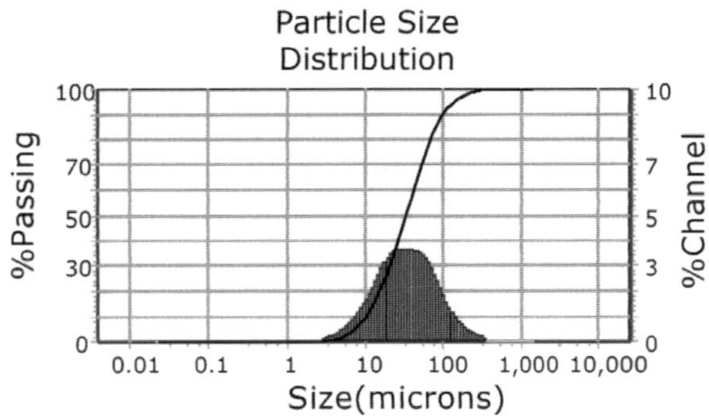

Figure 4. Particle size distribution of as-received BNi-2 powder used in this study

3.3. DSC Analysis and Braze Filler Flow Assessment

Differential Scanning Calorimetry (DSC) using a Netzsch model STA 449 C instrument was carried out under argon gas flow. Samples of the as-received powder and finished wire were analyzed through a complete heat ↔ cool cycle. DSC of the as-received powder was performed to ensure the melting range of the powder meets the requirement mentioned in section 2.2 (See Fig. 5). DSC analysis of the finished wire product was carried out to establish its melting behavior and determine its melting range. Stainless steel T samples were used for braze filler flow assessment.

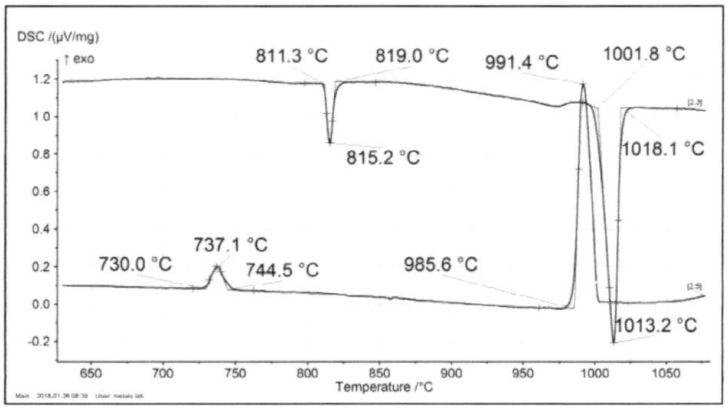

Figure 5. DSC trace of the wire powder component used in this study

3.4. Brazing Test and Braze Joint Strength Measurement

All furnace brazing experiments were performed in vacuum in the low 10^{-5} Torr range. The findings from DSC data analysis and filler flow experimental assessment were applied to optimize brazing cycle parameters. Joint strength measurements were performed by testing in tension the tensile specimen shown in Fig. 6. In this configuration, two tubular sections of an Inconel 625 alloy substrate material were braze joined along a ID/OD overlap 1/8" wide with a joint clearance of 0.005".

Figure 6. Tensile sample designed for brazed joint strength measurement. Top: 3D model; Bottom: Brazed specimen

Tensile samples brazed with BNi-2 flexible wire rings were tested and compared to similarly prepared and tested samples brazed with BNi-2 paste described in section 3.2. For the sample geometry shown in Fig. 6, a 0.046" wire size ring made to fit the OD of the small tube of the specimen was found adequate for providing sufficient amount of braze alloy to fill the gap between the tube sections. Similarly, when applying the brazing paste to the joint, a syringe of appropriate size was used to dispense the required amount of brazing paste. In both cases, it was verified that indeed, after brazing, an emerging brazing fillet could be seen on the opposite end of the joint as an indication of complete filling of the joint.

4. Results and Discussion

4.1. Wire Chemical Consistency, Alloy Melting Behavior and Flow

Wire product chemical consistency at the finished wire size was determined by ICP/OS analysis. Table 1 summarizes the chemical analysis data for a set of three samples randomly taken from a 700 ft spool of wire at 0.030" finished size. As shown in Table 1, all alloy major elements are within the limits given for BNi-2 alloy.

Table 1. Finished wire chemical analysis

	Element Contents wt.%		
Cr	6.7	7.1	6.9
B	3.2	3.0	3.1
Si	4.4	4.7	4.5
Fe	3.1	3.2	3.4

Figure 7 shows the heating and cooling curves of a DSC scan performed on the as-drawn wire at controlled heating and cooling rates of 10°C/min. The heating curve displays two broad endothermic peaks at onset temperatures of 973°C and 998°C which appear to match the published melting range 971°C-999°C for BNi-2. The slightly higher range detected upon heating is ascribed to a kinetic delay associated with the inter-diffusion process taking place between the atomic species of the sheath and powder components. Upon cooling the corresponding solidification range is shifted to 985°C-966°C which is indicative of some degree of undercooling, a very common effect which metallic alloys undergo upon solidification.

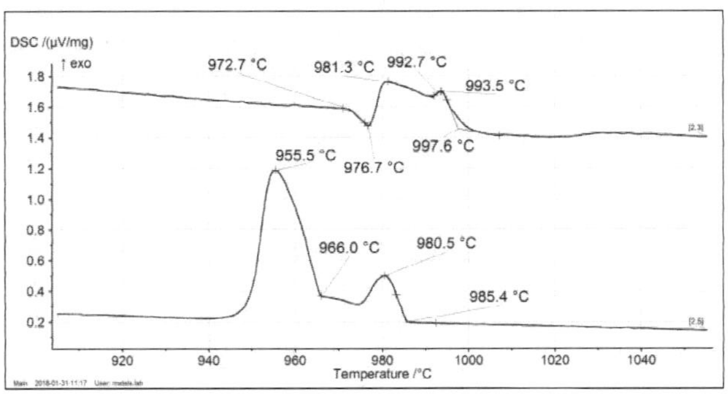

Figure 7. DSC trace of the as-drawn wire

The characteristics of the heating curve indicates that for achieving uniform melting of the wire, the brazing cycle must include a hold at a temperature below the brazing temperature to allow for the inter-diffusion reaction described earlier to reach its equilibrium. A closer examination of the DSC trace of the powder component of the wire shown in Fig. 5 suggests that a fast ramp to ~1010 °C during brazing is necessary in order to reach complete and instantaneous incipient melting of the powder. Complete wetting of the surrounding sheath along with enhanced atomic transport in liquid state was anticipated to uniform and simultaneous melting of the remaining sheath material.

Several brazing trials using T-shaped samples were performed for defining practical heating rates required to reach uniform and complete melting of the wire. Four brazing cycle tests consisting of fast ramps at 10, 15, 20 and 25 °C/min. to 1010 °C, a hold for 20 min., a slow ramp at 5°C/min. to 1038 °C, a hold for 10 min. followed by a cool down furnace power off. As anticipated, fast cooling at 15 °C/min. produced the best result in terms of uniformity and completeness of wire melting as well as capillary flow along the T-samples. Fig. 8 illustrate the outcome of the brazing test performed at a heating rate of 15 °C/min.

Figure 8. Wire brazing test using T-sample. Cycle: Ramp at 15 °C/min. to 1010 °C, 20 min. hold at 1010 °C, ramp at 5°C/min. to 1038°C, 10 min. hold at 1038 °C, cool down with furnace power off.

4.2. Joint Strength

The brazing cycle described in Fig. 8 was used to prepare the tensile samples. Testing in tension was performed on an Instron machine model 5985 at a ram speed of 0.1 inch/min.. The loads at rupture along with the corresponding stresses for the given joint area (0.285 in^2) are reported in table 2.

Table 2. Inconel 625 brazed joint strengths. BNi-2 wire versus BNi-2 paste

BNi-2 Wire		BNi-2 Paste	
Load at Rupture (lbs)	Stress at Rupture (psi)	Load at Rupture (lbs)	Stress at Rupture (psi)
14123	49554	10337	36270
13653	47905	10098	35432

From a statistical stand point, the set of two data point presented in this comparison may appear not sufficiently representative, nevertheless, there is evidence of a higher strength achieved using this wire form of BNi-2 compared to the strength attained using BNi-2 paste. There are many explanations to this finding, one of them is presumably the presence of organic compounds associated with the binder of the paste material. Binders burn off usually produces ash residues that could alter the mechanical strength of the joint. In other instances outgassing of the binder during heating could lead to porosity formation, however, SEM/EDS examination of the joint did not reveal noticeable porosity that could be ascribed to the lower joint strength produced by paste.

In the as-polished condition, joint microstructures produced by BNi-2 wire and BNi-2 paste looked fairly similar (See Fig. 9). Besides the joint centerline that showed no visible difference, the adjacent layers both revealed the presence of fine discrete (Mo, Nb)-borides and a network of (Mo, Ni)-borides. No Cr–borides such as those shown in Fig. 1 could be seen except in the fillet re-cast areas.

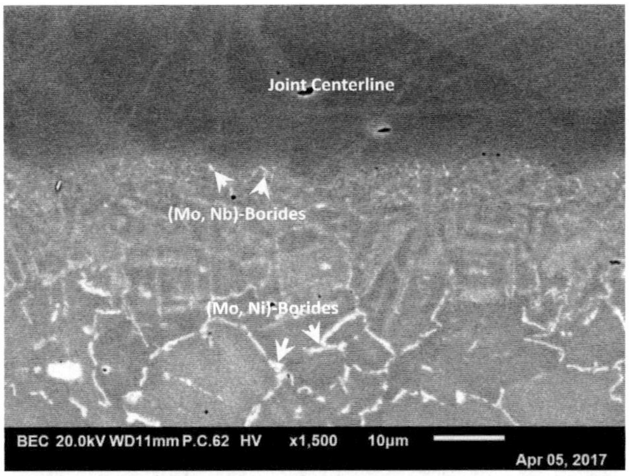

(a)

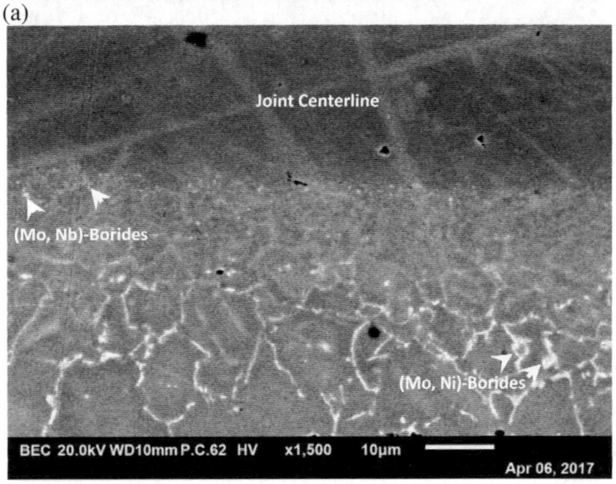

(b)

Figure 9. Inconel 625 brazed joint, as-polished microstructure. BNi-2 wire in (a), BNi-2 paste in (b)

In the absence of distinguishable features, we used Kalling's etch reagent mentioned in [4] at a di-ionized water to reagent dilution of 10:1 for further examination of the microstructures. The etched sample microstructures are shown in Fig. 10. As it can be seen, the sample brazed with BNi-2 paste exhibited the presence of micro-scale chromium boride precipitates in the centerline joint area (Fig. 10 (b)). Whereas no Cr-borides formation can be noticed in the sample brazed with BNi-2 wire (Fig. 10 (a)).

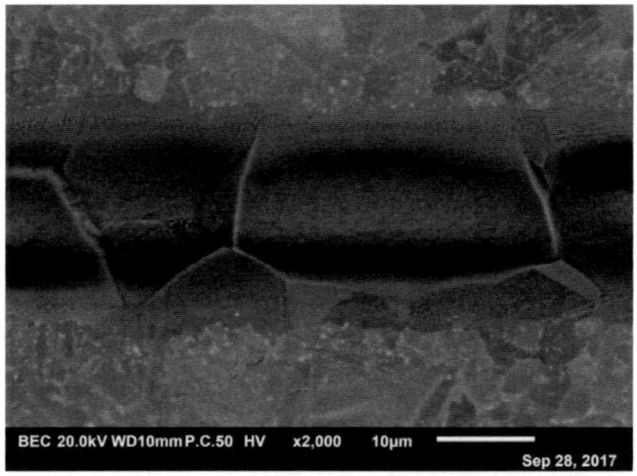

(a)

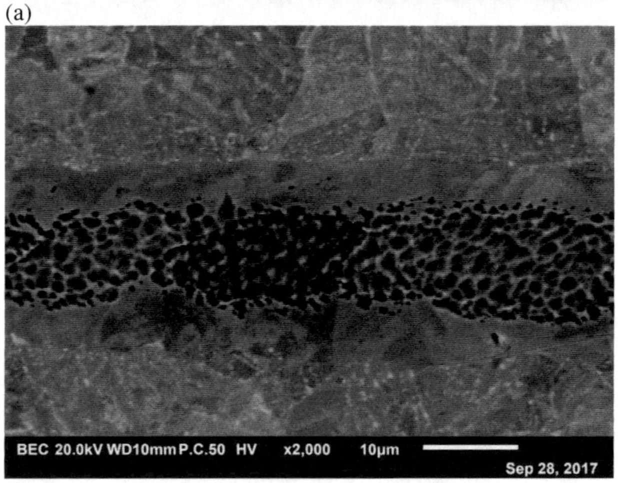

(b)

Figure 10. Inconel 625 brazed joint, Kalling's reagent etched microstructure. BNi-2 wire in (a), BNi-2 paste in (b)

Cr-borides, like other transition metals with high melting temperatures are very stable and often produce even higher melting point compounds. It is presumably the presence of Cr–borides in the as-atomized BNi-2 powder that is responsible for their re-precipitation in the brazed joint. One can speculate on the fact that even upon full melting of the paste at brazing temperature, there might still be incompletely molten Cr-boride crystallites which could have acted later upon cool down during the brazing cycle as potential nucleation sites for the seen Cr-borides of Fig. 10 (b). Unlike the case with BNi-2 paste, the powder used for BNi-2 wire production is Cr-free and therefore, Cr-boride-free as well. Upon melting of the wire at the brazing temperature, boron distribution among Inconel 625 alloy elements seems to follow a preferential reaction path leading primarily to the formation of Mo- and Nb-borides. Those preferential reactions being thermodynamically driven by the free energy of formation values of the respective competing borides, the predominance of Mo- and Nb-borides in BNi-2 wire joint is potentially related to their lower energy of formation [5]. As generally known, Cr-borides are detrimental for ductility in many alloys, particularly in Ni-based alloys such as those listed in the BNi series. The lower joint strength produced with paste is most likely the result of precipitated Cr-borides revealed in Fig. 10 (b). Further fractographic examination of the fracture surfaces of the tensile specimen would be a useful investigation for substantiating these findings.

5. Conclusion

A production route for manufacturing flexible BNi-2 wire was discussed. The critical aspects for successful production of a usable wire were presented in details.

A dedicated brazing cycle designed to accommodate the particular structure of the wire was established and demonstrated to achieve optimal brazing in terms of flow and melting within BNi-2 alloy melting range.

Furthermore, this wire form of BNi-2 alloy was found to produce superior joint strength with Inconel 625 substrates
Among the numerous advantages of this wire form of BNi-2 alloy, the following can be recognized:

1. Unlike paste products, the wire disclosed in this study is binder free, which could be very advantageous for maintaining clean furnace operation

2. The wire can be tack welded for precise positioning of brazing preforms around the joint area. Other products such as sintered BNi-2 preforms or boron coated nickel alloy rings are more than likely not weldable.

3. Low variation of preform-to-preform weight for high brazed joint repeatability

References

[1] R. E. Ballentine *et al.,"*Extruded Phosphorus-Silver-Copper Brazing Alloys", US Patent 2,795,520
[2] T. Masumoto *et al.,* "Process for the production of Fine Amorphous Metallic Wires", US Patent 4,495,691
[3] M. Puschner, "Method of Making a Filled Electrode for Arc Welding", US Patent 4,396,820
[4] ASM Specialty Handbook, *Ed. by J. R. Davis, Davis & Assoc.,* "Nickel, Cobalt, and Their Alloys", p. 316 ASM International, 2000
[5] M. Dahlqvist *et al.,* "Influence of boron vacancies on phase stability, bonding and structure of MB_2 (M= Ti, Zr, Hf, V, Nb, Ta, Cr, Mo, W) with AlB_2 structure", *J. of Phy.: Condensed Matter*, Vol. 27, No. 43 (2015), pp. 43-56

Filler Metal Control in Sinter-Brazing

Kyle H. Bear and Glenn Rishel
Pennsylvania State University, DuBois, PA

Brian Smith and Stephen L. Feldbauer, Ph.D.
Abbott Furnace Company, St. Marys, PA

ABSTRACT

Sinter Brazing has become a significant tool in the manufacture of complex powder metal components. The primary issue associated with this technology is holding the filler metal at the joint while sintering. The capillary forces of the porous base material and the braze joint, both compete for the filler metal. In cases where the filler metal is pulled into the compact and away from the joint, weak or no bonding may result in a defective component. Only by understanding the variables that influence the control of the filler metal, can a defined process be implemented to ensure optimal quality. The variables investigated in this study are: compact density, atmosphere chemistry, and the presence of flux in the filler metal.

INTRODUCTION

Sinter Brazing has become a necessary means of manufacturing many complex powder metal shapes. Although the compaction technology has advanced substantially over the years, the ability to bond two components together while sintering has enabled the industry to broaden its capabilities and remain competitive in many markets.

The primary issue with the brazing of two powder metal components, in either the green or sintered states, is the competition between the capillary effect that draws the filler metal into the gap and the capillary effect produced by the porosity of the compact which results in the filler metal being pulled away from the gap and into the part. If this filler metal is pulled into the part, the bond may be compromised or not form at all.

In this work, we review the key fundamentals of brazing and apply them in such a way as to better control the flow and retention of the filler metal at the sinter braze joint. This is achieved by reviewing the key variables and their effect on the wetting of the filler metal. The sintering atmosphere, flux content of the filler metal, and the density of the components are all important to the control of the filler metal.

BRAZING FUNDAMENTALS

The fundamental driving forces for the formation of a good braze joint are the wetting of the base metal by the filler metal and the capillary force that results and pulls the filler metal through the braze gap.

$$\text{Capillary Force} = \frac{2\gamma\cos\Theta}{r}$$

Where γ is the surface tension, Θ is the contact angle, and r is the radius of the capillary.

There are a few key steps that must be followed to produce a good braze joint.

1. Surface Preparation
   - When the filler metal melts, the wetting of the filler metal is strongly affected by the cleanliness of the surface.

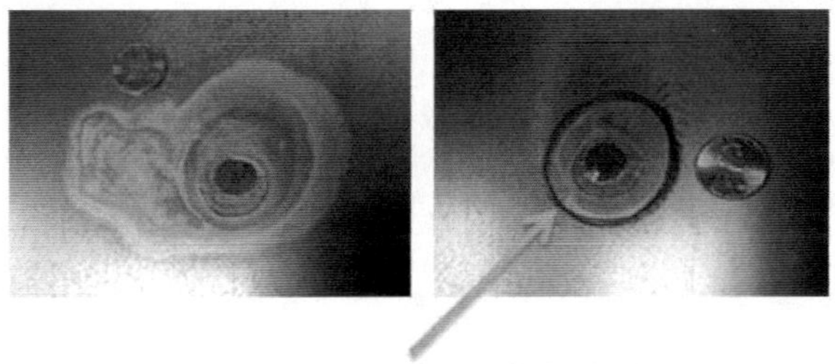

Figure 1. Effect of Surface Contamination on Wetting

The photo above shows the effect of a carbon ring on the flow of filler metal (AB-72) on a carbon steel plate. If the surface is not free of dirt, oxides, and oils, the filler metal will not wet the base metal and will not be pulled into the braze joint.

2. Braze Joint Design
   - Proper braze joint design is critical to achieving the desired strength in the component. The total length of the braze joint should be three (3) times the thickness of the thinnest cross section being joined. This will ensure that a good braze will result in a joint that is stronger than the base metal.

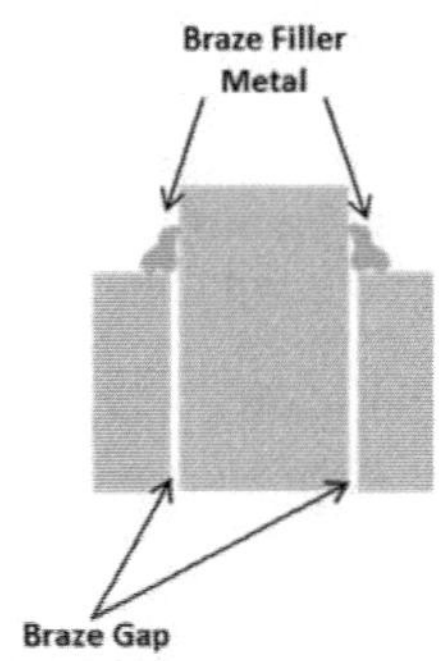

Figure 2. Proper Braze Joint Design

   - The gap, through which the filler metal is pulled via the capillary force and is a strong function of the gap r, must be 0.05 mm (0.002") to 0.20 mm (0.008") wide at the brazing temperature. This means that the thermal expansion of the components must be considered when designing the joint.

3. Furnace Atmosphere
    - All metals contain oxides on their surface. If these are not removed, the filler metal cannot alloy with the base metal. Hence, the oxides must be removed from the surfaces of the base metal and filler metal prior to the filler metal melting. This can be accomplished by using a flux (Typically an acid or other material that will remove/react with the oxide layer to produce a clean metal surface.) or a reducing furnace atmosphere. Unlike a flux, the reducing furnace atmosphere will not leave a residue that must be removed post brazing. Some typical dew points of reducing atmospheres are -51°C (-60°F) for stainless steels and -35°C (-30°F) for carbon steels.

4. Time and Temperature
    - The time for a filler metal to flow into a clean, properly designed braze joint, and form a good braze is three (3) to five (5) minutes at the brazing temperature that is above the liquidous temperature of the filler metal.

SINTER BRAZING

Sinter brazing is a manufacturing technique used to produce complex shapes from powder metal that cannot be produced directly from the compaction press. Multiple shapes are produced using current pressing technology. These shapes are then assembled in the green state, or some may be presintered, and a filler metal is placed in or near the adjoining surfaces of the shapes to allow for brazing to occur simultaneously with the powder metal sintering process.

Common to sintering and brazing, the metal particles and surfaces must be free of dirt and oils. The Hydrogen in the furnace atmosphere then reduces the oxides on the surface of the metals to produce clean particles and surfaces for sintering and brazing. As the components continue to travel through the sintering furnace and heat up to the sintering temperature, the filler metal melts and begins to wet the surface of the component surfaces; however, this is where the most common issue related to sinter brazing arises. Not only is the filler metal able to wet the surfaces that are to be brazed together but the pores within the compacted components also compete for the filler metal. The surface tension between the filler metal and the base metal, along with the capillary effect produced by the small pores of the compact, pull the filler metal away from the braze joint and compromise the bond. This compromise may be a reduced amount of joint filling to no bonding at all.

If the wetting characteristics of the filler metal can be controlled, the retention of the filler metal at the brazing joint can be maintained and an improvement in the bonding and overall strength of the assembly may be maximized. For many years, manufacturers have added Sulfur to the compact mix and Manganese to the filler metal to retard the wicking of the filler metal into the pores of the compact. Unfortunately, Sulphur is not always a desired additive. For this reason, current techniques use the addition of carbon monoxide containing atmospheres as a sintering atmosphere or as an addition to the sintering atmosphere. The impact of and mechanisms associated with the carbon monoxide addition were not clear; however, many swear by this approach. In this work, along with the influence of flux and density, the impact of and the mechanism associated with the addition of the carbon monoxide as it relates to controlling the wetting of the filler metal and its retention at the braze joint is investigated.

EXPERIMENTAL PROCEDURE

Discs of F-0000 were compacted to densities of 6.0 g/cc, 6.2 g/cc, 6.8 g/cc, and 7.0 g/cc. The discs had a dimension of approximately 7.62 cm (3 inches) in diameter by 2.54 cm (1 inch) thick.

Two filler metal chemistries were pressed into 1.5 gram pellets. One of the chemistries was a standard AB-72 material. The other filler metal chemistry did not contain a fluxing agent.

The pellets were then placed on top of the discs and sintered in two different furnace atmospheres, one atmosphere contained Hydrogen, Nitrogen, and carbon monoxide, and the other contained only the Hydrogen and Nitrogen.

The discs were then sectioned, polished and etch with 2% Nital etchant to reveal the infiltration of the filler metal into the compact. Image J software was used to measure the infiltration area.

RESULTS

As can be seen in Figure 3, increasing compact green density results in a decrease in the amount of filler metal that flows into the powder metal compact. This is to be expected since there is a 16.7% difference in the relative pore fraction when comparing a 7.2 g/cc compact and a 6.0 g/cc compact.

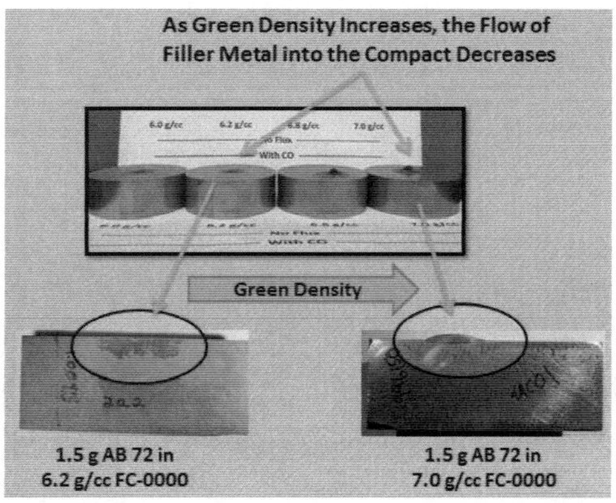

Figure 3. Effect of Green Density on the Flow of Filler Metal into the Compact

Flux is often added to the filler metal to help to clean the surface of the base metal and enhance wetting; however, in this application, the flux not only cleans the braze joint surface but also the particle surfaces inside of the pores of the compact. The result is an enhance wetting of the pore surfaces and more filler metal being pulled from the braze joint (Figure 4).

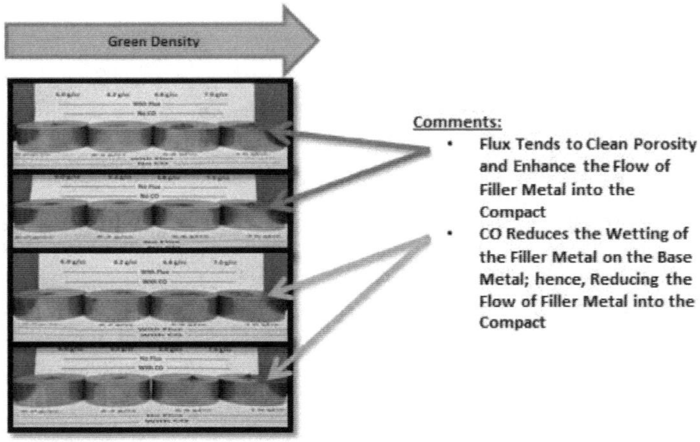

Figure 4. Effect of Flux and CO on Filler Metal Infiltration into the Compact

As the powder metal compact exits the preheat section of the furnace and enters the high heat section of the furnace, the compact temperature is in the range of 900°C (1650°F) and 1035°C (1900°F), carbon monoxide results in the carburization of iron and steel at these temperature[1]. Since the amount of time that the product is in this condition is small, the carburization will be a small layer on the surface of the exposed particles on the surface of the material and inside of the pores. This surface contamination reduces the wetting of the pores by the filler metal and results in less filler metal being pulled from the braze joint (Figure 4). Figure 5 shows that, although this material is a F-0000 and contains no carbon, the surface of the particles in along the very edge of the compact were carburized. The particle hardness on the surface is higher and the microstructure shows the presence of carbon on the surface of the particles.

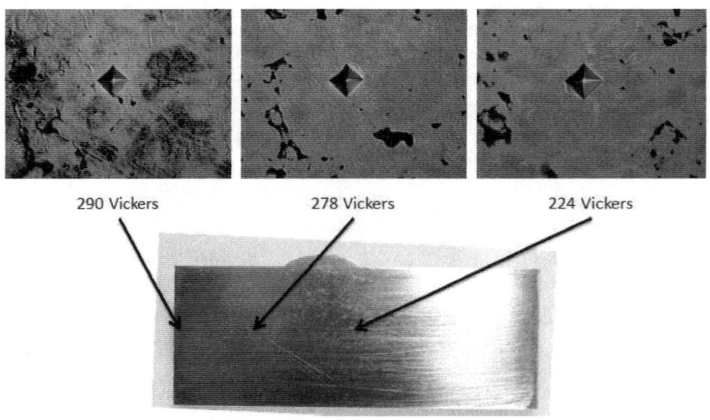

Figure 5. Edge to Center hardness and carburization variation due to the presence of carbon monoxide.

In comparing the overall impact of the variables (Figure 6), an interesting phenomenon was observed. In the high density region of graph (purple), the effects of the flux, carbon monoxide, and density were consistent with what was expected. In the lower density section of the graph (green), the control of the filler metal wetting was not influence by the wettability of the filler metal on the base metal.

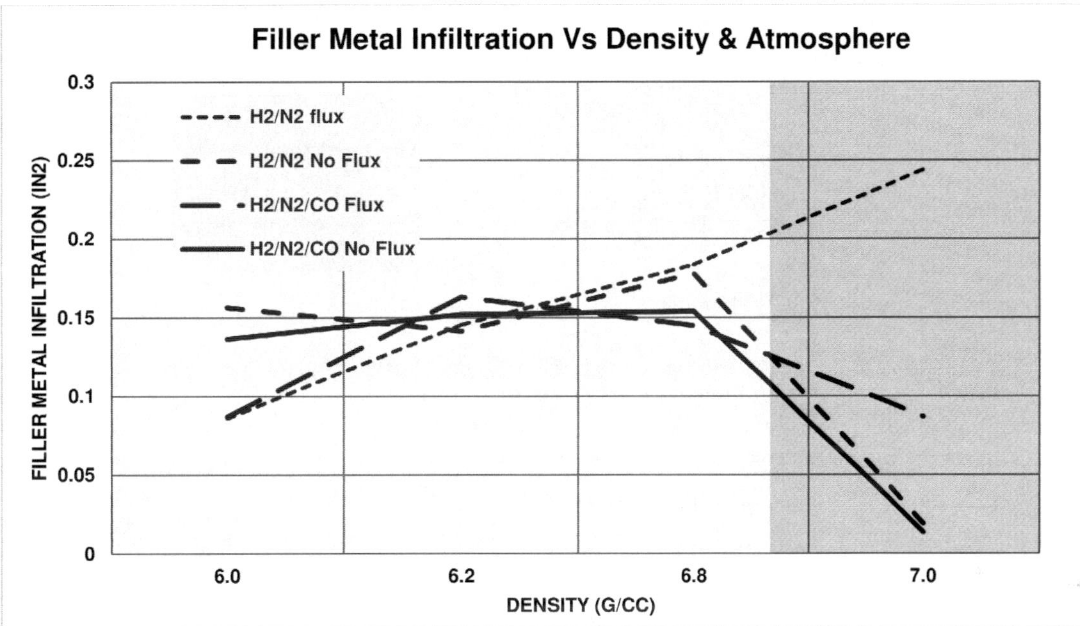

Figure 6. Influence of Density, Flux, and Carbon Monoxide on the Infiltration of the Filler Metal into the Compact

To better understand what is shown here, one must consider the relationship between the force due to surface tension, that would want to pull the filler metal into the pores of the compact, and the force due to gravity, resisting the movement of the filler metal flow. Figure 7 illustrates that as the pore size becomes larger, lower density of compact, there reaches a point where the force due to gravity will cause the filler metal to fall and not completely fill the pore. The result is that the influence of the altered surface tension is reduced and limits the ability of the producer to influence the wetting and flow of the filler metal with flux or carbon monoxide. For this reason, many producers find that they have more consistent results and better quality of sinter brazed products that are 6.9 g/cc and above.

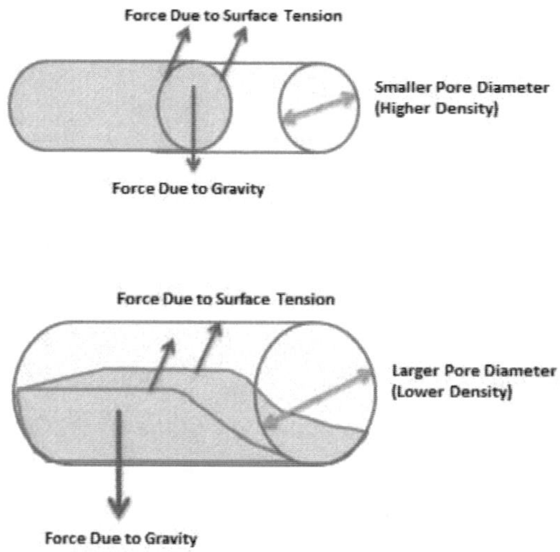

Figure 7. Surface Tension versus Gravitational Force as a Function of Pore Size

CONCLUSION

The control of the filler metal at the braze joint of a sinter brazed product has significant impact on the quality and yield of complex shapes that cannot be produced by current conventional compaction technology. Maintaining the presence of the filler metal at the sinter braze joint is necessary, or the components will not be bonded. However, the porosity of the powder metal compact results in capillary forces that tend to wick the filler metal away. By pressing the compacts to a density greater than 6.9 g/cc, removing the flux from the filler metal, and adding carbon monoxide to the sintering furnace atmosphere, the capillary action of the pore can be reduced and the flow of the filler metal at the joint can be controlled to produce a better bond.

REFERENCES

1. *Heat Treater's Guide*, 2nd Edition, pg. 47, ASM International, 1995
2. Bear, et al, Applying Brazing Fundamentals to the Sinter Brazing Process, May, 2017, American Welding Journal, pg. 44.
3. *Brazing Handbook*, 5th ed. 2007, Miami, Fla.: American Welding Society
4. Bear, K., and Feldbauer, S., Fundamentals of Brazing Applied to Sinter Brazing, Presentation, Special Interest Session, PowderMet 2016, Boston, MA.

Influence of Electric Current Passing Through Interface on Wetting of Zirconia with Metals

Durov O.V., Sydorenko T.V., Koval O.Y.
Frantsevich Institute for Problems of Materials Science National Academy of Science of Ukraine, Kyiv, Ukraine

Abstract

The influence of current passing through the interface on the wetting of zirconia by metal melts (copper, nickel, alloys) in vacuum was investigated. It has been found that electric current can improve wetting. The observed effects were used for brazing of zirconia to metal.

Introduction

Investigations of wetting and interaction in systems contained solid oxide and metal melt are interesting both in fundamental and practical aspects. Transition metal oxides, in particular zirconia, have the ability to pass electric current at high temperatures, it can influence the processes on ceramic to metal contact. This aspect was studied less then influence of other factors (composition of contact phases, temperature, medium [1,2,3]). One can note works [4] where aluminum to zirconia interaction under voltage applying studied, [5] where was found that voltage influence the interaction in ZrO_2-Cu system and [6] where Ag-Cu-O melt and ZrO_2 contact under electrical current was investigated. So the aim of current study was to investigate the contact phenomena in zirconia — metal melt systems and the possibility of ZrO_2-ceramic with metal filler brazing under electrical current pass.

Materials and methods

Ceramic based on 3.5 at. % Y_2O_3 stabilized ZrO_2, high pure metals (Cu, Ni, Ga, Cr), heat resistant steel and construction molybdenum were used. Cu-17.5Ga, Cu-45Ni and Ni-20Cr alloys were prepared by vacuum melting. For wetting experiments ceramic was machined to plates 1.3 mm thickness. 5×5×3 mm ZrO_2-ceramic blocks, steel disks 10 mm diameter and 5 mm thickness, 10×8×3 mm molybdenum blocks were used as parts of brazing joints. Molybdenum wire and 0.2 mm plate for current leads and clamps, Al_2O_3-ceramic and sapphire as insulator were used in samples for wetting experiments. The sample consists of two plates, both of ceramic or one of ceramic and other of molybdenum, between which aluminia spacers are inserted. Molybdenum plates as current leads were pressed down to ZrO_2-ceramnic plate by 1.2 mm molybdenum wire clamps with Al_2O_3 plates as insulators (Fig. 1). In experiments with copper and its alloys copper foil was situated between molybdenum current lead and ZrO_2-ceramic to provide tight contact. In gap between plates a piece of the studied metal was inserted, the sample was placed in the vacuum furnace, current leads were connected to wires leading outside the vacuum chamber. After 3 × 10^{-3} Pa vacuum creation the sample was heated to experimental temperature and current was supplied. Metal drops were photographed by the digital camera. Scheme of experiment is represented on Fig. 2.

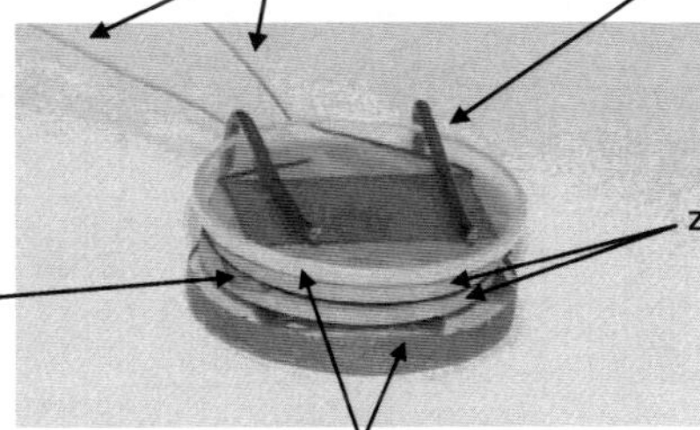

Fig. 1. Experimental sample

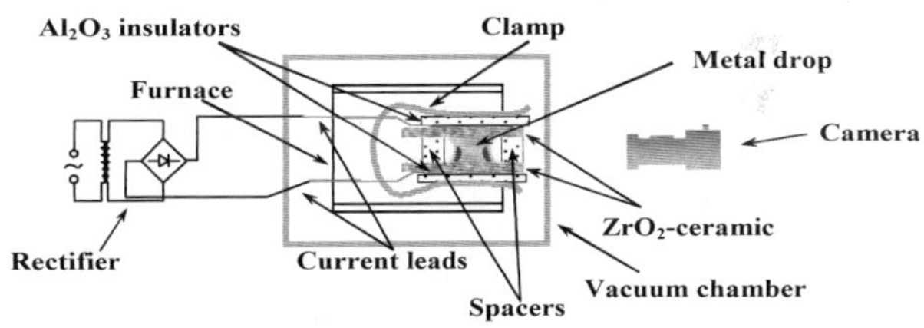

Fig. 2. Scheme of wetting experiment

For brazing procedure filler foil was situated between ceramic and metal details, this assembly and applied current leads were pressed out with the molybdenum clamp through Al_2O_3 insulators. Sample was put in the vacuum furnace, connected to electrodes, after heating current was loaded.

Results and discussion

Experiments where copper drop was situated between ZrO_2-ceramic plates (ZrO_2/Cu/ZrO_2) were carried out firstly. Electrical current passing leaded to wetting of both plates by melt up to full spreading (contact angle near 0°) and need to note that contact angles on both plates was close (Fig. 3). Kinetic of wetting is represented on Fig. 4 as a diagram, one can see that process depends on the amperage, full spreading reached faster for higher amperage.

In next experiments copper drops were situated between ZrO_2-ceramic and molybdenum plates (ZrO_2/Cu/Mo). When the positive electrode was connected to the ceramic plate copper wetted the ZrO_2 with kinetic similar to that for experiment with two ceramic plates (Fig. 5).

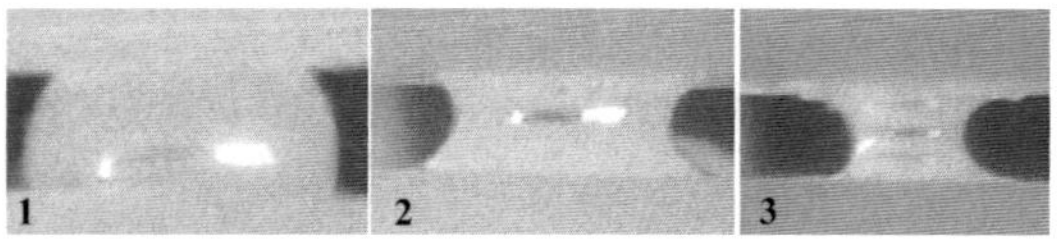

Fig. 3. Copper drop between ZrO_2 plates in the experiment of current through the interface passing, 1100 °C: 1 — without current; 2 — 1 A, 5 min; 3 — 1 A, 16 min

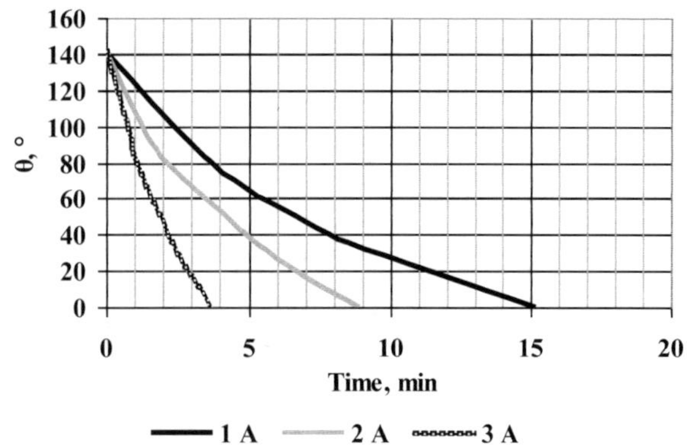

Fig. 4. Kinetic of copper melt on ZrO_2-ceramic spreading for drop situated between two ceramic plates at different amperage

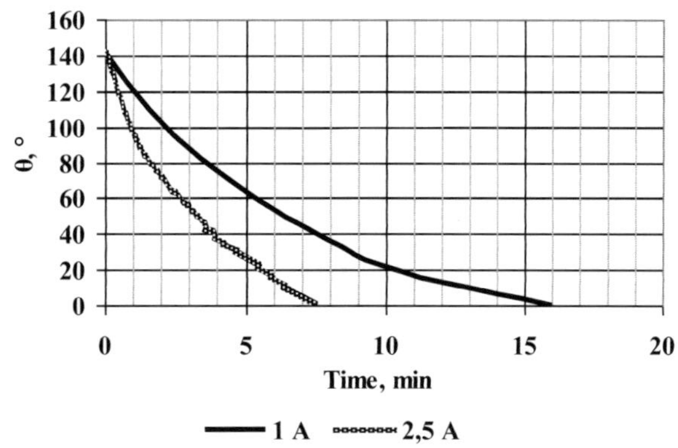

Fig. 5. Kinetic of copper melt on ZrO_2-ceramic spreading for drop situated between ceramic and molybdenum plates at different amperages

In case when the positive electrode was connected to the molybdenum plate and negative to the ceramic some improving of the wetting also observed but the minimal reached contact angle was just near 90° (fig. 6). At amperage 2 A or higher the formation of bubbles in the drop observed and the vacuum in the chamber notably decreased. If replace the electrode connected to ceramic plate from negative on positive, copper wets the ZrO_2 as in experiments where such connection was used initially.

A Cu-17.5Ga alloy was also studied, its wetting behavior is similar to that for pure copper (Fig. 7, 8).

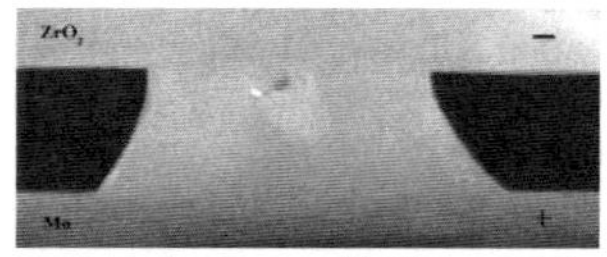

Fig. 6. Copper drop between ZrO_2 and Mo plates in the experiment of current through the interface passing, negative electrode connected to ZrO_2, 1.5 A, 15 min holding, 1100 °C

Fig. 7. Cu-17.5Ga drop between ZrO_2 plates in the experiment of current through the interface passing at 1.5 A amperage, 1100 °C: 1 — 1 min; 2 — 5 min; 3 — 8 min; 4 — 11 min

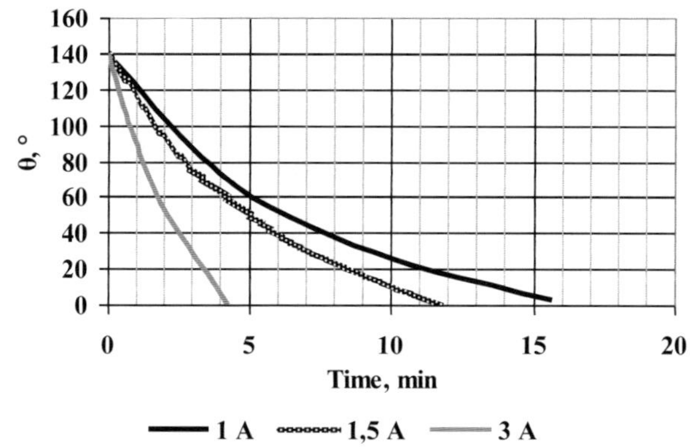

Fig. 8. Kinetic of Cu-17.5Ga melt on ZrO_2-ceramic spreading for drop situated between two ceramic plates at different amperage

Nickel and Ni-20Cr alloy was also studied. In both cases drops situated between ZrO_2-ceramic plates after voltage loading spread on oxide surface very fast. In the experiment with nickel drops disrupt during one minute (Fig. 9).

Fig. 9. Nickel drop between ZrO_2-ceramic plates in the experiment of current through the interface passing, 1 A, 56 s holding, 1470 °C

Ni-20Cr melt also spreads fast enough, although drops were not broke (Fig. 10).

The obvious reason of faster spreading of nickel and nickel-chromium is that the experimental temperatures (1470 °C for

Ni and 1450 °C for Ni-20Cr) were significantly higher then that for copper and Cu-17.5Ga (1100 °C) so all processes occurred faster.

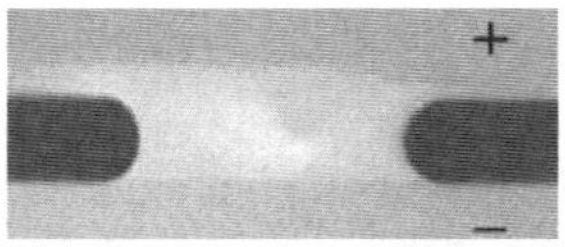

Fig. 10. Ni-20Cr drop between ZrO_2-ceramic plates in the experiment of current through the interface passing, 1 A, 3 min holding, 1450 °C

The following assumptions may be proposed to explain effects observed. For used experimental samples when a positive electrode connected to the zirconia plate oxygen anions move out of oxide — melt interface so the zirconia surface near the contact to metal is depleted of oxygen. The surplus of zirconium dissolves in the metal melt and in such a way improves the wetting for both ―negative" and ―positive" plates because zirconium is well known adhesive active element [1]. Similar effect was observed in [7] where deficit of oxygen in zirconia was formed by vacuum annealing or contact to reducing agent, it also provided improving of the oxide by inert metal melts wetting.

When copper drop situated between ―negative" zirconia and ―positive" molybdenum plates oxygen moves in the copper melt and it also improve the wetting, because oxygen is an adhesive active element also [1]. Solubility of oxygen in copper at high temperature in vacuum is relatively low so the bubbles formation observed and effect on wetting not as significant as for the zirconium dissolution.

Cross-sections were fabricated to verify these assumptions. On Fig. 11 are represented microphotographs of a cross-section for the copper drop situates between ZrO_2-ceramic plates (ZrO_2/Cu/ZrO_2), holding time at 3 A was 22 min.

1
2
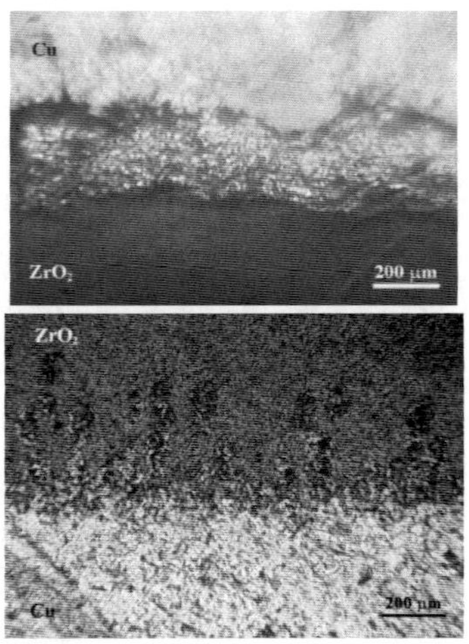

Fig. 11. Microphotographs of interfaces for the copper drop situates between ZrO_2-ceramic plates, 1 A, 22 min holding: 1 – for the plate connected to the negative electrode; 2 – for the plate connected to the positive electrode

As one can see on the ―negative" plate thick transition layer formed probably because zirconium dissolved from ―positive" zirconia plate has adsorbed on ―negative" plate and was oxidized by oxygen from it. ―Positive" plate significantly destroyed near interface due to dissolution in melt, also melt infiltrate in the body of ceramic. The reason of the infiltration may be both capillary and electrical.

On the Fig. 12 SEM microphotography of infiltrate melt ―peak" and elements distribution on it are represented. Zirconium presents in the copper-rich phase, it confirms the dissolution of the zirconia in the copper melt.

1
2
3
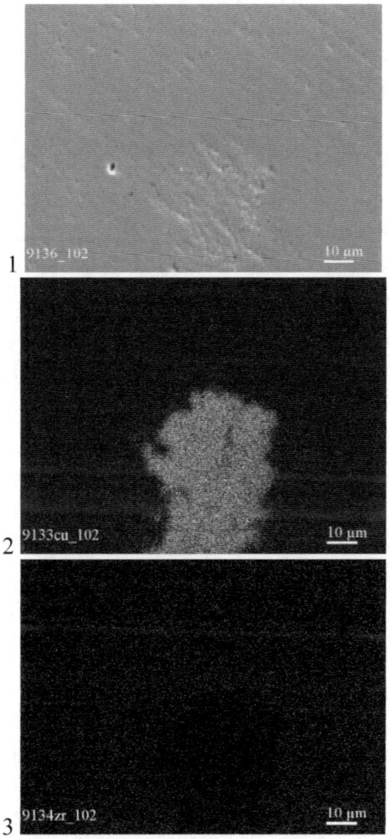

Fig. 12. SEM picture of infiltrate melt ―peak": 1 – microphotograph; 2 – copper distribution; 3 – zirconium distribution

On Fig. 13 microphotographs of metal-oxide interfaces for copper drops between ZrO_2 and Mo plates with different amperage and holding time are represented, ceramic plates were connected to positive electrodes.

As can be seen the interface structure depends on amperage and holding time: the stronger the current and the longer the holding the more destroyed the zirconia surface. Also it needs to note that the interface is not uniform probably because local differences of ceramic conductivity.

On Fig. 14 the microphotograph of the cross section for the nickel drop situates between ZrO_2-ceramic plates (ZrO_2/Ni/ZrO_2) is represented. Some dissolution of zirconia takes place but interface layer not so developed and much thinner than that for copper because the time of contact before drop disruption was significantly shorter.

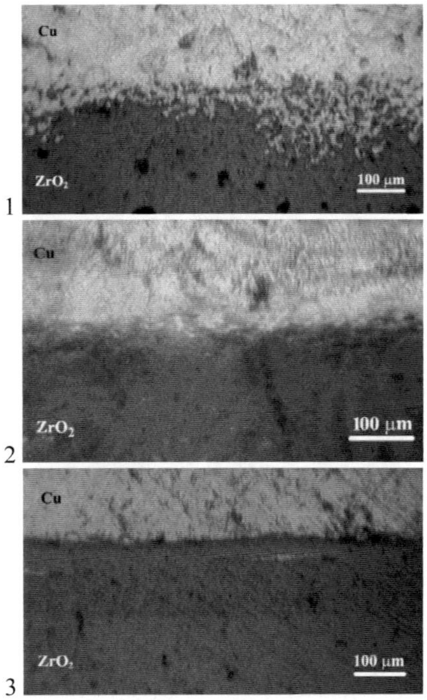

Fig. 13. Microphotographs of interfaces for copper drops situated between ZrO_2-ceramic and Mo plates, ZrO_2 plates were connected to positive electrode: 1 – 3 A, 16 min; 2 – 3 A, 5 min; 3 – 1 A, 5 min

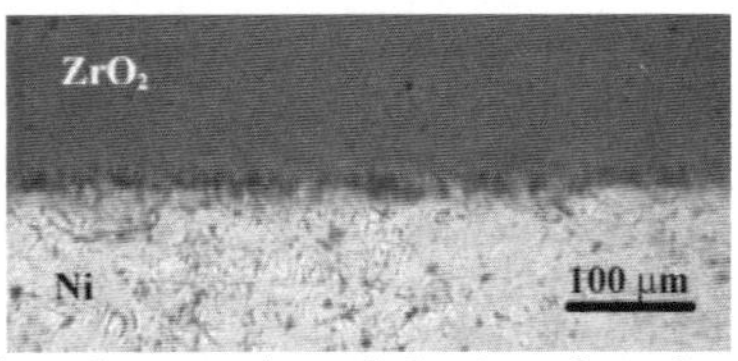

Fig. 14. Microphotography of the interface for nickel drop situated between ZrO_2-ceramic plates, 1 A, 1 min holding before drop disruption, “positive” plate

For the Ni-20Cr drop the interface layers are significantly different (Fig. 15) especially at the “negative” and “positive” plate: at “negative” interface layer is thicker and has another structure than that at “positive”. Reasons of the difference probably the same as for copper case: on the “positive” side dissolution of zirconium occurs and on the “negative” side interface is formed by adsorption and oxidization of this zirconium.

Effects founded were used for brazing of ZrO_2-ceramic to metal. When heating of brazing assemblies current was loaded before the filler melting because used as fillers inert metals do not wet the zirconia therefore the process providing the adhesion of metal to ceramic have to be started immediately at filler melting.

Brazing of ZrO_2-ceramic to molybdenum by copper was tested firstly, amperage 2, 3 and 5 A was used. For all samples filler did not fill brazing gaps fully because even for highest amperage too long time is needed to reach a sufficient wetting of zirconia. So a significant quantity of the filler is squeezed out of the brazing gap immediately after melting. It is necessary to note that a point superheat of ceramic was observed: in some small areas near contact to filler or current lead the glow of ceramic was significantly intensive probably due to non-uniform conductivity of interfaces or the ceramic body. Sometimes superheat has leaded to cracking of ceramic.

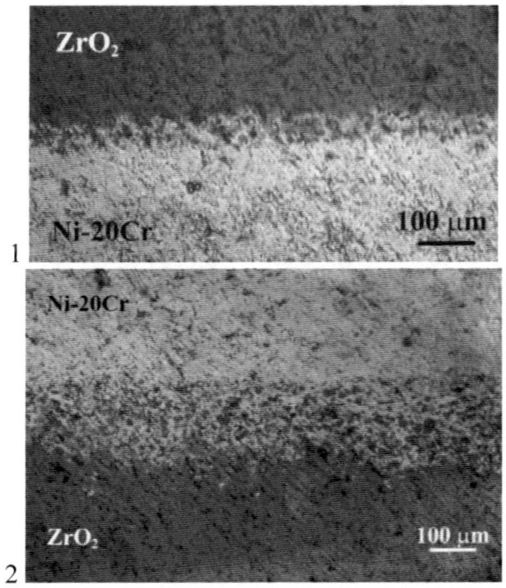

Fig. 15. Microphotographs of interfaces for the Ni-20Cr drop situated between ZrO_2-ceramic plates, 1 A, 3 min holding: 1 – “positive” plate; 2 – “negative” plate

Nickel and Ni-20Cr alloy were tested as fillers to braze ZrO_2-ceramic and molybdenum. At 1 A brazing gap was filled fully during one minute after filler melting. Superheat was not observed. Some dissolution in filler of molybdenum parts of joints is observed, so gaps were filled fast not only because intensive dissolution of zirconium from ceramic in filler but also due to some interaction of melt and molybdenum providing fast filler on molybdenum spreading.

Shear strength of joints was measured. It was 125-163 MPa for nickel filler and 133-176 MPa for Ni-20Cr, destruction occurred fully through ceramic or through ceramic and filler. Weibull plots for joints with nickel and Ni-20Cr filler are represented on Fig. 16 and Fig. 17 respectively.

As the improving of brazing joints quality for cases of nickel and Ni-20Cr in comparison to copper was explained by intensification of chemical processes at higher temperatures and by filler to molybdenum interaction a Cu-45Ni composition with liquidus point 1295 °C [8] was tested as filler for electrical brazing. The results were better than for copper but worse than for nickel and nickel-chromium: in batch of 9 samples for 3 brazing gap was not filled fully, 2 of them ceramic has cracked due to overheat. For last 6 samples shear strength was 110-181 MPa, Weibull plot is represented on Fig. 18.

Brazing of zirconia to heat resistant steel was also studied because such joints are needed in some technical application. Ni-20Cr filler and 1 A were used. Some excess dissolution of steel in filler occurred (Fig. 19). Shear strength was 102-139 MPa, Weibull plot is represented on Fig. 20.

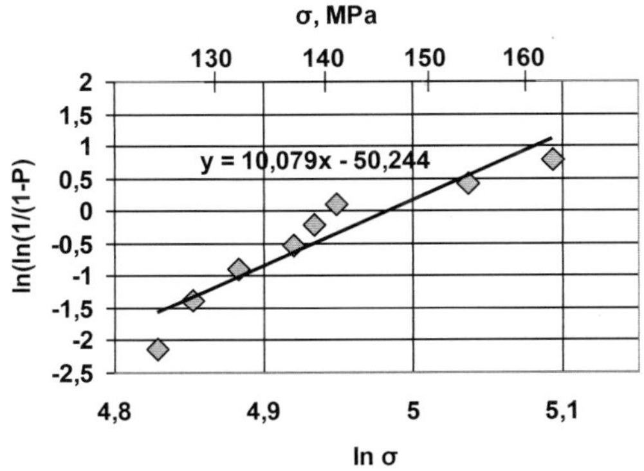

Fig. 16. Weibull plot for shear strength of ZrO_2-ceramic to molybdenum joints brazed using electric current passing with nickel filler

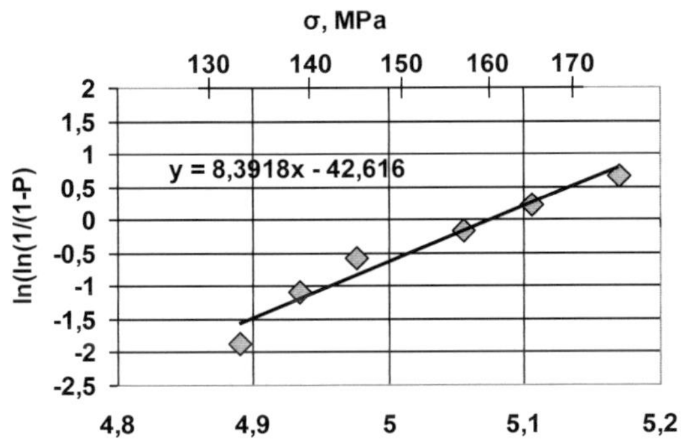

Fig. 17. Weibull plot for shear strength of ZrO_2-ceramic to molybdenum joints brazed using electric current passing with Ni-20Cr filler

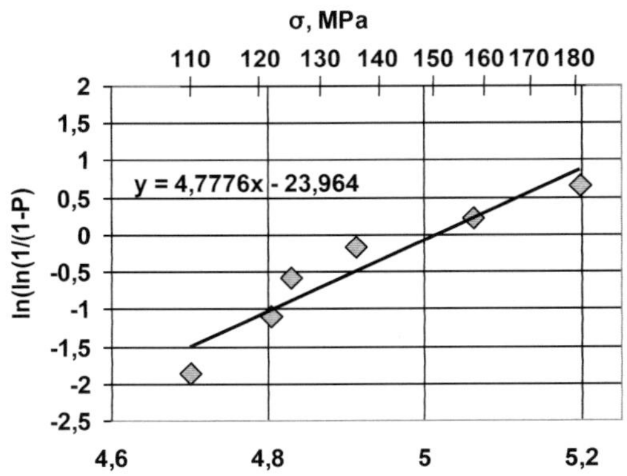

Fig. 18. Weibull plot for shear strength of ZrO_2-ceramic to molybdenum joints brazed using electric current passing with Cu-45Ni filler

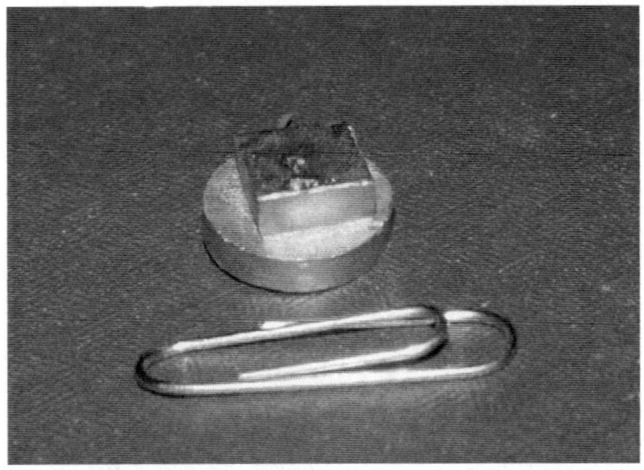

Fig. 19. Sample of ZrO_2-ceramic to heat resistant steel with Ni-20Cr filler using electric current

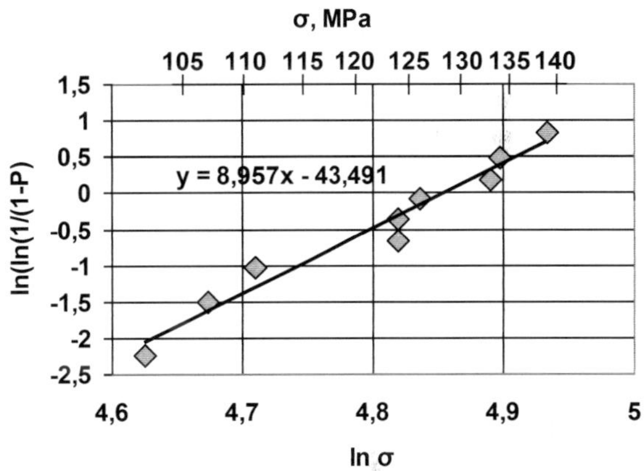

Fig. 20. Weibull plot for shear strength of ZrO_2-ceramic to heat resistant steel joints brazed using electric current passing with Ni-20Cr filler

Conclusions

The effect of electrical current passing through the interface on wetting of zirconia ceramic by metal melts was investigated for copper, copper-gallium, nickel, nickel-chromium melts. When negative electrode is connected to ceramic wetting significantly improves. It was explained by oxygen depleting of zirconia under the current and subsequent dissolution of surplus zirconium in the melt. For Ni and Ni-Cr melt spreading on zirconia occurs much faster than for Cu and Cu-Ga because of the intensification of processes at higher temperatures. When negative electrode connected to zirconia oxygen pass to melt (copper) it also somewhat improves the wetting. Microstructure investigations confirm the dissolution of ceramic in melt and its dependence on amperage.

Improvement of wetting at current passing was used for brazing of zirconia to molybdenum. Cu, Ni, Ni-20Cr, Cu-45Ni were used as fillers. Most comprehensive results were obtained for nickel and nickel-chromium fillers, for copper and copper-nickel incomplete filling of brazing gap and point superheat of ceramic take place. Electrical current was used to braze ZrO_2-ceramic and heat resistant steel with Ni-20Cr filler.

So electric current passing through the interface improve wetting of zirconia by metal melts and may be used for brazing.

References

[1] Naidich, Yu.V. ―The wettability of solids by liquid metals" *Progress in Surface and Membrane Science*, Vol. 14. (1981), pp. 353-487.

[2] Saiz, E., Cannon, R. M., Tomsia, A. P. ―High-Temperature Wetting and the Work of Adhesion in Metal–Oxide Systems" *Annual Rev. Mat. Res.*, Vol. 38. (2008), pp. 197–226.

[3] Peden, C., Kidd, K. B., Shinn, N. D. ―Metal–Metal-Oxide Interfaces: a Surface Science Approach to the Study of Adhesion" *J. Vacuum Sci. Technol.*, Vol. 9. (1991), pp. 1518–1524.

[4] Nogi, Kiyoshi at al. ―Effect of Applying a DC-voltage on the Reaction at the Al–ZrO_2 Interface" *J. Japan Inst. Met.*, Vol. 55, No. 11. (1991), pp. 1269-1273.

[5] Nogi, Kiyoshi ―Effect of applying a DC-voltage on the wettability of zirconia by liquid copper and on copper-circonia joining" *Materials Transactions*, Vol. 31, No. 1 (1990), pp. 83-90.

[6] Durov, O.V. ―Effect of electric current passing through the interface on the wetting of ZrO_2 ceramics with silver–copper–oxygen melt in air" *Powder Metallurgy and Metal Ceramics*, vol. 54, No 3–4 (2015), pp. 201-203.

[7] Durov, A. V., Karpets, M. V., Sydorenko, T. V., Kostyuk, B. D., Naidich, Yu. V., ―The Role of Stoichiometry in Contact Interaction of Zirconia with Metal Melts", *Powder Metallurgy and Metal Ceramics*, Vol. 55 No 9–10 (2017), pp. 612-616.

[8] Cu - Ni, *data from SGTE 2014 alloy database* (Internet resource), http://www.crct.polymtl.ca/fact/phase_diagram.php?file=Cu-Ni.jpg&dir=SGTE2014

Brazing of ferritic stainless steels - A new look

M.H. Biglari, L.C.P. Krassenburg, J.H.G. Brom, T. Hejazi, N.J.A. van Veen, A.A. Kodentsov
Mat-Tech BV, Ekkersrijt 4605, 5692 DM Son, The Netherlands
(www.mat-tech.com)

Abstract

The challenges associated with brazing of ferritic stainless steels span a wide range from recrystallization of the base material to reaction behaviour of a liquid brazing alloy at the contact interface of the joint. To shed some light into this matter, brazing experiments with a variable brazing gap joints were carried out using several Ni-based filler alloys. Wetting behaviour and gap filling capability of the brazing materials on mating surfaces of the ferritic stainless steels AISI 436 and AISI 441 have been evaluated for a set of well-defined experimental conditions.
The product brazing seam microstructure and interfacial reactions occurring during brazing procedures have been examined by means of optical microscopy, Scanning Electron Microscopy (SEM) and Electron-probe Microanalysis (EPMA). From the selected Boron-free Ni-based filler alloys only a (near-) eutectic Ni-based alloy containing 14 wt.% Cr and 10.1 wt. % of P is worthy of further serious consideration for brazing of the AISI 436 and AISI 441 ferritic stainless steels. No recrystallization in the ferritic stainless is commenced upon high-temperature exposure corresponding to the brazing conditions, and no brittle phosphide phases have been detected at the brazing seam/stainless steel substrate interfaces. Instead, all phosphides are found exclusively inside the brazing seam imbedded into (Ni,Fe,Cr)-based solid solution.

Introduction

Relatively high and fluctuating raw material costs for Nickel-containing alloys resulted in the current drive for the replacement of austenitic (Ni-bearing) stainless steels by lower cost ferritic ones for applications in corrosive environments. Obviously, any reduction of Nickel-content in stainless steel components is very attractive because it also significantly reduces cost of the product and provides clear competitive advantage.

Brazing remains an essential step in design and fabrication of various stainless steel structures. However, contrary to austenitic stainless steels, where vacuum brazing with Ni-based filler alloys are already well-developed procedure [1], information about the use of this enabling technology for joining ferritic stainless steel parts is very scarce [2, 3].

To alleviate, at least partially, this situation, the present study is focused on brazing experiments with two different grades of (fully stabilized) ferritic stainless steel, namely AISI 436 or AISI 441 as a base material (see Table 1 [4]), and a number of Boron-free Ni-based filler alloys (Table 2).

Table 1: Composition of the ferritic stainless steels (wt. %) used in the present study as a base material

Grade	Fe	C	Cr	Mn	Si	Nb
AISI 436	Bal.	0.02	17.5	0.25	0.4	0.5
AISI 441	Bal.	≤ 0.011	17.6	0.42	0.5	0.39

	N	P	S	Mo	Ti	
AISI 436	≤ 0.012	0.040	0.030	1.25	-	
AISI 441	≤ 0.011	0.023	0.001	-	0.14	

Table 2: Boron-free filler metals used in the present investigation

Composition, wt.%	Filler Alloy 1	Filler Alloy 2	Filler Alloy 3
Ni	Bal.	Bal.	Bal.
Cr	14	25	29
Si	-	-	4
P	10.1	10	6
Solidus, °C	888	882	
Liquidus, °C	888	949	
Brazing Temperature (°C)	940	1000	1080

The choice of base (substrate) materials is dictated by mainly three factors:

1. To reduce risk for intergranular corrosion (i.e. decrease propensity of the steel experiencing heat-treatment to be "sensitized" to intergranular corrosion), the carbon-content in the base stainless steels is very low, and both materials are alloyed ("stabilized") by Titanium and/or Niobium [5].

2. Higher level of Chromium and absence of Nickel in the steels also results in good corrosion properties at elevated temperatures.

3. Since the product joints are intended for applications in aggressive environments, the presence of Molybdenum in the ferritic AISI 436 grade provides extra resistance of the stainless steel against electrochemical corrosion.

On the other hand, compositions of the brazing alloys listed in Table 2 have been selected because of the following considerations:

- All these materials are Boron-free and Chromium-content in the fillers is relatively high (≥ 14 wt.%), which is an important factor given the corrosion resistance requirements.

- Since ferritic stainless steels are generally much more prone to recrystallization than austenitic ones [6], the brazing temperature, and hence, liquidus temperature of the filler alloy should not be too high.

The present work addresses three main issues:

1. Wetting behavior and gap filling capability of the selected Boron-free Ni-based brazing alloys on the AISI 441 and AISI 436 stainless steel mating surfaces.

2. Examination of the product brazing seam microstructure and interfacial reactions occurring during brazing procedures.

3. Evaluation of merits of the selected filler alloys for brazing of stabilized ferritic stainless steels for application in corrosive environments.

Experimental procedure

Sheets (0.8-3 mm thick) of AISI 436 and AISI 441 steel were used as substrate materials. The cold-rolled steels were annealed at 870-930 °C, pickled and finally cold (temper) rolled (a so-called "skin pass"). The grain size of the initial materials is below 50 μm. As an example, a typical microstructure of the "as-received" AISI 436 ferritic stainless steel is shown in Fig. 1.

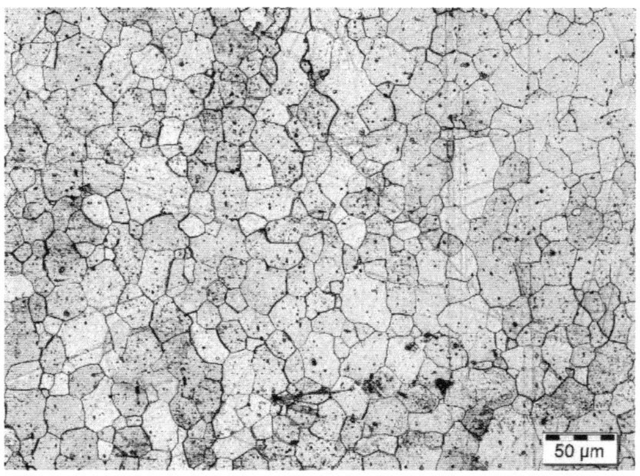

Fig. 1: Microstructure (bright-field optical image) of the "as-received" AISI 436 grade stainless steel ($FeCl_3$+HCl containing aqueous etch).

For evaluation of the merits of the selected brazing alloys (spreading on and wetting of steel surfaces by the filler, gap filling in the brazing assembly, interaction of filler metal with the steel substrate and recrystallization of the base steel) by means of examining brazement cross-sections, the initial steel coupons with dimensions of 40 × 20 mm and 30 × 10 were cut from the "as-received" stainless steel sheets. Variable brazing gap joints with a maximum distance between the mating surfaces in sample design of approximately 150 μm were used in the present investigation.

Prior to the assembling, steel part surfaces were cleaned in isopropanol and air dried. The Boron-free brazing alloys, listed in Table 2, were used in the form of paste. In order to ensure proper positioning of the steel components within the brazed assembly, the end-members were fixed before the applications of brazing paste by means of laser point welding.

Brazing was carried out in vacuum for varying times at the peak brazing temperature, ranging from 10 to 90 minutes. The relatively long holding time at peak temperature (90 minutes) was used to evaluate the merits of the suggested brazing technology for a large-batch fabrication.

For the sake of comparison, several additional experiments have also been made using a Boron-containing brazing alloy with composition 14 Cr; 4.5 Si; 4.5 Fe; 3.25 B; Bal. Ni.

After visual inspection, product test joints were cross-sectioned and after standard metallographic preparations, the samples ("side-views") were examined with Scanning Electron Microscopy (SEM) and Electron-probe Microanalysis (EPMA) using Energy Dispersive Spectroscopy (EDS). In order to study grain structure of the ferritic stainless steel components, the samples were etched in a specially developed aqueous $FeCl_3$+HCl containing reagent, and examined with optical microscopy.

Results and Discussion

Visual inspection of the brazed samples revealed no obvious (macro-)defects: wetting of the steel surfaces by the molten brazing alloys is good and fillet formation has been observed in all brazements. It is also interesting to note that after exposure to the brazing conditions, surface of the AISI 441 steel becomes "yellowish". This discoloration can be attributed to the presence of Titanium (as a "stabilizing elements") in this stainless grade. Typical microstructures of transition zones developed in the stainless steel brazements are shown in a series of Secondary Electron Images presented in Fig. 2.

It is important to realize that all Ni-based filler alloys used in the present investigation have been initially developed for brazing of Ni-based alloys, but not "Ni-free" ferritic stainless steels. Therefore, before proceeding further, it is necessary to give some definitions used in the text. In the context of the present discussion, a space between mating steel surfaces in the initial assembly is termed as "a brazing tolerance" or "a brazing gap". On the other hand, a domain filled with brazing alloy or with the products of the reaction between the molten brazing alloy and the initial steel is defined as "a brazing seam". According to these definitions, the apparent "brazing seam" in the case of the ferritic stainless steel joints brazed with Ni-based fillers is larger than the "brazing gap" set within the initial assembly. This is due mainly to the extensive (interdiffusion) intermixing between constitutes of the molten filler alloy and the steel substrates occurring during the brazing process.

Close inspection of the transition zone revealed that no phosphide formation takes place at the brazing seam/ferritic stainless steel interface in the brazements fabricated with Boron-free filler alloys. Instead, all phosphide phases (which are rather brittle) are found exclusively inside the brazing seam (Fig. 2a, b, c) imbedded into (Ni,Fe,Cr)-based solid solution. It was determined that the phosphide phases are based on the $(Cr,Fe,Ni)_2P$ compound.

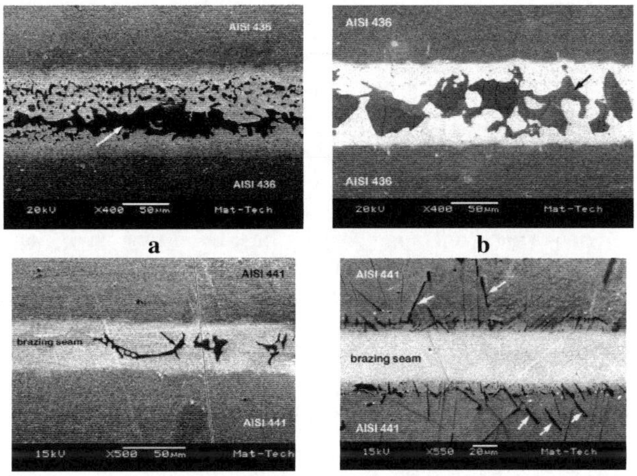

a b

c d

Fig. 2: Secondary Electron Images (SEIs) of the transition zone in the ferritic stainless steel brazements fabricated using different filler alloys: a) AISI 436-based joint brazed with the Filler Alloy 1 for 90 min; b) AISI 436-based joint brazed with the Filler Alloy 2 for 90 min; c) AISI 441-based joint brazed with the Filler Alloy 3 for 10 min and d) AISI 441-based joint brazed with Boron-containing brazing alloy with composition 14 Cr; 4.5 Si; 4.5 Fe; 3.25 B; Bal. Ni at 1135 °C for 10 min. (Intermetallic phases formed during brazing are indicated by arrows.)

All brazing alloys used in this work contain Phosphorous as a liquidus temperature depressant. According to the available phase diagram information (Fig. 3 [7]), solid-state solubility of Phosphorous in Ni and Cr is negligible, but in α-Fe it can reach up to 4.52 at. % at eutectic temperature (Fig. 3c). That is why, some phosphorous was also measured in the (Ni,Fe,Cr)-based solid solution matrix of the brazing seam. However, no harmful Phosphorous segregation along grain boundaries has been detected.

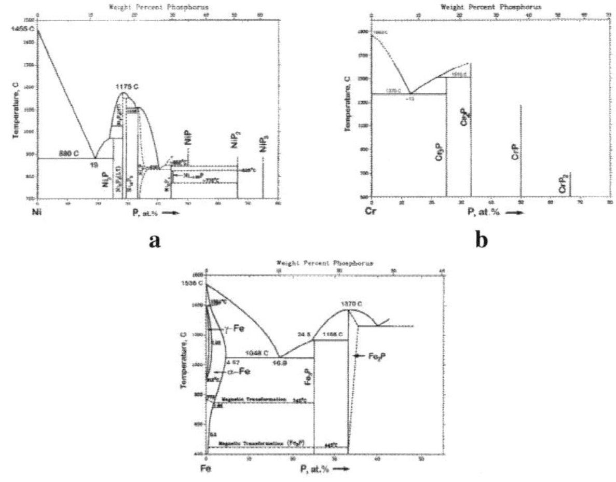

a b

c

Fig. 3 The binary phase diagrams: a) Ni-P; b) Cr-P and c) Fe-P [7].

It is also to be noticed that relative amount of phosphide phases within the product brazing seam is different depending on the Ni-based filler used. The Filler Alloy 3 contains less Phosphorous, viz. 6 wt.% (against 10 wt.% in the Filler Alloy 1 and Filler Alloy 2), and hence, it has lower propensity to form phosphides, which can be appreciated just by comparing micrographs shown in Figs. 2a, b with that given in Fig. 2c.

In contrast, when the Boron-containing filler alloy (14 Cr; 4.5 Si; 4.5 Fe; 3.25 B; Bal. Ni) is used for brazing ferritic stainless steels, the intermediate phases (Cr-based borides) are found in the transition zone in the form of "needles" along the ferritic stainless steel/brazing seem interface. Some boride formation is also observed inside grains of the ferritic stainless steel substrate (Fig. 2d). Obviously, the presence of the brittle borides in the vicinity of the base steel/ brazing seem interface may have a deleterious effect on thermo-mechanical stability of the brazed assembly.

During high-temperature exposures corresponding to the brazing with the Filler Alloy 2 and Filler Alloy 3, recrystallization and grain growth in the base ferritic stainless steels has been commenced. The latter can be best understood by looking at the microstructures of the base stainless steel before and after brazing presented in Fig. 1 and Fig. 4a.

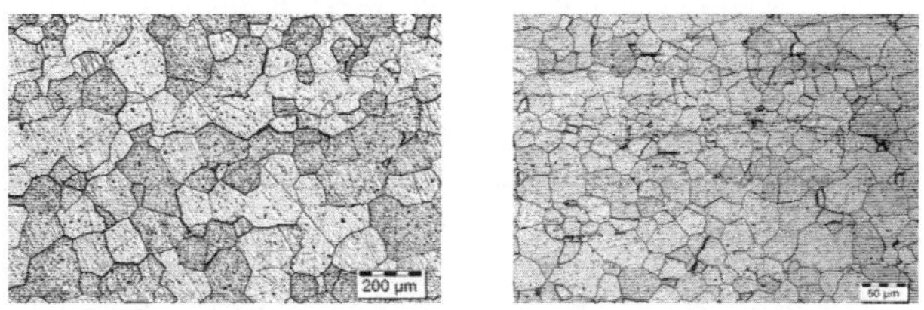

a b

Fig. 4 Microstructure (bright-field optical image) of the AISI 436 stainless steel component after brazing a) with Filler Alloy 2 at 1000 °C for 90 min and b) with Filler Alloy 1 at 940 °C for 90 min ($FeCl_3$+HCl containing aqueous etch).

It can be seen that a significant grain coarsening of the base stainless steel occurs during the brazing at 1000 °C using Filler Alloy 2. The grain size of the ferritic stainless steel component in the brazed assembly is one of the key parameter affecting overall performance of the product. The excessive grain growth in the base steel upon brazing has a detrimental effect on mechanical and structural integrity of the assembly, especially when thin-walled items are involved in the product design. Therefore, from this point of view, one can state that Filler Alloy 2 and Filler Alloy 3 (because of even a higher, viz. 1080 °C, brazing temperature is required) are not suitable filler metals for brazing of AISI 436 and AISI 441 stainless steel parts. The processing temperature is too high, and as a result, recrystallization and subsequent grain growth in the steel occurs.

On the contrary, no recrystallization of base ferritic stainless steel occurs under conditions corresponding to the brazing with the Filler Alloy 1. This can be appreciated by comparing microstructure of the stainless steel in "as-received" conditions (Fig. 1) with that that developed during brazing using this (near-) eutectic filler alloy (Fig. 4b).

Concluding Remarks

- Brazing of ferritic stainless steels for applications in corrosion-prone environments represents a number of specific features with regard to the filler alloy selection and the possibility to alter microstructure of the base material.

- There are no well-established brazing technologies allowing for joining below "re-crystallization" temperature of the fully stabilized ferritic stainless steels AISI 436 and AISI 441.

- No phosphide formation takes place at the brazing seam/ferritic stainless steel interface in the AISI 436- or AISI 441-base brazements fabricated using the selected Boron-free filler alloys with compositions listed in Table 2. Instead, all (brittle) phosphides are found exclusively inside the brazing seam imbedded into (Ni,Fe,Cr)-based solid solution, and no harmful Phosphorous segregation along grain boundaries has been detected.

- From the selected Boron-free Ni-based fillers only the (near-) eutectic Filler Alloy 1 is worthy of further serious consideration for brazing of the AISI 436 and AISI 441 ferritic stainless steels. No recrystallization in the ferritic stainless occurs upon high-temperature exposure corresponding to the brazing conditions.

- On the contrary, high-temperature exposure of the ferritic stainless steels AISI 436 and AISI 441 to the conditions of brazing with Filler Alloy 2 or Filler Alloy 3 results in recrystallization and grain growth in the steel.

- Brazing of the AISI 436 and AISI 441 ferritic stainless steels with the (near-) eutectic Filler Alloy 1, can be accomplished in vacuum at 940 °C. However, in order to avoid an excessive ingress of Iron from the steel substrate into molten filler alloy, brazing time should be kept as short as possible.

- Width of the brazing tolerance (brazing gap) suitable for brazing with the (near-) eutectic Filler Alloy 1 was found to be 0-40 μm. For the joining of components with larger clearances between the mating surfaces, a wide-gap brazing using Ni-Cr based alloy powder as a filler can be suggested [8].

References

[1] *Brazing Handbook*, 4th Edition, American Welding Society, Florida, 1991

[2] Y. Miyazawa, K. Saito, T. Fukikoshi, M. Abe, T. Ariga, J. Hiratsuka and T. Takizawa; Brazing of ferritic stainless steel to austenitic stainless steel, Proc. 10th Int. Conf. LÖT 2013, Aachen, Germany, DVS-Berichte Band 293, pp. 83-86

[3] M. Marsilius and T. Hartmann, Influence of boron and phosphor containing nickel based brazing alloy on different base materials, Proc. 11th Int. Conf. LÖT 2016, Aachen, Germany, DVS-Berichte Band 325, pp. 227-232

[4] J.R. Davis (Ed.), ASM *Speciality Handbook: Stainless Steels,* ASM International, Materials Park, Ohio, 1994

[5] D.A. Jones, *Principles and Prevention of Corrosion*, Macmillan Publishing Company, 1992

[6] G. Krauss, *Steels: Heat Treatment and Processing Principles*, ASM International, Materials Park, Ohio, 1990

[7] T.B. Massalski, *Binary Alloy Phase Diagrams*, ASM, Metal Park, Ohio, 1986

[8] E. Lugscheider, V. Dietrich and J. Mittendorff, Wide Joint Clearance Brazing with Nickel- Base Filler Metals, *Welding Research Supplement*, February (1988) 47s-51s

Residual Stress Reduction of Brazed WC-Co/Steel

Tatsuya Sasaki, Takumi Goto, Yasuyuki Miyazawa,
Tokai University, Hiratsuka-shi, Kanagawa-ken, Japan
E-mail: yamiyazawa@tokai-u.jp
TEL: +81-463-58-1211

Tatsuo Takato,
MMC RYOTEC, Godo-cho, Gifu-ken, Japan

Abstract

Brazing of WC-Co and steel material with Ni based brazing filler metal is necessary in fabricating fields of tools. However, crack in WC-Co caused by residual stress due to difference of thermal expansion coefficient between WC-Co and the steel.
In this research, size of unjointed area at the brazed layer is important to reduce the residual stress. The residual stress becomes smaller with the decrease in joining area size [2]. Therefore, we attempted reduce the residual stress by intentionally creating the unjointed area at the joint interface. WC-6mass%Co(Grain size:1.5μm) and S45C steel were employed as base materials. Each shape is round bar of φ9.2×20 mm. MBF-20 was employed as a filler metal. Filler metal was processed like a donuts shape with inner diameter φ6 mm. After brazing, the specimen was analyzed by metallographic observation, hardness test, and interfacial analysis with electron probe micro analyzer. The result shown that, it was possible to control the shape of unjointed part by applying the stop-off, WC-Co brazed joint was obtained with no crack due to the residual stress was reduced. Furthermore, W diffused and concentrated to vicinity the joint interface. The diffused W formed compounds with Cr and B. In addition, we compared to WC-Co of grain size 1.5 μm and 0.8 μm for obtain knowledge about interface reaction.

Introduction

WC-Co has good wear resistance and hardness, it has been mainly used for cutting tools. On the other hand, it is relatively expensive, so it is expected to be used with inexpensive steel materials. In this case, joining by brazing is effective, but that causes a residual stress due to difference in thermal expansion coefficient between the base materials. Residual stress reduction has been trying using Ag based filler metal having low liquidus temperature, but processing temperature after the brazing is limited. However, when WC-Co is brazed with Ni based filler metal having high liquidus, crack occurs in WC-Co.
Nevertheless, there are almost no report on brazing of WC-Co with Ni based filler metal. Therefore, we aimed to reduce the residual stress in brazed WC-Co/steel with Ni based filler metal, and obtain the joint without cracks.
Hence, we paid attention to area at the joint interface as a means of residual stress reduction. We attempted to reduce the residual stress by intentionally created unjointed area at the joint interface because residual stress becomes smaller with decrease in joining area of the brazed material.
We processed filler metal to square and cruciform in our previous research. These has succeeded to form unjointed part slightly but there was a problem that micro cracks occurred, filler metal was not filled in end portions.
In this research, filler metal was processed to like a donuts shape. Further, stop-off was applied on base material.

Experimental

Specimen
WC-6mass%Co(grain size:1.5μm) and S45C steel were employed as base material. Each shapes were φ9.2×20 mm round bar. S45C was polished with sandpaper(2000 Grit). MBF-20 was employed as Ni based brazing filler metal. Table 1 shows chemical composition, liquidus, and solidus. Filler metal was processed to shape of 9.2 mm diameter and 6mm internal diameter like a donuts shape. Stop-off was applied on base materials after ultrasonic cleaning with acetone, and which was set to jig that applies pressure to longitudinal even at high temperature. Schematic diagram of filler metal shape and specimen is shown in Fig. 1. We used only usual circle shape of 9.2 mm diameter in experiment comparing WC-Co with different grain sizes because we only focused on observing interface reaction.

Table 1 Chemical Composition(mass%), Solidus and Liquidus of the filler metal

Filler metal	Chemical Composition (mass%)					Temperture (°C)	
	Cr	Fe	Si	B	Ni	Solidus	Liquidus
MBF-20	6.00~ 8.00	2.5 ~ 3.5	4.00~ 5.00	2.75~ 3.5	Bal.	969	1024

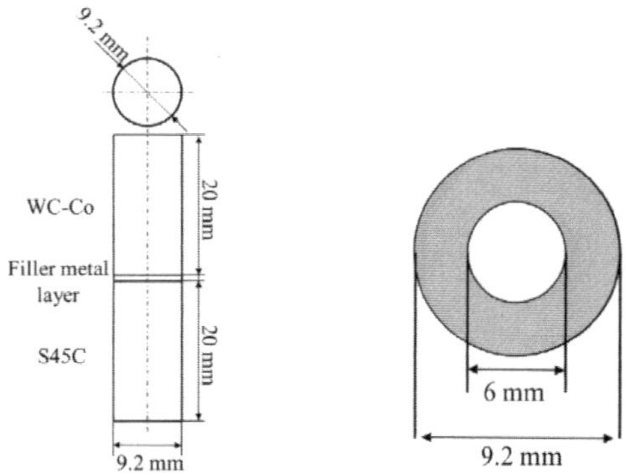

Fig. 1 Schematic diagram of specimen(left) and filler metal(right).

Fig. 3 SEM image of brazed WC-Co(1.5μm)/S45C

Brazing process

Specimen was brazed at 1024 °C with vacuum furnace after camber was evacuated until degree of vacuum reached under 3×10^{-3} Pa. The heating rate was 10 K/min and the cooling rate was 1 K/min. Brazing time was 1minute. After brazing, specimen was cut and polished. Then the cross-sectional microstructure was observed by optical microscope and scanning electron microscope. Hardness in WC-Co, reaction phase, and filler metal layer were examined by Vickers test machine.

Results and discussions

Microstructure and hardness distribution

The typical cross-sectional microstructure of brazed WC-Co(1.5μm)/S45C was shown in Fig. 2. The unjointed part was formed at center of the joint interface that introduced by stop-off. In previous experiments, crack occurred in the WC-Co when Ni based brazing filler metal was used. However, in this experiment, it was confirmed that crack did not occur as a result of formation of unjointed part. Fig. 3 shows cross-sectional micro structure of brazed WC-Co(1.5μm)/S45C observed by SEM. Observation was made at edge part because it tends to be the origin of crack. The dark area at the interface was observed due to dropout of the structure.

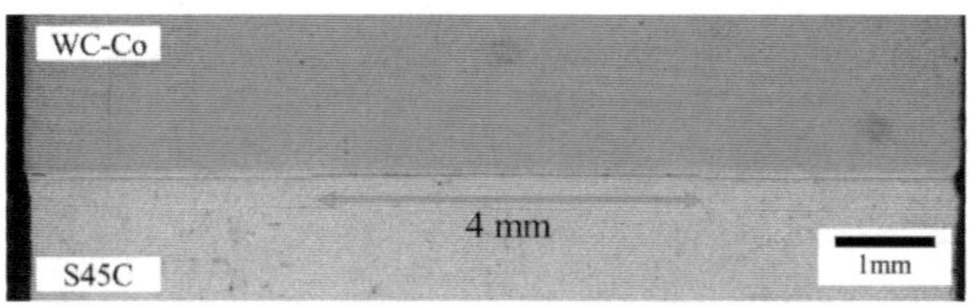

Fig. 2 Cross-sectional microstructure brazed WC-Co(1.5μm) /S45C

Degree of residual stress reduction was examined using the Vickers hardness machine. In this Vickers hardness test method, crack may occur from the edge of the indentation as the load increases; we considered the measurement of brittleness is possible using the crack length. Vickers hardness and average crack length of WC-Co is shown in Fig. 4. There was no much difference in hardness depending on the test points. Crack length also did not much differ in test point of 1 to 6, but it was larger in vicinity of joint interface(test point of 7 to 9). Especially, crack length at center of joint interface, It is assumed that the residual stress concentrates center part, not in the edge part where it tends to be the starting point of the cracks.

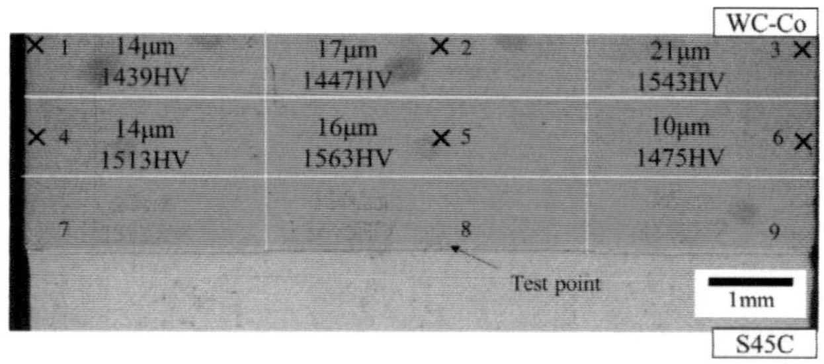

Fig. 4 Vickers hardness and average crack length of WC-Co

Elemental distributions and interface reaction

The EPMA result of brazed WC-Co(1.5μm)/S45C is shown in Fig. 5. B contained in the filler metal generally tends to diffuse to the base material. In addition, Co diffused to brazed layer side; therefore, low Co layer was formed.

In a part of the low Co layer, W also diffused and concentrated to vicinity the interface. It is assumed that Ni and Si which are difficult to make solid solution with W diffused and formed a compound. Furthermore, B and Cr were distributed in the same place as W. According to the W-Cr-B ternary phase diagram shown in Fig. 6, it is assumed that the compound composed of these three elements is a solid solution that Cr dissolved to BW_2.

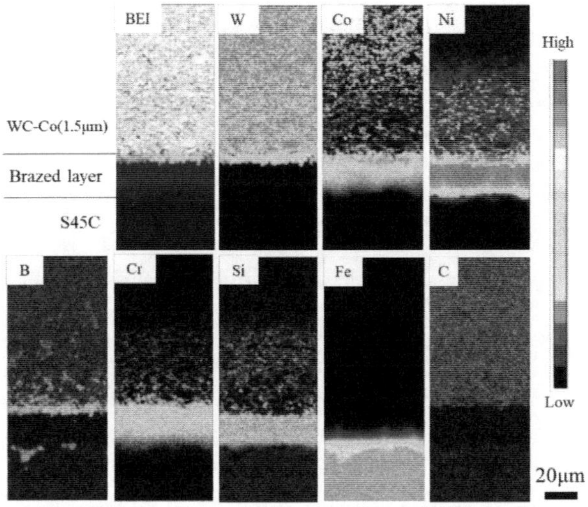

Fig. 5 Result of brazed WC-Co(1.5μm) /S45C by EPMA

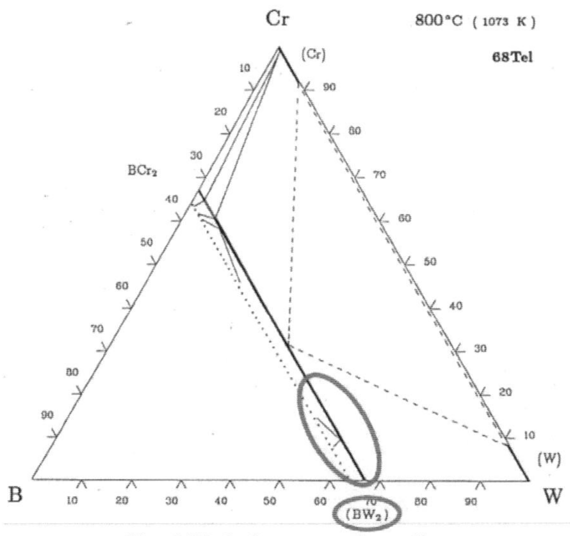

Fig. 6 W-Cr-B ternary phase diagram

Hardness of compound layer and brazed layer were examined by micro vickers hardness test. Result of micro vickers hardness test is shown in Fig. 7. Hardness of WC-Co was about 1500 HV. On the other hand, hardness of compound layer was 1143 HV, and hardness of brazed layer were 313 and 252 HV. We considered the values of hardness were obtained due to the different amounts of diffused Co with the result of EPMA. Compound layer was hard as compared to filler metal layer. The hardness of the brazed layer was harder as closer to WC-Co.

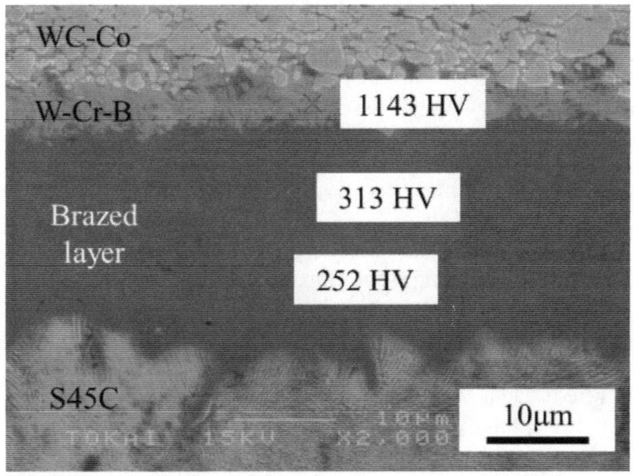

Fig. 7 Result of micro vickers hardness test on W-Cr-B composition and brazed layer.

We describe the reaction of brazed WC-Co(0.8μm)/S45C. The EPMA result of brazed WC-Co(0.8μm)/S45C is shown in Fig. 8. The diffusion length of Ni in WC-Co(1.5μm) was 60μm, WC-Co(0.8μm) was 116μm. B and Cr were also distributed in a wider range than WC-Co(1.5μm). It was established that diffusion length longer as the grain size decrease. Furthermore, W-Cr-B compound phase is a factor of lowering the bonding strength. We confirmed the compound phase behave as origin of cracks several times during the observation of cross-sectional microstructure. B and Cr diffuse to some extent toward the base material, hence we speculated that the bonding strength was relatively high at WC-Co(0.8μm).

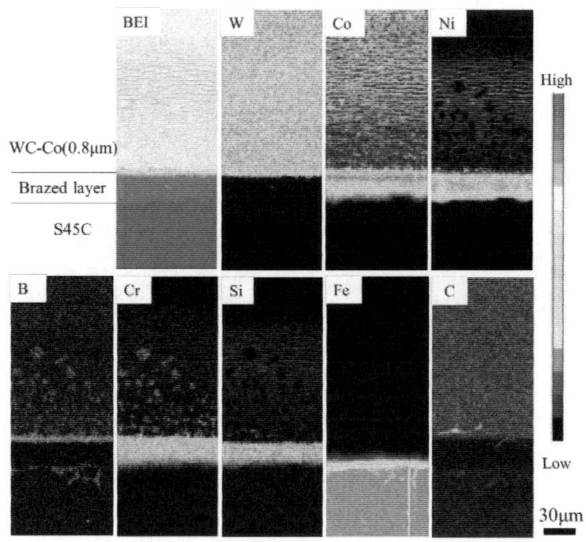

Fig. 8 Result of brazed WC-Co(0.8μm)/S45C by EPMA

Conclusions

This research attempted residual stress reduction as introduce unjointed part at center of joint interface. From results of vickers hardness test and analysis of the interface reaction between Ni filler metal and WC-Co by EPMA, this research can be concluded as follows:

(1) Unjointed parts was formed by processing the filler metal shape and applying stop-off to the joint interface.
(2) Residual stress was reducted by forming the unjointed part, and cracks were suppressed.
(3) Residual stress concentrates at the center, not at the end where it tends to be the origin of the crack.
(4) In the vicinity of the interface, the W-Cr-B compound phase is formed, which may adversely affect the bonding strength.

References

1) Rintaro Sakashita, Takumi Goto, Yasuyuki Miyazawa and Tatsuo Takato, TokaiUniversity, Effect of brazed layer thickness on interfacial reaction between WC-Co
2) Katsuaki Suganuma, etc,Joining of metal and ceramics,Uchidarokakuho (1990)p.1-15 p95-105

Braze Powder Characteristics and Their Effects on Brazing

Michael Graham
Prince and Izant Companies, Cleveland, Ohio, United States
mgraham@princeizant.com

Abstract

Braze powder is one of the most common forms of brazing filler metal used across all industries, yet the powder characteristics which impact brazing are rarely examined and published for end-users to optimize their applications. This paper identifies key characteristics of powder (morphology, particle size distribution, homogenous or discrete phase(s) composition, rheology, melt-range), how to measure these characteristics, and their impact on the behavior of the product. Two braze alloys from two manufactures were evaluated using high magnification, elemental mapping by x-ray energy dispersive spectroscopy (EDS), scanning electron microscope (SEM), laser-diffraction, differential thermal analysis (DTA), inductively coupled plasma (ICP) and combustion analysis (Leco) to draw correlations between behavior and characteristics. The results indicate no compositional difference, or noticeable difference in melt-temperature due to particle size. However, oxygen content increased by weight-percent as the particle size decreases. The level of oxygen increase varies by manufacturer, and by braze alloy. These results can be used to optimize the braze powder parameters which provide the highest quality and most economical results for a desired application.

Introduction

The intent of this paper is to review the characteristics of braze powder to educate users on the factors that influence the behavior of their powder and paste in different braze applications. The characteristics being evaluated include the Particle Size Distribution (PSD), morphology (shape), rheology (flow), melt-range and homogeneity of composition (phases). Many of these characteristics are well known within the powder production community as drivers of performance and costs for braze powders and paste, but are not widely known by the end-users. In this experiment, the technical requirements of AWS and AMS will be evaluated in attempt to determine their validity, and impending influence on the manufacturing, pricing structure, and application of braze powder.

Powder Specification Requirements and Experiment Setup

The AWS & AMS specifications which provide the technical requirements for braze powder are shaped around several key components:

1. Uniform composition throughout an entire melt or furnace-charge(s) of powder, regardless of particle size.
2. The impact of PSD on brazing; either through application or performance. This is derived from the tables listing the PSD requirements that limit the weight percent of particles that pass through a No. 325 screen for designations larger than 325 mesh. The 325 mesh allows for unlimited particles through a No. 325 sieve, with no restriction on smaller sizes.

Prince & Izant (P&I) evaluated powder from several manufactures from 2012 to 2018 including nickel-based (BNi), silver-based (BAg), and gold-based (BAu) powders. The results showed that not all gas atomized powders should be treated equally. Through evaluations of AWS BNi-5 (AMS 4782), PAL36, and pure gold powders in a sintering project conducted by Prince & Izant in 2015 it was discovered that the following properties varied between alloys and manufacturers:

1. Particle Size Distribution (PSD): Although material can be ordered to a mesh designation identified in both AWS A5.8M/A5.8:2011 (Specification for Filler Metals for Brazing and Braze Welding), and AMS (Aerospace Materials Specification), the PSD within the allowed limits still varies, with the widest variation occurring below the No. 325 sieve.

2. Homogeneity of particle compositions: It was observed that while the bulk composition always met the requirements listed by AWS & AMS when analyzed by approved testing methods (most commonly ICP), individual particles often exhibited variations of the overall composition. These variations were identified as discrete-phases. When these individual particles were lumped together to produce a mass of substantial weight for compositional analysis by ICP and ICP-MS their composition was within the specification limits.

3. Morphology: The most common process of braze powder production is gas-atomization. This generally produces spherical particles. However, the nozzle and overall reactor design can influence the roundness of particles leading to variation in shape between manufacturers.

These properties are important to brazing. Particle Size Distribution (PSD) is one of the primary cost drivers for any powder. The AWS & AMS specifications provide some standard mesh designations, but ultimately leave the PSD to the discretion of the end-user. The lack of sufficient data and wide spread knowledge on powder lead most end-users to select a standard mesh designation from one of the specifications. It has been the experience of Prince & Izant that the coarse distributions, (140C, 140F, 200 Mesh) are the most commonly selected from the AMS & AWS specifications. Each of these mesh designations have a maximum limit by weight percent on the amount of particles which pass through a No. 325 sieve. This reduces the overall sellable yield of the lot and thereby increases cost. This experiment focused on gold-based (BAu) alloys due to the availability of internally manufactured powders by P&I, as well as limited availability of published data on these material sets. Two gold-based braze powders from two manufacturers were tested to determine if there is a difference between composition, compositional homogeneity, morphology, melt-temperature, and oxygen levels based on the PSD which justifies the limitation of fine particles using rationale other than application or end-user preference. Table 1 shows how each characteristic was evaluated.

Characteristic	Tests	Conducted By:
PSD Classification	ASTM B214	Prince & Izant
Composition	According to AWS & AMS Standards	NSL Analytical
Composition Homogeneity	EDS (X-Ray Energy Dispersive Spectroscopy)	CWRU - Center for Surface Analysis of Materials
Morphology	Visual by SEM (Scanning Electron Microscopy)	CWRU - Center for Surface Analysis of Materials
Melt Temperature	DTA (Differential Thermal Analysis)	Prince & Izant
Oxygen Levels	EDS & Combustion (LECO)	NSL Analytical

Table 1: Testing of Powders

Table 2 shows the test matrix which details the tests that were performed on each of the PSDs by manufacturer.

SAMPLE	Alloy	Manufacturer	PSD	DTA (A)	Composition + Oxygen (B)	SEM + EDS (C)
1	Au82/Ni18	P&I	US 140F	X	X	X
4	Au82/Ni18	P&I	-100 / +140	X	X	X
5	Au82/Ni18	P&I	-140 / +200	X		
6	Au82/Ni18	P&I	-200 / +325	X		
7	Au82/Ni18	P&I	-325 / +400	X	X	X
8	Au82/Ni18	P&I	-400 / +500	X		
9	Au82/Ni18	P&I	-500 / +635	X		
10	Au82/Ni18	P&I	-635 / +850	X		
11	Au82/Ni18	P&I	-850	X	X	X
12	Au82/Ni18	Vendor B	US 140F	X	X	X
13	Au82/Ni18	Vendor B	-100 / +140	X	X	X
14	Au82/Ni18	Vendor B	-325/ +400	X	X	X
15	Au82/Ni18	Vendor B	-850	X	X	X
SAMPLE	Alloy	Manufacturer	PSD	DTA (A)	Composition + Oxygen (B)	SEM + EDS (C)
16	Au70/Ni22/Pd8	P&I	US 140F	X	X	X
19	Au70/Ni22/Pd8	P&I	-100 / +140	X	X	X
22	Au70/Ni22/Pd8	P&I	-325 / +400	X	X	X
26	Au70/Ni22/Pd8	P&I	-850	X	X	X
27	Au70/Ni22/Pd8	Vendor B	US 140F	X	X	X
28	Au70/Ni22/Pd8	Vendor B	-100 / +140	X	X	X
29	Au70/Ni22/Pd8	Vendor B	-325/ +400	X	X	X
30	Au70/Ni22/Pd8	Vendor B	-850	X	X	X

Table 2: Testing Matrix

Powder was evaluated from two separate manufactures to determine if small differences in process or manufacturing impact the behavior of the powders being tested. The manufacturers include P&I, and a second company identified as Vendor B. The braze filler metals evaluated were BAu-4 (AMS 4787) and BAu-6 (AMS 4786). The compositional requirements are shown in Table 3.

AWS	AMS	Au	Ni	Pd	*TAO
BAu-4	4787	81.5-82.5	Remainder	-	0.15
BAu-6	4786	69.5-70.5	21.5-22.5	7.5-8.5	0.15

*TAO = Total of All Other Elements

Table 3: Composition Requirements

All vendor samples by alloy were collected from the same Heat/Lot of powder with starting compositions and mesh designations (140F) that are fully compliant to the AWS & AMS specifications. The starting PSD was then classified using a mechanical sieve shaker by settings compliant with ASTM B214. Starting sieve analysis of each lot of powder is listed in Table 4.

Starting Powder				Starting Sieve Analysis (wt. %)			
Manufacturer	AWS	LOT No.	Mesh Designation	+100	-100/+140	-140/+325	-325
P&I	BAu-4	BAu-4-PI	140F	0.0%	10.0%	35.0%	55.0%
Vendor B	BAu-4	BAu-4-VB	140F	0.0%	4.0%	55.4%	40.6%
P&I	BAu-6	BAu-6-PI	140F	0.0%	10.0%	35.0%	55.0%
Vendor B	BAu-6	BAu-6-VB	140F	0.5%	4.3%	56.5%	38.7%

Table 4: Start Sieve Analysis

P&I analyzes the PSD of each atomization run utilizing laser-diffraction, prior to classification by size to evaluate the overall yield and make process improvements as required to optimize the sellable yield. Images 1 and 2 show the full PSD of Lot No. BAu-4-PI prior to sieving. These images show that the yield which should pass through a No. 325 Mesh screen (roughly 45 microns) is around 70%. If the entire lot of powder is to be sold as mesh designation 140F which has a maximum of 55% allowed to pass through a No. 325 mesh, the sellable yield is reduced, and must carry the fabrication costs of the remaining un-sold powder.

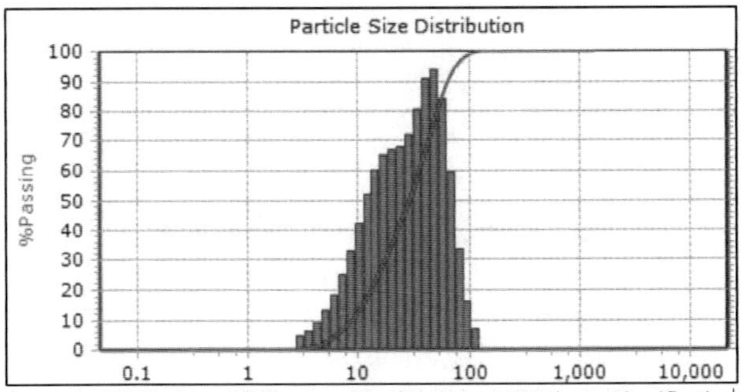

Image 1: PSD by Laser Diffraction of Lot No. BAu-4-PI Prior to Classification

As can be seen in Image 2, nearly 11% of the powder produced from gas-atomization in Lot No. BAu-4-PI is smaller than 10 microns (through a No. 850 sieve). After reviewing PSD data for 20 atomization runs by P&I, it was determined that each production run yields particle sizes from 150 microns to below 10 microns and therefore these tests focused on particles in these size ranges. The powders were classified using Humboldt U.S.A Standard Sieves, with the appropriate quantities being collected for each test performed.

Summary Data | Tabular Data | Notes/Warnings | Raw Data | Setzero

Distribution: Volume   Progression: Standard

Size(um)	%Chan	%Pass	Size(um)	%Chan	%Pass	Size(um)
1408	0.00	100.00	3.27	0.43	0.43	
1184	0.00	100.00	2.750	0.00	0.00	
995.6	0.00	100.00	2.312	0.00	0.00	
837.2	0.00	100.00	1.945	0.00	0.00	
704.0	0.00	100.00	1.635	0.00	0.00	
592.0	0.00	100.00	1.375	0.00	0.00	
497.8	0.00	100.00	1.156	0.00	0.00	
418.6	0.00	100.00	0.972	0.00	0.00	
352.0	0.00	100.00	0.818	0.00	0.00	
296.0	0.00	100.00	0.688	0.00	0.00	
248.9	0.00	100.00	0.578	0.00	0.00	
209.3	0.00	100.00	0.486	0.00	0.00	
176.0	0.00	100.00	0.409	0.00	0.00	
148.0	0.00	100.00	0.344	0.00	0.00	
124.5	0.70	100.00	0.2890	0.00	0.00	
104.7	1.62	99.30				
88.00	3.39	97.68				
74.00	5.96	94.29				
62.23	8.39	88.33				
52.33	9.39	79.94				
44.00	9.08	70.55				
37.00	8.04	61.47				
31.11	7.19	53.43				
26.16	6.80	46.24				
22.00	6.67	39.44				
18.50	6.49	32.77				
15.56	5.99	26.28				
13.08	5.18	20.29				
11.00	4.23	15.11				
9.25	3.32	10.88				
7.78	2.51	7.56				
6.54	1.83	5.05				
5.50	1.29	3.22				
4.62	0.89	1.93				
3.89	0.61	1.04				

No. 325 Sieve has openings equal to 44 microns

No. 850 Sieve has openings equal to 10 microns

Image 2: PSD by Laser Diffraction of Lot No. BAu-4-PI Prior to Classification

Compositional Analysis by Particle Size

All lots being examined have been tested and certified to conform to the AWS and AMS compositional requirements. After the lots were sieved, three PSDs were selected for compositional testing:

1. -100/+140
2. -325/+400
3. -850

All PSDs were evaluated by NSL Analytical using ICP, ICP-MS, and Leco, and found to conform to the AWS & AMS compositional requirements, with little to no variation within a lot of powder. The results of compositional analysis are shown in Table 6.

		Weight %			
AWS BAu-4 / AMS 4787		Au	Ni	Pd	TAO
Lot	PSD	81.5-82.5	Remainder	-	0.15
BAu-4-PI	140F	82.30	Bal.	-	<.10
BAu-4-PI	-100/+140	82.10	Bal.	-	<.10
BAu-4-PI	-325/+400	82.10	Bal.	-	<.10
BAu-4-PI	-850	82.20	Bal.	-	<.10
BAu-4-VB	140F	81.66	Bal.	-	<.10
BAu-4-VB	-100/+140	81.80	Bal.	-	<.10
BAu-4-VB	-325/+400	81.70	Bal.	-	<.10
BAu-4-VB	-850	81.80	Bal.	-	<.10
AWS BAu-6 / AMS 4786		Au	Ni	Pd	TAO
Lot	PSD	69.5-70.5	21.5-22.5	7.5-8.5	0.15
BAu-6-PI	140F	70.07	22.00	7.92	
BAu-6-PI	-100/+140	70.18	21.86	7.95	<.10
BAu-6-PI	-325/+400	70.18	21.88	7.92	<.10
BAu-6-PI	-850	70.20	21.86	7.90	<.10
BAu-6-VB	140F	69.69	22.49	8.02	<.10
BAu-6-VB	-100/+140	69.87	22.19	7.92	<.10
BAu-6-VB	-325/+400	69.88	22.14	7.95	<.10
BAu-6-VB	-850	69.84	22.16	7.91	<.10

Table 6: Compositional Analysis Results by PSD

Compositional Homogeneity by Particle Size

Samples classified by PSD were evaluated at the Center for Surface Analysis at Case Western Reserve University using EDS (x-ray energy dispersive spectroscopy carried out mostly at 15kV) to determine if there are noticeable trends of discrete phases (variation in composition) from particle to particle. The EDS compositions are not exact, but sufficient to identify noticeable differences and trends. The bulk compositional analysis performed by NSL cannot target the composition of a single particle due to the lack of mass. Therefore, EDS was used to examine composition on a particle level. The results show that discrete phases do not exist from one particle to the next on either alloy for all PSDs examined.

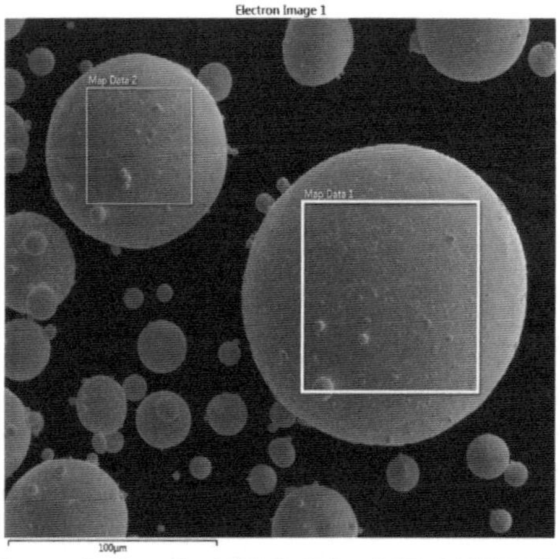

Image 3: Sample 1 EDS: BAu-4-PI: 140F PSD

Image 7: Sample 4 EDS: BAu-4-PI: -100/+140 PSD

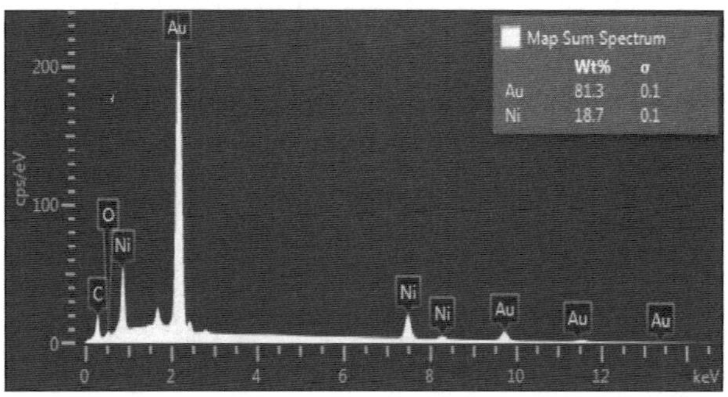

Image 4: EDS Results for Sample 1: BAu-4-PI: 140F PSD

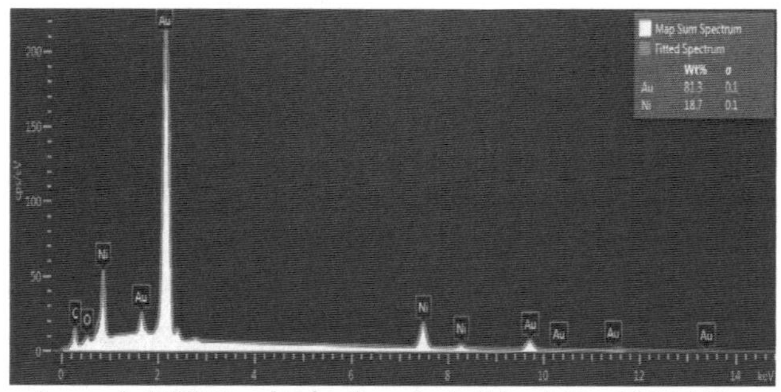

Image 8: EDS Results for Sample 4: BAu-4-PI: -100/+140 PSD

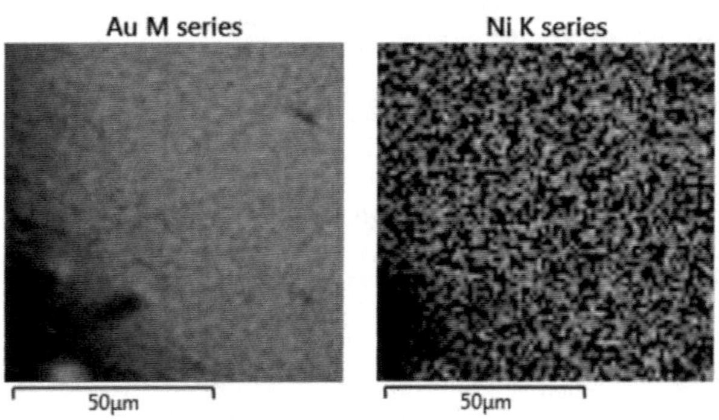

Image 5 & 6: EDS Results for Sample 1: BAu-4-PI: 140F PSD

Image 9: Particles for Sample 11 EDS: BAu-4-PI: -850 PSD

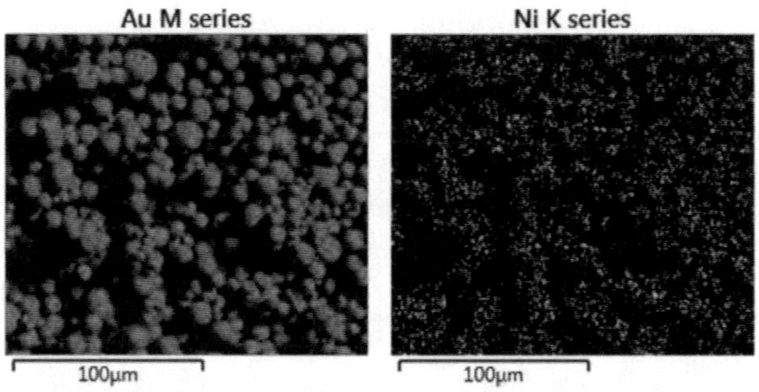

Image 10 & 11: EDS Sample 11: BAu-4-PI: -850 PSD

When examining on a single particle level, each particle demonstrates the same composition. When comparing a group of particles side by side, all particles show a uniform composition which matches the requirements of AWS & AMS. When comparing the coarse particles (-100/+140) to the fine particles (-850), there is not a difference in compositional homogeneity or in bulk compositional results. EDS was performed on several individual particles which were then analyzed as a collective group. The images below show the particles analyzed and their results.

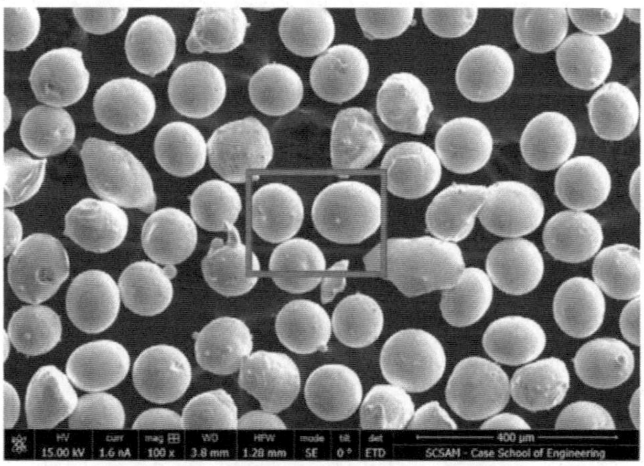

Image 12: Particles for Sample 13 EDS:
BAu-4-VB: -100/+140 PSD

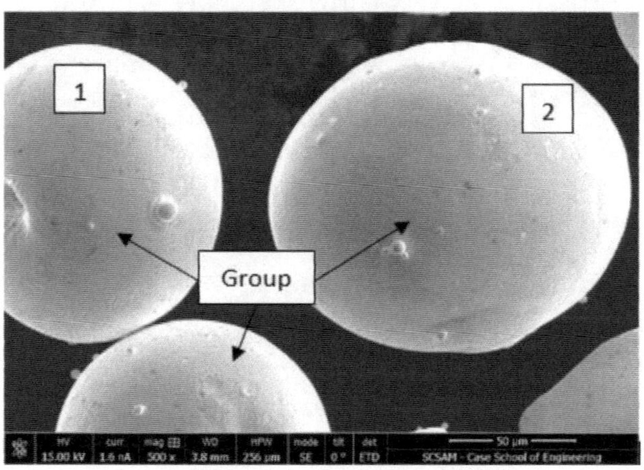

Image 13: Particles for Sample 13 EDS:
BAu-4-VB: -100/+140 PSD

Particle	Au	Ni
1	80.5	19.5
2	81.7	18.3
group	81.9	18.1

Table 7: EDS Quantification of Sample 13:
BAu-4-VB: -100/+140 PSD

The BAu-4 powder from Vendor B also showed uniform composition from particle to particle across all PSDs. This was not surprising since BAu-4 has a single melt-point of 955 °C. However, BAu-6 could more readily form discrete phases through formation of the Au82/Ni18 composition, with other particles making up the difference of the bulk chemistry (Ni/Pd). The results of the BAu-6 (AMS 4786) were the same as BAu-4. All particles demonstrate compliance to the bulk composition, with no variation in results based on PSD.

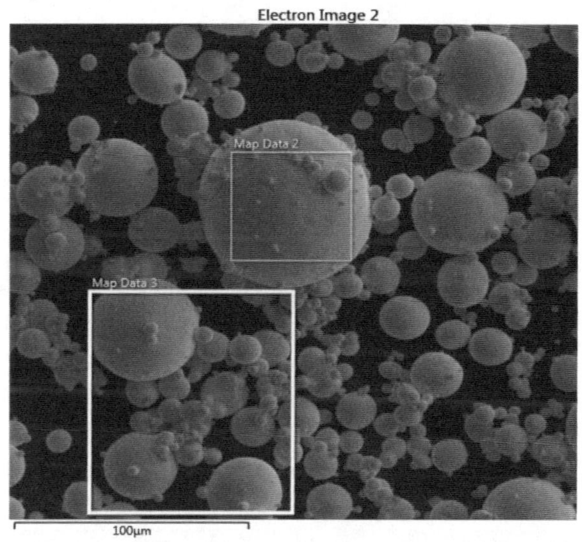

Image 14: Particles for Sample 16 EDS: BAu-6-PI: 140F PSD

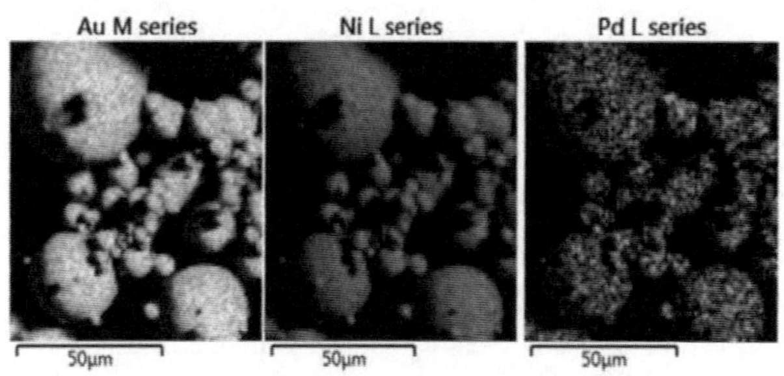

Image 15: EDS Results of Sample 16: BAu-6-PI: 140F PSD

Particle	Au	Ni	Pd
1	70.5	23.0	6.6
2	70.6	23.0	6.4
group	70.8	23.3	6.0

Table 8: EDS Quantification of Sample 16:
BAu-6-PI: 140F PSD

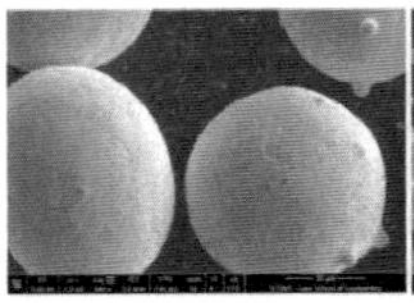

Image 16 (Left): Particles for Sample 19 EDS: BAu-6-PI: -100/+140 PSD
Image 17 (Middle): Particles for Sample 22 EDS: BAu-6-PI: -325/+400 PSD
Image 18 (Right): Particles for Sample 26 EDS: BAu-6-PI: -850 PSD

Sample	Particle	Au	Ni	Pd
19	1	69.2	23.8	7.0
	2	69.1	23.4	7.5
	group	70.8	23.3	6.0
22	1	69.6	23.0	7.4
	2	70.5	22.1	7.4
	group	68.0	24.8	7.2
26	1	69.8	23.0	6.8
	2	69.6	23.2	7.2
	group	67.7	25.3	6.9

Table 9: EDS Quantification Summary of Samples 19, 22 & 26 for BAu-6 produced by P&I

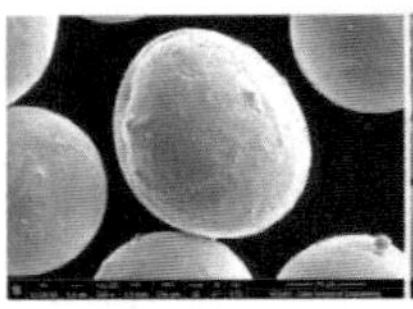
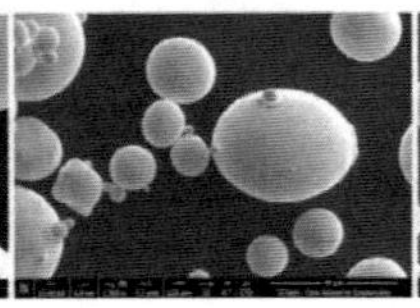
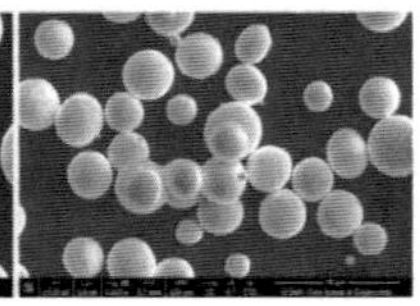

Image 19 (Left): Particles for Sample 28 EDS: BAu-6-VB: -100/+140 PSD
Image 20 (Middle): Particles for Sample 29 EDS: BAu-6-VB: -325/+400 PSD
Image 21 (Right): Particles for Sample 30 EDS: BAu-6-VB: -850 PSD

Sample	Particle	Au	Ni	Pd
28	1	70.9	22.2	6.8
	2	69.4	23.3	7.0
	group	69.2	23.4	6.4
29	1	70.6	22.2	7.3
	2	67.4	25.4	7.3
	group	68.8	24.2	7.0
30	1	69.5	23.3	7.3
	2	68.9	24.0	7.1
	group	68.0	25.1	6.9

Table 10: EDS Quantification Summary of Samples 28, 29 & 30 for BAu-6 produced by Vendor B

The compositional variation in EDS results is within the expected tolerance of a non-quantitative analytical method. This analysis is used for comparison, not conformance. The results show that there is minimal variation between particles, particle size, and vendors tested.

Examination of Morphology

The morphology or shape of the powder impacts the application and behavior of brazing in several ways. Relative to other geometries, spherical powder offers a lower surface area to volume ratio which in turn reduces the allowable surface area for oxide formation. The spherical shape also impacts the rheology or flow of powder. This is critical for applications in which powder is applied direct to a braze joint or surface by dry-pouring. The flow of dry powder is also critical in a bed of powder commonly used to apply the filler metal for honeycomb brazing. Braze powder is also mixed into extrudable (syringe dispensing), and printable braze pastes. In these slurries, the morphology of the powder impacts the ability to dispense paste in a controlled, accurate and repeatable manner. Oblong shapes or jagged edges do not flow as well or as consistently as spheres. Therefore, the preferred particle geometry for braze powder is spherical. The morphology of braze powder is impacted by the gas atomization equipment and settings such as superheat, nozzle design, gas pressure & temperature, quenchant, and gas flow within the atomization chamber. The morphology of product supplied to end-users is also impacted by the classification process which can be designed to remove undesirable particles. There are quantitative methods for measuring the rheology of powder. However, for this experiment the shape was inspected visually under high magnification and found to be generally spherical for all particle sizes, alloys and manufacturers. Below are some images showing the morphology of different size classifications:

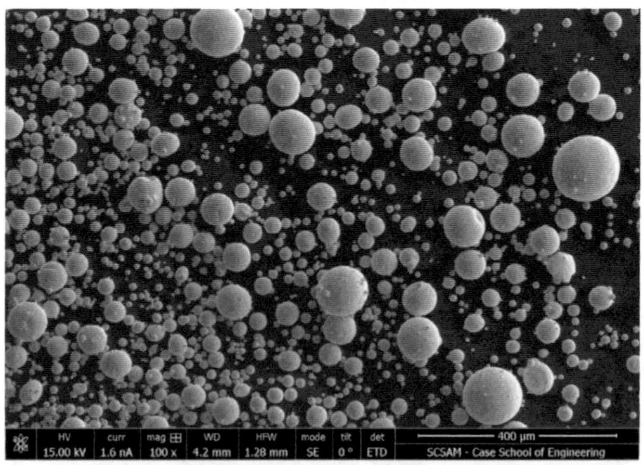

Image 22: Sample 1: BAu-4-PI: 140F PSD

Image 23 (Left): Sample 4: BAu-4-PI: -100/+140 PSD
Image 24 (Middle): Sample 7: BAu-4-PI: -325/+400 PSD
Image 25 (Right): Sample 11: BAu-4-PI: -850 PSD

Below are some images captured by Prince & Izant of BNi-5 (AMS 4782) braze powder which exhibits defects in what is supposed to be spherical powder.

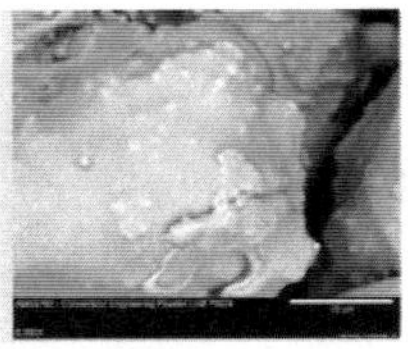

Image 26 (Left): BNi-5 at 500X zoom
Image 27 (Middle): BNi-5 at 1000X zoom
Image 28 (Right): BNi-5 at 5000X zoom

Melt Temperature

Limited testing has demonstrated that nano-particles of the same composition can have a reduced melt temperature [5]. The reduction in melt temperature is as much as 200 °C for some braze alloys [4]. The melt properties are still being studied to distinguish between surface melting and full-melting of particles. This experiment investigates potential correlations between melt temperature and particle size of powder manufactured by gas-atomization. Although gas atomized braze powders are generally considered coarse for metallic powders, the particle size analysis by laser-diffraction indicates that there are particles as small as 5 microns. This testing was performed by P&I on a Differential Thermal Analyzer with the conditions listed in Table 11:

Parameter	Condition
Protective Gas	Ar95/H5
Cup Medium	Alumina
Reference Material	Pt99.99
Ramp Rate	10 °C/min
Sample Size	200 mg

Table 11: DTA Conditions

Image 29 (Left): Differential Thermal Analyzer
Image 30 (Right): Test Sample loaded in ceramic liner on left – Pt reference in ceramic liner on right

The testing results show no difference in melt-temperature between particle sizes for both alloys of both manufacturers. Graph 1 shows the 3 coarsest PSD of P&I BAu-4 plotted with the 4 smallest PSD. The results are very uniform with minimal variation between PSD indicating that a 5-micron diameter particle melts at the same temperature as a 100 micron diameter particle.

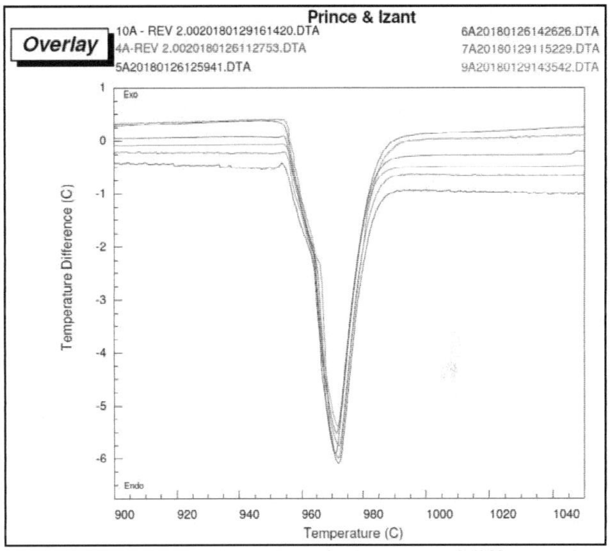

Graph 1: DTA Graph Overlay of 6 Samples of different PSD tested from Lot No. BAu-4-PI

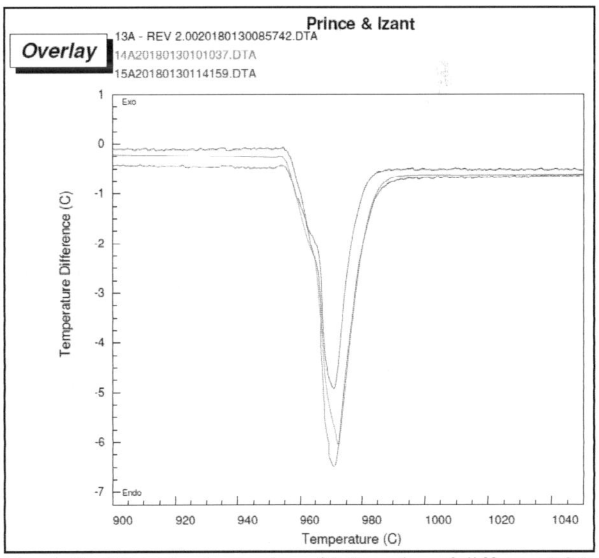

Graph 2: DTA Graph Overlay of 3 Samples of different PSD tested from Lot No. BAu-4-VB

Only the onset temperature was recorded for all BAu-4 (AMS 4787) samples since the solidus & liquidus are both published as 955 °C. Table 12 shows all onset temperatures recorded.

SAMPLE	AWS	Manufacturer	PSD	Melt °C
1	BAu-4	P&I	US 140F	954.77
4	BAu-4	P&I	-100 / +140	954.98
5	BAu-4	P&I	-140 / +200	954.58
6	BAu-4	P&I	-200 / +325	954.95
7	BAu-4	P&I	-325 / +400	955.35
8	BAu-4	P&I	-400 / +500	954.90
9	BAu-4	P&I	-500 / +635	955.05
10	BAu-4	P&I	-635 / +850	954.37
11	BAu-4	P&I	-850	955.01
13	BAu-4	Vendor B	-100 / +140	954.99
14	BAu-4	Vendor B	-325/ +400	954.85
15	BAu-4	Vendor B	-850	954.80

Table 12: Melt Temperatures of all PSD evaluated for BAu-4 (AMS 4787)

The solidus and liquidus were measured for BAu-6 (AMS 4786) with similar results. No variation in melt temperature by PSD was displayed for all samples tested.

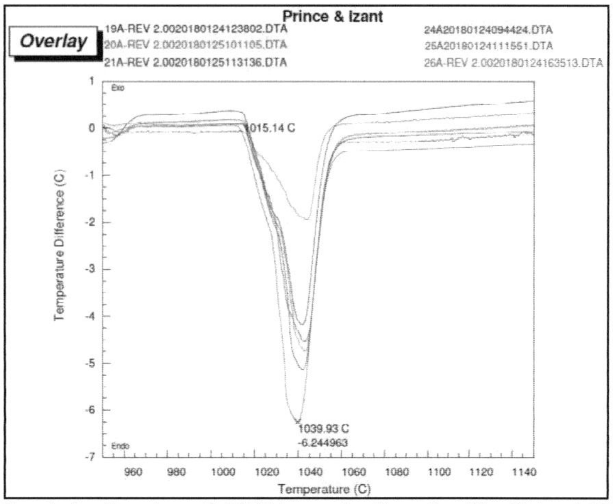

Graph 3: Overlay of 6 DTA graphs from Lot No. BAu-6-PI

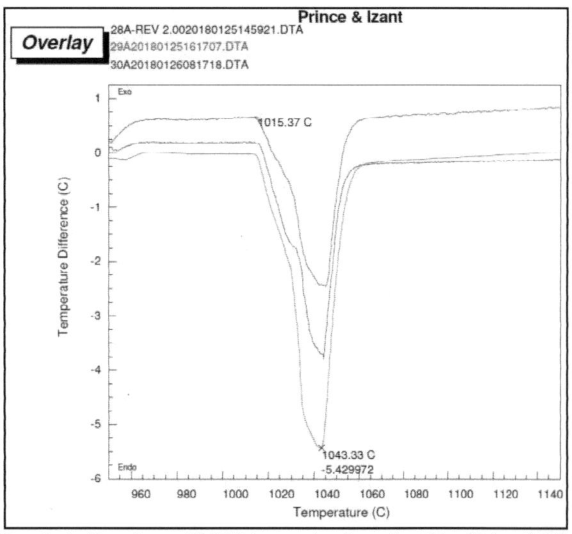

Graph 4: Overlay of 3 DTA graphs from Lot No. BAu-6-VB

SAMPLE	AWS	Manufacturer	PSD	Solidus °C	Liquidus °C
16	BAu-6	P&I	US 140F	1015.32	1042.97
19	BAu-6	P&I	-100 / +140	1014.88	1041.88
20	BAu-6	P&I	-140 / +200	1015.17	1042.86
21	BAu-6	P&I	-200 / +325	1015.50	1042.00
22	BAu-6	P&I	-325 / +400	1015.17	1042.61
23	BAu-6	P&I	-400 / +500	1014.94	1040.87
24	BAu-6	P&I	-500 / +635	1016.08	1042.79
25	BAu-6	P&I	-635 / +850	1013.18	1039.76
26	BAu-6	P&I	-850	1014.64	1043.36
28	BAu-6	Vendor B	-100 / +140	1015.66	1044.14
29	BAu-6	Vendor B	-325/ +400	1015.09	1042.99
30	BAu-6	Vendor B	-850	1017.75	1045.25

Table 13: Melt Temperatures of all PSD evaluated for BAu-6 (AMS 4786)

Oxygen Levels

As discussed above in the examination of morphology, the surface area to volume ratio can impact the level of oxidation for particles. In this experiment, several PSDs of each alloy were tested at NSL Analytical using Leco to determine the oxygen content by weight percent. Bulk compositional analysis, EDS, morphology, and DTA all found no difference in characteristics between PSDs. However, the Leco results demonstrate a trend that smaller particles that have a higher surface area to volume ratio have increased levels of oxygen. Three PSDs were evaluated for each alloy produced by each manufacturer. The PSDs evaluated include: -100/+140 mesh, -325/+400 mesh, -850 mesh. The results can be viewed in Graph 5 & 6.

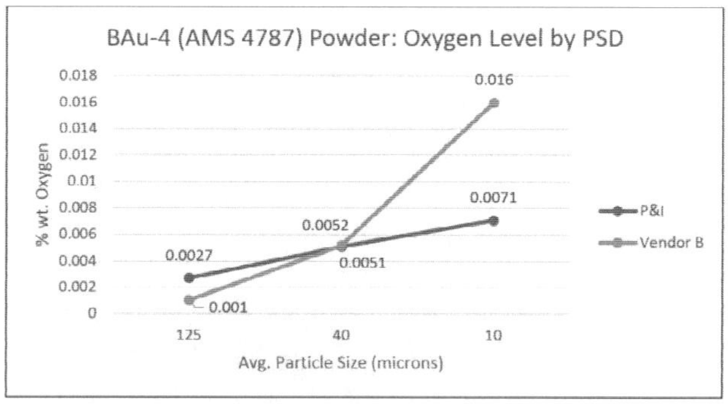

Graph 5: Oxygen Content of BAu-4 by PSD

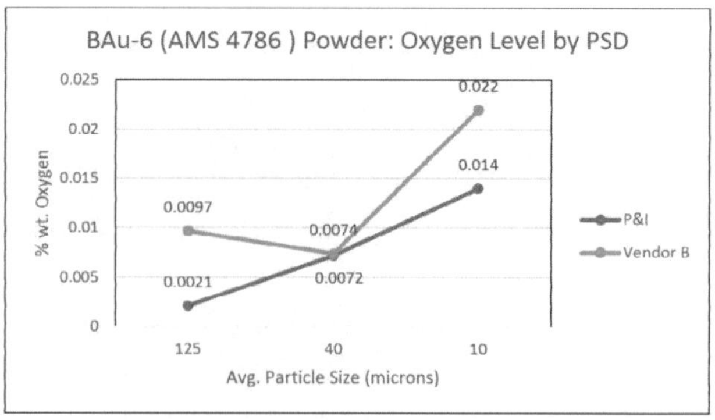

Graph 6: Oxygen Content of BAu-6 by PSD

The elevated levels of oxygen measured in BAu-6 as compared to BAu-4 are a result of the palladium content which has increased reactivity with oxygen. Although the oxygen content of Vendor B's material decreased from the -100/+140 mesh to the -325/+400 mesh, the difference between the oxygen levels was minimal. Whereas the smallest particles have significantly higher oxygen than either of the other two PSDs evaluated.

Conclusions

For the powders manufactured by P&I and Vendor B, the only measurable difference between particles of different size is the oxygen content. This indicates that the gas-atomization process utilized by P&I and Vendor B are tightly controlled processes designed to produce high quality powder. This calls into question the validity of using standard mesh designations to restrict the PSD. Although both alloys examined demonstrated increased oxygen content on the smallest particles, all levels of oxygen measured were relatively low. Any oxides present on the surface of these particles reduces by weight percent when blended with a larger distribution such as -140 (100% through Sieve No. 140). Increasing the -325 particles to 70 or 80% versus the 140F limit of 55% max, will not increase the oxygen ppm enough to impact most vacuum brazing applications. Furthermore, any oxides present on Au/Ni and Au/Ni/Pd commonly reduce before the melt temperature of these filler-metals is reached. Many end-users who vacuum braze BAu-4 (AMS 4787) and BAu-6 (AMS 4786) would see no reduction in braze quality or yield by changing their required PSD to -140 or -200, instead of using a standard mesh designation of 140C, 140F or 200. However, end-users could see a reduction in costs due to the increased sellable yield and decreased classification labor (manufacturer dependent). End-users should use the data and testing methods of this experiment to work with powder manufacturers to identify the optimal balance of powder quality and cost to meet their needs.

Acknowledgments

Danqi Wang – Case Western Reserve University Engineering Department: Performed all SEM & EDS

NSL Analytical (Cleveland, OH) performed all quantitative compositional analysis of bulk chemistry and oxygen by Leco

James Bush, Carl Fechko & Grigory Raykhtsaum – Prince & Izant Engineering

References

[1] AWS A5.8M/A5.8:2011 Specification for Filler Metals for Brazing and Braze Welding
[2] AMS 4787 (Aerospace Materials Specification)
[3] AMS 4786 (Aerospace Materials Specification)
[4] Anming Hu, Denzel Bridges, Suhong Zhang, Nanobrazing for Turbine Blade and Vane Repair, Advanced Material Process 2017
[5] R. John Bosco Balaguru, B.G. Jeyaprakash, Melting Points, Mechanical Properties of Nanoparticles and Hall Petch Relationship for Nanostructured Materials

Mechanical properties of stainless steel joints brazed with nickel filler metals

M. Penyaz, A. Ivannikov, O. Sevryukov, D. Bachurina and B. Kalin
National Research Nuclear University MEPhI, Moscow, Russia
amilenna@icloud.com

Abstract

At present, boron-containing rapidly quenched nickel filler metals are actively used for joining corrosion-resistant steels of various classes. However, in a number of cases a heterogeneous boride structure is formed in the seam zone during brazing. It can lead to a decrease in the mechanical properties and corrosion resistance of the joints. The formation of the structural-phase state of steel joints brazed with rapidly quenched boron-containing filler metals and with new Ni-Cr-Si-Be alloys was studied. The austenitic (12Cr18Ni10Ti) and ferritic-martensitic (16Cr12MoSiWNiVNb) steels were chosen as materials for research. The mechanical tests of brazed samples for the determination of strength characteristics were carried out. The fatigue processes and structural-phase transformations occurring in the brazed joint under cyclic temperature change are studied using the EBSD method. It is established on the basis of 50 cycles that the obtained joints have a resistance to thermal fatigue in the temperature range from room temperature to 450 ^{o}C. This indicates the possibility of exploiting these permanent connections in a significant fluctuations operating temperature.

Introduction

Recently, the use of rapidly quenched boron-containing nickel filler metals for high temperature brazing corrosion-resistance steels of different classes is perspective. The use of these alloys leads to the formation of a complex heterogeneous structure in the diffusion zone that contains separations of intermediate phases such as silicides and borides [1]. This structure negatively affects the strength characteristics of the joint, especially under dynamic loads and in corrosive environment. Fig. 1 shows images in back-scattered electrons (BSE) of brazed joints containing heterogeneities in the diffusion zone. In a number of cases products are subjected to extreme operating conditions, for example, intensive thermal effects. Therefore, serious requirements are imposed on permanent joints obtained by brazing and, accordingly, on filler metals. In this work filler metal based on Ni-Cr-Si-Be system was considered in addition to industrial boron filler metals. It is worth noting that in the joint obtained with the filler metals containing beryllium, there are no heterogeneities in the diffusion zone because of its significant solubility in steel.

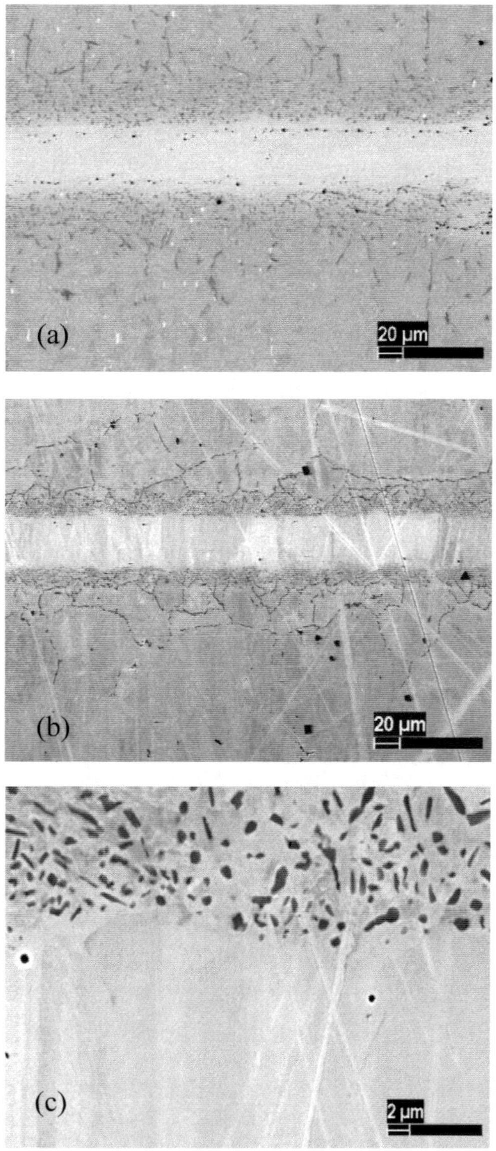

Figure 1. Brazed joints obtained with the filler metal Ni-7.5Cr-3.5Fe-4.5Si-2.6B wt.%: (a) ferritic-martensitic steel 16Cr12MoSiWNiVNb (EP-823), (b), (c) austenitic steel 12Cr18Ni10Ti (different magnifications)

In previously works [2, 3] the development of filler metal based on the Ni-Si-Be system and the method for obtaining a joint by high-temperature brazing are described. Vacuum high-temperature brazing of austenitic (12Cr18Ni10Ti) and ferritic-martensitic (16Cr12MoSiWNiVNb) steels was carried out with boron-containing nickel-based filler metals (type BNi-2) and developed alloys. The strength properties and the operating characteristics of brazed joint obtained with boron and developed non-boron filler metals are compared. The effect of thermocycling on the structure and properties of brazed steel joints are studied.

Methods

SSHVE-1.2.5 resistance furnace with tungsten heaters, which allows heating in a vacuum of 10-2 Pa, is used for brazing samples. The developed filler metal with 50 μm thickness and the boron filler metal with 35 μm thickness were used in one layer and placed directly in the brazing gap. Brazing was carried out in a special fixing conductor. The selected filler metals, base materials and their brazing parameters are listed in table 1. The chemical composition of used steels is shown in table 2. Brazing of the samples was carried out at the temperature 1160 °C [4]. The holding time for ferritic-martensitic steel is reduced to 15 minutes, compared to austenitic steel, in order to reduce the effect of structural-phase state changes at the brazing temperature that occurs due to the partial dissolution of Nb, V, W and Mo carbides.

The method of indentation is used to study the degree of brazed joint heterogeneity. Determination of the microhardness values by Vickers was performed on the installation HVS-1000. The measurement is carried out by pressing a diamond pyramid into the surface of the test sample under the action of a load of 50-100 g, which is applied for 10-15 seconds. Microhardness is measured directly in the seam and at a certain distance from the seam in steps of 50-100 μm. Several measurements are made for each characteristic point.
The strength properties of the samples under uniaxial tension were determined using a universal two-column QUASAR 50 testing machine, which has the maximum force of 50 kN. The tested specimens are listed in table 1. They were obtained with the same parameters of cylindrical samples manufacturing in accordance with State Standard 28830-90 with working part with a diameter of 6 mm (Fig. 2) [5]. The fracture zones of the tested samples were analyzed by scanning electron microscopy (SEM) after uniaxial tensile tests. Rectangular samples with dimensions of 10 × 10 × 55 mm without stress concentrators (notches) were made in accordance with the State Standard 23046-78 to determine the impact toughness of brazed joints [6]. The tests were carried out using special device with pendulum. A study of the seam structure before mechanical tests using optical microscopy was carried out to control the quality of the test samples.

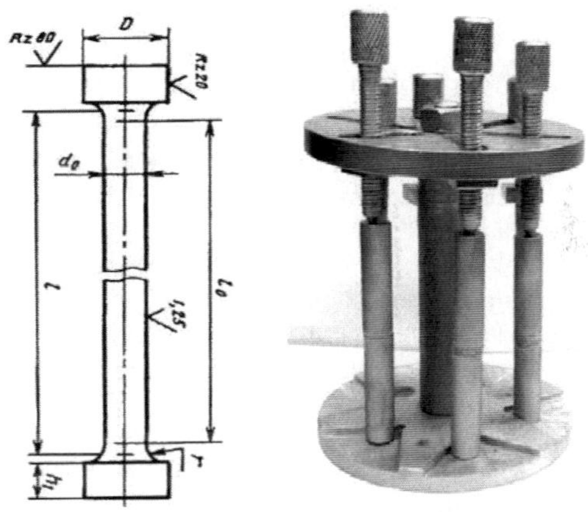

(a) (b)

Figure 2. Specimen for uniaxial tension test, (a) scheme of State Standard 28830-90, (b) assembly for brazing

Thermocycling of samples (cylinders 10 mm in diameter and 10 mm in height) was carried out in a metal flask which was placed in a muffle furnace. It was heated to 450 ° C and then cooled to 60 ° C by immersing the flask in cold water. The 50 cycles were performed. The graph of the heating and cooling dependence on time is shown in Fig. 3. The temperature was controlled with a thermocouple placed in the flask with samples.

Table 1. Brazed specimens for uniaxial tension tests

№	Steel	Filler metal	Brazing mode
1	12Cr18Ni10Ti	Ni-7.5Cr-3.5Fe-4.5Si-2.6B	1160 °C 40 min
2	12Cr18Ni10Ti	Ni-7Cr-5Si-3Be	1160 °C 40 min
3	16Cr12MoSiWNiVNb	Ni-7.5Cr-3.5Fe-4.5Si-2.6B	1160 °C 15 min
4	16Cr12MoSiWNiVNb	Ni-7Cr-5Si-3Be	1160 °C 15 min

Table 2. Chemical composition of used steels.

Material	Alloying elements, % wt									
	C	Cr	Si	Mn	Ni	Mo	V	Nb	W	Ti
12Cr18Ni10Ti	0,12	18	<0,8	<2	10	-	-	-	-	<1
16Cr12MoSiWNiVNb	0,15	11,2	1,1	0,7	0,5	0,6	0,2	0,3	0,7	

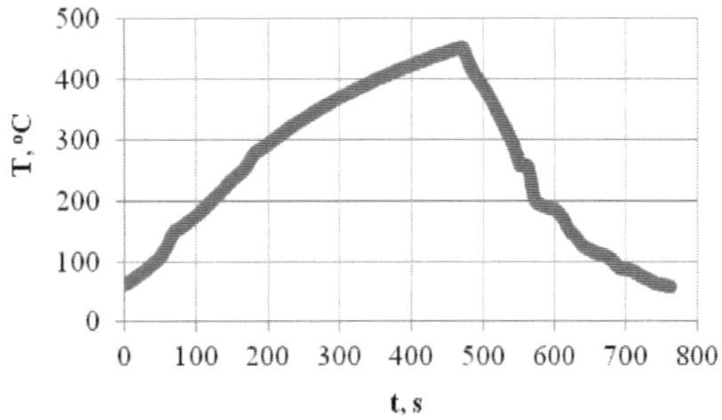

Figure 3. Dependence of heating and cooling on time (1 thermal cycle)

Results and discussion

The effect of the structural components microhardness on the brazed joints strength is analyzed [7]. The results for austenitic steel are shown in Fig. 4. The metallographic studies of the brazed joint showed the different structure of the diffusion zone after brazing with boron and beryllium filler metals. The diffusion zone with a size of 100-120 μm is characteristic for 12Cr18Ni10Ti steel, brazed with boron filler metal. However, recrystallization in this zone does not occur. Boron forms intermediate phases with chromium along the grain boundaries [8]. In both cases microhardness falls at a distance from the center of seam to values of annealed steel. The difference in the values of microhardness in these regions will lead to the appearance of microstrains. Isothermal crystallization is not completed fully in some places and an eutectic structure is crystallized in the center of the seam in the case of using the Ni-7Cr-5Si-3Be filler metal. The microhardness values from 240±10 $HV_{0.1}$ in the seam fall to 160±10 $HV_{0.1}$ in the diffusion zone and steel. The structural difference for these two filler metals, as will be seen below, will lead to different results of tests on strength and impact toughness.

Fig. 5 shows the results of the tests. As a result, recrystallization (in the case of using boron filler metal) leads to decrease of the mechanical properties. Therefore, the brazed joint 16Cr12MoSiWNiVNb / Ni-7Cr-3.5Fe-4.5Si-2.6B showed low values of the tensile strength (Fig. 5a). The joint obtained with the filler metal Ni-7Cr-5Si-3Be showed the best strength properties. It was also taken into account that steel after processing by the brazing mode changes its properties. It is strengthened and has microhardness higher than in the initial state. This is due to changes in the structural-phase state at the brazing temperature. The study of the brazed joint fractography showed the following: in the case of boron-containing filler metal, the destruction occurs mainly along the boundary of the seam-base material interface with elements of destruction along the seam body. This indicates its low cohesive strength. Destruction over the body of the seam is predominantly brittle with numerous secondary cracks [9]. In the case of a beryllium-containing filler metal, the destruction of the joint occurs along the seam, which indicates a

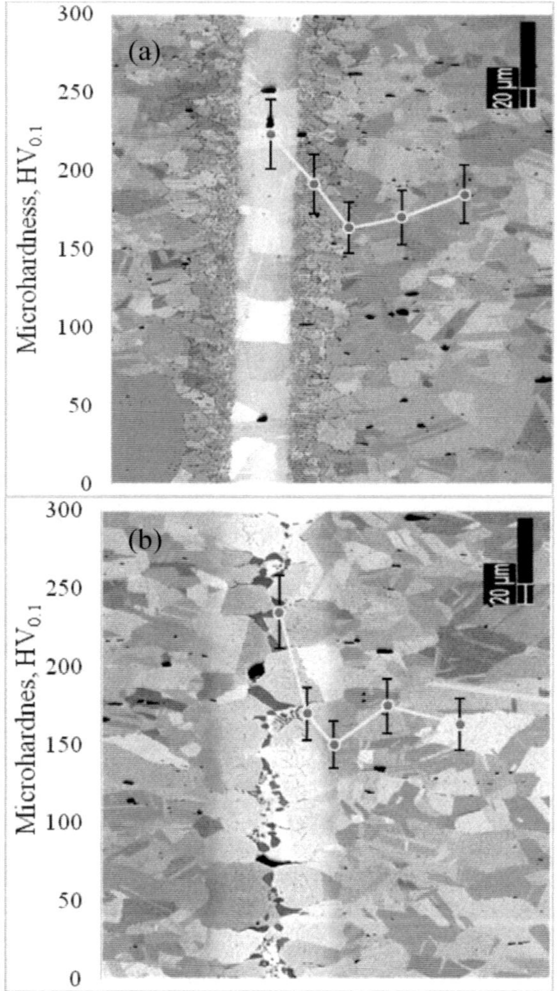

Figure 4. Microhardness of brazed joint 1160 °C 40 min
(a) 12Cr18Ni10Ti/Ni-7Cr-3.5Fe-4.5Si-2.6B
(b) 12Cr18Ni10Ti / Ni-7Cr-5Si-3Be

significantly higher cohesive strength of the seam-base material interface. The nature of the fracture is mainly viscous. The viscous type of fracture of the sample brazed with beryllium-containing filler metal indicates a high plastic deformation and a significantly higher energy of fracture of this sample compared to the sample, brazed with boron-containing filler metal and brittle destroyed. It can be seen that the grain is growing strongly due to recrystallization during brazing with boron filler metal. Moreover, intermediate phases $(Cr,Fe)_2B$ are formed along the boundaries, weakening them boundaries, therefore, destruction occurs subsequently on them. Diffusion of beryllium, unlike boron, does not lead to recrystallization in steel. It, significantly dissolving, is found both along the boundaries and in the body of grain.

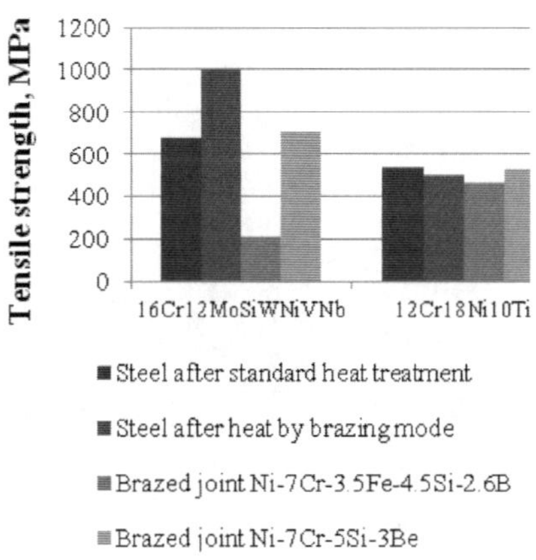

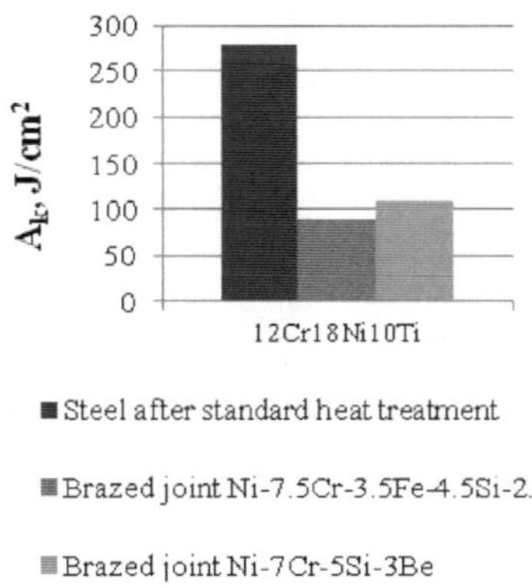

Figure 5. The results of tests on a) uniaxial tensile; б) impact toughness for steel 12Cr18Ni10Ti

It can be concluded that the Ni-7Cr-5Si-3Be alloy demonstrates high values of strength and plasticity. The structure of brazed joints (according to table 1) was visually examined by metallography method before and after thermocycling. It was found that after 50 cycles the structure was not changed at the macrolevel, no degradation was observed. A microstructure study was carried out using SEM. Fig. 6 shows the combined images (BSE) of the brazed joints. Comparative analysis of thermocycling results of joints 12Cr18Ni10Ti/Ni-7Cr-3.5Fe-4.5Si-2.6B and 16Cr12MoSiWNiVNb/Ni-7Cr-3.5Fe-4.5Si-2.6B 1160°C 15 and 40 minutes respectively, is based on it.

In the case of steel 12Cr18Ni10Ti, an increase in the number of borides after 50 cycles was not detected. The appearance of any defects was not detected. In the case of steel 16Cr12MoSiWNiVNb visually the size of borides increased and there was a decrease in their number. During more detailed examination of the heterogeneous structure in the border region an increase in the number of microcracks in borides was revealed (Fig. 7, 8).

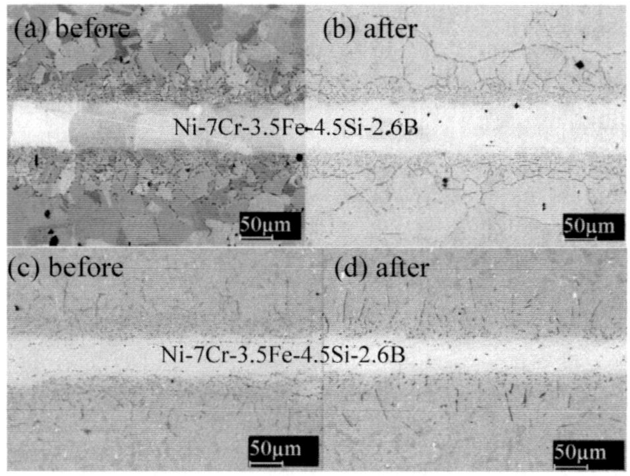

Figure 6. Microstructure of brazed joint (a), (b) 12Cr18Ni10Ti / Ni-7Cr-3.5Fe-4.5Si-2.6B before and after thermocycling, (c), (d) 16Cr12MoSiWNiVNb / Ni-7Cr-3.5Fe-4.5Si-2.6B before and after thermocycling

The contrast on the Fig. 8(a) is different because the sample is etched for EBSD analysis. Their further development can lead to their entry into the matrix and the formation of cracks in the steel. The occurrence of these microcracks may be due to the different values of the coefficient of thermal expansion of steels and borides.

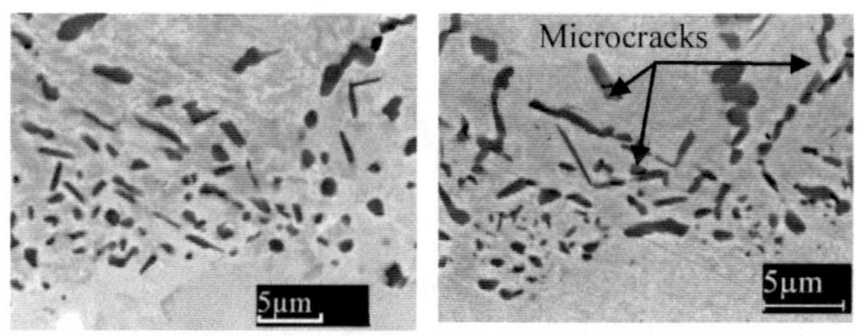

(a) (b)

Figure 7. Microstructure of borides in the border region of brazed joint 12Cr18Ni10Ti / Ni-7Cr-3.5Fe-4.5Si-2.6B, 1160 °C 40 min (a) before thermocycling, (b) after thermocycling

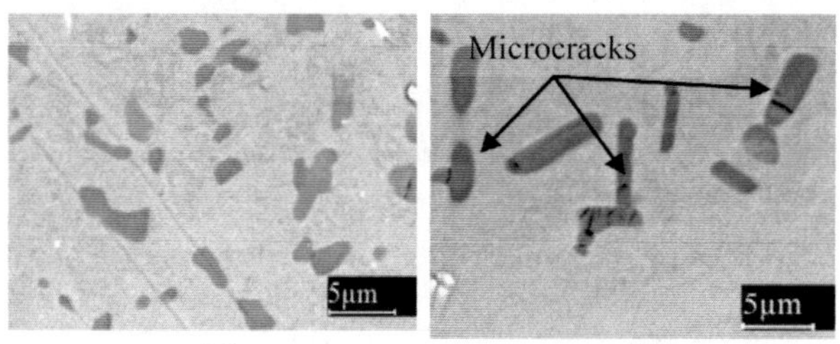

(a) (b)

Figure 8. Microstructure of borides in the border region of brazed joint 16Cr12MoSiWNiVNb / Ni-7Cr-3.5Fe-4.5Si-2.6B, 1160 °C 15 min (a) before thermocycling, (b) after thermocycling

Friability of borides with fatigue loads is a factor that reduces the performance of the joint. Such a heterogeneous structure helps initiate the onset of failure under aggressive operating conditions, such as, for example, periodic temperature changes. A noticeable change in the chemical composition in the seam region after thermocycling is not observed, since such such temperature (450 °C) is insufficient to activate the diffusion processes and form new phases. As indicated earlier, the properties of steel change both after brazing and after thermalcycling. Using the EBSD method, steel and an area of the brazed joint were investigated before and after 50 thermal cycles. The average percentage of identification is 91% due to the presence of small components.

Fig. 9 shows the structural-phase state analysis of the austenitic steel before the thermocycling test. It can be seen that the structure is an austenite and mainly consists of twins. There is a separate grain of ferrite, which can be found in such steels in very small quantities.

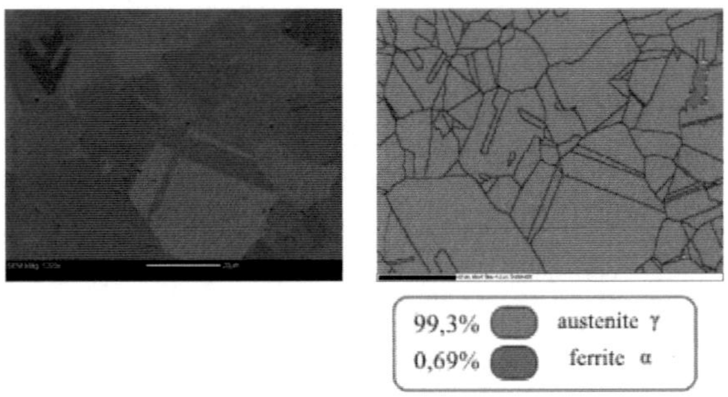

(a) (b)

Figure 9. EBSD analysis of the structural-phase state of steel 12Cr18Ni10Ti after heat treatment during brazing 1160 °C 40 min and before thermocycling, (a) BSE image, (b) phase map

Fig. 10 shows the structural-phase state analysis of the austenitic steel after the thermocycling test. It can be seen that a considerable amount of martensite was formed under the action of plastic deformation arising in the grains of steel.

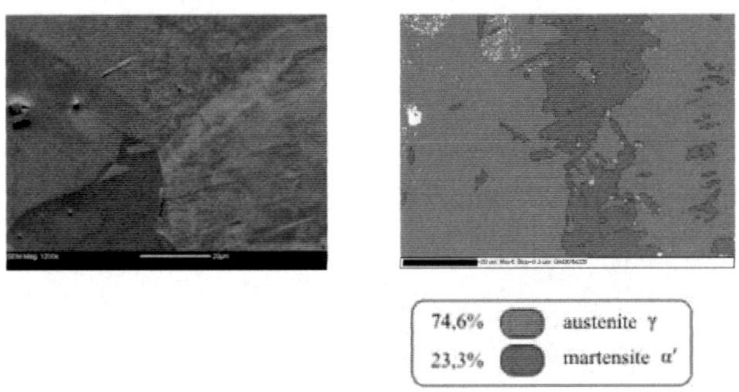

(a) (b)

Figure 10. EBSD analysis of the structural-phase state of steel 12Cr18Ni10Ti after heat treatment during brazing 1160 °C 40 min and after thermocycling, (a) BSE image, (b) phase map

It was found that the proportion of the formed martensite in the seam zone is higher. This is consistent with the fact that the degree of anisotropy of the structure contributing to the accumulation of stresses, plastic deformation and martensitic transformation is much higher here. Fig. 11 shows EBSD analysis of the structural-phase state of the brazed joint 12Cr18Ni10Ti / Ni-7Cr-3.5Fe-4.5Si-2.6B after brazing 1160 °C 40 minutes before and after 50 thermal cycles.

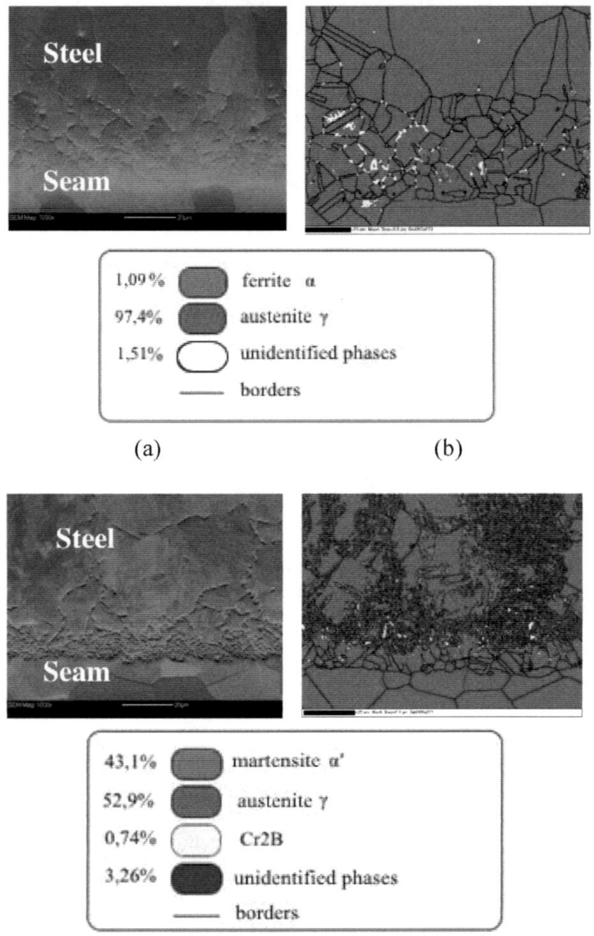

(c) (d)

Figure 11. Brazed joint 12Cr18Ni10Ti / Ni-7Cr-3.5Fe-4.5Si-2.6B (a) BSE image before thermocycling, (b) phase map before thermocycling, (c) BSE image after thermocycling, (d) phase map after thermocycling

In the case of fig. 11b, chromium borides have not been identified due to the capabilities of this method, however, they are present on the electronic image. If 23.3% of martensite was formed in the steel away from the seam, then in the joint zone it is 43.1% (almost twice as much), which indicates the production of a higher value of plastic deformation in this region. This also contributes to an increase in the degree of heterogeneity of the structure in this area.

Conclusions

High-temperature brazing of austenitic (12Cr18Ni10Ti) and ferritic-martensitic (16Cr12MoSiWNiVNb) steels with classical boron filler metal Ni-7Cr-3.5Fe-4.5Si-2.6B and developed filler metal based on the system Ni-(Cr)-Si-Be was carried out. The mechanical tests of brazed samples for the determination of strength characteristics were carried out. Joints were tested for uniaxial tensile strength and impact toughness. Comparative mechanical tests of brazed joints showed that the use of non-boron filler metal provides for the austenitic steel an impact toughness 110 ± 10 J/cm^2, a tensile strength 530 ± 30 MPa and for ferritic-martensitic steel a tensile strength is 710 ± 30 MPa. The values obtained for the boron filler metal type BNi-2 are lower: for the austenitic steel an impact toughness 90 ± 10 J/cm^2, a tensile strength 460 ± 30 MPa and for ferritic-martensitic steel a tensile strength is 210 $\pm$ 30 MPa.

Analyzing the obtained data, it can be concluded that the developed filler metal Ni-7Cr-5Si-3Be demonstrate high strength characteristics of brazed joints. The fatigue processes and structural-phase transformations occurring in the brazed joint under cyclic temperature change are studied using the EBSD method. It was determined that the heterogeneous structure in the form of a boride network promotes more intensive formation of martensite in the near-seam zone.

It is established on the basis of 50 cycles that the obtained joints have a resistance to thermal fatigue in the temperature range from room temperature to 450 °C. This indicates the possibility of exploiting these permanent connections in a significant fluctuations in operating temperature.

References

[1] Maksimova, S. V. "Amorphous filler metals for brazing stainless steel and titanium and the structure of brazed joints" *Adhesion of melts and brazing of materials* (2007), pp. 70-81.

[2] Ivannikov, A. and Kalin, B. "Study of the Ni–Si–Be system as a base to create boron-free brazing filler metals" *Science and Technology of Welding and Joining* Vol. 23, Is. 3 (2017), pp. 187-197.

[3] Ivannikov, A. and Fedotov, V. "The use of Ni-Cr-Si-Be filler metals for brazing of stainless steels" *IOP Conference Series: Materials Science and Engineering* Vol. 130, (2016), 012045.

[4] Ivannikov, A. and Tukhbatov, A. "Rapid-quenched nickel-based solder for high-temperature brazing of various constructive elements" *Tsvetnye Metally* (2014) Vol. 12, pp. 27-31.

[5] GOST (State Standart) 28830-90 Brazed joint. Test methods for tensile strength and long-term strength.

[6] GOST (State Standart) 23046-78 Brazed joint. Method of impact tests.

[7] Binesh, B. and Jazayeri Gharehbagh, A. "Transient Liquid Phase Bonding of IN738LC/MBF-15/IN738LC: Solidification Behavior and Mechanical Properties" *Journal of Materials Science and Technology* (2016), Vol. 32, pp. 1137-51.

[8] Bai, K. and Ng, F. "Understanding non-parabolic solidification kinetics in Ni-based alloys during TLP bonding via thermo-kinetic modeling" *Journal of Alloys and Compounds* (2017), Vol. 699, pp. 1084-94.

[9] Fellows, J. Fractografy and Atlas of Fracture Patterns [Reference Book: Moscow: Metallurgiya] (1982).

Investigation of CuTi, CuNiSi and CuNiSiB alloys for brazing applications

Tillmann, W.; Lehmert, B.; Manka, M.; Wojarski L.; Trelenberg P.
Institute of Materials Engineering – TU Dortmund, Dortmund, Germany

Abstract

Hybrid compounds of cemented carbide and tool steel are commonly used for machining equipment. In order to exploit the potential of both materials, high-quality joints with high strengths are desired. The main challenges when brazing these materials are considered to be the mismatch in the Coefficient of Thermal Expansion (CTE) and the poor wettability of cemented carbides with conventional filler alloys. For different brazing processes, copper-based alloys offer a cost-efficient alternative to the well-established silver-based alloys. Nevertheless, their application leads to a higher stress state due to their higher melting point.

This work concerns with the production and characterization of copper-based alloys (CuTi3, CuNi12Si5, and CuNi12Si5B4) for hybrid joints of cemented carbides and steel. The selected alloys offer the possibility of adjusting their mechanical properties by precipitation hardening procedures. This is especially interesting for reducing stresses within the final joints. It could be observed that melting, a multiple re-melting, and an inhomogeneous solidification during the production has no effect on the element distribution of the alloys and their properties such as the melting temperature or hardness.

Introduction

Tools for drilling, cutting, and sawing are exposed to exceptional mechanical and thermal loads during use. For this purpose, a hard and wear-resistant cutting material like cemented carbide is joined onto a ductile carrier material made of steel [1, 2]. Cemented carbides consist of very hard particles, mostly carbides, which are embedded in a ductile metal matrix such as cobalt [3]. Joining of composite materials like cemented carbides to steel requires high demands on the joining technology due to the different material properties [4, 5]. Brazing has the advantage that a material bond is achieved at significantly lower working temperatures when compared to welding. Thus, the thermal influence on the base materials is much smaller, which results in lower induced residual stresses in the composite part [5]. For brazing processes with cemented carbides, the challenges result from the weak wetting of the carbides and the carbon diffusion that leads to the formation of brittle η-carbides and negatively influence the properties of the joint [6]. For hybrid steel and cemented carbide joints, a further main challenge is the mismatch of the Coefficient of Thermal Expansion (CTE) and the poor wettability of cemented carbides [7]. Regarding the state of stress of the joining area besides the CTE, both the geometry and modulus of elasticity of the base materials as well as the melting point, the thickness, and plasticity of the filler metal are of importance [5]. Furthermore, the interactions between the braze material and the respective base material are fundamental for a successful braze joint [8, 9]. To improve the quality of the joint, there are usually two different approaches, which, however, also have disadvantages. A moderate cooling after the brazing process leads, on the one hand, to low stresses within the joint but, on the other hand, due to being exposed for a longer period to high temperatures, to a brittle microstructure and thus to low strengths. The opposite approach – fast cooling – induce a more ductile joint microstructure but leads to high internal stresses, which in turn can lead to fractures in the base materials, especially of the cemented carbide. A concept to optimize the stresses of the joint is the combination of both approaches. A fast cooling process after the brazing process and a subsequent precipitation hardening by means of a heat treatment should lead to a ductile microstructure and low residual stresses.

Therefore, it is necessary to deviate from usual brazing materials and to develop novel brazing alloys that, in addition to being cost saving, further combine positive properties such as a good wetting behavior, reasonable melting temperature, high strength, and the possibility of precipitation hardening.

This work is concerned with the production and characterization of three different copper-based filler alloys. They possess the ability to compensate residual stresses caused by the different thermal expansion coefficients to create a high-quality connection [10, 11]. These filler materials were melted and re-melted numerous times in an arc furnace to examine the influence of the increased melting factor on the element distribution and selected material properties such as the melting temperature and hardness. Subsequently their mechanical as well as thermal properties were analyzed before and after a heat treatment that leads to a precipitation hardening of the alloys.

Materials

As braze material, three different Cu-based alloys were chosen and melted with the following chemical composition: Cu-Ti (3 wt.%), Cu-Ni (12 wt.%)-Si (5 wt.%), and Cu-Ni (12 wt.%)-Si (5 wt.%)-B (4 wt.%). Compared to the well-established Ag brazes, the favorable price, the good flow ability, the gap filling capacity as well as the suspected ductility, and the resulting good applicability are the main arguments to use copper-based alloys for carbide steel brazing [12].

Experimental procedures

Melting and re-melting of alloys
Initially, the elements copper, nickel, titanium, silicon, and boron were found in granule form (Fig. 1). The pure elements (purity 99.9%) were weighed on a fine scale with a 0.001 g precision. The composition for a braze alloy was based on the total weight 5 g. Three samples were prepared for each alloy composition. In a subsequent step, the compositions were mechanically pressed together with a hydraulic press. For this purpose, the starting materials were collected and compressed with a force of >20 tons by means of the top stamp of a hydraulic press. This process was repeated several times to form a single unit from the material mixture. This pre-pressing of the granules was necessary to prevent that individual particles are blown away when touched with the arc.

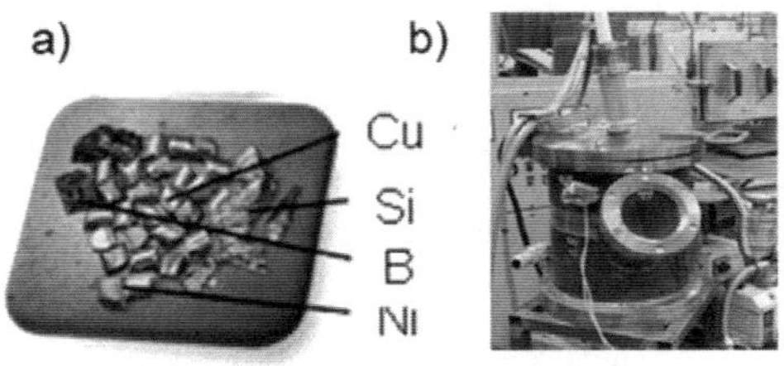

Figure 1: a) Granules of components; b) Used arc furnace

A metallurgical melt process was used to produce a final, fused unit of the compacted composites. The melting of the alloys took place in an arc furnace (Fig. 1) with a coupled drive unit of a TIG welding system, serving as the energy source for the arc. The weighed and pressed samples were successively placed on the water-cooled copper plate (anode) in the furnace. As a cathode, a commercially available water-cooled tungsten electrode was used, which was attached to the handle and allowed freedom of movement during the whole process. The melting was carried out under a pure argon atmosphere with a slight overpressure of 1200 to 1300 mbar in the working chamber. To ensure an oxygen-free environment the chamber was evacuated and rinsed with argon three times. The operation in the arc furnace was initiated by a contact ignition at 120 A and 1.6 V with a direct current (DC). With small circular movements, the arc was passed over the sample until a complete fusion took place. After the braze alloy was completely melted by the high temperature, the arc was removed and the material could solidify. In order to analyze the influence of re-melting, the braze alloys were re-melted two, four, and six times.

Thermal analysis
To identify the solidus and liquidus temperatures of the copper-based alloys, the phase transformation temperatures and changes in the physical conditions were determined by means of differential thermal analysis (DTA) with a STA PT 1600 by Linseis. For this purpose, a 20 mg sample was taken from each alloy and twice heated in a crucible up to a temperature of 1200 °C within an inert argon atmosphere. The heating and cooling rate was kept constant at 5 K/min.

Mechanical analysis
To determine the Young's modulus and hardness of the produced alloys, two different measurement methods were used before and after a heat treatment. In addition to the conventional hardness test according to Vickers (ISO 6507) [13], nanoindentation was performed with a "Nanoindenter G200" of the company Agilent. This device allows measuring an array of 5 rows of 5 measuring points, each with a distance of 25 µm and a penetration depth of 1 µm (Fig. 2). Therefore, it is possible to determine existing slight differences caused by an inhomogeneous element distribution within the microstructure. To calculate the elasticity module, the part of the relief curve, in which a purely elastic reaction of the counter-body occurs on the decreasing penetration force of the indenter, was used. Since an enlarged area is scanned at the same time, the evaluation of the result from the core to the edge area of the samples is also more accurate [14].

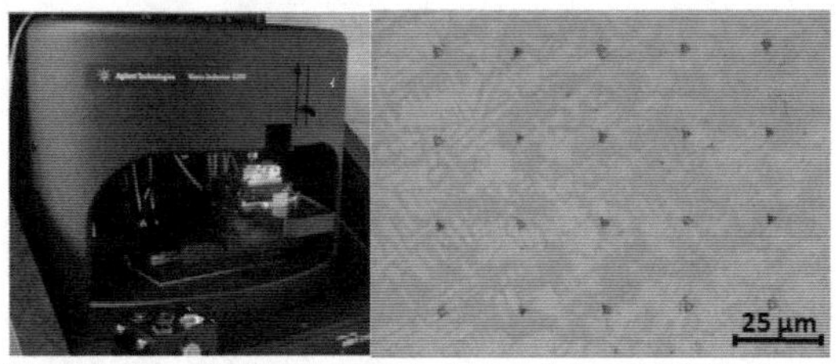

Figure 2: Nanoindenter and indents of the CuNiSi alloy

Heat treatment
The precipitation hardening forms very finely distributed precipitates in the microstructure. Depending on the treatment temperature and time, the structure of the alloys can be uniform after the heat treatment. Possible effects are an increase in the hardness, a strength increase, and lowering the self-tensioning condition [15]. For this purpose, the samples were exposed to temperatures of 295 °C, 325 °C, 425 °C, and 475 °C for 30 min and 60 min in small ceramic vessels. Each process was carried out under a vacuum atmosphere of 10^{-5} mbar at a heating and cooling rate of 25 K/min [13] in a high temperature vacuum furnace type TORVAC.

Microstructural observation
All produced samples were cold-mounted, ground, and polished. The cross-sections were examined by means of light microscope and scanning electron microscope (SEM) with a JSM 7001F by Joel, equipped with an energy dispersive X-ray (EDX) for chemical analyses. To carry out an examination of the homogeneity in the selected brazes, several point- and area-analyses were performed and correlated with EDX mappings. The investigations were performed both in the center and at the edge of the melted droplets. The chosen parameters were an acceleration voltage of 20 kV at a sample current of 14 V with a working distance of 10 mm.

Results and discussions

Cross-sections of all droplets of the molten alloys were analyzed by means of SEM and EDX. The structure as well as the element distribution were investigated. Both, the center and the fringe area of each individual droplet were compared to each other. It can be concluded that there are only marginal difference between the region at the edge and the center.

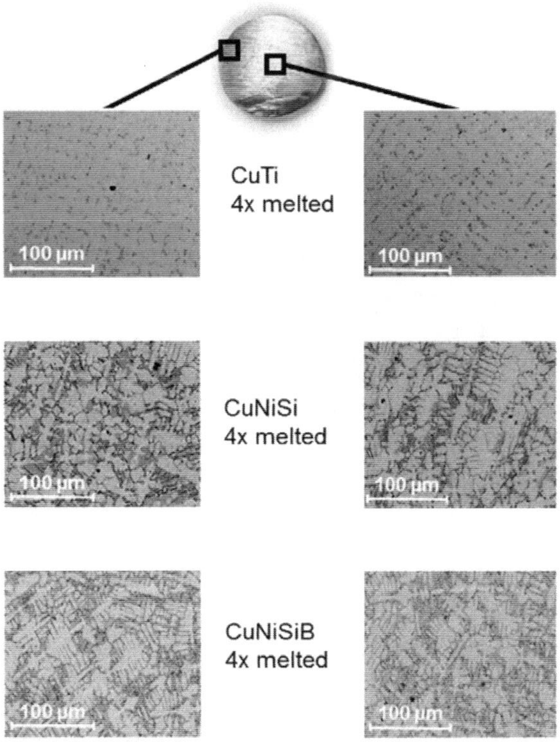

Figure 3: Comparison of the microstructure of 4 times melted samples

Figure 3 shows exemplarily the comparison of the cross-sections of the alloys that were melted 4 times. For all samples, the microstructure in the center of the droplet and at the edge do not differ significantly. The dimensions as well as the distribution of the phases seems identical. Thus, the inhomogeneous solidification caused by the water-cooled copper plate of the arc furnace does not seem to have any influence on the element distribution. This result could be further verified by hardness tests. Fig. 4 shows the average hardness for the alloys that were melted 4 times measured by means of nano-identification. A slightly higher hardness was obtained at the fringe area due to higher cooling rates. But the hardness differences are less then 7 HV and can therefore be neglected.

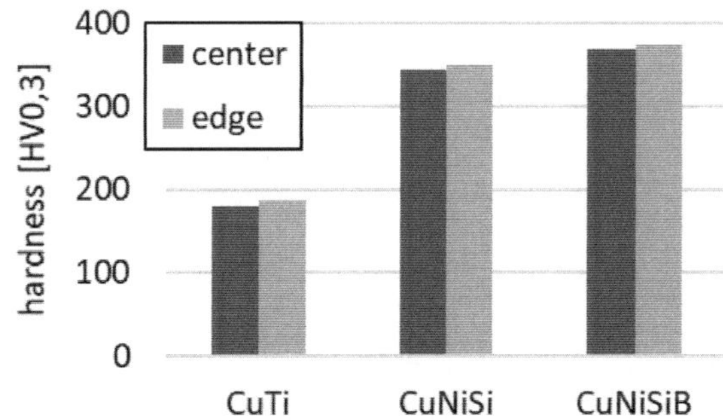

Figure 4: Comparison of the hardness at the center and the edge for 4 times melted samples measured by means of nano-identification.

In addition, the Young's moduli of all three alloys were determined and the values of the center and the edge regions were compared. For the CuTi the average values for the samples that were melted 2 times were 132 GPa while the samples of 4 and 6 times melting reached 131 GPa and a deviation of only 3 GPa could be determined. In comparison to pure copper with a Young's modulus in the range of 110-130 GPa, the doping of titanium (110 GPa) into copper has only a minor impact on the Young's modulus.

For the CuNiSi and CuNiSiB alloys, all values were between 145 and 157 GPa. However, a trend depending on the number of re-melting operations could not be recognized. In addition, the element boron did not seem to have any measurable influence on the Young's modulus. In this case, the doping of Ni and Si into copper tend to increase the Young's modulus of the alloys.

For the investigations of possible influences of the re-melting, SEM images and EDX spectra of the microstructure of all samples were compared. In Fig. 5, the recorded images and spectra of a CuNiSi alloy are shown exemplarily for re-melting twice, four, and six times.

By comparison, the changes of re-melting all samples show a slightly finer microstructure with a higher number of re-melts. While the structure of the brighter phase in the sample with a double melting is still more spherical, it becomes more dendritic and finely branched with an increasing number of melting processes. The element distribution in the EDX images shows a very homogeneous distribution. The microstructure can also be identified and the finer distribution with an increasing number of melting processes indicates a better homogenization.

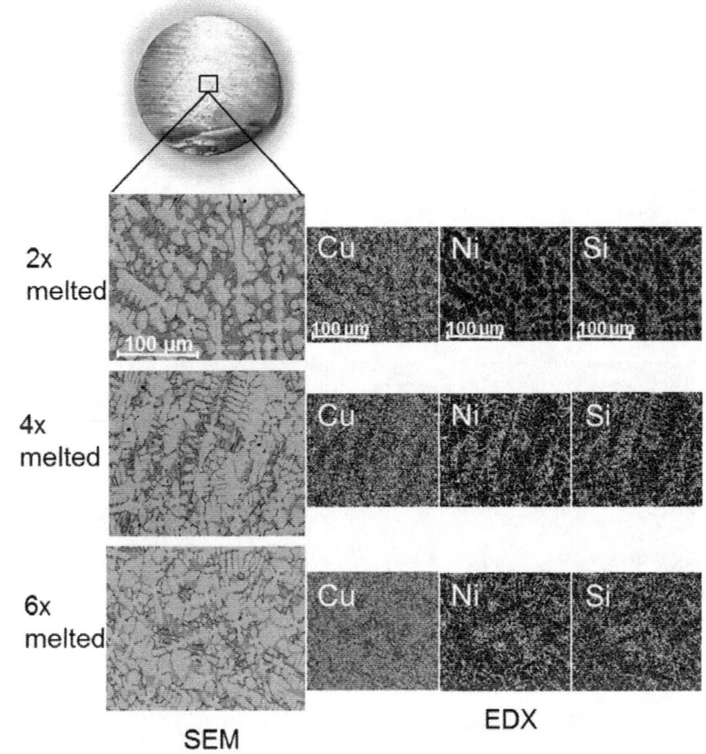

Figure 5: SEM and EDX images of a CuNiSi alloy with different numbers of melting

To determine possible thermal changes due to the re-melting, the liquidus as well as the solidus temperature were determined for all samples by means of DTA. The overview of all values is given in Fig. 6. The temperatures of all three identical samples were averaged and illustrated. It could be determined that re-melting has almost no influence on the melting temperatures. For the CuTi alloy, the liquidus temperature is identical to that of pure copper, while the solidus temperature is approx. 50 K below. In the case of the other two alloys, a melting point depression could be observed. The deviation between the solidus and liquidus temperature is approx. 80 K.

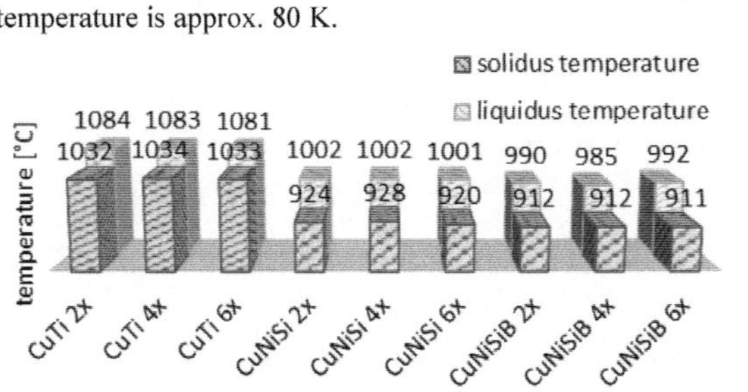

Figure 6: Overview of the melting temperature for all filler metals

Subsequently, the hardness values of all samples were determined as well. Besides illustrating effects of re-melting, these results were used in order to compare the values in a next step with the hardness of the alloys after the heat treatment. Figure 7 shows the mean hardness values with the corresponding deviations determined by means of Vickers.

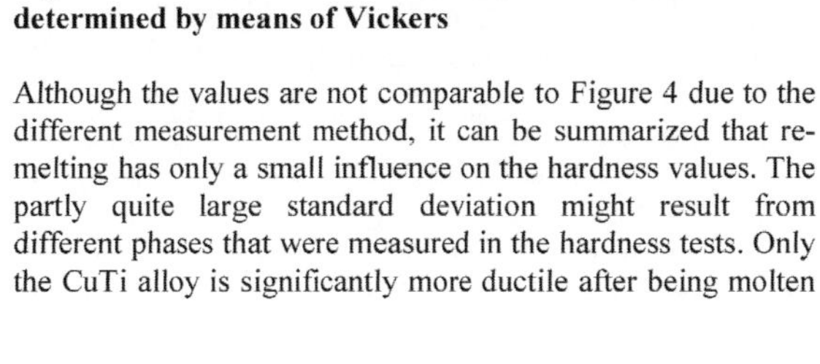

Figure 7: Overview of the hardness for all filler metals determined by means of Vickers

Although the values are not comparable to Figure 4 due to the different measurement method, it can be summarized that re-melting has only a small influence on the hardness values. The partly quite large standard deviation might result from different phases that were measured in the hardness tests. Only the CuTi alloy is significantly more ductile after being molten

4 times. Further investigations have to show if re-melting six times will already lead to precipitation hardening.
Based on these results, the further investigations concerning the heat treatments where focused on the samples, which underwent four melting processes.
In order to investigate a possible precipitation hardening, all alloys were heat treated for 30 and 60 min, respectively. Fig. 8 shows the comparison of the SEM images of the 4 times melted CuNiSi alloy after melting and after the heat treatment in each case for 30 min at 425 °C and at 475 °C. Essentially, three different phases could be identified: A dominant light gray phase (A) with a rounded appearance and an average composition of 87 at.% of Cu, 7.5 at.% of Si and 5.5 at.% of Ni that indicates a solid solution of Cu and Ni with solved Si. The slightly darker phase (B) partly has an dendritic structure and consists of 56 at.% of Cu, 25 at.% of Si, and 17 at.% Ni. This could be a $Cu_{56}Si_{11}$ phase. The dark gray phase (C) occurs partially with a dendritic structure and has the same composition of approximately 42 at.% Ni and 33 at.% Si, balanced with Cu. While this phase is rather lamellar in the untreated sample, the heat treatment leads to a more spherical appearance. In particular, the grain boundaries are increasingly surrounded by precipitations as can be observed in the sample treated at 475 °C for 30 min.

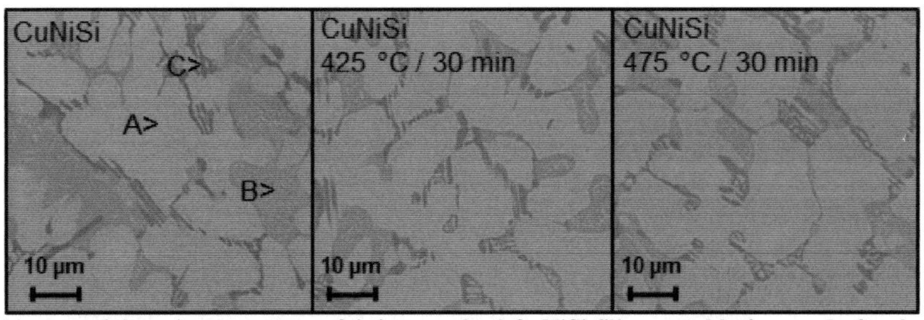

Figure 8: SEM image of the microstructure of 4 times melted CuNiSi filler metal before and after heat treatment

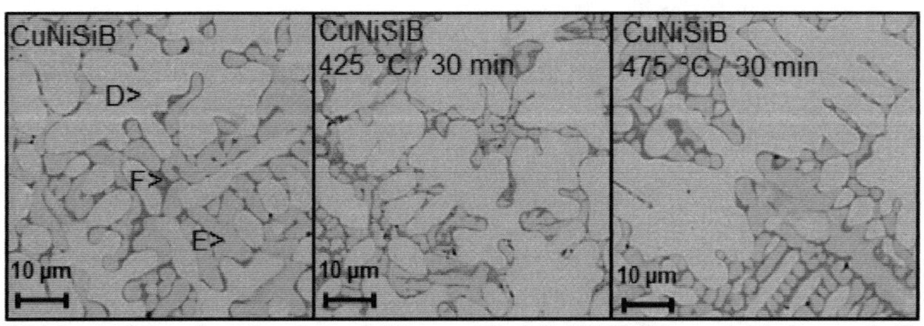

Figure 9: SEM image of the microstructure of 4 times melted CuNiSiB filler metal before and after heat treatment

The microstructure of the CuNiSiB alloy is different. Although the element boron is difficult to detect by means of EDS, the composition consisting of the elements Cu, Ni, and Si was investigated as well.
As shown in Fig. 9, the heat treatment for 30 min at 425 °C and for 30 min at 475 °C has only minor effects on the microstructure. While the various characteristics do not change, there are small variations in the size and the frequency of occurrence of each phase. Similar to the previous system, three phases are clearly visible.
The brightest phase (D) still has the highest Cu content with 88 at.%. The Si content is 9 at.%, and the Ni content is approximately 3 at.%. Concerning the area, this phase grows due to the heat treatment and displays, in particular after 475 °C and 30 minutes, elongated, rounded shapes.
The slightly darker phase (E) has no dendritic structure but with the same composition of 56 at.% Cu, 25 at.% Si, and 17 at.% Ni, it seems to be the same phase as in the CuNiSi alloy. Due to the heat treatment, the total area of this phase decreases slightly. However, it is clearly visible that the overall expansion is finer. It seems that individual areas after 425 °C and 30 min merge together and even form nest-like areas at 475 °C and 30 min.
The third and darkest phase (F) occurs more frequently after the heat treatment. It seems that it is formed preferably between the phases described above. With higher temperatures, the areas further seem to be growing together to form larger areas. With a higher magnification, these darkest regions exhibit a eutectic structure of a dark Ni-rich phase (F1) and a bright Cu-rich phase (F2) in which the Ni and Si content are fairly balanced (Fig. 10). These two phases are finely branched and only approximately 100 nm thick.

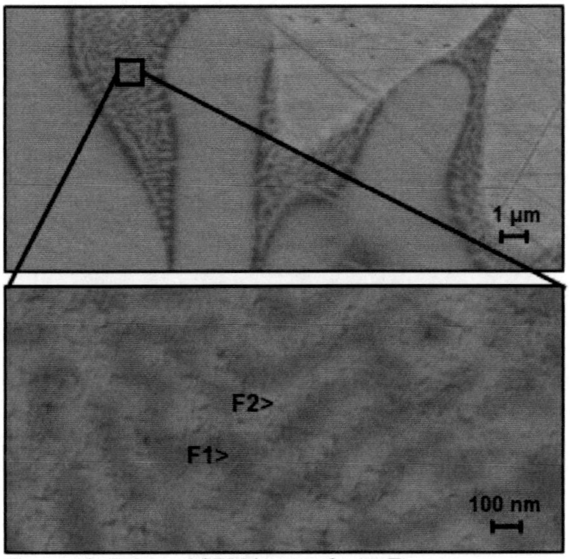

Figure 10: Enlarged SEM image of area F

In contrast to the CuNiSi and the CuNiSiB alloy, the most significant influence of the heat treatment for the CuTi alloy occurred at a lower temperature level. Therefore, Fig. 11 shows the SEM images of the microstructure of the 4 times melted CuTi alloy after melting, prior to and after the heat treatment, in each case for 30 min at 295 °C and at 325 °C. All samples differ only slightly. The EDS measurements verify a concentration ratio of approximately 2 at.% of Ti and 98 at.% of Cu in the brighter phase (G).

The darker phase (H) consists in average of 20 at.% Ti and 80 at.% Cu what indicates to be a Cu_4Ti phase. At higher temperatures, these areas appear to be more locally concentrated. Nevertheless, a possible precipitation hardening is not indicated at the cross-sections.

Furthermore, the hardness of the alloys was subsequently measured after the heat treatment. An overview of the hardness values before the heat treatment (0 min) as well as after 30 and 60 min of the respective alloys and the corresponding temperatures are summarized in Fig. 12.

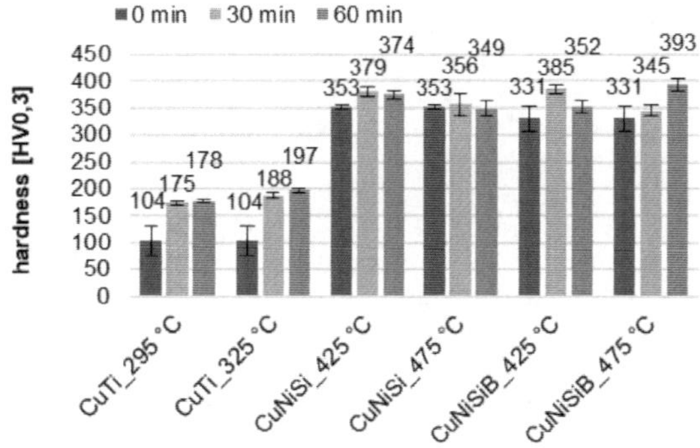

Figure 12: Overview of the hardness values for 4 times melted filler metals before and after the heat treatment determined by means of Vickers

An increase of almost 50% was observed for the CuTi alloy for both heat treatment temperatures. The dwell time of 60 min shows slightly better results compared to 30 min. The CuNiSi alloy displays only a limited hardness increase. The investigations revealed that the treatment temperature of 425 °C results in slightly better values than with 495 °C. Even a short dwell time seems to have a positive effect on the hardness degree.

The third alloy CuNiSiB shows a clear increase in the hardness when compared to the CuNiSi alloy, even if it is not as striking as for CuTi. The best results with a treatment temperature of 425 °C were achieved with a dwell time of 30 min. With a temperature of 495 °C, however, the dwell time of 60 min lead to better results.

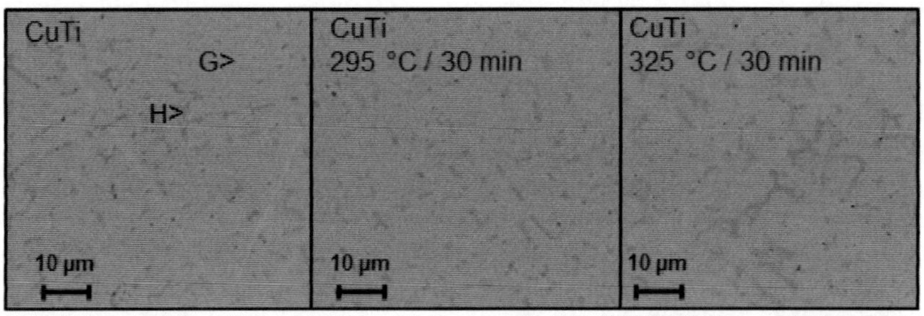

Figure 11: SEM image of the microstructure of 4 times melted CuTi filler before and after heat treatment

Conclusions

Within this paper, copper-based alloys were melted by means of an arc-melting. Further, it could be shown that the inhomogeneous solidification has no effect on the element distribution. All samples exhibit a very good homogeneity in the microstructure. Furthermore, minor dependencies on re-melting processes could be observed. The corresponding alloys showed similar solidus temperatures as well as similar hardness. The DTA investigations showed that the 4 times melted CuNiSiB alloy has the lowest solidus as well as liquidus temperature of 912 °C and 985 °C. The highest hardness before and after precipitation hardening could be documented for these samples. All filler metals showed the effect of precipitation hardening whereby the highest hardness increase (~50%) was measured for CuTi.

In further investigations the predicted phases by means of EDS have to be proven by XRD measurements. Furthermore, the fraction of phases formed in the microstructure have to be determined and set into relation with the hardness in order to develop a model for the explanation of the precipitation hardening mechanism.

Acknowledgments

The authors gratefully acknowledge the financial support from the German Research Foundation (DFG) with the project No. TI 343/112.

References

[1] Carsten P.: Herstellung und Einsatzverhalten von Keramik-Hartmetall-Verbundbohrwerkzeugen (Production and application behavior of ceramic-carbide composite drilling tools), Essen: Vulkan-Verl (2005).

[2] Schnee, D., Zenk, C., Magin, M., Rassbach, S., Hafner, T., Meyer, U.: Optimized Brazing Gap Width for Cemented Carbide Tipped Saw Blades, Proceedings of the 5th International Brazing and Soldering Conference, 2002.

[3] Hohenrainer, C.; Kny, E.; Schmid, L.; Bader, T.: Korrosionsresistente, hochverschleißfeste Hartmetalle (Corrosion-resistant, highly wear-resistant hard metals), Materials and Corrosion/Werkstoffe und Korrosion, Heft 5, S. 230 – 235.

[4] Herr, M.: Tribologisches Verhalten von ultra-feinkörnigen Hartmetallen mit verschiedenen Binderlegierungen (Tribological behavior of ultra-fine-grained carbides with different binder alloys), Erlangen, Nürnberg, 2002.

[5] Bronny M.; Schimpfermann M.: Hartlöten von HW-Schneidwerkzeugen (Brazing of HW cutting tools), Brazing is BrazeTec, Umicore AG & Co. KG (2011).

[6] Schimpfermann, M.; Schnee, D.; Umicore AG & Co. KG et al.: Löten in der Werkzeugindustrie (Brazing in tool industry) Umicore Brazing Technology (2015).

[7] Zhang, J.; Lin, L. Y.; Xu, J. C. et al.: Microstructure and Properties of Brazing Joint between YG8 Cemented Carbide and A3 Steel, Solid State Phenomena (2007), S. 265–270.

[8] Sievers, N.: Die elektrische Widerstandsmessung zur zerstörungsfreien Prüfung von Lötnähten am Beispiel von Hartmetall-Stahl-Verbunden (Electrical resistance measurement for non-destructive testing of brazing seams using the example of carbide-steel composites), Dissertation TU-Dortmund, 2017

[9] Tillmann W., Sievers N., Schnee D.; et. al.: „Comparison between the Electrical Resistance and the Shear Strength of Brazed Cemented Carbides", China Welding, ISSN: 1004-5341, 2015

[10] Hoyer, I. M.: Beitrag zur Entwicklung von Hochtemperaturloten auf Eisenbasis (Contribution to the development of high-temperature brazing alloys based on iron), Chemnitz, Techn. Univ., Diss., 2005, TU Lehrstuhl für Verbundwerkstoffe, Chemnitz, 2005.

[11] Mahler M., Zimmermann, K.F.: Löten von Hartmetallen (Brazing of cemented carbide), Technik die verbindet, Degussa AG, 1975.

[12] Müller, W., Müller J.U.: Löttechnik, Leitfaden für die Praxis (Brazing, guide to practice), Dt. Verl. für Schweißtechnik, Düsseldorf, 1995.

[13] Jin C.; Plucknett K.: The effects of microstructure on Vickers indentation damage in TiC-316L stainless steel cermets, International Journal of Refractory Metals and Hard Materials, Volume 61, December 2016, Pages 151-161.

[14] Mukhiya, R.; Kumar, A.; Kamat S.V.; Bhattacharyya T.K.: Static characterization of bulk micromachined accelerometer structure using a nanoindenter, Materials Science in Semiconductor Processing, Volume 15, Issue 4, August 2012, Pages 353-358

[15] Deutsches Kupferinstitut: Niedriglegierte Kupferwerkstoffe, Eigenschaften-Verarbeitung-Verwendung (Alloyed copper materials, properties processing and use), Deutsches Kupferinstitut Berufsverband e.V (2012).

Design approach for development of novel nickel-based brazing alloys: Initial CALPHAD Investigation

Liam Hardwick & Russell Goodall
Dept. Materials Science & Engineering, University of Sheffield, Sheffield, United Kingdom
lhardwick1@sheffield.ac.uk, r.goodall@sheffield.ac.uk

Ed Pickering
University of Manchester, Oxford Road, Manchester, United Kingdom
ed.pickering@macnhester.ac.uk

Pat Rodgers
VBC Group, Castle Business Park, Loughborough, United Kingdom
pat.rodgers@vbcgroup.com

Abstract

Nickel-based brazing filler metals are favoured in high-temperature brazing applications due to their high melting temperature, mechanical properties and corrosion resistance, for example making them an attractive option when joining superalloys. Recent advances in alloy design approaches can promote and benefit development of such filler metals. The present work demonstrates part of an ongoing alloy design approach, with the aim of developing new compositions of nickel-based alloys, with an evaluation of the ability to accurately predict the properties of such alloys. This approach starts with predictive computer modelling with Thermo-Calc software, which utilises the CALPHAD method along with databases of experimental and theoretical data to predict thermodynamic information. This was used to predict the phases produced during brazing superalloy Inconel-718 with nickel-based filler metal AWS BNi-2. By comparing the predictions with experimental results, an assessment is made as to the usefulness of Thermo-Calc as an alloy development tool. Predictions for different compositions based on AWS BNi-2 are presented.

Introduction

Nickel-based brazing alloys are often employed as filler metals for the joining and repair of nickel superalloys used in aerospace applications, such as jet engine components. Their properties, such as high melting temperature, mechanical properties and corrosion resistance, make them ideal for coping with the demanding conditions experienced. To bring down the liquidus temperature of these mostly nickel-containing alloys, melting point depressants (MPDs) such as boron and silicon are added. During the brazing process, these light elements can diffuse quickly into the base metal [1], and upon solidification of the filler metal, strong joints consisting of a central eutectic region and an isothermally-solidified region at the filler metal-base metal interface are produced [2]. The presence of MPDs such as boron can result in formation of brittle boride intermetallics that can have a detrimental effect on the mechanical properties of the joint. It is therefore important to consider the effects of the amounts of elements like boron in such systems, both in terms of their potency in suppressing the liquidus of an alloy, and the production of brittle phases in the joint region.

Development of brazing alloys has often based around already existing alloys, with compositions dating back several decades [3-5] and has been achieved by altering composition in an ad-hoc manner. Recently, thermodynamic modelling using software such as Thermo-Calc has seen use in predicting phases produced in brazed joints [6, 7]. Thermo-Calc utilises the semi-empirical CALPHAD method (CALculation of PHAse Diagrams) and databases comprising both experimental and theoretical data to predict stable phases, their abundance, as well as solidus/liquidus temperatures in alloys systems. Examples of such databases include the TCNI8 database, which is a database for nickel superalloys based on critically assessed binary and ternary systems and allowing for predictions of multi-component systems. Similarly, the SSOL4 database possesses a large number of critically assessed binary and ternary systems to allow predictions for multi-component systems and non-ideal multi-component solution phases. The present work details microstructural characterisation and compositional analysis of an Inconel-718 joint brazed using AWS BNi-2, and discusses the accuracy of predictions made using the Thermo-Calc software. Further predictions investigating alternative compositions based on AWS BNi-2 are made, and their implications explored.

Experimental Procedure

Table 1: Chemical composition if Inconel-718 base metal

Ni	Cr	Fe	Nb+Ta	Mo	Ti	Al	Co
53.37	18.73	17.74	5.07	2.93	0.97	0.55	0.23
Mn	**Cu**	**Si**	**C**	**P**	**B**	**S**	
0.14	0.10	0.08	0.03	0.011	0.002	0.0004	

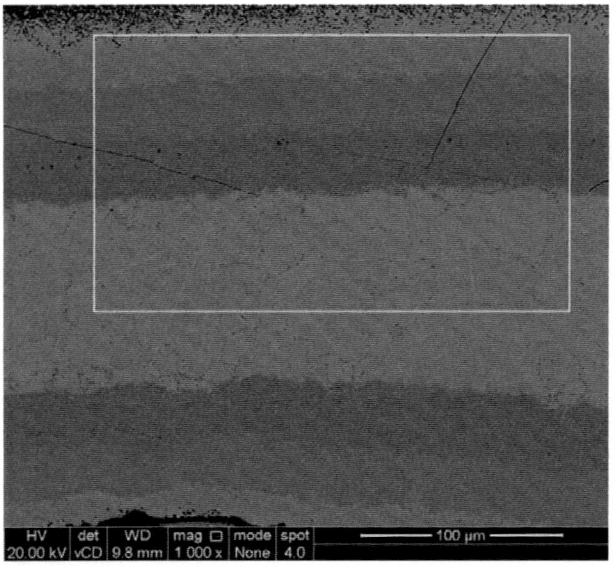

Figure 1: SEM BSE image of a cross section through AWS BNi-2 foil (plane of foil horizontal in the image). Representative area of EDS line scans shown.

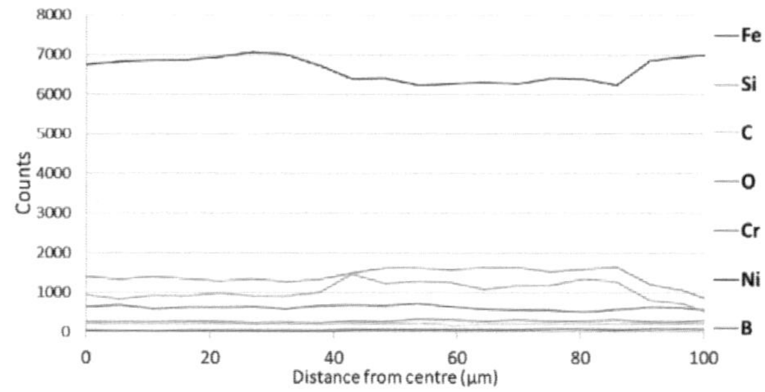

Figure 2: EDS linescan of BNi-2 foil, from central band outwards. Relative levels of the elements indicated at each point are shown.

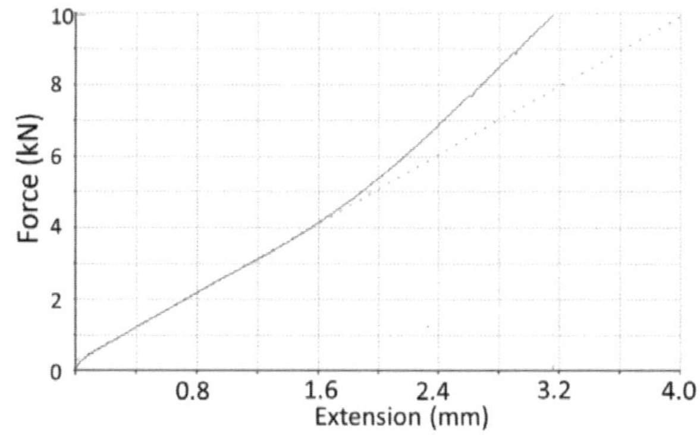

Figure 3: Force-extension curve produced from lap shear test of joint.

Materials

In the present study, the filler metal used was the nickel-based alloy AWS BNi-2, in foil form with average thickness of 48 µm. The base metal joined was nickel-based superalloy Inconel-718 (chemical composition shown in Table 1). The joint was prepared using the AWS C3.2 standard method for evaluating the strength of brazed joints. The Inconel-718 base metal was sectioned into tokens with dimensions 130 × 30 mm, and a thickness of 2 mm (which differed from the 3 mm thickness specified in AWS C3.2). Two such tokens were arranged in a lap joint configuration, with an overlap area of 30 × 12 mm. Post-brazing, this joint was electron discharge machined into a dog-bone shape for lap shear testing, with a width of 12 mm at the narrowest point, and a gauge length of 50 mm. Imaging and element mapping of the failed joint was achieved by scanning electron microscopy (SEM), energy-dispersive x-ray spectroscopy (EDS) and electron probe micro-analysis (EPMA). Alongside this joint, smaller tokens with dimensions of approximately 10 × 20 mm were joined in the same braze cycle to represent the joint pre-tensile test. The joint was sectioned, mounted, ground and polished with standard silicon carbide papers and diamond suspension. Microstructural and mechanical characterisation was conducted by optical microscopy, SEM imaging and nano-indentation, and compositional analysis was achieved by EDS. In addition, a cross-section of AWS BNi-2 foil was imaged by SEM and its composition analysed by EDS.

Brazing Procedure

Brazing was performed under a vacuum of less than 20 mTorr. The temperature ramp rate was 5°C min^{-1} up to 950°C, and then held for 10 minutes. The temperature was then raised to the brazing temperature of 1050°C, again with a ramp rate of 5°C min^{-1}, then held for 10 minutes. The joint was then allowed to vacuum cool.

Thermo-Calc

Thermo-Calc software, using the TCNI8 and SSOL4 databases, was used to produce property diagrams, predicting the phases that form and their molar fraction throughout cooling from 1600°C for a system representative of filler metal AWS BNi-2.

Results

AWS BNi-2 foil

Backscattered electron (BSE) images show the cross-section of a AWS BNi-2 foil sample (for example see Fig.1), revealing a banded structure throughout the thickness of the foil. EDS analysis (Fig.2) revealed mostly even element distribution, with slight chromium and silicon depletion in the outside layer, and a depletion of nickel in the bands exhibiting black plate-like particles which are thought to be boride precipitates (though detection of boron in EDS is challenging and was not detected directly through this). It is the size and concentration of these particles that appear to define the different bands observed. The outermost light grey regions are relatively free of these particles, and moving toward the centre the particles become coarser, elongated plates, which become coarsest at the boundary with the central band. Probable borides are also observed along the grain boundaries in the central band. The cracks seen in the foil in (Fig.1) appear to propagate straight through the plate-rich regions, and stop abruptly when at the central region, highlighting the brittleness of the regions with

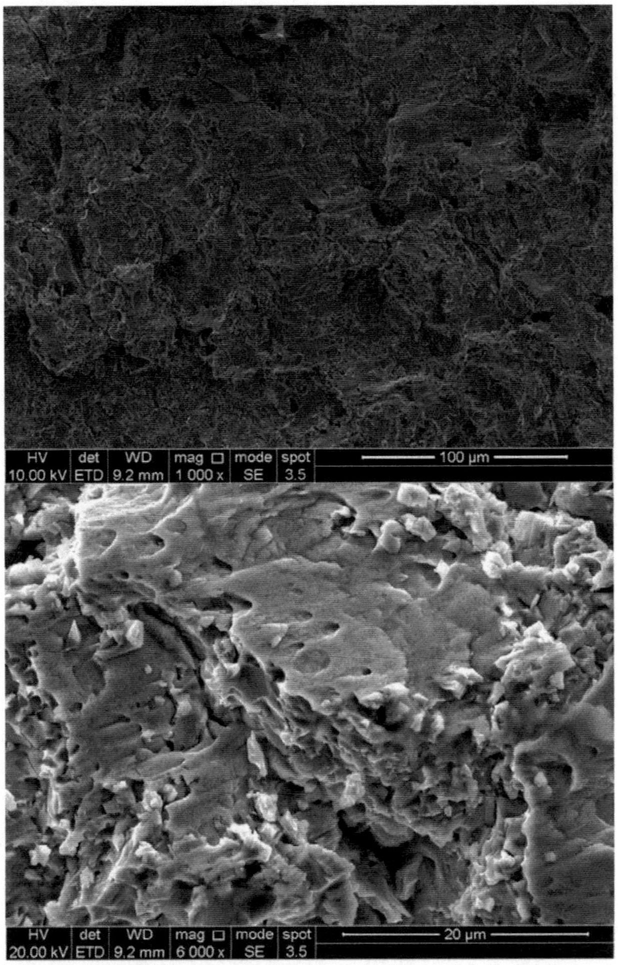

Figure 4: SEM image of fracture surface of failed joint, at 1000× and 6000× magnification.

these structures in high quantity, and further suggesting they are likely to be borides. This apparent banded structure of the foil is likely a result of its manufacture; boron was added to the material in a final step as a surface boriding process, and what we observe is due to incomplete diffusion of boron across the relatively thin foil cross section.

Joint Strength and Microstructure

Post-brazing, in order to verify a suitable joint had been achieved, the joint was lap-shear tested until failure at 9.96 kN, giving a shear strength of 199 MPa (Fig.3). Albeit low compared to achievable strengths [8], this is of comparable strength to other BNi-2 joints seen [9]. The fracture surface of the failed specimen was imaged by SEM (Fig.4), which indicated a brittle failure of the joint.

A typical section of the joint produced is shown in (Fig.5). The joint consists of a central eutectic region containing boride and silicide intermetallics, an isothermally solidified zone (ISZ), a diffusion affected zone (DAZ) and the base metal. EDS point scans and mapping allowed for an indication of the phases present in the joint to be obtained (Fig.5). The dark grey

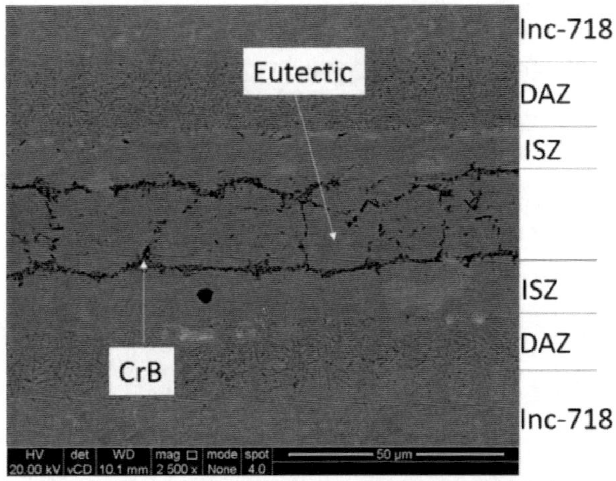

Figure 5: SEM BSE image of the brazed joint.

particles are likely CrB phase, and the lighter grey regions, rich in nickel and silicon, are potentially Ni_3Si phase. These are present in a nickel-rich matrix containing other elements found in AWS BNi-2. The CrB in particular appears to have precipitated in high concentration in two bands, and potentially along grain boundaries still present in the central region of the filler metal. The ISZ was also found to be nickel-rich with other BNi-2 elements. The DAZ was enriched in chromium, iron and other base metal elements such as niobium. Nano-indentation was performed at the CrB structure, the ISZ, and the DAZ, as shown in the AFM image (Fig. 6). The CrB structure was found to be extremely hard with an average hardness of 17.7 GPa (1805 HV), while the ISZ and DAZ were much softer with an average hardness of 7.0 GPa (714 HV) 8.2 GPa and (832 HV) respectively. These hardness values are greater than often produced by micro-indentation, possibly a result of the smaller contact area, difference in analysis, and the assumption of purely elastic contact during indentation as pointed out by [10].

Figure 7 shows element composition profiles produced by EPMA line scan from the filler metal into the base metal on the failed lap shear sample. High boron concentration is observed at the extreme left edge of the image, corresponding to the location where the sample failed in shear. This high concentration likely corresponds to the presence of intermetallic boride phases observed to have precipitated in continuous bands. Indeed, it is possible that the joint sheared along these bands of brittle borides, and such structures could clearly be detrimental to joint integrity. A marked decrease is then observed through the ISZ, before a sharp increase at the DAZ, indicating preferential diffusion into and reaction with the base metal. Enhanced boron concentration is also observed past the DAZ, again indicating diffusion into the base metal. The concentration abruptly falls and stays minimal after this, and so the boron diffused only as far as 60-70 µm into the base metal. Following this pattern at the DAZ region, an enhanced niobium concentration is apparent coinciding with the increased boron concentration. An enhancement is again observed coinciding with the second boron concentration increase just past the DAZ. This suggests reaction with the

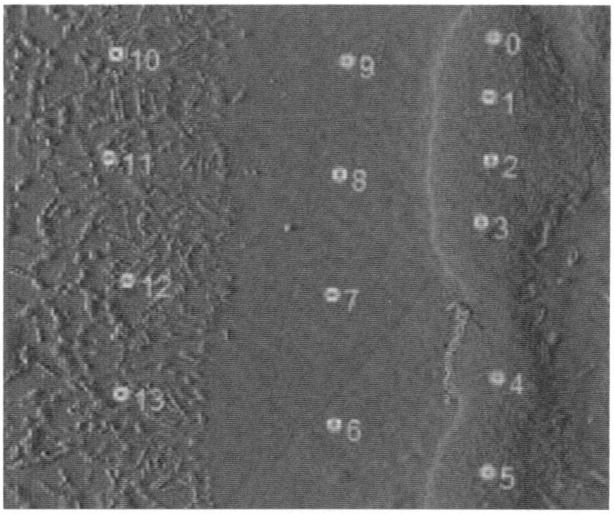

Figure 6: AFM image showing positions of nano-indents made at the CrB phase (0-5), ISZ (6-9), and DAZ (10-13).

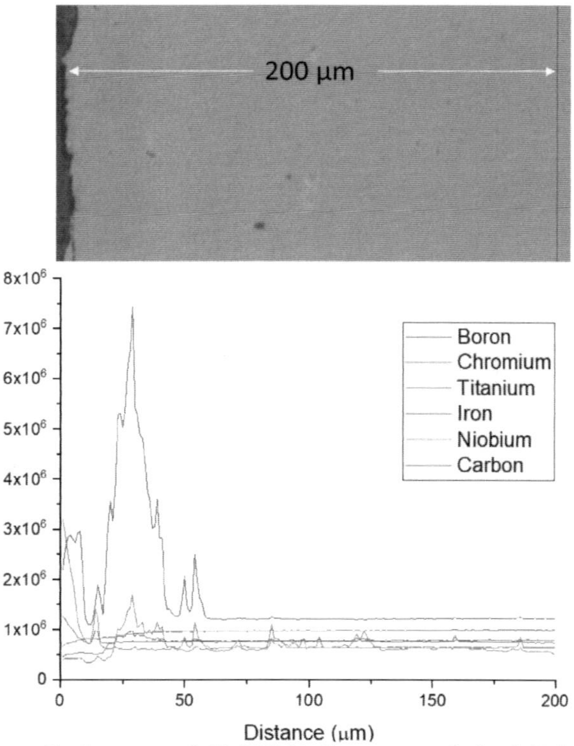

Figure 7: Average of 10 EPMA line scans of the failed joint, spanning 200 microns into the base metal, showing the relative levels of and variations in the content of the elements indicated.

niobium contained in the Inconel-718 base metal. While boron appears to have diffused into the base metal, there was no significant evidence of silicon diffusion, which remained in the filler metal. Chromium is as expected most prevalent in the filler metal. Carbon was also detected predominantly in the filler metal and is also somewhat co-located with the niobium. Figure 8 shows a region in the DAZ mapped via EPMA. The

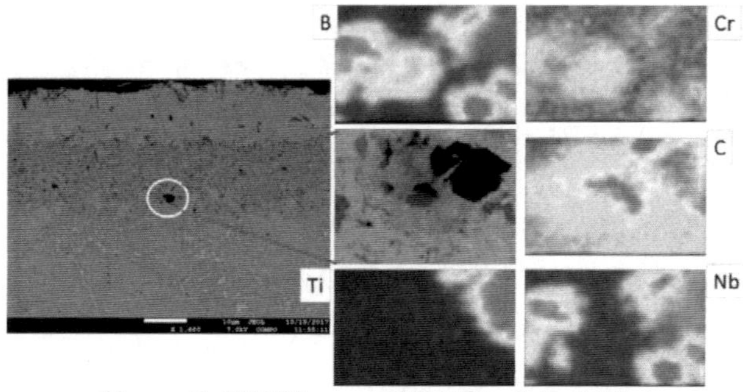

Figure 8: EPMA map of DAZ in failed joint.

maps show a CrB phase as a dark grey particle in the centre of the image, while the black phase is TiC. The lighter grey regions appear to be a NbB phase, which seems to be to quite prevalent in the microstructure.

Thermo-Calc Predictions

Figures 9 and 10 show the property diagrams produced in Thermo-Calc for a system with the composition of AWS BNi-2, using the TCNI8 and SSOL4 databases, respectively. Some notable differences are apparent. Using the TCNI8 database, five phases are predicted to form during solidification, the onset of which is predicted to be at the liquidus of 1288°C, where CrB (labelled MB_B33) precipitates. Following this, a nickel-rich phase labelled FCC_L12 and a phase Ni_3B labelled NI3B_D011, are predicted to precipitate at 1153°C and 1079°C respectively. Solidus temperature is predicted as 1053°C. At this temperature, a Ni_2B phase labelled M2B_TETR precipitates, and a probable Ni_3Si phase (labelled FCC_L12#2) precipitates at 976°C. In comparison, the SSOL4 database predicts a liquidus of 1165°C. Solidification starts with the precipitation of a CrB phase (labelled as CRB), followed by nickel-rich phase (labelled FCC_A1) at 1079°C. Two further phases precipitate before solidus is reached, a Ni_3B phase (labelled NI3B) and a Ni_2B phase (labelled M2B_TETR) at 995°C and 916°C respectively. Below the solidus of 897°C, an FeB phase (labelled FEB) is predicted to precipitate at 837°C, and a Ni_3Si phase (labelled NI3SI_L) precipitates at 669°C.

Discussion

The joints produced in this study possessed a largely typical microstructure consisting of a central eutectic region appearing to contain CrB and Ni_3B intermetallics, and a nickel-rich ISZ. The TCNI8 database consistently predicts the precipitation of all like-for-like phases at significantly higher temperatures than in the SSOL4 database. Nonetheless, from the Thermo-Calc predictions we can infer the solidification path in the joint. The brittle CrB intermetallics, observed as dark grey regions in Fig.5, formed first centrally in the joint. This phase appears to have precipitated preferentially into two bands parallel to (and positioned in between the original positions of) the base and filler metals and possibly along grain boundaries present in the central region of the filler metal. As the central band of the foil shown in Fig.1 appears to be depleted of

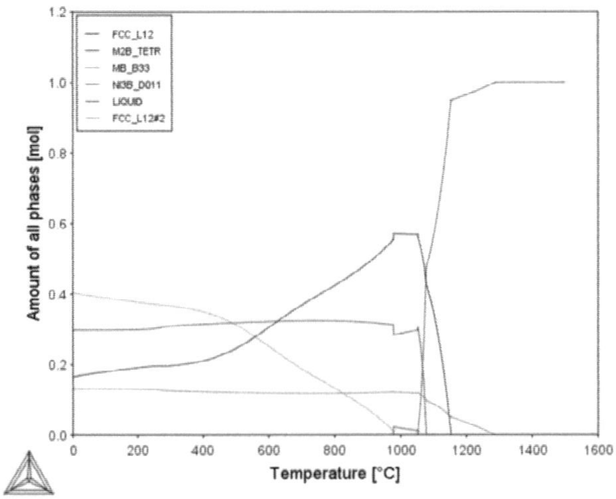

Figure 9: Thermo-Calc (TCNI8) property diagram displaying molar fraction of phases predicted during solidification of BNi-2. Labelled phases are described in the text.

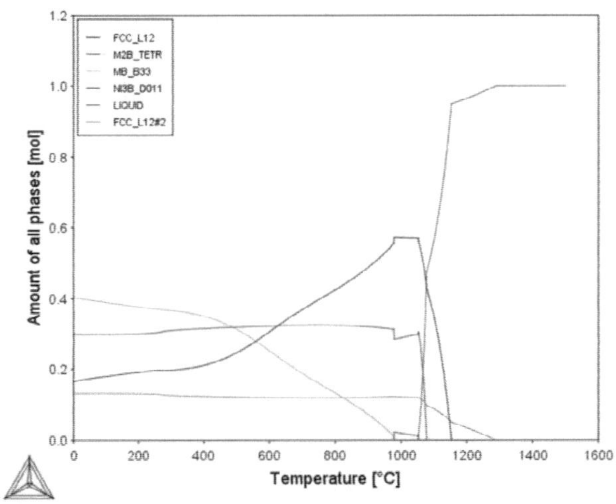

Figure 10: Thermo-Calc (SSOL4) property diagram displaying molar fraction of different phases predicted during solidification of BNi-2. Labelled phases are described in the text.

boron-containing phases, it is possible this region did not undergo melting during the brazing process, restricting boride formation to outside this region and along grain boundaries. These were then followed by the solidification of the nickel-rich ISZ, induced by boron diffusion, at the interface with the base metal and solidifying inwards, possibly contributing to the band-like distribution of the CrB phase. This corresponds to the FCC_L12 and FCC_A1 phases predicted in the TCNI8 and SSOL4 databases respectively. Nickel borides of the type Ni_3B, corresponding to the NI3B_D011 (TCNI8) and NI3B (SSOL4), then solidified, followed by Ni_2B (represented by M2B_TETR in both databases). Both again formed in the central region. The FeB phase predicted by the SSOL4 database was not observed experimentally, however, the low molar fraction predicted by Thermo-Calc (≤5 mol.%) could explain this. Both databases then predict the formation of Ni_3Si phases, corresponding to FCC_L12#2 (TCNI8) and NI3SI_L (SSOL4). These are also observed centrally and are likely the lighter grey regions in Fig.5. In terms of the phases predicted and the order of phases solidified, both databases agree well with experimental results. The TCNI8 database results in particular are in agreement with [7, 11]. It should be noted that the lack of any kind of niobium boride phase predicted is due to the base metal elements not being used in this simulation.

The major discrepancy between both databases arises when predicting solidus and liquidus temperatures. The TCNI8 and SSOL4 databases predicted solidus/liquidus temperatures of 1053°C/1288°C and 897°C/1165°C respectively. Compared to the accepted solidus/liquidus temperatures of 970°C/1000°C, the TCNI8 predictions are substantially higher for both solidus and liquidus. The SSOL4 liquidus prediction is somewhat closer, but the solidus prediction is substantially underestimated.

These predictions were investigated further, by calculating the solidus and liquidus temperatures for an AWS BNi-2-like system, over a range of boron and chromium weight percentages. For both databases, an approximately linear decrease in liquidus temperature is predicted as the boron weight percentage in the system is increased from 1.0 wt.%, until a minimum is reached at 2.1 wt.% boron, which yields a liquidus temperature of 1195°C (Fig.11). This is 93°C lower than the liquidus predicted by Thermo-Calc at the nominal BNi-2 boron weight percentage of 3.1%. After 2.1 wt.% boron, the addition of more is predicted to again increase the liquidus, albeit at a slower rate as more boron is added. At all boron weight percentages modelled, the TCNI8 database consistently predicts greater liquidus temperature, and above the 2.1 wt.% minimum, this gap increases. The solidus temperatures predicted by both databases show more immediately obvious differences. The TCNI8 database predicts a relatively stable solidus temperature across the whole range of boron weight percentages, with a slow decrease until 3.1 wt.%. The solidus the remains constant until another small decrease at around 4 wt.%. The SSOL4 database, however, predicts a sharp decrease in solidus temperature between 1.0 and 1.4 wt.%, and a slower decrease thereafter until 2.7 wt.%. The solidus then remains roughly constant across the rest of the boron weight percentages modelled. This prediction, while unexpected, and perhaps a spurious prediction arising from the incomplete nature of the databases used, could merit investigation as if it were borne out by experimentation it would be of significant use. It is however possible that the location of the minima is inaccurate, and should coincide with the 3.1% of current alloys.

To explore the potential application of the method to alloy design, chromium content was altered for a BNi-2-like system, over the range 0.5 to 10 wt.%, while the composition of other elements was the same as found in BNi-2 normally. Rather than seeking a significant effect on melting point, such a

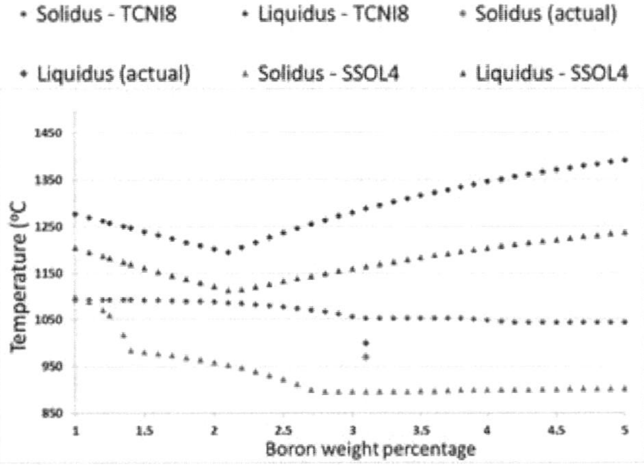

Figure 11: Solidus and liquidus temperature vs. boron weight percentage predicted by Thermo-Calc (both databases). Also plotted are the accepted values at the normal 3.1 wt.% found in BNi-2.

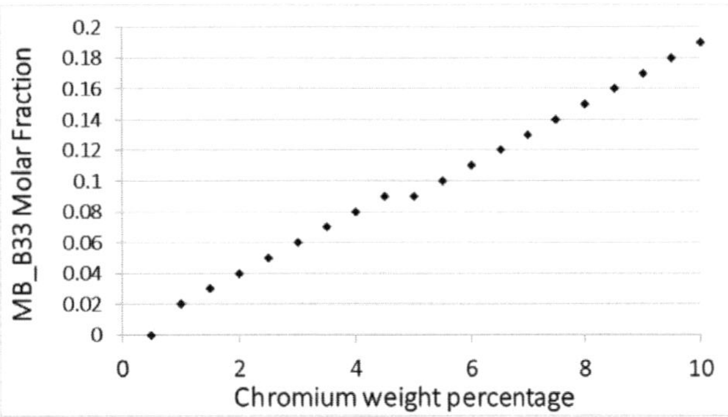

Figure 12: MB-B33 (CrB) molar fraction vs. chromium weight percentage predicted by Thermo-Calc (TCNI8 database).

variation may allow a reduction on some of the borides formed, through the removal of some chromium boride. Between 0.5 and 3.5 wt.%, both solidus and liquidus were relatively stable at around 1047 and 1125°C respectively. The addition of more chromium beyond this had a noticeable yet unsurprising effect on the liquidus temperature, which rose approximately linearly to a maximum of 1386°C at 10 wt.%. The solidus remained stable at 1047°C until 8 wt.% chromium, after which it rose approximately linearly, also reaching its maximum of 1073°C at 10 wt.%. The maximum molar fraction of the CrB phase, labelled MB_B33 in the TCNI8 database, was also recorded as a function of chromium content (Fig.12). Thermo-Calc predicted an approximately linear increase in molar fraction of the phase, from a negligible amount at 0.5 wt.% chromium, to 0.19 at 10 wt.% chromium. Again, while an unsurprising finding, this suggests refinement to the current BNi-2 composition in order to result in lower production of brittle boride phases such as CrB could be possible. While the predictions here give solidus and liquidus temperatures substantially different to actual values, they could nevertheless warrant exploring experimentally.

Combining these two sets of predictions, modifications to the current AWS BNi-2 composition for further investigation could therefore be foreseen exploring reductions in boron and chromium content. This would hypothetically result in a lower liquidus temperature, giving obvious benefits in costs of heating joints, as well as lower molar fraction of CrB in the final joint microstructure. Of course, considerations must be made as to the impact on corrosion resistance that a reduction in chromium might induce. Further investigations on this, and other compositional changes, led by the modelling predictions, will form the next phase of the work.

Conclusions

- Inconel-718 was vacuum brazed with nickel-based filler metal AWS BNi-2, producing a joint consisting of central eutectic region containing very hard boride phases and silicide phases, softer isothermally solidified zone and slightly harder diffusion affected zone. Lap shear testing demonstrated a joint strength of 199 MPa.
- Boron was found to have diffused into the base metal by approximately 60-70μm, resulting in a diffusion affected zone containing CrB phases and niobium-boron type phases.
- Thermo-Calc, using the TCNI8 and SSOL4 databases, made predictions of phases produced upon solidification in good agreement with experimental observations, suggesting usefulness for designing novel filler metal compositions. Discrepancies arose in both databases, however, when predicting solidus and liquidus temperatures.
- Predictions suggest there may be value in exploring reduced boron content alloys of the AWS BNi-2 type further. In addition, reduced chromium content could be exploited to subdue formation of brittle CrB phases in the joint post-brazing. Further investigation into this will form the basis for future work.

Acknowledgements

The author gratefully acknowledges the financial support of the Engineering and Physical Sciences Research Council UK (EPSRC) through the Centre for Doctoral Training in Advanced Metallic Systems. The author acknowledges the use of the EPSRC-funded JEOL JXA-8530F FEG-EPMA, award EP/M028097/1, and thanks Dr. Jonathan Fellowes for assistance with this. The author also thanks Dr. Claire Utton at the University of Sheffield for technical help with Thermo-Calc.

References

[1] MacDonald, W.D. and Eagar, T.W. "Transient liquid phase bonding processes". Annual Review of Materials Science. (1992). Vol. 1, no. 1, p. 23–46.

[2] Gale, W.F & Butts, D.A (2004) "Transient liquid phase bonding", Science and Technology of Welding and Joining (2004), 9:4, 283-300
[3] Lugscheider, E. *et al.,* "Development of Nickel-Chromium-Silicon base filler metals", 9th International AWS-WRC Brazing Conference, New Orleans, LA, (1978)
[4] Lugscheider, E. & Cosack, T., "High-temperature brazing of stainless steel with low -phosphorous nickel-based filler metal", 17th International AWS-WRC Brazing Conference, Atlanta, GA, (1986)
[5] Lugscheider, E. & Partz, K.-D., "High-temperature brazing of stainless steel with nickel-based filler metals BNi-2, BNi-5 and BNi-7", 12th International AWS-WRC Brazing Conference, Cleveland, OH, (1981).
[6] Piegert, S., *et al.*, "Design of a braze alloy for fast epitaxial brazing of superalloys" IOP Conf. Ser.: Mater. Sci. Eng. 33, (2012)
[7] Ruiz-Vargas, J., *et al.*, "Microstructure development during isothermal brazing of Ni/BNi-2 couples", Journal of Material Processing Technology 213, 20-39, (2013)
[8] Pouranvari, M., *et al.*, "Diffusion brazing of cast INCONEL 718 superalloy utilising standard heat treatment cycle", Materials Science and Technology, 30:1, 109-115, (2014)
[9] Khorram, A., *et al.* "Laser brazing of Inconel 718 and Inconel 600 with BNi-2 nickel-based filler metal", International Journal of Advanced Manufacturing Technology 88, 2075-2084, (2016)
[10] Qian, L., *et al.*, "Comparison of nano-indentation hardness to microhardness", Surface & Coatings Technology 195, 264 – 271, (2005)
[11] Riggs, B.B *et al.*, "Thermodynamic and kinetic simulations of high temperature brazing: microstructure evolution in CMSX-4 joints", Science and Technology of Welding and Joining (2017), 22:5, 428-437

Strength of Steel, Copper, and Brass Lap Joints Brazed by Silver-Based and Silver-Free Filler Metals

Boutilier J.[2], Brown N.[2], Klidas N.[2], Alexandrov B. T.[2], Shapiro A. E.[1]
[1] *Titanium Brazing, Inc., Columbus, OH*
[2] *The Ohio State University, Welding Engineering Program, Columbus, OH*

Abstract

The results of shear strength testing and metallurgical characterization of four Silver-based brazing filler metals BAg-7, BAg-22, BAg-24, and BAg-34 are reported and compared with two Silver-free filler metals P-14 (Cu-6P-4Sn wt.%), LOK59 (Cu-40Zn-1Sn-0.3Si wt.%), designated for brazing copper, brass, low-carbon steel, and stainless steel. All studied silver-free filler metals are suitable to replace some silver-based filler metals in brazing applications. Shear strength of brazed joints manufactured by both classes of filler metals (silver-based vs. silver-free) that are currently used in the US industry are discussed in order to find out their area of application and saving production costs.

1. Introduction

Deep technical and commercial interest in silver-free brazing filler metals (BFM) relates directly to their low cost. Compared to the cost of silver-based BFM, the range of silver-free alloy's cost can be as low as 1/4 to 1/8 of the silver BFM price. While cost is an important factor for filler metal selection, silver alloys are still used in brazing filler metals. That due to their low melting temperature and good capillary action, which contribute to their ability to easily create acceptable braze joints that meet various given requirements.

Many attempts have been made in the last 40 years to substitute silver to reduce the expenses in brazing production.

All of these various substitutes were developed and then produced using (a) classic technology: casting ingot and hot extrusion of rods, or of wires, or of strips, or else (b) by way of rapid solidification directly from the melt to obtain rods or foils [1-9].

Some conventional silver-free alloys that could provide a comparable strength have brazing temperature ranges that are significantly higher than silver alloys.

Only the phosphorus-containing silver-free braze metals have a brazing temperature in the range of 600 to 700℃ (1110 to 1380℉) that is comparable to Ag-Cu-Zn-Cd silver alloys. However, these phosphorous-containing alloys produce much lower strength brazed joints, especially in steel, due to the formation of brittle iron phosphides at the interface. On the other hand, silver-free filler metals do not contain cadmium, while some widely-used silver alloys (e.g., AWS BAg-1 and BAg-1a) contain 15 to 24% of cadmium. Cadmium is still used to depress the brazing temperature of some standard silver alloys, but this element is not permitted in new developments due to health concerns.

Silver based alloys' ductility in wire and strip forms is an important property that allows fabrication of different preforms such as wire, rings, shims, and more. Productivity during assembly and brazing is increased with the use of these preforms, and is why silver brazing alloys are preferred by most customers.

Hot deformation is needed to produce preform shapes of existing silver-free alloys due to insufficient ductility at room temperature. Hot deformation makes the manufacturing of silver-free preforms so expensive that they lose a significant part of their cost effectiveness.

In the last few years, some silver-free alloys have been made available in the form of powder or pastes [6, 9]. Other were made as rapidly-solidified foils [1-5].

High-strength silver-free BFM such as P211 (Cu-34Zn-7.3Sn-1.4Ni wt.%), LOK59-03 (Cu-40Zn-1Sn-0.3Si wt.%), and BCuP-9 (Cu-6Sn-6P-0.4Si wt.%) are manufactured as hot-extruded rods 2.2 mm and 3.1 mm in diameter that significantly improves their application for torch or induction brazing [4,6,7]. Silver-free alloys supplied in the form of powder, wires, or rods do not require any additional treatment or shaping. Therefore, silver-free alloys keep their low-cost effectiveness. Considering positive interest and a growing demand for silver-free BFM, this paper reports compared strength of brazed joints made by both competitive classes BFM (silver-based vs. silver-free).

2. Experimental

The strength and microstructure of brazed joint manufactured by silver-based filler metals BAg-7 (55% silver), BAg-22 (49% silver), BAg-24 (50% silver), and BAg-34 (38% silver) were evaluated in comparison to silver-free P-14 and LOK59-03 brazed joints that were tested earlier [6, 7]. All silver BFM were used in the form of 1.6mm wire. Compositions and melting temperature ranges of all tested filler metals are presented in Table 1. Base materials are 1018 carbon steel, 304 stainless steel, copper, and yellow brass.

Brazing was performed in air by heating with propane torch. The black, boron-modified, fluoroborate flux 601B/3411 (Superior Flux Mfg. Co., Cleveland, OH) was deposited inside the joints before heating.

Standard lap specimens (Fig. 1, on the left) were used for brazing and testing according to AWS Specification C3.2M/C3.2:2008 "Standard methods for evaluating the strength of brazed joints". Base metal bars with dimensions of 60x12.5x3.175 mm were brazed. A specimen and tooling for shear test by applying compression are shown in Fig. 1-2. Three to five specimens of every base metal / braze metal combination were fabricated and subjected to mechanical testing in order to determine shear strength of brazed joints and evaluate their ductility or brittle behavior.

Polished cross-sections of brazed joints were etched using the following etchant solutions: (a) 75% hydrochloric acid (6%) + 25% ethanol or (b) 10% acetic acid in ethanol for 10-30 s, and then were examined using an optical microscope to reveal solidification microstructures.

Table 1 Compositions and melting ranges of tested filler metals

Brazing filler metal	Composition, wt.%	Melting temperature range	
		°C	°F
Silver-based filler metals			
BAg-7	56Ag-22Cu-17Zn-5Sn	618-652	1145-1205
BAg-22	49Ag-16Cu-23Zn-7.5Mn-4.5Ni	680-699	1260-1290
BAg-24	50Ag-20Cu-28Zn-2Ni	660-707	1220-1305
BAg-34	38Ag-32Cu-28Zn-2Sn	649-721	1200-1330
Silver-free filler metals			
P-14	Cu-(3.5-4)Sn-(5.3-6)P	635-680	1175-1256
LOK59-03	Cu-39.7Zn-1Sn-0.3Si	880-905	1616-1662

3. Results and discussion

Each lap joint was mechanically tested in either tension or compression to find the shear strength of the brazed structure. The brazed joints in stronger base metals, 1018 carbon steel and 304 stainless steel, produced adequate results of shear strength in tension. The shear strength of brazed joints in more ductile materials, brass and copper, was tested in compression due to the inconsistency during shear testing in tension. Once the tests were completed, a ratio R between the tension and compression tests was calculated to compare the values in tension. The compressed test was normalized by the ratio R in order to make all measurements comparable. The average shear strengths, in terms of tension, for each base material are shown in the figures below.

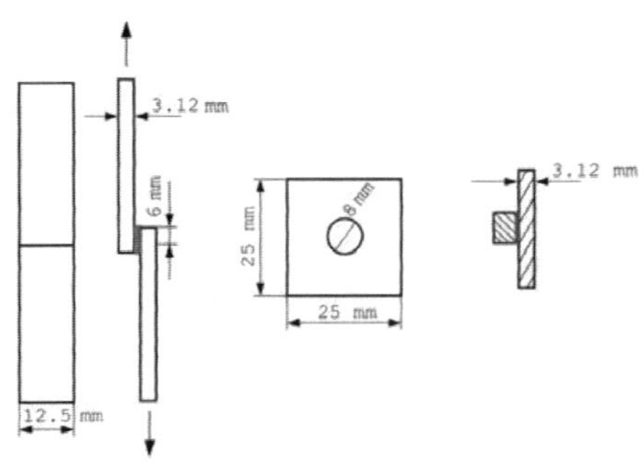

Fig. 1 Specimens used for measuring shear strength of brazed joints: standard design of a lap joint (on the left) - for tension tests, and a special design [10] for the compression test (on the right). Position of a brazed joint to be tested is colored in red.

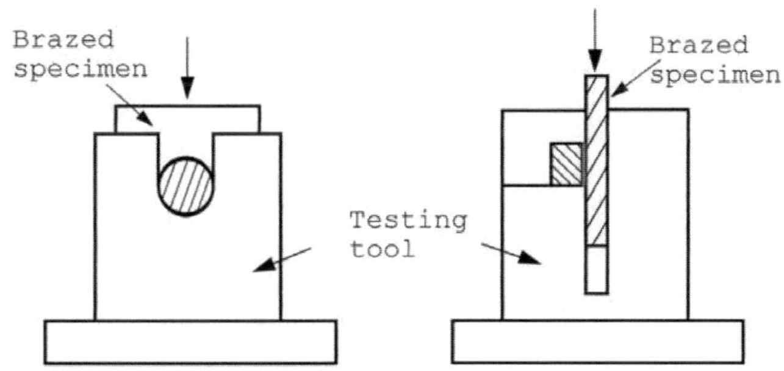

Fig. 2 Scheme of testing and a tool for measuring shear strength of brazed joints by applying compression [10]

The average shear strength of 1018 carbon steel brazed joints are shown in Figure 3. The carbon steel joints had high average shear strength: 283.4 MPa (41.07 ksi) for BAg-7, 324.16 MPa (46.98 ksi) for BAg-22, and 344.24 MPa (49.89 ksi) for BAg-24. P-14 had the lowest average shear strength at 137.31 MPa (19.90 ksi). LOK59 had an average shear strength between BAg-34 and BAg-7 at 266.34 MPa (38.60 ksi).

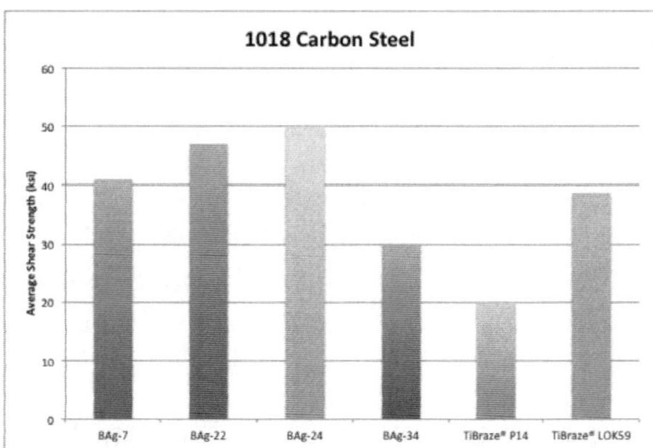

Fig. 3 - 1018 carbon steel joints, average shear strength for each braze metal.

These results mean that LOK59 can be substituted for BAg-34 when brazing carbon steel, while the strength data of 55% silver BAg-7 and silver-free LOK59 are comparable. Therefore, some users may be willing to accept a small decrease in strength in order to gain production costs that are at least five times lower than silver BAg-7.

Figure 4 shows that brazed joints of the 304 stainless steel exhibited high average shear strengths: 262.48 MPa (38.04 ksi) for BAg-22 and 265.17 MPa (38.43 ksi) for BAg-24. The LOK59 exhibited the highest average shear strength of 322.92 MPa (46.80 ksi), while BAg-34 had the lowest - at 170.15 MPa (24.66 ksi). Therefore, brazed joints of stainless steel can be manufactured with higher strength at a lower cost by using silver-free alloy LOK59-03 instead of any of the tested silver-based filler metals.

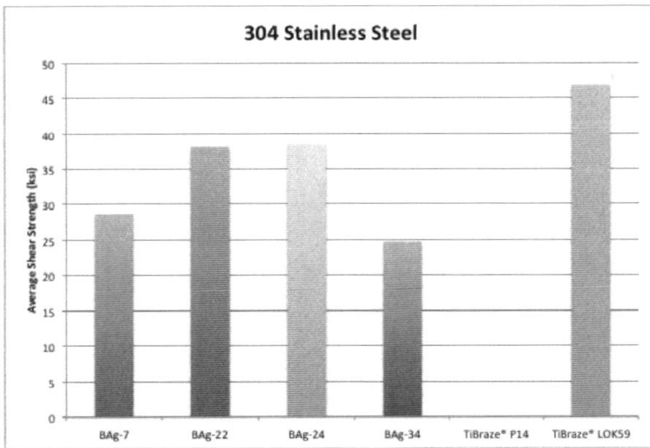

Fig. 4 - 304 stainless steel joints, average shear strength for each braze metal.

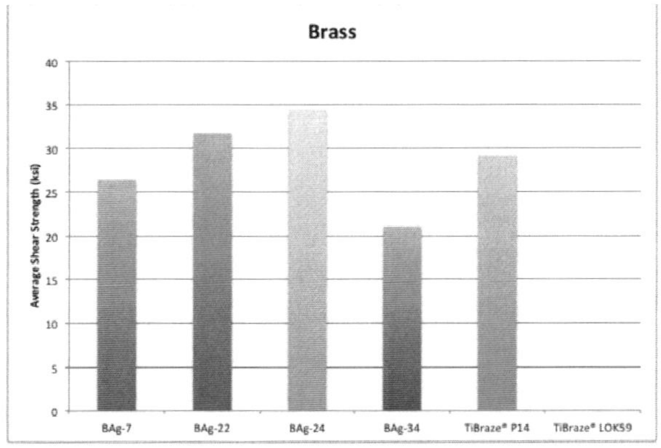

Fig. 5 - Brass brazed joints - average shear strength for each braze metal.

The average shear strengths for brass brazed joints are shown in Figure 5. BAg-24 exhibited the highest average shear strength of 237.29 MPa (34.39 ksi), while BAg-34 had the lowest strength of 145.38 MPa (21.07 ksi). BAg-22 had a shear strength of 218.80 MPa (31.71 ksi), P-14 had a shear strength of 200.79 (29.10 ksi), and BAg-7 had a shear strength of 182.57 MPa (26.46 ksi). The silver-free P-14 paste is an optimal solution to braze brass joints at a low cost, instead of BAg-7 or BAg-34, which contain 55% and 38% of silver, respectively.

BAg-7 exhibited the highest average shear strength for copper at 197.34 MPa (28.60 ksi) (Fig. 6). LOK59 showed the lowest strength of 124.20 MPa (18.0 ksi). BAg-22 and BAg-24 had shear strengths of 149.18 MPa (21.62 ksi) and 170.08 MPa (24.65 ksi), respectively. BAg-34 had a shear strength of 148.42 MPa (21.51 ksi) and P-14 had a shear strength of 153.18 MPa (22.20 ksi).

Thus, the silver-free braze alloy LOK59 exhibited the highest shear strength of stainless steel joints than that of all tested silver-based filler metals (Fig. 4).

The previous tests [6] projected that LOK59 is better to braze carbon steel in comparison to BAg-1. However, Figure 2 shows that LOK59 is only stronger than BAg-34 and P-14 for brazing carbon steels. For the copper samples, LOK59 has the weakest strength (Fig. 6).The silver-free alloy P-14 is better suited to braze brass than BAg-7 and BAg-34, based on the average shear strength, but does not match the strength of BAg-22 and BAg-24 (Fig. 5). P-14 is slightly stronger than BAg-22 and BAg-34 with copper base metal (Fig. 6). P-14 is the weakest filler metal for joining carbon steels (Fig. 3), however, the strength is unexpectedly high ~137 MPa (~20 ksi) for joints containing brittle iron phosphide layer at the interface of steel. The silver-free paste P-14 is a fair competitor to BAg-22 and BAg-34 in the manufacturing of copper brazed joints. In addition to the higher shear strength and low cost, P-14 has a lower brazing temperature than that of BAg-22, BAg-24, or BAg-34.

Generally, BAg-22 and BAg-24 have the higher shear strengths for all base metal-filler metal combinations compared to BAg-7 and BAg-34. This difference could be attributed to the alloying with nickel: BAg-22 and BAg-

24 have 4.0-5.0 wt% and 1.5-2.5 wt% of nickel respectively, while BAg-7 and BAg-34 are alloyed with tin instead of nickel. It was noted that the exception to this comparison is the shear strength of the copper BAg-7 sample.

The BAg-7 provided the highest shear strength (Fig. 6) among all copper specimens that were brazed. The copper BAg-7 sample was predicted to form FCC phase matrix with eutectic constituents, while the predictions for the copper BAg-24 samples included only eutectic constituents [11]. The eutectic constituents are known to have a lower strength than the FCC phase. The copper BAg-7 sample has higher strength possibly due to the FCC phase that formed in comparison to the samples with no FCC phase.

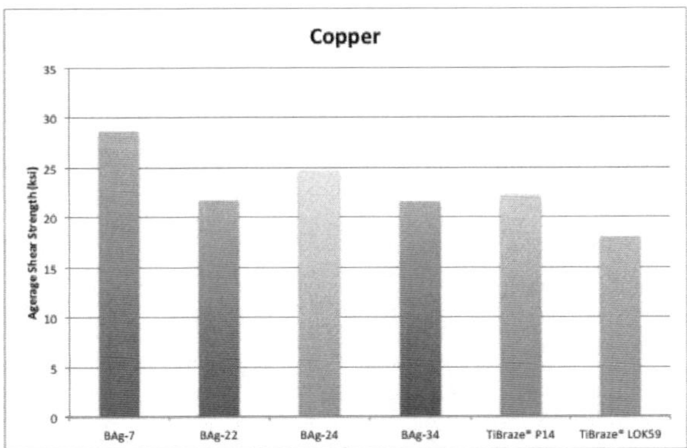

Fig. 6 - Copper joints, average shear strength for each braze metal.

4. Conclusions

4.1 Silver-based filler metals containing up to 50% of silver can be substituted by a silver-free LOK59-03 alloy for brazing 304 stainless steel.

4.2 LOK59 can substitute for BAg-34 when brazing carbon steel, while the strength data of 55% silver BAg-7 and silver-free LOK59 are comparable.

4.3 The phosphorus-contained, silver-free P-14 filler metal provided moderate shear strength brazed joints compared to silver-based braze metal when brazing brass and copper. Silver-free P-14 can be used to replace silver-based filler metals containing less than 50% silver for joining copper and brass.

4.4 Generally, silver-based filler metal alloyed by nickel create stronger brazed joints than those without nickel. The addition of nickel probably improves ductility but more research is needed to confirm this hypothesis. An exception to this is when brazing copper with Bag-7.

5. References:

1. Datta A., Rabinkin A., Bose D. 1984. Rapidly solidified copper-phosphorus base brazing foils, *Welding Journal, # 10, 14-21.*

2. Rabinkin A., Pounds S. 1998. Effects of load on brazing with Metglas MBF-2005 filler metal *Welding Journal, # 5, 33-45*

3. Rabinkin A. 1984. Resistance of copper/copper joints brazed with silver-base and amorphous Metglas alloys, *Materials Letters, vol. 2, # 6, 487-491*

4. Ilina I. I., Pashkov I. N, Rodin I. B. 2001, Interaction of some non-silver brazing metals with base materials, *Proc. of the 6th Int. Conf. on Brazing and diffusion bonding, Aachen, 7-9 June, 2001, Aachen, 326-328*

5. Pashkov I. N., Ilina I. I., Shapiro A. E. 2006. Properties and applications of Cu-based silver-free brazing filler metals made by rapid solidification technique,

Proc. of 3rd Int. Brazing and Soldering Conf., April 24-26, 2006, San Antonio, TX, 157-166

6. Duffey M. J., Marchal J. T., Loney M. R., Alexandrov B. T., Shapiro A. E. 2015. Evaluation of new silver-free brazing filler metals, *Welding Journal, 2015, vol. 94, #3, 40-46*

7. Marchal J. T., Duffey M. R., Loney M. R., Alexandrov B. T., Shapiro A. E. 2015. Silver-free filler metals meet strength requirements in brazed joints, *Welding Journal, 2015, vol. 94, #11, 62-66*

8. Duffey M. J., Marchal J. T., Loney M. R., Alexandrov B. T., Shapiro A. E. 2015. Strength of Brazed Joints and Metallurgical Testing of Cu-4Sn-6P and Cu-40Zn-1Sn-0.3Si Silver-free Filler Metals, *Proc. of 6th Int. Brazing and Soldering Conf. IBSC-2015, Long Beach, CA, April 16-19, 2015, paper #021*

9. Shapiro A. E., Alexandrov B. T., and Duffey M. J. 2016. Evaluation of new silver-free brazing filler metal in the form of wire or paste. *DVS Berichte, vol 325, 2016, Proc. of 11th Int. Conf. on Brazing and diffusion bonding, Aachen, 7-9 June, 2016, 338-340*

10. Matsu K., Miyazawa Y., Nishi Y., Ariga T. 2007. Brazing and interfacial reaction of CP titanium with Ti-Zr-based filler metals, Materials Transactions, vol. 48, No.5, 1055-1059

11. Boutilier J., Brown N., Klidas N., Alexandrov B. T., Shapiro A. E. 2016. Evaluation of silver-based filler metals for joining steels, copper, and brass, Report of the Capstone-2016 Project, Welding Engineering program, The Ohio State University, pp. 1-33